GEOLOGICAL DISPOSAL OF RADIOACTIVE WASTES AND NATURAL ANALOGUES

Lessons from Nature and Archaeology

Waste Management Series

1. Waste Materials in Construction. The Science and Engineering of Recycling for Environmental Protection
   Edited by G.R. Woolley, J.J.J.M. Goumans and P.J. Wainwright

Other relevant titles from Elsevier/Pergamon

Municipal Solid Waste Incinerator Residues 1997 By: A.J. Chandler, T.T. Eighmy, J. Hartlén, O. Hjelmar, D.S. Kosson, S.E. Sawell, H.A. van der Sloot, J. Vehlow

Harmonization of Leaching/Extraction Tests 1997 Edited by H.A. van der Sloot, L. Heasman, Ph. Quevauviller

Waste Materials in Construction: Putting Theory into Practice 1997 Edited by J.J.J.M. Goumans, G.J. Senden, H.A. van der Sloot

Waste Management Series, Volume 2

GEOLOGICAL DISPOSAL OF RADIOACTIVE WASTES AND NATURAL ANALOGUES

Lessons from Nature and Archaeology

William Miller
QuantiSci Ltd UK

Russell Alexander
Nagra, Switzerland

Neil Chapman
QuantiSci Ltd UK and Nagra, Switzerland

Ian McKinley
Nagra, Switzerland

John Smellie
Conterra AB, Sweden

2000
PERGAMON
An imprint of Elsevier Science

Amsterdam – Lausanne – New York – Oxford – Shannon – Singapore – Tokyo

ELSEVIER SCIENCE Ltd
The Boulevard, Langford Lane
Kidlington, Oxford OX5 1GB, UK

First edition 2000

Library of Congress Cataloging in Publication Data
A catalog record from the Library of Congress has been applied for.

ISBN: 0 08 043852 0 (Hardbound)
0 08 043853 9 (Paperback)

♾ The paper used in this publication meets the requirements of ANSI/NISO Z39.48-1992 (Permanence of Paper).
Printed in The Netherlands.

"In examining things present, we have data from which to reason with regard to what has been; and, from what has actually been, we have data for concluding with regard to that which is to happen hereafter. Therefore, upon the supposition that the operations of nature are equable and steady, we find, in natural appearances, means for concluding a certain portion of time to have necessarily elapsed, in the production of those events of which we see the effects."

James Hutton (1788) Transactions of the Royal Society of Edinburgh.

Acknowledgements

This book is an expanded and revised version of our earlier volume entitled *"Natural Analogue Studies in the Geological Disposal of Radioactive Wastes"* which was published by Elsevier in 1994. In the six years which have elapsed since that book was written, the application of natural analogues to radioactive waste disposal has matured and it is encouraging to see more discussion in safety assessment documents acknowledging that observations from nature and archaeology can be used to increase our confidence in the geological disposal of radioactive wastes. We hope this trend will continue. Also, in the last six years, a number of large-scale analogue studies have been completed and a few new ones undertaken. A relatively recent new use of analogues is their application in other environmental and waste issues. We hope that this book adequately reflects these developments.

In writing this book, the authors have drawn on their experience of many natural analogue studies over the last two decades, both large and small. These studies have involved researchers and data users from a wide range of scientific disciplines, and the authors wish to express their thanks to all their colleagues in many countries for fruitful, demanding and protracted discussion sessions in meeting rooms and watering holes around the world. Without these interactions the opinions expressed here would never have been developed or tested.

The writing of this book was funded by the following six organisations from around the world who, collectively, have been responsible for the promotion of many natural analogue studies: Enresa, Spain; Environment Agency, UK; JNC, Japan; Nagra, Switzerland; Posiva, Finland; SKB, Sweden.

We thank all of these organisations for their support and tolerance in bringing this book to press.

Contents

Case histories

Chapter 1: The issue of radioactive waste disposal

Throughout history, people have disposed of most types of solid wastes by either burning them or burying them. All too often this has resulted in a hasty and convenient shallow grave for all kinds of environmentally unfriendly materials. As a consequence, the perception of waste burial is often of dirty, old-fashioned landfill sites, strewn with garbage, and of contaminated lakes and rivers. So, when people hear of plans to bury radioactive wastes underground, they are understandably concerned for the safety of local inhabitants and for the environment.

However, the reality of radioactive waste disposal is so far removed from the images of common waste tips as to bear no comparison. Indeed, during the last two decades, the concept of underground radioactive waste disposal using purpose-built, engineered facilities has been developed to a degree far in advance of any other disposal practice adopted in any other industry, reflecting the high standards of safety which the nuclear industry is expected and legally required to achieve.

Most industrialised countries have some radioactive wastes which require disposal, although the volumes and types of these wastes varies considerably from country to country. However, in almost all cases, these countries have opted to dispose of their wastes underground, in radioactive waste repositories, rather than to store them indefinitely on the surface. Several repositories around the world have already been built, or are under construction, to contain wastes with low levels of radioactivity. Repositories for wastes with the highest levels of radioactivity are still in the design stages, although sites for the first of these are currently being identified, and construction is likely to begin within the decade. An example design is shown in Figure 1.1.

The location of a repository, its design and the depth of burial depend very much on the type of waste it is intended to contain, in terms of its level of radioactivity and physical and chemical properties. The waste materials, and the engineered barriers which initially contain them within the repository, are expected eventually to degrade and it is anticipated that some residual waste radionuclides might return to the surface in very low concentrations at some time in the distant future as part of the natural processes of groundwater movement and environmental change. One of the challenges facing the nuclear industry is to demonstrate confidently that a repository will contain wastes for so long that any releases that might take place in the future will pose no significant health or environmental risk. In this regard, the very fact that these wastes are radioactive can be considered helpful because natural radioactive decay will reduce their radioactivity down to levels similar to those of the surrounding rocks or of natural ore deposits. Ultimately, radioactive decay will convert the wastes to stable, non-radioactive materials. The time required for this conversion is defined by the half-lives of the particular radionuclides in the waste and, for most radioactive wastes, will range from a few hundred years to many thousands of years. In contrast, many other types of wastes which also present an immediate environmental risk in most industrialised countries, such as toxic chemicals and poisonous metals (e.g. arsenic and cadmium) will remain hazardous for ever. It is an irony then, that the same safety standards applied to radioactive wastes are generally not applied to these other dangerous wastes types.

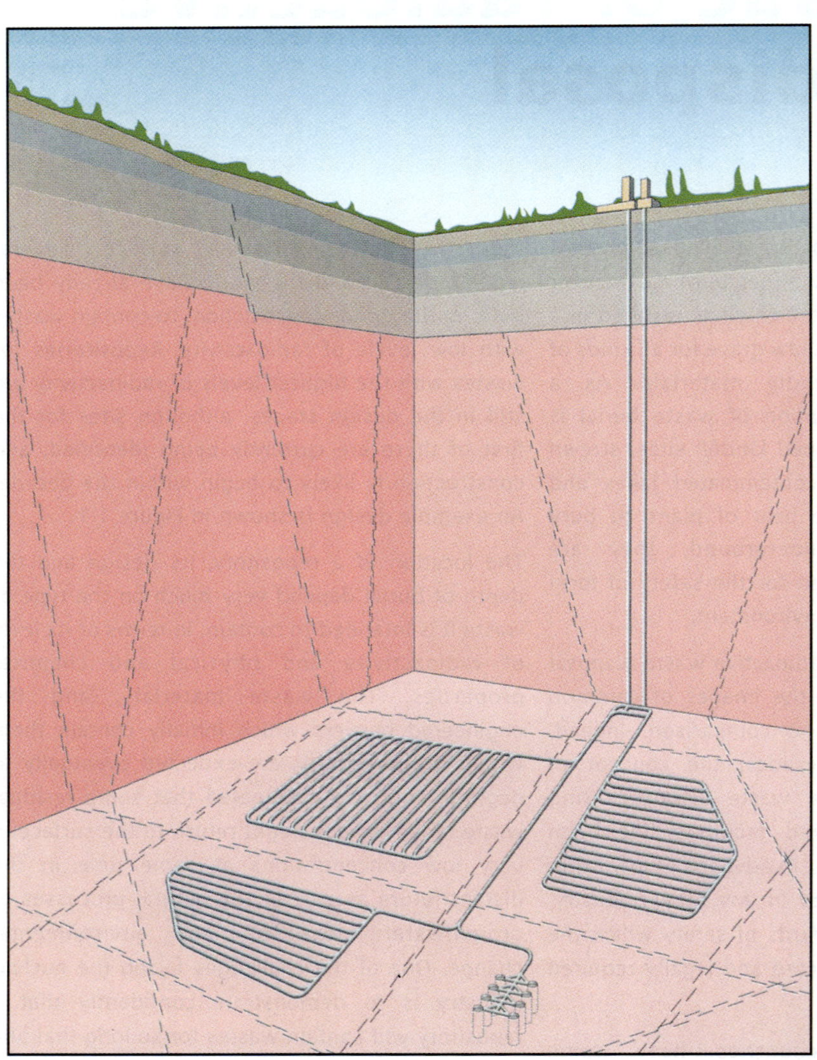

Figure 1.1: An example of a deep geological repository design to contain highly radioactive wastes. In this case, the design is from Switzerland. Radioactive waste would be packaged in massive metal canisters and placed in the horizontal tunnels, which would typically be between 500 and 1000 m below ground, as discussed in Chapter 2. Such purpose-built repositories are far in advance of the facilities used for the disposal of any other waste type and reflect the high standards of safety which the nuclear industry is expected and legally required to achieve. Illustration courtesy of Nagra.

As part of understanding how a radioactive waste repository will behave over many thousands of years, it is useful to look at processes which operate in nature and to draw appropriate parallels between geological systems and the repository. For example, there are many radioactive materials which occur naturally and which can be found in common rocks, sediments and waters around the world. Uranium, which is the principle radioactive component in the fuel used in nuclear power stations, is a naturally-occurring element which can be found in trace amounts in the soils in most peoples' gardens. By careful examination of the distribution of uranium in the natural environment and by learning about the processes which control the movement of uranium in rocks and groundwaters, we can begin to understand how it and other radionuclides might behave in a repository. Learning from nature in this way is using nature as an analogue for the processes which will determine the behaviour of a radioactive waste repository and which, therefore, will control its safety. *Natural analogue studies* are what this book is all about.

One aim of this book is to review and summarise the natural analogue studies which have been undertaken to date around the world, and to highlight how natural analogue information can be used to increase the scientific understanding of repository evolution and, thus, to make better technical predictions for repository safety. However, a second aim of this book is to consider how natural analogues can best be used for presenting the concept of geological disposal to various interested audiences in a simple, coherent and scientifically legitimate manner.

Anxiety about radioactive waste disposal is entirely understandable if people lack facts concerning the real risks presented by radioactive wastes and a repository. Reducing anxiety is a difficult problem and can only be achieved if repository developers are perceived to be making a serious effort to provide their scientific peers, decision-makers and the public with demonstrations of repository safety which can be readily understood. Complex mathematical demonstrations of safety are required by law before an operating licence for a repository will be granted but these are not easily comprehended by non-technical readers. In addition to their mathematical basis, the definitions of words such as *safe* and *risk* used in these formal safety demonstrations are hard to evaluate and to put into context with other normal, every-day uses of these words and measures of risks.

Natural analogues provide a sensible means of conveying the fundamental principles of a repository and its safety to wider audiences. Among the concepts which can be presented using analogues are the slow degradation of materials and radionuclide containment in deep rock. Suitably chosen analogue examples can relay immediate visual impact and understanding (however qualitative) and relate to objects and processes which people are familiar with, even if the quantitative similarity to the repository environment is sometimes limited. By providing some of these illustrations, it is hoped that this book will help advance understanding of the issues associated with radioactive waste disposal and allay some of the fears.

1.1 The nature of radioactive wastes

The majority of the radioactive wastes created around the world are the unwanted by-products of electricity generation using nuclear power, and of military activities. However, there is also a large number of industrial, medical and scientific research activities which use radioactive materials and create radioactive wastes, albeit in relatively small amounts. These last uses of radioactive materials mean that many more countries than just those with nuclear power have a waste problem to address, although the magnitude of the problem is much smaller in countries without a nuclear power programme.

Most radioactive wastes exist in solid form (or are solidified) but a small proportion, by waste and by radioactivity, arise in liquid or gaseous waste streams and may be discharged to the environment. Tightening of environmental controls and regulations over the last few decades has meant that the released proportion has been constantly shrinking as improved methods of waste treatment have been employed, such as better filtering of gaseous effluents.

Solid radioactive wastes may be categorised in many different ways, for example by the manner in which they were created, their physical or chemical form and their potential for reprocessing or recycling. However, the most commonly used categorisation is by level of radioactivity because it is this feature which sets radioactive wastes apart from any other form of waste. The different types of radioactive wastes and their origins are discussed in more detail in Chapter 2.

At the present time, the majority of wastes with high levels of radioactivity are held in surface stores at the nuclear facilities which created them (such as nuclear power stations) and are kept safe and separated from people and the environment by active institutional controls. Containment of the wastes in these stores depends on constant surveillance and maintenance which cannot be ensured indefinitely. Stored wastes could present a hazard if institutional controls were to fail through civil unrest or natural disaster, or if they became the target of deliberate military or terrorist activity. As a consequence, it is generally believed that a more permanent disposal option is required which can ensure these wastes are safely isolated from people (and from malicious tampering) and which does not, necessarily, depend upon long-term institutional control.

The duration of the required isolation period is defined largely by the level of radioactivity of the waste and the half-lives of the component radionuclides. In simple terms, the wastes with the highest activity and longest half-lives are considered to be potentially most dangerous and will therefore need to be disposed of in robust facilities designed to last for many thousands of years or longer. Lower activity wastes with shorter half-lives will require less robust facilities capable of isolating the waste for up to a few hundred years.

The very lowest activity waste may not require any special treatment if its activity is below designated exemption (radiation) levels. These exemption levels vary from country to country but are roughly equivalent to the levels of radiation found in many natural materials such as soils and rocks.

1.2 The concept of geological disposal

Many different options for the disposal of radioactive wastes have been proposed, several have been investigated in detail, and a few have been practiced to some extent in the past. The most commonly discussed alternatives include:

- storage until activity levels decay to below exemption limits;
- disposal into space;
- disposal in the polar icecaps;
- disposal on or beneath the seabed;
- nuclear transmutation;
- shallow land burial; and
- deep geological disposal.

When dealing with radioactive wastes, storage generally refers to a system which requires further management action before institutional control over the waste is given up, whereas disposal does not. In this regard, the first option is often promoted by individuals and environmental groups opposed in principle to the disposal of radioactive wastes by any means. However, consideration of the very long half-lives of some radionuclides means that institutionally-controlled storage is not a practical option for the longer-lived wastes because control cannot be guaranteed for the required very long (thousands of years) isolation period. In this case, there is no alternative to disposal. Institutional control can be predicted with confidence only for wastes with short-lived isotopes (IAEA, 1992).

Neither disposal to space nor to the icecaps has been investigated in any great detail: for a discussion on space disposal see Rice and Priest (1981) and for a discussion on icecap disposal see USDOE (1980). In contrast, disposal on the seabed has been carried out in the past in deep waters under international agreement. Furthermore, comprehensive international research programmes to develop and assess methods for disposal under the seabed (sub-seabed disposal) were undertaken in the 1970s and 1980s (e.g. NEA, 1988; Mobbs et al., 1988). However, seabed disposal by any method is now prohibited by international agreement (Sjoeblom and Linsley, 1994), although the International Atomic Energy Agency (IAEA) has been investigating seabed disposals made outside the international agreement by the former Soviet Union (Linsley and Sjoeblom, 1994). It is interesting to note, however, that several studies have shown that sub-seabed disposal may actually be, in radiological terms, a very safe option (de Marsily et al., 1988).

Nuclear transmutation is a process whereby some types of long-lived wastes are irradiated such that some of the longest-lived radionuclides they contain are converted to shorter-lived nuclides. Current opinion is that transmutation can provide a solution only for reducing the quantities of some particularly long-lived radionuclides and that it is not a feasible proposition for the large volumes of radioactive wastes which now exist. More significantly, transmutation would not avoid the need for the disposal of the shorter-lived but still highly radioactive nuclides it would generate.

Given the current situation, the majority of effort in most countries is focussed on land based disposal and several methods have been proposed, as follows:

- deep (> 100 metres) injection of liquid wastes into porous rock formations;
- ultra-deep (> several kilometres) burial of solid wastes in boreholes drilled from the surface;
- shallow (few metres) burial of solid wastes in surface or near-surface trenches or bunkers covered with engineered barriers; and
- deep (50 to 1000 metres) underground burial of solid wastes in excavations containing an engineered barrier system.

Of these four options, the first, deep injection of liquid radioactive wastes, has been carried out in Russia at several locations since the 1960s (Rybalchenko, 1998). However, these sites are planned to be decommissioned in the next decade and no new injection facilities are planned. Ultra-deep boreholes for solid waste disposal have been designed on paper (e.g. SKB, 1992: Gibb, 1999) but never put into practice because the latter two options, shallow and deep burial in repositories, have generally found favour in most countries. Shallow land burial is already practiced for low activity wastes and deep burial is planned for the higher activity wastes. For a comprehensive description of the background to geological disposal of radioactive wastes, see Chapman and McKinley (1987) and Savage (1995).

To ensure adequate levels of safety, the design of any shallow or deep repository for solid radioactive wastes must pay attention to the nature of the waste and to the geological environment. To be feasible, a repository must meet a number of fundamental design requirements, the most basic being that the repository:

- ensures the waste will not be released to the surface environment in concentrations which would represent an unacceptable hazard;
- effectively isolates the waste from the effects of human activities undertaken at the surface and from the effects of climate change;
- is capable of being built using available technology and achievable at a reasonable cost; and
- is sufficiently simple such that its future behaviour and safety can be assessed quantitatively and reliably.

In addition, certain other requirements might be placed on some repository designs, such as to ensure retrieval of the waste is both technologically and economically feasible, if so desired by future generations.

Given these basic requirements, two approaches to geological disposal are possible (Savage, 1995). The first is based on designs to contain the waste in one place for as long as possible or for as long as necessary to avoid unsafe radionuclide releases to the environment. While this objective appears sensible, it is very difficult to demonstrate convincingly for radioactive wastes containing very long-lived radionuclides. The second approach is based on designs which allow some progressive natural releases to the environment and use well-understood natural processes to dilute and disperse radionuclides so that concentrations are maintained below designated safe levels.

Both the shallow and deep land-based disposal schemes rely on elements of both the containment and controlled release approaches. Containment generally requires that the radioactive waste is solidified into a stable matrix which is isolated from the environment by a set of physical barriers. During containment, radionuclides within the waste undergo radioactive decay so that, for all nuclides, containment for a sufficient number of half-lives will effectively reduce their inventory to extremely low levels. For example, a containment time period equal to 10 half-lives leads to a reduction in inventory by around three orders of magnitude.

The optimal duration for a containment period depends on the type of waste and its radionuclide content. For low activity wastes containing only small amounts of long-lived radionuclides, containment for a period of around 300 years would be sufficient to reduce the activity in the waste to acceptably safe levels. Assuming proper institutional controls for this time period, these wastes could be safely placed in a near-surface facility.

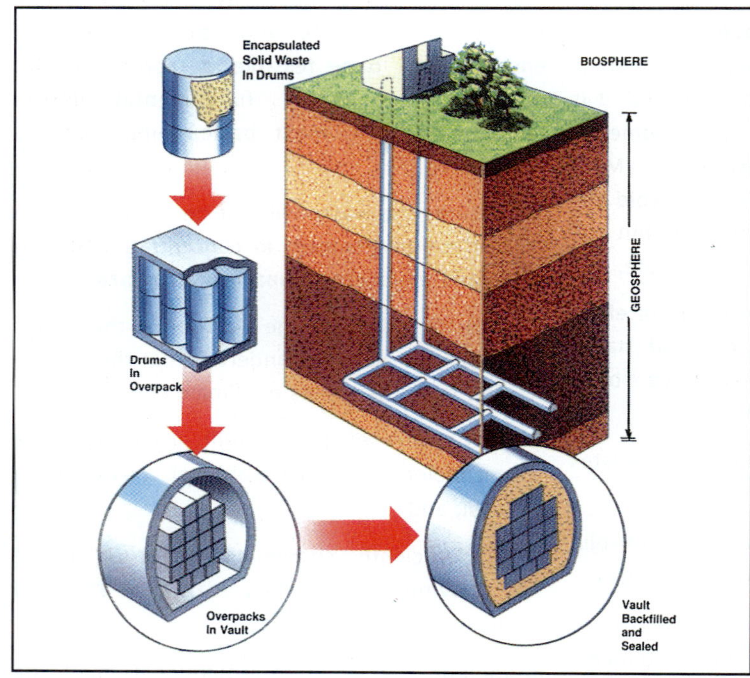

Figure 1.2: Diagrammatic illustration of the multibarrier concept, in this case for a generic repository design for wastes with intermediate levels of radioactivity. The waste is packaged in a drum or canister, then placed in a concrete overpack to create a stable waste package. This is then placed in the repository disposal tunnels and surrounded by an envelope of suitable buffer or backfill material. Physical containment is assured by the solid wasteform, the waste package, the buffer and the rock. Chemical containment is assured by low groundwater flows in a chemically reducing environment. Other multibarrier system designs are discussed in Chapter 2.

Ensuring safe disposal of the higher activity wastes containing very long-lived radionuclides requires more robust repository designs. Therefore, most of the higher activity wastes will be emplaced in deep repositories which utilise the *multibarrier concept*, whereby the wastes are emplaced inside a series of nested engineered structures and natural barriers which act in concert to restrict the rate of release of radionuclides over long periods. The multibarrier concept is shown diagrammatically in Figure 1.2 and is discussed in more detail in Section 2.3.

In a simple multibarrier repository, solidified wastes are packed into containers which may themselves be placed inside a thick overpack. Normally, both containers and overpacks are made of metal or cement and, together, they comprise the waste package. The waste packages are then emplaced at some depth in the repository excavations. The spaces around the waste package are filled with some suitable buffer material to provide long-term structural, hydraulic and chemical stability for the package.

After repository operations have been completed, access tunnels and shafts would be backfilled and sealed to the surface. A variety of materials and mixtures, often making use of crushed rock from the repository excavations, is available for backfilling and sealing operations. The host rock in which the repository is excavated is generally selected to be adequately stable for the construction and operation of the facility and, most importantly, to provide a stable environment where groundwater flows slowly through the repository zone, and other natural geological and geochemical processes are also slow and predictable.

Demonstrating complete containment of longer-lived wastes in a deep repository is considerably more difficult than for the short-lived wastes due to the inevitable process of degradation of the engineered barriers within the repository. For this reason, repositories for higher activity wastes are designed to allow for progressive release and dispersal, after containment failure, into the rock without adversely affecting the surface environment, as shown in Figures 1.3 and 1.4.

1.3 Evaluating repository safety

Before a repository can be built and operated, its safety must be adequately evaluated and shown to comply with various regulatory targets associated with radionuclide releases or movement in the environment. This formal process of evaluating repository safety is known as *performance assessment* and requires detailed mathematical analysis of all aspects of the repository system and its evolution.

In fact, performance assessment calculations will be undertaken long before a proposed repository site is finally selected. Various performance assessments will be undertaken throughout a complete repository design programme, which may last several decades, at different stages, e.g. to help identify suitable host rock formations, to evaluate alternative barrier materials and to assess the consequence of unusual, low probability events, such as accidental human intrusion into the repository. Some performance assessments may only focus on a certain aspect of the repository system, such as the behaviour of different types of multibarrier systems, and are likely to vary in complexity and realism. Since most of these performance assessment calculations will be undertaken at the conceptual repository design stage, they will use a variety of generic geological and geochemical information as input data.

Once a final repository site has been chosen and characterised (see Section 1.5.1), additional performance assessments will be performed in order to obtain a licence to build and to operate the repository. These performance assessments will benefit from the use of as much site specific data as possible and will examine the repository systems as a whole, from degradation of the waste through to future releases to the surface. Considerable effort has been expended over the last couple of decades to develop reliable assessment methodologies for all stages in a repository development programme. In addition, other work has been ongoing to define acceptable, independent criteria against which the safety of a repository can be judged.

Formal, legally-defined safety (or licensing) criteria are usually set nationally for specific waste and repository types, and international organisations such as the IAEA also provide general guidelines. It is common for licensing organisations to define the post-closure safety criteria in terms of radiological dose or risk to humans, where risk is defined as risk of death due to cancer arising from exposure to repository releases. Other forms of target criteria will be set for the pre-closure (operational) phase of repository development. Typical post-closure safety criteria for a specific target individual or group of individuals are:

- a radiological dose of < 1 mSv/yr (often < 0.1 mSv/yr), and
- a risk target of $< 10^{-6}$/yr.

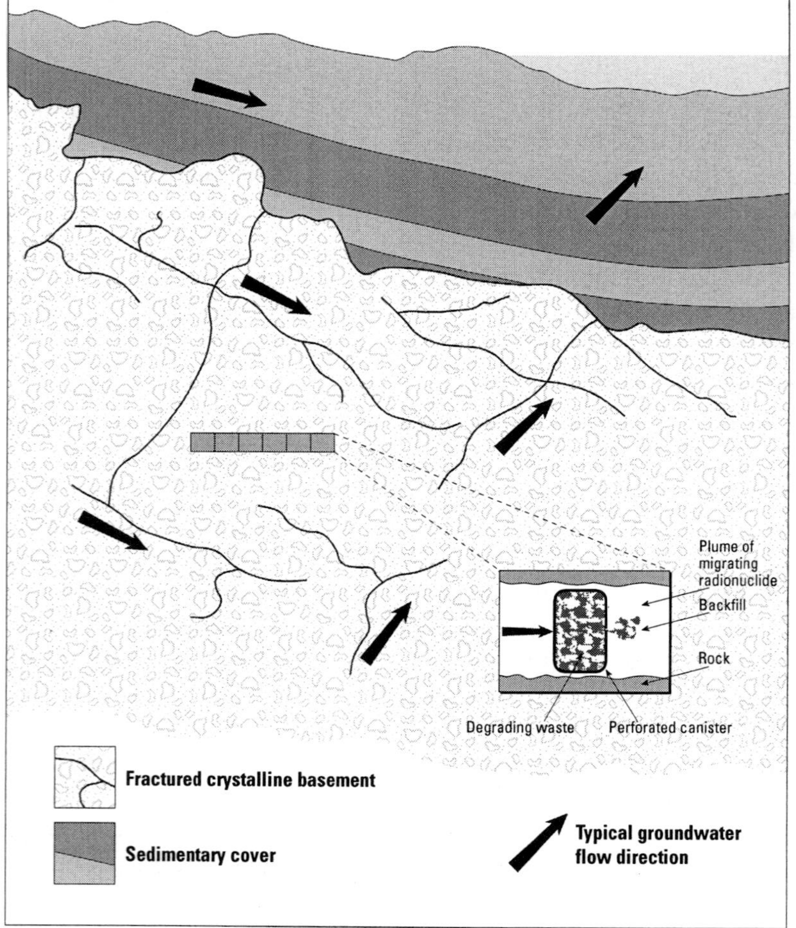

Figure 1.3: Over a long period of time the engineered components of the repository will degrade and, eventually, the waste will be exposed to the groundwater and will begin to degrade. Some radionuclides may dissolve in the very slow moving groundwater but their transport away from the repository will be retarded by interactions with the rock and the corrosion products of the engineered barriers. Dilution of radionuclides in the groundwater will also occur.

Most countries have adopted safety criteria similar to these. Depending on the type of the waste, quantitative assessments of dose and risk may be required to be made for some specified future post-closure time period, this might be 10^4, 10^5 or 10^6 years, although some current regulations give no time limit. Many authorities now believe that, after this time, more qualitative assessments of safety may be most appropriate to demonstrate the continued safe operation of the repository. In this case, qualitative predictions may be required for up to one million years or more. Again, these time periods are indicative and different systems may be adopted in different countries.

In order for a performance assessment to be able to calculate the expected dose and risk for a repository, it is necessary to represent understanding of the system behaviour by a series of conceptual models which can be converted into simple mathematical models for computational purposes. This simplification of the real system into something that can be dealt with in a computer model is shown illustratively in Figure 1.5. At the most basic, a performance assessment will need to include conceptual and mathematical models for:

- the degradation of the engineered barriers which contain the waste;

- dissolution of the solid wasteform;
- the solubility of the radionuclides in the groundwater;
- transport of groundwaters and dissolved radionuclides through the engineered barriers and the surrounding rock formations (natural barriers) to the surface;
- various processes which may retard radionuclide transport in the engineered barriers and the rock; and
- radionuclide release to the surface, incorporation of radionuclides into the food chain and eventual uptake by humans.

In reality, a separate model will be constructed to represent each of these processes. These will then be linked together, with the output from one model fed, as input, to the next, as shown diagrammatically in Figure 1.6. As well as the models in the formal performance assessment, other research codes may be used during performance assessment to perform scoping calculations to ascertain the significance of some processes. These research models are a fundamental part of safety assessment. If they indicate that a particular process is significant for safety, then that process may be explicitly included within the performance assessment calculations.

Undertaking a performance assessment is a complex operation which requires much more than simply writing a number of computer codes. From beginning to end, the development of a model to examine a particular feature of a repository (say, dissolution of the waste) is undertaken in a number of stages, as indicated in Figure 1.7, which roughly can be described as:

1) construction of a conceptual model which describes the system and includes all of the important processes and their couplings;

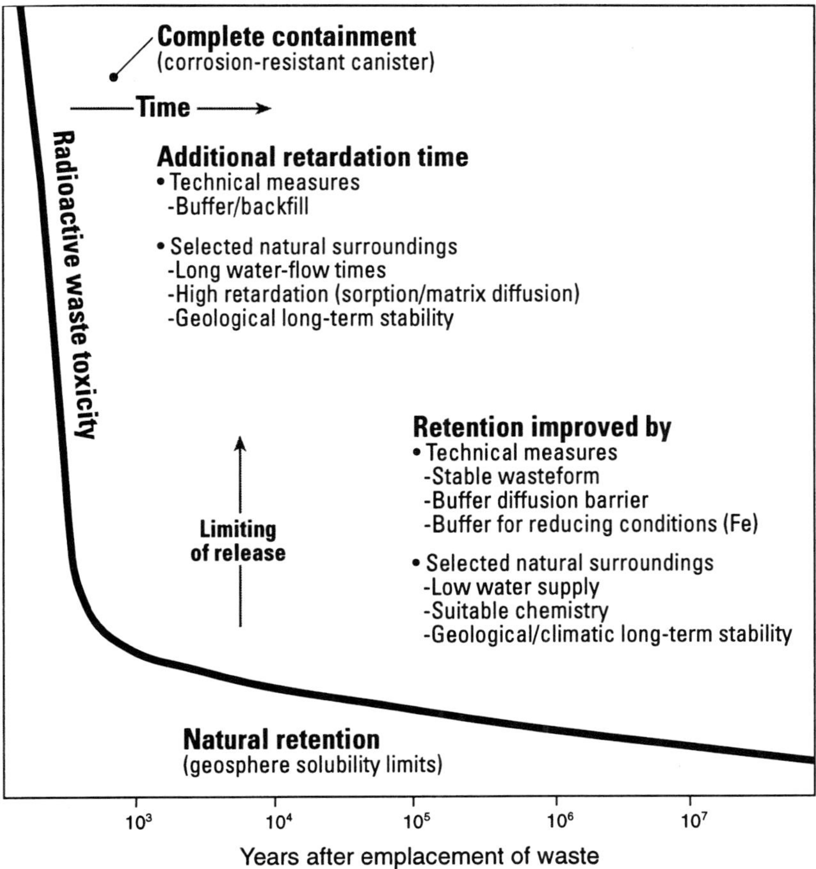

Figure 1.4: The waste canister will fail after some 10^3 to 10^4 years in the repository after which time continued retention of radionuclides in the repository will be due to the physical behaviour of the other barriers and chemical containment processes. While the canister is intact, the total radioactivity in the waste drops substantially due to decay of the shorter-lived radionuclides.

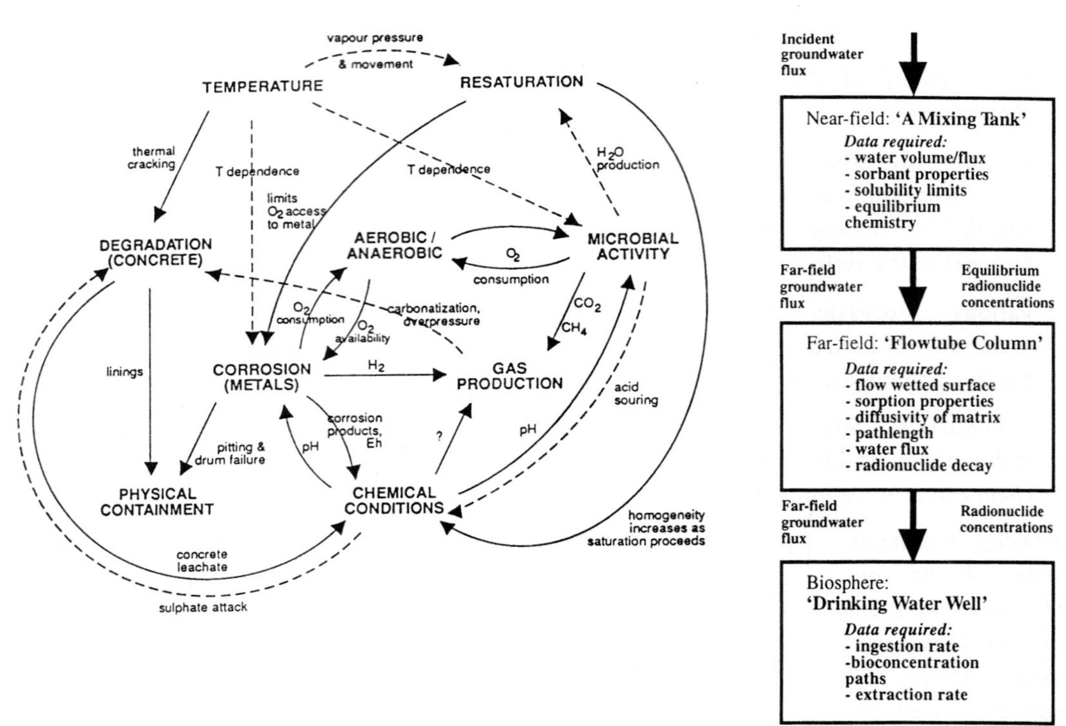

Figure 1.5: Comparison between the complex interactions of physical, chemical and microbial processes in one part of a real repository system (in this case for some low activity wastes) which may need to be considered in a performance assessment (right) and a typical, simplified model used in performance assessment modelling (left). From Chapman (1994).

2) translation of the conceptual model into a mathematical model and coding in the form of a computer program;

3) acquisition of quantitative input data for all the variable and constant parameter values included in the code;

4) verification of the numerical 'correctness' of the computer code; and

5) validation of the code's 'applicability' to the repository system to assess its predictive capabilities.

Stages 2 and 4 are purely desk-based activities. However, stages 1, 3 and 5 all require considerable research and development support to allow them to be completed successfully and much of this research and development will involve laboratory and field-based scientific investigations. For example, stage 1 (building the conceptual model) requires that the process to be modelled is investigated in detail to understand fully how it operates and this demands 'hands-on' practical science, either in a laboratory or in the field.

The output of a performance assessment will generally be expressed graphically, typically showing calculated radiological dose (or risk) and its variation with time after repository closure. An example of a performance assessment dose/time curve is given in Figure 1.8. In this case, the performance assessment was undertaken as part of an exercise to evaluate different types of potential host rocks in Europe (Cadelli et al., 1988). It can be seen that, for all cases, the calculated doses are below the safety criteria limit (1 mSv/yr, in this case) for all calculated times. Furthermore, the graph also shows that the calculated repository releases are at least 3 orders of magnitude less than the radiation exposure from natural background sources. This puts the repository radiological hazard into some natural perspective.

Further examples of performance assessment dose/time curves are shown in Figure 1.9. This graph was constructed by Neall et al. (1995) and includes the calculated individual doses from a number of performance assessments for proposed or generic repositories for the most radioactive wastes. The dose curves for each performance assessment and repository are different to each other. This reflects the various waste types, repository designs and modelling assumptions made during the performance assessment calculations. However, regardless of these differences, all of the performance assessment results indicate that calculated radiological doses due to repository releases will be at least two orders of magnitude below the safety limit (1 mSv/yr, in this case) for all calculated times. A further observation is that no releases to the surface are likely to occur for at least 1000 years, and generally for much longer, because of the long expected lifetimes of the waste containers.

1.3.1 Key uncertainties in safety calculations

In general, all performance assessment models and calculations will be hampered by an incomplete understanding of the processes which actually will occur in the real repository system and by the limited quantity and quality of input data. Together, these reflect modelling uncertainties in the performance assessment calculations and they can be categorised rather broadly as uncertainties regarding:

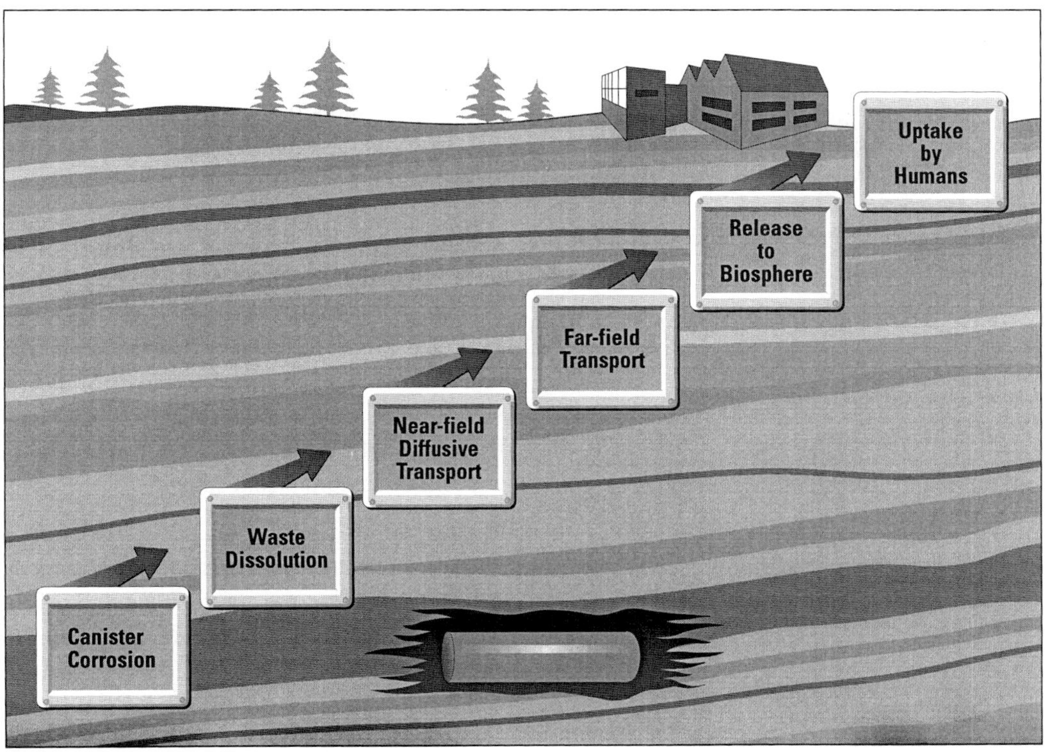

Figure 1.6. Diagrammatic scheme of a performance assessment model chain, starting with models of degradation of the engineered barriers and the waste, leading to models of releases to the surface environment and uptake by humans. The end-point of most performance assessment calculations are predictions of radiological dose and risk to humans living in the vicinity of the repository.

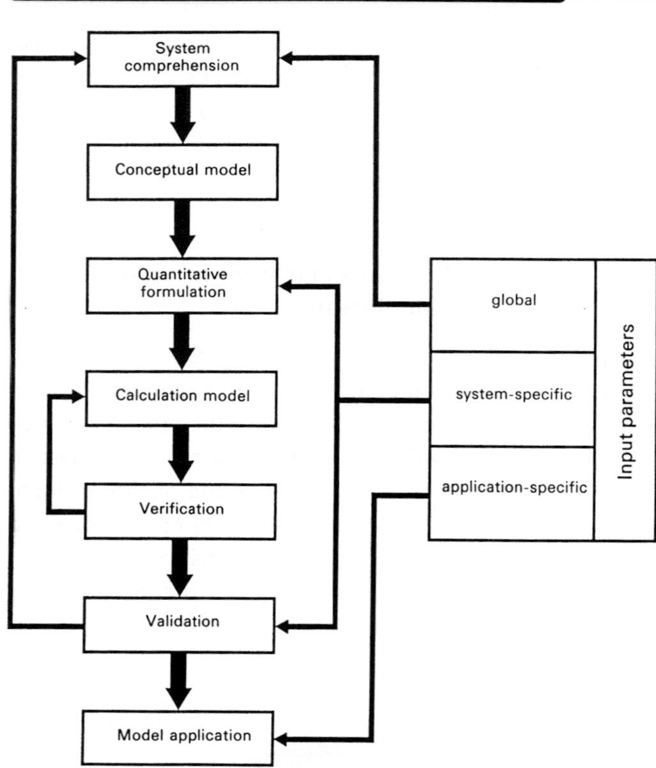

Figure 1.7: Stages in modelling for performance assessment. The system is iterative during development of repository designs and modelling approaches. Illustration courtesy of Nagra.

- which processes occur and how they are initiated;
- which processes are dominant over long periods;
- the critical interactions between many inter-linked processes;
- the rates of processes and whether these are constant;
- the most appropriate models to use to describe the processes; and
- the values of some of the parameters which are used in these models.

Such uncertainties are a fact of life for all natural sciences, and repository performance assessment is no exception. The fact that uncertainties exist is not an insurmountable problem and various approaches have been developed to deal with them. The most commonly adopted approach is to make both repository designs and performance assessment models robust and conservative which, in effect, means some element of over-engineering combined with pessimistic assumptions about repository performance. For a detailed discussion of uncertainty in performance assessment and various methods for dealing with it, see Savage (1995).

Conservatism in performance assessment can make allowances for a lack of conceptual understanding and for inadequate data. As an example of data (parameter) uncertainty, inadequate or imprecise data for the rate of metal canister corrosion will be expressed by a possible range of corrosion rates. The conservative modelling approach is to assume the fastest corrosion rate from the possible range. When faced with conceptual uncertainty, the conservative modelling approach is to exclude a process from the performance assessment calculations if it is not well characterised, if it acts to increase repository safety. So, for example, processes which slow down or retard radionuclide transport (such as matrix diffusion, see Section 5.3) can be conservatively omitted from the performance assessment model without any risk of compromising safety.

If, after adopting pessimistic parameter values and omitting uncertain processes which act to increase repository safety, the performance assessment output for a calculational case is within safety limits, then it can be assumed that the repository would be safe for more likely and realistic conditions. Encouraging as this is, the conservative modelling approach has to be used sensibly and it should not be taken to extremes

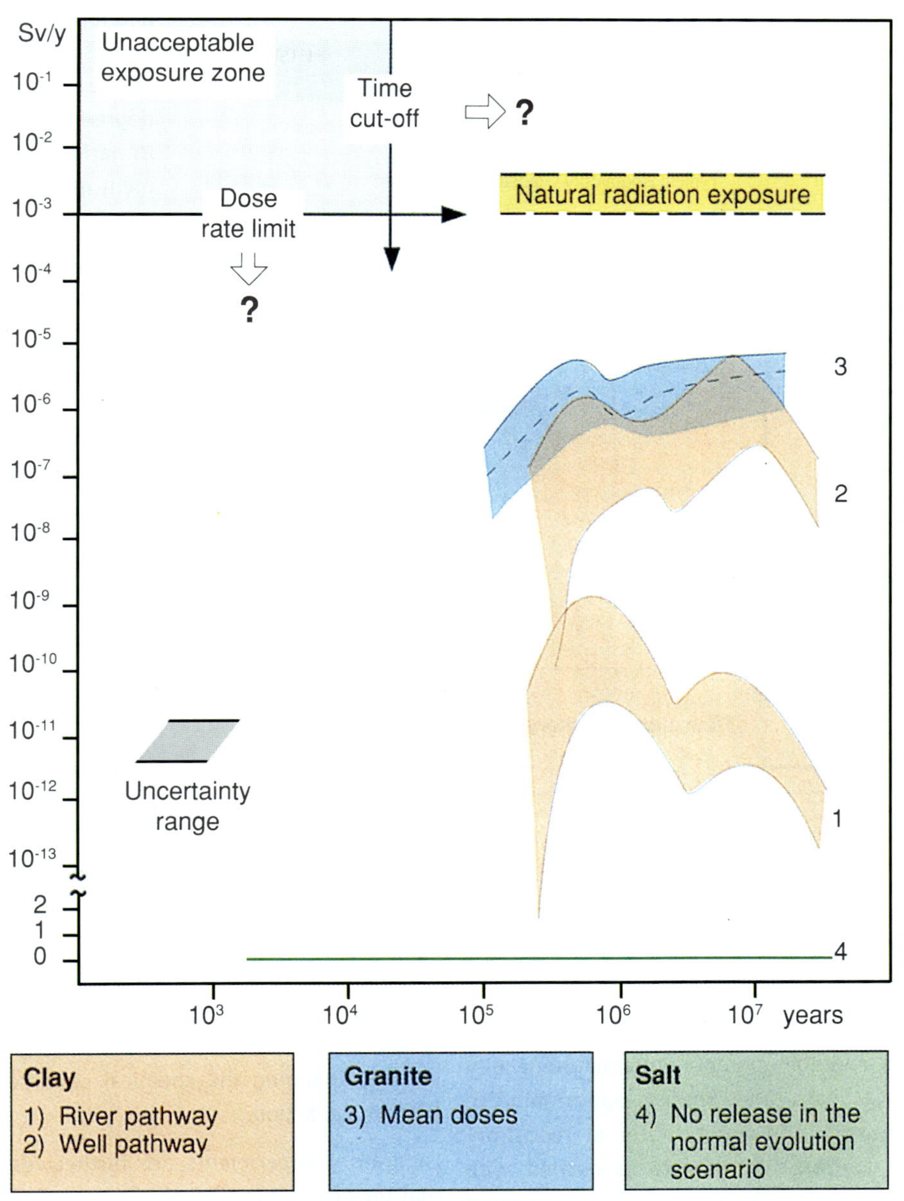

Figure 1.8. The performance assessment output expressed as a dose/time curve from an assessment of some different host rock options (clay, granite and salt) in Europe. This graph shows that all the host rocks examined would be safe compared to the safety criteria (1 mSv/yr) and natural background radiation. After Cadelli et al. (1988).

such that the performance assessment modelling assumptions become unrealistic.

The degree of conceptual and data uncertainty varies considerably for different parts of the repository system and its evolution. It is generally acknowledged that the greatest uncertainty is associated with the natural barriers and with radionuclide migration through the host rock, in terms of groundwater flow, elemental solubility and speciation, and processes which act to retard radionuclide transport. This uncertainty stems

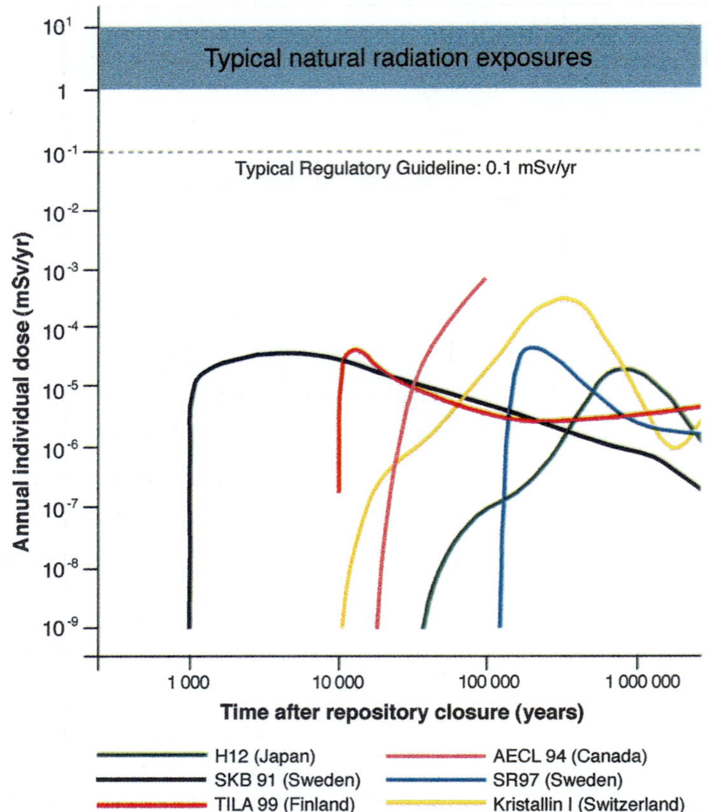

Figure 1.9. A range of performance assessment outputs expressed as individual dose/time curves for generic or proposed repositories for the highest activity wastes. In all cases, calculated doses fall several orders of magnitude below typical regulatory limits (0.1 mSv/yr) and below typical natural background radiation levels. The different outputs partly reflect different modelling assumptions and the degrees of conservatism in the assessments. Updated from Neall et al. (1995).

partly from a combination of natural heterogeneity in the rock mass and limited opportunity to sample the rock and groundwater at repository depth. In contrast, uncertainty associated with degradation of the waste and the engineered barriers is believed to be less. This uncertainty is reduced because these degradation processes are more amenable to laboratory study than processes which control radionuclide migration through deep rocks.

One of the objectives of the research undertaken to support performance assessment model development will be to reduce both conceptual and parameter uncertainties and, thus, to define an appropriate degree of conservatism in performance assessment models.

Much of this research will be undertaken in the laboratory or in field-based experiments. Although such studies can provide much useful information, they always suffer from the problem of short-term observation. Even the longest experiment is likely to be run for only a few years and the majority would be considerably shorter, many lasting only a few days. Due to the time issue, many laboratory experiments are artificially accelerated, to enable some reaction to be measured in the time available, by either raising the temperature, changing the chemical conditions or by mechanical action.

Laboratory experiments are further restricted in that they cannot reproduce the spatial scale of the repository system and the host rock. Combined, these limitations in time and space mean that laboratory studies are unable to replicate the true complexity of a repository, with all the various process-couplings and feedbacks, and changes in the external driving forces which control their effects. Thus, laboratory experiments bear limited resemblance to real repository evolution and, consequently, there is the risk that observations of

Table 1.1: The advantages and disadvantages of laboratory and analogue studies. In reality, both types of investigation are required to support performance assessment model development and should be viewed as complementary to each other.

Field and laboratory experiments	Analogue studies
▪ Short-term experiments lasting weeks to a few months, at most.	▪ Operate over very long time periods, typically thousands or millions of years
▪ Well defined boundary conditions for the experiment which are set by the researcher.	▪ The boundary conditions of the analogue system are often poorly constrained.
▪ Can use the technological materials which will actually be used in the repository.	▪ The materials in analogue systems only approximate the nature of repository materials.
▪ Very simple experimental systems which facilitate modelling of the results.	▪ Natural systems are complex and involve coupled processes, so are more realistic but hard to model.
▪ Reactions are often accelerated by raising the temperature or by using aggressive reactants.	▪ Processes take place at natural reaction rates and under natural conditions in the analogue systems.
▪ Thermodynamic assumptions allow little consideration of reaction kinetics.	▪ Reactions in analogue systems demonstrate inherent kinetic constraints.

processes and measured rates from the laboratory will not be representative of the actual long-term behaviour of the repository system.

1.3.2 Requirements for supporting natural data

As a direct consequence of the limitations of laboratory and field-based experiments to provide information which represents both the complexity of the repository system and the time-periods of relevance to performance assessment, these studies need to be complemented with natural analogue studies.

It is important that natural analogue studies should been seen as complementary to laboratory studies, rather than as a replacement. This is because both types of investigation have their advantages and disadvantages. Once of the largest drawbacks of natural analogue studies is that the original boundary conditions to the natural system are generally poorly characterised and, thus, interpretation of the results can be restricted. The primary advantages and disadvantages of both laboratory and analogue investigations are shown in Table 1.1.

Combining natural analogue studies with field and laboratory investigations provides a powerful tool for investigating the natural processes which will occur in the repository environment because the disadvantages of one method are balanced by the advantages of the other. However, it is extremely important that clearly appropriate analogue environments are chosen for study, in terms of physical and chemical conditions, materials present and time periods. As a consequence, considerable care was taken during the early development of natural analogue studies to explain the requirements for suitable analogue systems, and this is discussed in the following section.

The demands for physico-chemical similarity between the repository and natural systems can be relaxed somewhat if the natural analogue is to be used for qualitative, non-technical illustrations of some aspect of a repository concept. However, caution must be maintained because inappropriate comparisons between nature and a repository can be damaging to the scientific credibility of a disposal programme.

1.4 Natural analogue studies

The accepted definition of the term *natural analogue* has evolved over the last decade to reflect the changing acceptance and application of analogue studies within the radioactive waste disposal industry. The term was first coined in the late 1970s and fairly rapidly after that was in common usage. An early definition of the term (Côme and Chapman, 1986a) was *"...an occurrence of materials or processes which resemble those expected in a proposed geological waste repository."*

This definition was subsequently refined to reflect the growing usage of natural analogues for the development and testing of models, both conceptual and mathematical. For example, McKinley (1989) described analogue studies by saying that *"...the essence of a natural analogue is the aspect of testing of models, whether conceptual or mathematical, and not a particular attribute of the system itself."* This view was reflected by the IAEA (1989) who said that *"...natural analogues are defined more by the methodology used to study and assess them than by any intrinsic physico-chemical properties they may possess."*

This definition has been the commonly accepted one over the last several years while the large effort in natural analogue studies has been in their use to support performance assessment directly, through the development and testing of models. However, quite recently, there has been a growing interest in the use of natural analogues in non-technical demonstrations of safety which do not rely on the use of performance assessment codes and models. It should be recognised that all types of safety demonstrations have valuable contributions to make in building up a picture of repository performance for those who have to make decisions on waste management practices. Any definition of natural analogues should, thus, reflect their dual usage and not be limited only to those studies planned as direct support to performance assessment (Miller, 2000).

Natural analogue studies have been performed, to date, on a wide range of phenomena, including ore deposits, natural fission reactors, marine sediments and man-made copper and iron artefacts to name but a few. The latter, archaeological and industrial artefacts, although clearly not natural systems, are studied in the same way as natural analogues and are generally classed with them. They provide a record which spans nearly 5 000 years and give an indication of how the robustness and longevity of 'technological' materials compare with 'natural' materials, as discussed in Section 3.3.

Natural analogue studies should not be confused with repository site characterisation studies, even though a large-scale natural analogue study, such as an investigation of radionuclide movement around an orebody, can provide a valuable 'dry run' of a site characterisation methodology, using similar investigation techniques at a similar spatial scale. Site characterisation provides a different type of input to repository performance assessment in defining the nature of the natural environment into which the repository is fitted, as discussed in Section 1.5.1. Natural analogue studies provide information on the subsequent behaviour and effect of the repository, and may use data from the site itself or, more likely, from other sites. These may be in diverse environments, but are linked back to the repository concept by some well-defined similarity of process or material. It follows that not every geochemical study is a natural analogue; the vital aspect is whether information from the study can be used sensibly to increase confidence in geological disposal, either through supporting performance assessment or through building understanding and awareness.

In a nutshell, a natural analogue study is one which provides information on the behaviour of a repository which is derived from one site but applicable to another: natural analogue derived information must, therefore, be portable (Miller, 1996a).

Reasoning by analogy

Perhaps spurred on by the discussion on the definition of the term natural analogue, a number of authors have been concerned with the philosophical aspects of reasoning, or proof, by analogy. Notable amongst these are Ewing and Jercinovic (1987) and Petit (1992a), from whose papers much of the following discussion is derived. Analogy is implicitly recognised as a scientific method. Root-Bernstein (1988) identified several important mental qualities in famous scientists of the past two centuries; foremost amongst these qualities is the 'facility to recognise patterns'. This is a process very similar in style to reasoning by analogy. Root-Berstein (1988) concludes that *"...any mental activity which contributes directly to scientific discoveries should be recognised as a scientific method."*

Since the time of the Greek philosophers, people have argued over whether the use of analogy is another form of inductive thought, or is a distinct type of thought process in its own right. A personal view on this issue was given by Mill (1874) when he argued that any distinction between induction and analogy is artificial because both require collection and interpretation of observations. The two processes differ only when it comes to the important ability to demonstrate the validity of the interpretations reached. Induction is based on a scientific understanding of causality between events and phenomena, while analogy temporarily accepts a probable theory without absolute proof. It should be noted that this is an approach very commonly used in the physical sciences, where a theory is upheld until a negative instance is discovered. An obvious example is the displacement of Newtonian mechanics by Einstein's theory of relativity. This concept of searching for a negative instance is crucial to the idea of natural analogues.

Standard performance assessments are based on process models and model chains which explain how modellers believe the waste and the repository will behave in the geosphere from an understanding of chemistry, physics and geology. It is not possible to prove conclusively that such models are a correct description of the natural environment and the evolution of a repository. However, by applying the models to processes operating in the geosphere (i.e. by studying natural analogues), their validity can be tested by either finding a negative instance (to disprove the model) or finding the fit is good in a particular instance (increasing our confidence in the model). It is common, when testing models which describe the natural world, to find that they are neither absolutely right nor wrong, but rather that they are only approximately correct. It is easy to understand how a model can be completely correct; the process occurs just as predicted. It is also easy to understand how a model can be completely wrong; the process it predicts just does not occur (the negative instance). However, it is less easy to understand how a model can be approximately correct until it is considered that, in the geological environment, a process does not merely occur but rather it occurs at a particular (sometimes changing) rate, sometimes intermittently, and is almost certainly coupled to other processes.

From this apparent complexity comes one of the strengths of the natural analogue approach: it can be used to determine the range of applicability of a model. For example, assume that a model predicts that radionuclide transport will occur at a given rate in a particular environment. A natural analogue study in a similar environment may reveal that radionuclide transport does occur, but at a somewhat faster rate than predicted. In this case the model has not been disproved (this is not a negative instance) but rather some indication of the model's application has been determined, i.e.

for this environment the model is 'non-conservative', that is to say the model would under-estimate the radiological risk due to this process. The model can then be further tested by natural analogue studies in other similar environments. In some cases the model will be found to be conservative, in others non-conservative, and in yet others the model may be found to be correct. In this way the range of applicability of the model will be built up.

Perception of natural analogue studies

The foundations for quantitative natural analogue studies, with application to performance assessment, were laid in the seminal work by Chapman et al. (1984) which reviewed what was, at that time, an incoherent and generally unfocussed range of investigations. Up until that time, most so-called natural analogue studies were merely extensions of straightforward geological and geochemical investigations of chemical or isotopic anomalies, where the principal reasons for study were never allied to the needs of repository design and model testing or validation. Some natural analogue investigations were initiated with the specific aim of learning more about the potential evolution of a repository but, most often, the philosophy was to try to interpret a single site as a global analogue for the entire repository system.

This global analogue approach was to prove unsuccessful because adequate natural analogues for a complete disposal system do not exist. Even in the natural system often quoted to be most similar to a repository environment, the fossil nuclear fission reactors at Oklo (see Box 4), the boundary conditions are both significantly different and poorly constrained in comparison to a repository system. As a direct consequence of the global analogue concept, the information yielded by these early natural analogue studies was generally more qualitative than quantitative and, with the boundary conditions in many studies poorly defined, performance assessment modellers tended to reject the data in favour of results from laboratory investigations.

In an attempt to define natural analogue studies more clearly, and to orient them towards individual processes for which good analogues can be found, Chapman et al. (1984) listed a set of guidelines for selecting natural analogues for investigation. The need for well-characterised, process-oriented natural analogue studies is reaffirmed in this report and thus these guidelines are repeated here:

1) The process involved should be clear-cut. Other processes which may have been involved in the geochemical system should be identifiable and amenable to quantitative assessment as well, so that their effects can be subtracted.
2) The chemical analogy should be good. It is not always possible to study the behaviour of a mineral system, chemical element or isotope identical to that whose behaviour requires assessing. The limitations of this should be fully understood.
3) The magnitude of the various physico-chemical parameters involved (pressure, temperature, pH, Eh, concentration etc.) should be determinable, preferably by independent means and should not differ greatly from those envisaged in a repository.
4) The boundaries of the system should be identifiable (whether it is open or closed, and consequently how much material has been involved in the process being studied).
5) The timescale of the process must be measurable, since this factor is of the greatest significance for a natural analogue.

The quantitative philosophy outlined by Chapman et al. (1984) proved to be the impetus for a greater interest in natural analogues which resulted in an international symposium (Smellie, 1984) and the formation of the Natural Analogue Working Group (NAWG) which was sponsored by the Commission of the European Communities (CEC), now the European Commission (EC). Since its inception, the NAWG has met eight times (Côme and Chapman, 1986a, 1986b, 1989, 1991; von Maravic and Smellie, 1994, 1996, 1997; von Maravic and Alexander, 2000) and held one major international symposium (Côme and Chapman, 1987).

NAWG reports are a particularly valuable record of the evolution and application of natural analogue studies. Most of the large and significant analogues have been represented at NAWG meetings. Furthermore, the NAWG reports attempt to develop the perception of natural analogues by giving agreed introductory statements on their development.

In general, since Chapman et al. (1984), more natural analogue studies have focussed on the requirements of process models and performance assessment. This development has been facilitated by the following guidelines issued by the NAWG (Côme and Chapman, 1986a) for ways of applying natural analogues to modelling and assessment processes:

1) As natural experiments which replicate a process, or a group of processes, which are being considered in a model. This is probably the most quantitative application of analogues, which allows confident constraints to be placed on, for example, extrapolations of laboratory experiments to larger time or space scales.

2) For determining the bounds of specific parameter values. This application would be most useful at the stage where a modeller needs limiting values on a parameter, but can obtain these from any or many geological systems. The origins of the data are not particularly important, and need not be linked to the process being modelled. Diverse sources may be used and a statistical approach adopted. An example of this is thermodynamic or kinetic data, which could be obtained from any system.

3) As simple 'signposts' indicating which phenomena can occur in the system being modelled by reference to a parallel natural system. This is a purely qualitative application which gives 'yes-no' answers, or indicates the 'direction' of long-term process. It would be the first means of application used when carrying out scoping exercises.

4) In an empirical sense to integrate the results of many processes at one site, over long time periods. Not all of the processes involved may be evident, nor may the manner in which they have been linked. Only the end result is important, and in this sense this application is the most directly useful to a safety assessment (as distinguished from the individual models which comprise it). An example might be to determine whether there is any surface radiological manifestation of a deeply buried uranium ore body.

Many of the more recent analogue studies have adhered to the two sets of guidelines given above, and a variety of specialists have been involved, including geochemists, geophysicists, archaeologists etc. Nonetheless, there are still studies purporting to be natural analogues that have no clear radioactive waste application and this is reflected by the fact that building a consensus view on the usefulness of natural analogues has been a slow process.

Partly due to badly planned, poorly focussed studies over the the last two decades, critics of natural analogue studies have expressed doubts about their value on the grounds that:

1) Natural analogue-derived information is inherently only qualitative. It is believed that only quantitative data, which can be used as direct input to performance assessment are relevant and that these data cannot be obtained from natural analogue studies.

2) Natural analogues are not true, hard science. It is suggested that information derived from natural analogue studies is inherently obscure

and ambiguous and its interpretation equivocal in contrast to the data produced from laboratory experiments and from inductive and deductive thought processes.

However, with the development of recent, better planned analogue studies which have followed the guidelines listed above, it is now possible to counter these assertions

Quantitative and well-constrained data have been obtained from several natural analogue studies, such as Poços de Caldas, Maqarin, Oman and Cigar Lake, which are all discussed in detail later. Furthermore, some quantitative data obtained from natural analogue studies could only have been obtained from such investigations: laboratory experiments being unable satisfactorily to simulate the conditions likely to be encountered in a repository. However, even if much of the information from analogue studies is qualitative, this does not limit its usefulness because quantitative data for input to a process model or performance assessment are only required (and of importance) once it has been decided that the process is significant for the evolution and safety of the repository.

The decision as to whether a process is significant or not is a qualitative judgement and must be based on an understanding of geology and geological processes. In other words, qualitative understanding of a process must lead to quantitative examination. This is implied in points 3 and 4 of the NAWG guidelines (Côme and Chapman, 1986a) listed earlier. This heuristic function of natural analogues should be given greater prominence in the planning and definition of future studies and is being increasingly recognised by performance assessment modellers.

Whilst it is true that quantitative data can be obtained from natural analogue studies, it is also true that, even in the best-conducted studies, the boundary conditions are never certain. The validity of such data could, therefore, be questioned. A parameter may be measured with a high level of accuracy and precision but its significance may still be questioned because the processes which influence that parameter are not sufficiently well-known. However, it is this very complexity of the system, which results in the uncertain boundary conditions, that the modellers are trying to simulate. Only by studying natural systems, and attempting to strip away the effects of superimposing multiple processes, is it possible to determine the rates and effects of single processes, to reveal the coupling of processes which are inevitable under repository conditions and over geological timescales, and to verify that all the processes which an assessment should consider have been included.

Two examples serve to indicate the importance of the role which natural analogue studies play in ensuring that all relevant processes have been identified for incorporation into mathematical models:

- Some processes are susceptible to changes in their rate-limiting parameters over long periods of time. An example of this is the rate of elemental diffusion within glasses (both natural and radioactive wasteform), which is thought to change with time as a result of the formation of an alteration layer, the formation of secondary minerals, or both (Magonthier et al., 1992). This phenomenon was highlighted by natural analogue studies of volcanic glasses.
- Some processes which are immeasurably slow in the simple, confined conditions of the laboratory may be catalysed under complicated geological conditions, possibly as the result of bacterial action or the presence of colloids. An example would be the enhanced corrosion rate of metal in the presence of microbes. Both these phenomena may have implications for repository safety, and both might have gone unidentified if investigations were performed solely in the laboratory.

As a consequence of the many natural analogue studies performed to date, there has been a gradual reassessment regarding which processes are perceived to be important for repository safety. At the time when the first natural analogue studies were performed, much work was focussed on determining the stability and longevity of potential wasteforms and packaging materials. As a consequence many of the first process-oriented natural analogue studies were examinations of volcanic glasses (as analogues for borosilicate glass wasteforms) and copper and iron ore deposits and archaeological artefacts (as analogues for waste packages).

The development of repository concepts, and of the understanding of the long-term evolution of repository systems, has shown that a range of additional effects, such as colloidal transport, gas migration and biosphere processes, could be very significant in terms of overall repository performance. These processes were not considered from the viewpoint of available natural analogues by Chapman et al. (1984). Over the last decade, natural analogues have tended to focus on far-field radionuclide transport and retardation processes in addition to the more materials-oriented near-field analogue studies of earlier years.

One unfortunate consequence of the expanding interest in natural analogues is the continued publication of studies with no clear objective in terms of end-use of the information. In some cases the term 'natural analogue' is simply mis-understood and used to label purely academic studies which could possibly be of relevance to waste disposal, but in which the application is not discussed. More seriously, some workers have misunderstood the models used in performance assessment and misapplied them to natural systems. This book serves to highlight the better and most relevant natural analogue examples.

1.5 Other field-based studies of natural systems

Other field-based studies of natural systems share with natural analogues the general objective of improving our confidence in geological disposal. These studies may be undertaken at either a generic level or for a specific site or repository design. They include:

- site characterisation,
- palaeohydrogeology,
- natural safety indicators, and
- biosphere studies.

When referred to collectively, these are sometime called *natural system studies* and, although they are not generally considered under the banner of natural analogue studies, they are briefly discussed here to show how the complete picture fits together.

1.5.1 Site characterisation

Once a potential repository site has been identified, it has to be investigated in detail to determine if it meets the necessary requirements for a host rock environment in terms of geological stability, groundwater flow, geochemistry etc. This detailed investigation of the rock is called site characterisation or site investigation. It generally proceeds in stages, with the first stage being a study of the surface features in the broad region around the site and progressive stages focus in on the site scale and the deeper rock at the planned repository depth using remote geophysical techniques, borehole drilling etc, as shown in Figure 1.10. Finally, if all preliminary indications are positive, an underground research laboratory (URL) or rock characterisation facility (RCF) may be excavated comprising one or more shafts and trial excavations which allows the rock to be directly investigated at repository depth. Details of site characterisation programmes and techniques are given in Savage (1995).

Figure 1.10. Site characterisation involves several techniques to learn about the subsurface rock and groundwater systems. In this picture boreholes are being drilled in the vicinity of the proposed US repository at Yucca Mountain, Nevada (see Section 2.3.1) to obtain samples of the rock from depth.

The nature of the information which will be obtained from the site characterisation is diverse and is likely to include, amongst other items:

- fracture, fault and joint orientations, spacings and apertures;
- rock stress measurements;
- bulk rock and fracture mineralogy and compositions;
- groundwater flow rates, pressure (head) gradients and pathways; and
- groundwater major, trace element and isotopic compositions.

Of course, by its very nature, this information is site specific. Some of these data would be used directly in performance assessment models to obtain a detailed understanding of the behaviour of the proposed repository in the potential host rock. Up until this point, all previous performance assessments for the repository would have had to use varying amounts of generic data because the site specific data would be unavailable.

Given that these characterisation data are obtained from the actual proposed repository site, they are not usually regarded as analogue information. Site characterisation provides a different type of input to repository performance assessment in defining the nature of the natural environment into which the repository will be fitted. Natural analogue studies usually provide information on the subsequent behaviour and effect of the repository on the site. Consequently, site characterisation data is not a replacement for analogue information because, normally, it would not be possible to investigate processes such as natural radionuclide transport during site characterisation because the repository would be sited away from features such as orebodies which are often the focus of transport analogue studies. However, certain site characterisation data may replace or supplement analogue data; for example, site specific matrix diffusion depths (see Section 5.3) may replace generic matrix diffusion depths obtained from analogue studies.

1.5.2 Palaeohydrogeology

One of the objectives of performance assessment is to demonstrate that the long-term future evolution and stability of the groundwater system can be predicted with confidence. One method of making this demonstration is to show that the past evolution of the system is understood. This requires making palaeohydrogeological reconstructions of the system.

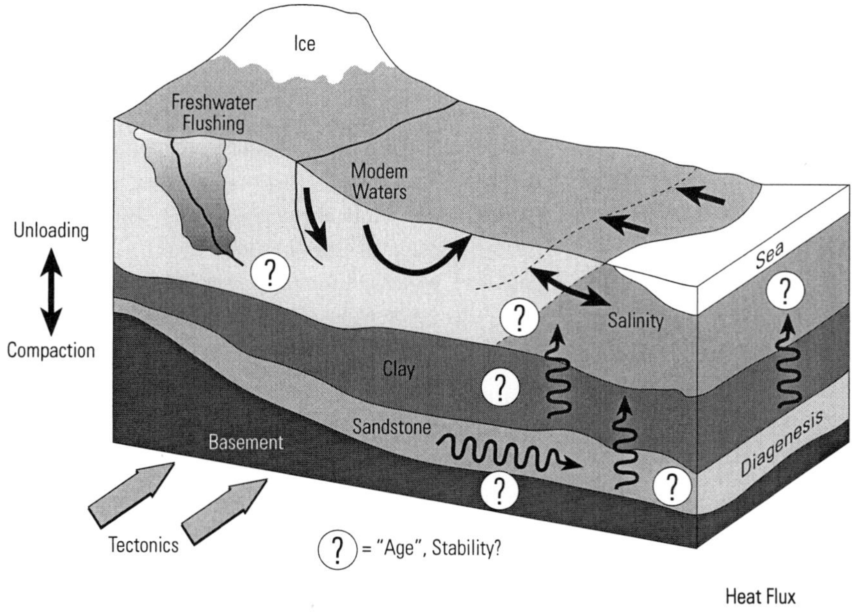

A further advantage of successful palaeohydrogeology is that it provides a potential method for testing or validating the groundwater flow models used in performance assessment (Chapman and McEwen, 1992).

Figure 1.11: Schematic illustration of some of the rock-water interaction processes and concepts to be accounted for or derived from a palaeohydrogeological study. The question marks indicate zones of the groundwater system where it would be important to establish the 'age' or degree of mixing of groundwaters. From Chapman and McEwen (1992).

Palaeohydrogeology applied to sedimentary sequences is fairly well developed, where it is used in studies of diagenesis and hydrocarbon reservoir modelling. However, it is considerably less well developed when applied to hard, fractured rock environments. Palaeohydrogeology is essentially a combination of observations on hydrochemical and isotopic differences in various groundwater zones or bodies, mineralogical data on the rock formations, and the hydraulic properties of the same formations, which are then compiled to allow interpretation of the evolution of the rock-water system over long time periods in the past, as indicated in Figure 1.11.

There are many difficulties in the interpretation of so much detailed and coupled information, and there are some well-known problems and complications which must be addressed. These relate particularly to the issue of resolving the different rock/water interaction events which will have occurred at a site over time, whose signatures are superimposed on each other in the geochemical and mineralogical data. Nonetheless, if sufficient data are available of adequate quality, then it is sometimes possible to derive information on:

- the number, location and type of different groundwater bodies which have been present in the rock;

- the stability, residence times and degree of mixing between the different water bodies;
- pathways for groundwater movement and how the pathways may have changed over time; and
- the external driving forces on the system, such as climate change, and secondary mechanisms which may have influenced the rock/water system.

Used most simply, palaeohydrogeology might provide indications of the suitability of a proposed site by indicating the presence or absence of fast flow paths from depth to the surface. However, the technique has more powerful applications in model testing and it should be able to be used to test or validate groundwater flow models (or coupled codes) in the same way that geochemical codes are tested 'blind' against well-characterised geochemical systems, as discussed in Section 5.1.

In terms of the physical and chemical processes investigated and the application to performance assessment model testing, there are many similarities between palaeohydrogeology and natural analogue studies. Nonetheless, palaeohydrogeology has generally been viewed as a subject in its own right, separate from analogues to the extent that they were rarely even discussed together in the same scientific meetings.

This separation is somewhat artificial and now considerable benefit is being gained from transfer of methodologies between natural analogues and palaeohydrogeology, and serious attempts are being made to use palaeohydrogeology as a mechanism for repository scenario development and for testing groundwater flow codes, e.g. at the Palmottu site in Finland (Blomqvist et al., 2000).

1.5.3 Natural safety indicators

As discussed earlier, the normal safety criteria against which repository performance is evaluated are radiological dose and risk. However, these are imperfect criteria, not least because the concepts underlying them are not readily understood by non-technical audiences but also because calculations of dose and risk become progressively less convincing as they extend into the far future particularly as they rely, in part, on assumptions for human behaviour. Consequently, assurances of safety cast only in terms of dose and risk are less credible the further into the future we look. As a result, there is a growing awareness that additional safety indicators are required which might be more readily understood by a broader range of audiences and which could also be used to place the hazard posed by the repository into a natural context.

Two indicators which have been suggested (e.g. IAEA, 1994) are natural elemental fluxes and abundances (concentrations). The idea of using some aspect of the natural system as a guide against which repository releases are evaluated has the advantage that future human behaviour can be removed from the assessment.

Natural fluxes are particularly useful as safety indicators because, over periods of 10^4 to 10^5 years, or longer, natural geological processes such as sediment diagenesis, deep and shallow rock-water interactions, erosion, weathering, sedimentation and mineralisation will be moving elements around within the same system which is being modelled in performance assessment, independently of the presence of the repository. This is shown illustratively in Figure 1.12.

Clearly, repository releases and natural elemental fluxes have a considerable degree of commonality and meaningful comparisons should be possible. Comparisons can be made in terms of fluxes measured as elemental mass (kg/yr) or in terms of total radioactivity (Bq/yr). When considering radioactivity, it can generally be assumed that if the flux to the surface from the repository is small compared with the natural flux from the rock, then its radiological significance should not be of great or priority concern (Miller et al., 1996). A number of authorities have begun to consider this approach and some specific proposals were

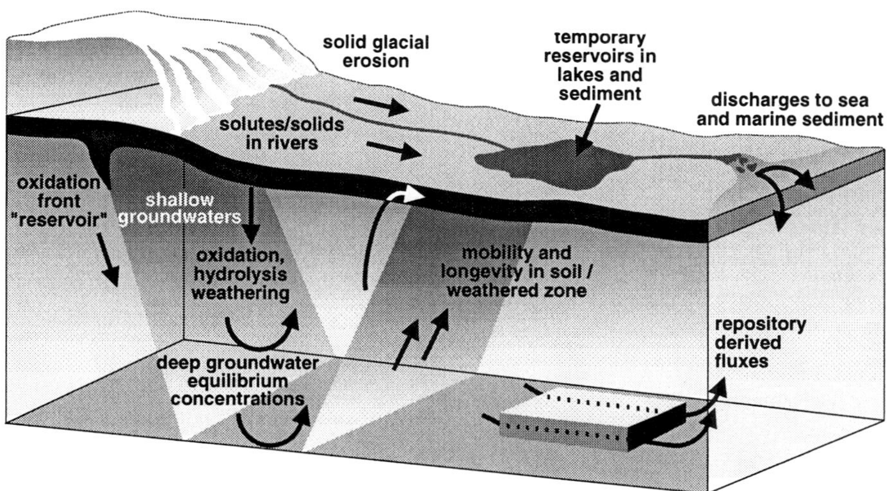

Figure 1.12: Natural geochemical fluxes will occur in the same system, over the same timescale and be driven by the same processes causing radionuclide releases from a repository. Therefore, looking at the natural geochemical fluxes which occur today may provide useful contextual information for evaluating performance assessment results.

included in the 'Nordic Flagbook' discussion document published by the radiation protection and nuclear safety authorities in Denmark, Finland, Iceland, Norway and Sweden (Nordic Radiation Protection and Nuclear Safety Authorities, 1993).

When considering elemental mass fluxes, there is a problem because many of the radionuclides present in radioactive wastes do not occur naturally and, as a consequence, direct comparisons cannot always be made. The most relevant comparisons are made for the natural uranium decay series radionuclides which are those which normally will contribute most to releases from a radioactive waste repository at times after about 10^5 years. However, other elemental fluxes are useful to consider because some elements in the repository present a chemotoxic hazard instead of a radiotoxic hazard, e.g. heavy metals. It is therefore interesting to consider the chemotoxic as well as the radiological hazards due to the repository derived flux and to compare these with risks from the natural flux. These additional comparisons can provide further perspective and act as a check that all the various hazards arising are being given an appropriate level of attention.

There is a growing interest and awareness of natural safety indicators and it is to be expected that many new studies in this field will be undertaken in the coming years. During this time, there is likely to be a transfer of methodologies from natural analogues to natural safety indicators, particularly with regards to the application of natural information to formal performance assessment and safety assessments in general.

1.5.4 Biosphere studies

The end-point of a performance assessment is generally the calculation of radiological dose and risk to individuals or groups of people living in the vicinity of the repository. These calculations require that the movement of radionuclides in the biosphere (the surface and near-surface zones containing living organisms) is understood and can be modelled.

Natural analogue studies are used to provide information on the transport of radionuclides through the repository and the rock mass. Biosphere studies essentially do the same thing but, instead of looking at radionuclide movement through rock, they look at radionuclide movement through surface sediments and waters, and ultimately the food-chain.

Biosphere studies employ laboratory experiments to obtain information on radionuclide uptake in plants and animals to support the biosphere codes in performance assessment calculations. However, there has recently been a move to use field-based studies at sites with elevated radionuclide concentrations to learn about the long-term behaviour of radionuclides in the biosphere. For example, studies are made of the distribution of radionuclides and stable elements in sediments, waters, plants and animals in the vicinity of old mine spoil tips (e.g. BIOMOVS, 1996a,b) and around the natural thorium deposit at Morro do Ferro, Brazil (Eisenbud et al., 1982, 1984).

This approach to learning from natural biosphere systems is essentially the same as that adopted in natural analogue studies of the geosphere. However, the division between natural analogues and biosphere studies is quite distinct in terms of the research and modelling groups involved. This largely is due to a division in scientific disciplines between the biological sciences and radiological assessment teams on the biosphere studies side, and the geologists and geochemists on the natural analogue side. Greater interaction between these two investigative approaches would clearly be valuable. This is especially important when considering the geosphere-biosphere interface, which is the critical zone at which radionuclides leave the geosphere and become accessible to humans. This is a complex zone, and proper understanding of it requires both geological, geochemical and biological input.

1.6 Toxic waste disposal

The treatment and disposal of toxic wastes is now beginning to be dealt with in a manner approximating the high level of care required for radioactive wastes. This is in response to increasingly tighter environmental legislation dealing with the management of these wastes.

In comparison with radioactive wastes, toxic wastes are considerably more heterogeneous in nature. A simple definition of a toxic waste would be any substance which presents a chemical hazard to humans and the environment, as either a poison or a carcinogen. Formal definitions and classifications of toxic wastes are made on a national basis and vary considerably from country to country, as do the legal controls on toxic waste disposal. Typical forms of toxic wastes arise in both liquid and solid form (gaseous wastes are not considered here) and include (Petts and Eduljee, 1994):

- inorganic acids;
- organic acids and related components;
- alkalis;
- heavy metals and compounds;
- fuels, oils and greases;
- polymeric materials and precursors;
- insecticides and other biocides;
- mining and quarrying wastes and spoils;
- medical and surgical wastes; and
- sludges including sewage.

Present-day management options for these materials strongly encourage waste reduction, reuse and recycling, rather than disposal, reflecting the current views on sustainable development. However, certain of these wastes are not suitable for reuse in any form and a final disposal option must be considered. Six options for disposal have been practiced in the past:

- incineration,

- specialist destruction,
- surface storage,
- export,
- discharge and dumping to sea,
- landfilling, and
- geological disposal.

Considerable amounts of toxic materials are disposed by incineration. This is an option for toxic wastes which is not a possibility for radioactive wastes because the hazard presented by toxic wastes can, in some cases, be eliminated by chemical reaction, including combustion. In contrast, radioactivity cannot, under any circumstances, be eliminated or reduced by chemical reaction. Although, incineration of some combustible wastes with low levels of radioactivity is undertaken for volume reduction purposes.

Less hazardous toxic wastes can be sent to landfill and even liquid wastes can be sent to landfill in small amounts. For the more hazardous toxic wastes which cannot be treated by incineration or other methods, the remaining options are surface storage and geological disposal. Deep-disposal in mines or purpose-built excavations is currently not a widely used technique because of cost considerations. However, geological disposal of solid toxic wastes is already practiced or planned in some countries, including Germany, Sweden, the Netherlands and the UK. In addition, several countries have injected liquid toxic wastes in deep boreholes as a final disposal technique.

The most extensive experience of deep geological disposal of solid toxic wastes is in Germany which has several operational and planned disposal facilities in salt and potash mines, such as those at Herfa-Neurode, Heilbronn and Zeilitz. The toxic waste types considered appropriate for geological disposal in Germany include incinerator ashes and slags, mineral sludges, non-ferrous metal wastes and halogenated mineral wastes all of which may contain high concentrations of heavy metals.

In most countries, the legislation governing toxic waste disposal does not require safety assessments which are equivalent in complexity and rigour to the performance assessments undertaken for radioactive wastes. However, the environmental impact assessments which are undertaken generally do require the potential for future migration away from the disposal facility to be investigated in a quantitative manner. It follows, therefore, that the natural analogue methodology developed for radioactive waste performance assessments can be equally well applied to environmental impact assessments for toxic wastes emplaced in shallow or deep disposal facilities. This use of natural analogues has recently been recognised and has been promoted by the NAWG (see Section 1.4) in its last few meetings (von Maravic and Smellie, 1997; von Maravic and Alexander, 2000).

This requirement for an analogue approach to toxic waste disposal may increase in the future if consideration is given to the co-disposal of toxic and radioactive wastes at the same site. At present, only the Netherlands is actively looking at co-disposal but there is no technical reason why the method should not be considered further. A key issue would be the need to evaluate the potential for chemical interaction between the two waste types and their engineered barriers.

Certain differences have to be considered when looking for analogues for toxic wastes because many of the waste types are entirely man-made and have no natural counterparts. This is particularly true for the refined organic compounds. Furthermore, when considering organic materials, it needs to be understood that their transport behaviour in the rocks, sediments and groundwaters may be controlled by processes which are of minimal importance for the inorganic species in radioactive wastes. For example, some complex organic compounds will degrade in the groundwaters and may form other hazardous species which have entirely different transport characteristics to the parent compound.

To date, the application of the analogue methodology to toxic wastes has been limited to possible transport of heavy metals in inorganic wastes such as mining wastes including waste rock, stock piles and tailings dams (Bowell et al., 1997) and to the durability of the different immobilisation matrices planned for toxic wastes (Côme et al., 1997). Although still very much in its infancy, the development of toxic waste analogue studies is very much to be encouraged. It is expected that this use of analogues will become much more in evidence over the next decade as environmental legislation regarding toxic waste disposal demands ever more rigorous safety assessments of future disposal facilities and begins to examine the hazard presented by existing toxic waste disposal facilities and sites of chemotoxic contamination.

Many of the materials in radioactive wastes also pose a chemical hazard as well as a radiological one. Therefore, there may potentially be a spin-off from using analogues for toxic wastes which improves the overall assessment of the total hazard (radiological and chemical) presented by any future releases to the surface environment from a radioactive waste repository.

Chapter 2: Radioactive waste types and repository designs

Radioactive wastes are characterised by their mode of formation, physico-chemical nature and level of radioactivity. Different repository designs are generally required to contain the individual waste types, although some repository concepts exist to house wastes with different levels of radioactivity. The generation of the different waste types and the repository designs developed to accommodate them are discussed in this chapter.

2.1 The nuclear fuel cycle and radioactive wastes

Radioactive wastes are the unwanted by-products of civilian nuclear power (electricity) generation programmes, military nuclear weapons construction and decommissioning strategies and, to a lesser extent, industrial, medical and scientific research activities. The majority of radioactive wastes from around the world are produced by nuclear power generation although, in countries with an extensive nuclear weapons programme such as the United States, the national inventory of weapons waste can be larger than the inventory from power generation.

The nuclear fuel cycle, shown in Figure 2.1, is the term used to encompass all activities associated with the production of nuclear power. Wastes are generated at each stage and these wastes are variously radioactive. At the front end of the fuel cycle (the part before fuel reaches the nuclear reactor), uranium mining and milling operations generate large masses of natural material, such as displaced rock and soils etc., and discarded waste, mainly in the form of mill tailings which are the residual materials from ore processing. On average, 86 000 tonnes of uranium-depleted tailings are produced every year for each reactor (US National Research Council, 1990). These wastes are not normally considered for disposal in a repository and are treated separately. However, they emit levels of radiation above average background and care is required to ensure safe management of this processed material.

Wastes are also produced during the uranium conversion and enrichment processes, and during fuel fabrication, and some of these wastes may be scheduled for disposal in a repository, depending on their levels of radioactivity.

The majority of radioactive wastes which will be sent for disposal in repositories are generated by the operation of nuclear power reactors. These wastes fall into three principal categories:

- used irradiated fuel and the wastes derived from its reprocessing;
- operational wastes created by the day-to-day running of nuclear reactors; and
- reactor decommissioning wastes.

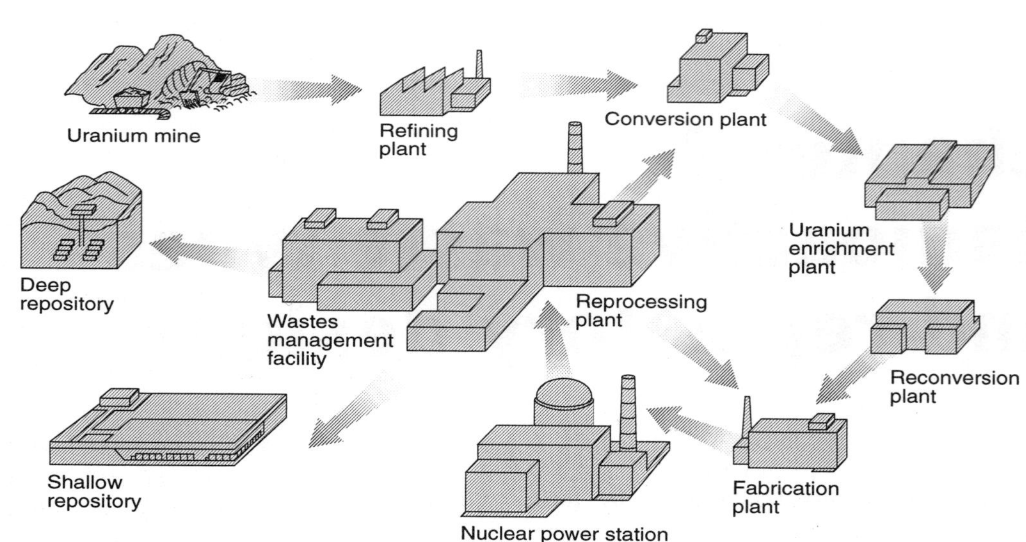

Figure 2.1: The nuclear fuel cycle from mining of uranium orebodies to the final disposal of reactor wastes.

These wastes are described below, together with some additional wastes from other sources that potentially may also be disposed of in repositories.

2.1.1 Used fuel and reprocessing wastes

Most nuclear reactors burn a uranium dioxide fuel (UO_2) enriched in ^{235}U that, after a few years in the reactor core, becomes poisoned with fission products and transuranic elements produced as a result of the nuclear reactions. These elements reduce the efficiency of the fuel and, consequently, it has to be replaced periodically. This used, irradiated fuel, known as *spent fuel*, once removed from the reactor, is stored for at least 6 months to allow the substantial radioactive decay heat to diminish to a level which allows safe handling and transportation.

After cooling, the spent fuel assemblies may be stored prior to direct disposal in a repository or they may be reprocessed to extract the unburnt uranium for incorporation in new fuel elements. Some countries, such as the United Kingdom and France, reprocess the majority of their spent fuel while others, such as Sweden, do not and are currently selecting sites for its direct disposal in deep geological repositories (see Box 1).

Spent fuel reprocessing is a complex process that involves dissolving fuel elements using aggressive inorganic solvents and then employing a sequence of organic and inorganic reactants to separate uranium and plutonium from the other unwanted radioactive products. The whole operation is performed on an industrial scale and produces large volumes of wastes with a broad spectrum of physical and chemical compositions, and varying levels of radioactivity. The most radioactive wastes are the liquids remaining after dissolution and separation of the uranium and plutonium. These liquids must be solidified, usually in a glass matrix, prior to final disposal.

Other less radioactive wastes generated from the reprocessing operation comprise a range of materials such as Zircaloy cladding from the fuel assemblies, ion exchange resins used in the chemical separation process, swabs used for cleaning laboratory benches and disposable clothing. These lower activity wastes contain a high proportion of solid, organic material and, to facilitate disposal, they are compacted and immobilised in either a cement, bitumen or resin matrix prior to disposal.

2.1.2 Operational wastes

The routine operation of a nuclear power plant generates many different types of waste in both liquid and solid form. The most significant waste, both in terms of volume and activity, are ion exchange resins which are used to recover radionuclides from liquid wastes. These liquid wastes are generated during routine decontamination procedures, either from chemical processes (e.g. primary coolant and fuel storage pond clean-up) or from detergent waste (e.g. laundry and personnel decontamination). Additional radioactive liquid streams are produced during solid waste treatment and volume reduction processes.

Within the nuclear plant, raw solid wastes are classed either as 'wet' (e.g. sludges, ion exchange resins) or 'dry' (e.g. rubber gloves, paper tissue). The ion exchange resins are generally dehydrated and powdered, and the final powder is solidified in either a bitumen or cement matrix. Other non-combustible solid wastes are compacted to reduce their volume and are then variously packaged in steel or cement containers before being solidified in cement or bitumen.

The low activity, combustible solid wastes may be incinerated to reduce their volume, creating an ash which is then immobilised in a cement matrix. Incineration itself generates secondary wastes such as the scrubber chemicals and particulate air filters which are used to clean the incinerator flue gases. These secondary wastes must also be processed and become part of the total inventory of wastes to be disposed.

2.1.3 Decommissioning wastes

At the end of a nuclear facility's operational life, it must be decommissioned. These facilities include nuclear power plants, and fuel fabrication and reprocessing systems. Various decommissioning options are possible but all involve partial or complete dismantlement of the facility and associated buildings. Some of the materials generated by this process will be radioactive. For example, in a nuclear power plant, the reactor vessel and the surrounding reinforced concrete structure (the biological shield) will become radioactive due to neutron irradiation by the reactor core, while other components, such as primary coolant piping, may contain residual radioactive liquids or solids.

Depending on the nature of the dismantled components and their levels of activity, some components will be scheduled for disposal in a deep geological repository after suitable packaging or will be routed for shallow burial. Many of the dismantled components will not, however, be radioactive and can be disposed or recycled as normal industrial wastes.

2.1.4 Other wastes

Radioactive wastes are generated outside the civilian nuclear fuel cycle in small amounts (by medical, industrial and research activities) but these still require careful handling and packaging, and most are scheduled for disposal in a repository. Larger volumes are also created by nuclear weapons programmes in some countries.

In medicine, the radioisotopes used, particularly for clinical and diagnostic purposes, tend to be short-lived with non-penetrating radiation and do not pose a significant radiological hazard. However, sources of high activity are used in radiotherapy leading to an important long-lived waste component. In addition, medical and biological research experiments sometimes make use of ^{3}H (tritium) and ^{14}C and these provide a significant contribution to the activity of medical radioactive wastes. Occasionally, in medical applications, radioisotopes may be associated with pathogenic substances which create special waste handling and treatment concerns (Savage, 1995).

In industry, radioactive sources are used for a range of applications such as radiography, material thickness and density gauging, well logging, moisture detection, and food sterilisation and preservation. These sources are relatively short-lived but may be intensely radioactive, posing a significant radiological hazard and requiring specialist disposal. They are increasingly located in less developed countries with no nuclear facilities and no experience in radioactive waste management. Studies are currently in hand to evaluate the disposal of these very small volumes in deep boreholes.

At nuclear research centres, various radioisotopes are produced in research reactors, particle accelerators and cyclotron facilities. In general, such facilities generate relatively small amounts of waste containing long-lived radioisotopes, especially accelerator and cyclotron facilities which do not possess nuclear fuel. For the latter, the main source of waste is derived from liquids produced during chemical processing or etching of target materials (Savage, 1995).

Lastly, military applications generate large volumes of radioactive materials. In broad detail, the reactors used to produce the nuclear component in weapons and to power some submarines and surface navel vessels are similar to civilian power plants. Thus, many of the radioactive materials produced as a result of military applications are similar to those from commercial nuclear power plants. However, many nuclear submarine reactors use very highly enriched fuel, which presents its own management issues.

Plutonium is being stockpiled due to the dismantling of large numbers of thermonuclear weapons. Plutonium disposal raises particular safeguards issues because of concerns regarding the hazards of theft and incorporation into new nuclear weapons (e.g. Garwin, 1996). Two options for plutonium treatment being considered are 'burning' in civilian nuclear power reactors in the form of a mixed uranium-plutonium oxide (MOX) fuel and direct disposal to a repository. However, not all stockpiles of plutonium and other military radioactive materials are currently classified as 'wastes' and thus are excluded from some inventories of radioactive wastes currently scheduled for disposal in repositories.

2.2 Classification of radioactive wastes

When defining a classification for radioactive wastes, it is important that the classification is linked to the planned method of disposal, including waste conditioning, to ensure adequate isolation from the surface environment. Although the classification of wastes tends to vary from one country to another, there is some commonality in approach. In most countries, the wastes described earlier are generally classified according to their levels of radioactivity and three categories of radioactive wastes are usually referred to:

- *High-level waste (HLW)* which includes spent fuel and the solidified forms of the liquid waste stream generated during reprocessing of spent fuel. These wastes are characterised by high-levels of radioactivity, have a component of very long-lived waste nuclides and are heat generating. The activity of these wastes generally ranges from 10^{16} to 10^{18} Bq/t.
- *Intermediate-level waste (ILW)* which includes a diverse range of materials such as ion exchange resins and metal wastes from normal reactor operations and spent fuel reprocessing. These wastes are generally solidified in either a cement or bitumen matrix. They are characterised by significant levels of radioactivity, have a component of long-lived waste nuclides but are not heat generating. ILW containing a high proportion of actinides is sometimes referred to as

transuranic containing (TRU) waste but is generally treated and packaged in the same manner as all other ILW. The activity of these wastes is generally higher than 10^9 Bq/t but below the limit for HLW.

- *Low-level waste (LLW)* which essential includes all other wastes that have a radioactivity content below the threshold for ILW coming from normal reactor operations, fuel fabrication and reprocessing. These typically includes the 'dry' operational wastes (e.g. paper tissues and disposable clothing) which are compacted and packaged in steel or cement containers.

Wastes from the nuclear industry with a radioactivity content below exemption limits (sometimes called very low-level wastes, VLLW) do not require disposal to a repository and can be treated as normal industrial wastes. These exemption levels vary from country to country but are generally around 1 Bq/g, which is broadly consistent with average background levels and the radioactivity contents of many natural materials such as soils and rocks.

Not all countries operate this basic three category system: for example, in the United States, commercial wastes from nuclear reactors generating electricity are designated only as HLW or LLW. However, irrespective of the terminology used, the principal objective of any categorisation system is to ensure that the timescale over which any waste is isolated from the surface environment is compatible with the radionuclide content of that waste. This generally means that the HLW and ILW are destined for deep geological disposal, while the LLW may be disposed of in near-surface facilities.

Most deep geological repository designs preclude the *co-disposal* of HLW with ILW, where co-disposal is taken to mean that the two waste types would be placed in the same or closely spaced underground excavations. This is to avoid complex interactions between wastes with very different physical and chemical characteristics. In particular, the ILW contains large volumes of cement as an immobilisation matrix which will cause the groundwater to become alkaline: interaction between this alkaline groundwater and HLW would be undesirable. An alternative engineering option is *co-location* which involves building two different, separated repositories at one site, which share common surface facilities, shafts and access tunnels. The distinction with co-disposal is that in co-location, the HLW and ILW are separated by a distance large enough to avoid any significant physical or chemical interactions between them.

LLW may be co-disposed with ILW because they both have similar characteristics and are generally both solidified in cement, as shown in Figure 2.2. However, separate disposal facilities are often chosen for these two waste categories due to the additional costs of building a deep repository to house both ILW and LLW. Nonetheless, some combined L/ILW repositories have been built: examples include the Finnish VLJ repository at Olkiluoto and SFR repository at Forsmark in Sweden (see Box 3). In many countries, however, two or more different repositories are planned to be built for the individual waste categories.

2.3 Repository designs

There are basically two types of engineered repositories which are either being built or planned for the disposal of radioactive wastes. These are deep, mined repositories in the subsurface rock which potentially could house any waste category and near-surface repositories which would be used only for LLW.

The exact design details of any repository will be dependent on the nature and volume of the waste it is planned to contain, the geological environment at the disposal site and engineering constraints imposed by the host rock. However, all repository designs are based on the *multibarrier concept* whereby the wastes are emplaced inside a

Figure 2.2: Most ILW and LLW will be immobilised in a cement matrix, as shown here in a demonstration waste drum. Several drums may be placed in larger reinforced concrete boxes to create a stable waste package which can be stacked in the disposal vaults in the repository.

series of nested engineered structures and natural barriers which act in concert to control the rate of release of radionuclides over long periods. These barriers are generally devised such that they have no common failure mechanism. This means that, if some event or process were to cause one barrier to fail, this should not lead to the other barriers also failing. In some sense, this implies a degree of redundancy in the disposal system but no individual barrier is actually designed in such a way that it, alone, would necessarily ensure the safe isolation of the waste.

The specific barriers employed in the repository designs for the different waste types vary from design to design but the fundamental multi-barrier concept is common to them all. A conceptual set of multiple barriers for a repository, starting innermost, includes the solid wasteform and the container (together referred to as the waste package), a low permeability backfill or buffer and finally the surrounding rock mass. A generic system showing these components was illustrated in Figure 1.2. The man-made barriers (wasteform, canister, buffer material) are generally referred to as the *engineered barrier system* and the rock as the *geosphere*. Clearly, in a deep repository for HLW, the extent of the geosphere will be considerably larger than that surrounding a surface or near-surface repository for LLW.

Disposal sites will be chosen such that the geosphere provides a stable physical and chemical environment to protect the engineered barrier system. The surface environment, populated by humans and other animals and plants, is generally referred to as the *biosphere*, see Section 1.5.4.

Construction of a deep repository is likely to employ either smooth-wall blasting techniques or tunnel boring machines. Either method will result in the development of fractures in the rock around the excavations, although the extent to which such fractures might form depends on the physical characteristics of the rock, the construction technique used and on the design of the facility. These newly formed fractures will extend a certain distance out into the rock and define the *engineered damaged zone* (EDZ) which will affect both the physical and hydrogeological conditions around the repository. According to standard terminology, the combination of the engineered damaged zone plus the engineered barrier system is referred to as the *near-field* and the physically undisturbed, intact rock between this and the surface is referred to as the *far-field*.

In addition to using deep engineered facilities, radioactive wastes can be disposed of in liquid form by injection into deep rock formations. This method of disposal has been carried out at three facilities (polygons) in Russia since the 1960s (Rybalchenko, 1998). This disposal method involves injection of liquid wastes into aquifers (permeable rock formations), known as 'collector layers', at depths of up to 1.5 km for HLW and up to 400 m for L/ILW. The wastes are assumed to be isolated from the surface by aquitards (low permeability rock formations) above and below the collector layer. Very large volumes of liquid waste are injected by this method: for example, 150 000 m^3/yr of liquid is injected at the Dimitrovgrad facility in Russia under an injection pressure of 5 MPa. This method of disposal is relatively cheap and conceptually easy, although practical problems related to the precipitation of solids in the injection wells have created operational difficulties. Due to the wastes being in liquid form, long-term safety cannot be guaranteed with confidence, although recent safety assessments of some of these injection facilities indicate that the wastes are well confined on a 300 year timescale (Hoek, 1998). Disposal by liquid injection is planned to be discontinued in Russia and is unlikely to be practiced in other countries, thus this method of disposal will not be considered any further here.

2.3.1 Deep repository designs for HLW

At the time of writing, no engineered repository for solid HLW (spent fuel and solidified reprocessing wastes) has yet been built. However, a number of conceptual repository designs have been developed and subjected to detailed safety analysis at a generic level. Furthermore, several countries are now involved in the site selection and site characterisation activities necessary to build such a repository. Examples include the Swedish and Finnish designs for spent fuel repositories, and the Swiss design for a repository for vitrified (glass) reprocessing wastes: these are presented in Boxes 1 and 2.

Most deep repository designs for HLW currently under development have a great deal of similarity, although alternative designs have been considered and evaluated over the last two decades. The basic design features for a HLW repository include a series of tunnels in which the waste will be emplaced, sometimes referred to as *galleries*, which are excavated at depths in excess of 100 m (typically between 500 and 1000 m) and a number of vertical or inclined shafts which connect the repository to the surface. An example is the proposed Swedish spent fuel repository design which is shown in Figure 2.3.

The waste itself will be placed in large metal canisters and these will be located in the disposal tunnels. All the spaces between the canister and the host rock will be filled with a buffer comprising compacted bentonite clay. Several geometric options for locating the canisters in the tunnel are under consideration. The simplest has the canisters arranged axially at intervals along the tunnel. This design has the advantage that a circular section tunnel can be driven easily into the rock with a tunnel boring machine but the disadvantage that the space around each canister must be backfilled at the time of emplacement before the next canister can be brought into position. Alternative designs locate the canisters in individual disposal holes drilled either into the floor or the walls of the tunnels. In this design, the tunnels generally have to be both wide and tall to allow long canisters to be manoeuvred into the disposal holes. This makes excavation of the tunnels a more difficult procedure. However, the advantage is that only the immediate space around each canister in the disposal hole needs to be backfilled at the time of emplacement. The main part of the tunnel can be left open, if so desired, for monitoring or other purposes until all canisters are in position.

The canisters will be massive, thick-walled metal structures designed to isolate the waste for long periods of time. The canister metal may either be iron or steel, or a less reactive metal such as copper or titanium, or a combination of both. Iron and steel corrode faster than other possible canister metals but, in so doing, will buffer the redox conditions by scavenging free oxygen from

Box 1: The proposed Swedish and Finnish spent fuel repositories

The proposed Swedish repository for spent fuel is based on a design presented in 1983 in an early performance assessment known as KBS-3 (KBS, 1983) which has been modified and optimised over the last two decades. The basic design of the repository is shown in Figure 2.3. Disposal would occur in several arrays of tunnels excavated at depths in excess of 500 m. A very similar design is also being developed in Finland and POSIVA, the Finnish implementing agency, are seeking approval to construct a repository in granitic rocks close to the Olkiluoto nuclear power plant.

Figure B1.1: Artist's impression of the proposed Swedish repository for spent-fuel. The repository will be excavated at depths in excess of 500 m. Illustration courtesy of SKB.

The repository will be sited in the fractured crystalline rocks which predominate in Sweden and Finland. These rocks frequently contain large-scale fractures and fracture zones which are the dominant control on the groundwater flow system. The repository will have to be sited and designed to fit within the largest, stable blocks of rock defined by these fractures.

The disposal tunnels have a distinctive 'key-hole' geometry in cross-section due to the individual disposal holes which are drilled into the floor of the galleries approximately 6 m apart, although this spacing will be modified to avoid smaller faults and fractures in the rock. It is not planned to emplace canisters along the axis of the tunnels. No tunnel liners are required because of the inherent strength of the crystalline rock. Repository designs with both one and two levels have been considered. The general design for a one level repository has tunnels with a diameter of 3.3 m placed approximately 25 m apart.

The spent fuel will be encapsulated in bimetallic canisters which have a copper outer shell surrounding an inner cast iron (or steel) vessel. Copper is chosen for the shell because it is essentially inert in most groundwater systems (Section 4.4) and, thus, will corrode slowly providing very long canister life-times: some estimates put the life-time of these canisters at 10^6 years or longer. The inner cast iron vessel provides mechanical support for the canister and a large redox buffering capacity if the copper outer shell is perforated.

Single canisters will be placed in the disposal holes with all the spaces around them filled with machined blocks of compacted bentonite. The space in the tunnels above the disposal holes will be backfilled with a mixture of sand and bentonite.

A number of recent performance assessments have been undertaken for this general repository design (SKB, 1992; SKI, 1994; SKB, 1999; Vieno and Nordman, 1999) which have used proxy data from real crystalline rock sites in Sweden and Finland to improve the level of realism in the assessments. These assessments and the research undertaken to support them highlighted a number of issues which have required more detailed investigation by natural analogues, these include:

- dissolution processes and rates for the spent fuel UO_2 matrix (Section 4.2);
- general and localised corrosion of copper and steel (Section 4.4);
- radionuclide transport and retardation in crystalline rock (Section 5.1);
- matrix diffusion in the rock adjacent to fractures (Section 5.3);
- radiolytic decomposition of groundwater (Section 5.4); and
- radionuclide behaviour at redox fronts (Section 5.5).

These performance assessments have all highlighted the potential very long life-time of the canister due to its unreactive copper outer shell. In the absence of any manufacturing defects or unexpected early perforation, the canister alone can provide adequate long-term isolation of the spent fuel assuming a stable disposal environment.

However, absence of defects cannot be guaranteed at the current pilot stage of canister manufacturing development and thus the performance assessments have investigated the consequences of various canister failure scenarios. The results of these calculations show that continued safety of the repository, after canister failure, is controlled largely by the groundwater flow barrier provided by the bentonite buffer and by radionuclide retardation in fractures in the crystalline host rock.

Figure B1.2: The cross-sectional geometry of the Swedish spent fuel repository tunnels. Individual canisters are placed in disposal holes drilled into the floor of the galleries with surrounding spaces filled with blocks of compacted bentonite. Canister spacing would be modified to avoid any large fractures or faults in the rock. Illustration courtesy of SKB.

For all probable scenarios, the performance assessment calculations indicate that no unsafe releases to the surface environment would occur. These scenarios include evaluation of the consequences of future glaciation and permafrost development which are expected to occur in Sweden and Finland over the next 10^5 years due to natural climate changes.

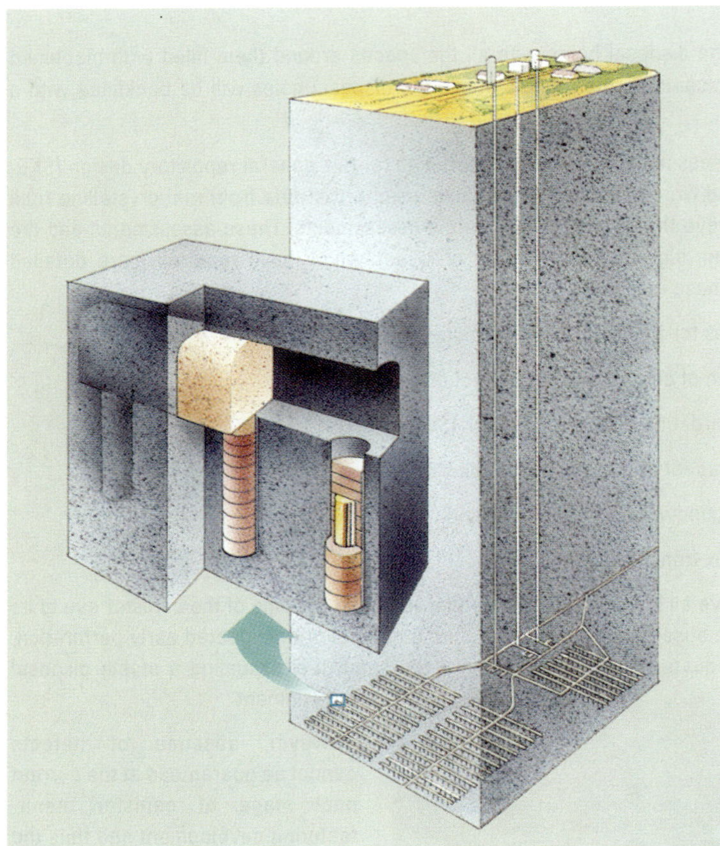

Figure 2.3: An example of a deep repository for spent fuel or HLW, showing an array of disposal tunnels at depth, linked to the surface by a number of shafts. In this case, the design is for a Swedish repository for spent fuel and shows the typical 'key-hole' near-field design described in Box 1. Illustration courtesy of SKB.

the near-field, thus maintaining a strongly chemically reducing environment (see Section 4.4). Chemically reducing environments are favourable because most of the radionuclides of concern are poorly soluble under such conditions. Copper and titanium, in contrast, will corrode much more slowly than steel, providing a potentially longer canister lifetime but less capacity for buffering the redox conditions. Both 'corrosion allowance' and 'corrosion resistant' metals offer different containment advantages.

The bentonite clay buffer placed around the canister plays a very important role in ensuring repository safety. Bentonite exhibits special properties of plasticity and expansion in the presence of water (Section 4.5). Once the repository has been closed and sealed, groundwater will begin to flow back into the excavations causing it to hydraulically resaturate. The bentonite will adsorb this water and expand to fill all remaining void spaces, creating a material with a very high swelling pressure, and very low porosity and hydraulic permeability. As a result, groundwater flow will not be able to occur through the buffer between the rock and the canister; movement of water, canister corrosion products and released radionuclides could only then occur by diffusive processes which are very slow. Other advantages of the bentonite are that it will sorb some proportion of the released radionuclides to the clay particle surfaces, act as a barrier to colloid movement (see Section 5.6) and evenly distribute water around the canister surface, lessening the chance of localised corrosion.

Sodium-bentonite is presently favoured as the buffer material in most repository designs due to its superior swelling capacity. This will react with the groundwater, buffering pH to mildly alkaline conditions (around pH 8). This buffering capacity provided by the large volume of bentonite means that the porewater composition close to the canisters is, to some extent, independent of the original groundwater composition in the far-field.

Sodium-bentonite can undergo ion exchange with any dissolved calcium or potassium in the groundwater, causing mineralogical transformation to either a calcium-bentonite or illite (see Section 4.5). These clays have a lower swelling capacity than sodium-bentonite and, thus, these transformations potentially may affect repository performance. However, given the very slow rate of these transformations, coupled with the almost stagnant groundwater flow conditions in the repository near-field and the massive amounts of bentonite present, these reactions are not expected to be significant.

One possible disadvantage of bentonite as a buffer is its load-bearing capacity and there is the possibility that the heavy canister may slowly sink through the bentonite if it were to act as a viscous fluid. If the canister did sink completely through the bentonite and come into contact with the rock, this would effectively negate the bentonite's barrier capacity. This is an issue which has not yet been fully resolved, although most assessments indicate that it is unlikely to be a serious problem. Another drawback of bentonite is its generally low thermal conductivity, in contrast to the surrounding bedrock, which means that the spacing of canisters and their waste loading must be carefully planned to avoid overheating in the near-field.

The host rock for a HLW repository could be hard crystalline rocks, various types of sedimentary rocks or evaporite (salt) deposits, all of which vary in their physical and chemical characteristics, in terms of groundwater flow, chemistry, thermal conductivity and physical strength. As a consequence, the design of a HLW repository will require optimisation for the particular host rock and geological environment chosen, although the basic elements of the near-field design would not change significantly. Potential host rocks and geological environments are discussed in Section 2.4.

Alternative HLW repository designs which have been proposed in the past include disposal of containerised solid waste in very deep boreholes, drilled to depths ranging from 3 to 4 km, or in very long boreholes with closely spaced containerised waste at depths of around 500 m (e.g. SKB, 1990; Gibb, 1999). Some of these options are shown in Figure 2.4. Although some alternatives have been investigated in detail, they generally have been found to be less suitable than the 'standard' HLW design for a number of reasons, such as the potential requirement for future retrievability of the waste.

A conceptually very different HLW repository design is being developed in the United States and is unique in that it is located above the water table in unsaturated rocks (volcanic tuffs). This is in contrast to all other current HLW and spent fuel repository designs which will both be located below the water table and, thus, will be water saturated, such as the Swedish and Swiss designs detailed in Boxes 1 and 2. The United States repository is planned to be built at Yucca Mountain in Nevada, and the preferred site is the focus of intense site characterisation (see Section 1.5.1). If an operation licence is obtained, this repository will house spent fuel and some military wastes.

The subsurface layout of the Yucca Mountain repository consists of a series of disposal tunnels connected to the surface by two inclined ramps leading to the lower slopes of the Yucca Mountain ridge. Waste will be encapsulated in large containers manufactured from an outer corrosion allowance metal, such as mild steel, and an inner corrosion resistant metal, such as Inconel. The containers will be located along the centre lines of the disposal tunnels. It has not yet been decided if the void spaces between the containers and the rock will be backfilled immediately after the containers are put in place. The repository could be left without a backfill for a considerable period of time to allow monitoring and waste retrieval, if required.

Although the volcanic tuffs at Yucca Mountain are technically hydraulically unsaturated, they do contain significant amounts of groundwater in the pore spaces and some perched water tables have been encountered in the tuffs during the site

Box 2: The proposed Swiss repository for vitrified HLW

The proposed Swiss repository for vitrified (glass) spent fuel reprocessing wastes is based on a design presented in 1985 in an early performance assessment known as Project Gewähr (Nagra, 1985) which has been modified and optimised over the last fifteen years. Alternative host rocks which have been considered for this repository include crystalline basement rocks in which the repository would be sited at a depth of around 1200 m, and argillaceous (clayey) sediments in which the repository would be sited at a depth of around 850 m. An artists impression of this repository design was shown in Figure 1.1.

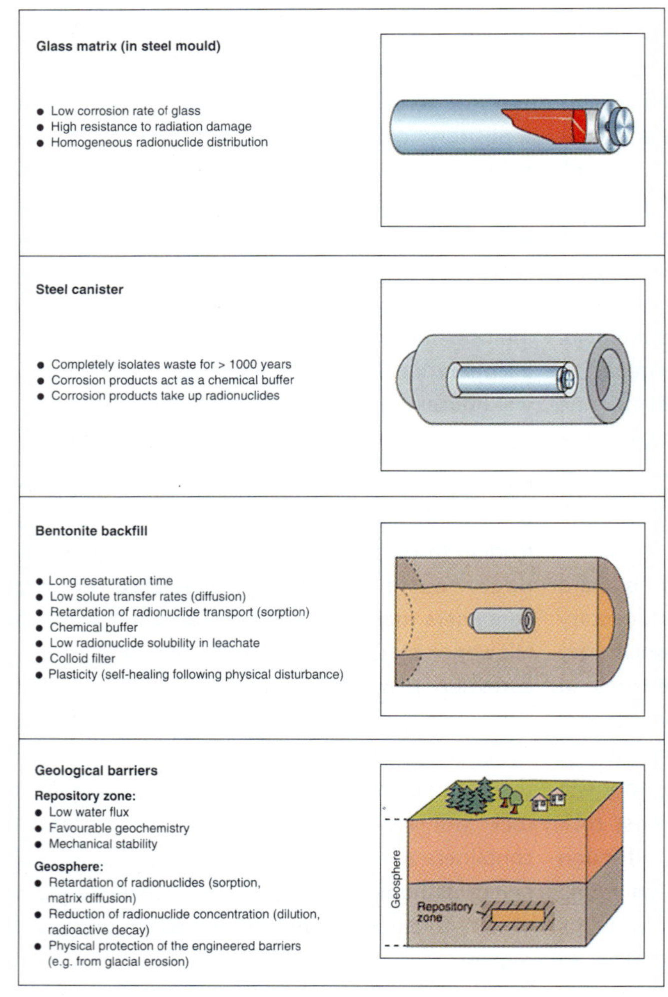

Figure B2.1: The series of engineered and natural barriers which isolate the waste from the surface environment in the Swiss HLW repository design. The repository design was shown in Figure 1.1. Illustration courtesy of Nagra.

The disposal tunnels have a circular cross-section and canisters will be emplaced axially along the tunnels. Due to the different physical and thermal characteristics of the two types of rocks, a repository in argillaceous sediments would have smaller tunnel diameters (2.5 m instead of 3.7 m) and tunnel spacings (25 m instead of 40 m). For stability, a tunnel liner would be required if the repository was built in sediments. For both the crystalline and sedimentary options, the vitrified HLW will be encapsulated in massive, 25 cm thick steel canisters. Canisters will be spaced along the tunnel axis and all the spaces around them filled with machined blocks of compacted bentonite.

Steel is reactive in groundwater and will corrode to form amorphous iron oxyhydroxides. However, the benefits of this reaction are that iron corrosion will buffer the geochemistry to maintain chemically reducing reactions and the iron oxyhydroxides will provide a large capacity for radionuclide sorption (see Section 4.4). In addition to the iron, the bentonite buffer will help to control the groundwater chemistry such that the near-field chemistry will be very similar in both the crystalline and sedimentary options. The largest differences between the two options relate to contrasts in the thermal conductivities and mechanical strengths of the rocks.

Both the thermal conductivities and heat capacities of the sediments are lower than for the crystalline rocks, leading to higher predicted near-field temperatures. However, the temperature can be controlled by operational measures (e.g. lower waste loadings or longer storage prior to disposal) or repository design (e.g. larger tunnel diameter, higher backfill conductivity or greater spacing between canisters).

The mechanical strength of the sediments is considerably less than that of the granite. Therefore, tunnels in the granite would be self-supporting while those in sediments may require steel liners. These liners have advantages (e.g. enhanced near-field redox buffering capacity) and disadvantages (e.g. delayed hydraulic resaturation) for repository safety and thus the consequences of leaving steel liners in place need to be evaluated.

A number of performance assessments (e.g. Nagra, 1994) and supporting research studies have been undertaken on these designs and they have highlighted a number of issues which have required more detailed investigation by natural analogues, these include:

- dissolution processes and rates for the glass wasteform (Section 4.1);
- corrosion rates and products from steel (Section 4.4);
- radionuclide transport and retardation in crystalline rock (Section 5.1);
- radionuclide transport and retardation in argillaceous rock (Section 5.1);
- matrix diffusion in the rock adjacent to fractures (Section 5.3); and
- radionuclide behaviour at redox fronts (Section 5.5).

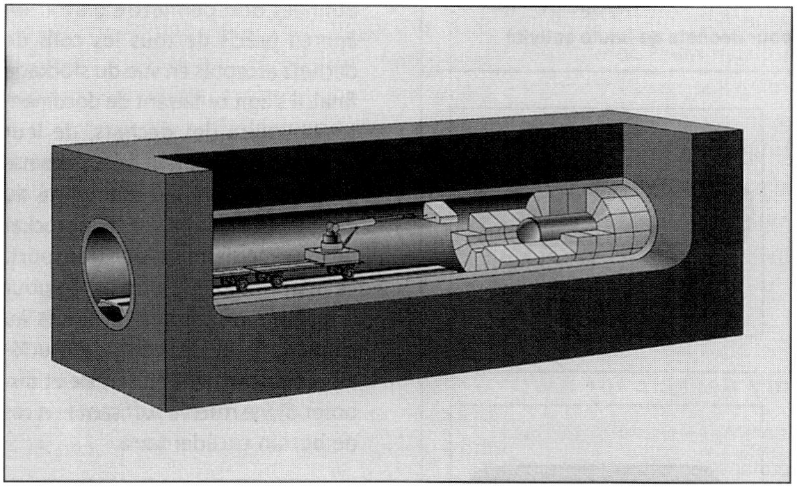

Figure B2.2: Canisters will be placed axially along the tunnels with a separation of 5 m. All spaces will be backfilled with machined blocks of compacted bentonite. All void spaces between the bentonite blocks will disappear as the bentonite resaturates and swells. Illustration courtesy of Nagra.

With conservative assumptions for canister lifetime, performance assessments for this repository design indicate that the most important issues for safety are the geochemical controls on radionuclide solubility in the near-field and retardation in the far-field. In particular, they show that the advective flow path must be specified in detail at the small-scale to calculate solute transport. From this point of view, homogeneous sediments have advantages over fractured, crystalline rock because the flow path will be easier to define. However, the performance assessments indicate that both potential host rocks will be adequate to guarantee repository safety

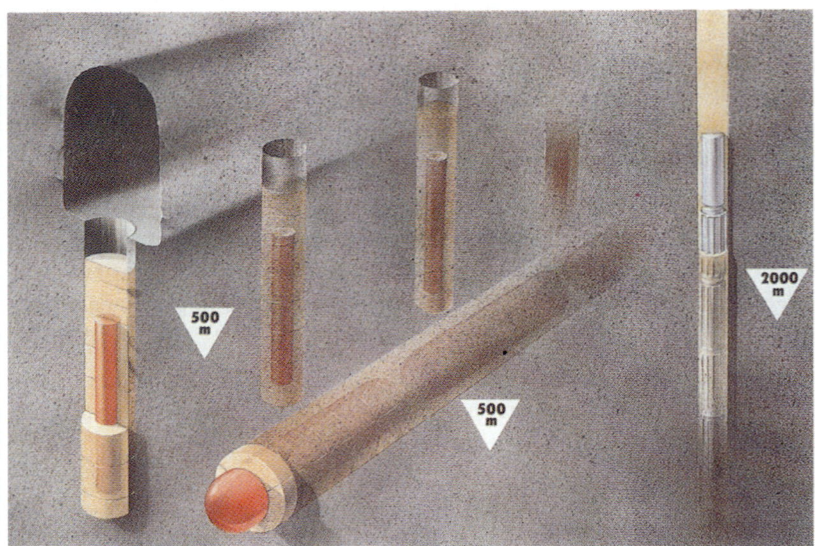

Figure 2.4: Artist's impression of the Swedish spent fuel repository concept (left) and two alternatives: the very long hole (middle) and the very deep hole (right). The size and spacing of canisters varies between concepts although, in each case, the canister is surrounded with compacted bentonite blocks. Illustration courtesy of SKB.

investigations. There are significant conceptual uncertainties about the mechanisms controlling the behaviour of water in unsaturated porous rock and the transport of radionuclides away from the repository. Consequently, natural analogues for the Yucca Mountain design are required to address a different series of transport issues to those of relevance to other HLW repository concepts, such as radionuclide migration in a two-phase (water and air) system.

2.3.2 Deep repository designs for ILW

Designs for deep ILW repositories depend on the nature of the waste to be disposed of, in terms of its volume, radioactivity and physical nature (i.e. if it is immobilised in cement, bitumen or other material). However, all designs for deep ILW repositories currently under development differ significantly from proposed designs for deep HLW repositories.

In contrast to HLW repositories, several deep ILW repositories have actually been built and are receiving waste. In Scandinavia, both the repositories at Forsmark in Sweden, and at Olkiluoto and Lovisa in Finland take both ILW and LLW. The Forsmark repository is described in Box 3. In the United States, the WIPP (Waste Isolation Pilot Plant) facility in New Mexico, which is excavated from bedded evaporite (salt) deposits, has recently begun receiving military derived TRU wastes. Other ILW repositories are planned in other countries.

Most ILW repositories are designed to be located at shallower depths than proposed HLW repositories. Typically, ILW repositories will be at depths of around 100 m below the ground surface, although some designs are deeper. For example, the proposed British ILW repository was to have been built at a depth of around 750 m below the Sellafield site in north-west England, although plans for this repository have now stalled.

The excavations for an ILW repository generally consist of a number of large disposal caverns (sometimes referred to as vaults), rather than the disposal tunnels planned for a HLW repository. In most designs, the waste will be placed in large reinforced concrete boxes and stacked in the caverns, with spaces between them backfilled with cement or concrete. The proposed Swiss repository design is shown in Figure 2.5.

The number and dimensions of caverns in an ILW will vary from design to design but can be very large. As an indication of the upper size range, the proposed British ILW repository was to have been based around a number of caverns 25 m wide,

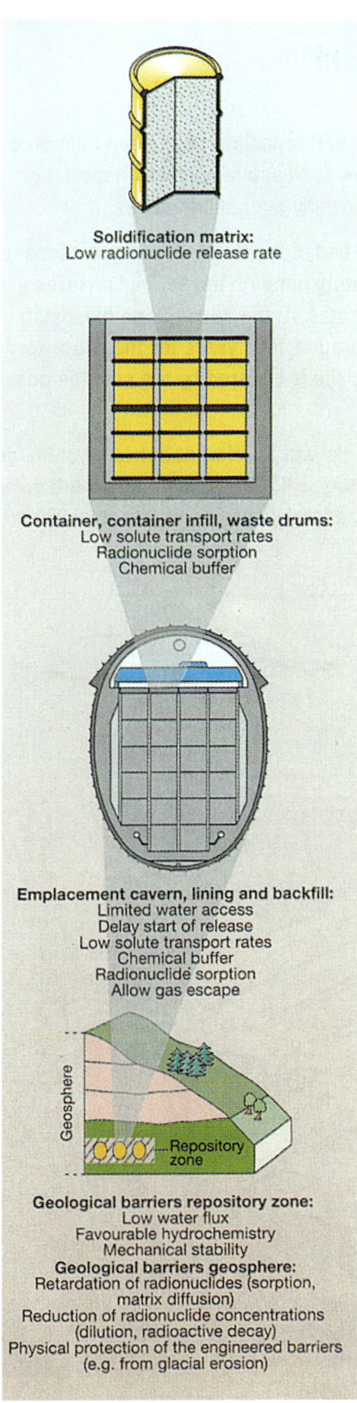

Figure 2.5: The series of engineered and natural barriers which isolate the waste from the surface environment in the Swiss ILW repository design. Illustration courtesy of Nagra.

16 m high and 265 m long. To reduce groundwater flow through the cavern walls and to support the excavations, some caverns may be lined with concrete and steel reinforcements.

In addition to the caverns, certain ILW repository designs also include a silo structure to contain the higher activity wastes. A silo is a reinforced concrete, cylindrical structure built within a tall cavern. These typically may have a diameter of 20 to 30 m and a height of several tens of metres. The spaces between the reinforced concrete silo shell and the rock may be backfilled with bentonite so that groundwater flow through the silo will be substantially less than through the other caverns. The silo structure at the Finnish ILW repository at Olkiluoto is shown in Figure 2.6.

As discussed in Section 2.2, ILW is heterogeneous, comprising a mixture of materials arising from the routine operation or decommissioning of nuclear power plants and fuel processing plants. These wastes will be solidified (immobilised) in either a cement, bitumen or resin matrix, although cement will be used most abundantly. The solidified wastes will be contained in either thin-walled metal drums or large concrete boxes, depending on the waste type and repository design, to provide a stable waste package. In most designs, these waste packages are largely designed to aid handling and transport to and within the repository, and are not intended to provide long-term isolation of the waste from the groundwaters in the repository.

In a typical repository, these waste packages will be stacked in the caverns and in the silo, and all residual spaces between them will be backfilled with cement or a cementitious mortar. Wastes with different levels of radioactivity or chemical form may be segregated and placed in separate caverns, with the highest activity wastes located in the silo (if present). If LLW is also emplaced in the repository, this will be located in separate caverns to the ILW.

Box 3: The Swedish L/ILW repository at Forsmark

The Swedish combined repository for ILW and LLW (known as the SFR repository) has been built on a coastal site at Forsmark, adjacent to an existing nuclear facility to minimise radioactive waste transportation. It is one of the few examples of an operating geological repository, currently accepting wastes.

The repository has been excavated at a depth of 60 m beneath the bed of the Baltic Sea in gneissose rock with a low hydraulic conductivity of 10^{-8} to 10^{-7} m/s. Locating the repository beneath the seabed ensures a very low hydraulic gradient and, as a consequence, low groundwater flow rates. At the current rate of isostatic uplift of the land of around 6 mm/yr (due to glacial rebound), it will take around 1000 years for the repository area to become dry land. In the short to medium-term, however, while the site is covered by the sea, the possibility of inadvertent human intrusion is clearly minimised.

Currently, the repository consists of one silo and four caverns, each with a different barrier system, designed to hold different waste streams. The disposal volume in the repository is 60 000 m^3, although there are plans to extend it with a second silo and two additional caverns to provide a total disposal volume of 90 000 m^3.

Figure B3.1: Artists impression of the Swedish L/ILW repository at Forsmark (the SFR repository). The silo houses the highest activity wastes, while lower activity wastes are placed in the four caverns. Illustration courtesy of SKB.

The silo is constructed from a reinforced concrete shell, 90 cm thick. It is 50 m high, 28 m in diameter and is located within a 70 m high cylindrical rock cavern. The void between the rock and the concrete silo is backfilled with bentonite, approximately 1.3 m thick. At the top and bottom of the silo, a sand/bentonite mixture (90% sand) is used as the buffer material to provide greater bearing strength (at the bottom) and gas permeability (at the top). The silo is receiving ILW in the form of ion exchange resins solidified in cement and bitumen matrices, and packaged in steel drums and concrete moulds. After this silo is filled, all spaces will be backfilled with cement.

The four caverns are each 160 m long and their walls are lined with shotcrete to provide mechanical support. The caverns hold LLW and lower activity ILW immobilised in cement and packed in steel or concrete containers. After they are filled with waste, the spaces above and between the waste packages will be backfilled with either cement or crushed rock. No low-permeability buffer material is used in the caverns.

Near-field groundwater chemistry in the repository will be both very alkaline and reducing. The large volumes of cement and concrete will buffer the pH, and the iron present in the canisters, reinforcing rods etc., will buffer the redox potential in the repository.

On closure of the repository (scheduled for 2010), drainage pumping will end, the repository will be sealed and hydraulic resaturation will commence. The present conceptual hydrogeological model anticipates groundwater flowing horizontally through the caverns, leaching radionuclides from the waste form, and then flowing vertically through fissures in the rock to reach the seabed, where the Baltic Sea will ensure large dilution of releases. Radionuclides leached from the waste in the silo would first diffuse through the bentonite buffer before being advectively transported with the slowly moving groundwater to the seabed. This situation will change with time due to ageing and weakening of the engineered barriers and continued uplift of the seabed.

Several performance assessments for the repository consider separately the period when the repository remains beneath the Baltic Sea when recharge waters are saline and the period, after uplift, when the repository area dries out and a freshwater system is created. These performance assessments have highlighted a number of issues which have required more detailed investigation by natural analogues, these include:

- degradation of cement (Section 4.6)
- degradation of bitumen and organic waste (Sections 4.7 and 4.8);
- radionuclide solubilities in high pH environments (Section 5.1);
- colloidal activity (Section 5.6);
- microbial activity (Section 5.7); and
- gas generation and transport (Section 5.8).

A particular issue which is more important for this repository than for HLW repository designs is gas generation from the degradation of organic materials. Preliminary assessments showed that a build-up of gas pressure potentially can affect the structural integrity of the engineered barriers and the rock. Consequently, the repository was designed to allow gas to escape freely from the silo and caverns.

For all probable scenarios, the performance assessment calculations indicate that no unsafe releases to the surface environment would occur, including changes that result from land uplift and retreat of the sea.

Figure 2.6: Artist's impression of the silos in the Olkiluoto L/ILW repository in Finland. These silos are constructed from reinforced concrete, surrounded by a bentonite based buffer between the concrete and the rock. Waste is packaged in large reinforced concrete or steel boxes and stacked in the silos. Illustration courtesy of Nagra.

The large volumes of cement in a typical ILW repository have an important safety role. As groundwaters infiltrate the near-field after closure, they will chemically equilibrate with the cement and concrete, buffering the pH of the porewaters to around pH 13 or 14. Such hyperalkaline conditions are expected to persist for 10^4 years or more, depending on the mass of the cement, and the chemistry and flow rate of the groundwaters.

Progressive corrosion of the steel canisters and steel reinforcing rods will occur, and the corrosion products (principally hydrogen and dissolved iron) will control the redox environment, creating and maintaining chemically reducing conditions. High pH, chemically reducing environments are favourable because most of the radionuclides of concern are poorly soluble under such conditions.

Over a long time period, the high pH groundwater will migrate from the repository into the host rock as a hyperalkaline 'plume' which may result in slow changes to the composition of the cement and the pH of the near-field. The high pH groundwater could influence the host rock and, depending on the nature of any reactions, may cause both physical and chemical changes, thus creating a disturbed zone around the repository. The expected slow groundwater flow rate will allow chemical equilibrium to be maintained in the near-field and along the migration path from the near to far-field.

The high initial pH in the near-field will gradually drop to 12 over a few thousand years but will then be maintained for some 10^5 years by the slow dissolution of the cement minerals (see Section 4.6). After about a million years the pH will have dropped to around 10, depending on the groundwater flow rate. This is much in excess of the expected life span of the engineered barriers and indicates that the conditions in the near-field should act as a chemical buffer even after the physical integrity of the concrete has been lost.

2.3.3 Near-surface repository designs for LLW

The disposal of LLW in near-surface or shallow repositories is practiced routinely in many countries. These repositories are located at or close to the land surface because the low activity and short half-lives of the waste they contain means that they do not require the very long isolation times necessary for HLW or long-lived ILW.

The earliest LLW disposal designs were little more than landfills, based on simple trenches into which the waste was tipped and then backfilled with earth. More recent designs employ a more robust, multi-barrier engineering approach and it is these that are considered here, e.g. Figure 2.7. The basic design involves a series of reinforced concrete bunkers or trenches into which packaged waste is emplaced. These structures may be located on the ground surface or they could be fully or partially set into the ground, with eventual closure beneath a thick earth cap.

Figure 2.7: Photograph of the Centre de l'Aube surface repository for LLW in France. Wastes packaged in steel or concrete boxes will be placed in the large engineered compartments which are each 24 m long. When the repository is filled, the compartments will be sealed with a concrete 'lid' and the entire facility will be covered with an earth and clay cap, to form a small hill. Illustration courtesy of Nagra.

As discussed in Section 2.2, LLW is heterogeneous, comprising a mixture of materials from normal reactor operations, fuel fabrication, reprocessing and decommissioning. These typically include items such as cleaning materials and disposable clothing but may also include large items from decommissioning activities. These materials will be compacted and solidified (immobilised) in cement. Most solidified wastes will be contained in large concrete boxes or steel containers, depending on the waste type and repository design. These waste packages are not intended primarily to provide any long-term isolation of the waste.

In a typical repository, these waste packages will be stacked in the bunkers and trenches and all residual spaces between them will be backfilled with earth, clay or cement. After each trench or bunker is filled, it may be closed with an 'anti-intrusion' slab of reinforced concrete designed to prevent inadvertent excavation into the wastes in the future. When all disposal activities are completed, the entire repository system will be covered with a low permeability earth and clay cap to provide an outer protective barrier. The earth cap may form a mound or hill. This cap will often comprise layers of low permeability clay or an impermeable membrane to reduce water percolation into the repository. Specially chosen vegetation is often planted on top of the earth cap, again to limit water percolation.

The various combinations of waste packages, concrete structures and earth caps used reflect the volume and characteristics of the LLW and the geology and climate at the repository site.

2.4 Geological disposal environments

The fundamental requirement of a suitable geological environment for a repository is that it should be relatively stable to protect the engineered barriers and its behaviour should be adequately predictable. The requirement for predictability arises from the need confidently to demonstrate the long-term behaviour of the repository system in performance assessment, as discussed in Section 1.3. In this regard, geological stability relates not only to the physical features of a site but also to the geochemical and hydrochemical aspects. This is particularly important for deep spent fuel and HLW repositories designed to contain the longest-lived radioactive wastes.

Put simply, the basic requirement of any suitable host rock is that it should provide a stable cocoon for the repository in which the rates of all physical and chemical processes which might disturb the engineered barrier system are very slow and are not likely to be subject to significant disturbances or modifications over the time period of containment.

The search for suitable host rocks for deep repositories has concentrated on four principal rock types:

- crystalline basement rocks;
- extrusive volcanic rocks (e.g. lavas and pyroclastics);
- low-permeability sedimentary sequences; and
- thick or diapiric evaporite (salt) deposits.

Some geological environments which have been considered for a repository combine one or more of these rock types, as with basement rocks under sedimentary cover, the so-called 'BUSC' environment. The fundamental characteristics of these different environments are described later, together with the transport and retardation issues specific to these environments which have been addressed by natural analogue studies.

In any geological environment, the greatest uncertainty with regard to natural conditions and processes will probably be associated with the flow of groundwater and radionuclide transport in the far-field rock. Much of the relevant information on the transport pathways at a site needed for performance assessment will be provided by site characterisation programmes (see Section 1.5.1) in an attempt to understand the nature of heterogeneity and spatial variability of the system, especially for fractured crystalline rocks in which the heterogeneity of the hydraulic system can be particularly marked (Hodgkinson and McEwen, 1991).

Nonetheless, natural analogues can provide useful generic information in this regard, by identifying and quantifying the most important processes which control transport and retardation in the rock types listed above. This is particularly so for natural analogue studies which look at sites on a scale large enough to examine an entire flow system and can show how radionuclide transport is affected along the flow paths. A good example is the investigation of colloidal transport at the Morro do Ferro site, part of the Poços de Caldas analogue study (see Box 14).

A lot of the information that comes from the natural analogue studies which is useful for characterising disposal environments is not strictly analogue information. Rather, it is geoscientific data that is collected about the analogue sites in order to understand the analogue system, and to constrain and interpret other data. For example, analogue information on colloidal populations cannot be usefully interpreted with regard to radionuclide transport, if the hydrogeological parameters of the system are not known. As a consequence, there is a considerable volume of field data obtained from analogue studies which can be used to improve our understanding of flow and transport in different rock types, which is generally not given an analogue label.

Most rocks considered as host rocks for a repository will have low hydraulic conductivities and low hydraulic gradients, and it is these features that are largely important for determining groundwater flow velocities. In rocks with plastic behaviour (e.g. argillaceous rocks and evaporites) the hydraulic conductivities may be so low

(< 10^{-10} m/s) that groundwater and contaminant movement occurs predominantly by diffusion, especially at depth. The hydraulic conductivities of the matrices of rocks with brittle behaviour (e.g. crystalline and volcanic rocks, and some hard sediments) may also be very low particularly in the rock mass chosen to locate the deposition tunnels. At these locations, some diffusion may also be expected to occur.

Contrastingly, in the far-field environment, movement of groundwater predominantly occurs through the fracture networks which are ubiquitous. Consequently, for these rock types, it is the hydraulic and transport properties of the fractures rather than of the matrix that are important and these can vary considerably. Crystalline rocks generally have very low total porosities which comprise the porosity of the fracture network and the porosity of the dead end pores which are connected to the flowing fractures.

Crystalline rocks

Crystalline rocks form at high temperatures and pressures deep beneath the surface either by cooling and crystallising from a magma (in the case of igneous rocks) or by solid state recrystallisation of existing rocks (in the case of metamorphic rocks). These rocks are very hard but tend to be brittle and, thus, are usually cut by faults and fractures. Several countries are considering crystalline basement rocks for deep repositories, including Canada, Czech Republic, Finland, France, Japan, Spain, Sweden, Switzerland and the UK.

The critical feature of these rocks is that groundwater flow and, thus, radionuclide transport will be concentrated in the fracture network, with essentially no advection occurring in the intact matrix of the rock, which can act to retain radionuclides which diffuse into it from mobile water in the fractures: a process known as *matrix diffusion* (Section 5.3). The detection and the measurement of representative transport properties of such pathways is inherently difficult. It is only by linking hydrogeological measurements and modelling with geochemical and structural geological studies that sufficient understanding is likely to be obtained (Nagra, 1994; Mazurek et al., 1992a) and natural analogue studies have helped in this regard.

Natural analogue studies have tended to focus on determining the retardation capacity of the fracture surfaces in these rocks because it is predominantly these surfaces which are in contact with mobile groundwater. The low effective porosity also means that the surface area available for sorption is much less than in permeable sedimentary sequences. In addition, in contrast to the potentially long pathways which are possible in sedimentary environments, pathways in crystalline rocks are invariably short owing to the frequent presence of near-vertical fractures or faults (SKB, 1992; Mazurek et al., 1992b).

Natural analogue studies in hard, fractured rocks have examined a number of processes which might affect flow and transport, and have provided important information on:

- the groundwater chemistry in crystalline rocks and radionuclide solubilities in these hydrochemical environments (Section 5.1);
- the retardation processes which operate in fractures and quantitative information on the sorption capacity of fracture lining and fracture filling minerals (Section 5.2);
- the potential for matrix diffusion and the depth of interconnected porosity adjacent to the fractures (Section 5.3);
- the development and movement of redox fronts in fractured systems (Section 5.5); and
- the populations and movement of colloids, organics and microbes through the fracture network (Sections 5.6 and 5.7).

Extrusive volcanic rocks

Extrusive igneous rocks are formed by the cooling and solidification of volcanic lavas and ash deposits. In some cases, the ash deposits are deposited at very high temperatures which 'glues' together the individual particles to form a massive rock formation known as a *welded tuff*. These rocks tend to have low matrix porosity and, after cooling, are hard and brittle, and thus can have physical characteristics similar to the crystalline rocks, although their mineral grain size tends to be much smaller. In geochemical terms, welded tuffs have similar ranges in bulk compositions to crystalline rocks, although they are often altered by hydrothermal fluids associated with their volcanic origin, leading to the formation of zeolite and clay minerals in fractures and pore spaces.

Geologically old volcanic rocks may be located below the surface under more recent sedimentary sequences and form potentially suitable repository host rocks. Some countries are considering these rocks for deep repositories, including Japan and the UK. Due to their hard, fractured characteristics, when these rocks occur below the water table they have quite similar transport characteristics to crystalline rocks and, hence, analogue study objectives for the two rock types would be similar.

Young welded tuffs are frequently found at or close to the surface in volcanic regions. Such rocks occur at the site of the proposed US repository at Yucca Mountain, described in Section 2.3.1. The US repository would be excavated in the upper parts of the welded tuffs, above the water table, and hence the host rock environment would be unsaturated. Although the basic geochemical and mineralogical characteristics of this rock would be similar to other welded tuffs, the unsaturated nature of the tuffs at Yucca Mountain mean that analogue studies would need to address specific issues, such as:

- the degradation of UO_2 (spent fuel) in an unsaturated, oxidising environment (Section 4.2);
- radionuclide speciation in an unsaturated, oxidising environment (Section 5.1); and
- the transport and retardation processes which operate in an unsaturated fractured rock (Section 5.2).

Sedimentary sequences

The sedimentary rocks which may provide potentially suitable repository host rocks are clays, mudstones, marls and shales. These are all fine grained rocks which form by the accumulation of sediment underwater in marine or freshwater environments. Due to their mode of formation, they can be laterally extensive and homogeneous although, in both the vertical and horizontal directions, sedimentary sequences can change rapidly from fine to coarse grained material.

As sediment accumulates, it becomes compressed and its porosity is reduced. In some cases, the individual grains become cemented together (lithified) making the sediment mass brittle and liable to fracture. The majority of shales and many mudstones are fractured and can have quite similar flow characteristics to fractured crystalline rocks and, hence, analogue study objectives for the two rock types can be similar.

In some clay-rich sedimentary rocks, the clay particles do not cement together and the sediment mass can retain its plasticity. Thick clay sequences are potentially suitable for deep disposal because they tend to have low hydraulic conductivities, are associated with low solute transport rates and can have discontinuities which are self-sealing. Transit times for water from within a relatively deep clay horizon to the surface can also be extremely long. Several countries are considering clay-rich sediments for deep repositories, including Belgium, France, Spain and Switzerland.

However, there is concern that the plastic nature of clay-rich sediments could be lost in some circumstances and that radionuclides might be able to move more freely through them than predicted (e.g. Gera et al, 1992). Natural analogue studies in sedimentary sequences have examined a number of processes which might affect flow and transport in these rocks types and have provide important information on:

- the groundwater chemistry in sediments and radionuclide solubilities in these hydrochemical environments (Section 5.1);
- the retardation processes that operate in sediments and quantitative information on the sorption capacity of clay minerals (Section 5.2);
- diffusion coefficients in compacted sediments (Section 5.2);
- matrix diffusion in sedimentary rocks (Section 5.3);
- the development and movement of redox fronts in sediments (Section 5.5); and
- the populations and movement of colloids through sediments (Section 5.6).

Evaporite deposits

Evaporite deposits form by the evaporation of closed water bodies to leave behind an accumulation of salts. They tend to be located in sedimentary basins, in association with sedimentary rocks and, hence, can be interstratified with clays and other sediments. A wide range of minerals can be found in evaporites, typically chlorides and sulphates of sodium, magnesium, calcium and strontium, although most interest for waste disposal is focussed on large masses of relatively pure rock salt (sodium chloride). Thick beds of evaporites (bedded evaporites) may remain in horizontal formation but, if they are covered by large thicknesses of sediment, they may become unstable, in terms of density and plasticity, and rise upwards to form salt domes.

Both bedded evaporites and salt domes appear to provide some of the most suitable environments for deep disposal because the transport rate of radionuclides within salt is very slow for most situations which can be envisaged. Salt has higher thermal diffusivity than crystalline or argillaceous rocks and, therefore, temperature rises due to heat from the waste can be lower. Due to their plastic behaviour, these rocks are also essentially self-sealing. Several countries are considering evaporates for deep repositories, including Belgium, France, Germany, Spain, Switzerland and the US.

However, despite the interest in evaporites, very few natural analogue studied have examined them in detail and relatively little analogue information has been obtained for them. Field studies at the WIPP repository site in New Mexico, now receiving military TRU wastes, have revealed that evaporites can be associated with high pressure brines and the presence of local deposits of both oil and gas.

Chapter 3: Varieties of analogue studies

In essence, a natural analogue study can be any form of investigation of any relevant system, provided that it results in qualitative or quantitative information which can be used to support and build confidence in geological disposal. This may mean that an investigation provides data which are directly applicable to performance assessment (for the development of models or the provision of parameter values) or, alternatively, it may provide illustrations of concepts or processes that allow non-technical demonstrations of safety to be made. On this basis, two important guidelines for selecting analogue studies are focussed on the end-user of the analogue information:

- the output of the study needs to be defined clearly, together with the intended use of the analogue information, and
- the end-user (performance assessor or information presenter) should be involved and informed at each stage of the study, including at the very beginning when the studied is first planned.

For example, performance assessment modellers are keen to test their conceptual and mathematical approaches against well-defined test cases, and the best way to define such cases is with the active participation of the modeller when selecting analogue sites and when designing projects. When using analogues in this way, much of the value of a study may stem from careful management and dissemination of data. For example, predictive model testing at its most rigorous should be done 'blind' to the data against which it is being tested, as discussed in Section 5.1. This requires tight planning and control of precisely which data need to be collected, and how these data are released to the modellers (Pate et al., 1994; Alexander et al., 1998).

Whilst each repository design will require unique information to assist in building and presenting a safety case, there are a number of broad areas where information is required generically in the geological disposal field. Historically, analogue studies have tended to focus on only a small number of types of natural system and, thus, analogue studies can be categorised into a few broad groups which are representative of some major components of a repository system or feature of its evolution. These groups are:

- natural geological and geochemical systems;
- archaeological systems; and
- sites of anthropogenic contamination.

Although this general categorisation is a convenient way of presenting the most obvious analogue systems which have the potential to provide relevant information, it is important to recognise that useful data may be obtained from many types of environment. The objective of the following discussion is, thus, briefly to describe the features of typical analogue systems and to discuss some of their limitations.

Within many analogue studies there is considerable emphasis on sites containing natural radionuclides or stable isotopes of elements of direct relevance to those in a repository. However, several studies have investigated the behaviour of other naturally occurring chemical species which are believed to behave in an analogous manner to chemical species in the waste. Consequently, as well as discussing the broad natural analogue types it is also worth considering this concept of *chemical analogues.*

3.1 Chemical analogues

The elements of relevance for radioactive waste management are widely spread throughout the periodic table. However, not all of these elements (or their individual nuclides) occur in nature. Depending on the natural availability of a particular radionuclide, a natural analogue study has three options:

- to examine the radionuclide directly when it occurs in nature, such as in the case of the natural decay series radionuclides (Ivanovich and Harmon, 1992);
- to examine naturally occurring isotopes of the same element, when the waste nuclide does not occur in nature, such as in the case of stable caesium or bomb fall-out ^{137}Cs as an analogue of waste ^{135}Cs;
- to examine a different naturally occurring element which behaves in a chemically similar manner to the waste element; and
- to examine the end fission products of parent radioisotopes no longer present, e.g. at the Oklo natural fission reactors (see Box 4) where the behaviour of in situ generated plutonium and iodine can be interpreted with respect to the location of decay products such as ^{235}U and ^{129}Xe.

In the first case, potential weaknesses of the analogy with the waste radionuclide mostly relate to differences between the natural and repository environments, in terms of water chemistry and source terms etc. However. differences in concentration also need to be considered; for example, naturally occurring radionuclides can now be measured at extremely low concentrations (detection limits of about 10^5 atoms) which are of no possible radiological significance and are far below the levels considered in performance assessment. It is questionable if the behaviour of plutonium, for example, at very low water concentrations is representative of its behaviour at more repository relevant concentrations. Hence, such observations need to be interpreted with care.

When the analogy is extended to other, naturally occurring radionuclides of the waste element of interest, differences due to the decay process must be added to the caveats above. This is not generally a problem for very long-lived (or stable) isotopes but there are well-known examples of decay-induced processes which can give rise to geochemical partitioning of shorter-lived nuclides, such as recoil, radiolysis, hot atom effect. For very short-lived nuclides with, say, a half-life of less than one year, the timescale of ingrowth and decay may be much shorter than that of the process to be studied (e.g. sorption, precipitation, matrix diffusion etc.) and, hence, the observed behaviour of a short-lived nuclide cannot be extrapolated to longer-lived isotopes of the same element. Probably the best known example of a decay induced processes is the case of $^{234}U/^{238}U$ disequilibrium in natural waters, which is discussed in detail by MacKenzie et al. (1990a).

When considering a naturally occurring element as an analogue for a different waste element, great care must be exercised. The concept of chemical analogue elements was first discussed by Chapman et al. (1984) who presented a list of possible chemical analogues for some of the long-lived radionuclides present in HLW. Experience

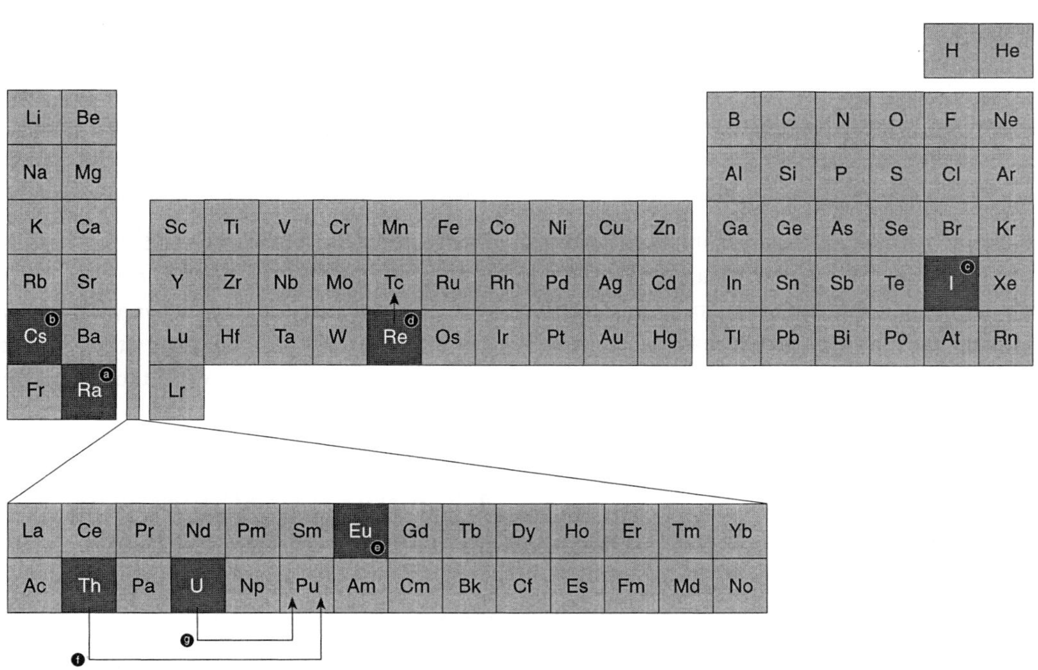

Figure 3.1: Periodic table of the elements. Some examples of chemical or isotopic analogues are illustrated: (a) natural ^{226}Ra for waste ^{226}Ra; (b) fallout ^{137}Cs for waste ^{135}Cs; (c) stable I for waste ^{129}I; (d) Re as an analogue for Tc; (e) Eu, or any lanthanide, as an analogue for a trivalent actinide, e.g. Am; (f) Th as an analogue for Pu(IV); (g) U, under oxidising conditions, as an analogue for Pu(VI).

since this time has shown that this and similar tabulations can be misleading, and may lead to over-interpretation of the extent of the analogy involved. The true extent to which one element can be considered a chemical analogue of another is very dependent on the system and process studied, and must be examined on a case by case basis.

In the simplest case, chemical analogues can be selected from the well established chemical periodicities. For example, an alkali metal such as rubidium would be expected to behave in a similar manner to its neighbour caesium, whilst barium might be considered as an analogue of strontium, and bromine as an analogue of iodine etc. These chemical analogues are shown in Figure 3.1. For elements of simple chemistry, such assumptions may be quite easy to justify but great care must be taken for more complex cases, especially if different valencies occur over the range of Eh/pH conditions encountered in the natural environment.

The limitations of the analogy must always be borne in mind, even in the 'simple' cases. For example, bromine may be a good analogue for iodine when examining transport processes in deep groundwaters but can be significantly less satisfactory when examining near-surface environments, due to the tendency of iodine to be associated with microbial processes in soils and sediments, and its variable redox behaviour.

A particular challenge is to select chemical analogues for the transuranic elements, because they often have complex chemistry and are not found in significant concentrations in most natural systems. Americium and curium are similar, and are not redox-sensitive in natural waters, being present only in the III oxidation state. Obvious chemical analogues of the actinides would be any of the lanthanides, which have similar chemistry, although curium and europium should generally be avoided because they may diverge from the III state (to IV and II, respectively) causing a chemical partitioning which has been observed in some cases, such as at Poços de Caldas (MacKenzie et al., 1990a; Miekeley et al., 1990a).

Plutonium has an extremely complex chemistry and may be found in natural waters (as anthropogenic contamination) in four oxidation states (III to VI), all of which may be present in measurable concentrations simultaneously. Under reducing conditions (e.g. in the presence of ferrous iron), plutonium is found predominantly in the III and IV oxidation states (e.g. Schweingruber, 1983). Some III-valent (lanthanides) or IV-valent (thorium or, possibly, zirconium or hafnium) elements can be considered to have similar behaviour but this analogy must be regarded with caution. The use of Th(IV) as a chemical analogue for Pu(IV) is extensively discussed by Eisenbud et al., (1984) in relation to the Morro do Ferro analogue site. Under oxidising conditions, the IV, V and VI states of plutonium may all be important, the higher states especially so in the presence of high carbonate concentrations. In such a case, the closest similarity to the PuO_2^+ and PuO_2^{2+} species may be uranium (found predominantly as UO_2^{2+} under oxidising conditions) but the analogy is not very close. In a system of varying redox conditions, uranium is probably the only reasonable analogue of plutonium but, especially here, the analogy should be considered as qualitative only.

Neptunium has slightly less complex chemistry, being found predominantly as Np(V) in oxidising conditions and Np(IV) in reducing waters. For the reducing case, thorium may be a good analogue, with protactinium as a possible analogue for oxidising conditions. Again, only uranium could be considered as a qualitative analogue for a case of varying redox conditions.

Technetium is not found in significant quantities in most natural systems. The chemistry of technetium is similar to rhenium but quite different from manganese, its other neighbour in the periodic table. Critical to the behaviour of technetium, however, is the transition from cationic species under reducing conditions to the anionic pertechnetate under oxidising conditions. The redox conditions under which this transition occurs will differ for technetium and rhenium.

Finally, it should be emphasised that there is no general recipe for assessing the relevance of a particular chemical analogue; every system must be evaluated separately and the processes occurring in nature compared to those expected in the repository.

3.2 Natural geological and geochemical systems

Various natural geological and geochemical systems may be investigated as analogues, provided they are appropriate to the repository system of interest. The natural systems that have attracted the most interest are uranium orebodies, naturally occurring high-pH systems and naturally occurring metals, glasses and bitumens.

3.2.1 Uranium orebodies

Economic and subeconomic primary and secondary concentrations of uranium occur in many different geological environments. Their principal interest as a natural analogue lies in the mechanisms which have been responsible for their original deposition, and any subsequent remobilisation of the natural series radionuclides. These processes are shown graphically in Figure 3.2 and are analogous to those which might be expected to occur in the near-field of a HLW or spent fuel repository. Secondary deposits and remobilised regions adjacent to orebodies are of most interest because they usually form at temperatures which are representative of conditions in a repository (i.e. $< 100°C$).

One limitation of many orebodies is that a number of the better known sites are at relatively shallow depths, where high fluxes of oxygenated groundwaters will be dominating both the current and recent transport processes.

For processes where it is interesting to extrapolate to the low-flux, chemically reducing repository conditions expected in a repository, this requires very careful characterisation of the hydrochemical history of the site.

In addition, ore-bodies which have been actively mined may be so perturbed that it becomes difficult to define the natural boundary conditions to the geochemical and hydrogeological systems.

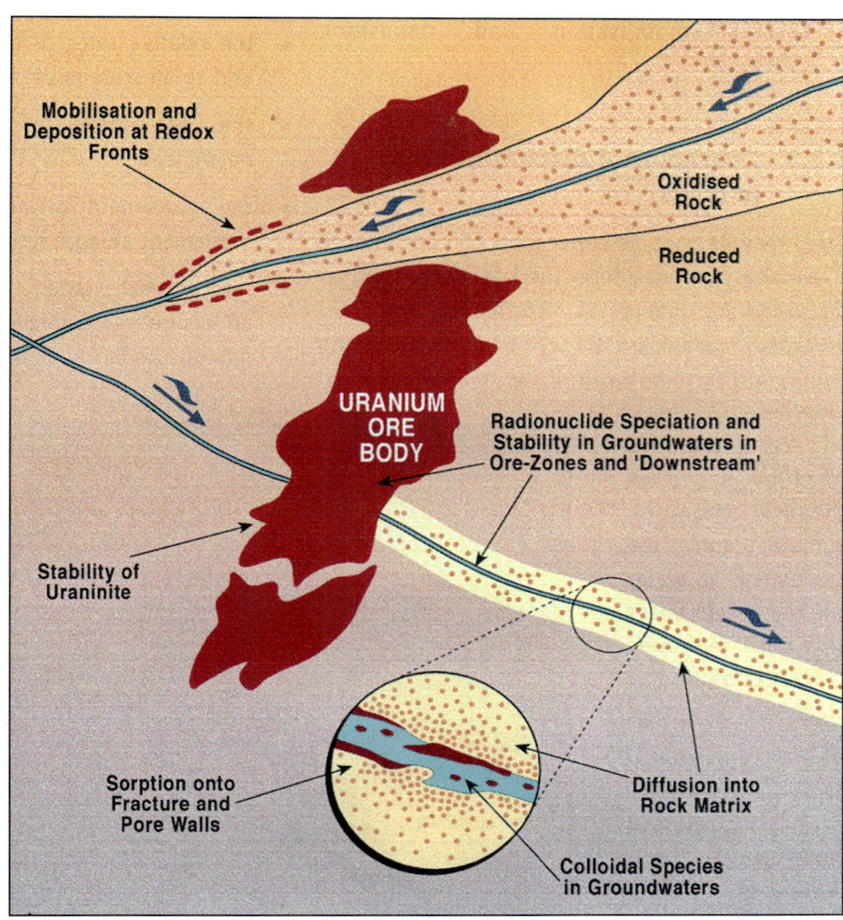

Figure 3.2: Diagrammatic view of a generic orebody showing the principal physical and chemical processes which potentially might be investigated in natural analogue studies in this type of geological environment.

Nonetheless, useful information can be obtained from mines and quarries, as indicated by the work on redox fronts performed at the Osamu Utsumi mine as part of the Poços de Caldas study (see Box 14). However, the results from these studies have yet to be explicitly included in performance assessment.

The main features of uranium orebodies of potential relevance as an analogue are:

- the composition, long-term stability and corrosion/dissolution behaviour of uraninite as an analogue to spent fuel;
- the role of redox processes in mobilising and retarding radionuclides, including redox fronts and other geochemical discontinuities;
- the speciation and solubility controls of radionuclides in groundwaters, including colloid formation and behaviour;
- the downstream retardation processes affecting remobilised radionuclides, including sorption phenomena on various surfaces and diffusion into the rock matrix porosity; and

- the ability to use natural decay series disequilibria to estimate the longevity of various mobilisation and deposition processes.

3.2.2 Geochemical discontinuities in clays

Clays may be used as either a backfill or buffer in a repository, or may form the host rock itself, as discussed in Chapter 2. Geochemical transport processes in clays are not well understood, owing largely to the complexity of the multiple coupled processes of solute and clay interactions which not only drive chemical migration but also control the movement of water and the development of the hydraulic properties of the clay medium.

Where geochemical discontinuities occur in clay as a result of heterogeneity in a sedimentary sequence or due to contact with a compositionally different material, such as an igneous intrusion or the burial of an archaeological artefact, they offer the opportunity to study small scale migration and other processes. These processes are shown graphically in Figure 3.3.

The main features in clay formations of potential relevance as an analogue are:

- the relative roles of diffusion and advection and small-scale physical heterogeneities;
- the estimation of elemental diffusion coefficients;
- the movement of redox fronts, including movement along fractures in clays;
- the thermal stability of clays and thermal effects on transport properties, in cases where igneous bodies have intruded into clays; and

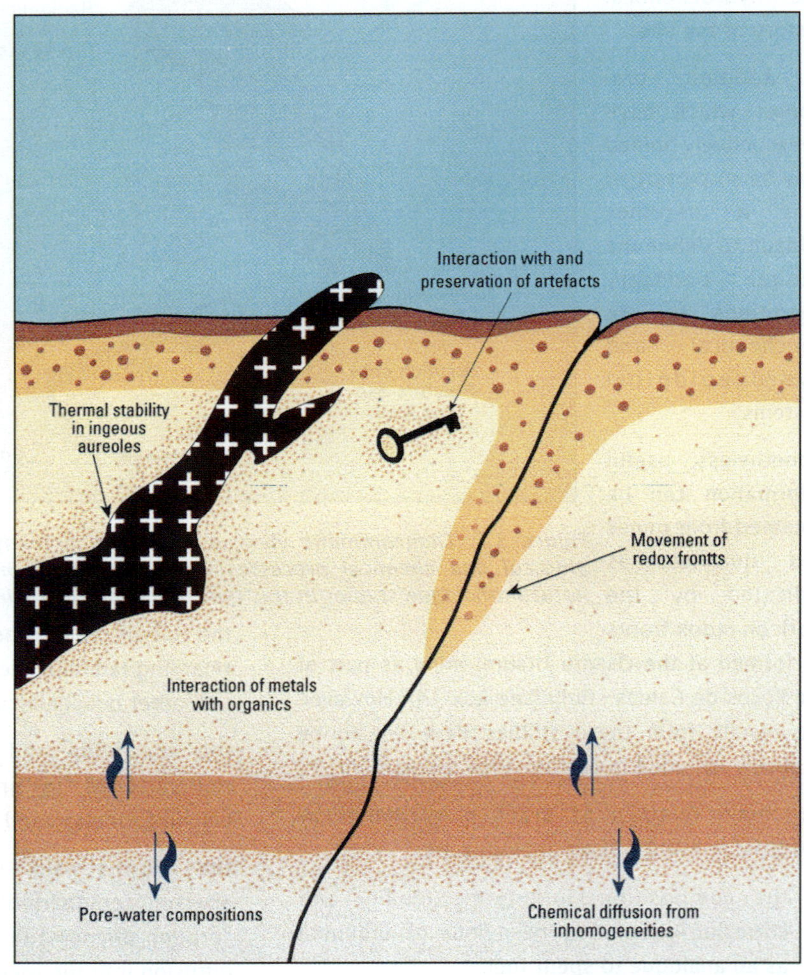

Figure 3.3: Diagrammatic view of a generic clay formation showing the principal physical and chemical processes which potentially might be investigated in natural analogue studies in this type of geological environment.

- the chemical complexing behaviour of radionuclides and other trace elements in pore fluids of different compositions, and in contact with organic material in clays.

3.2.3 Hyperalkaline environments

Natural waters with unusually high alkalinities may be analogous to the porewaters found in cements and concretes. These offer the opportunity to study the hydrochemical behaviour of radionuclides and trace elements in conditions similar to those of some cementitious repository near-fields.

Hyperalkaline waters are relatively rare in nature and have their origin in uncommon systems such as complex rock/water interactions involving the alteration of ultramafic rocks (e.g. at the Oman analogue site) or thermally metamorphosed limestones and marls (e.g. at the Maqarin analogue site). To be reasonably representative of cementitious porewaters, the pH must be greater than 12.

The main features in natural high-pH environments of potential relevance as an analogue are:

- the solubilities and speciation of radionuclides, and other relevant elements in high pH conditions where solubility calculations suggest rates of mobilisation into solution will be very slow;
- the interaction of high pH fluids with surrounding rocks which may be analogous to near-field porewaters which migrate from a repository into the host rock;
- the nature and viability of microbially mediated geochemical processes at high pH, with respect to their potential significance for wasteform breakdown and near-field radionuclide mobilisation; and
- the nature and stability of colloidal species formed in high pH waters, and at the interface between high pH waters and neutral waters.

3.2.4 Hydrothermal systems

Natural hydrothermal systems have sometimes been advanced as being analogous to the near-fields of HLW or spent fuel repositories during the early period after disposal when temperatures around the waste are high. It has been suggested that they provide the opportunity to study geochemical transport processes in warm, near-field fluids.

Unfortunately, the analogy is generally weak, because repository host rock temperatures in most designs (with the possible exception of the proposed US repository at Yucca Mountain, see Section 2.3.1) generally never reach more than about 80 to 100°C, which is lower than the temperature of most hydrothermal systems. In any event, most disposal concepts aim at complete containment of the wastes within the waste package during the early thermal peak. In addition, most natural hydrothermal systems have been flushed with much larger volumes of fluids than would occur around a repository, although the duration and depth of water circulation in the two systems might be comparable (Cathles and Shea, 1992).

Combined, these differences between the hydrothermal and repository systems mean that the supposed near-field rock-water interactions and mass transport analogy may be exaggerated both in terms of the scales and amounts of materials involved, and the nature and kinetics of the reactions observed. In addition, attempts to extract pertinent information for performance assessment from hydrothermally altered granites are usually made difficult by the superimposition of several hydrothermal events at various temperatures.

Nonetheless, a few studies (e.g. Ménager et al., 1992a; Parneix, 1992; Shea, 1998) have been described which claim to have extracted some useful information from hydrothermal or fossil hydrothermal systems. Thus, it may be possible, although very difficult, to obtain usable information on elevated temperature near-field processes from these geological systems but the limitations of the analogy must be borne in mind, and the guideline to 'define the end-use' must be remembered.

The main features in hydrothermal systems of potential relevance as an analogue are:

- alteration of physical properties (e.g. pore and fracture enlargement or blocking) and chemical properties (e.g. fracture and pore surface mineralogy) of near-field rocks which could occur during a thermal transient and which may affect transport and sorption of radionuclides released at a later time;
- the solubility and speciation of radionuclides, and relevant trace elements in geothermal fluids where hydrothermal processes have affected an existing orebody or region of rock with elevated concentrations of relevant elements; and
- elemental and isotopic matrix diffusion profiles associated with veins containing hydrothermally transported and deposited minerals containing relevant trace elements, from which apparent diffusivity values can be determined.

These analogue applications are only really useful to high temperature repository concepts or to scenarios where early canister failure during the initial thermal peak are considered. Otherwise, these applications appear tenuous for most repository designs and normal evolution scenarios, which perhaps explains the lack of interest by performance assessment modellers in using information derived from hydrothermal systems.

3.2.5 Natural occurrences of repository materials

This category includes a wide range of naturally occurring materials analogous to wastes or other repository components. These are generally found as isolated and often uncommon occurrences in nature. Typical examples are glassy igneous rocks, tektites, natural bitumens, iron meteorites and native metals. These materials are appealing analogues, but frequently they are compositionally inappropriate when compared to repository materials, to an extent which may make their properties and behaviour fundamentally different.

Some good analogues can, however, be found, although careful evaluation is essential if data are to be used in performance assessment. As with archaeological analogues (see below), a critical aspect is the physical and chemical nature of the environment in which they are found, and its relevance to that of a repository. The range of studies which can be envisaged is, thus, very dependent on the exact characteristics of the material and of the host system. However, natural materials also have considerable potential for illustrating the basic concepts of repository design and performance for non-technical audiences and, in this case, the demands on the analogue may be relaxed somewhat. The types of studies that have been undertaken include:

- natural volcanic glass corrosion as an analogue for borosilicate glass wasteform dissolution;
- natural metal corrosion and pitting as an analogue for metal HLW and spent fuel canister corrosion;
- decomposition of naturally occurring bitumen as an analogue for the degradation of the bitumen L/ILW immobilisation matrix; and
- stability of naturally occurring cement minerals as an analogue for the long-term behaviour of cement in a L/ILW repository.

3.3 Archaeological analogues

The use of archaeological analogues has grown in importance over the last few years as their potential for performance assessment support and, particularly, for providing illustrations for non-technical demonstrations of safety has been more widely acknowledged (Miller and Chapman, 1995; Miller, 1996b).

The progressive decay of man-made artefacts can provide a direct analogy to the long-term behaviour of repository materials and can be studied at carefully chosen archaeological sites (e.g. Figure 3.4). The types of artefacts useful as analogues ranges from jewellery to buildings.

The time period of study is, naturally, constrained to a few hundreds or thousands of years at most for the majority of relevant materials. However, although shorter in comparison to geological analogue systems, the timescales applicable to archaeological analogues may be more tightly constrained. Although the term 'archaeological' is often used, some man-made systems which are only a few decades old also yield useful information on the rates of processes relevant to the early life of repositories. In some cases, the terms *anthropogenic analogue* or *industrial analogue* are used instead.

Due to the fact that archaeological materials and the environments in which they occur can be quite dissimilar to a repository situation, it is important that care is taken when selecting archaeological systems for investigation. Not all well-preserved, old artefacts can provide quantitative analogue information for input to performance assessment. However, archaeological analogues are also very important as providers of non-technical information and as illustrations of the basic

Figure 3.4. Archaeological artefacts can be used as analogues for the long-term behaviour of engineered barrier materials. In this case, an excavation of a Roman settlement in Colchester, England provided a range of iron and bronze artefacts (rods and rings which might have been used for divination). Qualitatively, this shows the much slower corrosion of iron over bronze (copper alloy). If the chemistry of the metals and the burial environment can be shown to be similar to the repository, then quantitative corrosion rates for materials may possibly be obtained. From Crummy (1997).

concepts of repository design and performance for non-technical audiences. This is partly due to the fact that archaeological systems can sometimes be associated directly with materials, artefacts and time periods with which people are familiar.

To aid the selection of suitable archaeological analogues for both quantitative and qualitative uses, Miller and Chapman (1995) drew-up a list of basic parameters which should be known about the artefact, these include:

- the age of the artefact;
- the composition of the material;
- the deposition and preservation history;
- the physico-chemical environment in which the artefact was preserved;
- any events or environmental changes which have affected the artefact when in situ; and
- the composition of the surrounding soil or sediment.

A key issue is that the burial environment in which an artefact is preserved is as important as the artefact's composition for defining its suitability as an analogue. However, characterising the burial environment can be difficult, especially determining if the environment has been constant over time.

Most artefacts come from the surface or from shallow depths, often in conditions which have been more chemically aggressive than those expected in a repository, e.g. in terms of chemical fluxes and redox environment. Clearly, if the burial environment has been highly chemically active, most artefacts will have been significantly degraded. The general approach to archaeological materials is, therefore, one of trying to estimate rates and mechanisms of degradation as a function of the chemistry of the environment of preservation, and to extrapolate or interpolate the data from one or more sites or objects to the chemical conditions expected in a repository.

However, this type of investigation is potentially prone to bias if focussed on artefacts obtained from museum collections, because museums will (naturally) tend to house the best preserved artefacts. Corrosion rates based solely on archaeological material could, thus, be non-conservative. This sample bias problem is likely to be less important if artefacts are collected in situ, rather than from a museum, for then it would be possible to see artefacts in all possible degradation states for that environment. Analogue studies based on archaeological artefacts must be considered carefully to determine if this type of bias has occurred. Nonetheless, considering all the caveats on the suitability of archaeological artefacts, the range of materials and issues for investigation is potentially wide and includes:

- corrosion of metallic or cementitious objects analogous to waste containers or waste matrices;
- degradation of glasses and cementitious or bituminous materials analogous to wasteforms;
- long-term evolution of the physico-chemical properties of cements and other building materials analogous to repository structures;
- decay and breakdown products of organic materials and complexation with trace elements, analogous to waste degradation; and
- chemical interaction of buried objects with surrounding rocks and soils which may be analogous to near-field processes.

3.4 Sites of anthropogenic contamination

There are many thousands of sites around the world where carelessly disposed wastes and spillages have resulted in the migration of contaminants through subsurface materials. Several of these contaminated sites have been studied as anthropogenic analogues. For example, investigations have been made of:

- the migration at the geosphere-biosphere interface of radionuclides released from the Chernobyl nuclear power plant explosion;
- the migration through volcanic rocks of radionuclides released from underground nuclear explosions at the nuclear weapons test site in Nevada; and
- the migration through rocks and sediments of radionuclides leaked from liquid radioactive waste storage tanks and shallow LLW disposal trenches.

Several other similar contamination systems could be imagined as potential analogues. For example, the injection facilities for the disposal of liquid radioactive wastes in Russia, described in Section 2.3, are planned to be investigated in further detail to understand how these wastes have behaved and migrated underground since injection, as an analogue of radionuclide migration away from a repository.

The problem with many of these anthropogenic contamination sites is that the environments are grossly unrepresentative of either the near or far-fields of any deep geological repository concept. In some cases the physical systems are greatly perturbed (such as around sites of nuclear weapons tests) or the chemistry is too far removed from repository conditions to warrant being considered as an analogue (as may be the case at the some liquid waste injection sites). As a result, the great majority of sites of anthropogenic contamination have little or no analogue relevance for a radioactive waste repository.

Nonetheless, there may be a few anthropogenic contamination sites from where useful analogue information could be obtained. Of particular interest would be locations where radioactive materials have migrated into otherwise undisturbed rocks at concentrations sufficiently dilute to be representative of the releases which may occur in a repository far-field. This is important because, if contaminant concentrations are too high, then they may display solubility limited transport behaviour. This would be in contrast to the repository system in which many elements will be released in such dilute concentrations that they would be expected to remain below solubility limits, especially in the far-field.

If suitable anthropogenic sites can be found with a demonstrable analogue potential, then it may be possible to investigate migration and retardation of transuranic elements in the subsurface environment and, hence, provide a means to support laboratory data for these non-natural nuclides. Nonetheless, particular consideration and care would need to be given to the extrapolation of data from any anthropogenic contamination site to the repository system.

Chapter 4: Analogues of repository materials

This chapter examines how natural analogue studies can provide information on the post-closure behaviour of the many different materials which can be found in the near-fields of the various repository designs discussed in Chapter 2. Most materials in a repository can be grouped into three categories which together comprise the engineered barriers, they are:

- the wasteform,
- the waste packaging, and
- buffers, backfills and seals.

Some materials may be used extensively in a repository; for example, metals may be present in both the wasteform (as decommissioning waste) and the waste packaging (as drums and canisters). Similarly, cement might be present as the wasteform, the immobilisation matrix and the backfill. In addition, some materials, such as cement and steel, may also be used in the construction and support of the repository excavations, and be left in place when the repository is closed.

Many of the materials which will be used in a repository are familiar from everyday experience. However, the requirement for predictable behaviour and longevity in a repository is quite unlike any other demands placed on these materials in the past. Consequently, it is not possible simply to apply the performance of these materials, under normal engineering conditions, to the repository situation. Other materials (e.g. ceramics used as waste matrices) are essentially new and have been developed specifically for use in repositories. With no previous data on the real life performance of these materials in any environment, they require full testing and characterisation.

The performance of many repository materials can be measured most simply by quantifying their rate of degradation or decomposition, such as the rate of dissolution of glass or spent fuel wasteforms. However, in reality, the issue is somewhat more complicated due to the fact that the products of material degradation may impact directly on the performance of the repository system by impeding or enhancing radionuclide transport, or by interacting with other materials in the near-field. For example, laboratory studies of glass leaching indicate that a corrosion layer builds up on the surface of the glass and that this protects the glass, lowering further leach rates.

However, for the most part, it is the stability of the different materials within the repository environment which controls the duration for which radionuclides are retained within the near-field. It is difficult to recreate, in a laboratory, the physico-chemical conditions expected in a repository and impossible to simulate the time scales involved without encountering kinetic effects which cannot be scaled back to natural process rates. It follows that a full appreciation of the stability and longevity of repository materials can only be reached from a combination of complementary field, laboratory and natural analogue studies.

The materials discussed in this chapter are those which are most likely to be present in the near-fields of the different repository designs discussed in Chapter 2, either as the wasteform, in the waste packaging or in the other engineered barriers. They are:

- silicate glass (vitrified waste),
- spent fuel (including mixed oxide fuel),
- mineral and ceramic wasteforms,
- metals,
- bentonite,
- cement and concrete,
- bitumen, and
- organic materials.

Some of these materials exist in nature or have similar natural counterparts. Examples include some metals, bentonite and bitumen. The long-term behaviour of these materials can be investigated in analogue studies of relevant geological systems and appropriate archaeological systems, if the materials have been used historically by man. Other materials do not have equivalent natural counterparts, such as metals which do not occur naturally in a pure metallic (native) form. In this case, the only possible analogue studies are on archaeological systems, and then only if the material has had an historical use. Relevant natural analogue studies of all these materials are discussed in the following sections.

4.1 Silicate glass

At the present time, silicate glass is the preferred matrix for the solidification and immobilisation of the liquid HLW produced from spent fuel reprocessing operations in Europe, the United States and Japan. Currently, full-scale vitrification plants are operating in France (the AVM process at Marcoule since 1978 and the AVH process at La Hague since 1989) and in the UK (at the Windscale Vitrification Facility at Sellafield since 1990). Similar plants will soon be operational in the United States and in Japan.

Vitrification plants incorporate liquid HLW into the glass via calcining and vitrification. The vitrified (glassified) wastes are encapsulated in metal containers for storage prior to final disposal, as shown in Figure 4.1. The attraction of glass as a wasteform stems from its apparent physical and chemical durability under repository conditions, its ease of production and the relative insensitivity of the vitrification process to waste stream composition.

Silicate glass was first proposed as a HLW matrix in the 1950s and, since then, has been investigated in depth in both laboratory and natural analogue studies.

Numerous vitreous compositions have been investigated including phosphate-based glasses, nepheline syenite glasses (derived from a silica poor igneous rock), high-silica glasses and boron containing (borosilicate) glasses. However, the latter, borosilicate glass, has universally been adopted as the favoured vitrified wasteform for HLW because of its stability and relatively low formation temperature (around 1100°C) which minimises waste losses due to volatilisation during manufacture. A number of variant borosilicate glass compositions are being developed in different countries to reflect individual waste streams and manufacturing methods, and a compilation of waste-free base glass compositions is given in Table 4.1.

Glass has also been proposed as a suitable matrix for the immobilisation of surplus plutonium and other actinides generated by the dismantlement of nuclear weapons and the clean-up of weapons production sites. Unlike vitrified HLW (reprocessing wastes) which may contain

Table 4.1: Compositions of four manufactured borosilicate HLW glasses (without a waste loading) and typical rhyolitic, basaltic and tektite natural glasses for comparison. On the basis of silica content, basaltic glasses are most similar to HLW glass, although compositional differences exist, and hence most analogue studies have looked at basaltic glass. Data are in weight %. From Lutze (1988).

	R7T7 France	**SRL131 USA**	**AVM France**	**UK209 UK**	**Rhyolite Natural**	**Basalt Natural**	**Tektite Natural**
SiO_2	54.9	58.7	56.1	68.5	73.1	50.3	71.2
B_2O_3	16.9	14.9	25.3	15.0	-	-	-
Li_2O	2.4	5.8	-	5.4	-	-	-
Na_2O	11.9	18.0	18.6	11.2	3.5	4.9	1.5
K_2O	-	-	-	-	4.5	1.9	1.9
TiO_2	-	1.04	-	-	0.2	2.8	0.8
CaO	4.9	-	-	-	2.6	7.1	3.1
MgO	-	2.1	-	-	1.0	3.9	2.9
Al_2O_3	5.9	-	-	-	11.9	17.3	12.5
ZnO	3.0	-	-	-	-	-	-
ZrO_2	-	0.45	-	-	-	-	-

around 0.01 % plutonium, the loading for these glasses is expected to be at least 5 % plutonium. This high actinide content places different constraints on glass behaviour compared to the standard HLW glasses; for example, the glass matrix must be designed to minimise the risk of a nuclear criticality event. Although the exact composition of the glass to be used for plutonium vitrification has not yet been identified, it will be somewhat different to the standard borosilicate glass used for reprocessing wastes. Laboratory studies are currently investigating alternatives such as alkali-tin-silicate (ATS) glasses which contain a small amount of gadolinium as a neutron adsorber (Bates et al., 1996) and lanthanide-rich glasses (Bibler et al., 1996). Since the plutonium wastes have a lower activity than HLW, the glass wasteforms for plutonium are likely to have additional radioactive components added to increase their activity to deter diversion of the plutonium and maintain nuclear safeguards (making them as radioactive as spent fuel, known as the 'spent fuel standard').

Figure 4.1: Example of the borosilicate glass waste form in a section of steel canister.

There is a large volume of literature concerned with the behaviour of borosilicate glass wasteforms under repository conditions, including laboratory studies and mechanistic modelling. There have also been many investigations of natural glasses as analogues of waste glasses, with particular emphasis placed on defining dissolution processes and rates, and on determining the nature of solid secondary alteration products, for example: Daux et al. (1991), Smith (1991), Magonthier et al. (1992), Crovisier et al. (1992) and Mazer et al. (1992).

The abundance of natural glass analogue studies is partly due to the fact that a considerable body of knowledge has been accumulated on natural glasses because they have been studied for many years by geologists and mineralogists, outwith the concerns of radioactive waste disposal. The transfer of knowledge regarding alteration and dissolution from natural glasses to glass wasteforms was thus simplified. However, whilst the analogue literature is at first sight extensive, in effect much is repetition of the same or similar studies and results.

The majority of natural glasses are formed when lavas and magmas cool too rapidly to allow mineral growth to take place in the melt. As such, natural glasses exhibit similar bulk compositions to those of extrusive igneous rocks, ranging from silica-poor (sideromelane: basaltic composition) to silica-rich (obsidian: rhyolitic composition). On the basis of silica content, the basaltic glasses are generally considered to be the most appropriate analogues for the borosilicate HLW glasses, as shown in Table 4.1.

However, significant compositional variations do occur between basaltic and borosilicate glasses and these must be borne in mind when interpreting the results of the analogue studies to avoid reaching inappropriate conclusions. In particular, the natural glasses obviously do not contain any waste component and, hence, the radiation induced effects (e.g. radiation damage to the glass matrix) which might occur in the repository will not be observed in natural glasses. In addition, boron occurs in most natural glasses in only trace amounts, so the impact of boron on glass durability cannot be addressed in studies of natural glass. Furthermore, nuclear magnetic resonance investigations of basaltic and borosilicate glasses suggest that the arrangement of certain atoms (such as sodium) in the glass can vary between natural and man-made glasses, and that this may have some impact on their relative long-term durabilities (Angeli et al., 1998).

Rhyolitic glasses and tektites (glassy meteorites) have been proposed as analogues for the high-silica, low-alkali formulations considered as a glass wasteform but these glass compositions are no longer actively being investigated for HLW reprocessing wastes. However, these silica-rich glasses may be suitable as analogues for some of the novel glass formulations being considered for the immobilisation of excess plutonium.

In addition to natural glasses, man-made (archaeological) glasses have also been considered as analogues to HLW glasses. Glasses are thought to have been first manufactured in Egypt around 2500 BC (Tomabechi, 1995) and there is some indication that glass had been discovered even earlier than that. The ability to produce glass soon spread to other civilisations and it has been used extensively ever since. A wide range of glass compositions has been produced throughout history but many are chemically too dissimilar to modern borosilicate glasses to be good analogues. Early glasses were usually soda-lime based until about 1000 AD when potassium-lime glasses were introduced and became widespread. Many other sorts of glass, such as lead-glass, have been produced in smaller amounts for specific purposes (Kaplan, 1980a). For a discussion of various historical glass compositions and their potential use as analogues, see Vandiver (1994).

An interesting consideration is the variety of metallic compounds used to colour archaeological glasses which may be considered loosely analogous to the waste incorporated into borosilicate glass. Different production techniques could potentially influence the chemical and thermal stability of glasses, and hence their longevity. Therefore, the basic similarity in the manufacturing process between archaeological and modern borosilicate glass (in terms of production temperatures and cooling rates), may allow relevant information to be obtained from the archaeological glasses that could not be obtained from natural glasses.

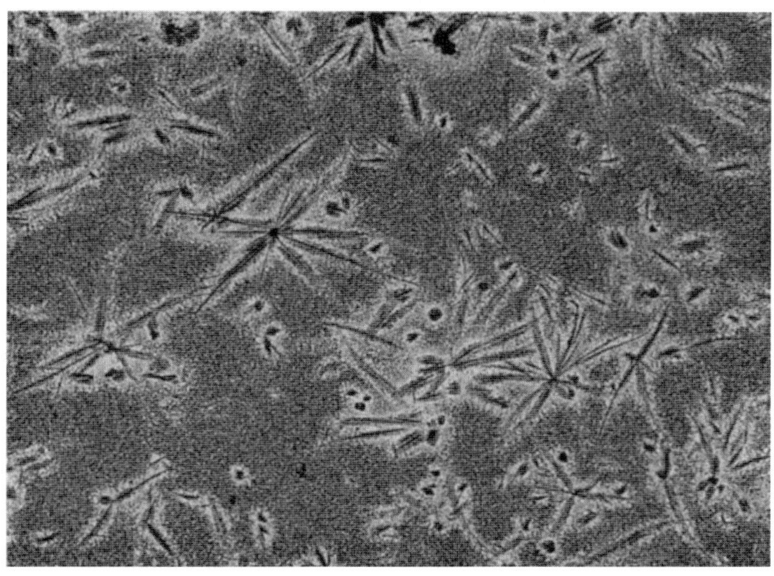

Figure 4.2: Photomicrograph of the devitrification of a natural volcanic glass from Scotland to form radiate clusters of crystals. The bulk of the glass contains many thousands of smaller crystals giving the glass its grey colour. The field of view is 1 cm.

Glass is a metastable substance and will, over a period of time, devitrify (nucleate and crystallise) to a more thermodynamically stable assemblage of mineral phases. In addition, glass can dissolve in the presence of water and react to form secondary phases. Thus, the processes of devitrification and dissolution of HLW glass in the repository potentially may allow radionuclides to be released from the glass matrix.

The issues of most relevance to the behaviour of glass in the repository which have been (or potentially could be) addressed in natural analogue studies are:

- devitrification;
- dissolution and alteration;
- radiation induced effects; and
- radionuclide retardation by secondary alteration products.

These issues are discussed in the following sections.

Devitrification

Devitrification (solid state crystallisation of the glass matrix) can have a number of deleterious effects on a glass wasteform. Laboratory experiments indicate that devitrification may cause a glass to be more prone to chemical degradation or dissolution, with measured increases in corrosion rate of a factor of three (see Lutze, 1988). Nonetheless, laboratory techniques are unable to predict devitrification rates at repository temperatures with any degree of confidence.

Most natural volcanic and other glasses show some signs of devitrification (Figure 4.2). However, the rate of devitrification of natural glasses is sufficiently slow such that the duration for which natural glasses survive before becoming totally devitrified varies from a few thousand years to tens of millions of years (Forsman, 1984), although most natural glasses are younger than the Miocene, i.e. less than 25 million years old (Marshall, 1961).

Ewing (1979) compiled statistics on the ages of 425 natural glasses from North America and found that more than half were younger than 2 million years, but some were as old as 40 million years. These statistics demonstrate the general point that all glasses ultimately will devitrify but that they can persist in a natural environment for very long periods of time. However, these data cannot be used quantitatively to predict the lifetime of borosilicate glass in the repository because information on the corresponding compositions of the natural glasses and co-existing waters, and the thermal histories etc., which are necessary fully to explain this range in ages is not available. Nonetheless, the average natural glass age of around 2 million years is significantly longer than the time periods relevant to radioactive waste disposal, so these data, together with other observations of natural glasses, can be used to support the qualitative conclusion that devitrification is unlikely to be a significant problem over repository timescales.

Localised occurrences of considerably older glasses (some Precambrian, i.e. more than 600 million years old) are occasionally reported (e.g. Philpotts and Miller, 1963; Lindqvist and Laitakari, 1980). In some individual exposures, old glass is fairly common, such as in the 80 million year old Cyprus Ophiolite (Robinson et al., 1983; Rautenschlein et al., 1985). An interesting Precambrian glass occurrence is reported by Palmer et al. (1988) who found small fragments of 1100 million year old rhyolitic glass at Lake Superior, Canada. The glass occurred as 25 to 35 μm sized, silica-rich, brown isotropic fragments. Although its composition is quite dissimilar to that of borosilicate glass, it is of relevance because its longevity has been ascribed to the physical state of the surrounding rock which was believed to have isolated the glass from the groundwaters.

This conclusion is supported by the abundance of extremely old natural glass collected from the Moon. The famous orange glass collected by the Apollo 17 astronauts in the vicinity of the Shorty Crater was dated radiometrically using the $^{40}Ar/^{39}Ar$ method by Husain and Schaeffer (1973) at 3710 ± 60 million years; some three orders of magnitude older than an average terrestrial glass. On the Moon, free water does not exist and, hence, devitrification proceeds exceedingly slowly. This observation is relevant to radioactive waste repositories because the glass wasteform will be isolated from groundwater for extended periods of time by the metal canister, as discussed in Section 2.3.

Numerous old glass tektites (Izett, 1991) have been discovered from the Cretaceous-Tertiary boundary (60 million years) and these, together with other tektites, may be useful if they approximate to the composition of some of the novel wasteforms being developed for the immobilisation of excess plutonium. An interesting observation of tektites is that they generally show very little devitrification or other alteration compared to basaltic glasses, which has been ascribed to their very high silica contents (Ewing and Haaker, 1979).

Devitrification can be observed in archaeological glasses but no known analogue studies have addressed this issue in detail. However, it is probably not worthwhile undertaking any specific studies because the devitrification of archaeological glass exposed to air with varying degrees of humidity and temperature is unlikely to proceed at the same rate or in the same manner as devitrification of borosilicate glass in a water saturated repository.

Dissolution and alteration

The related processes of dissolution and formation of secondary alteration products are crucial for understanding the release of radionuclides from borosilicate glass in the repository environment.

After canister failure, when the glass wasteform first comes into contact with groundwater, the glass will begin to dissolve. Predicting the rate of glass dissolution is problematic because the dissolution rate is not thermodynamically controlled due to the fact that the glass is a metastable material. This means that the glass and the co-existing groundwater can never reach true thermodynamic equilibrium and, thus, the dissolution process is not solubility limited. In contrast, the dissolution of most crystalline substances is solubility limited, meaning that dissolution stops when the concentration of the principal elemental constituent(s) reach a thermodynamically controlled saturation point (solubility limit) in the water. The lack of solubility limited dissolution for glass means that the durability of the glass wasteform is, instead, controlled by the kinetics of the alteration process which proceeds by ion exchange and hydrolysis reactions.

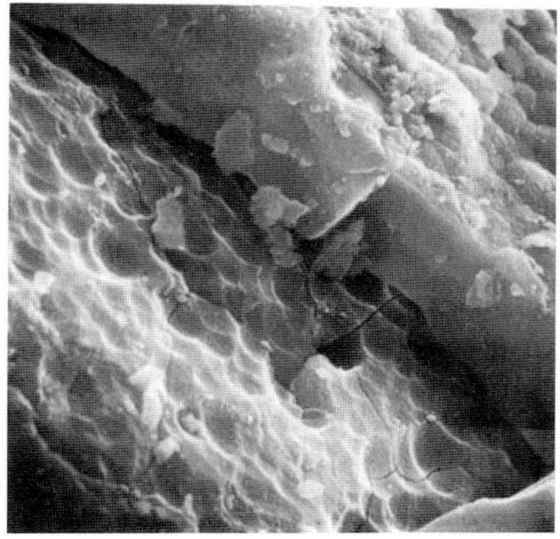

Figure 4.3: Electron microscope photomicrographs of alteration products on the surfaces of samples of basaltic glasses from British Columbia. Top: A palagonite layer forming over a dissolution-pitted glass surface (× 4000). Bottom: Zeolites and a mat of smectite growing over the alteration zone (× 1500). From Byers et al. (1987).

As glass corrosion proceeds, waste elements released from the glass to solution can become supersaturated with respect to certain solid phases, leading to the precipitation of secondary alteration phases on the glass surface, as shown in Figure 4.3. These secondary phases can incorporate some of the waste elements released from the glass, limiting their mobility.

The combination of ion exchange, hydrolysis and precipitation reactions causes an alteration rim to form on the glass surface which comprises three distinct layers, as shown in Figure 4.4. The innermost layer, adjacent to the pristine, unaltered glass, is a thin reaction (diffusion) zone in which the glass is depleted in soluble elements (e.g. boron, lithium and sodium) and where compositional variations and pitting can occur. Immediately outside the reaction zone is the altered layer where amorphous, gel-like phases exist. This layer is generally porous and allows reaction products and water to diffuse through it. The third layer occurs on the surface and is formed by growth of an authigenic cement composed of crystalline phases, such as clays and zeolites, which form either by direct precipitation or by replacement of the amorphous gel layer.

These reactions are observable and measurable in laboratory studies, and the data suggest that the formation of the altered layers provides a kinetic control on the bulk glass corrosion rate, causing the rate to slow with time. Consequently, there is uncertainty over the validity of extrapolating the results of short-term laboratory studies to the long time periods relevant to radioactive waste disposal. Therefore, numerous analogue studies have attempted to quantify long-term natural

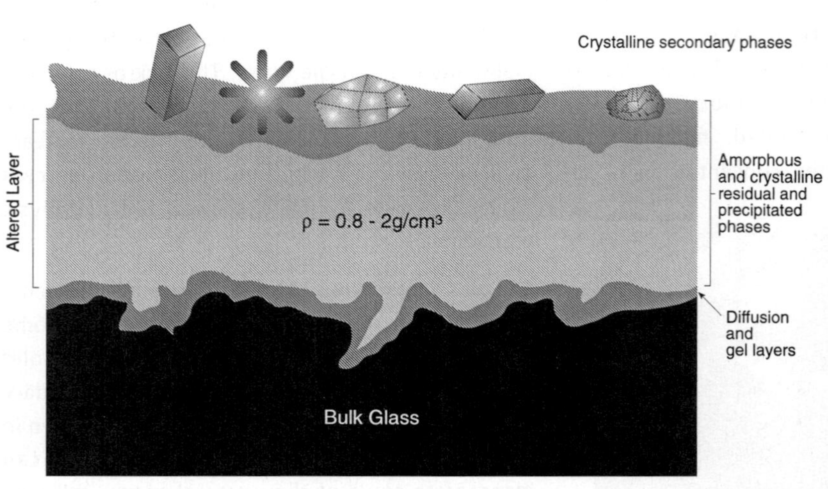

Figure 4.4: Schematic illustration of the three alteration layers that form when glass corrodes in water. The crystalline precipitate layer may become sufficiently thick that further alteration of the glass is inhibited. 1 = distributed amorphous precipitate; 2 = highly structured layered region, generally amorphous with species migration; 3 = glass reaction zone, exhibits extensive pitting, depletion of soluble elements. From Lutze (1988).

glass dissolution rates and the nature of the secondary solid alteration products which form on the glass surface. However, few of these analogue studies have been able to investigate alteration of natural glass under conditions that are truly representative of the repository near-field, in terms of temperature, pH, Eh and groundwater composition. Nonetheless, these studies have indicated that the general degradation mechanism observed to affect natural glasses is essentially the same as that observed in laboratory studies on borosilicate waste glasses (e.g. Mazer et al., 1992).

Many analogue studies have also indicated that the nature of the alteration materials identified on natural basaltic glasses is the same or similar to those identified in laboratory experiments of HLW glass alteration (e.g. Crovisier et al., 1992; Mazer, 1994). Commonly, it is found that the early, gel-like material which forms on natural and borosilicate glass is palagonite, which is replaced over time by more crystalline zeolite phases.

Many attempts have been made to establish the time dependence of basaltic glass dissolution by measuring the thickness of alteration rims on glasses of known ages. A good example of one such analogue study is that by Grambow et al. (1986) who examined basalts from subaqueous environments, such as pillow lavas from the sea floor at Iceland and from deep sea drill cores. The alteration rims on these samples varied from 0 to more than 1000 µm, with the thicker rims being found on deep-sea dredged samples.

Unfortunately, meaningful calculation of the corrosion rate on these samples was frustrated by two problems; the first was establishing the age of the glass, which cannot easily be dated by normal radiometric methods, and the second was that an assumption had to be made for the duration of contact with water. The result is that a wide range of apparent dissolution rates was determined. However, the data did appear to suggest that there was a mechanistic control on the alteration rim thicknesses which Grambow et al. (1986) attributed to a two-stage dissolution process:

a) The first stage process occurs in glasses on the open sea floor. In this environment, rapid hydration of the glass occurs and the amorphous palagonite forms, but growth of the authigenic cement is restricted due to the low silica concentrations in the seawater. At this point dissolution rates are relatively fast, ranging from 3 to 20 µm/1000 years.

b) The second stage process operates when the glass has become buried by pelagic sediments. In this environment, the porewater rapidly becomes supersaturated with silica and, consequently, the growth of authigenic cements on the glass surface is enhanced. Eventually the authigenic cement will completely cover the glass surface. Further dissolution is then diffusion-controlled because the cement layer is relatively impermeable. As a consequence, the dissolution rate is significantly reduced, typically about 0.1 µm/1000 years.

This two-stage nature of glass dissolution means that it is inappropriate to calculate a uniform dissolution rate simply from the thicknesses of reaction rims and the ages of the samples. Instead, dissolution rates are strongly controlled by the physico-chemical environment. This conclusion has implications for extrapolating data from short-term laboratory experiments to the long time periods of relevance to performance assessment calculations. It is most likely that the slower second process is more applicable to the repository situation than the rapid first process because groundwater movement through the near-field will be restricted by the bentonite buffer, allowing silica saturation to be approached

A unique experiment designed to measure palagonitisation rates was performed in the basaltic tephra on the volcanic island of Surtsey over the time period 1963 to 1967 (Jakobsson and Moore, 1986). Palagonite formation was monitored via a borehole penetrating into a hydrothermal system. The palagonite formation was observed to be accompanied by precipitation of calcite, chabazite, phillipsite, analcime, tobermorite and smectite clays. The rate of palagonite formation was temperature-dependent, doubling with every 12°C increase. At 100°C the palagonite formation rate was 3 µm/year.

In the repository, glass corrosion should be much slower than this because the temperatures at which glass-water contact occurs will be much lower than at Surtsey. The canisters should not fail until long after the thermal peak has subsided and the availability of water will be restricted by the low permeability buffer around the canisters. Furthermore, the composition of the Surtsey basalts is significantly different to the borosilicate glass wasteform. However, the Surtsey corrosion data could, perhaps, be viewed as an extreme upper bounding limit to borosilicate glass corrosion rates.

In addition to temperature and the flux of water, the dissolution rate must also depend on the composition of the water, especially pH, Eh and concentration of silica. In the repository environment, no palagonite formation will occur when the canister is intact and, after canister failure, palagonitisation will occur in an environment where solute transport occurs predominantly by diffusion. The effect of chemically reducing, mildly alkaline conditions on the long-term rate of palagonite formation is unknown because this issue has not been addressed in natural analogue studies. These restrictions are generally not taken into consideration in natural analogue studies where, it would seem, access of water to natural glass is often not restricted, other than by the formation of the authigenic cement on the glass surfaces, and the co-existing waters are normally oxidising and sometimes slightly acidic. These factors make it highly problematic to draw firm conclusions regarding borosilicate glass corrosion rates from natural analogue data or to derive quantitative data for direct input to performance assessment codes. However, the qualitative evidence from natural analogues should not be underestimated for providing confidence that the glass degradation processes are well understood and for providing upper bounding limits to the degradation rates.

Archaeological glasses also exhibit degradation and formation of secondary alteration products, although sometimes the degree of alteration appears limited. It has previously been remarked upon (Kaplan, 1980b) that some archaeological glasses have survived for 3500 years with only minor degradation. This is despite the high alkali content of some of these glasses and the aggressive environments they have endured, all of which could be expected to decrease their life spans. In contrast, the repository environment may be considered less hostile. Kaplan (1980b) identified five decomposition processes in archaeological glasses:

- *crizzling*, the formation of a network of small cracks over the artefact's surface (Figure 4.5);
- *weeping*, the 'sweating' of water droplets when excavated glasses initially come in contact with the air;
- *pitting*, formation of pits and scars on the surface filled with weathering products, which result from abrasion or chemical dissolution;
- *layering*, development of a filmy iridescent surface formed from multiple layers of mica-like minerals; and
- *crusting*, development of amorphous residues by leaching of the glass.

Figure 4.5: An example of extensive crizzling on a Venetian goblet from the collection at the Victoria and Albert Museum in London. Photographs courtesy of Dr. David McPhail.

All of these decomposition processes relate to the same glass corrosion mechanism described earlier (comprising ion exchange, hydrolysis and precipitation reactions), although they reflect the fact that the glass is reacting in a humid atmosphere rather than in a water saturated environment. Crizzling would appear to be a direct consequence of glass exposure to a humid atmosphere (Brill, 1975; Ryan et al., 1993) when an ion exchange reaction takes place between alkali ions in the glass and H^+ or H_3O^+ ions from atmospheric water. This replacement results in the surface of the glass being placed under tensile stress and in the formation of a network of microcracks in the surface layers. Crizzling is exacerbated if a glass artefact is exposed to changing levels of humidity, which suggests that it may only become relevant for HLW glass in an hydraulically unsaturated environment, such as the proposed Yucca Mountain repository in the USA (see Section 2.3.1), where changing temperatures and water content may cause considerable humidity variations in the repository. However, at Yucca Mountain, the glass wasteform would be isolated from the humid atmosphere for the lifetime of the canister which will be longer than the initial high temperature period.

Radiation induced effects

In HLW glass, the radionuclide waste content is homogeneously dispersed throughout the glass matrix and acts as an internal radiation source whose strength is time (decay) dependent. This radiation may cause structural damage to the glass and to the secondary alteration products, and this might potentially increase the radionuclide release rate by accelerating glass-water interactions.

Radiation induced effects cannot easily be investigated in natural glasses because they generally contain low (natural background level) radionuclide contents. In some cases, sporadic radiation damage effects could be observed in natural glasses if they are in close proximity to radionuclide bearing minerals (such as zircon or apatite) but it would not be sensible to upscale these observed effects to the very high radiation doses that are present in HLW glass.

A more interesting and productive method to examine radiation effects might be to examine archaeological glasses coloured with uranium compounds (Figure 4.6). These uranium glasses are extensively described by Tomabechi (1995). In 1789 the German chemist Klaproth first extracted pure uranium compounds and later published recipes for mixing these compounds into glass to colour them. However, it was not until the first decades of the 1800s that uranium glasses became popular for making decorative items. The oldest existing item of uranium glass with a known fabrication date was made in 1841. Some of the best known uranium glasses were manufactured at this time in Bohemia by the Reidel company. They perfected uranium glasses with either a yellow colour, known as

Figure 4.6: A goblet and decanter made from uranium glass. Such pieces can contain up to 5 % uranium and offer a potential archaeological analogue to the borosilicate glass wasteform. Photographs courtesy of Ken Tomabechi.

Annagelb, produced by adding sodium diuranate ($Na_2U_2O_7{\cdot}6H_2O$) to glass or a green colour, known as *Annagrün*, by adding copper sulphate to Annagelb. Similar uranium compounds were used by other manufacturers.

These uranium glasses have an attractive surface sheen effect due to the fluorescence of the uranium and are commonly referred to as *vaseline glass*. They are known commonly to contain up to 5 % uranium, with some pieces possibly containing more. These uranium glass pieces essentially provide a century long glass doping experiment, in which the affect of radiation damage on the glass may be measurable. However, as far as is known to the authors, these uranium glasses have never been examined as analogues to HLW glasses.

Radionuclide retardation by secondary alteration products

As discussed earlier, when glass corrodes, elements released to solution can become supersaturated with respect to certain solid phases, leading to the precipitation of secondary alteration phases on the glass surface and in colloidal form in the groundwater.

The secondary phases forming on the glass surface can directly incorporate radionuclides released from the glass matrix, effectively immobilising them. Understanding the formation of these radionuclide-bearing solid phases is important because these phases can control radionuclide solubilities in the near-field system.

Secondary alteration phases may also retard radionuclide migration even if radionuclides are not directly incorporated into the mineral by providing a high surface sorption capacity. In the later stages of dissolution, a more crystalline cement is formed which may include zeolites. The potential retardation capacity of these zeolites is likely to be much higher than for the initial amorphous alteration products. However, compared to the retention of radionuclides directly incorporated into the alteration phases, sorption onto their surfaces may be less important.

This view has been supported by short-term laboratory hydration and leaching experiments performed on radioactive waste glasses which have suggested that alteration products (typically palagonite) do act to retain a range of elements, including iron, rare earth elements and actinides (see, for example, Petit et al., 1989) but, in most cases, this appears to be due to direct incorporation rather than by surface sorption.

Despite the large number of analogue studies performed on natural glass to investigate dissolution and alteration processes, few have looked in any detail at the retardation capacity of the secondary alteration products. One analogue study which did was performed on Icelandic basalts to investigate the behaviour of rare earth elements during basaltic glass dissolution (Daux et al., 1991). The rare earth elements are considered as possible chemical analogues of the trivalent transuranic elements americium and curium (see discussion in Section 3.1). Mass balance calculations based on major and trace element analyses of the glass and palagonite showed that dissolution of the glass resulted in a net release of rare earth elements. However, it would appear that, for these elements, retardation by the alteration phases via sorption processes may be insignificant compared to the retention of radionuclides directly incorporated into alteration phases as they form. This analogue observation supports the laboratory data described above.

An alternative method of investigating radionuclide retardation by secondary alteration products might be further examination of the uranium glasses mentioned earlier. Investigation of these archaeological glasses might reveal whether uranium was incorporated into, or sorbed onto, the secondary alteration layer, or whether it was lost entirely from the glass. If old, discarded uranium glass fragments from a water-saturated burial environment could be investigated, this could potentially yield interesting results.

Conclusions

Natural glasses several millions of years old are relatively common, which provides qualitative evidence of their stability and durability in natural systems. The fact that glasses millions of years old are not completely devitrified suggests that devitrification may not be a problem for HLW glass. Furthermore, the existence of much older glasses from water-free environments suggests that, in the repository, the onset of devitrification will be delayed until after canister failure. However, natural glasses cannot be used to provide a quantitative estimate for the time at which devitrification will begin, or the rate at which it will proceed, in the repository environment.

Examination of natural glasses from water saturated environments indicates that they corrode by the same mechanism observed in laboratory experiments on HLW glasses. Archaeological glasses also appear to degrade by the same mechanism. The dissolution rates measured on natural glasses are variable but always very slow. The dissolution rate is kinetically limited and is likely to be controlled by the formation of layers of solid secondary alteration minerals, such as palagonite, which restrict further water contact with the glass.

Radionuclides released from the glass matrix may be incorporated into the secondary alteration minerals. These minerals then provide thermodynamic controls on the solubility of radionuclides in the system. The initial alteration products from glass decomposition (amorphous palagonite) have some potential for retarding radionuclide release. In the later stages of dissolution, a more crystalline cement is formed that may include zeolites. The potential retardation capacity of this cement is likely to be much higher than for the initial amorphous alteration products. However, compared to the retention of radionuclides directly incorporated into the alteration phases, sorption onto their surfaces may be less important.

Natural analogues have not yet provided any useful information regarding radiation induced effects on glass durability because suitable glasses have not yet been examined. However, it might be possible to obtain some relevant data from examination of archaeological glasses containing uranium oxides as colourants.

In all cases, the differences in the chemistry between HLW glasses, and natural and archaeological glasses needs to be considered when interpreting analogue data to avoid drawing inappropriate conclusions from the studies. Nonetheless, the qualitative evidence from analogues on natural and archaeological glasses has added to confidence that the glass degradation processes are well understood and has provided upper bounding limits to the degradation rates.

4.2 Spent fuel

The fuel burnt in most nuclear power plants is crystalline UO_2, although some reactor designs such as the Magnox reactors developed in the UK burn metallic uranium fuel. Metallic uranium does not occur in nature and consequently there are no natural analogues for it; hence metallic uranium is not considered further. In the following discussion, spent fuel refers to used UO_2 fuel unless otherwise stated.

The UO_2 fuel is manufactured in the form of pellets which are stacked into Zircaloy fuel rods, as shown in Figure 4.7. This uranium oxide is specifically chosen for the fuel owing to its stability at high temperatures. Although the fuel is typically > 95 % UO_2, it is notoriously non-stoichiometric and, once irradiated, can contain a number of additional characteristic components (Curtis, 1996):

- inclusions of metallic minerals and oxides containing non-volatile fission products in the spent fuel matrix and along grain boundaries;
- solid solutions in the UO_2 matrix containing non-volatile nuclear products;
- solid phases containing volatiles in some parts of the fuel; and
- gaseous bubbles within grains, grain boundaries and structural defects, and in void spaces in the fuel assemblies.

Figure 4.7: Top: Nuclear reactor fuel pellets. One pellet produces the same energy as 800 litres of oil; about five pellets are needed to supply the average house for one year. Bottom: The pellets are encased in fuel rods which are bundled together to form fuel assemblies. The photograph shows the fabrication of one fuel assembly.

The radioactive components present in the spent fuel include the remaining natural series radionuclides (predominantly ^{235}U), fission products, actinides and actinide daughters. In addition, nuclear activation products will be present in the non-fuel parts of the fuel assemblies.

Analogues for spent fuel are the naturally occurring uranium minerals uraninite and, to a lesser extent, pitchblende. These natural minerals have a nominal composition of UO_2 but, in reality, are a mixture of $UO_{(2.00-2.07)}$ and $UO_{(2.23-2.25)}$ (Johnson and Shoesmith, 1988). Pitchblende refers to the fine grained, poorly crystalline variety. Crystallographically, spent fuel and uraninite are essentially identical; both are cubic, having the same structure as fluorite.

However, whilst there are obvious chemical and structural reasons for using uraninite as a natural analogue for spent fuel, it must be borne in mind that there are also important differences between the two phases. Most importantly, spent fuel is artificially enriched in ^{235}U and contains nuclear reaction products. In contrast, uraninite contains a higher proportion of other, non-radiogenic, impurities. In addition, the thermal history of spent fuel is unlike that of natural minerals. This thermal history, particularly the high thermal gradient present across the fuel in the reactor, may cause the spent fuel to exhibit lattice and crystallisation structures not evident in uraninite, although the high temperatures may rapidly anneal any such defects.

Other lattice defects may form in spent fuel as a result of more extensive radiation damage. In turn, these lattice defects may influence the reactivity of the phase by, for instance, increasing reactive surface areas due to thermal expansion cracks. Such phenomena will be specific to each fuel batch and controlled by its burn-up history in the reactor. It is important, therefore, for the fuel to be well-characterised and for these characteristics to be borne in mind when interpreting analogue information obtained from naturally-occurring uraninite and pitchblende.

During reprocessing of standard UO_2 reactor fuel, significant amounts of plutonium are removed and are stored for future use. One application of this plutonium is to mix it with uranium during fuel fabrication to create a mixed oxide fuel (MOX) which typically comprises some 5 % PuO_2 in the mix. MOX fuel is currently being burnt in more than 30 reactors around the world and several more are licensed to burn MOX (NEA, 1997).

Once used, it is possible that MOX fuel will not be reprocessed but will be stored prior to geological disposal. Little specific information is available regarding disposal concepts for spent MOX fuel, but it is likely that it will be treated in a similar manner to 'normal' spent fuel. Due to the different chemistries of plutonium and uranium, the behaviour of the PuO_2 component of MOX fuel may be expected to differ from the UO_2 component in the repository environment. Plutonium exists naturally in only very low concentrations of two main isotopes; as the last remaining traces of the long-lived, primordial ^{244}Pu and as shorter-lived ^{239}Pu generated by neutron capture by naturally occurring ^{238}U. Only extremely small quantities of these isotopes of plutonium can be found in uranium minerals (Katz et al., 1986). Consequently, no natural minerals with sufficiently high concentrations of plutonium exist to be suitable as analogues for MOX fuel.

The issues of most relevance to the behaviour of spent fuel in the repository that have been (or potentially could be) addressed in natural analogue studies are:

- dissolution and radionuclide release; and
- radionuclide retardation by secondary alteration products.

These issues are discussed in the following sections.

Dissolution and radionuclide release

It is clear that the stability of spent fuel is critical if it is to retain radionuclides in the repository environment. In an attempt to quantify the stability of UO_2 and its dissolution rate, many laboratory experiments have exposed spent fuel or non-irradiated UO_2 to a wide range of solutions with different pH, Eh, temperature and electrolyte concentration, only a few of which may be considered similar to natural groundwaters. Many of these experiments can be criticised because they were performed on powdered UO_2 or sintered pellets in an attempt to accelerate the dissolution process to a measurable rate. Nonetheless, these

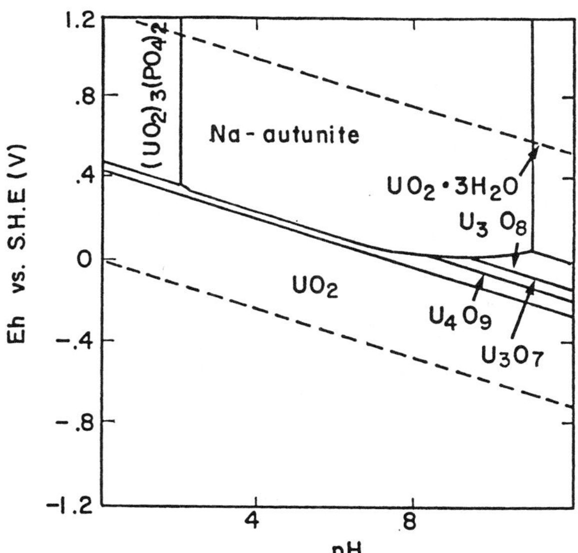

Figure 4.8: Stability (Eh-pH) diagram for uranium species as defined in laboratory experiments using synthetic groundwater at 150°C. The solid solution boundaries are drawn at [U] = 10^{-9} mol/kg. From Finch and Ewing (1991).

experiments strongly indicate that UO_2 dissolution is extremely slow under the chemically reducing conditions which are expected to dominate in the repository near-field. The stability of uraninite and other uranium phases in different chemical environments is shown in Figure 4.8.

These laboratory studies further show that the dissolution rate of UO_2 becomes faster as conditions becomes increasingly oxidising but that it does not become significant until the fuel is oxidised beyond the U_3O_7 state (Johnson and Shoesmith, 1988). This has relevance since, although a spent fuel repository will be sited where groundwaters are reducing, radiolytic oxidant production may cause near-field condition to become locally oxidising, enhancing the dissolution rate of the spent fuel. Radiolysis is discussed in more detail in Section 5.4. However, even if the matrix does not dissolve completely, radionuclides may still be released if the spent fuel is oxidised to the U_3O_8 state because the crystal lattice structure of the material is significantly rearranged at this oxidation state. Liberation of radionuclides would be possible during lattice rearrangement and, thus, oxidation of the fuel can lead to radionuclide loss by both dissolution and lattice rearrangement processes (Brandberg et al., 1993). Loss due to lattice rearrangement could potentially be the most important release mechanism for very soluble elements, such as strontium.

As with all laboratory dissolution experiments, there is uncertainty over the validity of extrapolating the results to the long time periods relevant to radioactive waste disposal. Therefore, numerous analogue studies have attempted to confirm the laboratory data and to quantify long-term natural uraninite dissolution rates and processes.

In a general sense, the existence of abundant uranium orebodies containing natural uraninites hundreds of millions of years old attests to the possibility of slow dissolution kinetics in natural systems. In an attempt fully to explain this apparent longevity, a number of natural analogue studies have studied the critical factors affecting the long-term stability of uraninite as an analogue to spent fuel under repository conditions. In addition, these analogue studies have been used to test current models for spent fuel dissolution (e.g. Bruno and Casas, 1994).

Rates of natural UO_2 dissolution can be quantified by measuring the amount of fission product released from the uraninite and using this as a tracer. Concentrations of this tracer in the rock or in the groundwater at, or close to, the uraninite are proportional to the dissolution rate, assuming that the tracer is released from the uraninite by dissolution (Curtis, 1996). The tracers used for this method are ^{99}Tc in rock (or its stable daughter ^{99}Ru when technetium has decayed to insignificant amounts) and ^{129}I in groundwater. There are several uncertainties in the modelling and assumptions made in this approach but some consistency is apparent in the results obtained from different uranium orebodies when using the same isotopic system. For example, using the ^{99}Tc tracer at Oklo (see Box 4) and at Cigar Lake (see Box 5) provided average release rates of 1.5×10^{-6} yr^{-1} and 1.1×10^{-6} yr^{-1}, respectively. However, different rates are obtained when using ^{129}I as a tracer. For example, applying this tracer at Cigar Lake provided release rates of between 9×10^{-9} and 3×10^{-10} yr^{-1} which are 2 to 4 orders of magnitude less than the values obtained using the ^{99}Tc tracer. While this method clearly has potential for quantifying UO_2 dissolution under natural conditions, the method has yet to be refined and differences between results for the two tracers explained.

Setting aside the fission product tracer method, no other technique yet exists for quantifying directly long-term uraninite dissolution rates in natural analogue studies. Most other analogue investigations of UO_2 dissolution are, at most, semi-quantitative. Uraninite stability has been investigated in many natural analogue studies but most of the relevant work has been done at the

Table 4.2: Disposition of a number of elements in the reactor zones at Oklo. Information based mostly on analysis from reactor zone 10. Presence of plutonium and technetium is inferred from decay products. Summarised from Blanc (1996).

Element	Uraninite	Inclusions	Clays	Migration
Cs				✓
Rb				✓
Sr				✓
Ba				✓
Mo		✓		✓
Tc		✓	✓	
Ru		✓	✓	?
Rh		✓	✓	?
Pd		✓	✓	?
Y	✓		✓	
Nb	?	?		
Zr	✓	✓	✓	
Te		✓		
REE	✓		✓	
Ce		✓		
Pb	✓	✓		✓
Bi		✓		
Th	✓		✓	
U	✓		✓	
Np	?	✓		
Pu	✓		✓	

Oklo natural fission reactors (see Box 4) and at the Cigar Lake uranium orebody (see Box 5). These analogues are described here in some detail because they reflect contrasting geochemical environments. Parts of the Oklo orebody are in a near-surface, oxidising environment while the Cigar Lake orebody is located in a deep, water saturated, chemically reducing environment which has many similarities to a deep geological repository environment.

The Oklo orebody (see Box 4) exhibits a unique geological evolution, during which parts of it (the so called 'reactor zones') achieved nuclear criticality as a result of exceptionally high concentrations of ^{235}U. The ore comprising the reactor zones is principally uraninite, together with some pitchblende and coffinite, and is very high grade (up to 70 % uranium oxides). Unlike Cigar Lake and all other known uranium deposits, Oklo uraninite contains significant quantities of fission products (or their stable daughters), directly equivalent to those present in spent fuel. Of note is the inferred presence of radiogenic plutonium (from location of its decay products). As a result, it is not surprising that the Oklo natural fission reactors continue to be put forward as one of the most interesting analogues. However, the analogy between Oklo uraninite and spent fuel has its limitations because:

- the Oklo uraninite contains lower concentrations of fission products than does spent fuel;
- the maximum temperatures (400 to 600°C) and the power density at Oklo were somewhat lower than those in a reactor; and
- the duration of criticality at Oklo was very much longer than the lifetime of reactor fuel.

However, despite these differences, several large-scale analogue investigations have been undertaken and much relevant information has been obtained, including some semi-quantitative information on the fate of radionuclides contained in the orebody. Recent papers include Brookins (1990), Blanc (1996) and Oversby (2000).

These analogue investigations have revealed that when the reactor zones were cooling after periods of criticality, some dissolution of the uraninite and elemental remobilisation occurred. However, the limited extent of this remobilisation is indicated by the uranium 'fuel', more than 90 % of which has remained in the same spatial configuration since criticality. This implies that uranium has been almost fully retained within the uraninite minerals. The disposition of some performance assessment relevant and other elements in the reactor zones at Oklo is summarised in Table 4.2. The transuranic elements neptunium, plutonium and americium were all formed in situ within the uraninite during

Box 4: The Oklo natural fission reactors

The uranium orebodies at Oklo contain the only known examples of natural fission reactors and are, therefore, unique to nature. Just as in a man-made nuclear power plant, the fission reactors at Oklo generated many waste radionuclides in the form of fission products and actinides, including transuranic nuclides. Oklo is located in the southeast part of the Republic of Gabon, in the Francevillian basin which forms an elliptical, elongated depression of some 2500 square kilometres.

Figure B4.1: Photograph of a natural fission reactor 'core' imprint at Oklo exposed in the quarry walls. Scale is indicated by the man standing by the reactor.

The reactor zones were not identified until 1972 when scientists from the Pierrelatte Diffusion Plant in France found that the ^{235}U content of the ore being processed to make nuclear fuel pellets was depleted from the normal 0.72 % to 0.62 %. Further investigations revealed that the isotopic ratios of some other elements were also perturbed, which confirmed that nuclear fission had occurred in the ore (IAEA, 1975, 1978).

Nuclear criticality at Oklo occurred approximately 2 billion years ago. Before that time, natural uranium had high isotopic abundances of ^{235}U. However, uranium was widely dispersed in rocks and sediments in mineral form, and critical masses could not accumulate because surface conditions were chemically reducing, meaning that uranium remained as a solid and so could not easily be mobilised and concentrated.

At the onset of the evolution of plants capable of photosynthesis, surface conditions gradually became chemically oxidising, allowing uranium in near surface environments to go into solution. At Oklo, this newly soluble uranium was mobilised from a primary source and accumulated in a sequence of organic rich sediments in sufficient mass to achieve criticality. The fission reactions were moderated by the sediment porewaters and operated intermittently for between 10^5 and 10^6 years.

Since criticality occurred so long ago, the majority of the transuranic elements and fission products have long since decayed to insignificant levels. However, their original presence can be inferred by the distributions and abundances of their decay daughters. In a simple manner, the natural fission reactors at Oklo can be considered as an analogue for a very old radioactive waste repository, although the analogue is not perfect. Oklo thus provides an opportunity to investigate several processes which cannot be observed in other uranium orebodies, such as the transport behaviour of transuranic radionuclides and the stability of uranium minerals which have undergone criticality.

Figure B4.2: General view of the open cast mine at Oklo.

A total of 16 separate reactor 'cores' have been identified at Oklo, with one found at Bangombé about 20 km to the south-east of the main group at Oklo/Okelobondo. The rocks in the Francevillian basin are sediments of Lower Proterozoic age which include a lower sequence of uraninite-bearing conglomerates and sandstones (the 'FA' Formation), and an upper sequence of organic-rich pelites associated with limestones (the 'FB' Formation).

The uraninite in the FA Formation was not sufficiently rich to allow criticality to occur. Instead, the reactor cores formed when uranium was leached from the uraninite in the FA Formation by oxidising groundwaters and transported upwards through the sedimentary pile. When these groundwaters contacted the organic-rich FB Formation, the local reducing conditions caused the uranium to precipitate. Continued mobilisation and precipitation over time established a uranium-rich layer (up to 60 % uranium) a few metres thick (the C1 layer) at the boundary between the FA and FB Formations.

Figure B4.3: Photograph of the Oklo mine. One of the reactors (Reactor 2) has been preserved as a 'museum' by enclosing it in concrete to protect it from further weathering and decay.

The reactors consist of an inner core and an outer aureole. The core contains only uraninite whilst the aureole contains uraninite with coffinite and pitchblende. The reactors themselves are surrounded by roughly concentric zones of clay material (illite and chlorite) which formed by hydrothermal alteration of the sandstone due to heat released by the nuclear reactions.

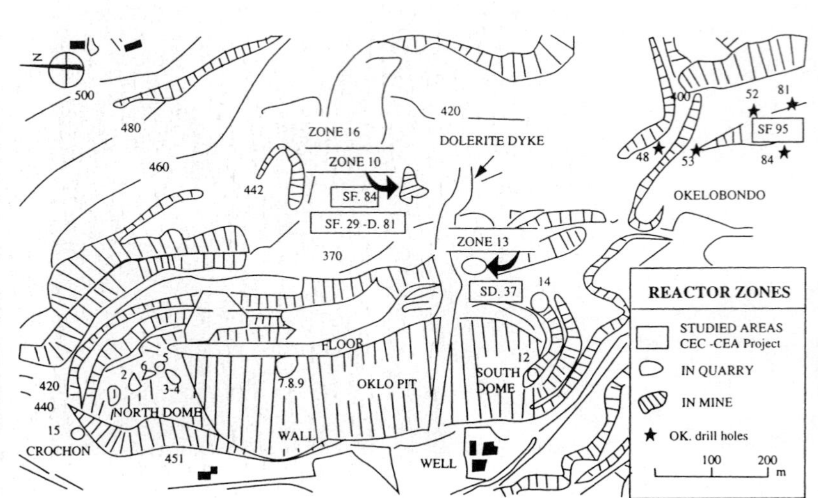

Figure B4.4: Map of the Oklo quarry showing the locations of some of the reactor zones which have been identified.

The reactors became critical about 2 billion years ago and operated intermittently for some 10^5 to 10^6 years afterwards. Roughly 1000 to 2000 tonnes of uranium was initially present as fuel, of which about 6 to 12 tonnes consisted of ^{235}U which underwent fission, producing some 4 tonnes of plutonium. The reactors operated at temperatures of up to 600°C and pressures of 30 to 40 MPa.

Nuclear criticality was initiated in the C1 layer due to the very high concentration of fissile nuclei, a low content of poison (neutron capturing) elements and the presence of water which acted as a neutron moderator. Criticality ceased when the ^{235}U content decreased and when water was isolated from the reactors due to the formation of the low-permeability clay halo. The continued existence of the reactors is due to the fact that the region has undergone very little regional metamorphism or tectonic activity.

The early studies at Oklo performed in the 1980s were restricted to basic examinations of the radiochemical system and were not aimed at direct application to performance assessment. However, since the early 1990s, more performance assessment relevant research has been undertaken which has been focussed primarily on:

- the stability and longevity of UO_2 (see Section 4.2);
- the transport and retardation of radionuclides (see Section 5.2);
- the degradation and radiolysis of bitumen formed by maturation of the organic material in the sediments (see Section 4.7); and
- the production of radiolytic oxidants (see Section 5.4).

Recent studies concentrated mainly at Bangombé and on reactor zones 10, 13 and 16 which more closely represent repository conditions because they are located at depth in reducing conditions. Unfortunately, field studies are no longer practicable at Oklo because now the mines are closed and flooded. Only Bangombé has been preserved for future study.

The volume of literature about the Oklo natural fission reactors is vast. More recent summaries of the investigations on radiolysis and the behaviour of the radionuclides in the reactors are given by Curtis et al. (1989), Loss et al. (1989), Brookins (1990) and Del Nero et al. (2000). The latest analogue programme is summarised by Blanc (1996), Gauthier-Lafaye (2000) and the most up-to-date performance assessment application of the Oklo data is given by Oversby (2000).

criticality, and their stable daughters have also been retained due to their compatibility with the crystal chemical structure of their host or in inclusions in the uraninite.

Other radiogenic elements, which are less compatible with the uraninite host, have been partially or totally lost by diffusion from the uraninite. Some elements, however, migrated only short distances and were totally retained within the clay matrix enclosing the reactors. Data for other elements indicate clear deficiencies in the noble gases, halides and lead (possibly due to volatilisation) and suggest that some (usually minor) loss from the system has occurred for other elements.

The loss of some radionuclides, particularly the large loss of lead, must be noted, although lead itself is of little radiological significance. Brookins (1990) suggests that lead loss was the result of volume diffusion subsequent to the reactor criticality and is a process common in all uranium ores.

It is important to note that most of the observed uraninite alteration at Oklo occurred under hydrothermal conditions. Little uraninite-groundwater interaction has taken place at present-day ambient temperatures, apart from at the Bangombé reactor zone. It is evident that the geological system at Oklo has retained radionuclides to a considerable extent but, not surprisingly, this retention has not been total. This is clear, if qualified, support for the predicted stability of uraninite in the repository environment. The evidence becomes even more impressive when the long timescales and the aggressive environmental conditions at Oklo are considered. In comparison, the repository environment is relatively stable.

In contrast to Oklo, the Cigar Lake orebody (see Box 5) is located deep underground and never experienced a nuclear criticality event. This orebody formed by precipitation from oxidising, uranium-rich aqueous fluids circulating in porous sandstones, when they encountered reducing fluids emanating from fractures in the lower basement rocks.

While demonstrating the mobility of uranium under hydrothermal conditions, the Cigar Lake ore body appears to be stable in the present lithological and hydrogeochemical environment. However, mineralogical investigations indicate that the uraninite at Cigar Lake has been subject to partial dissolution, lead loss and slight alteration to coffinite. These processes are believed to have occurred under reducing conditions, as witnessed by the presence of sulphides, while the uranium and lead isotopic data indicate that several isotopic fractionation events have also occurred. Despite the very long history of interaction between the groundwater and the ore, secondary uranium mineralisation has been limited, and restricted to migration along some fractures. However, these processes most probably occurred under hydrothermal conditions and, as such, are not directly relevant to the repository environment.

Comparison of the observed and calculated uranium concentrations in the groundwater in the vicinity of the ore suggests that the main uranium solubility limiting phase in the ore is U_3O_7. This is supported by studies of the uranium ore mineralogy which indicate stoichiometries in the range U_4O_9 to U_3O_7 (Janeczek and Ewing, 1992).

It is well established that the redox state of groundwater largely determines the stability of uranium ores. Bruno and Casas (1994) have found good agreement between measured and calculated redox potentials for the Cigar Lake deposit, based on pyrite/siderite stabilities. Their study concludes that oxidative dissolution of uraninite does not occur at Cigar Lake, even where relatively high redox potentials (between 100 and 200 mV) have been measured. Combined mineralogical and geochemical information indicates that the oxidative dissolution threshold of uraninite in the ore zone occurs at redox potentials exceeding 200 mV. Comparison with

Box 5: The Cigar Lake uranium mine

The Cigar Lake uranium deposit, in northern Saskatchewan in Canada, is the second largest and richest uranium orebody known in the world and is notable because it is located entirely below the surface at repository depths, which makes it a particularly useful analogue site.

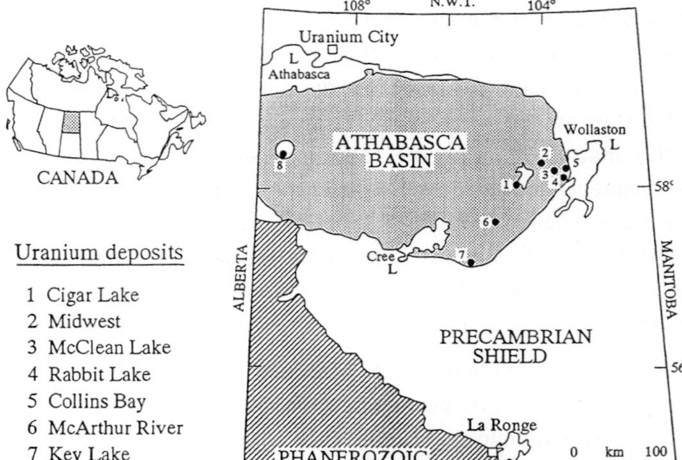

Figure B5.1: Location of the Cigar Lake uranium deposit, northern Saskatchewan, Canada. From Cramer and Smellie (1994b).

The uranium mineralisation is hosted by the Proterozoic Athabasca Sandstone Formation, just above the contact with the underlying high-grade metamorphic basement rocks of the Archean Shield, in an area where several other uranium orebodies have been discovered. The Cigar lake orebody lies at a depth of around 430 m and is lensoid shaped, about

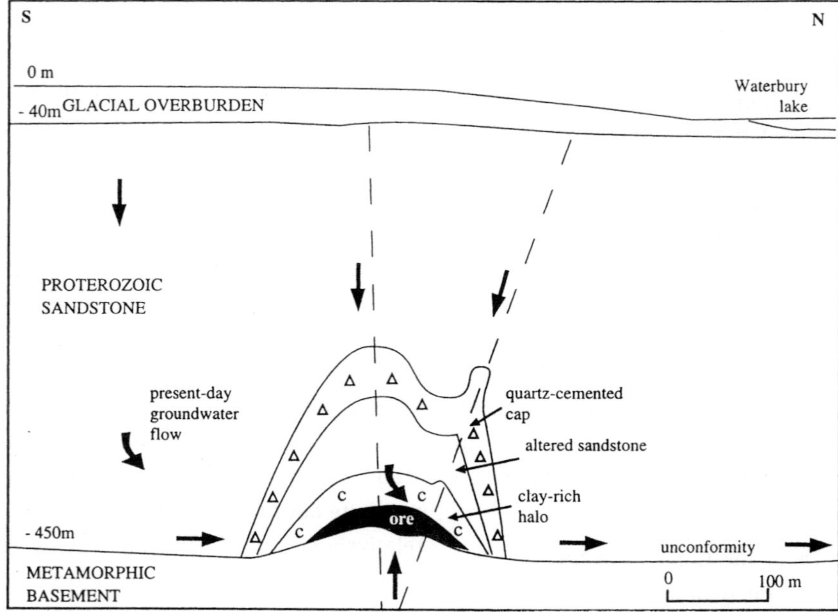

Figure B5.2: Simple cross-section of the Cigar Lake orebody, showing the clay envelope that helps to retain radionuclides in the orebody. After Cramer and Smellie (1994b).

Figure B5.3: Core sections representing the various materials at Cigar Lake from the clay halo (white), through the iron oxyhydroxide rich zone (red) to the ore (black). Illustration courtesy of SKB.

2 km long, 50 to 100 m wide but only 1 to 20 m thick. The uranium ore is uraninite (UO_2) and coffinite ($USiO_4$), with an average grade of 14 % but reaching 55 % in some areas. The ore formed around 1300 million years ago by precipitation from oxidising, uranium-rich aqueous fluids circulating in the sandstones, when they encountered reducing fluids emanating from fractures in the crystalline basement rock. The uranium-rich fluid has been calculated to have been at a temperature of 150 to 200°C.

Although the Cigar Lake deposit is more enriched in uranium than Oklo (see Box 4), spontaneous nuclear fission was not possible because of its younger age; i.e. by the time the orebody formed, the natural $^{235}U/^{238}U$ ratio had decayed to too low a level to allow a nuclear criticality event to occur. Furthermore, there were too many potential neutron-capturing nuclei present and insufficient water (porosity) to moderate a nuclear reaction.

The orebody is typically polymetallic in nature, with the uranium accompanied by a suite of elements including nickel, cobalt, zinc, manganese, iron, vanadium and radiogenic lead. The ore zone is surrounded by a 10 to 50 m thick clay-rich halo (illite, kaolinite and quartz) which formed by hydrothermal alteration of the host sandstone. At the contact between the ore and the clay halo there is often an iron oxyhydroxide rich zone which becomes less concentrated away from the contact. Several major fractures cut the orebody and the clay envelope; these are originally features of the basement rocks but were reactivated after deposition of the sandstone and formation of the ore. Most of the fractures were sealed during early clay formation.

Many features of the Cigar Lake orebody make it a particularly useful analogue site (Cramer and Smellie, 1994a). Although the idea that any natural system could be used as a 'global' natural analogue for a whole repository has been dismissed (see Section 1.4), the Cigar Lake uranium deposit has clear similarities to proposed spent fuel repositories: the host rock and geometry match potential repository designs, the ore is similar to the gross structure and composition of spent fuel, and the clay envelope is somewhat analogous to a compacted bentonite buffer. The main difference between the orebody and a repository is that there is no analogue for a metal canister at Cigar Lake.

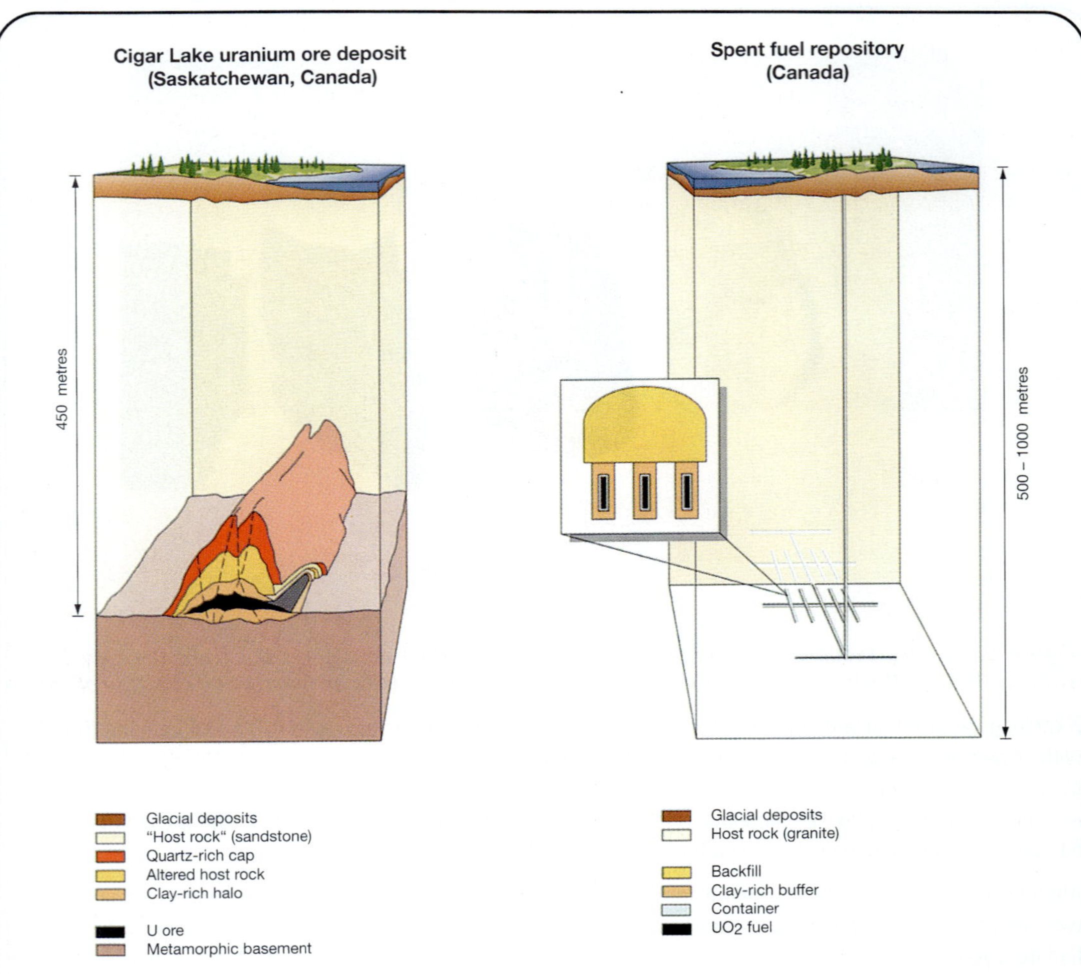

Figure B5.4: Comparison of the Cigar Lake orebody and the structure of a spent fuel repository, showing the clear similarities between the two systems. Illustration courtesy of Nagra.

A large multinational analogue study was started in the 1984 and the results were published by Cramer and Smellie (1994b). This study had the objective of performing detailed investigation of a number of performance assessment relevant issues, including:

- the stability and longevity of UO_2 (see Section 4.2);
- the transport and retardation of radionuclides in the clay halo (see Section 5.2);
- the production of radiolytic oxidants (see Section 5.4); and
- colloidal transport (see Section 5.6).

In addition to its use as an analogue for supporting performance assessment, Cigar Lake also has considerable value as an illustrative analogue for non-technical demonstrations of safety. This is because, despite its high uranium content, great age and presence in a porous sandstone, no significant radionuclide migration from the ore to the surface has occurred. In fact, there is no direct chemical or isotopic signature in the surface waters, soils, rock or vegetation to indicate the presence of the orebody below.

similar data from Poços de Caldas (see Box 14) indicates potentials greater than 300 mV.

Currently, the laboratory evidence for spent fuel stability suggests that the dissolution of the waste matrix (UO_2) is controlled by the low solubility of UO_2 under reducing conditions. Based on laboratory studies, a threshold value of around Eh = 120 mV has been calculated for the switch between reductive and oxidative dissolution of UO_2 (Shoesmith and Sunder, 1991). This is a conservative value when compared with the value of 200 mV derived from pyrite/siderite stability calculations at Cigar Lake. However, the fact that the localised oxidative effects appear to be small, as shown from mineralogical and geochemical evidence, indicates that this 200 mV redox threshold value has not been exceeded at Cigar Lake.

Theoretically, the Cigar Lake orebody could contain oxidising species due to radiolytic decomposition of the groundwaters (see Section 5.4) and this could cause oxidation of the uraninite. According to some predictive radiolysis models, this process should have caused the Cigar Lake orebody to have been totally oxidised within 200 million years after formation. Clearly, this is not the case and, thus, these radiolysis models are grossly over-conservative (Karlsson et al., 1994). In an attempt to improve radiolysis models used in the performance assessment of spent fuel repositories, some recent studies have used data from Cigar Lake to establish a more theoretical basis for radiolytic models to understand the radiation fields around uraninite grains and to establish what fraction of the radiation reaches the groundwater to initiate radiolysis. These studies have found that only a very small fraction of the total radiation impacts on the groundwater to cause radiolysis and revised models now more closely replicate the extent of oxidation observed in the Cigar Lake ore (Smellie and Karlsson, 1996). This is a good example of the use of natural analogue data to develop and improve performance assessment models.

Most studies have examined uraninite stability under conditions analogous to a water saturated repository environment. However, the recent study at the Peña Blanca uranium orebody in Mexico was undertaken to examine the behaviour of UO_2 in unsaturated volcanic tuffs which have a great deal of similarity with the tuffs at the Yucca Mountain proposed repository site (see Section 2.3.1).

The uraninite at Peña Blanca (see Box 6) formed some 8 million years ago but has had a long history of oxidation and leaching in a near-surface oxidising environment. This has resulted in the transformation of the primary uraninite minerals to form secondary uranophane with subsidiary soddyite and other uranyl minerals. Some remnant uraninite crystals remain and detailed mineralogical observation of these has shown a characteristic pattern of alteration on mineral surfaces and at grain boundaries. Comparison with the results from laboratory studies of spent fuel degradation, designed to simulate Yucca Mountain conditions, has indicated that both the natural and laboratory UO_2 systems degrade broadly by the same processes to form the same alteration products (Pearcy et al., 1994). These similarities provide very good support for the performance assessment models used to predict the future evolution of spent fuel in the Yucca Mountain repository.

The host volcanic tuffs at Peña Blanca are commonly fractured, particularly in the welded zones, and these fracture porosities are the key factor controlling unsaturated flow through the rock. Field studies clearly show that uranium transport and water flow occurred predominantly through the fractures but have also suggested that matrix diffusion is not an important retardation mechanism in the volcanic rocks at Peña Blanca or, by analogy, at Yucca Mountain.

Mobilised uranium has been transported away from the ore into the surrounding tuffs and has been partly redeposited. Association of the redeposited uranium with minor phases such as iron oxyhydroxide in the fractures and the rock

Box 6: The Peña Blanca uranium mine

Peña Blanca is located in the desert near Chihuahua, Mexico and the local geology comprises unsaturated, silicic volcanic tuffs that are very similar to the volcanic rocks at the site of the proposed US repository at Yucca Mountain, Nevada (see Section 2.3.1). Numerous uranium mineralisations occur in the Peña Blanca region, and one particular orebody, the Nopal I deposit, was chosen as the site for detailed analogue investigation.

The Nopal I deposit is exposed at the surface in both the walls and the floor of the mine, allowing a three-dimensional investigation to be performed of the distribution of primary and secondary mineralisation. Uranium-lead dating of the ore at Nopal I suggested that the primary uranium mineralisation formed approximately 8 million years ago.

Originally, the uranium would have been present in the form of uraninite, however, 3 million years ago, the deposit was largely oxidised and, as a result, the mineralisation is now in the form of uranophane, weeksite, soddyite and schoepite, although a few centimetre sized remnant pods of uraninite still remain.

Natural, long-term alteration of the uraninite at Peña Blanca has thus occurred largely in near-surface oxidising environments, and the observed alteration patterns on mineral surface and at grain boundaries were found closely to mimic those observed on spent fuel in laboratory experiments designed to simulate Yucca Mountain conditions.

Figure B6.1: Photograph of the Nopal I uranium orebody. The uranium mineralisation can be seen as the darker area in the foreground, with the valley in the distance.

Preservation of the secondary mineral assemblage indicates that the rate of uraninite oxidation must exceed the dissolution rate of the secondary phases and uranium transport out of the system.

The rate limiting step for uranium migration appears to be advective transport in groundwaters, due to the low water flow rates in the dry, desert conditions (Murphy and Pearcy, 1994). Similar rate limits are expected to control the source term in the Yucca Mountain repository.

The host volcanic tuffs are commonly fractured, particularly in the welded zones and these fracture porosities are the key factor controlling episodic unsaturated flow through the rock. Uranium transport and water flow occurred predominantly through the fractures but field observations indicate that migration through the rock matrix perpendicular to the fractures also occurred, suggesting that matrix diffusion is not an important retardation mechanism in the rocks at Peña Blanca or, by analogy, at Yucca Mountain. Mobilised uranium has been transported away from the ore into the surrounding tuffs and have been partly redeposited.

Figure B6.2: A smaller uranium mineralisation in the Peña Blanca region in unsaturated volcanic rocks. The dry, desert environment of this region is very similar to conditions at Yucca Mountain.

An important conclusion from the uranium isotopic measurements on fracture coating minerals is that uranium transport in the system at Peña Blanca was episodic rather than continuous, and assumed to relate to preferential mobilisation during periods of high groundwater flow. This observation has clear implications for modelling groundwater flow and transport at Yucca Mountain.

The age of the mineralisation and information on its alteration were used to calculate a radionuclide release rate which has been applied in a performance assessment for Yucca Mountain to compare with an experimentally derived base case value. The analogue-derived value gave lower calculated doses than the base case, providing confidence in the conservatism of the base case value (Pickett and Murphy, 2000; Murphy et al., 2000).

matrix is a significant retardation process. An important conclusion from the uranium isotopic measurements on fracture coating minerals is that uranium transport in the system at Peña Blanca was episodic rather than continuous, and assumed to relate to preferential mobilisation during periods of high groundwater flow (Pickett and Murphy, 2000). This observation has clear implications for modelling groundwater flow and transport at Yucca Mountain.

The age of the primary and secondary mineralisation at Peña Blanca, and information on its alteration were used to calculate a radionuclide release rate which has been applied in a performance assessment for Yucca Mountain to compare with an experimentally derived base case value. The analogue-derived value gave lower calculated doses than the base case, providing confidence in the conservatism of the base case value (Pickett and Murphy, 2000; Murphy et al., 2000).

Other studies which have examined the dissolution of uraninites under oxidising conditions found that the dissolution rate was diminished by the presence of thorium, lead and rare earth element impurities in the uraninite (Grandstaff, 1976). This observation is supported by later work by Finch and Ewing (1992) and it is important because spent fuel has a lower content of these impurities than uraninite (less than 5 % compared to up to 20 %) and, on the basis of this study, might be considered to dissolve more rapidly. Also, Grandstaff (1976) discovered no relationship between the age of uraninite and its dissolution rate. This is encouraging, because it indicates that the cubic structure is resistant to radiation-induced damage to the crystalline lattice, which otherwise may be expected to promote dissolution. Again, care must be taken in extrapolating this result to spent fuel due to the dissimilarity in the fission product content.

In summary, both laboratory and natural analogue investigations indicate that the kinetics of UO_2 dissolution, either as spent fuel or uraninite, is exceptionally slow under the reducing conditions expected in the near-field of a spent fuel repository. While dissolution rates cannot be quantified readily from natural analogue data, the abundance of naturally occurring uraninite some 10^9 years old indicates its stability in the geological environment. Extrapolation of this apparent longevity to spent fuel in the repository environment should, however, be done cautiously, due to the uncertain long-term effects of the high levels of radioactivity and thermal history of spent fuel.

Interestingly, the natural analogue which has supplied some of the most useful data to performance assessment is the Peña Blanca study which is due to the very close similarity between the geological environment and history at the site and the conditions expected in the proposed Yucca Mountain repository (as discussed in Box 6).

Radionuclide retardation by secondary alteration products

There is a large body of literature on laboratory investigations of the surface structure of spent fuel during dissolution (for review, see Johnson and Shoesmith, 1988). During these laboratory experiments it has been noticed that UO_2 dissolution is accompanied by the formation of secondary phases on the fuel surface and that these corrosion products can passivate further dissolution. However, limited attention has been given to the nature of these surface phases due to the amorphous nature of products formed under relevant temperature conditions in laboratory timescale experiments.

Natural analogue studies provide an obvious alternative to these types of experiment. There is a large number of sites where uraninite accumulations occur and where long term UO_2 dissolution products and processes can be studied. However, natural analogue studies need to contend with the problems associated with the differences in chemistry and structure between spent fuel and uraninite noted earlier.

A comprehensive investigation of the corrosion products of uraninite was undertaken by Finch and Ewing (1989) who studied material from the Shinkolobwe orebody in Zaire. While most studies have focussed on UO_2 dissolution under reducing conditions, the Shinkolobwe orebody is in an oxidising environment which could be somewhat representative of a near-field made oxidising by the build-up of radiolytic oxidants (see Section 5.4).

The Shinkolobwe deposit weathers under oxidising conditions in a monsoonal-type environment where rainfall is above 1 m/year. As cautioned in the previous chapter, the limitations of observing processes active in this type of hydrochemical environment must constantly be borne in mind. At Shinkolobwe, the uraninite is coarsely crystalline and lacks many of the impurities (e.g. thorium and rare earth elements) found in other uranium deposits. This lack of impurities led Finch and Ewing (1989) to suggest that the thermodynamic stability of the Shinkolobwe uraninite may closely approximate spent fuel. Over 50 secondary uranyl phases were identified from the alteration of uraninite at this site. It was concluded that uraninite transforms to Pb-U oxide hydrates and then to uranyl silicates if sufficient silica is present in the system.

The conditions at Shinkolobwe are very different to the reducing environment expected in most deep, water saturated repository near-fields. As such, the Shinkolobwe natural analogue is more relevant to either the US Yucca Mountain repository or a spent fuel repository that has become oxidising due to groundwater radiolysis (see Section 5.4). In an expansion of the Shinkolobwe study, Finch and Ewing (1991) listed the most important uranium deposits around the world and also provided a comprehensive catalogue of the uranyl minerals that have been identified at each locality. The extent to which this information can be of direct use in performance assessment is yet to be established.

Conclusions

Uraninite is a good analogue mineral for spent UO_2 fuel, although important differences exist which need to be considered when evaluating the analogue information. Qualitatively, the abundance of uraninite in chemically reducing conditions from a range of geological environments is a strong indication for the potential stability of spent fuel in the repository environment. Detailed studies of natural uraninites indicate that their rate of dissolution is extremely slow when temperatures and groundwater fluxes are low, and when oxidation does not progress beyond U_3O_7.

The concentration of fission products as a tracer in rock and groundwater surrounding uraninite provides a satisfactory approach to estimating natural dissolution rates, although the technique needs to be refined and additional data acquired from other analogue sites.

Dissolution of spent fuel in laboratory experiments shows that a passivating layer of uranyl phases may form on the spent fuel surface. Understanding corrosion behaviour and the nature and reactivity of the corrosion products is important for modelling purposes and natural analogue studies have contributed to the testing of different types of model for spent fuel corrosion (e.g. leaching versus oxidative corrosion). Dissolution experiments on spent fuel are not easily performed in the laboratory due to the very slow kinetics of the process. Natural analogue studies of uraninite corrosion when subjected to conditions similar to those expected in a repository would be a useful aid to such experiments.

In all cases, the differences in the chemistry between spent fuel and uraninite needs to be considered when interpreting analogue data to avoid drawing inappropriate conclusions from the studies. Nonetheless, the qualitative evidence from these analogue studies has added to confidence that the spent fuel degradation processes are well understood and has provided upper bounding limits to the degradation rates.

Spent mixed oxide fuel (MOX) may be disposed of in the same manner as ordinary spent fuel. However, no natural analogue of MOX fuel can be found with minerals with sufficiently high concentrations of plutonium to be particularly useful.

4.3 Mineral and ceramic wasteforms

Naturally occurring radioactive elements can be included in minerals in two basic ways. They can either substitute for other elements in a mineral structure to form a solid solution or they can form minerals in their own right. In many cases, when a radioactive element forms its own mineral, that mineral is both physically and chemically stable in subsurface, geological environments. For example, uraninite (in which uranium is an essential component) is extremely stable in deep, chemically reducing conditions and may remain unaltered for many millions of years, as discussed in Section 4.2. This is a principal reason why direct disposal of spent fuel in a geological repository is considered to be safe.

Besides uraninite, there are many other minerals which contain radioactive elements, such as zirconolite, although few are as abundant as uraninite. However, it was the recognition that these minerals do exist in nature, that they are stable and that they can naturally contain high concentrations of radionuclides which led to the development of mineral and ceramic wasteforms. For this reason, it is true to say that the entire concept of a mineral wasteform can be said to be natural analogue led.

In essence, mineral wasteforms are an alternative to borosilicate glass for the solidification and immobilisation of the liquid HLW produced from spent fuel reprocessing operations and military wastes. The fundamental difference is that, in contrast to glass, radionuclides in a mineral wasteform are tightly bound in a mineral lattice.

The best known mineral wasteform is SYNROC (Ringwood et al., 1979, 1988), although a number of other forms of ceramic and crystalline materials have been investigated over the last two decades. SYNROC is the generic name for a number of multi-phase, titania-based wasteforms which are synthetic mineral assemblages consisting of several oxide structure types, including fluorites, perovskites, hollandites, reduced rutile and magnetoplumbite (Savage, 1995). The three basic phases in SYNROC are designed to include all of the chemical species in HLW by isomorphic substitution (Hart et al., 1996):

- *zirconolite*, ideally $CaZrTi_2O_7$, designed to incorporate uranium, zirconium, neptunium, neptunium, and the rare earth elements;
- *hollandite*, ideally $Ba(Al,Ti)_6O_8$, designed to incorporate caesium, rubidium and barium; and
- *perovskite*, ideally $CaTiO_3$, designed to incorporate strontium, neptunium, neptunium, and the rare earth elements.

Although the majority of the radionuclides in HLW do enter these phases, a number of different SYNROC formulations have been developed to take account of variations in the composition of wastes and radionuclide contents. As a consequence of the possible variations in phase proportions, SYNROC has considerable flexibility to deal with many different radioactive waste streams.

In comparison to borosilicate glass, SYNROC has better mechanical properties and excellent thermodynamic and chemical durability. Its increased resistance to elevated temperatures allows higher waste loading, and its fracture toughness and thermal conductivity minimises the number of thermally-induced fractures. Long-term radionuclide release from SYNROC would be expected to occur by either base-catalysed hydrolysis of the titanate structure, diffusion controlled ion exchange or precipitation and layer formation.

Despite its appealing physical and chemical characteristics, no countries are currently planning to use SYNROC technology to immobilise liquid HLW from commercial reactor and reprocessing operations. This is largely because borosilicate glass immobilisation (vitrification) technology has already been developed and is now in industrial-scale use. In contrast, SYNROC research involving 'active' testing using actinide-doped materials under repository conditions has yet to be completed. Nonetheless, there is some potential interest in using SYNROC or variations to immobilise certain types of military radioactive wastes and, perhaps, for the final disposal of excess weapons plutonium (e.g. Vance et al., 1996).

SYNROC is not the only mineral wasteform that has been under development. In Canada, a sphene-based glass ceramic has been considered as an immobilisation matrix for waste from CANDU reactor fuel recycling wastes (Savage, 1995). Glass-ceramics based on crystalline sphene ($CaTiSiO_5$) were selected for development because of the well-documented persistence of sphene as a naturally occurring mineral, with an ability to take a wide variety of ions into solid solution. The aluminosilicate glass matrix remaining after sphene crystallisation is a highly durable material for immobilising those waste ions which do not partition into the sphene phase, such as ^{135}Cs, ^{137}Cs and ^{79}Se. However, despite early interest in this wasteform, no current research is known to be underway.

Zircon has also been proposed as a mineral wasteform for the immobilisation of excess waste plutonium (Burakov et al., 1996). This mineral was first suggested as a suitable wasteform after it was recognised to have formed by crystallisation from the reactor core melt generated in the course of the accident at the Chernobyl nuclear power plant (Burakov et al., 1996).

The majority of the research into mineral and ceramic wasteforms has been undertaken in laboratory studies, despite the original natural system inspiration for SYNROC. Nonetheless, a few relevant analogue studies have been performed. However, since the component minerals in SYNROC are relatively rare in nature, the issues of most relevance to the behaviour of mineral wasteforms which have been (or potentially could be) addressed in natural analogue studies are limited to:

- the long-term stability of SYNROC component minerals.

This issue is discussed in the following section.

Long-term stability of SYNROC component minerals

Much of the analogue studies on SYNROC component minerals have focussed on zirconolite because actinides partition into this mineral, rather than perovskite, by a factor of between 5 and 10 (Hart et al., 1996), although some analogue studies have looked at other natural minerals as well (e.g. Lumpkin et al., 1998).

Zirconolite is a relatively rare mineral in nature, crystallising in a range of geological environments but normally associated with silica-poor rocks. To date, there are only 55 recorded occurrences of zirconolite (Hart et al., 1997).

The range of compositions of these natural zirconolites is wide and includes up to 30 elemental components with concentrations greater than 0.1 %. Natural compositions deviate substantially from the ideal structure due to extensive substitutions involving rare earth elements, actinides, niobium, iron and other elements (Gieré et al., 1998). Almost 80 % of the calcium site and up to 65 % of the titanium site may be substituted. However, in natural

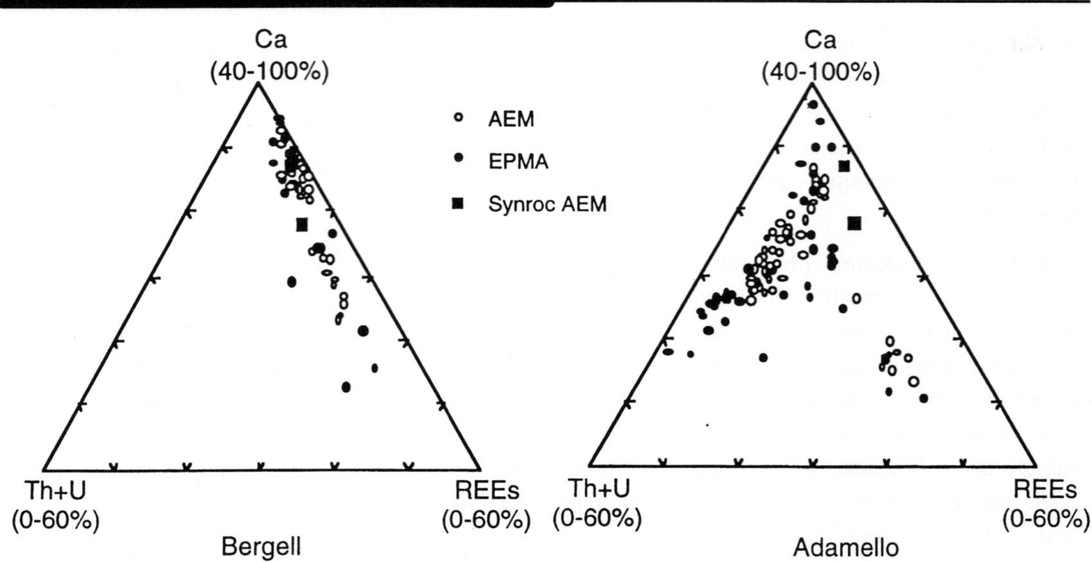

Figure 4.9: Comparison of the phase chemistry of natural zirconolites from two locations (Bergell and Adamello) measured using electron microscopy techniques (AEM and EPMA) with the average compositions of synthetic zirconolite in SYNROC. From Hart et al. (1996).

zirconolites, only about 3 % of the zirconium site is substituted, in contrast to synthetic zirconolite in SYNROC in which up to 50 % of the zirconium site may be substituted by actinides. In terms of radionuclides, natural zirconolite can contain up to 25 % UO_2 and 18 % ThO_2. Furthermore, the abundances of substituted components in the mineral may be zoned. Despite this range of compositions, there is very little apparent variation in the stability of compositionally different zirconolites in nature, which provides good qualitative evidence for the stability of waste-loaded synthetic zirconolite in SYNROC. The measured compositions of zirconolites from two locations, Adamello in Italy and Bergell in Switzerland, are shown on Figure 4.9, together with average compositions of synthetic zirconolite in SYNROC containing 10 and 20 % waste loading.

Leaching studies on natural zirconolite samples (Ringwood et al., 1988) indicate that these minerals are closed systems with respect to uranium, thorium and lead isotopes for periods of up to 5.5×10^8 years, which is far longer than the time periods normally considered in performance assessment. However, there is some uncertainty regarding the interpretation of these leaching experiments and, in addition, new experiments examining the leaching of actinides are required.

There is geochemical evidence that 5 samples from the 55 known zirconolite occurrences have suffered from interaction with infiltrating ground or porewaters. However, most of these samples were collected as detrital grains and consequently no information on environmental conditions pertaining to alteration could be obtained. However, in the case of one zirconolite sample from Adamello in Italy, this was collected in situ and, thus, the conditions under which it corroded could be determined. This proved to be at temperatures in excess of 500°C. On the basis of this single sample, the only general statement that can be made is that alteration is likely to occur only at relatively high temperatures exceeding those expected in a repository.

Some studies have also been performed on natural pyrochlores (e.g. Lumpkin and Ewing, 1989; Lumpkin et al., 1994; Lumpkin and Mariano, 1996). Pyrochlores are a group of minerals related to zirconolite which have the general formula $A_2B_2O_6(O,OH,F)$, where the A and B sites can naturally contain a wide range of elements, including uranium, thorium, lead etc. Examination of pyrochlores from a range of alkali-rich igneous rocks, such as carbonatite, which have experienced hydrothermal alteration over a range of temperatures and time periods, suggest that they can be extremely stable for hundreds of millions of years. Chemical evidence suggests that corrosion of natural pyrochlores does take place but only under hydrothermal conditions which far exceed those experienced in a repository near-field. Laboratory tests on these minerals also indicate that they undergo a crystalline aperiodic transformation due to alpha-decay. The principal effects of this transformation are volume expansion and microfracturing which might lead to enhanced radionuclide loss.

Conclusions

Natural zirconolite and pyrochlores are good analogue minerals for synthetic component minerals in SYNROC. However, these natural minerals are very rare and generally are acquired as detrital grains rather than from their place of formation. As a consequence, it is difficult to relate their observed stability to in situ geological conditions. A few natural samples have provided geochemical evidence to suggest that natural zirconolites can corrode but probably only under hydrothermal conditions which are not relevant to repository systems.

The limited natural analogue information on synthetic component minerals in SYNROC suggests that they are very stable, long-lived and suitable solidification and immobilisation matrices for the liquid HLW produced from spent fuel reprocessing operations. However, little quantitative information can be gained from the analogue studies which would be appropriate for input to performance assessment code development.

Laboratory studies will probably remain the best means of investigating the stability of these mineral phases. However, additional natural analogue studies would be welcome, particularly if more samples could be found in situ from sites that are known to have experienced long rock-water interaction events at temperatures approximating repository near-field conditions.

4.4 Metals

Metals will be used extensively in the near-fields of all repository designs, as discussed in Chapter 2. They will be present in engineered barrier systems as:

- canisters and overpacks for vitrified HLW and spent fuel;
- metal canisters and containers for solidified and compacted L/ILW;
- reinforced concrete structures such as silos in ILW repositories; and
- rock supports and reinforcements in all repository excavations.

The majority of these components will be made from steel. However, certain other metals might be used for waste canisters to hold vitrified HLW or spent fuel. For example, both copper and titanium have been proposed for spent fuel canisters. Also, lead has been suggested as a filler material for the void spaces in some spent fuel canister designs.

In addition to the engineered barriers, metal might be present in the repository near-fields as a component of the wastes. For spent fuel disposal, it is possible that entire fuel elements will be placed in canisters for disposal rather than only the fuel pellets. A typical fuel element comprises fuel rods made from Zircaloy (zirconium alloy with 98 % zirconium) with welded Zircaloy end plugs, fixed together by stainless steel and Inconel (nickel alloy with 73 % nickel) spacer grids and upper and lower tie-plates. If complete fuel elements are loaded in a canister, it means that significant masses of various metallic non-fuel components will be present in the waste: approximately 300 kg/tU. If the fuel pellets are removed from the fuel elements (or if the spent fuel is reprocessed), the metallic components will be routed to an ILW repository instead.

In an ILW repository, metal is likely to be present as a waste generated from reactor refuelling and maintenance operations (including metallic components from spent fuel elements), and from reactor decommissioning.

In the post-closure repository environment, these metals will begin to corrode. The significance of metal corrosion for repository evolution will depend, in part, on the nature of the metallic objects themselves and their importance for radionuclide containment and structural support. The most significant consequences of metal corrosion will be:

- degradation of canisters and waste packages leading eventually to complete failure and contact between the wasteform and the groundwater;
- formation of large volumes of solid secondary alteration products;
- direct release of radionuclides (activation products) from corroding metallic waste components; and
- mechanical failure of reinforcements and supports leading to stress readjustments in the near-field rock and engineered barriers, especially in ILW caverns which may not all be backfilled.

In addition, corrosion of steel or iron will generated large volumes of gas, as discussed in Section 5.8. The discussion in this section is focussed mainly on the mechanisms and rates of metal corrosion.

Metal corrosion proceeds by a series of complex process which may involve several coupled reactions. As corrosion progresses, a metallic compound may be formed on the metal surface, although some metals (such as copper) may corrode under certain conditions without the formation of any compound at the surface. This layer of corrosion product may limit the flow of metal ions to solution, thus slowing the corrosion rate. This property is called *passivity*, and is exhibited by most proposed container metals (van Orden, 1989). The formation of this passive film may slow down further corrosion. A consequence of passivity is that the corrosion rate of the metal becomes time-dependent which complicates predictions of the operating lifetimes of metallic components of the engineered barriers. In the case of a steel canister, the presence of alloying metals in the steel (e.g. nickel and chromium) may increase the passivity of corrosion (Smellie et al., 1997).

When a passivating layer is disrupted, or suffers chemical breakdown, then localised corrosion may occur. Localised corrosion may also occur as a result of stress cracking, particularly at welds. Localised corrosion is known as *pitting* and this is seen in Figure 4.10. Pitting is relevant because any small perforation may allow radionuclide release to occur, even if the canister maintains its

Figure 4.10: Pitting or localised corrosion seen on a Saxon iron helmet. Pitting on a steel canister can cause small perforations to form, allowing radionuclides to escape, before the canister fails mechanically.

mechanical integrity. This is most significant for canisters for HLW and spent fuel because these are designed to isolate the waste for long periods of time. However, in the case of L/ILW repositories, most performance assessments do not assume that the waste package provides any significant physical containment capacity, although this is usually a very conservative assumption.

Metal corrosion can be exacerbated by the action of microbes by a number of different processes. Microbes can produce a corrosive species as a by-product of their metabolic cycle, they may enhance the electron transfers involved in the electrochemical reactions (Iverson, 1987), or they may be able to ionise the metal surface itself (Miller, 1981). There are microbes which can attack most metals and, in the case of steel, microbes have been found particularly to attack welds (Dexter, 1986).

Numerous laboratory studies have investigated the corrosion rates and mechanisms of many metals and alloys (e.g. Simpson 1983, 1984, 1989; Simpson and Vallotton, 1986; Beavers and Durr, 1991). Whilst these experiments may provide useful indications of the relative durability of metals, few are conducted in conditions that adequately simulate those expected in a repository. Of particular concern are the very short timescales involved in the laboratory experiments and, in this case, natural analogues may be useful for providing bounding limits to the long-term corrosion rates for different metals. The time dependent corrosion rates exhibited by some metals (e.g. iron and steel) as a consequence of the formation of a passivating layer are difficult to establish adequately in the short-term laboratory experiments.

A number of analogue studies have examined metal corrosion and these are comprehensively reviewed by Vira (1996). Some of these analogue studies have examined naturally-occurring metal deposits but most have focussed on archaeological artefacts. Since many of the metals which will be used in repositories have been produced only in recent times (such as stainless steel, metallic titanium, Inconel and Zircaloy) they have no counterparts in nature or in archaeology. Consequently, the natural analogue studies on metals described in this book have been limited to copper, iron and steel, and other alloys using these metals. Archaeological analogue studies on lead have been performed but, because no lead structure is ascribed any isolation capacity in current repository designs, the degradation of lead will not be discussed here. Readers interested in the long-term stability of lead are referred to Tylecote (1983) who examined a number of archaeological artefacts in order to investigate the durability of the metal in different environmental conditions.

Archaeological analogue studies can provide useful information on general corrosion rates and pitting. However, this type of investigation is potentially prone to bias if they focus on metal samples from museum collections because museums will (naturally) tend to house the best preserved artefacts. Corrosion rates based solely on archaeological material could, thus, be non-conservative. This sample bias problem is likely to be less important if artefacts are collected in situ, rather than from a museum, for then it would be possible to see artefacts in all possible corrosion states for that environment.

Corrosion studies based on archaeological artefacts must be considered carefully to determine if this type of bias is evident. Indeed, the most extreme bias may be the fact that only 'unusual' samples survive, the rest having corroded away. To guard against this, it is important to examine artefacts from sites where historical documents back up the finds, so allowing some independent assessment of the 'survivability' of the entire collection of material which was buried. In such a case, bias could be carefully avoided or assessed. The issues of most relevance to the behaviour of metals in the repository which have been (or potentially could be) addressed in natural analogue studies are:

- the durability and longevity of iron and steel;
- the durability and longevity of copper; and
- the properties of secondary alteration products.

These issues are discussed in the following sections.

The durability and longevity of iron and steel

Natural occurrences of metallic iron occur very rarely in the geological record due to the high reactivity of this metal. In most cases, iron is found in mineral form as a silicate, oxide or sulphide. When metallic iron does occur, it is usually alloyed with small amounts of nickel. In fact, native (pure metallic) iron can generally form in nature by only two processes; first from cooling of an iron-rich magma and, second, from hydrothermal alteration (serpentinisation) of ultrabasic rocks. For a detailed discussion of these processes, see Hellmuth (1991a).

Iron formation due to cooling from a magma only occurs very rarely because the necessary conditions for formation are both complex and unusual; they are roughly similar to those which take place during technical iron smelting. Only two large occurrences of iron formed by this process are known; one at Disko Island, Greenland (Ulff-Möller, 1990) and the other at Bühl, Germany (Hellmuth, 1991b, 1994). Native iron formed by serpentinisation is more common but, because it is usually very fine grained and disseminated in the rock mass, it is often overlooked. However, large iron accumulations do occasionally form by this process and a number of well known iron ore bodies are of this type, e.g. the Muskox intrusion in Canada.

Iron is generally rapidly oxidised in near-surface environments and, consequently, it is interesting to understand why the large native iron occurrences at Disko Island and Bühl have not corroded. This issue has been addressed by Hellmuth (1991b) who examined samples from both localities.

The Bühl iron is enclosed in a basalt matrix which has been exposed to oxidising groundwater for more than 1 million years. Groundwater advection through the basalt is limited to fractures because the basalt matrix is almost impermeable. As a result, corrosion of the large mass of iron has been limited to diffusive mass transport. Furthermore, dissolved oxygen in the groundwater has been scavenged by the FeO in the basalt matrix, actively buffering the redox conditions. The high FeO content of the basalt has restricted the migration rate of the redox front into the rock to only a few centimetres every million years. Results from this investigation also suggest that the groundwater conditions (high pH and low Eh) ensure passivity of the iron.

Interpretation of the geochemical results for samples from Disko Island is more ambiguous. The iron here is in the form of interconnected inclusions, which extend to the weathered surface. Generally, oxidation of the iron has occurred to a depth of a few millimetres only, despite surface exposure for 10^3 to 10^4 years. The role of the permafrost in this region is difficult to assess but it may have helped to limit corrosion by restricting surface water access to the iron.

The results of these studies have led to suggestions that olivine-bearing rock (rich in FeO) should be considered as an additional redox buffer in the near-fields of repositories for vitrified HLW (Hellmuth, 1991b; Hellmuth et al., 1994). However, most performance assessments of HLW repositories containing steel canisters, such as the Swiss Kristalline I assessment (Nagra, 1994), show that the massive steel canisters provide sufficient redox buffering capacity for the period of concern. Nevertheless, this suggestion may bear further consideration for repository designs to contain spent fuel or MOX fuel due to the larger radiolytic oxidising capacities of these wastes (see Section 5.4).

Extraterrestrial sources of metallic iron have also been considered as analogues for the iron and steel components in the engineered barrier system. Metallic iron is sometimes found in certain types of meteorites but is generally alloyed with other metals which, together with the extreme conditions they have endured, means that no conclusive results can be drawn from their investigation. As a result, it is not recommended that iron meteorites are considered in future analogue studies.

The most useful analogue studies on iron and steel are probably those that investigate the corrosion of archaeological artefacts. However, the standard analogue caveats of sample bias, similarity in materials, processes and conditions between the analogue and repository systems must be borne in mind. Iron has been used extensively throughout much of recorded history. Iron beads were worn in Egypt as early as 4000 BC but these were of meteoric iron. The use of smelted iron ornaments and ceremonial weapons became common during the period extending from 1900 to 1400 BC. About this time, tempering was first carried out in the Hittite empire and, subsequently, knowledge of iron smelting was passed throughout Europe, Asia and, later, to the Americas.

Due to the widespread use of iron by early civilisations, iron archaeological artefacts are very abundant and can be found in a wide range of burial environments. These provide the opportunity to examine materials subject to long-term degradation processes and rates in various subsurface environments, and a number of useful analogue studies have been undertaken.

The most comprehensive study is that of Johnson and Francis (1980) who examined a couple of iron meteorites and over forty archaeological artefacts composed of iron or alloys of iron. Disregarding the data from the meteorites, the remaining artefacts provided surprisingly similar corrosion rates of between 0.1 and 10 μm/year, as shown in Figure 4.11.

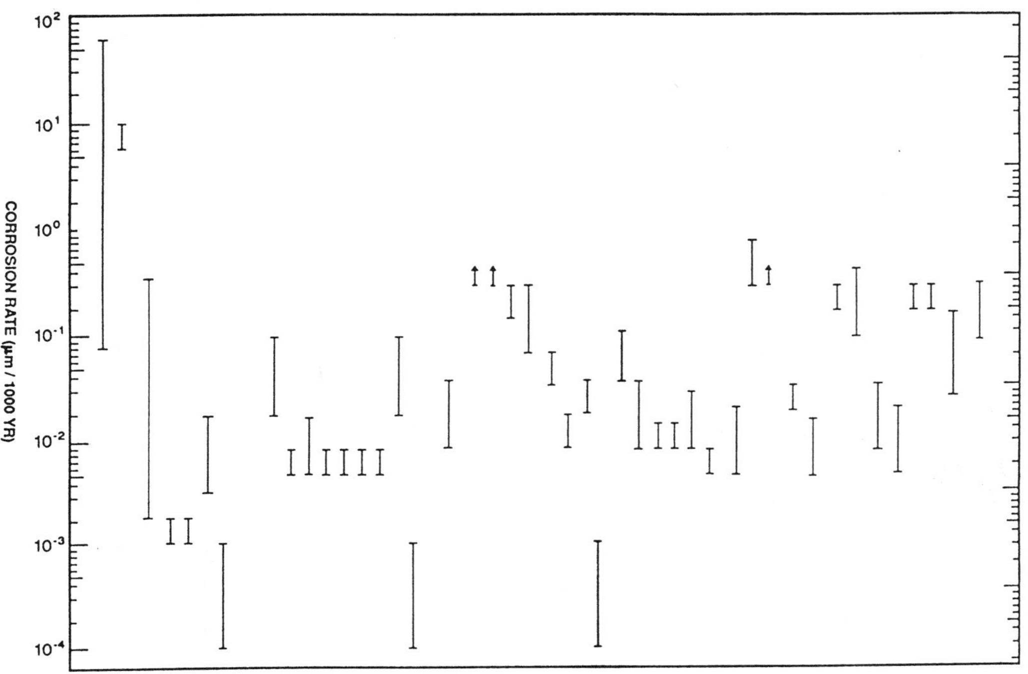

Figure 4.11: Corrosion rate data for archaeological artefacts composed of iron or iron alloys and iron meteorites. The corrosion rates for the archaeological artefacts range from 0.1 to 10 μm/year. After Johnson and Francis (1980).

Box 7: The Inchtuthil Roman nails

Inchtuthil, in Perthshire, Scotland is the site of the most northerly legionary fortress in the Roman Empire (Angus et al., 1962; Pitts and St. Joseph, 1985). The fortress was rapidly abandoned by the Romans in 87 AD, soon after construction, when they retreated south. In a successful attempt to hide from the local Pictish tribes metal objects which potentially could be fashioned into weapons, the Romans buried over one million nails in a 5 m deep pit and covered them with 3 m of compacted earth.

Figure B7.1: Artist's impression of the scene when nearly one million nails were buried at the Roman legionary fortress at Inchtuthil, Scotland. Illustration courtesy of Nagra.

These nails remained buried until the 1950s, when the site of the fortress was excavated, and the nails unearthed.

After excavation, it was found that the nails on the outside of the hoard, particularly those on the top, were very badly corroded and had formed a solid iron oxide crust. However, nails inside the hoard showed only very minimal corrosion, limited to the formation of a thin passivating layer on the nail surfaces.

Although all the nails were composed of iron, they were heterogeneous in composition, with regions of high and low carbon content. Nonetheless, neither the nail composition or size showed any significant correlation with the degree of corrosion.

The survival of the central nails was attributed to the redox buffering capacity of the outer nails which removed free oxygen from the infiltrating groundwaters such that the waters were chemically reducing (less corrosive) by the time they came into contact with the nails in the centre of the hoard.

This is directly analogous to the way in which the large volumes of iron in waste canisters are expected to buffer redox conditions in the repository near-field and so maintain canister integrity and chemically reducing conditions, despite the production of oxygen due to radiolytic decomposition of the water.

The excavation at Inchtuthil was undertaken as an archaeological study and not as a natural analogue. Consequently, no information is available on the chemistry of the soils or porewaters in the burial environment, which means that no quantitative information on corrosion rates or processes can be derived from the nails. However, in qualitative terms, the fact that the central nails survived for over 2000 years in a river flood-plain provides very strong supporting evidence for the lifetime of steel containers in a repository.

It is probable that other buried hoards of nails and other iron artefacts exist and could be excavated in future. If a relevant site could be found, it would be important to combine the archaeological excavation with a detail geochemical survey of the burial environment and of the nails (and corrosion products) to obtain the most complete understanding of the corrosion process.

Archaeological analogues such as Inchtuthil are particularly useful as providers of non-technical demonstrations or illustrations of the performance of a repository. They can deal with materials, processes and timescales with which many people are familiar, making the analogue and its message much clearer.

Figure B7.2: Photograph of one of the nails from the central part of the hoard at Inchtuthil. This photograph shows the excellent state of preservation, despite being buried in the soil for over 2000 years. The nail is approximately 35 cm long.

This uniformity is surprising, considering the various environmental conditions involved. Only two artefacts showed significantly higher corrosion rates; these were cannon balls which had lain in highly corrosive, oxidising seawater and hence are unrepresentative of repository conditions.

The average corrosion rate from Johnson and Francis (1980) is supported by work presented by Yusa et al. (1991) who investigated corrosion of buried gas and water pipes made from various steels and located in a clay-rich burial environment. They reported maximum corrosion rates of up to 10 μm/year and noted that the principal corrosion products were $FeCO_3$ and iron oxyhydroxides.

An archaeological analogue which exhibits a very similar low iron corrosion rate is that of a hoard of iron nails discovered at the most northerly legionary fortress in the Roman Empire, at Inchtuthil, Scotland (see Box 7). These nails remained buried for over 2000 years under 3 m of earth. When they were excavated, the surfaces of all the nails exhibited some corrosion but the degree of corrosion was quite variable and was found to be greatly controlled by the location in the hoard. The nails on the inside of the hoard show minimal corrosion, limited to the formation of a thin corroded layer, whilst those on the outside of the hoard, and in particular those at the top, were corroded to such an extent that they formed a solid iron oxide crust. This crust would presumably have had a low hydraulic permeability and this, combined with the oxygen consumption of the outer nails, ensured that anaerobic conditions were maintained at the centre of the hoard, regardless of its position in the flood plain of the River Tay.

It was noticed that, on a few limited areas on some larger nails, appreciable localised corrosion (pitting) occurred. There is no information given as to the location in the hoard of the nails exhibiting pitting and no quantification of the pitting factor. It may be that the pitting corrosion is influenced by the iron composition. It has been noted that the larger nails have higher carbon contents (Angus et al., 1962), but without further investigation this must remain speculation. In conclusion, it is clear that the corrosion rate of the outer nails was fast in these oxidising conditions but that the excess of iron probably ensured a negligible corrosion rate at the centre of the hoard for over 2000 years.

This Inchtuthil situation can be considered broadly analogous to the behaviour of steel and iron in a HLW or spent fuel repository. However, the analogue is not complete because the conditions at Inchtuthil, being strongly oxidising with a high water flux, are much more aggressive in comparison to the reducing, low flow conditions expected in a repository. Nonetheless, qualitatively it may be concluded from this analogue that, where large volumes of steel are present in a repository, much of the steel will be unaffected by corrosion for a duration in excess of that at Inchtuthil. More quantitative information would have been useful (e.g. the Eh, pH and chemistry of the groundwater), but it should be remembered that the excavation was not performed as a natural analogue study. However, it is illuminating to note that finds of Roman nails are quite common; so common in fact that archaeologists rarely discuss them in any detail (Angus et al., 1962). It would be worthwhile for archaeologists to be made aware of the potential importance of such finds and, if further large hoards are discovered, for the archaeological excavation to be accompanied by detailed geochemical and hydrological investigations of the burial system.

In summary, archaeological examination of iron-based archaeological artefacts indicates consistent corrosion rates which suggest a typical HLW canister could have a lifetime of over 100 000 years (Alexander and McKinley, 1999). However, none of this information can be used to quantify the time to initial canister failure (perforation) because these studies do not give any quantitative information on the pitting factor, which is a parameter commonly required in performance assessment.

In a cementitious environment, such as the near-field of an ILW repository, where the pH will be hyperalkaline (initially > 13), steel corrosion will occur by different reactions and at different rates than in a pH neutral repository without a cement-based backfill. The effect of hyperalkaline environments on iron corrosion and the solid corrosion products has been investigated in laboratory studies (for a review, see Grauer, 1988). It would appear that, in this environment, corrosion rates are negligible since the corrosion product, magnetite, is stable in alkaline solutions and passivates further corrosion.

No detailed natural analogue studies have examined steel corrosion in a high pH environment and this is an area where further studies are recommended. Relevant analogue

Figure 4.12: Photograph of a copper nugget from Michigan in the USA. The copper is millions of years old and can be seen to be in an excellent state of preservation. This provides useful qualitative information for the possible lifetime of a copper disposal canister.

information would be very useful because gas production during iron corrosion could be significant in a L/ILW repository due to the very large volume of steel present (see Section 5.8) and this has to be taken account of in the repository design. More reliable corrosion rates could lead to improved engineered barrier designs if it could be shown conclusively that passivation greatly reduced gas production to rates below those conservatively assumed in current performance assessments.

Potentially suitable analogue systems would include early reinforced concrete structural components, where the steel reinforcing rods have been in contact with cement and high pH pore fluids for several decades. However, it would need to be possible to show that the concrete has been in a reducing environment since construction to be of direct relevance to a cementitious repository.

The durability and longevity of copper

Copper is one of the few metals commonly found in its native state in the geological environment. Occasionally samples are found with purity in excess of 99 %. Copper deposits are widespread, with the largest known deposit being found on the Keweenaw Peninsula, Michigan, which has been studied as a natural analogue of copper longevity (Crissman and Jacobs, 1982). Nuggets of native copper with only thin oxide layers have frequently been found in glacial outwash plains in Canada, as seen in Figure 4.12.

Most large copper deposits, such as that at the Keweenaw Peninsula, were formed by the action of extensive hydrothermal activity on primary host rocks causing wide scale copper mobility and concentration (Jacobs, 1984). This indicates that copper can be very reactive and mobile in the geosphere (Apted, 1992) under extreme hydrothermal conditions but not under the conditions expected in the near-field of a repository.

Meaningful natural analogue studies of native copper need to focus on low temperature environments remote from hydrothermal systems. One such investigation was undertaken by Marcos (1989) who investigated copper deposits from a range of different geological associations. Although not all of these deposits occurred in environments chemically similar to the near-field of a repository (in terms of Eh and pH), all of the samples studied indicated that the copper had remained stable since formation. Other natural occurrences of copper are currently under investigation as analogues (Marcos, 1996; Blomqvist et al., 1997; Marcos and Ahonen, 1999). In these studies, native copper aggregates present on a number of fracture surfaces from the Cu-U mineralisation at Hyrkkölä, Finland are being examined. These aggregates occur at depths of up to 150 m and are in association with copper sulphides, gummite and calcite, and in contact with oxidising groundwaters.

Archaeological materials made of copper and copper alloys have also been investigated as analogues of copper in the repository. Abundant archaeological material is available because copper and bronze (a common alloy of copper and tin) have been used extensively by man since the bronze age. However, few studies have examined copper-based archaeological materials from burial environments with known chemistry, which means that only limited quantitative data on copper corrosion can be obtained. Nonetheless, the broad qualitative evidence is that copper artefacts corrode very slowly and this gives useful illustrative evidence for the long-term stability of copper canisters in a repository environment.

One early investigation of copper durability was performed by Tylecote (1977) who studied many archaeological artefacts made of copper, together with some made of lead, tin and tin-bronzes. This study was, in the main, qualitative with the investigation centering on the different sources of ore and methods of smelting, and how these differences affected the durability of the artefacts and contributed to their survival. Nonetheless, the study directly considered the suitability of the different metals for encapsulating radioactive wastes, albeit for disposal at sea. Tylecote (1977) concluded that copper and copper alloys were by far the most suitable material, having shown the greatest resistance to corrosion of all the materials examined.

In another comprehensive investigation, Johnson and Francis (1980) examined 34 archaeological artefacts composed of copper or alloys of copper. This study is more quantitative than that of Tylecote (1977) and an attempt was made to calculate the corrosion rate directly from the age of the artefact. This approach is useful when an artefact can be dated accurately but this is not often the case. However, the relevance of this type of archaeological analogue can be limited due to

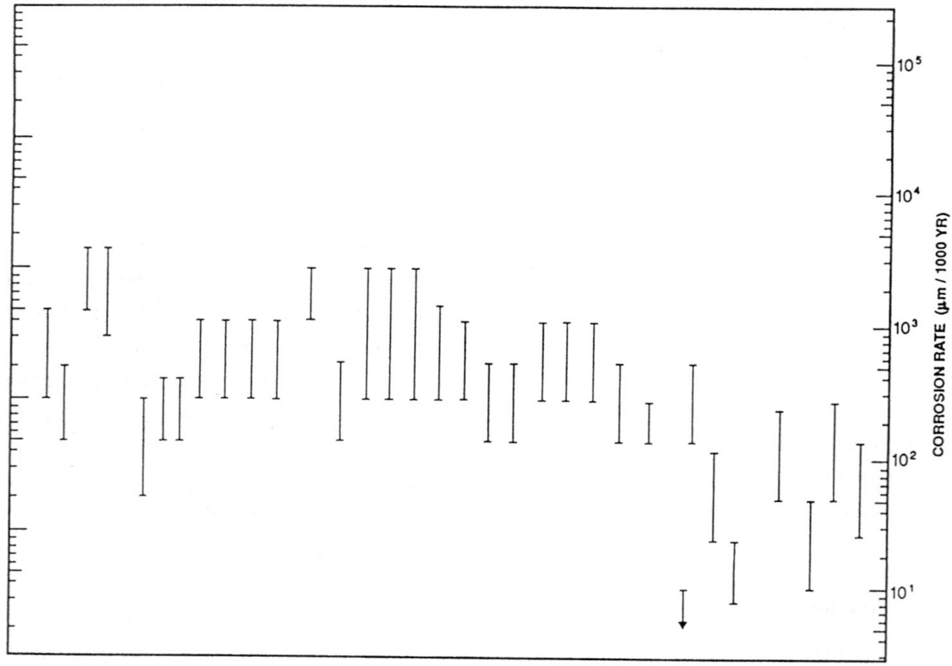

Figure 4.13: Corrosion rate data for 33 archaeological artefacts composed of copper or copper alloys. The range of corrosion rates is from 0.025 to 1.27 μm/year, with an average of around 0.3 μm/year. After Johnson and Francis (1980).

often dissimilar copper compositions when alloyed and environmental conditions which are unlikely to have been constant during the artefact's burial. The problem of sample bias towards better preserved samples may have occurred in the Johnson and Francis (1980) study. Nonetheless, this study revealed a range of copper corrosion rates of 0.025 to 1.27 µm/year (Figure 4.13) which, due to the harsh environmental conditions, may be considered an upper limit to that expected in a repository.

The problem of biased sampling due to rapid corrosion of artefacts located in aggressive conditions has been investigated by Tylecote (1979). The object of this study was to relate the corrosion of copper and tin-bronzes to the chemical condition of the soils in which they were buried. Artefacts from a total of 53 sites were examined and, for each location, the soil pH together with the content of organic matter, CO_2, P_2O_5 and SO_3 was given. Each artefact was examined in detail and its composition and state of corrosion assessed. It was concluded that the most corrosion-resistant metals were the tin-bronzes, but in no case was corrosion excessive. In a typical mildly alkaline soil, the corrosion rate was found to average 0.225 µm/year, which is similar to that calculated by Johnson and Francis (1980).

Whilst this agreement is very encouraging, it must be borne in mind that the composition of the archaeological artefacts is somewhat dissimilar to modern copper, and the environments in which the artefacts were found were generally oxidising whilst a repository is reducing. Furthermore, the artefacts would have been manufactured by cold-wrought processes whereas a copper canister may be fabricated using some form of heat treatment, and this may affect the corrosion rates.

It seems likely that archaeologists might have much more data on corrosion rates and processes that could be very useful in further bounding the durability of copper (and other metals), particularly for samples from reducing environments. This information would have been acquired without any thought for radioactive waste disposal. A thorough literature search, specifically for artefacts discovered in reducing environments, may be informative.

Pitting factors in copper were determined by Bresle et al. (1983) and Mattsson (1983) for use in the Swedish KBS-3 performance assessment for spent fuel disposal. The copper and copper alloy materials used in these studies included several archaeological artefacts such as Roman coins and vases, objects from the Swedish Bronze Age, 17th Century coins, buried lightening conductor plates and lumps of native copper. The objects were all of different age, came from a wide range of environments and had differing compositions. All samples (excluding the native metal) indicated pitting factors of less than 3 (see Figure 4.14); the native metal had a pitting factor of 2 to 6. Comparison with pitting factors used in performance assessment (a factor of 25 was used in the KBS-3 assessment) would suggest that the performance assessment values are overly conservative.

Microbially induced pitting corrosion has been observed on copper water supply pipes in oxidising conditions (Bremer and Geesey, 1991). In this study, the formation of biofilms on the surface

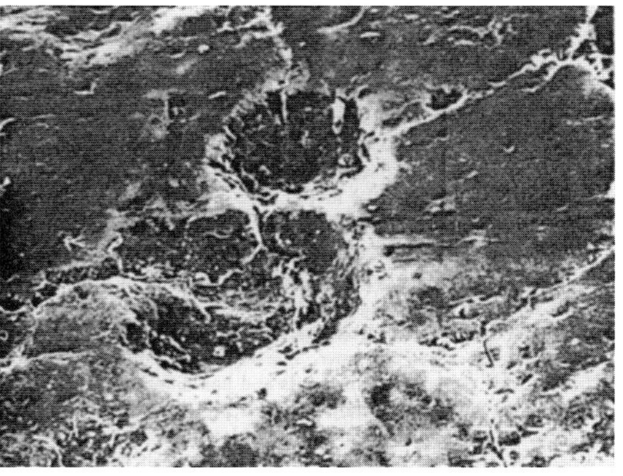

Figure 4.14: Corrosion pits seen in an SEM photomicrograph of a copper alloy razor from the Bronze Age. Magnification X 300. From Bresle et al. (1983).

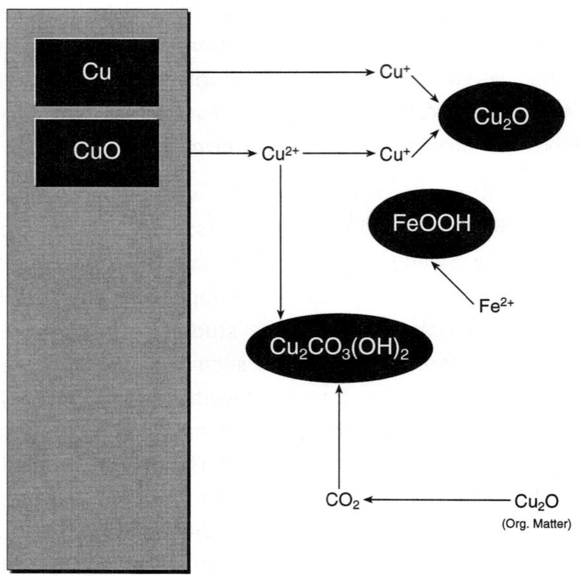

Figure 4.15: The main processes involved in the corrosion of the copper cannon from the Swedish warship 'Kronan'. The corrosion products indicate a generally oxidising environment; the copper was oxidised predominantly to Cu_2O while the CuO slag altered to a hydrated copper carbonate. After Hallberg et al. (1987).

of the copper were found to be important controlling factors in the corrosion process. This process may relate to corrosion of copper canisters in the near-field of a repository which has become oxidising due to the build-up of oxidants from groundwater radiolysis, although the viability of microbes in a spent fuel repository would be limited, as discussed in Section 5.7.

A widely quoted study of copper corrosion is that of the cannon recovered from the wreck of the 300 year old Swedish warship *Kronan* (see Box 8). When the Kronan sank, some of the cannon became partly buried in the marine clays. One cannon in particular was investigated in detail as an analogue for the corrosion of the copper canisters planned to be used in the Swedish spent fuel repository design (see Box 1). This cannon had remained partly buried in a vertical position, muzzle down, in clay sediments since the ship sank (Neretnieks, 1986a; Hallberg et al., 1987). This is generally a good analogue system, although there are important differences between it and the repository, such as the composition and compaction of the clay, and the porewater composition. Nonetheless, other factors, such as the known age of the cannon's burial make this a valuable analogue investigation.

The Kronan cannon had a high copper content and the clay, composed partly of montmorillonite, was tightly packed and water-saturated. The sediment pore waters around the Kronan cannon had neutral pH with variable Eh. The pore waters nearest the top of the sediments were more strongly oxidising due to ingress of oxidising seawater. The change in redox potential with depth was evident from a change in sediment colour from brown in the top 10 cm to grey at depth. Identified corrosion products include Cu_2O and Fe_3O_4, which confirm a generally oxidising environment, as shown in Figure 4.15. A corrosion rate of 0.15 µm/year was calculated, and was constant over the bronze surface. However, inclusions of CuO slag in the bronze weathered more rapidly. Hallberg et al. (1987) concluded that oxygen may be disregarded as the principal oxidising agent because the corrosion products were the same all over the cannon surface. In other words, there is no change in oxidation product with change in redox potential due to limited seawater penetration at depth. That the cannon has suffered only minor corrosion increases confidence in the suitability of copper as a canister material. Indeed, the corrosion of a copper disposal canister should be slower than that of the cannon due to the less harsh repository environment (reducing rather than oxidising) and higher quality copper production (no inclusions).

Metal canister corrosion in the oxidising, unsaturated Yucca Mountain repository (see Section 2.3.1) is likely to proceed at different rates to metal corrosion in other water saturated repository near-fields. One analogue study to investigate metal corrosion in volcanic ash deposits was undertaken by Murphy and Pearcy (1994) at Santorini, Greece. At Santorini, a large volcanic eruption approximately 3600 years ago covered the island with a thick deposit of silicic volcanic ash, which is quite similar to the volcanic rock at Yucca Mountain. This ash inundated the local Minoan settlements causing them to be abandoned.

Ongoing archaeological excavations at Akrotiri on the island have uncovered artefacts made of many different materials in contact with the ash deposit. Various metal artefacts, especially bronzes, in the upper unsaturated layers were considered somewhat analogous to metal canisters in the Yucca Mountain repository. The bronze artefacts were found to be fairly heavily corroded and estimated to have lost approximately one third of their original mass. A quantitative corrosion rate was not given by Murphy and Pearcy (1994) but it is evident, from comparison with other analogue observations, that metal corrosion at Santorini occurred considerably faster than corrosion in other burial environments. The metal lost by corrosion was observed to form a contaminant plume in the adjacent volcanic ash and this information has been used to test trace elemental transport codes (Murphy et al., 1997), as discussed in Section 5.2.

Properties of secondary alteration products

When metal engineered barrier components corrode in the repository, the metal surfaces generally will become coated with solid secondary alteration products. The nature of these alteration products and their rate of formation will be controlled by the type of metal and the chemistry of the reacting porewaters.

As mentioned earlier, the formation of this layer may passivate further corrosion, which is a positive feature. However, when the canister is ultimately perforated, these secondary alteration products can play an additional important role in retarding radionuclides by providing sites for surface sorption or incorporation. Some radionuclides might also be taken-up directly into the structure of the alteration products by precipitation and mineralisation reactions. If this occurs, understanding the formation of these radionuclide-bearing solid phases is important because these phases can control radionuclide solubilities in the near-field system.

The most important iron alteration product for immobilising radionuclides is expected to be iron oxyhydroxide, $Fe(OH)_3$, because very large volumes of iron will occur in most repository near-fields and because this oxyhydroxide is a very efficient scavenger of aqueous phases. Radionuclides may be either sorbed to the surface of the iron oxyhydroxide or may co-precipitate with it, as discussed in Section 5.2.

The exact nature and behaviour of solid metal corrosion products will be determined by the near-field environment. Regardless of repository type, initial iron corrosion will be aerobic as trapped oxygen in the near-field is consumed and iron oxyhydroxides are formed. Aerobic corrosion will occur for a short time only. As an example, for the proposed UK ILW repository concept it has been estimated that this process will operate only for between 50 and 100 years (Atkinson et al., 1988a, 1988b). Once all the free oxygen has been consumed, anaerobic corrosion will begin and magnetite may become the dominant solid phase corrosion product. Anaerobic corrosion is likely to proceed at a significantly slower rate than aerobic corrosion, even when catalysed by microbial activity.

Despite the likely positive behaviour of the solid corrosion products, recent performance assessments, such as the Swiss Kristalline I assessment (Nagra, 1994), have taken a conservative stance and ignored the likely uptake of radionuclides, mainly because relevant examples of trace element association with anaerobically produced magnetite are rare. In the current Swiss

Box 8: The Kronan cannon

The *Kronan* (the 'Crown') was a Swedish warship built in 1668 when she was one of the world's largest ships. She exploded and sank in June 1676 during the Battle of Öland, fighting a combined Danish-Dutch fleet. At the time, she was the most powerful warship in the Swedish navy and was armed with 126 bronze cannon. The total weight of the cannon has been estimated to be around 230 tons. Several of the cannon were captured from other ships during the Thirty Year War (1618-48). Consequently, there would have been a range of cannon types, including Swedish, Spanish, Danish and German.

Figure B8.1: Photograph of the cannon from the Kronan being examined in situ in the marine sediments prior to recovery. Photograph courtesy of the Kalmar County Museum/The Kronan Project.

In the period from 1680 to 1686, around 60 cannon are known to have been salvaged from the wreck using primitive dive bells. In recent times, between 1980 and 1987, a further 32 cannon were salvaged. These ranged in weight from 300 to almost 5000 kg and were cast between the years 1514 and 1661. Some of them may well have come from the Vasa which sank in 1628. The oldest cannon recovered is a German 30-pounder, cast in 1514, the youngest is a Swedish 36-pounder cast in 1661.

In addition to the cannon, more than 22 000 objects, such as highly sophisticated musical instruments, rare coins and elaborate decorations have been recovered from the wreck. The hull of the Kronan is broken apart but a large section of the port ship side remains intact and is lying with the outside facing the bottom clay. This will possibly be salvaged in the future.

Figure B8.2: Professor Rolf Hallberg and Dr Per Östlund on board a Swedish coastguard ship examining the cannon soon after its recovery from the wreck of the Kronan. Photograph courtesy of the Kalmar County Museum/The Kronan Project.

The analogue study focussed on a particular bronze cannon which had remained partly buried in a vertical position, muzzle down in clay sediments since the ship sank (Neretnieks, 1986a; Hallberg et al., 1987). The cannon is a good analogue for the canisters planned to be used in the Swedish and Finnish spent fuel repository designs (see Box 1) which have a copper outer shell because the cannon had a very high copper content (96.3%). The marine clay is also an approximate analogue to the bentonite buffer which will surround a canister because the marine clay is composed partly of montmorillonite, was tightly packed and water-saturated.

Analysis of the cannon surface showed that corrosion had progressed at a rate of 0.15 μm/year since the Kronan sank and that this was constant over the whole cannon, with the exception of some inclusions in the metal which corroded faster. At this rate of corrosion, it would take some 70 000 years to corrode away 1 cm thickness of copper, which provides very strong supporting evidence for the predicted very long life of the copper spent fuel canisters in the repository.

Chemical analysis of the clay from around the cannon showed that copper leached from the cannon metal had diffused 4 cm into the clay, causing a reduction in the copper content at the surface of the cannon from 96.3 to 95.2 %. The sediment pore waters around the cannon had neutral pH with variable Eh; pore waters nearest the top of the sediments were more strongly oxidising due to ingress of seawater. The change in redox potential with depth, was evident from a change in sediment colour from brown in the top several centimetres to grey at depth. Identified corrosion products include Cu_2O and Fe_3O_4 which confirm a generally oxidising environment.

assessment, this issue is being re-examined in an attempt to produce more realistic estimates of radionuclide retardation around the steel canisters but, to date, little relevant laboratory and no relevant analogue data have been found. It is even more difficult to find relevant analogues for iron corrosion in alkaline conditions representative of cementitious ILW repository near-fields. One possible natural analogue could be the corrosion of reinforcing rods in old cements, as discussed earlier.

It was suggested that metallic materials at the Maqarin site in Jordan (see Box 11) could be used to evaluate metal corrosion in high pH conditions. In the adit and at the Maqarin railway station, iron nails and rails have been in contact with hyperalkaline waters for some tens of years (Alexander et al., 1992a). However, because the majority of the groundwaters at Maqarin are oxidising, in comparison to the reducing conditions in a L/ILW repository near-field, no useful information could be obtained from investigation of the corrosion of these materials (Smellie et al., 1997; Smellie, 1998).

An additional source of information on the reactivity of corrosion products comes from uranium ore deposits which have been subject to oxidative weathering. Secondary iron oxy-hydroxide alteration products of iron sulphides and other ferrous iron minerals present in rocks can be observed strongly to sorb a wide spectrum of trace elements mobilised in solution. In the orebody at Poços de Caldas (see Box 14), trace elements were measurably sorbed onto amorphous iron oxyhydroxide phases at the redox fronts. Similar behaviour is seen on a smaller scale in redox halos in sedimentary rocks (Hofmann, 1990a, 1990b).

However, in a recent review of the trace element retardation around redox fronts, including uptake on secondary alteration products, Hofmann (1999) noted that not only are quantitative data rare, in no case studied to date is it possible quantitatively to assess the efficiency of uptake by the secondary products. This being the case, performance assessment treatment of this phenomena will remain conservative for the foreseeable future by continuing to discount radionuclide uptake.

When a copper canister corrodes in a repository, the canister will also become coated with reaction products. The nature and chemical reactivity of these corrosion products should be known so that their influence, if any, on the release and transport of radionuclides can be assessed. None of the natural analogues discussed so far is relevant to this issue because the geochemical environments are so dissimilar. Even in the case of the Kronan cannon, the environment was generally oxidising. If copper artefacts could be found in the vicinity of reducing, neutral to slightly alkaline ground-waters, then some meaningful investigations might be performed. Unfortunately, no such sites are known as yet. It seems probable that investigation of this issue is most efficiently performed in the laboratory if it is deemed necessary to use such data in a performance assessment.

Conclusions

Natural occurrences of iron are rare, which testifies to its generally reactive nature in near-surface rocks. The few large occurrences of native iron which do exist show remarkably low corrosion rates due to buffering of the redox conditions by FeO in the host material (ultrabasic igneous rock) and restricted water access by either low-permeability host material or extended permafrost conditions.

Examination of iron-based archaeological artefacts indicates consistent corrosion rates that suggest a typical HLW canister could have a lifetime of over 100 000 years (Alexander and McKinley, 1999). Little quantitative natural analogue information is available on localised corrosion (pitting) of iron or steel and, although laboratory data clearly indicate that pitting of a steel canister should constitute no significant problem, qualitative support of this conclusion by analogue data would clearly increase confidence in the conclusion. A rare, but unquantified, reference to pitting of iron artefacts comes from the excavation of the Inchtuthil nails.

The reactivity of iron and steel corrosion products has, to date, played no real role in the calculated performance of the near-field of some repositories but is currently being considered in more realistic assessments of this issue. Most of the information on the sorptive properties of secondary iron minerals comes from qualitative studies of redox fronts and concerns iron oxyhydroxides. More information on the development and reactivity of iron corrosion products formed under anaerobic conditions would be valuable. Investigation of the reinforcing rods in old concrete from a chemically reducing environment may provide a situation where steel corrosion products can be examined in a reducing, alkaline environment.

Native copper is relatively abundant in certain geological environments which points to its stability under the conditions expected in a repository. The very low corrosion rates obtained from analogue investigation of both native copper samples and archaeological artefacts suggest that a copper canister could have a lifetime in the order of several hundreds of thousands of years.

Localised corrosion (pitting) has been identified in copper archaeological artefacts, with pitting factors of below 3. Similar studies of native metal indicate a pitting factor of 2 to 6. However, further data on localised corrosion would be useful.

It would be useful to determine the nature and reactivity of copper corrosion products. However, laboratory investigations are, perhaps, the best way to approach this issue due to the lack of native copper or archaeological artefacts found in repository-relevant environments to date.

Further analogue studies on metal corrosion processes and rates are probably only warranted if more relevant systems can be identified. In particular, metal compositions should be as close as possible to canister metals. This means investigating copper not bronze, and steel not iron. This clearly has an impact on the age of artefacts that would be relevant for each metal.

4.5 Bentonite

Bentonite is the name given to naturally occurring clays that comprise mixtures of the minerals montmorillonite and beidellite, both of which are members of the smectite group of clays. Bentonites generally form by alteration of volcanic ash or tuff, but they can also form by hydrothermal alteration of igneous rocks, such as trachyte, as in the case of the bentonites at Almeria, Spain. The characteristic component of bentonite are the smectite clays, of which montmorillonite is the most common. The smectites are swelling clays, that is they show the ability to take up water or organic liquids between their structural layers, causing an increase in volume, as can be clearly seen in Figure 4.16. This ability is most pronounced in the sodium-rich smectites which can expand to up to 15 times their dry volume, if unconfined. The smectites also have significant cation exchange properties and may act as a pH and Eh buffer.

The first of these attributes has led to their incorporation into most HLW and spent fuel repository designs as a buffer to surround disposal canisters. According to current design concepts, bentonite will be placed into these repositories in the form of heavily compacted, machined blocks. A demonstration emplacement is shown in Figure 4.17. Bentonite may also be placed in some L/ILW repository designs as a backfill around disposal vaults and silos, when it might be mixed with sand or crushed rock aggregate.

Once a repository has been sealed, groundwater will flow into the near-field and will be adsorbed by the bentonite, causing it to expand. This expansion will be restricted by the enclosing rock mass and, as a consequence, the swelling pressure will reach some tens of MPa. Water and dissolved radionuclides will be able to pass through the bentonite only by diffusion rather than by advection, therefore limiting radionuclide mobility.

Colloid and microbe migration in the microporous fabric of compacted bentonite is also negligible.

Figure 4.16: Demonstration of the very high swelling capacity of compacted bentonite. Right: Perforated tube containing pellets of dry, compacted bentonite. Left: The same tube and bentonite after immersion in water for 24 hours. This particular tube and bentonite construction is used for sealing boreholes, larger cut blocks of compacted bentonite will be used in the near-fields of HLW and spent fuel repositories. Illustration courtesy of SKB.

Other physical properties of bentonite which are important are a high level of plasticity which allows the bentonite to flow into and seal any void spaces, a reasonable load-bearing capacity so that the waste canister should not sink through the bentonite and a relatively high thermal conductivity so that radiogenic heat generated within the canister can be dissipated.

One commercially available bentonite which has been investigated in detail is MX-80. This material has physico-chemical characteristics that are representative of the type of bentonites which will eventually be used in repositories. The mineralogical composition of MX-80 is: montmorillonite, 75 %; quartz, 15 %; feldspar 5 to 8 % and the remainder comprised of mica, carbonate, kaolinite, pyrite and organic carbon (Müller-Vonmoos and Kahr, 1983). It is unfortunate that some studies do not provide the compositions of the clays they are investigating in comparison to bentonites such as MX-80. This makes assessing the relevance of any quantitative analogue conclusions difficult.

Smectite clays may transform progressively into illites at elevated temperatures, when the sodium in the clay can be exchanged for potassium in groundwater. This happens naturally in a number of geological processes, including diagenesis, contact metasomatism, regional metasomatism and hydrothermal alteration. Illites have lower swelling capacities and higher permeabilities than smectites and it follows that the illitisation of bentonite may be of concern for HLW or spent fuel repository behaviour. However, as the illitisation reaction is very temperature-dependent and the repository thermal period will be relatively short, significant illitisation is not expected. In contrast, smectite alteration potentially may be a greater problem in repositories where potassium is abundant in the host rock groundwater or in the cement leachate.

In the case of the Swiss HLW repository design, McKinley (1985) showed that, for a repository in crystalline basement rocks, the groundwater potassium flux to the repository is so low that the bentonite should survive significant alteration for 10 to 100 million years, depending on the actual conditions of the site.

In the case of cementitious repositories, the high pH of the system is likely to give rise to alteration products other than illite. The interaction of bentonite with other engineered barrier materials (such as metal and cement) may also locally affect its properties.

Regional metamorphic processes are not normally suited to natural analogue study due to the fact that both maximum temperatures and duration of heating are much more extreme than would be encountered in a repository environment. Hydrothermal alteration is also often of limited value for natural analogue studies because the boundary conditions (temperatures, fluid-rock ratios etc.) are usually highly variable and not easily determined (see Section 3.2.4). However, diagenesis and contact metamorphism have been studied as natural analogues of illitisation. These processes have the advantage that the maximum temperature, duration of heating and fluid-rock ratio can sometimes be determined, allowing the system to be better characterised than in some other cases. Previous studies of these two processes are discussed below.

In addition to promoting illitisation, high temperatures may cause compacted bentonite to become cemented when only partially resaturated, if a steam phase forms (Couture, 1985).

Figure 4.17: Demonstration emplacement of a canister in a disposal tunnel. The machined blocks of compacted bentonite are clearly seen lining the tunnel. Once groundwater resaturates the tunnel, the bentonite will swell to close all voids between the bentonite blocks, the canister and the tunnel walls. Illustration courtesy of Nagra.

The issues of most relevance to the behaviour of bentonite in the repository which have been (or potentially could be) addressed in natural analogue studies are:

- longevity and rate of alteration;
- physico-chemical changes due to heating;
- canister sinking;
- interaction with other repository materials; and
- hydraulic barrier and colloid filter functions.

These issues are discussed in the following sections.

Longevity of bentonite and the rate of alteration

A number of natural analogue studies have examined diagenetic illitisation in the Gulf of Mexico, USA (e.g. Eberl and Hower, 1976; Roberson and Lahann, 1981), and elsewhere (Pusch and Karnland, 1988). These studies do not really represent the repository

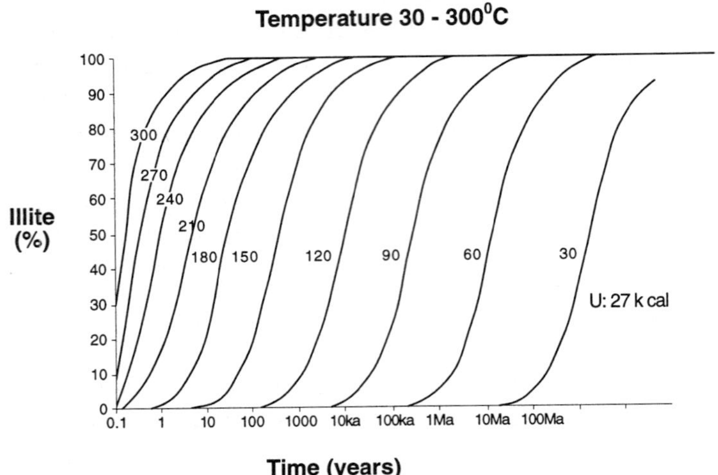

Figure 4.18: Graph showing that the illitisation rate is strongly dependent on the temperature. The data here were obtained from laboratory studies in a closed system containing abundant potassium. These data imply that at repository temperatures, the illitisation rate would not be significant. In the real repository environment, illitisation might be slower than predicted here because the supply of potassium is likely to be restricted. After Pusch (1985).

environment because the duration of heating is several orders of magnitude longer than would be the case in a HLW or spent fuel repository. The temperature is an important factor controlling the illitisation rate, as can be seen in Figure 4.18. Nonetheless, some useful information was obtained in that these studies suggesting the illitisation rate in the natural environment is considerably slower than that predicted by kinetic models (Anderson, 1983). However, this is due to the fact that the process depends on the rate of supply of potassium, which may often be limited. This is evidenced by the fact that natural bentonite deposits which are millions of years old but have suffered only partial alteration are quite common.

A number of natural analogue studies have also examined contact metamorphic illitisation (e.g. Benvegnú et al., 1988; Pusch and Karnland, 1988; Yusa et al., 1991). In most cases, such studies poorly represent the repository environment because contact temperatures may reach up to 900°C, which is much higher than temperatures expected in a repository. To avoid this problem, it is best to examine thick bentonite sequences where samples may be collected some distance from the contact. In the study performed by Yusa et al. (1991), the process was well-constrained, consisting of an intrusive rhyolitic rock in contact with an homogeneous bentonite sequence. Cooling rates of 60 to 70°C per million years were determined from radiometric mineral ages. The analogy with a repository is not ideal since this cooling rate is much slower than that expected for a HLW repository. However, these values, together with the illite-smectite ratios at different locations, allowed an activation energy for the illitisation process of 27 kcal/mol to be calculated. This calculated activation energy is close to the 30 kcal/mol value obtained by Roberson and Lahann (1981) from laboratory experiments, providing some measure of validation of the short-term laboratory data.

Pusch and Karnland (1988) investigated several natural bentonites and their data gave important constraints on the conditions required for bentonite alteration. Bentonite from Sardinia gave definite proof that significant heat-induced dissolution of smectite occurred at 150 to 200°C and precipitation of siliceous material occurred during cooling. This siliceous cementation was found to have measurably affected the rheological properties of the bentonite, in a manner which might adversely affect the containment of radionuclides if it occurred in a repository environment. The duration of heating above 100°C was estimated to be only three months. The authors concluded that there is a critical temperature (about 150°C) at which montmorillonite converts to beidellite. Further alteration to mixed-layer illite/smectite clays and separate illite depends entirely on the access of potassium to the system.

There are many other natural sites for which data exist on smectite/illite transformation, many of which were examined for purely academic mineralogical research and not for any particular analogue objective. In total, these natural data cover pressure and temperature conditions which are more extreme than the conditions expected to be achieved in most HLW and spent fuel repository near-fields, and none of these studies indicate that the rate of illitisation is ever significantly fast. Thus, it can be concluded that illitisation is unlikely to be a serious problem in a HLW or spent fuel repository.

However, bentonite alteration in a cementitious environment, such as the near-field of an ILW repository, could be more problematic than for HLW or spent fuel concepts. High potassium concentrations are likely to occur when the groundwater leaches the large volumes of cement present (see Section 4.6). However, increased bentonite alteration due to this potassium may be offset by the lower temperatures expected in a L/ILW repository which would slow the rate of the process. The hyperalkaline environment in the near-field of a L/ILW repository would also cause alteration products other than illite to form in this situation, although at high pH the potassium concentration is less relevant to the alteration process. Bentonite alteration in a high pH environment is not easy to investigate by natural analogue studies since relevant analogue sites are rare and no suitable sites have yet been identified.

Bentonite will help to buffer the Eh and pH of the porewaters in the near-fields of spent fuel and HLW repositories. As well as modifications resulting from interaction of the clay with the engineered barriers, the chemical buffering capacity might also be affected by changes to the clay mineralogy resulting from interaction with the groundwater. This process may be amenable to analogue study; the interaction of low ionic strength groundwaters of various compositions from rocks overlying bentonite deposits with the clay porewaters being the most obvious possibility. Such information could be used to test thermodynamic model predictions or to compare with laboratory experiments. However no analogue study has yet provided useful information on the changes in bentonite porewater chemistry and laboratory studies probably will remain the primary source of information on this issue.

Physico-chemical changes due to heating

A detailed investigation of the changing isolation properties of clays was performed by Pusch et al. (1987) who studied seven natural clays with smectite contents ranging up to 25 %. Although these clays ranged in age up to several hundred million years old and had experienced slow mineralogical alteration processes in relatively near-surface active environments, most still possessed swelling and rheological properties which would be adequate for buffer performance. As such, this qualitatively suggests that the bentonite buffer in the relatively stable, low flow conditions of a repository near-field would be likely to retain its barrier functions for at least similar periods of time.

Two of the clays examined by Pusch et al. (1987) had been cemented. The likelihood and performance implications of cementation occurring in the buffer of a HLW or spent fuel repository have not yet been adequately investigated. If cementation did occur, then it is possible that the cemented bentonite could then fracture, which would allow groundwater to flow through the fractures and, thus, the bentonite's barrier functions would be diminished. Furthermore, if the bentonite were cemented, its ability to sorb radionuclides might be lessened because the available bentonite surface area would be reduced, although this may be a minor effect. However, it is unlikely that the whole bentonite buffer mass would become cemented, so a considerable capability to seal-heal may be retained.

Since this is a potentially significant process, a number of natural analogue studies have attempted to investigate bentonite cementation, particularly with the intention of determining whether a cemented and fractured bentonite mass is capable of self-healing. Such studies have been attempted at sites where clays close to igneous intrusions or lavas have been cemented as a consequence of heating in Scotland, France and Italy (e.g. Bouchet et al., 2000). Once such study was undertaken at Orciatico, Italy (Gera et al., 1994; Pellegrini et al. 1999, 2000) at a site where a small igneous intrusion penetrated Pliocene argillaceous sediments. The intrusion had an initial temperature of around 800°C which is far higher than near-field temperatures and, thus, the behaviour of the clays closest to the intrusion has no analogue significance.

The study, therefore, focussed on clays at varying distances away from the contact with the intrusion where maximum temperatures would have been lower. A number of mineralogical changes were observed in these samples, including crystallisation of potassium feldspar and smectite causing cementation. The mineralogical changes away from the intrusion were associated with subsequent fracturing of the cemented clays as the interstitial waters were driven out. Convective water circulation was suggested as the reason why smectisation occurred. Unfortunately, these observations cannot directly be applied to performance assessment because the thermal, chemical and pressure histories experienced by these clays have not yet been adequately characterised. More detailed information may acquired in further stages of this analogue study.

Canister sinking

If a HLW or spent fuel canister were to sink through the surrounding bentonite and come to rest on the floor of the tunnel or disposal hole, then the buffering capacity of the bentonite would be effectively short-circuited, allowing more rapid transport of radionuclides from the canister to the host rock to occur. The colloid transport barrier function of the buffer would also be lost.

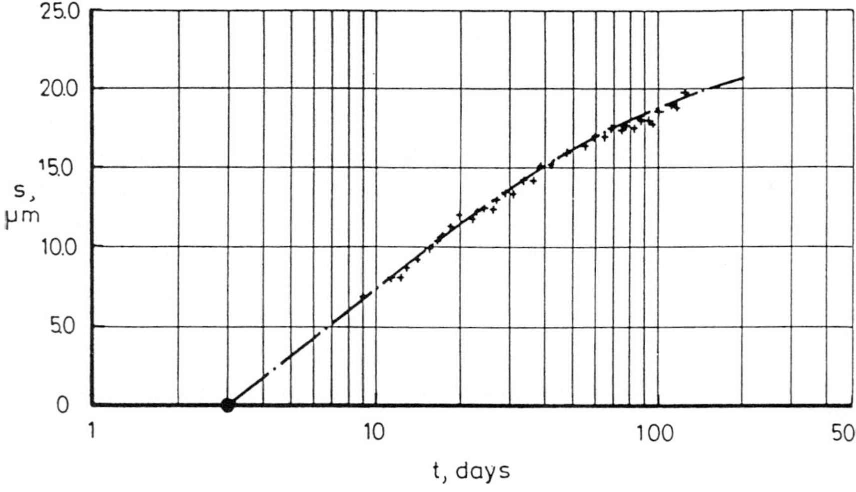

Figure 4.19: Canister sinking has been investigated in laboratory experiments. In this case, a model (not full size) canister was found to sink at a rate of about 1 µm/day at a system temperature of 70°C. However, such laboratory data are inappropriate to scale-up to the long time periods of relevance to performance assessment and supporting analogue data would be useful to supplement these data. From Pusch (1986).

There is some disagreement over the possibility and mechanisms of sinking. One suggestion is that consolidation of the bentonite would occur beneath the canister causing it to settle, but only to a limited depth. Another suggestion is that the water saturated bentonite clay would exhibit viscous flow behaviour and be squeezed out from beneath the heavy canister. In this case, the canister might conceivably sink right through the bentonite until it comes into contact with the near-field rock.

Canister sinking has been investigated in laboratory experiments (Pusch, 1986) and on a larger scale at the Stripa mine (Börgesson and Pusch, 1989). In both experiments measurable movement of the canister was observed, although not on a scale sufficient to cause concern, see Figure 4.19. It must be noted, however, that neither investigation was performed on full-size canisters and, of course, the experiments were of limited duration, three to four years at most.

In an assessment of the probability and performance consequences of canister sinking in the Swiss HLW repository design, Smith and Curti (1995) concluded that it was very improbable that the canister could sink completely through the bentonite buffer. In the more realistic case of the canister settling part way through the buffer, the radiological impact was found to be negligible, mainly due to the over-engineered design of the full engineered barrier system. However, recent moves to optimise repository designs may necessitate a re-analysis of this process.

Canister sinking is clearly a process which requires investigation although, to the best of the authors' knowledge, no natural analogue study has yet adequately addressed this issue. There are some natural processes which may give rise to dense objects resting on bentonite; for example meteorites, barite nodules or glacial erratics, and study of these might provide some qualitative information. However, almost certainly, there will be differences in the size, shape and density of any analogue object compared to a disposal canister, and the rheological behaviour of the clay may not be the same as compacted, water saturated bentonite. Also, in many cases (such as the meteorite situation) the manner of emplacement will be very dissimilar to that of a canister in a repository. This means that it is very unlikely that quantitative data could be obtained but, nonetheless, any relevant qualitative information would still be useful as supporting evidence for the modelling predictions.

One study which has recently been initiated may in the future provide some useful information on this issue. The bentonite deposits near Almeria, Spain are overlain by a limestone horizon of varying thickness (Hernán and Astudillo, 2000). This limestone would have been deposited slowly, without exerting sudden impacts on the bentonite. The deposition of the differing thicknesses of limestone would have exerted variable loading pressures on the bentonite layer beneath. There is a possibility, therefore, that some qualitative information might be obtained from observations of the way in which the bentonite has responded to this differential loading which might be useful for understanding the behaviour of bentonite beneath a dense waste canister. However, to date, no detailed results from this study have yet been published.

Archaeological environments may provide the best chance for finding a good analogue. Burial chambers lined with clay have been discovered in China: (see later). If one of these, or similar chambers elsewhere, contained heavy metal items and the clay was water-saturated then these may provide semi-quantitative information. It would be possible to establish the duration of sinking of any object whilst, at the same time, being certain that no other disturbances had occurred if it could be shown that the chamber had remained sealed. Unfortunately, no suitable burial chambers containing such items are known to the authors.

Interaction with other repository materials

The elevated temperatures in the near-field of a HLW or spent fuel repository may cause chemical interaction between the bentonite and other engineered barrier materials. For example, the bentonite could react with the canister metal, once the bentonite has become water-saturated during the initial high temperature period. This could,

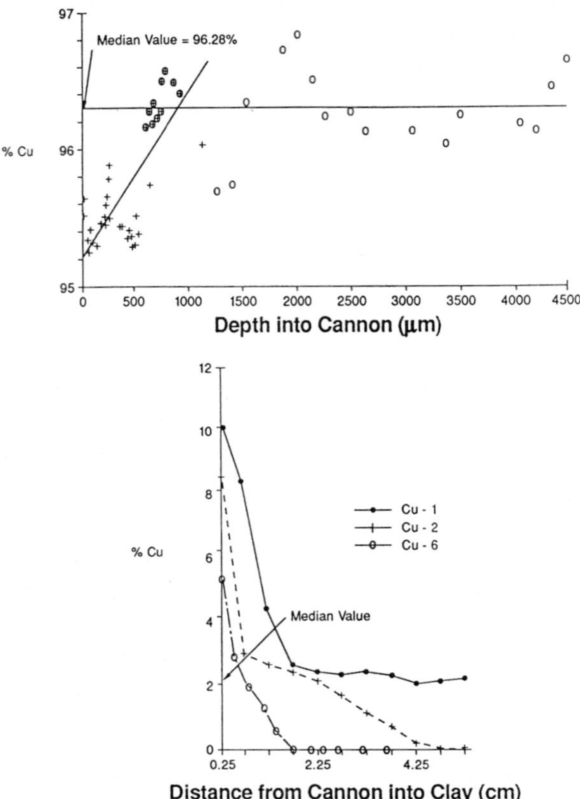

Figure 4.20: During burial in marine sediments of a copper cannon from the Swedish warship 'Kronan', migration of copper occurred from the surface of the cannon into the clay. Top: The copper content of the cannon is reduced near the surface from an average of 96.28 % to 95.2 %. Bottom: Analysis of the clay shows the copper has migrated about 4 cm from the cannon; from three profiles. From Neretnieks (1986a).

potentially, have deleterious consequences because leached metal ions from the canister might exchange with cations in the bentonite clay, with a corresponding loss of isolating properties.

This phenomenon has been recognised in experiments where copper, in contact with bentonite for 3 to 6 months, was leached and the copper ions replaced sodium in the bentonite (Pusch, 1982a). Extrapolating from these data, it was thought that, over the lifetime of a copper spent fuel canister, ion exchange and clay particle rearrangement may occur in the entire buffer. As a consequence, permeability could be increased by around a factor of 2 or 5 but this would not be significant due to the very low initial permeability. However, the experiment performed by Pusch (1982a) was not conducted under repository-relevant conditions and, consequently, the observed rate of ion exchange may not be realistic as the rate will be dependent on ionic strength and pH.

One natural analogue study that has a bearing on this matter is that of the copper cannon from the Swedish warship Kronan which was buried in marine clays (see Box 8). The cannon is somewhat similar to the copper canisters planned to be used in the Swedish spent fuel repository design, in that the cannon had a high copper content and the clay was tightly packed and water-saturated. Whilst this study was more concerned with the rate of copper corrosion (as discussed in Section 4.4), the study did reveal that copper from the cannon had diffused 4 cm into the clay over a 300 year period, as can be seen in Figure 4.20. Unfortunately, there is little information regarding the actual products of the copper interaction with the clay. If possible, this clay should be re-examined with a view to identifying any mineralogical changes resulting from interaction with the copper, whilst bearing in mind the limitations of the analogy.

In the case of iron canisters, a few practical studies have been performed (Simpson, 1983; 1984) which suggest that some reaction takes place. Grauer (1990) indicates that insufficient thermodynamic data are available to fully assess the reactions between magnetite and bentonite. However, he does suggest that the formation of iron phases such as chamosite, greenalite or nontronite is likely and, given the mass ratios, some 20 % of the bentonite may become involved. These minerals will be microcrystalline and will not, therefore, impair the sorption behaviour of the buffer but they have no swelling capacity and, hence, the physical barrier properties of the clay may be reduced. No natural analogue studies are known which have investigated this issue.

Other locations where metal-clay interaction may be studied include the contacts between ore bodies and host sediments, and buried metallic archaeological artefacts. If such contacts could be found at depth, where groundwaters are reducing and neutral to alkaline, a useful natural analogue study could possibly be undertaken.

Laboratory experiments have investigated the reaction between bentonite and Portland cement (e.g. Pusch, 1982b; Milodowski et al., 1990). These investigations reveal that clay in contact with cement alters initially with the exchange of sodium and magnesium in the clay for calcium from the cement-derived fluids. If alteration continues, the clay can be completely degraded and can form zeolites. This process leads to an increase in volume and, therefore, a pressure increase in a confined system, combined with a loss of plasticity. In the repository environment, the extent and rate of this reaction would be controlled by mass ratios and by the aqueous diffusion of cement pore waters into the bentonite. In locations where old cement or concrete foundations have been laid in bentonite, it may be possible to investigate this process. Alternatively, it may be possible to find instances where bentonite is in contact with hyperalkaline groundwaters at a site similar to Maqarin in northern Jordan (see Box 11) where hyperalkaline waters react with clay-rich limestones. However, no analogue study has yet investigated this process in any detail.

Hydraulic barrier and colloid filter functions

A key function of the bentonite buffer is to restrict water movement to diffusion. This is achieved by a very high swelling pressure which causes the confined bentonite to develop a very low hydraulic conductivity. The bentonite will swell into small fissures in the surrounding rock, further limiting water flow and reducing the potential for radionuclide release. On swelling, the bentonite pore spaces become very compressed, to the point that colloids and microbes are unable to move through them. Combined, all these processes act to isolate the canister from the groundwaters in the near-field rock, and the maintenance of this hydraulic isolation capacity is important for the repository performance.

There are several impressive examples of organic materials being preserved in clay environments. The best known is the 2 million year old preserved forest at Dunarobba, Italy but there are several others, such as 2000 year old cadavers buried in excavated tombs in China (Lee, 1986). These examples of organic material preservation indicate significant isolation capacity for clays and, furthermore, have great potential for illustrations for non-technical audiences, a matter which is discussed in Chapter 6.

The preserved trees at Dunarobba (Ambrosetti et al., 1992; Benvegnú et al., 1988) are particularly interesting because they are still in their original vertical positions and, unlike most other examples of buried forests, these trees are still composed of wood (see Box 9). In normal circumstances, wood alters to lignite and is subsequently lithified when buried. The trees at Dunarobba are enveloped in a lacustrine clay, above which are sand deposits with freely circulating, oxidising water. There is a large difference in hydraulic conductivity between the two materials, with the clay layer having a hydraulic conductivity in the range 2×10^{-13} to 2×10^{-10} m/s compared to an estimated hydraulic conductivity for the sand of 10^{-4} m/s (Lombardi and Valentini, 1996). The wood has been protected from active degradation processes largely by the clay envelope which restricted ingress of the oxygenated water which would have allowed aerobic decomposition to take place. Unfortunately, there is currently no detailed geochemical or mineralogical information on the clay and,

Box 9: The Dunarobba forest

The Dunarobba forest, near Todi in central Italy, is an exceptional case of the natural preservation of wood. In the Dunarobba and Cava Topetti quarries, dead tree trunks can be found still in their original, upright position. The soils and sediment they were growing in have been dated to the Upper Pliocene, making the trees approximately 2 million years old (Ambrosetti et al., 1992; Benvegnú et al., 1988).

Figure B9.1: Photograph of the Dunarobba forest from a distance showing the trees standing in their original growing positions. The clay horizon which allowed the preservation of the trees is the cream coloured material in the middle of the photograph.

The most unexpected feature of these trees, which separates them from typical fossilised trees is that they are still composed of wood. In normal circumstances, wood alters to lignite and is subsequently lithified when buried, if it does not decompose completely.

Hydrogen, oxygen and carbon isotopic ratios measured on material sampled from the Dunarobba trees are very similar to those of modern plants, indicating that very little alteration to the organic material has taken place. This is supported by the fact that the wood can be sawn, polished and even burned.

Preservation of the wood has been attributed to the very low hydraulic conductivity of the clay which has surrounded the trees. In the sedimentary sequence at Dunarobba, groundwater preferentially flows through more permeable sandy layers that lie above the clay horizon containing the tree trunks (Lombardi and Valentini, 1996). The clay stopped oxygenated waters from contacting the wood, limiting aerobic decomposition processes. The stable burial environment also meant that little mechanical disturbance to the trees occurred.

Figure B9.2: Photograph of one of the trees from the Dunarobba forest being cut with a chain-saw, showing clearly that they are still comprised of wood.

To date, there has been very little detailed mineralogical, geochemical or hydrological information published on the clays. Therefore, the Dunarobba tress are generally used only as qualitative illustrations of the isolating capacity of clay, in terms of the compacted bentonite buffer found in HLW and spent fuel repository designs. However, in this respect, they are very convincing, especially as non-technical demonstrations for the general public. The Dunarobba trees also have particular relevance for L/ILW repository concepts, because the wood can be considered to be somewhat analogous to the organic/cellulosic materials which will comprise a large part of the waste.

This analogue thus suggests that if anaerobic conditions can be maintained, decomposition and thus gas generation could be minimised in the repository near-field. The impact of this on L/ILW repository design and performance is worth considering further.

consequently, the natural analogue is poorly characterised. Nevertheless, it is a clear illustration of the potential isolating capacity clay-type materials can possess, which qualitatively supports the isolation role of the bentonite buffer in the repository. Further detailed studies at the site could be beneficial to allow more quantitative conclusions to be reached.

It is worth noting that, despite the fact that the wood is organic, the clay at Dunarobba has clearly had a role in retarding microbial degradation. This issue would be worth examining further because the wood is somewhat analogous to the organic and cellulosic materials which may be placed in some L/ILW repositories (see Section 4.8).

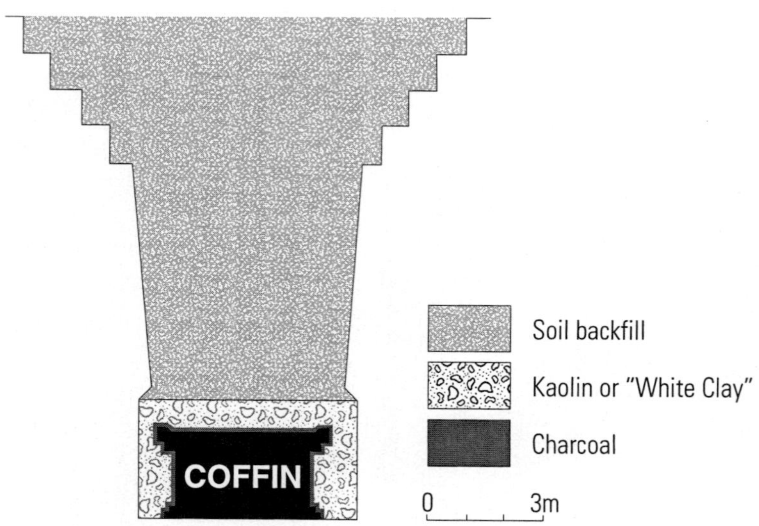

Figure 4.21: Cross-section of the 2 100 year old Chinese tomb in which a well preserved cadaver was found which was not mummified before burial. In addition to the body, wooden artefacts and animal and vegetable foodstuffs were also found within the tomb in a good state of preservation. The preservation of the body and the artefacts was due to the thick layers of clay and charcoal that were surrounding the coffin which restricted ingress of oxygenated water which would have allowed aerobic decomposition to take place. After Lee (1986).

The 2100 year old cadaver from the Chinese tomb (Lee, 1986) also provides a good illustrative example of the isolation capacity of clays. Here, the body was placed in a wood coffin together with various burial artefacts, including silk, wood, meat and vegetables. After exhumation, the cadaver was found to be well-preserved; the skin was complete and retained some of its elasticity whilst the abdominal organs were intact and some of the joints were partially movable. The burial objects were in an equally good state of preservation, the meat and vegetables showing only partial decomposition.

This high degree of preservation is attributed to the clay lining (a few metres thick) placed deliberately in the burial chamber and totally enclosing the coffin, see Figure 4.21. In addition to the clay, a layer of charcoal surrounded the coffin; this may also have helped to preserve the cadaver by absorbing any moisture present. The clay liner effectively provided an air-tight seal and putrefaction proceeded only until the oxygen trapped in the burial chamber had been exhausted. Again, the geochemical and mineralogical composition of the clay is not well known; however, the principal clay mineral is believed to be kaolinite which is abundant in the region.

Other similar burial chambers are thought to exist in the region and it would be useful to undertake further quantitative studies if new tombs are excavated, particularly to focus on the geochemical and hydraulic properties of the clay.

The hydraulic barrier function of clay has also been well demonstrated at the Cigar Lake uranium orebody (see Box 5). The orebody at Cigar Lake is largely surrounded by a 10 to 50 m thick illite/kaolinite clay halo, locally isolating it from the overlying sandstone host rocks. This clay halo has provided an effective, long-term seal for the orebody for most of its existence (1300 million years) which demonstrates the stability of the clay under these hydrochemical conditions.

A considerable conductivity contrast exists at Cigar Lake between the host sandstones which have a hydraulic conductivity of 10^{-6} m/s and the clay halo which has a hydraulic conductivity of 10^{-9} m/s (Winberg and Stevenson, 1994). In a repository near-field, bentonite, rather than illite, will be used as the buffer material and this can be expected to provide an even more efficient hydraulic barrier than the illite at Cigar Lake, because of the much high swelling capacity of bentonite compared to illite.

In addition to limiting water flow (advection), the clay halo at Cigar Lake also acts to reduce the movement of colloids. This is an important safety role for the repository bentonite buffer and laboratory studies indicate that the porespaces within the compacted bentonite will be too small to allow transport of colloids (e.g. Torstenfelt et al., 1982a; Eriksen and Jacobsson, 1982). At Cigar Lake, the populations and geochemical characteristics of colloids inside the clay halo adjacent to the orebody are distinct from those outside the clay halo in the sandstone, clearly indicating that the clay has effectively acted as a colloid barrier (Vilks et al., 1991; Vilks and Bachinski, 1994). Again, the much higher bentonite swelling pressures in the repository near-field should mean that the buffer should be at least as effective as filtering colloids as the clay halo at Cigar Lake.

Conclusions

The alteration of bentonite to form illite, in diagenetic and contact metamorphic environments, has proved amenable to natural analogue study. It is clear from such studies, that the rate of illitisation is so slow that complete illitisation of the bentonite buffer in a HLW or spent fuel repository could take some 10^8 years which is much longer than the time period of relevance to performance assessment. Analogue data suggest there is a critical temperature (about 150°C) at which montmorillonite converts to beidellite. Further alteration to mixed-layer illite/smectite clays and separate illite depends entirely on the access of potassium to the system. No further natural analogue studies of this process seem to be needed as long as repository designs ensure that temperatures remain below 150°C.

The evolution of bentonite pore water chemistry might be amenable to natural analogue study. Although this is not of primary concern to performance assessment, analogue data on this issue could help further to build confidence to demonstrate the long-term stability of the engineered barrier system.

The sorption of radionuclides onto clay minerals has not been investigated in detail in natural analogue studies and this is an area where further research could be undertaken. Potential analogue systems would be the clay haloes that surround both the Cigar Lake and Oklo uranium deposits.

Cementation of bentonites may restrict their ability to self-heal following mechanical displacement, thus causing a significant rise in hydraulic conductivity. In extreme cases this mechanism may possibly cause fissuring, allowing direct radionuclide transport through the buffer by advection. This issue has not yet been adequately resolved in natural analogue studies and further investigations at relevant sites are warranted.

There remains a possibility that canisters might sink through the buffer depending on the rheological behaviour of the bentonite. While this issue does not appear to be critical to performance assessment, a good natural analogue would provide confidence in this conclusion. Unfortunately, no appropriate natural or archaeological analogue systems are known at present. If none come to light, long-term laboratory and in situ experiments in an underground research laboratory probably represent the best approach to address this issue.

Leached metal ions from the canister or other metal components of the barrier system may exchange with sodium or calcium in the bentonite, with a subsequent loss in permeability. This phenomenon has been recognised in laboratory studies where it was concluded not to present a serious problem, but has not been addressed in natural analogue studies. A number of natural analogue studies have investigated metal cation migration in clays (e.g. the Kronan); it may be worthwhile extending these investigations to examine any ion exchange reactions that occur.

Further natural analogue studies could usefully address the issues of bentonite interaction with other repository materials (e.g. cement and wasteform) and host rocks. Particular emphasis should be placed on examining any mineralogical changes (e.g. zeolite formation) and subsequent lowering of isolation capacity. There are a number of both natural and archaeological analogues that may be examined in which bentonite is in contact with a dissimilar material. However, it may prove difficult to find such a system in a relevant environment.

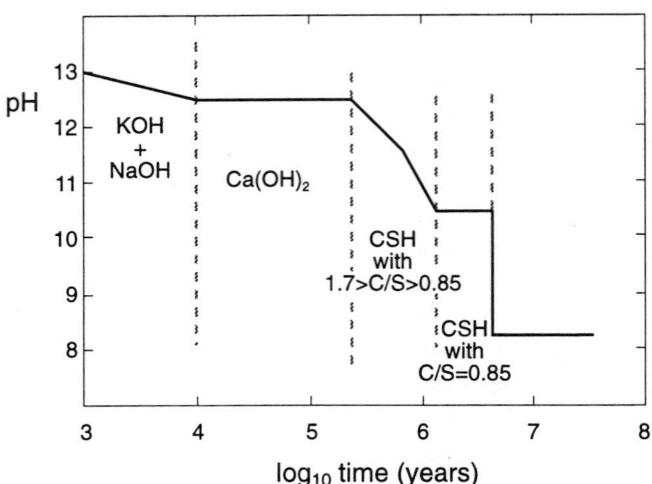

Figure 4.22: Predicted evolution of the pH within the near-field of the proposed UK ILW repository with an average cement content of 185 kg/m³. After 1 000 years a pH of 13 will be attained which will gradually decline with time but remain above pH 10 for, at least, the first one million years. After Atkinson (1985).

Good isolating capacities for clays are qualitatively indicated by the preservation of buried organic material, including trees, cadavers, foodstuffs etc. Unfortunately, little quantitative information has been gained from these studies because they have been performed for archaeological reasons, not as natural analogue studies. Consequently, important information is missing, such as physical, chemical and mineralogical data on the clays themselves, the exact nature of alteration products on the buried objects etc. Nonetheless, the qualitative evidence should not be underestimated because it has added to confidence that bentonite will provide an effective isolation capacity in the repository environment.

4.6 Concretes and cements

Concretes and cements will be used in the near-fields of all repository designs, as discussed in Chapter 2, although by far the largest volumes will be found in LLW and ILW repositories. They will be present as:

- rock supports and reinforcements in most repository excavations;
- as plugs and seals in shafts and tunnels;
- a solidification matrix for some LLW and ILW;
- reinforced concrete structures such as silos in ILW repositories, e.g. in the Swedish SFR repository (Box 3);
- concrete waste packages and 'tanks' for certain L/ILWs; and
- as a cementitious buffer and backfill in some L/ILW repositories.

The actual quantities of cement to be used in some repositories are huge; for example, the near-field of the current Swiss L/ILW repository design will contain up to 1.5 million tonnes of cement, which will be approximately 90 % of the total mass of all materials emplaced in the repository (Alexander, 1995).

These very large quantities of cement will strongly buffer the pH conditions. According to models of cement evolution (e.g. Atkinson, 1985; Berner, 1990; Neall, 1994), the hydration by groundwater of the cementitious materials will produce an initial stage of hyperalkaline leachates dominated by alkali hydroxides, with a pH of around 13.5, followed by a longer period of portlandite buffered leachates with a pH of around 12, as indicated in Figure 4.22. Under such hyperalkaline conditions, most radionuclides will exhibit very low solubilities. Furthermore, the cementitious minerals and gels will provide a very large surface area for sorption (Hodgkinson and Robinson, 1987). As a consequence of both the pH buffering and the sorption capacity, a cementitious repository design should ensure considerable radionuclide retention in the near-field.

In addition to the concrete and cement emplaced in the near-field as part of the engineered barriers, it is likely that concrete will also comprise part of the waste itself in L/ILW repositories, particularly waste from reactor decommissioning operations.

Cement will also be used in some HLW and spent fuel repository designs, as structural supports, tunnel and shaft seals etc., although in much smaller amounts than will be used in L/ILW repositories. Although this cement will also hydrate in the groundwaters, it is not assigned any pH buffering capacity role in performance assessments.

Modern concretes of the type that would be used in a repository are based on Portland cement, of which calcium silicate hydrate (CSH) compounds are the main hydration products. These CSH compounds form an amorphous gel which provides the bonding strength between aggregate particles. The CSH gels are thermodynamically unstable and transform spontaneously into stable crystalline forms (Steadman, 1986).

The rate of this process is too slow to be measured experimentally and cannot easily be calculated. It is likely that any conversion product would still bond together the aggregate, but there is no possible way of predicting the resultant bond strength. It follows that the most appropriate way to study the long-term stability of CSH-bearing cements and concretes is through combined laboratory and analogue studies.

There are two approaches to analogue studies on concretes and cements. The first is to study archaeological building cements (hundreds to thousands of years old) and industrial building cements (tens to hundreds of years old). The second approach is to study natural occurrences of minerals which are analogous to compounds found during hydration of Portland cement. Natural CSH compounds have been identified in a small number of formations (e.g. McConnell, 1954, 1955).

There are about 40 naturally occurring minerals which can be considered as counterparts to phases found in Portland cement; these are listed by Milodowski et al. (1989b) together with their chemical composition. Most of these minerals are extremely rare, forming only in restricted environments such as:

- high-temperature, low-pressure metamorphism of organic rich marls and limestones, or of coal bearing strata as a result of natural in situ combustion, followed by subsequent retrograde alteration;
- high-temperature, low-pressure contact metamorphism and metasomatism of siliceous limestones and calcareous rocks; and
- zeolitic alteration of basaltic lavas.

Neither archaeological artefacts nor occurrences of natural cement minerals provides a complete analogue for the repository environment for reasons that will be discussed below, but it is generally accepted that valuable information can be gained from such studies if care is taken when interpreting and extrapolating data (Rassineux et al., 1989; McKinley and Alexander, 1992a).

The issues of most relevance to the behaviour of cement in the repository which have been (or potentially could be) addressed in natural analogue studies are:

- durability of cement;
- cement-rock-groundwater interactions;
- radionuclide sorption;
- colloid production and filtration;
- gas and water permeability; and

- bonding properties.

These issues are discussed in the following sections.

Durability of cement

Concretes have recently been found to have been used even earlier in history than first thought. The use of a lime-based concrete in the floors of a Neolithic construction in Galilee has been described by Malinowski and Garfinkel (1991). Carbon dating of organic materials at the site indicate that they were constructed around 7000 BC. The type of material used at this site indicates that Neolithic man had technology for the burning and calcining of limestone. Prior to this discovery, it was thought that the first inorganic cements were made from gypsum ($CaSO_4 \cdot 2H_2O$). It is generally agreed that Egyptians used gypsum in the construction of the Great Pyramid of Cheops built between 2613 and 2494 BC (Lea, 1970).

From these early times, cements and concretes have been widely used in most civilisations. Lime based mortars were used by the Greeks who sometimes added sand to produce a mortar. This type of material had a major problem in that the calcium hydroxide was easily washed away if subjected to high water flows. The problem was solved by the Romans who developed the use of pozzuolanic lime concrete. Pozzuolans are fine siliceous and aluminous materials (usually powdered volcanic tuff) that cause the cement to set and harden without drying, and this characteristic led to their application in underwater construction.

The history of the development of concretes and cements is set out in several good text books, such as those by by Gani (1997), Stanley (1979) and Francis (1977). Modern concretes, including those that would be used in a repository, are composed of Portland cement which largely comprises calcium silicates with little free lime. The physical and chemical properties of modern Portland cements are somewhat different to those of the older lime cements so, to a large extent, the archaeological cements and concretes therefore appear to be rather poor analogues for repository materials.

Thomassin and Rassineux (1992) reviewed some of the literature on Gallo-Roman cement-based materials. One of the most impressive examples is the 1700 year old Roman mortar used in Hadrian's Wall (see Box 10) which still contains substantial amounts of CSH compounds, see Figure 4.23. These mortars were studied specifically with the behaviour of an ILW repository in mind (Rayment and Pettifer, 1987; Jull and Lees, 1990). The origin of the CSH compounds in Hadrian's Wall is thought to be from calcining of siliceous limestones to produce lime or by the inclusion in the mortars of larnite from metamorphosed cherts found locally in limestones. The formation of the CSH compounds reduced the porosity and permeability of the cement which helped to ensure the wall's preservation for the last 2000 years.

Several other studies have been performed to determine the chemistry and mineralogy, and to test the physical properties of archaeological lime based concretes and cements with and without pozzuolans (Mallinson and Davies, 1987; Rassineux et al., 1989; Jull and Lees, 1990). As with the Hadrian's Wall mortars, other archaeological concretes were also found to contain CSH compounds with compositions and structures similar to those in modern cements. Generally, those concretes that did have CSH compounds all contained crushed vitreous fireclay as a pozzuolan, as opposed to the usual volcanic material. These CSH compounds are thought to have formed by reaction between the vitreous pozzuolans and free lime: the vitreous fireclay is more reactive than normal pozzuolans. Thus, a compositionally wide variety of ancient (greater than 1500 year old) cements show similar results for their durability under quite varied conditions. As mentioned earlier, modern Portland cements

Figure 4.23: Photograph of a section of 1700 year old cementitious mortar from Hadrian's Wall. This mortar contains the same calcium silicate hydrate phases which characterise modern Portland cement.

have a much greater mechanical strength, and are more resistant to chemical corrosion than lime-based cements. It follows that CSH compound-bearing concretes may exhibit durabilities equivalent to, or greater than, that of the archaeological examples.

Perhaps the most striking example of the use of Roman cement is the Pantheon in Rome. Here, the circular temple was built in about 120 BC during the reign of Emperor Hadrian. The dome's 43.4 m span was cast solid in a lightweight concrete containing pumice and pozzuolana (Lea, 1970). This concrete was, however, different from modern materials because the aggregate and mortar were not pre-mixed but laid in horizontal courses (Harries, 1995). Nonetheless, despite the differences in manufacturing technique, the many examples of Roman concrete structures still in existence provides very strong qualitative evidence of the durability of cement-based materials. The writings of Marcus Vitruvius Pollio recorded in his treatise *De Architectura* (27 BC) make clear that the Romans exercised effective quality control during cement manufacture and this will have helped their preservation.

Modern Portland cement was first manufactured in 1824 by Joseph Aspdin, so the oldest technical concretes to include CSH compounds are only 150 years old. One of the first large-scale industrial uses of Portland cement was in the construction of a brick-lined tunnel under the River Thames at London. This tunnel was built by Marc Isambard Brunel and work started in 1825 (Young, 1995). Several studies have examined the durability of Portland cement based concretes (e.g. Idorn and Thaulow, 1983; Steadman, 1986; Mallinson and Davies, 1987; Yusa et al., 1991). Steadman (1986) discusses a 60 year old concrete from a sea-wall and a concrete formed from a cement paste consignment found in a ship wreck from 1848. In the latter case, the paste hardened in its barrels and was later retrieved and the barrel shaped concrete blocks used to form a sea defence. When examined, the CSH compounds were found to be indistinguishable from those found in modern Portland cements, despite several decades of exposure to harsh oxidising, aqueous conditions.

The study of Mallinson and Davies (1987) examined samples of concrete from Britain's first multi-storey reinforced concrete framed building, Weaver's Mill, Swansea (1897-8) and its first reinforced concrete marine structure, Woolston Quay, Southampton (1899). The CSH compounds in concrete from these two structures were found not to exhibit any form of degradation, although hydration of the cement was incomplete. It might be expected that further curing and strength gain could occur if hydration proceeds to completion.

Box 10: Hadrian's Wall

The Romans occupied Britain from the middle of the 1st century to the beginning of the 5th century and for much of this time northern England was the edge of their empire. In AD 122 Emperor Hadrian ordered the building of a wall across England to separate the land of the Britons from the land of the Picts to the north. The wall was manned continuously until it was abandoned in AD 383. This wall is now referred to as *Hadrian's Wall.*

Figure B10.1: Section of the remains of Hadrian's Wall showing the stone block construction bound together with Roman cement.

Hadrian's Wall was built from stone blocks cemented together. When it was built, the wall was around 100 km long and 5 metres high. Along the wall are milecastles (small forts), watchtowers and larger fortresses.

After the 1745 Jacobite uprising, sections of the wall were destroyed in order to use the stone for a military road which would allow the King's troops to move quickly from east to west. Today, the best remaining sections of the wall, at Housesteads in Northumberland, are only about 1 metre high The wall is now officially recognised and protected as a World Heritage Site, which means it is now very difficult to gain permission to collect concrete samples in situ from the wall for investigation.

Hadrian's Wall is of interest as an analogue because of the longevity of the Roman cement used to bind together the stone blocks. Roman cement has some similarity with modern Portland cement because it contains calcium silicate hydrate (CSH) compounds which provide Portland cement with its strength and bonding properties. In Roman cement, these CSH compounds derived from calcining of siliceous limestones to produce lime or by the inclusion in the mortars of metamorphosed cherts found locally in limestones.

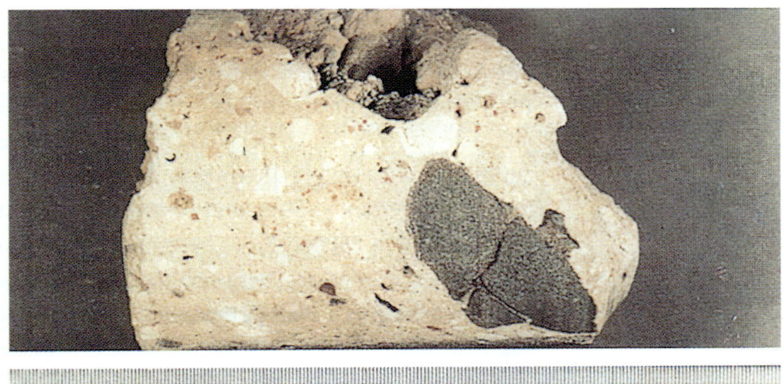

Figure B10.2: Photograph of part of Hadrian's Wall showing the Roman mortar and, below, a core through a section of the mortar showing its excellent strength and stability after 1700 years.

CSH compounds can also be found in other Roman cements used at other localities throughout the range of the Empire. This occurs because of the Roman's favoured use of vitreous pozzuolans in the cement mixture. In this sense, the inclusion of the CSH compounds in the cement used in Hadrian's Wall and in other Roman cements was more by accident than design.

The surface environment in northern England will be dissimilar to the conditions in a deep repository but, nonetheless, the chemical and mineralogical similarities between the Roman cement and modern Portland cement allows some qualitative conclusions to be drawn regarding the potential stability and longevity of modern cements in a repository. Since Hadrian's Wall is also a well known structure, it also provides a very understandable non-technical demonstration of cement durability for general audiences.

Milodowski et al. (1989b) obtained samples of naturally occurring CSH compounds from metamorphosed flints, close to the contact with a dolerite plug, from a locality in County Antrim, Northern Ireland, as seen in Figure 4.24. It was evident that the CSH compounds had formed during retrograde hydrothermal alteration subsequent to intrusion of the dolerite. This implies that the CSH compounds have been stable for 58 million years, the radiometric age of the dolerite intrusion.

In summary, the examination of industrial concretes composed of Portland cement indicates that the CSH compounds are sufficiently stable to survive in a variety of environments for up to

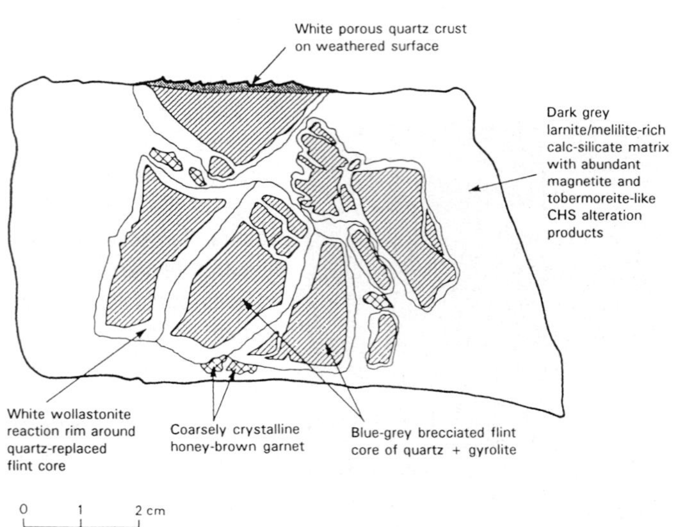

Figure 4.24: Drawing of a photomicrograph showing naturally occurring CSH compounds. The shaded areas are brecciated flint which are surrounded by wollastonite reaction rims in a metasomatic matrix of CSH minerals and opaque minerals: magnetite, perovskite and sulphides. From Milodowski et al. (1989b).

150 years. Archaeological materials indicate these compounds may survive for thousands of years. Geological materials considerably extend the known longevity of CSH compounds to tens of millions of years. Whilst this information is clearly very encouraging, it is must be understood that all of the analogue studies outlined above have been performed on material from environments that are quite dissimilar to the conditions expected in a repository.

Nonetheless, although no quantitative information can be derived from these archaeological examples for direct use in performance assessment, their illustrative value should not be underestimated for use in non-technical demonstrations of repository performance to a wide range of audiences.

Cement-rock-groundwater interactions

The emplacement of a large mass of concrete and cement into the deep geological environment will clearly cause a massive perturbation to the geochemical conditions in the rock at depth. Concrete in a repository will come into contact with a wide range of other materials, both man-made and natural, and it is important to be able to predict any chemical interactions that may occur. Indeed, it has been predicted (e.g. Haworth et al., 1987; Steefel and Lichtner, 1994) that, as the hyperalkaline porewater leaches out of the near-field, significant interaction with the repository host rock may occur, possibly leading to deterioration of those characteristics for which the formation was originally chosen (e.g. low groundwater flux, high radionuclide retardation capacity etc).

These model predictions must be tested to assess the true significance of these predicted interactions, not least because the geochemical codes use incomplete thermodynamic databases for the numerous species and phases of interest, as discussed in Section 5.1. These codes may also represent the interface between the near and far fields in a less than perfect manner. Even the most sophisticated codes currently available are as yet unable fully to couple flow, reaction and evolution of physical properties, such as porosity.

To date, few laboratory data of relevance have been produced against which to test the model predictions of cement evolution and associated host rock degradation (see McKinley and Alexander, 1992a; Steefel and Lichtner, 1994). Simple, open system, column experiments (Bateman et al., 1995, 2000) and closed system, batch reaction experiments (Chermak, 1992, 1993; Adler et al., 1999) are currently ongoing. Although providing some insight into the problem, such laboratory data require additional support because they cannot replicate the complexity or timescales of the repository system. Natural analogue investigations of old cement structures are one way to investigate these issues.

An interesting study has been performed on an old water tank installed in the towers of Uppsala Castle, Sweden (Trägårdh and Lagerblad, 1998). In this case, the steel tank was lined with a 20 mm thick layer of cement mortar. The tank was installed in 1906 and was demolished in 1991. In the intervening 85 years, the tank was regularly refilled with fresh water and, because equilibrium could never be reached between the cement and the water in the time available, the cement mortar was continuously being leached. The concrete mortar was investigated by chemical, physical and optical methods.

The results showed the mortar to be covered with a thin layer of carbonates that are believed to have formed by reaction between the cement and bicarbonate in the water. Behind the bicarbonate layer was a 5 to 8 mm thick zone with an enhanced porosity, reduced calcium content but a relative increase in sulphate and iron. This zone represents a region of complex leaching, elemental redistribution and recrystallisation. Portlandite was not apparently depleted in the porous zone, although it had recrystallised to coarser aggregates and the CSH compounds had reorganised to a lower calcium-silicon ratio. Cement leaching can be modelled in performance assessment using the shrinking core model which assumes instantaneous release of leachate to the water (Höglund and Bengtsson, 1991). Applying this model to the water tank, leads to a prediction that the leaching depth should be 6 cm and that all the portlandite should be dissolved. Clearly, the analogue data indicate that diffusion controlled leaching proceeds at a slower pace that predicted by the performance assessment model and, thus, that the model is conservative.

However, although 85 years is a much longer time than any laboratory study could be operated, it is still very short compared to repository lifetimes. As such, the water tank analogue falls short of being able to validate the slow and progressive decrease in pH predicted in the models described earlier. One possible way to circumvent this problem is to examine the evolution of groundwaters which are naturally highly alkaline.

The natural analogue site most suitable to constrain hyperalkaline groundwater evolution and interaction with the host rock is Maqarin in Northern Jordan, Figure 4.25. This site appears to be unique in that the hyperalkaline groundwaters in the area are the product of leaching of an assemblage of natural cement minerals produced as a result of high temperature and low pressure metamorphism of marls and limestones.

Hyperalkaline groundwaters exist elsewhere, for example in ophiolite environments such as in Oman (Bath et al., 1987a,b) and in rift valley terrains in Kenya (Jenkins, 1936). However, these natural systems are produced by processes which are of little relevance to a repository environment and generate waters with a pH of only around 11. Nonetheless, they have proved useful for investigating certain issues, such as trace element speciation and microbial populations in alkaline environments.

The Maqarin site has been investigated as a natural analogue over the last decade (Alexander, 1992; Alexander et al., 1992a; Linklater, 1998; Smellie, 1998). The main areas of interest for this analogue study have been investigation of the overall hyperalkaline groundwater evolution, including the question of the evolution of the cement leachates, interaction of the hyperalkaline leachates with the host rock and the testing of a variety of geochemical, transport and biological codes (see Box 11).

The hydrogeology of the Maqarin site is rather complex and it appears that at least two geochemically distinct flow systems have been identified. In the eastern part of the area, the groundwater pH is 12.5 and is buffered by abundant portlandite, $Ca(OH)_2$, in the source rock (Alexander et al., 1992a). In the western part of

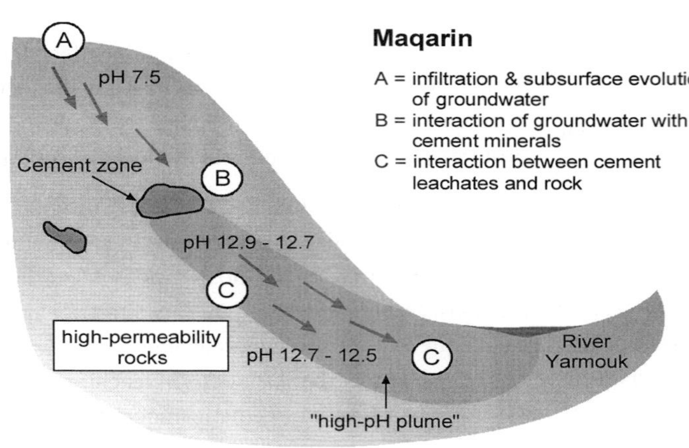

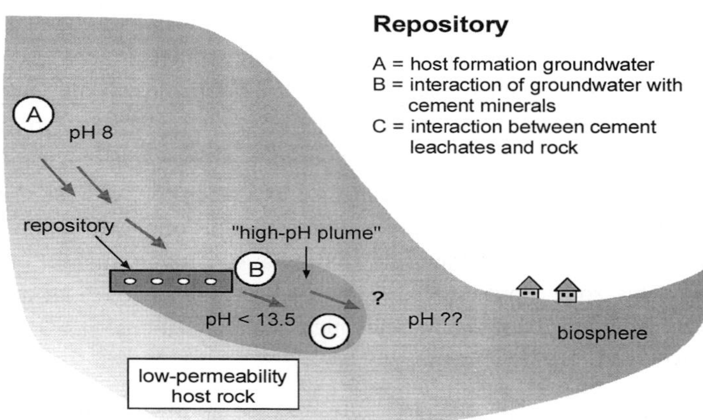

Figure 4.25: The basis of the analogy between the Maqarin site and a cementitious repository. At Maqarin, groundwaters (A) percolate through the rock until they meet the zone containing the natural cement minerals where rock/water interaction produces hyperalkaline groundwaters (B). In a repository, it is assumed that the same course of events will be followed, with the local groundwaters leaching the cementitious L/ILW repository producing hyperalkaline leachates (or groundwaters). At Maqarin, following interaction with the cement zone, the hyperalkaline groundwaters have been observed to continue their percolation through the fractured rock, interacting with the fractures and rock matrix as they migrate (C). In a cementitious repository, a similar sequence of events is also likely and thus the observations from Maqarin can be used to provide a guide to the possible effects of the leachates on the host formation.

the site, the groundwater contains much higher levels of sodium and potassium, and appears to be a younger system. Allied to this is the fact that new in situ measurements indicate pH levels up to 12.9 which are nearer to KOH and NaOH controlled values.

At Maqarin, an unusual assemblage of secondary minerals has been observed which are the result of interactions between the hyperalkaline groundwaters and the rock. As part of the Maqarin project, these analogue observations were used to help develop and constrain a conceptual model to explain the possible interactions which might occur between a hyperalkaline plume migrating away from a repository through the host rock (Savage, 1998), as shown in Figure 4.26.

This conceptual model assumes that cement leachates rich in sodium, potassium and calcium flow outward from the repository, driven by the groundwater flow system. As the plume begins to interact with the host rock, a complex sequence of reactions can occur, involving dissolution of the aluminosilicate minerals in the rock, and precipitation of CSH compounds and, eventually, zeolites as the pH decreases and the aluminium concentration increases in the groundwater.

This conceptual model developed and changed throughout the course of the Maqarin project as more mineralogical data became available from the site and from supporting laboratory studies, and as improved coupled codes were developed. This is a clear example of the necessity to involve analogues in performance assessment model development (as discussed in Section 6.1) because this model could not have been derived solely on the basis of short-term laboratory results.

Cement-rock interactions have also been investigated in archaeological analogue studies. One early example of a study of this type was the examination of profiles through a seventy year old concrete-clay interface at the base of the Washington Ship Canal (Andersson and Fontain, 1981). Here, ion exchange reactions were clearly identified but there is no evidence of mineralogical or physical alteration of the clay. Another, more recent study has examined the cement/rock interactions which may occur during the operational phase of a repository constructed with concrete tunnel liners. Here, groundwater will drain into the tunnels through the liner and oxygen, and carbon dioxide will diffuse out into the liner and adjacent rock. Although the tunnel liner itself is not assigned a containment role in repository safety assessment (i.e. the liners are only there for tunnel engineering reasons, not repository safety), a study of the effects of the above two mechanisms is useful for two reasons. First, examination of the groundwater interaction with the cement phases in the liner will provide information on alteration likely to occur at the upstream side of a L/ILW repository, i.e. that part affected by fresh groundwaters entering the repository and reacting with the cement. Second, as has been shown elsewhere (Granger and Warren, 1969; Rainey and Rosenbaum, 1989), that oxygen entering the rock can oxidise pyrite, thus producing highly acidic groundwaters which then attack both the rock and the tunnel liners. This can produce a damaged zone of much higher permeability around the repository, inducing greater groundwater flow in the vicinity of the repository after closure.

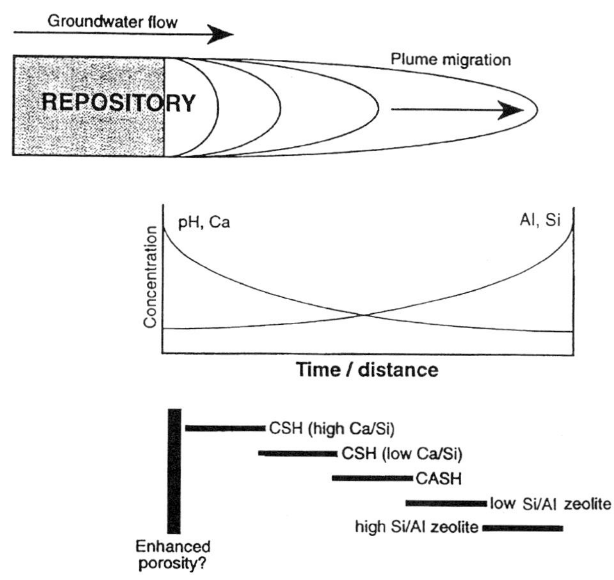

Figure 4.26: The complex sequence of precipitation and dissolution reactions which can occur in repository host rocks as a hyperalkaline leachate plume flows through and reacts with them. From Savage (1998).

An attempt was made by Mazurek (1990) to study these effects where samples were taken through the concrete liner of a 15 year old ventilation tunnel and into the host (Valanginian marl) formation. It was found that reactions at the liner/rock contact were confined wholly to the concrete. In the zone immediately adjacent to the marl, the cement was completely recrystallised to form a mass of fine-grained calcite, and portlandite was altered to an optically isotropic phase, thought to be a gel. The combined alteration resulted in a significant increase in porosity, up to 20 to 40 % in places. Deeper into the cement, newly formed calcite occurs in lesser abundance and is associated with microfractures. It seems likely that the groundwater carbon dioxide content is high enough to produce the large amount of calcite in the cement (c.f. Baeyens and Bradbury, 1991), thus leading to a much more porous zone behind the liner, although the permeability is not known. Interestingly, there was

Box 11: The Maqarin hyperalkaline system

The Maqarin analogue study site is located at the natural springs area of north-west Jordan, at the Jordanian-Syrian border by the Yarmouk river. The rocks at Maqarin have undergone an unusual evolution that has led to the natural spring waters becoming hyperalkaline (pH 12 to 13) and a number of naturally formed cement minerals have precipitated from the groundwaters. As a result, Maqarin is a good analogue for a cementitious L/ILW repository, although the majority of groundwaters at Maqarin are oxidising in contrast to the reducing conditions in a repository near-field.

Figure B11.1: One of the springs at Maqarin showing the hyperalkaline waters, with a geological hammer for scale. From Smellie (1998).

The rocks at Maqarin are late Cretaceous marls and bituminous limestones, known locally as the Bituminous Marl Formation, which are overlain by Tertiary chalks and limestones. The bituminous rocks are biomicrites, composed essentially of calcite with accessory quartz, dolomite, apatite, pyrite and clay minerals, and have a high organic content, up to 20 %, and an SO_3 content of up to 12 %. Trace elements, including uranium, are mostly adsorbed by the organic materials. The organic material itself, despite the rock name, is not actually bitumen but is similar to kerogen.

The Bituminous Marl Formation is the focus of the investigations because it contains a rare assemblage of naturally formed cement minerals, including portlandite. These minerals formed by a two-stage process:

1) The Bituminous Marl Formation underwent spontaneous combustion (temperature > 1000°C) to form a high temperature mineral assemblage of graphite, apatite, diopside, wollastonite and anorthite.

2) The high-temperature minerals hydrated by interaction with the normal aquifer waters (pH ≈ 8) to form a low-temperature mineral assemblage of gypsum, ettringite, tobermorite and portlandite.

The resulting high pH of the groundwaters is controlled by the solubility of portlandite and the other cement phases. This is exactly analogous to the situation in a cementitious L/ILW repository but is different to the system at the Oman natural analogue site (see Section 5.1) where the hyperalkaline conditions relate to the alteration of ultramafic minerals.

Figure B11.2: Photograph of cementitious precipitates in the vicinity of one of the springs discharging hyperalkaline groundwaters, with lens cap for scale. The cement minerals at Maqarin include portlandite and are essentially the same as those found in modern Portland cement. From Smellie (1998).

The geomorphology of the site is important for the development for the high pH waters. It has been suggested that spontaneous combustion was initiated by earthquake activity which caused landslips on the valley side that allowed air (oxygen) to come into intimate contact with the organic materials.

The focus of the investigations was on the valley, where the contact between the lower Bituminous Marl Formation and the upper limestones crops out. Groundwater flowing down through the sequence tends to discharge in springs in the hillside at the contact between the two rock formations because the marls are less permeable than the fractured limestones. The cement zone occurs at the contact between the two formations.

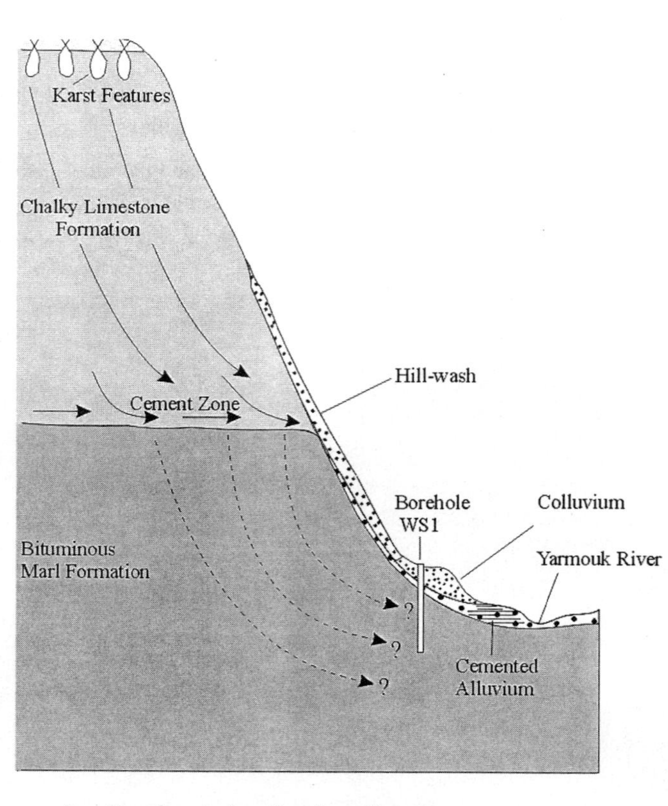

Figure B11.3: Diagrammatic cross-section through the Maqarin site showing the groundwater flow paths, and the locations of the cement zone at the contact between the limestones and the bituminous marls. From Smellie (1998).

One objective of the Maqarin study is to test the applicability of geochemical codes to high pH environments and to improve the relevant mineral data in the associated databases (see Section 5.1). Consequently, a great deal of effort has gone into careful groundwater sampling and analysis of the hyperalkaline springs and other groundwater recharge and discharge zones to establish the input groundwater source composition.

In addition to the testing of geochemical codes, other studies that have been undertaken at Maqarin include:

- understanding the evolution of the hyperalkaline groundwater and the cementitious mineral assemblage (see Section 4.6);
- interactions between the hyperalkaline spring waters and marl (see Section 4.6);
- colloidal populations in the hyperalkaline spring waters (see Section 5.6); and
- microbiological populations in the hyperalkaline spring waters (see Section 5.7).

The project has been undertaken in three main phases, and a comprehensive summary report has been written at the end of each stage. The project has also been described in other publications (e.g. Alexander, 1992; Alexander et al., 1992a; Linklater, 1998; Smellie, 1998; Alexander and Smellie, 1998, 2000).

no oxidation of the 1 to 2 % pyrite in the rock (Mazurek, 1990), but this may be a site-specific feature due to the armouring of the pyrite by clays or organics, or both (Bradbury et al., 1990). In addition, the possibly dominant chemical transport effect of radial flow of water towards the tunnel is likely to have affected the scale and nature of reactions in this interface zone. Thus, although this small project has shown the potential of such work, a much more detailed study is necessary before any conclusions can be reached.

No similar natural analogue studies have been reported for cement leachate interactions with crystalline rock, although a number of laboratory studies have been performed (Fritz et al., 1984; 1985; 1988). In addition, a large field experiment is underway at Nagra's underground test site at Grimsel in Switzerland. This study, the HPF (Hyperalkaline Plume in Fractured Rock) experiment aims to integrate data from Maqarin, laboratory studies and modelling work with the field results (Kickmaier et al., 2000). As noted by Alexander et al. (1998a), well designed, realistic field experiments can bridge the gap between laboratory and natural analogues by offering repository relevant natural conditions with some of the constraints of the laboratory and intermediate timescales. Combining information from the three sources (natural analogues, in situ field experiments and laboratory studies) can provide greater confidence in the extrapolation of laboratory derived data to repository relevant timescales and conditions.

Radionuclide sorption

A number of laboratory experiments have been performed to investigate radionuclide sorption on concretes and cements (e.g. Allard et al., 1984, 1985a,b; Kindness et al., 1994; Bradbury and van Loon, 1997). These laboratory experiments indicate that generally the highly hydrolysed actinides, as well as cobalt, iodine, technetium and nickel, are more strongly sorbed onto concrete than onto common minerals, whilst alkali metals and alkali-earth metals, especially caesium and strontium, are very poorly sorbed, see Figure 4.27.

Carbon, as carbonate, is removed from groundwaters by precipitation as $CaCO_3$. Whilst this laboratory derived information is interesting, it may not accurately represent the repository system because experimental work is generally carried out with 'young' cements and it is known that the mineralogy of such materials changes considerably with ageing. Reducing and carbon dioxide free conditions are also difficult to maintain in the laboratory.

No corresponding natural analogue studies which back up these laboratory studies are known to have been performed, despite the obvious importance of this issue. An attempt was made to study ^{14}C uptake in earlier phases of the Maqarin project (see, for example, Alexander, 1992) and this is continuing in the current phase of the project, although no data are yet available. In addition, it may be possible to study sorption on archaeological cements in places such as Roman baths which have held thermal springwaters with above average concentrations of naturally occurring radionuclides.

Other industrial analogue investigations could be usefully performed, for example at shallow LLW or toxic waste disposal sites from which uncontrolled leakage has occurred. One such example is the Oak Ridge site in Tennessee in the United States, where cement grouts have been injected into the ground in an attempt physically to contain toxic wastes by blocking groundwater flow. It may be possible to retrieve samples of grouts which have been in contact with the contaminated groundwaters for some years. Such studies must,

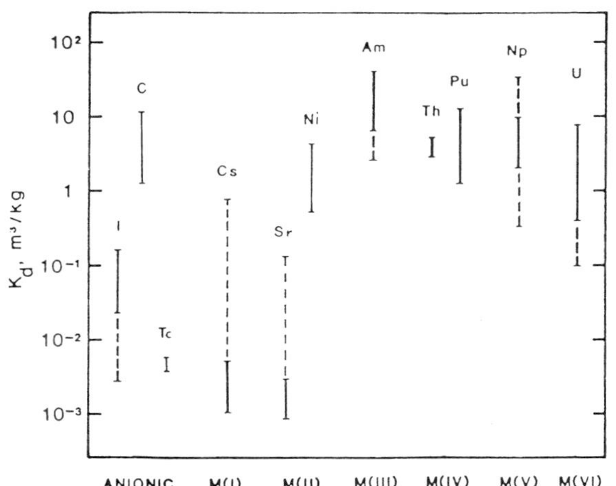

Figure 4.27: One of the functions of the large volumes of cement placed in L/ILW repositories is to sorb migrating radionuclides. Laboratory studies indicate that some elements (e.g. uranium, thorium, plutonium and americium) strongly sorb to cement while other elements do not (e.g. caesium and strontium). The dashed lines indicate the maximum range; the solid line indicates the probable range for a cementitious near-field. From Allard et al. (1985b).

however, carefully distinguish between sorption and precipitation processes if they are to be useful (McKinley and Alexander, 1992a,b).

Alternatively, it may be possible to study sorption on naturally occurring CSH minerals, for example at locations where hydraulically-active fractures cut through rock formations containing these minerals. Ongoing work in the Maqarin project has shown that uranium appears to have been preferentially taken up by secondary jennite, presumably by solid-solution, from fracture waters. It is hoped that the new data can be combined with new laboratory work on radionuclide uptake by secondary cement phases (e.g. Bradbury and Baeyens, 1997) and with the HPF experiment (mentioned above) which will include the injection of a suite of safety relevant radionuclides along with the hyperalkaline fluids.

Colloid production and filtration

The fine-grained nature of cement may result in it forming colloids and suspended particles, particularly when the cement degrades, with the possibility of enhancing radionuclide transport (see Section 5.6). As pH slowly decreases with time, the solubility of silica will also drop, increasing the possibility of the formation of colloidal silica species. In addition, the presence of steep chemical gradients at the interface of cementitious regions and the host-rock provides a suitable environment for colloid formation. If significant colloid formation were to occur, it may negate the benefit of low radionuclide solubility brought about by the hyperalkaline environment. Alternatively, the pore spaces in the cement may be sufficiently small that the body of cement will act as a colloid filter. It seems likely, therefore, that only colloids forming on the outermost edge of the cement mass may travel any distance from the repository. However, the potential of cement to form and filter colloids appears not to have been thoroughly investigated.

Only a small number of laboratory experiments examining colloid production during cement degradation have been reported (e.g. Wieland, 1997 and Gardner et al., 1998). In all laboratory studies the production of colloids by experimental artefacts are difficult to overcome and it is not possible to reproduce the conditions of the near-field/far-field interface in a realistic manner. Additional, supporting data from analogues are therefore required to put the laboratory results in a more realistic perspective.

Some relevant information has been supplied by the Maqarin project. In the first phase of work at the site, a preliminary assessment of the colloidal population was carried out. The results indicated maximum colloidal populations were around 1 ppm. These colloids consisted mostly of $Ca(OH)_2$ and $Fe(OH)_3$ and contained insignificant quantities of uranium (West et al., 1992). Further colloid studies were undertaken in a later phase of the project (Wetton et al., 1998), which indicated populations of 10^7 colloids per litre. Both these results indicate that the colloidal populations at Maqarin are low in comparison to other near-surface waters (see Section 5.6).

Comparison with other data on near-field cementitious colloids is difficult due to the significantly different methodologies employed by researchers. Frequently, there is no clear relationship between reported colloid populations and mass concentrations from the laboratory experiments, making it difficult accurately to obtain data for consideration in performance assessment.

Arguably, the data on colloids from Maqarin are the most realistic in so far that the Maqarin site is a better representation of a cementitious repository than any laboratory experiment. However, further hydrogeological information from Maqarin will be required fully to evaluate the significance of colloid transport. In particular, it is necessary to sample colloids both upstream of the cement zone (for input populations) and downstream, some distance from the interface between the cement zone and the host rock, and from both oxidising and reducing groundwaters.

It is recommended that future work in analogue studies and in the laboratory would benefit from a common approach to colloid characterisation, which should try to minimise method inherent differences, so producing a more compatible data set for use in performance assessment.

Gas and water permeability

The gas permeability of cement is an important issue for repository engineering, especially for designs with large volumes of steel or organic material, which degrade to produce large volumes of hydrogen, methane and other gases (see Section 5.8). It follows that this issue is most important for L/ILW repositories. There is some concern that a build up of gas may cause structural damage to the near-field and, consequently, from this point of view a high gas permeability is desirable. This is contrary to the requirement to maintain a low hydraulic conductivity. A compromise is necessary to resolve this dichotomy and this will be specific to individual repository designs.

If a high permeability concrete is desirable, then this can be achieved by using a porous or uniform sized aggregate. Some qualitative information may be obtained from studying old concrete constructions, in particular, reinforced concrete for signs of damage resulting from gas production due to corrosion of reinforcing rods. It is expected, however, that any such investigations may be frustrated by gas leakage from microcracks rather than from diffusive migration in the cement. In addition there will be the usual problem of finding an analogue in repository-relevant conditions. One possible location may be the foundations of piers which are embedded in sediments, where conditions within the concrete may be mildly reducing and hyperalkaline.

Bonding properties of cement and concrete

The bonding ability of cements is due to CSH compounds. These not only bond the aggregate particles together but, in the case of a repository, will be required to bond to the host rock walls when used as a buffer or seal. All of the archaeological cements and concretes indicate that the bonding capacity of the CSH compounds is maintained for as long as they are protected from degradation. It is noted, however, that the bonding behaviour was not explicitly evaluated in the archaeological cements described earlier. It is possible that these materials could be re-examined with this point in mind.

Conclusions

The physical stability of cements and concretes depends on the binding properties of CSH compounds. Such compounds, in Portland and pozzuolanic Roman cements, have been shown to be stable for up to 2000 years. Naturally formed CSH compounds have been identified in hydrothermally altered igneous rocks some tens of millions of years old. They appear to have remained stable over this time period largely because they have been physically isolated by the host rock mass.

Cement-rock-groundwater interactions require further investigation. Modelling studies indicate that cement may be adversely affected by interactions with porewaters from clay formations, while hyperalkaline groundwaters in the far-field may affect the porosity and sorptive capacity of the rock. Maqarin has gone some way to characterising the complex dissolution and precipitation reactions that result from reaction between high pH leachates and rocks, and this information has been used to develop conceptual models for this process. Further studies of this type are recommended, linking analogue, laboratory and modelling work.

Radionuclide sorption on cement has been poorly addressed in natural analogue studies. A number of possible analogue systems could provide information on sorption on archaeological cements and naturally occurring CSH compounds.

The permeability of cements and concretes to gas produced from steel corrosion is an issue which may potentially affect repository performance. The relative importance of this issue has not been resolved by modellers or by laboratory investigation. It is possible that natural analogue studies of old reinforced concrete structures may indicate if concern over this issue is justified.

The bonding properties of cements have not been explicitly examined in natural analogue studies of ancient cements and concretes. It is recommended that some early concretes be examined to determine if the bonding ability changes significantly with time.

One study to date on colloids in a relevant hyperalkaline system (Maqarin) has shown relatively low colloid populations and this observation suggests that colloid production during cement leaching may not be a problem in the repository. However, further analogue data are required from a range of different groundwater environments before such a conclusions can be confirmed. The Maqarin data also show that the colloids present show only minimal uptake of dissolved uranium. This observation also requires further detailed study.

4.7 Bitumen

Bitumen is used in a number of countries as an immobilisation matrix for some L/ILW because it provides a wasteform which is both sufficiently physically and chemically stable for disposal (IAEA, 1993). Typical materials immobilised in bitumen are dehydrated and powdered ion exchange resins, and reactor wastes such as filters from reactor clean-up systems. Bitumenised wasteforms are generally encapsulated in either steel drums or large concrete containers before being emplaced in the repository.

The technological bitumen used as an immobilisation matrix is derived from natural organic materials. It consists of mixtures of mainly aliphatic and aromatic hydrocarbons of high molecular weight obtained from the heaviest petroleum fractions. There is some confusion in the scientific literature over the use of the terms bitumen and asphalt. Due to limited knowledge of the chemical structures of many complex hydrocarbons, classification schemes are normally based on physical parameters, such as the one shown in Figure 4.28. A widely accepted classification scheme is that of Abraham (1960). According to this scheme, a bitumen is a carbon based substance which can be extracted with an organic solvent. Thus, crude oils are bitumens, whilst kerogen and coal are not. Bitumen is further classified into liquids and solids; the solids may be fusible or not. Solid, fusible bitumens according to Abraham (1960) are asphalts.

Bitumens consist predominantly of carbon and hydrogen, but also contain oxygen, sulphur and nitrogen as minor constituents. The average composition of this class of material is (Savage, 1995):

- carbon, 80 to 88 %;
- hydrogen, 8 to 11 %;
- oxygen, 1 to 12 %;
- sulphur, 1 to 7 %; and
- nitrogen, trace to 1.5 %.

Strictly, the bitumens which are used for the immobilisation of L/ILW are technological asphalts distilled from crude oil at several hundred Celsius for a few hours. The bitumen composition is controlled both by the composition of the crude oil and the precise distillation process. Consequently, the technical bitumens have a range of compositions, all of which are well characterised.

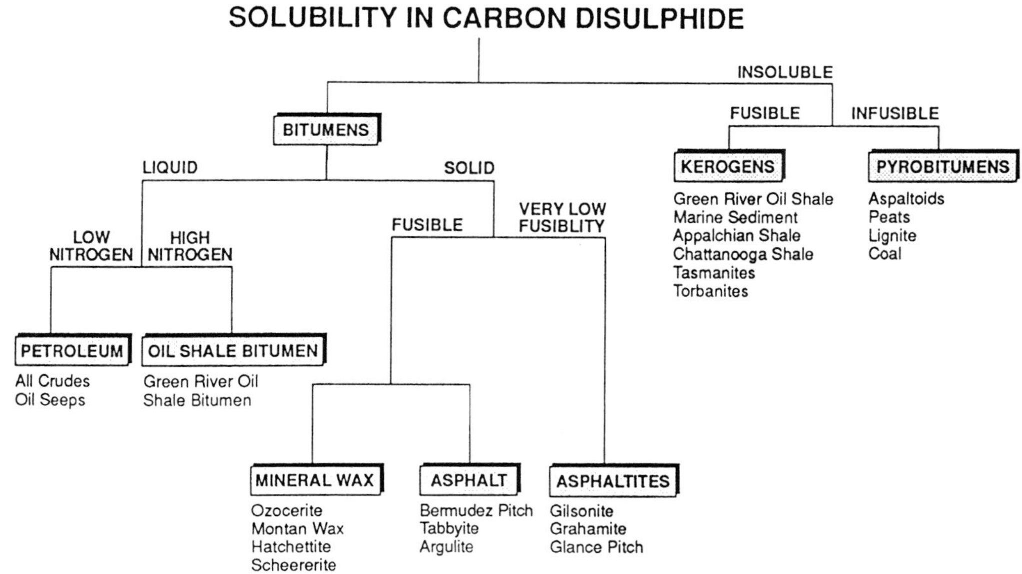

Figure 4.28: The classification of natural organic materials based on their physical state and their reaction with an organic solvent. The natural materials that are closest to the technological bitumens which may be used as an immobilisation matrix are the asphalts. From Hellmuth (1989a).

Natural bitumens (and asphalts) are also distilled from crude oils, but by geological processes over much longer time periods and over widely variable temperatures. In general then, natural bitumens are the product of a complex, usually unknown history which produces an end-product of a highly variable nature. Consequently, the similarity between technical and natural bitumens is limited but this has not prevented natural bitumens from being used as analogue materials. For a review of the characteristics and genesis of bitumens in geological processes, see Mossman and Nagy (1996).

For the most part, the behaviour of bitumen has been addressed by laboratory studies, which appears to be the best approach. However, a few natural analogue studies of bitumen have been performed, most of which have lumped all the various bitumen decomposition processes together as simple 'degradation' and tried to quantify the longevity of the materials (for a review see Alexander and Miller, 1994).

The issues of most relevance to the behaviour of bitumen in the repository which have been (or potentially could be) addressed in natural analogue studies are:

- durability and longevity;
- groundwater leaching;
- microbial degradation;
- radiation induced degradation; and
- interaction with saline water.

These issues are discussed in the following sections.

Durability and longevity

Natural bitumens, including asphalts, are found in a number of geological environments and in all climatic zones from tropical to permafrost, in arid to water-logged conditions. The most impressive are probably the asphalt lakes of Trinidad and Guanoco, Venezuela; other forms include the impregnated sandstones and limestones (e.g. Athabasca, Canada; Utah, USA; Val de Travers,

Switzerland and Hannover, Germany) and those in hydrothermal veins (e.g. Derbyshire, UK). Large asphalt blocks (up to several hundred cubic metres) have frequently been found floating in the Dead Sea. This asphalt has leaked from hydrocarbon reservoirs, which are abundant in the region, to the sea floor where it floats to the surface of the water.

In some cases, natural bitumens have been found to have preserved organic material for long periods of time. Perhaps the best examples are the asphalt pits at Rancho la Brea, California and Talara, Peru which contain abundant fossil bone and wood remains (Behrensmeyer, 1980). The Rancho la Brea asphalt pit is the site of the largest find of Pleistocene fossils (Rolfe and Brett, 1969). The mechanisms that led to the preservation of these organic materials, including the fossil remains, are complex but basically they act to limit microbial decomposition by isolating the organic material from air and water (see Hellmuth, 1989a for review).

Due to their good water-proofing ability, natural bitumens have been used by man for more than 5000 years (Hellmuth, 1989a). Babylonian buildings, from 1300 BC, have been found to have used asphalt to provide an impermeable coating in floor constructions and also as a building material in river banks and piers where it was also used as a cement. Bitumen was also used to preserve organic materials by impregnating wooden cases and baskets made of palm leaves. Since then, tars and bitumens have been abundantly used to water-proof a range of items such as the hulls of ships and the roofs of houses. In almost all cases where archaeological artefacts have been found coated in bitumen, they have been well preserved when mechanical disruption of the bitumen has not occurred.

Qualitatively, the many geological and archaeological occurrences of bitumens point to their stability under a wide range of physico-chemical conditions as well as their ability to isolate materials from water for long periods of time. However, although there are many documented geological and archaeological occurrences of bitumen, there is usually little discussion in the literature concerning the actual composition of the bitumen or of any weathering and alteration processes that have been operating.

Furthermore, the majority of natural bitumens that are recorded come from near-surface environments which are very different to the chemically-reducing, high-pH conditions which will dominate the near-fields of L/ILW repositories where bitumenised wastes will be emplaced. Thus, it cannot be concluded on the basis of these analogue studies that bitumen durability in the hyperalkaline repository environment will be the same as for bitumens at the surface. Since very few natural or archaeological systems are known in which bitumen is in contact with cement or high pH waters, the long-term stability of bitumen in cementitious ILW repositories is best examined in laboratory studies.

Groundwater leaching

An obvious issue for bitumen is leaching by groundwaters. In theory, this process should be easy to study in a natural system but, in practice, it is difficult to characterise natural systems well enough to avoid ambiguous results. For example, any observed alteration to a natural bitumen may have been on-going for so long that no original bitumen substances remain. Alternatively, any apparent alteration (e.g. loss of volatiles) may be a reflection of high-temperature maturation rather than low-temperature groundwater leaching.

One notable attempt has been made to characterise a bitumen-groundwater system in sufficient detail to describe the processes involved and rates of alteration. Hellmuth (1989b) investigated a bitumen impregnated limestone in Germany. This limestone is exposed at the surface but dips beneath permeable sedimentary cover. Variations in the chemical structure, composition and physical properties of the bitumen were measured in samples taken from a range of locations and depths. It was discovered that degradation (oxidation) results in a bleaching of the bitumen as volatile substances are leached from the bitumen but that degradation was confined to only a very shallow near-surface zone. The limited extent of leaching is a consequence of the breakdown process: degradation of the bitumen by oxygen and water is fastest where it is exposed to visible and ultraviolet light, i.e. at the surface only. Deeper penetration is prevented as long as the weathered layers are not mechanically destroyed and removed.

This could be taken to indicate that leaching of bitumen in the repository will be very slow. However, it would not be sensible to reach this conclusion only on the basis of these analogue results because of the gross differences between the chemical conditions of the analogue site and the hyperalkaline near-field of a L/ILW repository.

In fact, as mentioned earlier, very little information exists on the leaching of bitumens by hyperalkaline waters. The review of Hellmuth (1989a), for example, cites over one hundred publications but not one deals with the long-term behaviour of bitumen under the hyperalkaline conditions expected in the near-field of a L/ILW repository. This is clearly an area worth further study but known relevant analogue sites are rare. To date, only the Maqarin region of northern Jordan (see Box 11) has been identified as containing both natural bitumens and hyperalkaline (portlandite-buffered) groundwaters in close association. Unfortunately, the bitumen has undergone post-depositional combustion and is, thus, a poor analogue of any technical bitumens of interest (Alexander, 1992).

The soluble breakdown products of bitumen (leachates) include organic molecules which can act as complexants for some of the poorly soluble radionuclides from the waste. Consequently, the products of bitumen degradation might enhance radionuclide transport in the repository. The effects of organic leachates of bitumen on radionuclide complexation have been examined under the controlled conditions of the laboratory (e.g. van Loon and Kopajtic, 1990) but few natural analogue studies have investigated this issue.

In the bitumen impregnated limestone study (Hellmuth, 1989b), there was some indication that humic and fulvic acids in the groundwater could be identified as bitumen decomposition products. However, studies such as this one are always faced with the problem of trying precisely to identify which organics in groundwaters were a product of bitumen leaching and which were 'background' and, as a consequence, the results are likely to be ambiguous. For this reason, it is probably not sensible to try to undertake detailed analogue studies of this type.

Microbial degradation

Bitumen could be subject to microbially-mediated degradation because bitumen, in common with most organic compounds, is susceptible to many microbes which have the ability to utilise hydrocarbons as sources of energy and nutrients (for discussion, see the review of Hellmuth, 1989a). It is generally accepted, however, that the rate of biodegradation of bitumen is highly dependent on the chemical nature of the hydrocarbons present, on the microbial community and on environmental factors that influence microbial activities.

The oil industry has highlighted a number of cases of microbial degradation of crude oil. Superficially, the genetic connection between crude oil and bitumen suggests that similar degradation processes may affect both hydrocarbons. One convincing example of microbial degradation of crude oil is that at the Saskatchewan oil fields (Bailey et al., 1973). In this case, samples ranging progressively from non-degraded to highly-degraded could be clearly related to the influx of microbe-bearing surface waters. Little work has been performed on the microbial degradation of natural bitumens, although some work has been performed on technical asphalts and this has been reviewed by Zobell and Molecke (1978) who concluded that the chemical durability of asphalts in a repository environment could potentially be compromised by microbial degradation. However, technical bitumens are normally located in environments dissimilar to that of a repository and extrapolating the results of studies on technical bitumens to the repository is, therefore, not as simple as suggested by Zobell and Molecke (1978).

Again, because of the gross differences between the chemical conditions surrounding most natural bitumens and the near-field of a L/ILW repository, it is probable that no unambiguous analogue studies for microbial degradation could be found.

Radiation induced degradation

One area where a natural analogue study could be of use is in the examination of radiation induced degradation of bitumen. The standard laboratory technique is to subject bitumen samples to a massive, externally applied radiation dose. This effectively gives the bitumen sample its predicted lifetime (i.e. several hundred years) dose in two or three months (for details, see Burnay, 1987; Kopajtic et al., 1989). This is clearly unrealistic and a natural analogue study of bitumen with associated radionuclides obviously recommends itself.

Uranium orebodies and other metallic orebodies with enhanced radionuclide concentrations are sometimes accompanied by bitumen or kerogen (Parnell et al., 1993). An example of this association is the Oklo natural fission reactors (Nagy, 1993) where natural graphitic bitumen and uraninite occur together in some of the reactor zones. This association was investigated by Nagy et al. (1991). The bitumen at Oklo (see Box 4) is derived from syngenetic kerogen through hydrothermal processes during criticality. Initially the bitumen was liquid, and its presence caused the reduction of uranium in aqueous solution to form uraninite. Some of these newly formed uraninite crystals were enveloped in the bitumen which subsequently hardened into a solid graphitic bitumen, as shown in Figure 4.29.

Comparison of the retention of fission products between uraninite crystals enclosed in this graphitic bitumen and those enclosed in clay minerals clearly demonstrates that containment was greater in the graphite. This is despite the irradiation of the graphitic bitumen which must have occurred around the uraninite crystals. In simple qualitative terms, this is an interesting observation but it should not be used to quantify the stability of bitumen in the repository because there are too many important differences between the repository and Oklo systems.

At Oklo the bitumen composition is far removed from that of the technological bitumens which will be used as an immobilisation matrix. Most

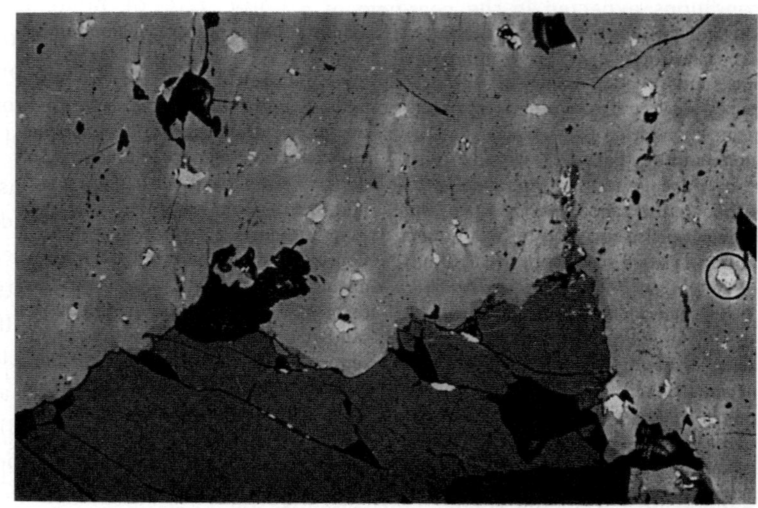

Figure 4.29: Reflected light photomicrograph of a sample of bituminous ore from Oklo showing uraninite grains surrounded by graphite with radiation haloes of increased reflectance around the uraninite. From Mossman and Nagy (1996).

importantly, the bitumen at Oklo has been substantially altered to form almost pure graphite due to the loss of volatiles. In addition, the uraninite exists as discrete crystals or aggregates of crystals in the graphitic bitumen, leading to highly localised radiation effects, whereas in the latter, the waste is homogeneously dispersed throughout the bitumenised waste package leading to low, non-localised doses. It is, therefore, very difficult to apply the results of such a natural analogue to the repository design.

Bitumen and other organic materials can be affected by radiolysis in the presence of a high radiation field. This can result in various radiolytic alteration and degradation processes, including aromatisation, dehydrogenisation, polymerisation and oxidation (Hofmann, 1996). These processes have been observed in a number of different orebody types, including roll-front deposits, uraninite placer deposits and black shales (e.g. Leventhal et al., 1986; Lewen and Buchardt, 1989; Meyer et al., 1991; Landais, 1993; Nagy et al., 1993). Radiolytic degradation of organic materials is discussed in more detail in Section 5.4.

Interaction with saline waters

One aspect of groundwater leaching of bitumen where even qualitative data would be useful is that of the interaction of saline waters with bitumen. Repositories for L/ILW may be sited in coastal regions, like the Swedish L/ILW repository at Forsmark which is constructed beneath the Baltic Sea (see Box 3), and hence saline waters may intrude into the repository near-field.

To date, no analogue information exists on the potential effects of such waters on technical bitumens. One possible study would be a comparative examination of tars and bitumens used to waterproof wooden-hulled ships. Where such ships have sunk into undisturbed anoxic muds, it might be possible to examine the bitumen for degradation in a similar manner to the the Swedish study of the Kronan cannon discussed earlier (see Box 8). In addition, information could be collected on the microbial attack of the bitumen. Nevertheless, it must be implicitly recognised that such a study could supply no more than qualitative data but, in this case, it represents a useful beginning.

The large asphalt blocks floating in the Dead Sea represent an obvious situation where bitumen is in contact with brines, and this could provide some information for sites where brines may enter the repository near-field, such as at depth in the Scandinavian Shield.

Conclusions

The long-term durability of bitumen in near-surface environments and its ability to isolate organic materials are illustrated by the many natural and archaeological occurrences of bitumen from many different locations. Particularly impressive examples are the preservation of fossil bone and wood in tar pits in California and Peru. Qualitatively, these examples suggest that bitumen can be stable and long-lasting. However, few analogue studies have attempted to characterise natural bitumen in sufficient detail to assess the extent of the analogy between natural and technical bitumens.

Most importantly, no natural or archaeological examples of bitumen have yet been examined from high pH systems which approximate to the chemistry of a L/ILW repository near-field. For this reason, few of the bitumen analogue studies have produced anything more than qualitative data and most studies provide no data of use to a repository performance assessment.

Areas of potential future application of more quantitative natural analogue data include the study of bitumen degradation in saline, brine and hyperalkaline environments. Further areas for examination include radiation effects on bitumen degradation, assuming that samples with well dispersed, non-crystalline, radionuclide bearing phases are found.

4.8 Organic materials

A significant proportion of the material which will be placed in L/ILW repositories will be organic. This includes organic wastes which are generated mostly by nuclear installations and smaller contributions from research, industry and health care activities.

The nature of these organic wastes is highly variable and includes, among other things, ion exchange resins and sludges; filter pulps; halogenated and non-halogenated rubbers and plastics; disposable clothing, paper and cleaning materials. Some of this material is compacted and directly encapsulated in cement. However, the lower activity combustible material may be incinerated to reduce waste volumes and the resulting ash immobilised in cement. Incineration will destroy most, but not all, of the organic constituents in the material.

In addition to the organic wastes, some organic material will be used in the engineered barrier systems of some repositories. The bitumen used as an immobilisation matrix is an organic substance and this material is discussed separately in Section 4.7. Organic additives may also be used in structural concrete and cementitious backfills to prevent cement-water separation and to control the hardening rate.

Other organic materials in the engineered barriers include polymers which may be used as a specialist immobilisation matrix for a small percentage of L/ILW. Different types of thermosetting and thermoplastic polymer matrices have been used to immobilise ion exchange resins as an alternative to bitumen and also liquid waste such as evaporator concentrate which is rich in borate and nitrate. These polymer matrices include polymethyl methacrylate (PMMA), phenolformaldehyde (PhF), styrene divinylbenzene (for evaporator concentrates), polyester, polyurethane and epoxy (for spent ion exchange resins).

The solidification technology generally involves thorough mixing of the binder with a catalyst, followed by the waste material, to the required viscosity. Addition of the initiator or promoter results in gelation within minutes and hardening in several hours. A cross-linking agent may also be added to the initial mixture if additional mechanical strength is desired.

The degradation of these organic materials may impact significantly on the performance of a L/ILW repository by a variety of processes, the most important being:

- the generation of large volumes of gas which may impact on the structural integrity of the engineered barriers and may alter groundwater flow patterns;
- the generation of significant populations of colloids and organic complexants which may act to increase radionuclide solubility; and
- the release of the radionuclides directly to the groundwaters by the degradation of radionuclide-bearing organic materials.

With regards to the first two processes, the material of greatest concern is cellulose which is readily biodegradable. Cellulose may occur in the waste in the form of wood and paper etc., although materials such as these may be incinerated prior to solidification. Other organic materials, such as waste ion exchange resins, are generally less biodegradable than cellulose having been designed to be inert in laboratory and industrial usage.

Concern regarding the implications of cellulose degradation for repository safety has increased in a number of countries to the point where the consequences of organic degradation are the focus of much of the current research into L/ILW disposal. This issue is clearly one that requires further investigation. For the most part, the degradation of organic materials has been addressed by laboratory studies, which appears to be the best approach. However, a few natural analogue studies of organic degradation have been performed. Further natural analogue studies of cellulose breakdown have the potential to increase understanding of the process and to provide quantitative information that may help define the actual magnitude of the problem and suggest measures to alleviate it. A summary of cellulose degradation processes and their significance for repository behaviour is given by Askarieh et al. (2000).

The issues of most relevance to the behaviour of cellulose in the repository which have been (or potentially could be) addressed in natural analogue studies are:

- cellulose degradation;
- cellulose degradation products; and
- natural resins

These issues are discussed in the following sections.

Cellulose degradation

Cellulose, in the form of everyday materials such as wood and paper, is abundant in the surface environment where it can be observed to decompose rapidly when exposed to water and the atmosphere. However, surface conditions are unlike the chemically reducing, hyperalkaline conditions of a L/ILW repository near-field. In the repository environment, cellulose may decompose to form large volumes of gas, principally methane and carbon dioxide, via the following reaction chain:

i) hydrolysis of bulk cellulose to form soluble polysaccharides;

ii) physical and microbially-mediated hydrolysis of the polysaccharides to form glucose; and

iii) microbially-mediated anaerobic and aerobic decomposition of the glucose by a number of reactions to form organic compounds plus CO_2, H_2 and CH_4.

The first reaction proceeds more rapidly under high pH conditions, a phenomenon used to advantage in the paper-making industry. This is significant considering the hyperalkaline conditions in a L/ILW repository because the low molecular weight hydrolysis products are easily used as microbial substrates. The second and third reactions are dependent on the populations of microbes in the near-field and the ambient redox conditions.

Conditions in a L/ILW repository are not thought to be sufficiently hostile to inhibit microbial activity (Grogan and McKinley, 1990; McKinley et al., 1998). This is supported by measurements of viable microbe populations in the hyperalkaline groundwaters of the Oman and Jordan analogue sites (Bath et al., 1987a; Alexander, 1992; West et al., 1995).

In the repository, cellulose degradation reactions may be coupled to metal corrosion reactions because some of the aerobic reactions will compete for the limited available oxidant, whilst hydrogen production from metal corrosion may inhibit some cellulose degradation reactions. The actual balance between metal corrosion and cellulose degradation reactions will depend on the relative volumes of materials and the physico-chemical conditions and will be, therefore, site and repository-specific. However, this coupling may not be important because the aerobic phase is relatively short. The complexity of this coupled system has meant that most studies consider cellulose degradation separately from metal corrosion.

There are few sites where analogue studies can investigate cellulose degradation in high pH systems. However, there are sites where cellulose degradation has been limited due to specific burial conditions. One particularly interesting natural analogue study which has examined cellulosic material from a reducing environment was an investigation of the long-term breakdown of cellulosic materials in the fossil trees at Dunarobba (see Box 9). These trees are unusual due to the fact that, after 2 million years of burial in clay, they are still composed of wood and have not decomposed, as can be seen in Figure 4.30. This preservation is ascribed to the isolation of the wood from oxidising groundwaters by the impermeable clay envelope.

In this natural analogue study, organic leachates from samples of the Dunarobba wood have been compared with leachates obtained from Roman oak, fresh oak and 150 million year old lignites (Chapman, 1990). The total organic carbon (TOC) content produced from the Dunarobba wood is comparable to the steady state values from cellulose breakdown in the fresh oak sample and contains significant polymeric material. In other words, the isolation of the wood from oxidising conditions and advective water flow has significantly reduced the rate at which the cellulose degraded.

Cellulose degradation products

The soluble breakdown products of cellulose (leachates) include organic molecules that can act as complexants for some of the poorly soluble radionuclides from the waste. Consequently, the products of cellulose degradation might increase radionuclide solubility in the repository. The effects of organic leachates of cellulose on radionuclide complexation have been examined under the controlled conditions of the laboratory but few natural analogue studies have investigated this issue.

Figure 4.30: Fossil wood from the 2 million year old trees from Dunarobba, Italy, embedded within clay (see Box 9). The woody nature is clearly seen in the fragments. This preservation is due to the lack of cellulose decomposition resulting from the protection given by the envelope of clay that formed when the trees were buried. From Benvegnú et al. (1988).

However, as part of the Dunarobba experiment, the effect of the TOC content on the solubility of plutonium was investigated. It was found that the solubility increased from 10^{-11} M in NaOH to about 10^{-8} M in the leachate from the Dunarobba wood. If the cellulose degradation process was inhibited in the repository environment to the same extent observed at Dunarobba, then neither gas production nor increased radionuclide solubility would be such significant problems. However, Dunarobba does not replicate the L/ILW repository in one important respect, namely the hyperalkaline environment.

It was mentioned earlier that the hydrolysis of bulk cellulose proceeds much more rapidly under alkaline conditions. It follows that the slow rate of decomposition of the trees at Dunarobba may not be representative of the behaviour of cellulose in the repository. If this is true, then the Dunarobba natural analogue study has limited significance for a repository safety assessment and, hence, it is not appropriate to use the Dunarobba observations as a basis to revise near-field evolution models.

Natural resins

The ion exchange resins will be an important component of the wastes in some L/ILW repositories in terms of volume. These are likely to be dehydrated and powdered, and may be immobilised in bitumen.

The only natural materials that might be used as an analogue for the ion exchange resins are natural resins which are normally derived from tree sap (e.g. amber and rubber). However, these natural resins are compositionally and structurally very dissimilar to the technological ion exchange resins and, consequently, are not very close natural analogues. The only known discussion on natural analogues of polymers and resins is in Hellmuth (1989a) and is limited to a brief discussion of resinous plant products.

Natural resins are known which range in age from a few hundreds of years to 140 million years old. The most common resinous plant product found in the geological environment is amber. The preservation of these natural resins occurs under aerobic and anaerobic sedimentary conditions, as long as the sedimentation rate is not too slow. These resins may also survive under the temperatures associated with the initial stages of coal formation; i.e. up to 200°C for up to a few million years, or up to 100°C for up to 20 million years.

The occurrence of resins in sedimentary rocks attests to their preservation during erosion and transport in water. However, resins found in less permeable formations (e.g. clays) demonstrate a better degree of preservation than those found in permeable formations (e.g. sandstones) which suggests that groundwater leaching can enhance their deterioration.

The preservation of archaeological artefacts made from amber has sometimes been discussed (e.g. Beck et al., 1978) although never with radioactive waste objectives in mind. As with the natural occurrences of amber, the archaeological artefacts often exhibit slow deterioration rates. Whilst these observations are interesting, they can be used to give only very limited qualitative assurance of the stability of resins in the repository environment because of the gross differences in the chemistry of the materials and the depositional environments between the natural systems and the repository near-field.

Conclusions

Natural analogue examination of the fossil trees at Dunarobba suggests that in such environments the rate of cellulose degradation is slow. However, Dunarobba is a poor analogue of a L/ILW repository because this analogue site does not replicate the hyperalkaline conditions of the repository.

Relevant examples of cellulose degradation in hyperalkaline environments are necessary before firm conclusions can be drawn on degradation rates. Wooden pit props exposed to hyperalkaline groundwaters at the Maqarin site might possibly provide some useful qualitative information.

Chapter 5: Analogues of transport and retardation

This chapter discusses a number of physical and chemical processes which have been identified as being potentially important for the transport and retardation of radionuclides in the near and far-field environments. The specific issues discussed are:

- elemental solubility and speciation;
- elemental retardation processes;
- matrix diffusion;
- radiolysis;
- redox fronts;
- colloids;
- microbial activity; and
- gas generation and migration.

Not all of these issues are relevant to all repository designs. For example, gas generation is principally a problem associated with L/ILW repositories, although some gas is likely to be produced in all repositories.

Whilst all of these processes may potentially affect the transport and retardation of radionuclides, the actual importance and significance of some of them has not yet been fully demonstrated, and natural analogues may have a role in resolving these issues.

5.1 Elemental solubility and speciation

Geochemical models incorporating chemical thermodynamic codes and databases are widely used in performance assessment studies for radioactive waste repositories. Applied to the near-field, such models and codes are used to evaluate the evolution of the engineered barriers and radionuclide release from the near-field. Given the major element composition of the near-field groundwater, the solubility and aqueous speciation of performance assessment relevant elements (which generally will occur in trace amounts) can be predicted, thus allowing their transport properties to be estimated. In addition, this information may be used to support the extrapolation of laboratory measurements to near-field conditions.

Applied to the far-field, where it may be difficult to obtain comprehensive and representative groundwater samples, these codes may be used to define the far-field water chemistry in equilibrium with the host rock and any fracture coating minerals. As in the near-field, the solubility and speciation of radionuclides in the groundwater may also be predicted.

As a repository evolves, a chemically disturbed zone may develop around the repository due to migration into the host rock of contaminants derived from the near-field, such as high pH leachates from cement. This zone may penetrate some way into the far-field rock causing an extensive region of alteration. A further use of geochemical models is thus to predict the movement of such chemical fronts and their consequences for repository performance.

Geochemical codes also form one component of coupled codes which attempt to link the geochemistry of rock-water interactions to models of groundwater flow and, in some cases, to thermomechanical models of rock behaviour, thus allowing the transport of solutes in a dynamic system to be predicted. Coupled codes are still in development and none are used routinely in performance assessment. However, it is necessary to ensure that the geochemical components of coupled codes are equally as capable as stand-alone geochemical codes.

There are a number of different approaches to modelling geochemical systems. A good summary of these and the development of geochemical modelling in performance assessment is given by Nordstrom (1996). The most commonly used modelling approach assumes that chemical equilibrium between the rock and the groundwater is rapidly reached and, hence, that the distribution of aqueous species and saturation indices for solids can be calculated from free energies of formation (or equilibrium constants) by solving a set of equilibrium distribution and mass balance equations. This approach has been applied to a wide range of problems but is known to break down, to some extent, in low-temperature groundwater systems which may not be in equilibrium due to the sluggishness of rock/water interactions under these conditions. In particular, redox pairs which involve multiple electron transfers (such as sulphate/sulphide, nitrate/ammonia, carbonate/methane) are usually far from equilibrium (Lindberg and Runnels, 1984). However, in such cases it may be possible to use the standard equilibrium codes by 'switching off' reactions known to be so slow as to be insignificant.

An alternative approach is to take explicit account of the kinetics of rock-water reactions and, hence, to calculate the distribution of aqueous species and solid phase dissolution/precipitation as a function of time from known rate constants. This kinetic modelling approach has the potential to represent more accurately low-temperature systems but suffers greatly from the paucity of appropriate data. Rate constants are very difficult to measure and must be extrapolated from measured conditions with great care. Nonetheless, kinetic models have been successfully used in a number of cases but they usually require considerable model simplification of the natural system.

It is important to distinguish between the application of geochemical codes to major elements (e.g. those which are abundant in most groundwaters and essential to mineral formation) and to the rather exotic trace elements of interest to performance assessment (e.g. those which occur in low concentrations in groundwaters and which may be included by elemental substitution in a wide range of minerals). In the case of major elements, there are extensive databases of relevant thermodynamic data which are well supported by many field and laboratory studies in a wide range of geochemical environments. However, even for the major elements, care must be exercised when predicting solid phases or redox couples assumed to be at equilibrium. The use of thermodynamic codes to interpret groundwater major element chemistry is well illustrated in several analogue studies, such as Poços de Caldas (Nordstrom et al., 1990a) and Cigar Lake (Cramer and Smellie, 1994b) but, as these studies involve fairly standard geochemical analysis and use of the codes, this application will not be considered further here. The reader can find more details in standard geochemistry texts, such as Langmuir (1997).

Most performance assessments for HLW repositories (and many for L/ILW) have identified the low solubility of some elements as a key factor contributing to the safety case. In such analyses, solubility is represented rather simplistically, generally as a time-independent *solubility limit*. Solubility may be defined as the maximum achievable equilibrium concentration, in a solution of defined chemistry, which can be reached by a specific element. This definition refers to concentrations in true solution and does not consider uptake on colloids (see Section 5.6). The definition of solubility limits for performance assessment is a fairly subjective procedure which involves integration of field, laboratory and theoretical information to select values which may be termed *realistic* or *conservative*. Realistic values are intended to provide a reasonable estimate of the maximum solubility expected for a defined reference water (which, in some cases, may still pessimistically overestimate actual solubility in real groundwaters by many orders of magnitude) while conservative values are even more pessimistic estimates considering all uncertainties in the chemistry of the elements involved (see McKinley et al., 1998 for further details).

A comparison of the solubility databases used in a number of recent performance assessments for HLW repositories has indicated that a rather large scatter in data exist for some elements, and identified a lack of transparency in the procedures used to derive such databases as a particular problem (McKinley and Savage, 1996). The comparison also noted that geochemical thermodynamic modelling was extensively used to develop solubility databases, with little mention of direct laboratory or field measurements.

However derived, the geochemical databases need to be *validated* (demonstrating that they are appropriate for the use envisaged). The validation process is controversial due to its inherently subjective component, resulting from the impossibility of quantifying all heterogeneity and uncertainty in a slowly evolving repository system. It is neither practical nor useful to reproduce here the debate on validation terminology, and the interested reader is referred to Pescatore (1995) and references therein. Nevertheless, a specific regulatory requirement for validation may be specified in some countries, as is the case in Switzerland.

Any validation or testing scheme should place particular emphasis on the role of natural analogues in assessing data applicability over the long timescales and inherent heterogeneity of the geological environment. The subjective nature of validation means that it is important that it is carried out in a transparent and logical manner. It is also important to note that validation applies to a specific defined use and, although the testing scheme may be generic, the detailed arguments used need to be developed on a case-by-case basis. For any performance assessment database, the testing scheme can be focused on answering the following questions:

- Is the application clearly defined and scientifically reasonable?
- Are the theoretical arguments and models used defensible?
- What extent of validation is required to demonstrate conservatism?
- Are the selected data consistent with laboratory, field and natural analogue studies?

As part of testing and validation, it is possible to compare geochemical model predictions with field observations in a natural system which has some geochemical similarity with repository conditions. For example, a uranium ore body in a reducing groundwater system could be analogous to a spent fuel repository. For a situation like this, the trace element solubilities and speciations can be predicted using as input only the major chemical variables (major element concentrations, Eh, pH and temperature). These predictions can then be compared to actual field measurements in a procedure shown diagrammatically in Figure 5.1. Agreement between model predictions and reality (elemental concentrations actually measured in the groundwaters) indicates that the databases used may be applicable to a repository. In contrast, poor agreement means that the extrapolation was somehow in error and the data require re-evaluation. This methodology is sometimes known as *blind predictive modelling* and represents a rigorous validation methodology which is fundamentally different to simulation exercises in which measured data are retrospectively fitted to the model (Pate et al., 1994). However, it should be emphasised that a truly rigorous validation of the code is not possible because there will always remain some differences between the analogue and repository systems. Nevertheless, reasonable agreement in a number of tests at different analogue study sites, exhibiting a range of chemical conditions, greatly increases confidence in the applicability of the model or code in performance assessment.

In a formal repository performance assessment, safety will be demonstrated by means of a series of model predictions of how the repository will behave. To be confident in the robustness of such predictions, it is important to use codes and databases which have been shown successfully to predict the behaviour of similar systems, be they natural analogues or laboratory experiments, and not simply have been shown to simulate the results. This is all the more important for a performance assessment where the modellers will have to use such codes and databases to predict, without prior knowledge, repository performance far into the future.

Very extensive databases of trace element concentrations in groundwaters exist (e.g. Coughtrey and Thorne, 1982) but little of this information is of use in geochemical code testing because the rock/water systems the data are from are often insufficiently characterised. Consequently, it is difficult to assess if the elements of interest are saturated in solution. Also, in many cases, the water chemistry is of little relevance to performance assessment, particularly due to the high concentration of organics in near-surface waters.

Simulation studies using performance assessment geochemical models have been carried out using field data acquired from several analogue studies, e.g. Oklo (see Box 4); Broubster (see Box 12); Needle's Eye (see Box 13); Poços de Caldas (see Box 14), Alligator Rivers (see Box 15) and Palmottu (see Box 16). However, such simulation studies provide only a weak test of the predictive capabilities of geochemical codes and are more of an aid to interpretation of the geochemical system under study. Consequently, simulation exercises will not be considered further. Likewise, many

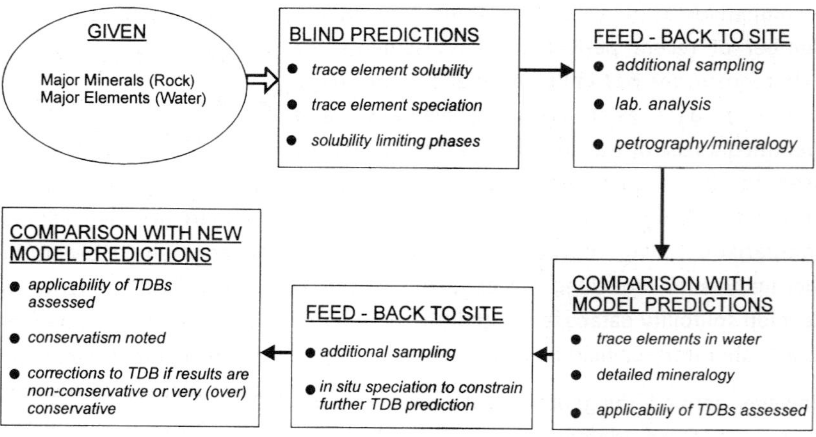

Figure 5.1: Procedure adopted for blind modelling tests, in this case as applied to the Maqarin study. The modellers are provided with reference water chemistry and the mineralogy. Predictions of the trace element solubilities, limiting solid phases and speciation are, thus, made blind. The modelled predictions are then compared to reality. TDB = thermodynamic database. From Smellie et al. (1997).

other model simulations of elemental solubility and speciation can be found in the geochemical literature but, as they do not have the aim of testing performance assessment codes and databases, they fall outwith the scope of this book.

The simplest blind predictive modelling exercises involve measuring the concentration of the elements of interest in a geochemical system in which they are greatly enriched (e.g. around an orebody). Assuming the ores are actually in contact with the groundwater analysed, and that their contact time is sufficiently long for equilibrium to have been reached, the measured aqueous concentration should approach the model prediction for the element's solubility limit in the system studied. An early example of such an approach is the work carried out on groundwaters from Morro do Ferro (Eisenbud et al., 1982; Miekeley et al., 1982) which, despite the presence of the thorium-rich orebody, have rather low concentrations of dissolved thorium. This observation is consistent with code predictions of low thorium solubility in most natural waters. However, such observations could be over-interpreted by inappropriate extrapolation from one element to another (e.g. the misuse of the chemical analogue approach discussed in Section 3.1). For example, using the measured concentrations of natural thorium to suggest that equally low concentrations of waste plutonium would occur in a repository environment would be dangerously inappropriate.

Figure 5.2: Photograph of one of the springs at Oman discharging hyperalkaline groundwaters which precipitate a range of unusual minerals including portlandite. These hyperalkaline waters and mineral assemblages make Oman a good analogue to a cementitious repository.

The blind predictive modelling approach has been applied in several analogue studies and some of these are described below. Two of these studies (Oman and Maqarin) are in hyperalkaline environments directly applicable to cementitious L/ILW repositories, where a number of specific problems arise when applying speciation models in high pH solutions. Most available databases do not adequately represent the solution species or range of minerals found in such environments and, thus, there is a pressing requirement to evaluate the application of the

Box 12: The Broubster uranium mineralisation

The Broubster natural analogue site in Caithness, Scotland comprises a quarry with exposed faulted sandstones, laminated bituminous limestones, boulder clay and a surrounding area of 5000 year old peats. A small quarry has been dug into the limestones and this has now largely been filled with spoil. This quarry has disturbed the near-surface hydrochemical system.

Young, shallow groundwaters flowing along the interface between the limestone and the clay, and in the peat, discharge into the quarry and then flow through the spoil to the surface sediments. These waters are acidic and oxidising and, thus, mobilise a variety of soluble elements from the in situ rock and the quarry spoil.

Examination of elemental abundances in the quarry spoil and the surface sediments along the flow path, show that uranium is accumulating in the peat between 100 and 200 m down-slope of the quarry. Maximum uranium concentrations are up to 0.1 wt.% in the peat. Chemical conditions in the peat are oxidising and acidic and, therefore, uranium would normally be expected to be very mobile in such an environment.

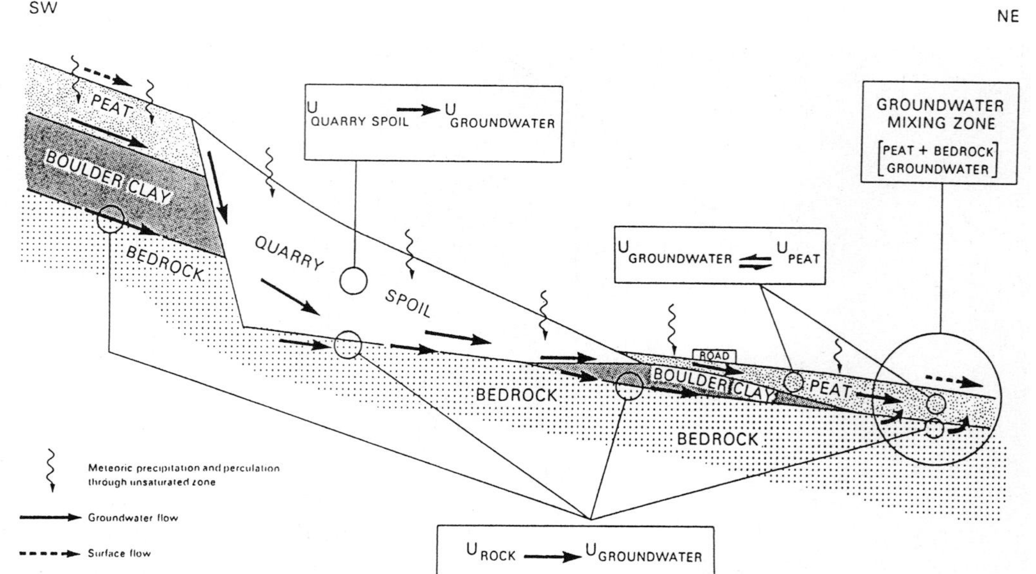

Figure B12.1: Diagrammatic cross-section of the Broubster natural analogue study site showing the movement of ground and surface waters, and the major uranium mobilisation and fixation processes. From Ball and Milodowski (1991).

It was found that, although both uranium and thorium originated from the same source, the quarry in the limestone, the migration pathways were quite distinct. Uranium is transported in true solution as carbonate or, possibly, phosphate aqueous complexes. At the peat bog, when the pH drops below 6, the organic ligands become more successful in complexing with the uranium and neutral inorganic complexes are replaced by positively charged humates which sorb onto the peat material and become fixed.

In the peat bog, the majority of soluble uranium was found to be bound to high molecular weight organic acids, and little to Fe/Mn oxyhydroxides. Reduction of U(VI) to U(IV) does not appear to occur and, therefore, cannot influence the speciation behaviour. The situation is, thus, different to a 'roll-front' deposit, or the mechanism expected to occur at a redox front in the repository environment.

In contrast, the thorium was transported as colloidal oxyhydroxides and not in solution. Retention by the peat is thought to be partly physical, with pores becoming clogged by the thorium-rich colloidal material. It is noted that neither thorium nor uranium are transported downwards into the clay beneath the peat, suggesting that the retention of both elements is connected with the presence of the humic substances in the bog.

As a consequence, this is not a good analogue for a repository near-field environment but the easily accessible and well-characterised source (quarry) and sink (peat) makes the site ideal for investigating radionuclide migration behaviour (processes, rates and distributions) and to use geochemical codes and databases in a simulation exercise.

This practical work and experimental results from the study are detailed in Read (1988), Milodowski et al. (1989a), Longworth et al. (1989a), Higgo (1989) and Ball and Milodowski (1991).

geochemical codes and databases in these conditions. The other analogue sites discussed here (Poços de Caldas, Cigar Lake and El Berrocal) are uranium ore bodies and are somewhat representative of HLW and spent fuel repositories.

Oman

The Oman natural analogue study site is located in the Semail Ophiolite Nappe of northern Oman. These rocks represent a complete cross section, some 15 km thick, of obducted oceanic crust and upper mantle rocks (Lippard et al., 1986). They now form a mountainous terrain rising to between 500 and 1800 m. The original mineralogy of the upper mantle rocks (harzburgites) was predominantly olivine with lesser amounts of pyroxenes, whilst in the oceanic crust (gabbros and basalts) the original mineralogy was predominantly pyroxenes with lesser amounts of olivine. The olivine in these rocks, which is magnesium and iron-rich, has been substantially hydrated to form serpentine and discrete iron oxide phases; with the effect of increasing the porosity. This process (serpentinisation) results in the groundwaters becoming both strongly alkaline and reducing. At the surface, the groundwaters degas large volumes of dissolved hydrogen.

The hyperalkaline groundwaters tend to flow in the mantle-derived rocks and reach the surface at springs at the contact between the crustal and mantle-derived rocks, as seen in Figure 5.2. Seven springs were sampled and the groundwaters were found to be Na-Cl-Ca-OH solutions with pH levels between 10 and 12. This unusual chemistry is reflected in the minerals precipitated at the springs, which include brucite, $Mg(OH)_2$, and portlandite, $Ca(OH)_2$. The measured concentration of trace elements in the groundwaters proved to be extremely low.

The approach the blind predictive modelling approach the geochemical teams adopted in the Oman study (Bath et al, 1987a,b) was simple and robust. Assuming that the site is a good analogue for a cementitious repository, it allows the behaviour of a suite of elements in the natural system to be predicted using the codes and databases which will be employed in an actual repository safety assessment.

The modelling teams first predicted the behaviour of the elements of interest in the system and only then was the measured data revealed for comparison. This was the first comprehensive blind predictive modelling exercise and the procedures developed for this study have formed the basis for most later studies.

Box 13: The Needle's Eye uranium mineralisation

This analogue site in south-west Scotland is on the northern shore of the Solway Firth, close to a natural rock arch known as the Needle's Eye. The area contains a suite of uranium and other metal-rich veins radiating out from the Criffel Granodiorite, the largest of which reaches the shoreline.

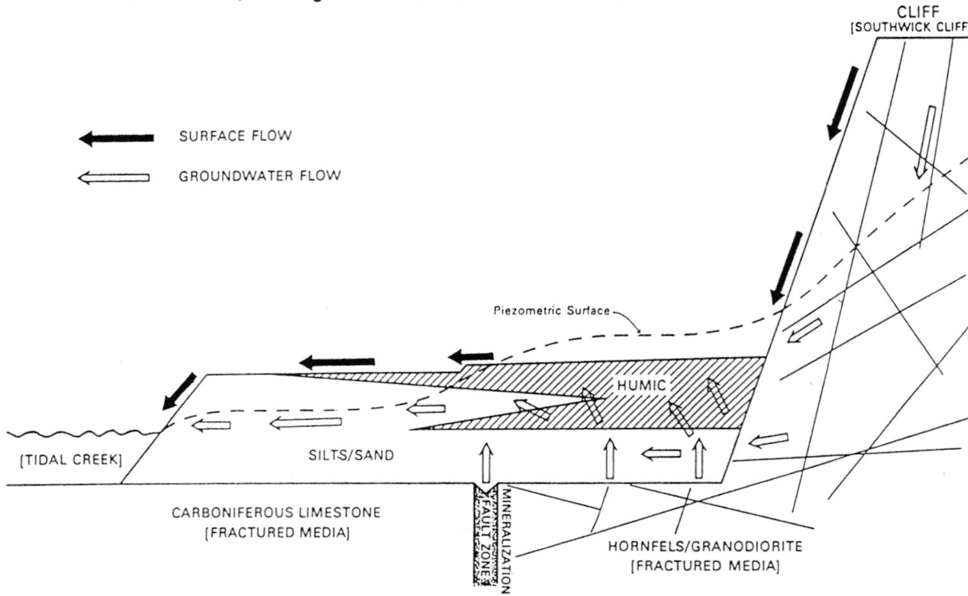

Figure B13.1: Diagrammatic cross-section of the Needle's Eye natural analogue study site. Radionuclides are leached from the mineralised fault zone and are carried towards the creek. From Hooker (1990).

The Needle's Eye study site comprises an ancient sea-cliff forming the edge of the granodiorite where the mineralised vein is partially exposed, below which is an area of anoxic, organic-rich soil which extends some 10 m from the foot of the cliff; beyond is a flood plain of fine-grained marine silty sediments with salt-resistant vegetation.

The generally simple site layout suggested that it would be fairly easy to identify the processes involving the transport and retardation of uranium and its daughter isotopes (Hooker, 1990). The analogue study focussed on the largest mineralised vein exposed in the sea-cliff and investigated radionuclide migration behaviour, and used geochemical codes and databases in a simulation exercise.

Uranium is present in the vein as pitchblende (UO_2) associated with secondary minerals such as uranophane (Ca-U-silicate). The pitchblende has undergone dissolution by two processes. The first is slow leaching over a long time period, probably by a reducing water, resulting in preferential loss of ^{234}U relative to ^{238}U; the second is rapid contemporary dissolution by oxidising waters. Some mobilised uranium is redeposited in close proximity to the vein as oxidised uranium minerals which are stable under present-day conditions and contribute little dissolved uranium to the groundwater.

However, the majority of the mobilised uranium (80 to 90 %) migrates with the shallow groundwater to the anoxic silty sediments where they are fixed and accumulate by sorption onto humic substances and iron oxyhydroxides.

In contrast to the uranium, the dissolution and transport of thorium is negligible. As with the Broubster site, the near-surface system does not make Needle's Eye a good analogue for a repository near-field environment but the easily accessible and well-characterised materials makes the site ideal for investigating radionuclide migration behaviour (processes, rates and distributions) and to use geochemical codes and databases in a simulation exercise.

The Needle's Eye natural analogue study has been discussed in a number of reports and is summarised in two volumes (MacKenzie et al., 1989a; 1991).

At Oman, two phases of modelling were carried out. In the first, literature data on major element concentrations in hyperalkaline springs were used to make initial predictions of trace element concentrations which were published prior to field sampling (McKinley et al., 1987) and also used to help plan the analytical programme. In the second phase (Bath et al., 1987a,b; McKinley et al., 1988) groundwaters from five springs were analysed and two thermodynamic speciation codes were run 'blind': MINEQL for uranium and thorium with two different databases, and PHREEQE for nickel, palladium, selenium, tin and zinc.

In most cases, the concentrations of all these elements were low, i.e. at or below the detection limits in many cases, which complicates the assessment of the modelling predictions. Generally, though, modelling using PHREEQE proved reasonably conservative, as shown in Table 5.1, although some predictions (e.g. for palladium) were well below measured concentrations.

For the MINEQL modelling of thorium, both databases gave consistent answers, and $Th(OH)_4$ was predicted to be the major species in solution, with ThO_2 as the solubility limiting solid. While this agreement is, at first sight, encouraging it probably reflects the fact that few thermodynamic data were available for this element at the time and both datasets consequently used the same sources of data. In the case of uranium, the predicted concentrations were not as consistent, and were above the measured values. In this case, the uranium databases do not appear to be able realistically to represent the uranium behaviour, although from a performance assessment perspective, the codes and databases are over-conservative. However, the measured uranium concentration in the most oxidising waters was surprisingly low, suggesting that uranium may be held in U(VI) minerals under these hyperalkaline conditions, although this was not confirmed. Other possible minerals have been proposed, including $Ca(OH)_2 \cdot UO_2(OH)_2$ or $UO_2(OH)_2 \cdot H_2O$. If these minerals did exist in hyperalkaline conditions, the solubility of uranium could be several orders of magnitude lower than predicted by the thermodynamic codes used. Unfortunately, the low uranium concentration in the groundwater means that it was difficult to draw firm conclusions.

Maqarin

The Maqarin site represents an excellent natural analogue of a cementitious repository because it contains natural cement phases such as portlandite, ettringite and tobermorite (see Box 11). These, and associated minerals, act as the source of both the hyperalkaline solutions and the relatively high concentrations of uranium,

*Table 5.1: Predictive and observed elemental concentrations from five hyperalkaline springs from the Oman blind predictive modelling exercise. *Predicted using the EIR database and **predicted using the NEA database. After Bath et al. (1987b).*

Element	Nizwa	Jebel	Bahla	Karku	Nidab
Selenium					
Predicted	5×10^{-3}	5×10^{-3}	5×10^{-3}	5×10^{-7}	5×10^{-3}
Observed	$< 3\times10^{-9}$	$< 3\times10^{-9}$	$< 3\times10^{-9}$	$< 3\times10^{-9}$	$< 3\times10^{-9}$
Palladium					
Predicted	10^{-16}	10^{-10}	10^{-22}	10^{-22}	10^{-23}
Observed	2.8×10^{-9}	6.6×10^{-9}	2.8×10^{-9}	2.8×10^{-9}	3.8×10^{-9}
Tin					
Predicted	10^{-19}	10^{-19}	10^{-18}	10^{-18}	10^{-18}
Observed	$< 2\times10^{-9}$	$< 2\times10^{-9}$	$< 2\times10^{-9}$	$< 2\times10^{-9}$	$< 2\times10^{-9}$
Zirconium					
Predicted	5×10^{-4}	5×10^{-4}	5×10^{-3}	2×10^{-3}	10^{-4}
Observed	$< 10^{-9}$	$< 10^{-9}$	1.1×10^{-9}	10^{-9}	2.2×10^{-9}
Nickel					
Predicted	3×10^{-7}	6×10^{-7}	3×10^{-7}	3×10^{-7}	3×10^{-7}
Observed	$< 1\times10^{-8}$	$< 1\times10^{-8}$	2.7×10^{-8}	2×10^{-8}	3.4×10^{-8}
Thorium					
Predicted*	5×10^{-10}	5×10^{-10}	5×10^{-10}	5×10^{-10}	5×10^{-10}
Predicted**	10^{-10}	10^{-10}	10^{-10}	10^{-10}	10^{-10}
Observed	$< 2\times10^{-10}$	$< 2\times10^{-10}$	2×10^{-10}	2×10^{-10}	2×10^{-10}
Uranium					
Predicted*	2×10^{-7}	8×10^{-4}	10^{-7}	10^{-6}	6×10^{-7}
Predicted**	2×10^{-4}	5×10^{-3}	3×10^{-9}	3×10^{-9}	2×10^{-9}
Observed	$< 4\times10^{-11}$	4.2×10^{-11}	$< 4\times10^{-11}$	$< 4\times10^{-11}$	$< 4\times10^{-11}$

selenium, tin, nickel and other trace elements of interest in the groundwaters.

Detailed mineralogical analyses have also identified a number of sink phases which are controlling elemental solubilities in this system. The blind predictive modelling study is discussed in Alexander et al. (1992a), Alexander (1992), Tweed and Milodowski (1994), Linklater (1998), Smellie (1998) and Alexander and Smellie (2000).

The aims of the thermodynamic database testing exercise at Maqarin were twofold (Alexander et al., 1998b). The first was to assess the applicability of the thermodynamic databases employed at the high pH conditions anticipated within a cementitious repository. In some countries the predictions derived from thermodynamic databases may be used directly as input to performance assessment and, in this case, it is important that results are either accurate or conservatively (safely) over-predict radionuclide releases from a repository.

The second aim was to evaluate how well different modelling groups (drawn from both performance assessors and geochemical modellers) can predict radionuclide solubilities for a range of trace elements in a relevant natural system. This is also important because the modellers themselves need to be trained in the use of the codes which require a good deal of geochemical knowledge in order to obtain appropriate and satisfactory results.

The geochemical modelling procedure employed at Maqarin was very detailed and it is explained here at some length because the validity of the databases relies on the rigour of the exercise. The system was modelled with a number of different thermodynamic databases used in different national radioactive waste management programmes. The testing procedure involved six stages, as follows.

1) Defining a groundwater major element chemistry, in this case three groundwater chemistries, based on measured data, were specified for five modelling groups. The modellers were allowed to comment on the appropriateness of the data (e.g. with respect to charge balance, degree of saturation with respect to major minerals etc.) but were required to carry out further evaluations without correction of, for example, charge imbalance.

2) Blind prediction of elemental speciation and degree of saturation of potential solubility controlling solids for uranium, thorium, radium, selenium, nickel, tin and lead using two thermodynamic codes and five different databases. Predictions were made using an arbitrary element concentration of 1 μg/l. In this phase of the test, no data suppression was permitted, such as ignoring mineral phases unlikely to form in the particular groundwater system, because this would introduce a bias to the procedure.

3) Comparison of predictions to determine major database discrepancies. This procedure is a simple intercomparison between the databases used (a type of verification) which can identify major inconsistencies in the databases.

4) Provision of measured concentrations of the seven trace elements of interest to the modelling groups.

5) Second phase of blind prediction of elemental speciation, solubility and appropriate solubility controlling solids.

6) Final comparison with the detailed analytical data on the site mineralogy and approximate groundwater speciation (separation of anionic, cationic and non-ionic, or neutral, components). Interpretation of anomalies between the code predictions and the analytical data provides information on the applicability of the databases to a cementitious repository environment and may additionally indicate areas for database improvement.

Although the testing procedure was detailed, there were still a number of assumptions or simplifications which had to be made, as described below. The significance of these had to be evaluated because the same assumptions and simplifications would probably have to be made in geochemical modelling as part of a performance assessment for a real repository system.

The blind predictive modelling exercise assumes that the concentration in the groundwater of any given trace element is controlled only by equilibration with the most stable phase present in the associated rock/groundwater system. Theoretically, this phase could be part of either the primary or secondary mineral assemblage. In the case of control by a primary mineral, it is assumed that saturation of the contacting water can be achieved before the mineral has dissolved completely. This may not be true, in which case the trace element would be *source-term limited* and the solution concentration controlled by the volume of the phase present in the primary mineral assemblage, and not equilibrium with the solution. It should be borne in mind that another

Box 14: Poços de Caldas

The Poços de Caldas plateau, which formed from a volcanic caldera, is located in the state of Minas Gerais in Brazil. The area has been the focus of diverse investigations for some twenty years owing to its high levels of natural radiation: Morro do Ferro in particular has been identified as amongst the most naturally radioactive places on Earth. A symposium to discuss the phenomenon of high natural radioactivity was held at Poços de Caldas (Cullen and Penna Franca, 1977).

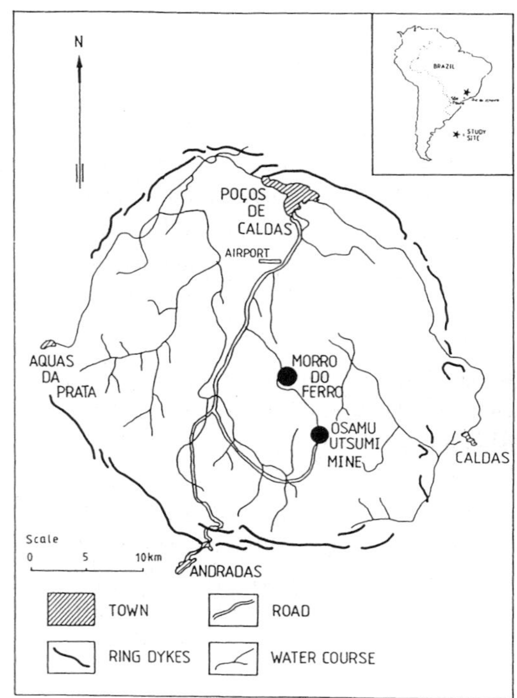

Figure B14.1: Location of the Osamu Utsumi mine and Morro do Ferro study sites that formed the focus of the Poços de Caldas natural analogue study. Both locations lie within the ring dykes marking the edge of a caldera. From Chapman et al. (1990).

The Poços de Caldas caldera is a ring structure about 35 km in diameter, composed of alkaline volcanic and plutonic rocks, mainly phonolites and nepheline syenites, of Mesozoic age. The caldera has suffered two major postmagmatic hydrothermal events which have led to widespread argillation and zeolite formation. The second hydrothermal event was more localised than the first and is believed to be responsible for the formation of the many ore bodies within the caldera complex.

Two of these ore bodies were the centre of the Poços de Caldas natural analogue study; the Osamu Utsumi mine, which is a uranium orebody with subsidiary thorium and REE enrichment, and Morro do Ferro which, in contrast, is a thorium and REE ore body with subsidiary uranium.

At the Osamu Utsumi mine, the rocks are predominantly phonolites. Primary mineralisation is mostly low-grade, disseminated throughout the rock, and is associated with hydrothermal alteration related to the intrusion of two breccia pipes.

Intense weathering in the semi-tropical environment has led to alteration of the upper exposed rock and secondary supergene enrichment of uranium along redox fronts, due to the downward migration of oxidising groundwaters.

The weathered rock is brownish-red due to the presence of iron oxyhydroxides, while the underlying rock is fresh and reduced, grey in colour and contains disseminated pyrite. The redox front is generally very sharp but irregular in profile as it follows the dips of hydraulically active fractures and faults, along which the oxidising waters have penetrated deeper into the rock mass. Uranium mineralisation occurs at the redox front itself, in the form of kidney-shaped accumulations of pitchblende several centimetres across and the mine has, consequently, been dug down to this level.

Morro do Ferro is a roundish hill, rising to 140 m above the surrounding plateau, and is some 5 km to the north of the Osamu Utsumi mine. The original rock is believed to have been a carbonatite but it has been heavily weathered to a depth of at least 100 m. Now the hill is composed of gibbsite, kaolinite and illite with additional iron and manganese oxyhydroxides that tend to form distinct layers.

There are also numerous magnetite veins that reach a few metres thick. Thorium and rare-earth element mineralisation extends from the summit down the south side of the hill and is very enriched; up to about 3 wt.% ThO_2 and up to 20 wt.% total rare-earth elements in some parts. No uranium mineralisation is known at Morro do Ferro despite many detailed investigations prior to the Poços de Caldas project.

Figure B14.2: Photograph of the Osamu Utsumi mine, showing the redox fronts which were the focus of much investigation. Photograph courtesy of Nagra.

The two sites were investigated in three-year (1986 to 1989), international natural analogue study of radionuclide transport in the geosphere (the Poços de Caldas natural analogue project). The focus of the analogue studies were concerned with radionuclide transport and retardation, with particular attention paid to:

- uranium mineral stability, degradation and dissolution (see Section 4.2);
- radionuclide transport and retardation processes at the redox fronts in the Osamu Utsumi mine (see Section 5.5);
- colloidal transport of radionuclides at Morro do Ferro (see Section 5.6); and
- testing thermodynamic solubility and speciation codes and databases in 'blind' predictions at both the Osamu Utsumi mine and Morro do Ferro (see Section 5.1).

Figure B14.3: Photograph of Morro do Ferro. This hill hosts a very enriched thorium and rare earth mineralisation, and is one of the most radioactive places known on Earth. Photograph courtesy of Nagra.

The Poços de Caldas natural analogue study has been comprehensively documented in a series of fifteen reports (the summary volume is Chapman et al., 1990) which discuss the data collected, their interpretation plus the predictive modelling and the relationship to radioactive waste disposal, particularly performance assessments. In addition the work has been published in a special issue of the Journal of Geochemical Exploration (Chapman et al., 1992).

possible control in natural systems might be leaching from the adjacent rock.

No equilibration with the identified solubility-limiting phase was carried out. For a low solubility trace element, the approach used is sufficiently accurate and simplifies comparison of calculated speciation. For a highly soluble trace element, this procedure is less appropriate as saturation with this element may significantly influence groundwater chemistry. However, for most trace elements of interest to a performance assessment, this approach is probably justified as they are unlikely to be present (in either a repository or a natural analogue site) at concentrations high enough to perturb the water chemistry to any great extent.

All the databases tested (and the majority of databases in present use) assume solubility control by pure mineral phases whereas, in reality, co-precipitation or solid solution will be the most likely control for trace elements. The detailed mineralogical data (Alexander, 1992; Milodowski et al., 1998) show this to be the case in the Maqarin system. However, as it seems unlikely that relevant thermodynamic databases including co-precipitation and solid solution will be available for the foreseeable future, it is necessary to understand any differences between prediction and observation using existing databases and measurement. Similar considerations can be made for the influence of colloids and organics on the model results.

No charge balancing was carried out during the calculations. This was intended to simplify intercomparison of results because differing methods of charge balancing would require back calculation to non-balanced conditions to assess whether differences in results between the databases were simply an artefact of the balancing procedure. In most groundwaters, the difference is not great anyway, but in highly mineralised systems, such as the Maqarin groundwaters, scoping calculations should be carried out to check that any charge imbalance does not skew the results. This has been done here and shown to produce differences of less than 1 % in the calculated solubility.

After undertaking the detailed practical exercise and then a comprehensive process of interpretation of the similarities and dissimilarities between model predictions and measured trace element concentrations, a number of conclusions were reached. The principal conclusion was that, with the exception of uranium, the observed elemental solution concentrations were two or three orders of magnitude less than the code predictions, when it was assumed that solubility was controlled by the simple stoichiometric oxide or hydroxide phases listed in the databases. This means that, although the databases do not contain mineral phases directly relevant to this hyperalkaline environment, the code still manages to provide results which are conservative from a performance assessment perspective. Clearly, more accurate, less conservative results may be obtained if the databases contained more relevant phases. A second limitation is that those minerals which are in the database are represented by pure end-members which do not commonly occur in nature. For example, the code underestimated the concentration of uranium owing to the choice of $CaUO_4$ as a solubility-limiting phase, even though this pure end-member was not observed in the study area. However, a similar amorphous phase does exist and may control the solubility of the uranium. Alternatively, uranium oxides of varying stoichiometry and degree of crystallinity are potential candidates for solubility limiting solids which would alter the predicted extent of oversaturation.

An attempt was also made to assess the in situ speciation of uranium in the groundwater using ion exchange columns. Unfortunately, the columns failed to function in these hyperalkaline conditions. Further work on the in situ measurement of trace element speciation in a range of geochemical conditions, including hyperalkaline environments, would be beneficial.

Poços de Caldas

For the geochemical part of the study, well characterised groundwater compositions were obtained from two separate sites at Poços de Caldas (see Box 14): the Osamu Utsumi mine and the Morro do Ferro thorium deposit (Nordstrom et al., 1990b). Participating organisations calculated, in blind tests, the solubility, speciation and solubility-limiting phases for a number of trace elements of relevance to radioactive waste disposal (uranium, thorium, lead, strontium and nickel) and some others which reflected the geochemistry of the sites (manganese, vanadium, aluminium and zinc). In a pilot study several additional elements were also evaluated (tin, selenium and radium).

The specific results of the modelling exercise for each element are summarised in Figure 5.3 and discussed in detail by Bruno et al. (1990). In general, the results from each participating organisation were comparable and fairly consistent with the measured geochemical data. Only the predictions from one group for two elements (nickel and zinc) were highly non-conservative. This was explained by a ferrite mineral being incorrectly taken as the solubility-limiting solid: these minerals do not form at low temperatures and are not relevant in this environment. As a consequence, it was concluded that ferrite minerals should be excluded from low-temperature repository assessment solubility limit calculations. This highlights the important lesson that training of the modellers is as important as testing of the geochemical codes.

For a few elements the predicted speciation was compatible with field measurements but this was not the case for uranium and thorium. However, it was clear that, even when different modelling groups predicted similar solubility limits, they often differed considerably in their predictions of aqueous speciation, not even agreeing on the charge of the dominant species. This is clearly an area where more work could be carried out using improved methods for determining chemical speciation in situ.

Cigar Lake

As part of the Cigar Lake study (see Box 5), a blind predictive modelling exercise was undertaken to compare measured trace element concentrations in the groundwater in contact with the uranium orebody with model predictions made using the PHREEQE code and the ZZ-Hatches v3.0 thermodynamic database (Casas and Bruno, 1994).

In general, quite reasonable agreement was found, particularly for those elements where suitable thermodynamic databases exist (e.g. barium, copper, thorium and, to a lesser extent,

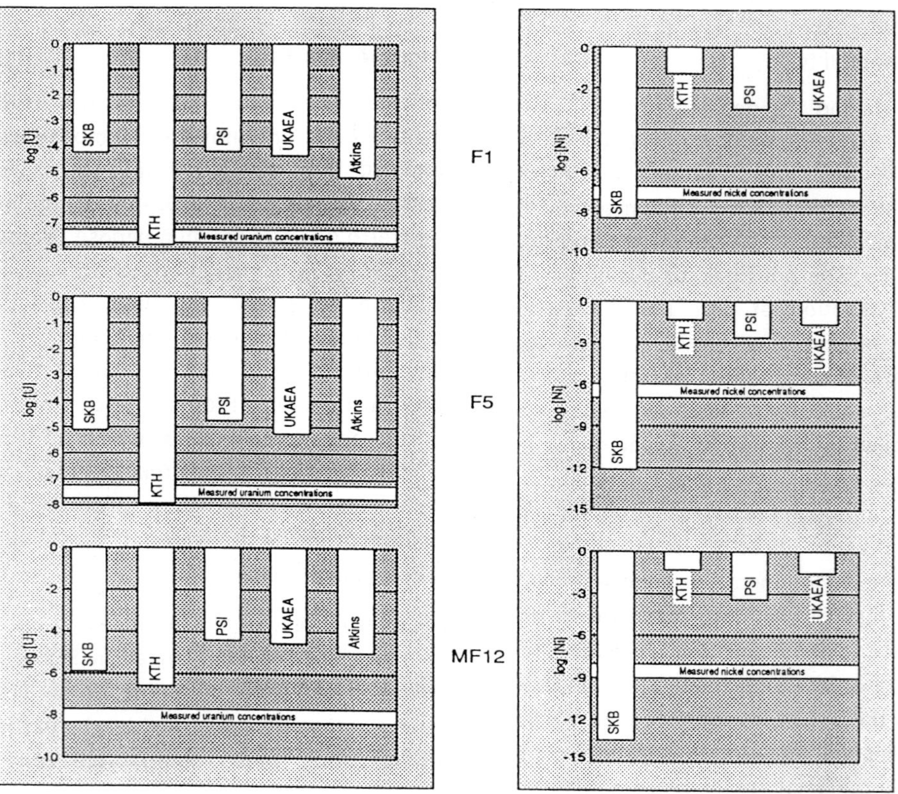

Figure 5.3: Calculated (columns) versus measured (horizontal lines) solubilities for uranium, on the left, and nickel, on the right, for selected groundwaters from the Osamu Utsumi mine (F1 and F5) and from Morro do Ferro (MF12). See text for details.

zinc). This indicated that the equilibrium approach is useful to describe the behaviour of these elements at Cigar Lake. In addition, the similarity of the groundwater, clay and ore system to the near-field of a spent fuel repository implies that the equilibrium approach will also be suitable for performance assessment in this type of chemical environment. Unfortunately, for nickel and lead, groundwater concentrations were below detection in most cases.

Inadequate thermodynamic data for critical mineral phases were a problem for predicting the concentrations of strontium, molybdenum and arsenic. The chromium prediction was largely non-conservative, believed to result from the uncertainty of the Cr^{6+}/Cr^{3+} redox potential, which is necessary for defining its solubility in groundwater.

Inadequate detailed mineralogical data meant that some of the predicted phases could not be confirmed, e.g. the solubility-limiting phase of thorianite for thorium, compared to monazite and brannerite identified, and native copper for copper (although some native copper was found associated with the ore, the dominant phase is chalcopyrite). In contrast, the predictions for nickel and lead were confirmed with the identification of bravoite and galena.

El Berrocal

An extensive blind predictive geochemical modelling exercise was performed at El Berrocal which was made possible by the very comprehensive geochemical and mineralogical investigations performed as part of the overall El Berrocal study (see Box 17). The groundwaters at the site exhibit significant compositional variations, particularly with regard to uranium, sulphate and pH. This range of groundwaters within the single study site allowed for a very demanding validation exercise for the geochemical codes and databases.

Eight groundwaters were chosen for the geochemical modelling exercise and four modelling terms participated using different combinations of geochemical codes (HARPHRQ, PHREEQE, EQ3NR) and thermodynamic databases (HATCHES, LLNL). A suite of 9 elements was identified for the exercise based on data availability and relevance to performance assessment, they were: barium, copper, manganese, nickel, lead, strontium, thorium, uranium and zinc. The exercise adhered strictly to the blind predictive modelling procedures outlined earlier, by ensuring the modelling teams had no previous knowledge of measured groundwater compositions beyond those provide as part of the exercise. The modelling work took place in two stages:

- Stage One required the modelling teams to predict solid solubility limiting phases, and aqueous elemental concentrations and speciations for the suite of elements in all the waters. The only input data provided to the modellers was the major component groundwater chemistry, pH and Eh.
- Stage Two allowed the modellers to recalculate their predictions on the basis of the known mineral assemblages in fractures (i.e. minerals in contact with the groundwaters) and considering solution solutions.

Within these constraints, modelling teams could adopt different modelling procedures and assumptions. Some of the Stage One solubility predictions are indicated graphically in Figure 5.4, for barium, nickel, thorium and uranium. It is clear from these graphs that the modelling groups sometimes agree and sometimes disagree. However, when they disagree there is no single reason why. In some cases, it is because different solubility limiting phases are chosen by the modelling teams. This is particularly noticeable in the case of nickel when three groups chose trevorite ($NiFe_2O_4$) as the solubility limiting phase in most waters. However, the fourth team decided that trevorite would be unlikely to form in these

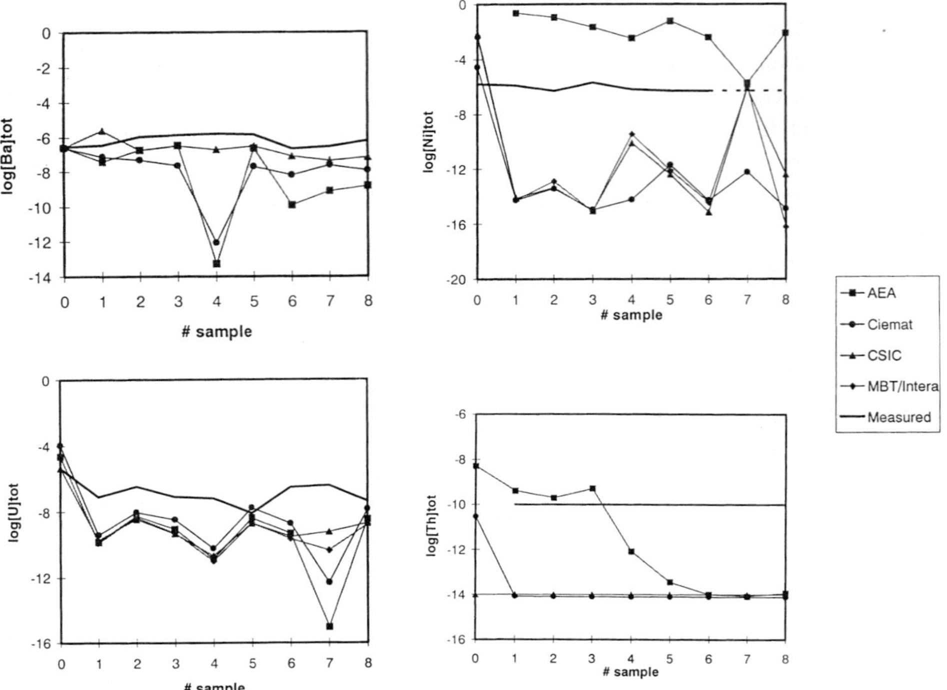

Figure 5.4: Stage One solubility predictions for barium, nickel, thorium and uranium from the El Berrocal study. These predictions were made assuming pure solids selected from the thermodynamic databases. From ENRESA (1996).

low temperature waters and chose the soluble phase $NiSO_4 \cdot 7H_2O$ instead. This raises an important general point regarding the use of geochemical codes and that is that the modellers themselves should be recognised as a key variable in predictive modelling together with the codes and the databases.

In the stage two predictions, use was made of the geochemical and mineralogical information from the site which indicated that many of the trace elements were included in solid solutions and not pure minerals, as is generally assumed in thermodynamic models. Attempts were made, therefore, to calculate elemental solubilities considering co-precipitation or co-dissolution (assuming present-day net mineral growth or loss in the system) and an assumed molar fraction trace element association in the solid solution minerals (Bruno et al., 1998).

The impact of considering solid solutions on the modelling is exemplified by the case of barium. In the Stage One calculations, barium concentrations were underpredicted in almost all cases when pure mineral forms were assumed. Under-prediction is non-conservative with respect to performance assessment calculations and is, therefore, potentially unsafe. However, by considering association of barium with carbonate, as Ba-$CaCO_3$, and assuming co-precipitation and co-dissolution, then the predictions changed. The co-precipitation assumptions continued to underpredict solubilities but this is because the solid phase in the system is undergoing net dissolution rather than net growth. Assuming congruent co-dissolution, however, the predictions slightly overestimated concentrations compared to the measured values, as shown in Figure 5.5. A prediction was also made for incongruent co-

Table 5.2: Summary of the results from assumptions of solid solution in the El Berrocal geochemical modelling exercise. From Rivas et al. (1997).

Element	Predicted solubility limiting phases
Barium	Good predictions assuming congruent co-dissolution of Ba_x-$Ca_{1-x}CO_3$.
Copper	Good predictions assuming co-precipitation of CuO-$Fe(OH)_3$.
Lead	Results not conclusive. Lead is likely to be source-term limited.
Manganese	Good predictions assuming congruent co-dissolution of Mn_x-$Ca_{1-x}CO_3$.
Nickel	Results not conclusive. Nickel is likely to be sorption controlled.
Strontium	Good predictions assuming congruent co-dissolution of Sr_x-$Ca_{1-x}CO_3$.
Thorium	Overprediction by 2 orders of magnitude assuming $Th(OH)_4$.
Uranium	Good predictions assuming co-precipitation of schoepite-$Fe(OH)_3$.
Zinc	Results not conclusive, possibly co-dissolution of Zn_x-$Ca_{1-x}CO_3$.

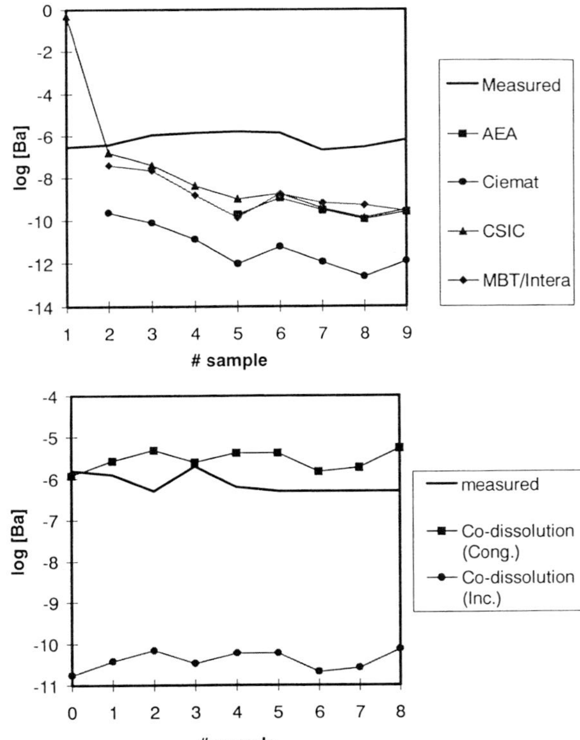

Figure 5.5: Stage Two solubility predictions for barium assuming co-precipitation and co-dissolution. These predictions were made assuming solid solution in Ba-$CaCO_3$. From ENRESA (1996).

dissolution but this proved not to be able to reproduce the system.

Overall, assumptions for solid solution rather than pure minerals tended to improve the predictions but the degree of success was variable for different minerals, as shown in Table 5.2. In a number of cases, trace elements were found to associate with carbonate, indicating the importance of this species in the El Berrocal system. Although carbonate would not necessarily control radionuclide precipitation in a repository far-field, the general principle holds true that, in real systems, trace elements only infrequently precipitate as pure minerals. The inability for geochemical models accurately to simulate solid solutions relates to the currently inadequate nature of the current generation of thermodynamic performance assessment codes and databases which are unrealistic compared with natural mineralogical systems. This is an area where natural analogues potentially have an important role in future performance assessment code development.

The work at El Berrocal has confirmed the observations from several analogues studies, such as Cigar Lake and Poços de Caldas, that may elements co-precipitate with other elements to form solid solutions, rather than as pure solids. This has important implications for performance assessment because current geochemical models used in performance assessment are unable to simulate this process due to a lack of relevant data. Neglecting co-precipitation may well be a conservative assumption but it can lead to overestimation of the solution concentrations by many orders of magnitude and, thus, sensible conservatism can give way to a gross lack of realism in the modelling. The observed magnitude of this problem was not recognised before it was highlighted in natural analogue studies (Smellie et al., 1997) and, thus, this is a good example of how information from analogue studies can be used to improved our conceptual understanding of natural processes, which can then hopefully lead to the development of improved performance assessment codes, as discussed in Chapter 6.

Conclusions

Testing of the geochemical codes used to predict radionuclide solubility and speciation has been one of the successes of natural analogue studies to date. Such tests have varied in rigour but have ranged from qualitative support of the predicted behaviour of particular elements to clear identification of errors or missing information in the codes used or their associated databases. In this regard, it should be noted that the natural analogue test cases described above can be usefully recycled for testing of new codes and thermodynamic databases (e.g. Pate et al., 1994).

Further work is recommended, particularly aimed at more rigorous blind predictive modelling tests in systems where saturation of a range of relevant trace elements may be expected in groundwater of appropriate chemistry, such as chemically reducing, low organic content waters. Thus, there is a need for further appropriate analogue sites, particularly for ore bodies containing a wide range of performance assessment relevant elements (e.g. selenium, palladium, nickel, tin) which are in contact with slowly moving, reducing groundwaters, where saturation can be expected to have been reached.

Further development and application of in situ speciation measurement techniques is also strongly recommended because this is a particular area where considerable discrepancy between codes and databases was identified.

In carrying out further studies, it is important that the difference between true blind predictions and the more common approach of simulating the geochemical behaviour of the system is appreciated. The former approach is a good approximation to the amount of information which will be available, and the manner in which predictions will be made, in a performance assessment exercise. The latter approach is essentially a model, code and database calibration exercise which is valuable for building experience in the application of geochemical codes, but does not comprise a rigorous test of performance assessment methodology.

A major limitation of the standard geochemical codes is their inability to treat solid solutions into which many radionuclides are included. Omitting solid solutions from the performance assessment models may be shown to be conservative but could involve overestimates of solubility by many orders of magnitude. Attempts to improve realism of codes are needed to improve this situation and some projects are now underway. It is encouraging to see that analogue studies are leading the way in this field.

5.2 Elemental retardation processes

The movement of radionuclides in solution through the near-field engineered barrier system materials and along migration pathways in the far-field rock is controlled by the processes of advection and diffusion in porewaters and groundwaters. A very small number of radionuclides, such as ^{3}H (tritium), ^{36}Cl and ^{129}I, interact so weakly with the solid materials through which they are passing that they can be considered effectively to move at the same rate as the individual molecules of groundwater, and are are said to migrate conservatively with respect to the water.

However, the majority of radionuclides released from the waste will interact with the repository materials and rock surfaces over which water passes, or are subject to changes in solution behaviour owing to modifications in rock-water interactions along the flow path. These processes can retard their movement relative to that of the water, not only slowing their progress through the system, but progressively reducing their concentration in solution to a point where equilibrium may be attained. These retardation processes are thus beneficial for repository safety and a great deal of effort has gone into characterising and quantifying them.

Retardation processes can be essentially chemical or physico-chemical in nature, and a number of mechanisms have been identified in natural systems, as shown diagrammatically in Figure 5.6:

- chemical retardation mechanisms:
  - adsorption,
  - ion-exchange,
  - precipitation, and
  - mineralisation;
- physico-chemical retardation mechanisms:
  - diffusion into the secondary, 'matrix' porosity of rocks (see Section 5.5),
  - molecular filtration, and
  - ion exclusion.

Adsorption and ion-exchange are often collectively termed sorption, a term used generically to encompass chemical interactions with solids that retard transport. Sorption is generally modelled as a reversible process whereby sorbed radionuclides may be released to solution if solution concentrations or composition change. Desorption kinetics are generally slower than sorption kinetics and there may be instances where sorption could be considered irreversible.

Strictly speaking and according to thermodynamics, no chemical process, such as the sorption of radionuclides by mineral surfaces, can ever be truly irreversible. All chemical reactions are reversible, the significance is in the time taken and the conditions necessary for the reaction to be reversed. When irreversible sorption is discussed in terms of performance assessment, it is generally meant that the kinetics of desorption are slow compared to the time period of interest to the assessment, assuming the physico-chemical conditions are constant.

Precipitation is not a sorption process but it can be difficult to discriminate between the two mechanisms in both field and laboratory studies. Generally, sorption would be expected to be the dominant process at low solution concentrations of radionuclides. As concentrations of radionuclides increase, precipitation of phases in which the radionuclide of interest is a stoichiometric component (i.e. an essential component of the mineral structure) may occur if saturation is reached in the groundwater. In circumstances where the total amount of dissolved solids in the groundwaters is high, a very complex chemical environment may develop in which radionuclides may co-precipitate as solid-solutions in a variety of mineral phases, or be scavenged by amorphous precipitates, such as iron oxyhydroxides, and also effectively co-precipitated. The stability of newly precipitated minerals and amorphous phases depends on high solution concentrations being maintained. If concentrations fall below saturation, perhaps after a pulse of contaminated groundwater has passed, or as the near-field is flushed with fresh groundwater, these minerals will begin to dissolve and release any radionuclides they contain back to solution.

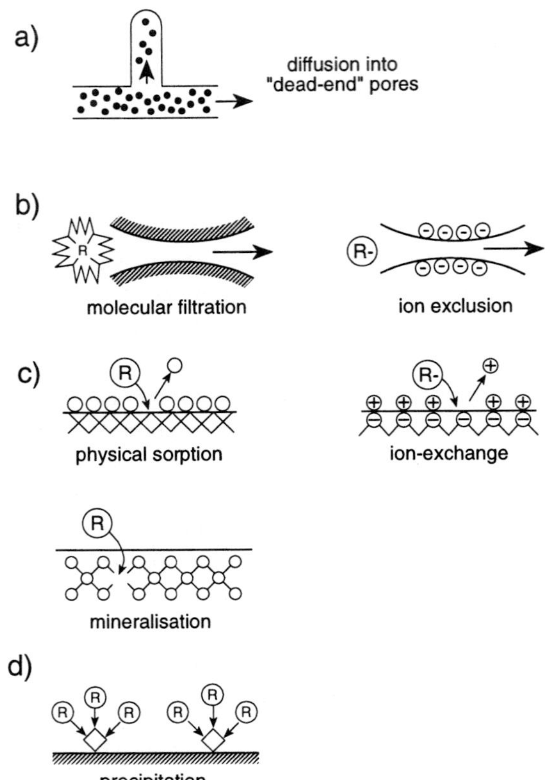

Figure 5.6: The retardation mechanisms that may affect radionuclide transport in groundwaters: (a) and (b) only occur in dynamic systems and retard solute transport, (c) and (d) are sorption processes that may occur in dynamic or static systems. After McKinley and Hadermann (1984).

The kinetics of sorption and dissolution, and precipitation processes are clearly important to long-term predictions of radionuclide behaviour. Irreversible sorption (slow desorption) and precipitation processes are obviously beneficial in performance assessment terms because they immobilise radionuclides very effectively. They are, however, difficult to demonstrate, and the most robust performance assessment models make the assumption that all chemical retardation processes are instantaneously reversible. A challenge for natural analogues would be to provide convincing evidence of irreversible sorption (slow desorption) processes.

In most cases, irreversible sorption would be a conservative process. However, in cases where sorption is irreversible but the sorption capacity of the rock surface is low, sorption sites could become saturated leading to a state where no further net sorption can occur. A further objective of analogue studies could be to establish sorption capacities of different rock surfaces as well as the kinetics of the sorption mechanisms.

Sorption processes generally occur sufficiently rapidly to allow their kinetics to be ignored in performance assessment models. In an essentially very sluggish groundwater flow environment, such as in the near-field of some repository concepts, the system is often modelled as a chemical mixing tank, where the presence of sorbing surfaces, such as cements and corrosion products, is a key part of the model generating steady-state solution concentrations of radionuclides for the far-field source term. In this environment, the kinetics of sorption processes may become important if any mechanisms are identified which can lead to gross chemical inhomogeneities in the system or rapid groundwater transit pathways through the system. In the far-field, sorption kinetics are not considered in performance assessment, although precipitation and mineralisation kinetics would clearly be relevant if these processes were to be included in an assessment. Where groundwater flows are relatively rapid (as may occur, for example, in a major fracture zone), then retardation mechanisms become increasingly less significant in affecting the rates of release of radionuclides and, in some performance assessments, are given no credit.

Retardation during transport was one of the first mechanisms to be investigated by natural analogue studies. Much of the early work was performed at the Oklo natural fission reactors (see Box 4), which could be considered as containing point sources or zones of fission products and actinides from which transport could be measured (e.g. Brookins, 1984). Similar investigations were performed at other locations where sources of natural series radionuclides could be found, such as at Morro do Ferro (Eisenbud et al., 1982; 1984) or at the contact between igneous intrusions and host rocks (Brookins, 1984; Laul and Papike, 1982; Laul et al., 1984; Wollenberg et al., 1984). Other investigations examined geological environments where there was no single point source of radionuclides, for example rock weathering profiles (e.g. Michel, 1984); aquifers with well defined flow rates (Pearson et al., 1983; Andrews and Pearson, 1984); and aquifers with poorly defined flow rates (Krishnaswami et al., 1982).

Since these early studies, many more natural analogues have investigated transport and retardation in a more comprehensive and quantitative fashion. It would not be possible to review all the work that has been performed on this issue and, consequently, only the most important and representative studies which have yielded information relevant to performance assessment will be described.

The issues of most relevance to radionuclide transport and retardation that have been (or potentially could be) addressed in natural analogue studies are:

- transport and retardation within fractured crystalline rocks;
- transport and retardation within argillaceous rocks;
- transport and retardation within volcanic ash deposits;
- transport and retardation within evaporites;
- transport and retardation at the geosphere-biosphere interface; and
- measurement of in situ distribution coefficients.

These issues are discussed in the following sections. Other related issues dealt with elsewhere are matrix diffusion in Section 5.3 and redox fronts in Section 5.5.

Transport and retardation within fractured crystalline rock

The principal mechanism for radionuclide transport in fractured crystalline rock is advection along hydraulically active channels within fracture networks, as seen in Figure 5.7. Hydraulically active fractures in crystalline rocks are usually coated by secondary mineralisation resulting from the hydrous or hydrothermal alteration of the rock. The fracture coating minerals that form depend upon the mineralogy of the rock mass and the groundwater chemistry and, as such, are site-specific. These minerals are extremely important in understanding radionuclide transport in the rock because it is the fracture coating minerals which dominate retardation and sorption in the fissures. In strongly altered fractures, these minerals may also comprise much of the higher porosity zone into which matrix diffusion can occur (see Section 5.3).

Unfortunately, many analogue studies do not pay sufficient attention to the exact mineralogy and chemistry of fracture surfaces, and so not enough data are made available to extract quantitative information for performance assessment purposes. An example of a more detailed study is that of the Klipperås study site (Landström and Tullborg, 1990) where a complex suite of fracture minerals was recorded, including chlorite, calcite, siderite, quartz, epidote, muscovite, illite, hematite, pyrite, goethite, kaolinite, mixed-layer clays, gibbsite, plagioclase and potassium feldspar.

In addition to the mineralogy, elemental analyses (including those for rare-earth elements, uranium and thorium) were performed on samples of fracture filling minerals, the host rock and the associated groundwater. These data indicated that redistribution of certain elements occurred in response to alteration (clay formation) of the host

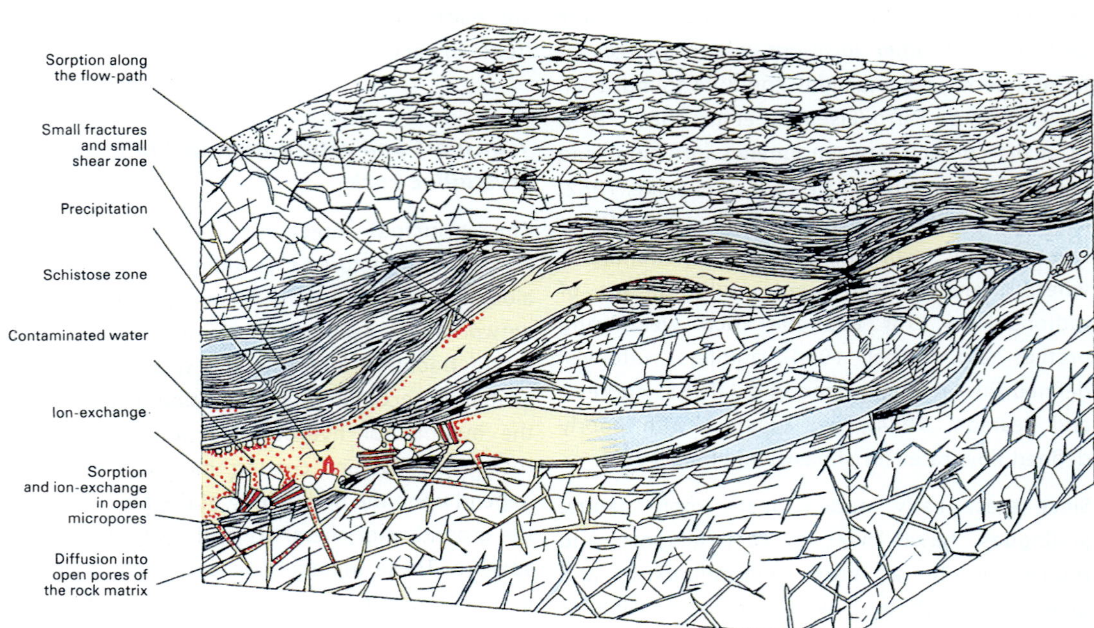

Figure 5.7: Some transport and retardation processes which may occur in fractured rocks. In characterising such systems, care must be taken to distinguish between low-temperature processes relevant to migration away from a waste repository and earlier higher-temperature events, the effects of which may dominate the present-day distribution of elements in the bulk rock and fracture minerals. Illustration courtesy of Nagra.

rock by groundwater. Uranium, for example, appears to have been very mobile and is associated with most fracture materials, preferentially with iron-rich phases. Similarly, thorium was found to have been mobile and strongly associated with iron-rich phases. Rare-earth element behaviour varied; the light rare-earths were preferentially associated with iron-rich phases, while heavy rare-earths showed some selectivity for carbonate material.

In general then, this study confirmed the association of remobilised elements with iron-rich phases (principally secondary iron oxyhydroxides) previously demonstrated elsewhere (e.g. Smellie et al., 1986a,b,c; Guthrie, 1989; Hofmann, 1989). However, the thorium data from this study show the problems associated with trying to separate the effects of recent low-temperature processes with earlier high-temperature events. Thorium is generally immobile under low temperature conditions (c.f. Langmuir and Herman, 1980) but may migrate in association with colloids (Miekeley et al., 1990b; 1992). The thorium mobility described from Klipperås is, therefore, most likely to be an artefact of either colloid transport or a previous high-temperature event, rather than recent low-temperature transport. Likewise, the association of the heavy rare-earth elements with calcite could also be a product of a hydrothermal overprint (see Taylor and Fryer, 1982, for discussion).

As part of the investigations at the Äspö Hard Rock Laboratory in Sweden, the sorption capacity of the fracture coating phases was investigated in detail (Landström and Tullborg, 1995). At Äspö, the host rocks are fractured granite and the most common fracture coating phases are clay gouge, calcite and iron oxyhydroxides. The distribution of a range of elements including uranium, thorium and rare-earth elements on these was determined using sequential extraction methods on the separated minerals. The results showed that many of the elements (rare-earth elements, scandium, thorium, uranium, radium and barium) were associated with mixed iron oxyhydroxide and calcite precipitates, although it was not clear whether sorption or coprecipitation was the dominant fixation mechanism. The clay minerals were also found to have strongly sorbed several different elements, including rare-earth elements, thorium and uranium.

In a further study at the Äspö Hard Rock Laboratory, biogenic mediation was found to be significant for radionuclide retardation processes (Tullborg et al., 1997). In this case, it was determined by chemical and isotopic analysis of fracture minerals, that the presence of sulphate reducing bacteria (SRB) had caused the production of HCO_3^- and subsequently the local formation of calcite. The mechanisms responsible for elemental incorporation in this biogenic calcite are unclear but geochemical data indicated favourable uptake of the light rare-earths in the calcite. The relevance to radionuclide retardation in the far-field from this study is unclear because of uncertainty regarding the population and significance of microbial processes in undisturbed, deep rocks, as discussed in Section 5.7.

A similar study of the sorption properties of the fracture coating phases was undertaken at El Berrocal site in Spain (see Box 17), which is also in a fractured granite host rock. At El Berrocal, the fracture minerals consist of quartz, potassium feldspar, clay minerals, carbonates and minor pyrite and iron oxyhydroxides. Furthermore, uranyl phosphates and silicates are present in many samples. The site has experienced a complex hydrothermal history, which makes interpretation of the geochemical data difficult. However, useful information on the association of a range of elements with the fracture coating minerals was obtained (Rivas et al., 1997; Pérez del Villar et al., 1997).

Comparison of the compositions of the granite matrix and the fracture coating minerals shows that the clay minerals in the fractures formed directly by granite alteration. Many of the trace elements, including uranium, were found to be bound in specific mineral phases. Uranium was mainly precipitated as mineral phases or coprecipitated with calcite: uranium sorption on iron oxyhydroxides was less common in comparison to other analogue sites.

Comparing the results from a number of analogue study sites shows that similar sorption behaviour is seen at many locations. For example, uranium and rare-earth elements are frequently associated with calcium and iron oxyhydroxides. Unfortunately, it is rarely possible to define in detail the palaeohydrogeological history of a site and, thus, it is generally the case that recent low-temperature retardation events cannot be uniquely identified.

This is possibly the reason for the apparently anomalous observations at El Berrocal related to the limited observed uranium sorption on iron oxyhydroxides. Therefore, much of the analogue sorption data can be taken as no more than qualitative and, to be conservative, would probably not be used in a formal performance assessment.

Furthermore, although several natural analogue studies have demonstrated the effects of sorption and precipitation processes on fracture surfaces, none have been able to distinguish clearly between these processes or to provide quantitative data on retardation with respect to transport of trace elements in natural waters.

Essentially, these studies provide useful observations of the net effects of interactions between solutes and rock surface, and highlight which phases are most active, but do not provide the type of sorption data that are required in performance assessment. To address this issue, mineralogical studies are being increasingly linked with laboratory experiments to try and quantify the processes of interest, e.g. as part of the Palmottu study (Box 16).

Transport and retardation within argillaceous rocks

A number of studies have examined radionuclide transport within unconsolidated sediments and some of the most significant are described below. Although unconsolidated sediments cannot be considered as a complete analogue to either argillaceous repository host rocks or clay buffers, the analogy is generally valid for argillaceous environments where transport is dominated by diffusion. The clay content and associated mineralogy of the sediments is also very similar to relevant argillaceous rocks, and the sediments have maintained reducing conditions similar to those in a repository for long time periods, although the pH conditions may be dissimilar.

One interesting example is the study which was performed on marine turbidites of the Madeira Abyssal Plain (Colley et al., 1984; Colley and Thomson, 1985; 1991). This study investigated the phenomenon of the formation of uranium-rich layers by redox front migration (Wilson et al., 1985; 1986; Thomson et al., 1987). This uranium enrichment mechanism occurs when an organic-rich sediment with high uranium content is emplaced by a turbidite into a deep-sea environment. Oxygen-rich seawater penetrates this turbidite deposit and oxidises and dissolves the uranium, which is then free to diffuse through the sediments. Uranium migrating downwards reaches more reducing conditions where it precipitates, enhancing the uranium concentration at that level (Figure 5.8). A strong redox front is, thus, established and, as more oxygenated seawater penetrates the turbidite, this redox front migrates further downwards, driving the progressively more enriched uranium with it. This redox front remains active until another turbidite is emplaced above the first, cutting off the supply of oxygenated water. Colley and Thomson (1991) examined the enriched uranium layers at the inactive redox fronts and measured the longer lived parent-daughter pairs of the ^{238}U decay series (^{238}U-^{234}U; ^{234}U-^{230}Th; ^{230}Th-^{226}Ra; ^{226}Ra-^{210}Po). It was discovered that the only element to exhibit migration since the front became inactive was ^{226}Ra, whose symmetrical concentration peak around its parent ^{230}Th implies transport occurred only by diffusion and not advection (Figure 5.9): if the advection had occurred in water flowing in any particular direction, the resulting ^{226}Ra distribution would be asymmetrical. The concentration profile for ^{226}Ra was used to

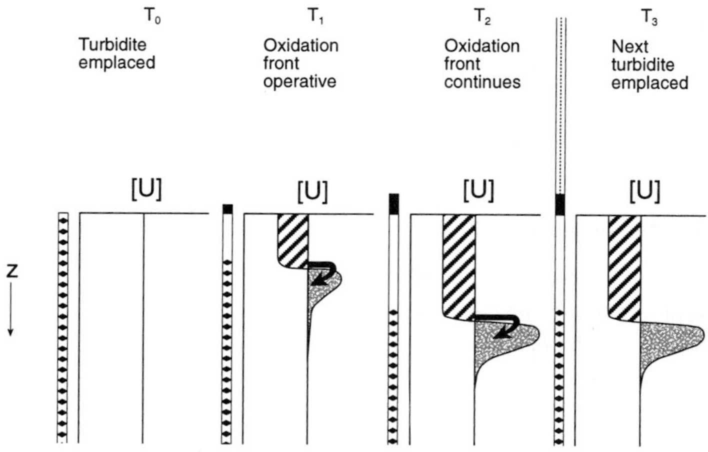

Figure 5.8: Formation of uranium-rich profiles in turbidite sequences. At T_0 a new turbidite is emplaced with uranium phases homogeneously dispersed throughout. At T_1 oxygenated seawater infiltrates the turbidite creating a redox front which oxidises and mobilises the uranium. The redox front moves downwards and entrains the uranium in the sediment as it progresses. At the same time a small quantity of pelagic sediment (the black bar) accumulates above the turbidite. At T_2 the processes continues, more seawater infiltrates the sediment and the redox front continues downwards and its uranium concentration progressively increases. Further pelagic sediment accumulates. At T_3 the next turbidite is emplaced (the dotted bar) which cuts off the supply of seawater to the redox front causing it to stop. The uranium accumulation at the redox front marks its location. A new redox front begins to form at the top of the newly emplaced turbidite and the sequence is repeated. From Colley and Thomson (1991).

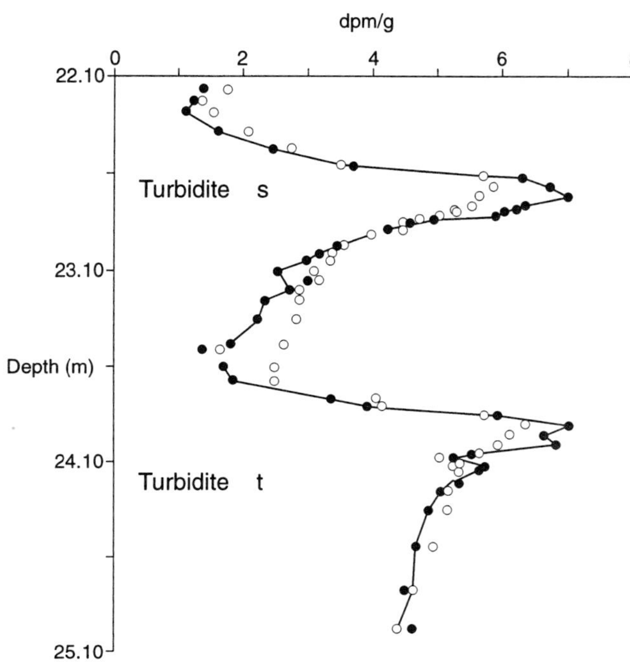

Figure 5.9: Profiles of ^{230}Th (filled circles) and ^{226}Ra (open circles) through two turbidites. The ^{230}Th data points are joined together to illustrate the profiles. The fossil redox fronts are marked by the peak isotopic concentrations. The ^{230}Th has been immobile since the redox front stopped moving but its daughter isotope (^{226}Ra) has migrated away from the redox front as shown by the lower concentrations of ^{226}Ra at the redox fronts and higher concentrations between fronts than its parent isotope. From Colley and Thomson (1991).

calculate effective diffusion coefficients of between 6×10^{-13} to 1×10^{-13} m^2/s.

Perhaps the best known natural analogue study of radionuclide transport within sediments is that performed at Loch Lomond in the southern Highlands of Scotland (MacKenzie et al., 1983; 1984; 1989b; 1990b; Hooker et al., 1985; Falck and Hooker, 1990). Although this study involved the examination of only a few sediment cores from the loch (lake), much valuable, quantitative information has been gained from this study. At present Loch Lomond is freshwater and landlocked but a marine (Flandrian) transgression from the Firth of Clyde resulted in incursion of seawater into the loch. This event is clearly recorded in the sediment by a one metre thick band of marine detritus which is overlain and underlain by freshwater sediments and, consequently, forms a geochemical discontinuity (Figure 5.10). All the sediments are clay-rich, containing up to 80 % clay in some horizons. The porewater chemistry also identifies the marine sediments by higher concentrations of chloride, bromide and iodide. Migration of these halogens into the freshwater sediments and porewaters, above and below the marine band, records a history of diffusive transport.

A range of concentration profiles was constructed for various elements, from analysis of Loch Lomond sediment core samples, and these were used to determine processes of mobility and retardation. The sedimentation of the marine band was found to have occurred between 6900 and 5400 years ago, using ^{14}C analysis, palaeomagnetic and palynological studies. Fixation of iodine, bromine, uranium and ^{226}Ra was clearly identifiable within the marine layer, a feature which correlated with the presence of organic carbon. Another core was taken from the Dubh Loch which lies just to the east of Loch Lomond, but at a higher elevation, and which did not experience the marine transgression. This second core has shown that the effects observed in the Loch Lomond cores are indeed due to the marine transgression and not due to a possible change in composition of eroded material supplied from the surrounding catchment. Comparative data are also available from Loch Long, which parallels Loch Lomond but has always been marine.

Unlike the study by Colley and Thomson (1991) in marine turbidites, no mobile redox front was established in the Loch Lomond sediments. The pore water concentrations of bromine and, to a lesser extent, iodine decrease with distance from

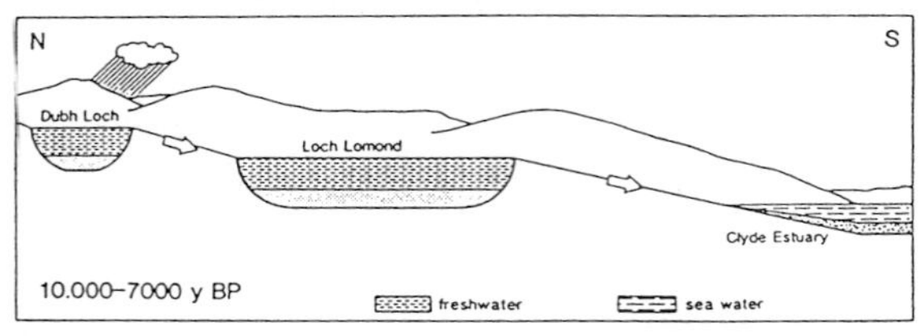

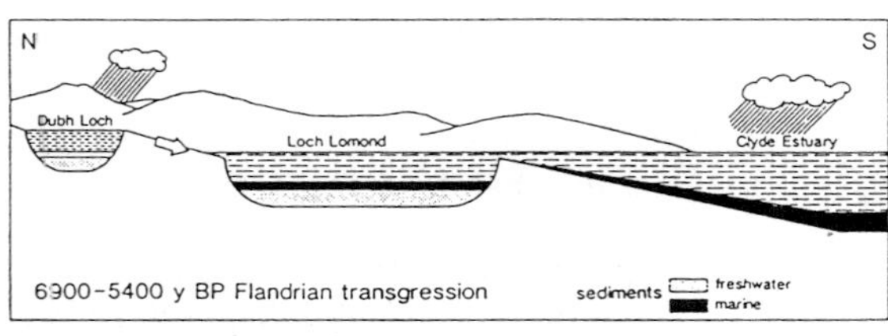

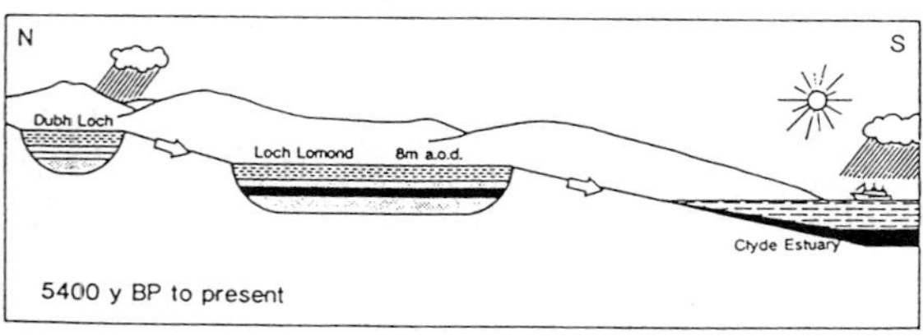

Figure 5.10: Representation of the formation of the marine band within Loch Lomond during the Flandrian transgression 6900 to 5400 years ago. From McKinley (1989).

the marine sediments, as shown in Figure 5.11. The bromine concentration profile was modelled according to simple diffusion with reversible sorption, and assuming no component of advective transport. When an initial bromine concentration of 60 ppm was assumed, the model produced an apparent diffusivity of 8×10^{-11} m^2/s. Batch sorption experiments were performed on the core material for iodine and bromine and it was discovered that, in both cases, the calculated apparent diffusivities were an order of magnitude less than measured values, possibly due to sample perturbation in the laboratory. This suggests that, although the laboratory studies are conservative, the calculated diffusivities are probably more realistic. In addition, the laboratory experiments indicated that some sorption processes were apparently irreversible for several radionuclides. It follows that the assumption made in many models of migration, that sorption is instantaneous and reversible, is inaccurate and the underlying kinetics of the process may need to be investigated further.

The results from these two studies are most useful when applied to diffusion-dominated argillaceous repository environments. Clearly, this would include transport within a bentonite buffer material, but the information would also be appropriate to matrix diffusion in a fractured clay host rock where advective flow occurs.

A number of other studies on clay-rich environments may provide semi-quantitative information. The fossil forest at Dunarobba in Italy (see Box 9) has some potential, but has not been studied effectively to date. The preservation of the cellulosic material in the trees, together with their degradation products, would allow the study of complexation of trace elements from the clays with organics from the wood, and their migration in an essentially undisturbed environment over a well quantified time period of around one million years.

At first sight, the clay haloes around uranium orebodies might be considered to provide a suitable means of studying transport in clay buffers around waste containers. However, there are inherent problems in the interpretation of data from these sites which may make them less useful. At Oklo, for example, many of the radionuclides leached from the uraninite were subsequently retained in the clay envelope around the reactor zones, including niobium, rhodium, ruthenium, tellurium and tin. Unfortunately, none of these elements is particularly relevant to performance assessment. In addition, because the clay envelope formed as a result of the high temperatures prevalent whilst the reactor zone was active, evolution of fission products, their mobilisation, and the formation of the clay may be contemporaneous processes, which makes interpretation difficult. In addition, present-day oxidising conditions around some of the studied reactor zones may explain the apparent loss of some other elements from the clay in these cases.

Similarly, at Cigar Lake (see Box 5), the clay halo around the ore represents the alteration products formed by hydrothermal dissolution and breakdown of the host sandstones. These residual clays, characterised mainly by illite and accumulations of accessory minerals, form an efficient hydraulic barrier to groundwater movement in and around the ore body. Radionuclide movement in the clay halo, when observed, is mainly diffusive and local, up to a few tens of centimetres, in extent (Cramer and Smellie, 1994b).

Other possible analogues might include the transport of trace elements from igneous intrusions emplaced in or adjacent to bentonite bodies, but in this situation high temperature processes, including the effects of enhanced illitisation, dominate the behaviour of the system and make them of very tenuous relevance to the repository environment.

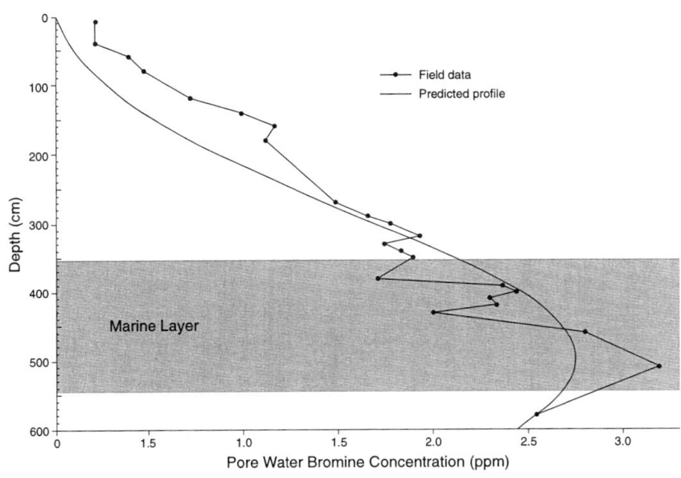

Figure 5.11: Measured and modelled (assuming an initial bromine concentration of 10 ppm) bromine profiles through the marine layer in Loch Lomond. The marine layer was laid down between 6 900 and 5 400 years ago. Since this time, bromine has migrated upwards and downwards into the freshwater sediments above and below the marine layer. The modelling results suggest an apparent diffusivity of 8×10^{-11} m^2/s. From Hooker et al. (1985).

Transport and retardation within volcanic ash deposits

The proposed US repository at Yucca Mountain is located in unsaturated volcanic ash deposits (tuffs), as described in Section 2.3.1. The transport behaviour for porewaters and dissolved radionuclides in these unsaturated rocks is expected to occur by processes and at rates that are quite different to those that will occur in saturated fractured rocks or in saturated argillaceous sediments.

One interesting study to investigate elemental migration in unsaturated volcanic ash was undertaken at Santorini (Murphy and Pearcy, 1994). At Santorini, a large volcanic eruption in 1628 BC covered the island with a thick deposit of silicic volcanic ash. This ash inundated the local Minoan settlements causing them to be abandoned. Recent archaeological excavations at Akrotiri on the island have uncovered artefacts made of many different materials in contact with the ash deposit. Various metal artefacts, especially bronzes, in the upper unsaturated layers were considered analogous to metal canisters in the Yucca Mountain repository.

The bronze artefacts were found to be fairly heavily corroded and investigation of the volcanic ash around them revealed a metallic contaminant plume. Since the boundary conditions for system initiation are well constrained, in terms of location, timescale, chemistry etc., this archaeological system provided a good opportunity to test the performance assessment models for trace metal migration in tuffs that would be applied to the Yucca Mountain repository. Comparison between measured contaminant distribution and model predictions gave some degree of success (Murphy and Pearcy, 1996; Murphy et al., 1997), raising confidence in the performance assessment models. However, some differences between the ash at Santorini and the tuffs at Yucca Mountain do exist, particularly in terms of porosity and permeability, and consequently interpretation of this analogue must be done with care.

Radionuclide mobility at the microscopic scale in microfractures in grains and along grain boundaries has been investigated in the volcanic sediments at the Tono analogue site (see Box 18). This work is discussed in Section 5.3.

Transport and retardation within evaporites

Anhydrite and other evaporite formations are generally considered to be effectively dry, containing only isolated pockets of brines. There is some debate as to whether certain evaporite formations may contain a mobile intergranular film of fluid which may allow radionuclide migration in solution. Otherwise, the only transport mechanism available for radionuclides would be solid-state diffusion, which is so slow that it poses no safety risk for a repository. This is demonstrated by Wollenberg et al. (1984), who discuss natural analogue investigations of locations where salt formations have been intruded by crystalline rock. In all the cases examined, the migration of uranium and thorium from the igneous rock to the salt is minimal after tens to hundreds of millions of years.

If brine accumulations containing dissolved radionuclides were able to migrate out of a repository host formation, or radionuclides were able to diffuse in an intercrystalline liquid phase into brines in surrounding formations, then it would be important to know how such brines would interact with other rock types, and the consequent fate of radionuclides within them.

A novel natural analogue-type investigation to understand the possible transport, precipitation and retardation behaviour of radionuclides in salt formations was outlined by van Luik (1987). The idea is quantitatively to study the fate of radionuclides, such as uranium and thorium, during evaporite crystal formation at the brine surface, and diagenesis at the bottoms, of saturated hypersaline lakes. Such lake environments occur throughout the world and have different geochemistries and input and evaporation rates etc. There is already a large volume of literature (e.g. Bell, 1956; Thurber, 1965; Simpson et al., 1984) that discusses the fate of radionuclides in hypersaline lakes. Although this type of study may provide useful information for scenarios involving mobile brine pockets in evaporites, or the dissolution and transport of contaminated zones of a salt repository host rock, it is not totally clear how such data could be used, and the concept requires further development.

Transport and retardation at the geosphere-biosphere interface

Although it is possible to apply natural analogue methodology to radionuclide transport in the biosphere, this is such a vast area of research that it cannot be adequately covered in this report. Nonetheless, it is worth considering the interface between the geosphere and the biosphere because this is a critical zone for controlling radionuclide releases to the surface environment, as discussed in Section 1.5.4.

By convention, the geosphere-biosphere interface is defined as the zone including the upper, weathered rock horizon and soil cover to the point at which radionuclides are taken up by the root system of plants or are incorporated into surface waters. Within this zone, there are concentrations of naturally occurring radionuclides which may be considered as chemical analogues and used as tracers for the possible movement of radionuclides from a repository. However, other useful information may be gained from investigating the behaviour of anthropogenic radionuclides dispersed into the environment either deliberately (e.g. from aerial, surface or subsurface atomic bomb tests or discharges from nuclear establishments) or accidentally (e.g. from the Chernobyl reactor explosion or the Windscale fire).

An example of the use of naturally occurring radionuclides to show the transport of radionuclides in surface sediments and their uptake by organic materials and peat is that of Landström and Sundbland (1986). In this study, thorium, uranium, radium and ^{137}Cs concentrations were determined. It was discovered that thorium and uranium were enriched in the organic material in peat bogs and peat horizons in soils, whilst the daughter radium was preferentially taken up into the roots of plants. In the strict sense, this is not a natural analogue study because no information has been gained on the mechanisms of transport or retardation of radionuclides in peat bogs but this study could easily be extended to investigate these matters. An extended study may prove be extremely useful for modelling radionuclide migration in organic material-rich soil horizons.

Several studies have examined the fate of radionuclides released from the Chernobyl explosion in April 1986. Gustafsson et al. (1987) monitored the concentrations of several radionuclides at the Gideå and Finnsjön study sites in Sweden, and established their transport behaviour over the twelve months following the accident. Radionuclide transport was indicated in soil profiles, groundwater, rock fissures and in surface waters. Radionuclides had been taken up by many different species of vegetation. Five months after the radionuclides were deposited, some had migrated to depths below 20 cm, indicating that transport was not diffusion controlled. Measurements in drill cores at this time showed that radionuclides had penetrated 2 to 3 cm into vertical fractures. At nine months after deposition, ^{106}Ru was detected in artesian wells around 100 m deep. Monitoring continued for five years after deposition, and radionuclide concentrations were recorded from packed-off sections of a deep artesian well. Transport through the rock took place through fractures and the total distance travelled was approximately 300 m.

The extent to which this type of information (about essentially the reverse situation to releases from depth) can be used to aid description of processes in the dynamic, high energy zone of the geosphere-biosphere interface is not clear. However, studies within the IAEA sponsored BIOMOVS and BIOMASS programmes have been aimed at evaluating the potential uses of such analogues in modelling the biosphere.

Measurement of in situ distribution coefficients

Most performance assessments and process models require quantitative expressions of the sorptive capacities of rock. This is usually expressed in terms of the distribution of an element between solid (sorbed) and liquid (groundwater) phases as a distribution coefficient (K_d). There are now many thousands of published laboratory K_d determinations on a wide range of water-rock pairs (see Sibley and Myttenaere, 1986, for example). Unfortunately, most may be criticised for their inadequate representation of the repository environment. For example, many used crushed fresh whole rock even though it has been demonstrated that it is mainly the fracture coating or surface alteration minerals which control sorption in a fractured rock.

Other problems include those of ensuring that in situ conditions are maintained, or that the water used in experiments is in equilibrium with the rock and representative of real groundwaters in the environment sampled, and the scale of the experiment. To circumvent these types of problem, some recent studies have attempted to measure K_d values in the field: i.e. to measure in situ K_d values. A recent discussion of the issue was given by Murphy (1996). The following three methods for measuring in situ K_d values are commonly used and are sometimes claimed to be applicable to a variety of different situations.

The first method uses *radionuclide concentrations in rock-water pairs*. This method basically applies laboratory batch style studies to the field situation: radionuclide concentrations are measured in a rock, taking care to measure only the sorbed concentration, and a groundwater from an appropriate location. Data obtained from such experiments can be meaningful if the water-rock pair are taken from the same location, the solid phase measurements are taken from material which was available for exchange with the water, precipitated phases are identified and discounted and that the entire system was in equilibrium.

Unfortunately, it is extremely difficult to prove that any system meets all these requirements and, due to this, most in situ K_d values must be considered qualitative at best (see McKinley and Alexander, 1992b for full discussion). The principal problem with this method for determining in situ K_d values is that the measured solid phase radionuclide concentration does not systematically distinguish between sorption on surfaces available for exchange and precipitation. In some cases, the measured concentrations do not even distinguish between surface deposits of radionuclides and those present in, and coeval with, the bulk rock mineralogy.

In an attempt to solve this problem, a number of studies have tried using selective leachants to distinguish radionuclide species held by sorption from those held in the crystalline lattice of minerals (Lowson et al., 1986; Ivanovich et al., 1988; Nightingale, 1988; Yanase and Isobe, 1991). Phase selective extractions are widely used in geochemical exploration for analysing elemental distributions in ore samples. The issue of selective leaching is discussed in Chao (1984) and Martin et al. (1987). There is abundant evidence that many leachants are non-specific; i.e. the relationship between the extraction solutions and the targeted mineral phases is not precisely identified (for discussion see Alexander and McKinley, 1992). Some effort has been put into testing the specificity of the leachants, particularly within the Alligators Rivers study (see Box 15). Yanase and Isobe (1991), for example, have used optical and X-ray diffraction identification of the mineral phases at each stage to ensure that specificity is maintained.

Whilst this is clearly a more rigorous approach to measuring in situ K_d values, two questions remain unresolved. First, the abundant evidence (see, for example, Sholkovitz, 1989) of readsorption during leaching and redistribution of the very elements of interest onto different minerals during the extraction process, thus distorting the apparent radionuclide distributions, has yet to be adequately addressed. Second, this method does not take into account the possibility of radionuclides existing in the colloidal state. If they do, then the basic assumption of a distribution between only the solid and solution states is unfounded and the resulting K_d value is invalid.

The second method uses *radionuclide concentration profiles*. This method is based on the principle that the redistribution of a well-defined and datable geochemical anomaly can be interpreted in terms of retardation, if the water transport velocity is known. The simplest case would be where no advective transport occurs and radionuclide migration occurs by diffusion only. For example, the redistribution of various elements from the marine band in Loch Lomond (discussed earlier) has been analysed to derive 'best fit' retardation factors (e.g. Hooker et al., 1985; Falck and Hooker, 1990). In the very simplest case, such studies can directly yield a retardation factor from the ratio of peak transit times of a sorbing and a 'non-sorbing' tracer. However, this only holds true under the assumption of fast, concentration independent, reversible sorption (i.e. K_d type). If the active sorption mechanism is more complex, the whole concept of deriving a K_d in such systems is inappropriate.

The third method uses *isotopic ratios*. This method for determining in situ K_d values from isotope ratios is based on the work of Krishnaswami et al. (1982) on closed aquifers in Connecticut in which the K_d values are calculated from the isotopic ratios of radionuclides in the groundwater and rock. There are a number of reasons why this approach cannot be applied directly to in situ K_d determinations within an environment analogous to that of a repository. Principally, this method allows for radionuclide input to solution by dissolution and recoil, but removal only by decay and sorption. Such assumptions are not generally applicable to a groundwater system because, in low flow situations or in a state of disequilibrium, addition by mineral dissolution and removal by precipitation or co-precipitation are also likely to occur. In addition, this approach contains numerous other unproven assumptions, many of which are criticised as being unrealistic by McKinley and Alexander (1993a, 1993b).

Other, alternative isotopic methods of in situ K_d measurement exist, one of the simplest being the measurement of the solution phase concentrations of members of the natural decay series chains (e.g. Laul et al., 1985, 1986; Laul and Smith, 1988). Here, a radionuclide retardation factor is defined as the reciprocal of the activity ratio of that radionuclide relative to ^{226}Ra, which is assumed to be non-sorbing in the particular brine system studied. Given the complex geochemistry of the natural decay series, along with signs of solubility limits for uranium and thorium in this system, this approach appears to be over-simplistic (see McKinley and Alexander, 1992b, for details).

In summary, it would appear that in situ K_d values cannot presently be regarded as transferable between natural and repository systems or directly applicable as conservative parameters for input to performance assessment models. Nonetheless, increasing improvements to analytical techniques and procedures provide insights to the rates and mechanisms for water-rock interaction and radionuclide transport (Murphy, 1996). It is hoped that future developments in this field will allow true analogue-derived in situ K_d values to be measured.

Conclusions

The uptake of radionuclides and other trace elements on fracture surfaces in crystalline rock is a phenomenon frequently observed in the field. The principal phases taking up elements from solution are iron oxyhydroxides, organic coatings and clay minerals, although many other fracture surface and bulk rock minerals have sorptive capacity. It is often very difficult to distinguish quantitatively between sorption and precipitation processes. Many studies resort to laboratory experiments to determine distribution coefficients (K_ds) for the purposes of modelling transport in the analogue system. Studies of natural decay series disequilibria can give useful information on the timescales of uranium mobilisation and deposition in some systems. The direct transfer of K_d data derived from natural analogue studies to performance assessments is not yet a practical proposition. However, the analogue data can be used in a semi-quantitative way to provide a 'reality check' to the magnitude of the values used in performance assessment, and to provide qualitative information on the overall effects of transport and retardation processes and to assist in identifying the processes which are most active in any given environment.

Elemental retardation during diffusive transport has been clearly identified in studies on unconsolidated sediments, and some quantitative diffusivity data have been obtained. Such studies are reasonable natural analogues of radionuclide behaviour in argillaceous rocks and clay backfills, although the analogy is not complete.

The issue of irreversible sorption of radionuclides over long time periods, a potentially enormously beneficial process in performance assessment terms which currently cannot be given any credit,

Box 15: The Alligator Rivers uranium orebody

The Alligator Rivers natural analogue project is an investigation of a secondary enriched uranium deposit in the Northern Territory of Australia. Four large uranium deposits are located in the region; Nabarlek, Jabiluka, Ranger and Koongarra, of which the latter is the focus of the project.

The Koongarra uranium deposit was discovered in the early 1970s by airborne geophysical measurements and over forty boreholes were sunk to characterise the orebody. The deposit was never exploited due to political restrictions imposed by the Australian government on uranium mining and hence is relatively undisturbed.

Figure B15.1: Photograph of the Alligator Rivers study site. Open boreholes can be seen in the foreground. Although dry in this photograph, the terrain becomes difficult in the wet season when torrential downpours occur. Photograph from Snelling (1992).

These uranium deposits are the most significant mineralisation in the Pine Creek Geosyncline which comprises about 14 km thickness of Lower Proterozoic metasediments (carbonaceous pelites, psammites and carbonates) together with interlayered tuffs which were deformed and metamorphosed 1800 million years ago. These Lower Proterozoic rocks rest on late Archean granites and are themselves overlain by the Kombolgie sandstone which is of Middle Proterozoic age. The uranium deposits lie at the base of the Cahill Formation schists and are located in extensively chloritised zones and adjacent to massive dolomites.

The Koongarra or body lies in two distinct parts separated by a barren zone. They are composed of uraninite and pitchblende-bearing veins within a zone of steeply dipping, sheared quartz-chlorite schists and a fault which brings the ore body in contact with the Kombolgie sandstone. The primary ore at depth is being leached by groundwater to form a secondary mineralisation that extends from the surface to the base of the weathered zone at about 30 m.

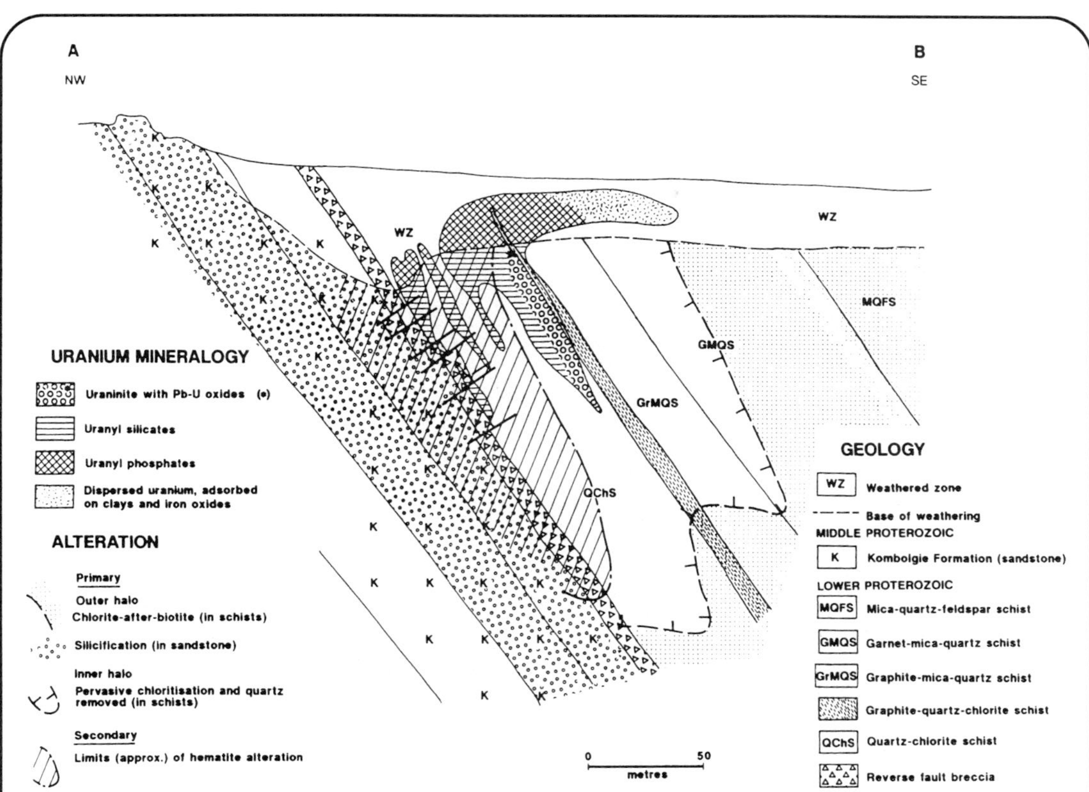

Figure B15.2: Simplified cross-section through the Koongarra number 12 orebody showing the distribution of the primary ore, secondary uranyl minerals and the dispersion fan which extends some 80 m downslope of the ore. From Duerden (1992).

This leaching process has resulted in the orebody, and the region above it, forming four zones: the primary ore of uraninite and pitchblende; a uranium-silicate zone formed by in situ alteration of the primary ore; a zone of secondary uranyl phosphate minerals which are currently being leached by the groundwater; and a shallow dispersed uranium zone with the uranium in association with clays and iron oxyhydroxides, which extends downslope some 80 m away from the ore.

Geochemical data show that the areas of greatest leaching are near the upstream zone of the orebody. Rapid accumulation occurs near the edge of the dispersion fan. During weathering, uranium distribution follows the formation of various iron phases, with iron and manganese controlling the fixation of various radionuclides. Radionuclide mobilisation and transport are processes are controlled by the climate. The upper zone is particularly seasonally affected, fluctuating from saturated in the rainy season to unsaturated in the dry season. Below 15 m depth, the rock is permanently water saturated.

The groundwaters are all oxidising (Eh of 100 to 400 mV) and mildly acidic. Therefore, the conditions at Koongarra are very unlike those found in most deep, water-saturated repository near-fields, which are chemically reducing, making it a poor analogue for a repository near-field. However, the well-characterised source makes the site ideal for investigating radionuclide migration behaviour (processes, rates and distributions) and to test geochemical codes and databases.

Figure B15.3: Photograph of the Kombolgie sandstone cliffs of Mount Brockman which lies close to the Koongarra No. 1 orebody. These sandstones overlie the mineralised Cahill Schists. Photograph from Snelling (1992).

The orebody is more relevant to proposed US Yucca Mountain repository design which is oxidising and unsaturated (Section 2.3.1), and it is probable that some of the information from the Alligator Rivers project could be applicable to the Yucca Mountain situation.

The natural analogue study had a number of broad objectives, including:

- to investigate the processes leading to the decomposition and leaching of the primary ore (Section 4.2),
- to investigate the processes of radionuclide transport and retardation (Section 5.2), and
- to investigate the effect of colloids on radionuclide mobility (Section 5.6).

As part of this programme of investigations, much effort was also put into improving techniques for in situ Kd measurements and also into testing thermodynamic solubility and speciation codes.

The results from the Alligator Rivers natural analogue project have been widely published (e.g. von Maravic and Smellie, 1993) and discussed in detail in a series of 16 project reports, the first of which is a summary of findings (Duerden et al., 1992).

can probably only be resolved by natural analogue studies.

Very few natural analogue studies have addressed the issue of radionuclide transport in anhydrite and other evaporite formations. The information that is available indicates that transport will probably be diffusive and so may be very slow. It is possible that natural analogues could help provide information on the behaviour of radionuclides in mobile brines, particularly in mixing zones with other formation waters, but this has not been pursued.

Radionuclide behaviour at the geosphere-biosphere interface can be investigated by natural analogue studies on the migration of radionuclides released from underground bomb tests and accidents such as Chernobyl. Quantitative data have been obtained from this type of investigation, but further work is necessary to define how such data can be applied in modelling this complex zone.

In situ K_d determinations are potentially extremely valuable in quantifying the sorptive capacity of different mineral assemblages. However, the process of obtaining these measurements is fraught with technical difficulties and interpretative ambiguities. Despite these problems, every effort should be made to try to improve the techniques before such investigations form part of future natural analogue studies. Most such data acquired to date must be treated as only qualitative due to various inherent problems with the different techniques but, if these problems can be rectified, the technique may yet have value for testing.

5.3 Matrix diffusion

The term matrix diffusion is applied to the process by which solutes, carried in groundwater flowing in fractures, penetrate the surrounding rock mass by diffusive processes. Diffusion into the rock occurs in a connected system of pores and microfractures, while diffusion through the solid phase is presumed to be insignificant by comparison (see the review of Valkiainen, 1992). The importance of matrix diffusion in the context of contaminant transport is that it greatly enlarges the area of rock surface in contact with solutes and can delay releases of contaminants (Neretnieks, 1980; Grisak and Pickens, 1980; Rasmuson and Neretnieks, 1981; Hadermann and Roesel, 1985).

The matrix diffusion theory proposes that dissolved radionuclides will diffuse from a water-filled fracture, through any porous fracture coating layer and into the rock matrix. If these radionuclides are reactive, they will sorb onto the inner surfaces of these pores or else will remain dissolved within the immobile pore water, as indicated in Figure 5.12. This process can be envisaged as an extreme case of a dual porosity medium, in which advective flow occurs entirely within the fracture system, the primary porosity, whilst all solute transport in the bulk rock, the secondary porosity, takes place by diffusion (for example, Barenblatt et al., 1960; Grisak and Pickens, 1980).

In hard rocks, the secondary porosity of the matrix can be very small (generally about 0.1 to 1 %) and, if the groundwater is flowing rapidly, the impact of matrix diffusion on contaminant transport will thus be limited because the rate of advection in the fracture is much greater than the rate of diffusion into the rock mass. However, for the slower flow paths likely in a repository host rock, matrix diffusion may result in a significant delay in release and a reduction in the maximum concentration of radionuclides in the groundwater entering the biosphere. This process is particularly significant if the resulting transport time to the biosphere is greater than the half-life of the radionuclide, since the total release can be reduced by several orders of magnitude. In addition, pulsed releases can be spread over longer time periods, thereby decreasing release concentrations by a process of temporal dilution. For non-sorbed radionuclides, this process

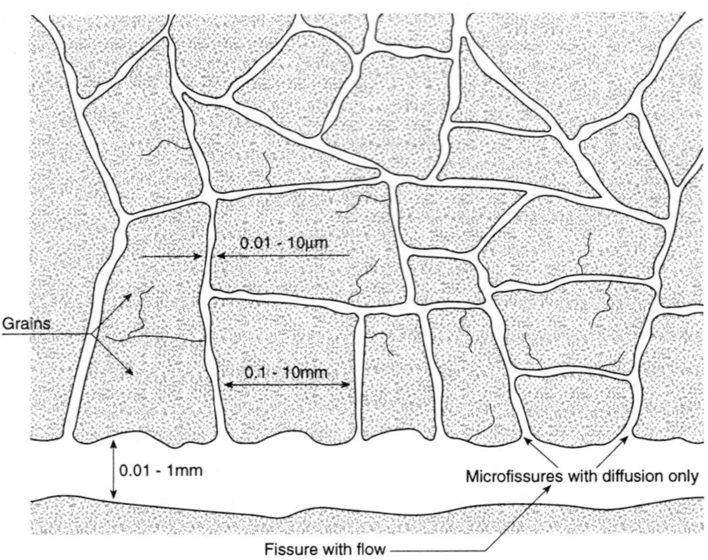

Figure 5.12: Representation of matrix diffusion theory in fractured crystalline rock. Advective transport of radionuclides occurs in water conducting fractures. Some radionuclides will enter the rock matrix via the microfractures that run between and through the grains and where transport occurs by diffusion only. The depth of rock which is available to the diffusing radionuclides will vary from rock to rock but is often between 10 and 40 mm for granites. From Birgersson and Neretnieks (1983).

represents an important retardation mechanism, because these species would otherwise be transported at the advection rate of the groundwater.

The theoretical basis for matrix diffusion is well established but there are, however, different views regarding the volume of rock that would be available for matrix diffusion, i.e. the extent of connected microporosity. In the early Swedish KBS-3 performance assessment (KBS, 1983), it was assumed that the entire volume of rock was available for matrix diffusion. In contrast, Glueckauf (1980) assumed that matrix diffusion is limited only to dead-end pores, whilst the early Swiss Project Gewähr performance assessment (Nagra, 1985) assumed that matrix diffusion was limited to a 1 mm thick, microfractured damage zone. Considerable laboratory and field work has thus been focussed on determining the depth of interconnected porosity adjacent to fractures.

Laboratory-scale studies on a range of hard rocks appear to support the matrix diffusion theory with evidence from experiments on limestone (e.g. Garrels et al., 1949), sandstone (e.g. Klinkenberg, 1951) and crystalline rocks (e.g. Skagius and Neretnieks, 1983, 1986; Bradbury and Stephen, 1986). One significant problem, however, is that laboratory experiments are conducted, out of necessity, on core samples which have probably been disturbed by the drilling and sub-coring processes. Disruption to the rock may occur only on a microscopic scale but, as this is the scale of interest to matrix diffusion, this suggests that the laboratory results must be treated with caution (Ohlsson and Neretnieks, 1995). In general, all of the changes induced by sampling tend to cause over-estimation of the rock diffusivity, leading to an overestimation of matrix diffusion (McKinley, 1989) which is, in turn, non-conservative in the performance assessment sense, because it leads to an apparently greater degree of radionuclide immobilisation in the far-field.

It should be noted that a petrological description of bulk matrix porosity does not indicate the actual porosity available to matrix diffusion because not all of the rock porosity will be interconnected. Furthermore, not all of the porosity seen in hand specimens may exist when the rock is in situ, due to compression and closure of pores at depth. To investigate this issue, a detailed attempt to identify pre-existing, original fractures and pores, and those induced by sampling was made by Chernis (1984) and by Mori and Alexander (2000) but the major problem remains that it is difficult to identify unambiguously the generation to which pores and fractures belong.

Attempts have also been made to verify matrix diffusion by more complex experiments which, by confining large rock samples under high pressures (e.g. Brace et al., 1968; Bischoff et al., 1987; Drew and Vandergraaf, 1989) hoped to recreate the in situ conditions more precisely. There are still many problems with such experiments: the low hydraulic conductivity means that unrealistically high pressure gradients have to be applied to the infiltrating fluids to produce breakthrough of radionuclides within reasonable time scales, plus there is no way to guarantee that the re-established confining pressures will recreate anything like the original pore geometry. In fact, in two studies which have attempted to quantify the degree of disturbance, it was estimated that laboratory produced data probably overestimated in situ diffusion coefficients by a factor of two to five (Skagius, 1986) and that in situ porosities were up to an order of magnitude lower than those measured in the laboratory (Mori and Alexander, 2000).

An alternative method to assess the extent of matrix diffusion is to conduct the type of in situ tracer tests carried out in the Stripa and Grimsel test sites. Generally, these are of too short a duration to allow matrix diffusion effects to be observed (SKI, 1991; Frick et al., 1992), and the 'proof' depends on a model interpretation of tracer breakthrough curves, not on direct measurement of tracer penetration into the rock matrix. Unfortunately, other processes such as dispersion and sorption could equally produce similar 'model fits' to the breakthrough data and, as such, the results cannot be uniquely attributed to matrix diffusion. Indeed, similar criticisms may be applied to the vast majority of laboratory experiments. In only a few cases has an attempt been made to measure directly the actual penetration profiles (for example, Ittner et al., 1988; Alexander et al., 2000a) and it is usually shown that the calculated diffusion coefficients measured on laboratory samples are several orders of magnitude greater than those measured in situ in the rock mass.

A detailed example of a field experiment which might reasonably be claimed to have indicated matrix diffusion on the field scale is the long-term caesium migration test conducted at the Grimsel Test Site (Smith et al., 2000; Alexander et al., 2000a). This test was so well constrained by the elimination of all unknown parameters, apart from matrix diffusion, that little doubt can exist as to the interpretation of the results. It is, however, worth noting that this particular experiment was the culmination of ten years of field, laboratory and modelling effort and is also, obviously, site-specific.

Measuring matrix diffusion by laboratory and field experiments is clearly problematic and, thus, several analogue studies have attempted to measure real matrix diffusion parameters that could be used in performance assessment: see the review by Neretnieks (1996) for details.

One analogue approach to the assessment of matrix diffusion is to carry out studies on natural tracers in rock, although care must be taken to exclude sampling-induced artefacts (Mazurek et al., 1992a). Here, perturbations to the average chemical composition of the bulk rock are used to indicate the presence of previous rock-water interaction. The distribution of a range of indicator elements is examined along a profile, usually away from a known (or suspected) water-conducting fracture in the bulk rock. In an ideal case, matrix diffusion would result in an smooth concentration profile away from the fracture but heterogeneity in the rock and the effects of different geological events on the system will complicate this picture.

Usually, the natural decay series, rare-earth elements and a suite of other redox-sensitive elements are studied with the intention of covering a range of geochemical behaviour, thus increasing the chances of obtaining evidence of rock-water interaction. In the case of the natural decay series, for example, uranium is more soluble than thorium and ^{234}U is more mobile than ^{238}U. Uranium is also redox-sensitive, while radium is more likely to be involved in exchange reactions or in solid-solution with a variety of carbonate phases. This range of geochemical behaviour is supplemented by the large range in the half-lives of the various members of the series (e.g. half life for ^{234}U is 2.5×10^5 years and for ^{226}Ra is 1.6×10^3 years), the combination of both features allowing a variety of types of rock-water interaction to be identified, along with some idea of the duration of any given event.

Two geological environments have been identified as suitable for this type of study. The first is where a water-conducting fracture zone transports measurable concentrations of radionuclides, or leaches them from the surrounding rock, and the second is in the marginal zones around hydrothermal vein deposits where radionuclides migrate from the vein minerals into the surrounding host rock. The first of these environments is the most relevant to performance assessment because of the similarity in the temperatures of the natural and repository systems. In contrast, the hydrothermal systems generally have inappropriate high temperatures and inapplicable rock types or water geochemistry, making interpretation of the results very difficult.

Other types of analogue system that have been used to examine matrix diffusion are where a rock has been flushed with new and chemically distinct waters (for example, where water changes from fresh to saline) and the penetration of the 'new' water into the rock can be measured. This situation can occur both in natural and anthropogenic systems.

Although the great majority of analogue studies which have investigated matrix diffusion have looked at fractured, crystalline rocks (igneous and metamorphic) a few have addressed matrix diffusion in sedimentary environments. This is important because many sedimentary rocks are brittle and may fracture, particularly shales as discussed in Section 2.4. As a consequence, sedimentary host rocks may exhibit dual porosity, allowing matrix diffusion to be an important retardation process. This suggestion has only recently been acknowledged as greater information on the physical structure of sedimentary rocks has become available.

The issues of most relevance to matrix diffusion that have been (or potentially could be) addressed in natural analogue studies are:

- depth and volume of interconnected porosity;
- bulk rock chemical buffering capacity;
- the extent of matrix diffusion in sedimentary formations; and
- estimation of diffusion coefficients.

These issues are discussed in the following sections.

Depth and volume of interconnected porosity

One interesting early study to measure the depth of interconnected porosity examined rock in contact with saline water, taking the saline water to be a tracer moving through the rock by matrix diffusion (Jefferies, 1987). In this investigation, granite blocks that had been submerged in seawater for 30 years, as part of a pier construction, were studied. Solute diffusion profiles of Cl^-, Br^-, F^- and SO_4^{2-} were determined in the granite and apparent diffusion coefficients and solute-accessible porosities calculated for Cl^- and Br^-. These anions were chosen because they are weakly sorbing or non-sorbing and are, therefore, most mobile. Laboratory investigations of the diffusion coefficients from these samples suggest that this parameter is enhanced by up to 200 times close to a fracture surface, due to mechanical disruption. This is, however, not a good example because the rock was drained before emplacement in the water and, therefore, capillary suction rather than matrix diffusion may explain these results. Certainly, in a study of uranium migration in intact granite, Baertschi et al. (1991) unambiguously identified capillary suction as the migration mechanism, rather than matrix diffusion.

In another study, Olin and Valkiainen (1990) measured the diffusion of saline waters out of granite into a water-conducting fracture. Owing to post-glacial uplift, the study site (Hästholmen Island) has risen out of the Baltic Sea, creating a layered groundwater regime with recent freshwater above the original, marine-derived saline water. The work was conducted on core samples of fresh, unfractured granite with saline porewaters from between water-conducting fractures carrying fresh water. In principle, it would be expected that the concentration gradient between the saline porewater and the freshwater in fractures would induce matrix diffusion towards the fractures, resulting in observable concentration profiles.

Here, as in the work of Jefferies (1987), the cores were immersed in distilled water and the rate of change of the concentration of chloride and sulphate in this water was observed. Calculations provided an apparent diffusion coefficient of 2×10^{-9} m^2/s, about an order of magnitude higher than measured in laboratory through-diffusion experiments. This discrepancy is not explained, but may be due to the presence of microfractures or the use of an inappropriate model to calculate the apparent diffusion coefficients (Valkiainen, 1992).

In a study of matrix diffusion in the vicinity of a mineralised vein, Pinto Coelho (1987) measured the depth to which percolating groundwaters had redistributed (leached) elements in the rock adjacent to the vein. A series of elements were studied, including uranium, thorium, lanthanum, neodymium, bromine, strontium and barium. It was generally found that the depth to which redistribution of the elements occurred was between 32 and 65 mm and this was interpreted to be a measure of the depth of matrix diffusion. Greater elemental redistribution occurred closer to the vein, and the depth of redistribution varied for each element; barium was affected to a depth of 32 mm, while uranium and zirconium were affected to 65 mm into the rock. All the major elements, measured as oxides, (except Fe_2O_3) were disturbed to a depth of 50 mm.

It is recommended that these be treated with caution because important boundary conditions to the system are unknown. Particularly, there is no information on the maximum fluid temperatures or the duration of fluid flow in the fracture. Fluid temperatures must have been significantly higher than would ever be reached in a repository to enable precipitation of quartz in the vein. It is likely that the higher fluid temperatures would have enhanced the diffusion penetration depth, and so these depths might be considered as maximum estimates but, without knowing the duration of the event, this must be speculative. As such, the results of this study should be considered as non-conservative.

Some of the most detailed natural analogue studies of matrix diffusion were performed on rock adjacent to well characterised, water-conducting fractures (e.g. Smellie et al., 1986a,b,c; Alexander et al., 1988, 1990a,b). In these studies, core samples from boreholes drilled perpendicular to fracture surfaces were analysed for a suite of elements (Figures 5.13 and 5.14). All of the core samples were crystalline rock; two from the Swiss underground test site at Grimsel, one from a Swiss borehole at Böttstein and the other from a Swedish borehole at Kråkemåla. Alexander et al. (1990a) concluded that interconnected porosity may extend 500 mm into the matrix of heavily altered granites, but would be more restricted in granites that had experienced less physical alteration or hydrothermal activity. With reference to the early Swiss Project Gewähr performance assessment, which assumed a limited connected pore depth, Alexander et al. (1990b) concluded that a depth of 50 mm would be more realistic, and 10 mm would

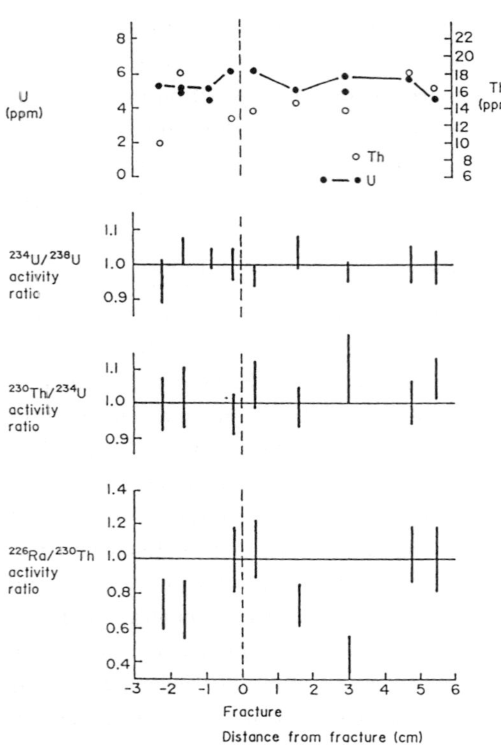

Figure 5.13: Natural decay series profiles measured in a core section perpendicular to a water conducting fracture from the Grimsel Test Site, Switzerland. Assuming immobility of thorium, the lowest profile indicates preferential mobility of ^{226}Ra in a region 30 to 40 mm deep either side of the fracture. This profile also suggests diffusion of ^{226}Ra towards the fracture over the last 1 600 to 8 000 years. From Alexander et al. (1990a).

be sufficiently conservative. In the Grimsel granites, Alexander et al. (1990b) saw no evidence that the connected porosity extends throughout the entire rock mass, as assumed in Swedish KBS-3 performance assessment (KBS, 1983). However, the data for the Kråkemåla (Smellie et al., 1986b) and the Böttstein (Alexander et al., 1988) cores indicate that physical and hydrothermal alteration are capable of opening the rock matrix to a significant depth, perhaps even enhancing advective transport well outside the obvious fracture zone (see also Alexander et al., 1990a; Bossart and Mazurek, 1991).

Mazurek et al. (1992a) report matrix diffusion depths of up to 40 mm in samples from the north Switzerland crystalline basement. Five cores from aplitic/pegmatitic fracture zones all displayed natural decay series disequilibria, indicating rock-water interactions at depth in the bulk rock matrix (c.f. also Alexander et al., 1988). It must be noted that the above studies assessed the depths penetrated by measuring disequilibria in the uranium-series radionuclides and, as a consequence, the actual penetration depths might be greater than those measured because the apparent penetration depth is inherently limited by the radionuclide half-lives. Alexander et al. (1990a) do not consider these penetration depths to be significantly influenced by physical disruption during sampling because they obtained consistency between elemental and radionuclide profiles and observed microporosity in thin sections.

Matrix diffusion was also investigated as part of the Palmottu natural analogue study (see Box 16). Radionuclide concentration profiles, from a fracture into the rock matrix, were examined from several different drill cores (Suksi and Ruskeeniemi, 1992). An indication of the radionuclide distribution was first obtained from alpha-autoradiography and then more detailed investigations proceeded with phase selective extractions. The alpha-autoradiography clearly showed the distribution of radionuclides decreasing from the fracture surface into the rock in an almost exponential fashion. Diffuse alpha could be identified some 80 mm into the rock matrix in one sample. The alpha-autoradiography also indicated that the radionuclides were associated with clay minerals associated with iron oxyhydroxides, formed from feldspar alteration, and biotite grains. Selective leaching experiments showed that most of the radionuclides that had diffused into the rock matrix were only loosely bound to the iron-rich phases. In some samples, activity correlated to microfractures indicating that fluid flow was channelled even at the microscopic

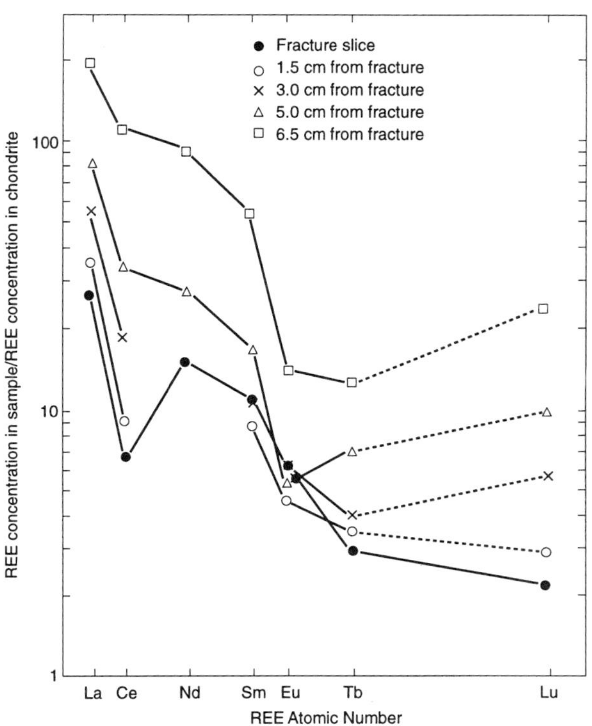

Figure 5.14: Chondrite-normalised rare earth element concentration profiles measured in core sections perpendicular to water conducting fractures from Böttstein, Switzerland; note there is a pegmatite adjacent to the fracture. These profiles clearly show that the fracture influences the concentration of many elements in the rock and, although the exact processes are not easily deduced from these data, such changes in elemental concentrations must be due to migration within the microfracture network associated with the main fracture. From Alexander et al. (1990a).

level, whilst, in another sample, diffusion was predominantly along grain boundaries. This strongly supports the conclusions noted above that physical disruption or chemical alteration of the rock around a fracture may control the mechanism of transport into the rock matrix.

Studies were also carried-out on fractures at Palmottu located only a few metres away from the mineralisation (Kumpulainen et al., 1992). A range of elemental concentration profiles were measured in rock adjacent to the fractures and these indicated significant alteration over a depth of around 25 mm in one sample but very little alteration in another, highlighting the spatial variability of matrix diffusion depths in natural systems that results from natural heterogeneity in the rock matrix structure and, possibly, the extent of groundwater flow in the fractures due to channelling.

A comprehensive investigation of matrix diffusion processes compared granite samples from a number of locations: the El Berrocal natural analogue study site, Spain; the Stripa test mine, Sweden; the Underground Research Laboratory (URL), Canada and the Grimsel Test Site, Switzerland (Montoto et al., 1991a,b; Heath, 1995). Each rock sample was taken from close to a hydraulically active fracture and then cut to provide a series of slices parallel to the fracture. All the slices were extensively characterised for mineralogy, chemistry and physical properties by a number of different techniques, including uranium series disequilibrium measurements. Concentrations of uranium, thorium and selected trace metals, the extent of iron oxidation, $^{234}U/^{238}U$ and $^{230}Th/^{234}U$ activity ratios were compared with porosity, water content, void index, longitudinal wave velocity and dry density for each slice. It has been found that, for all samples, the region of enhanced uranium mobility correlates with the zone of microstructural alteration in the rock adjacent to the fractures. This zone of microstructural alteration often penetrates deeper into the rock than does the region of enhanced uranium mobility but, in any case, is itself confined to a zone extending only a few tens of millimetres from the fracture. The actual depth of enhanced uranium mobility varies from site to site and are: 35 mm in the shallow El Berrocal granite; 80 mm in the deep El Berrocal granite; 25 mm in the Stripa granite; 50 mm in the altered URL granite; 50 mm in the unaltered URL granite.

Box 16: The Palmottu uranium orebody

The Palmottu natural analogue study is based on a small uranium-thorium deposit located at Nummi-Pusula in southwestern Finland. The orebody is hosted by Precambrian gneisses and migmatites whose protoliths were arkoses and greywackes. These high-grade metamorphic rocks are part of the Svecofennian fold belt that stretches from south west Finland into central Sweden.

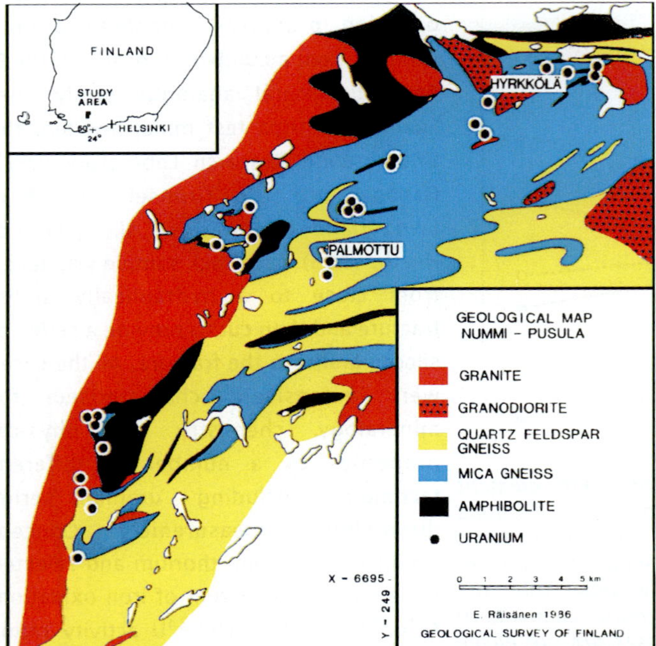

Figure B16.1: Geological map of the Palmottu region showing the location of the Palmottu analogue study and other uranium orebodies in the region. The Hyrkkölä site is under investigation for its copper deposits which are described in Section 4.4. From Blomqvist et al. (1995).

The orebody was discovered in the late 1970s during routine airborne geophysical investigations. In the subsequent characterisation, 62 boreholes were drilled with a total length of over 8 km. Some of these boreholes and several new holes, which were drilled perpendicular to the strike direction, were investigated as part of the natural analogue study.

The ore body is up to 15 m thick and some 400 m long, but is discontinuous in the form of uraniferous pegmatites and veins. The principal ore mineral is disseminated uraninite, thinly coated with the uranium silicate coffinite, and was formed in the latest stages of metamorphism, some 1800 to 1700 million years ago.

The average grade of the ore reaches up to 0.1 % uranium. The uranium is thought to be derived from the late-stage granitic fluids of the nearby late-kinematic Perniö granite.

The groundwaters, measured in open boreholes, show a distinct layered zonation, with the upper waters being fresh, oxidising Ca-Na-HCO_3 type and the lower waters being slightly saline, reducing Na-Cl-SO_4-HCO_3 type. The change in composition and the redox front occur at about 125 m depth. High dissolved uranium concentrations (up to 500 ppb) are associated with the groundwaters in the vicinity of the mineralisation down to the depth of the redox front. At greater depths, uranium has very low concentrations (< 10 ppb) in the reducing waters.

The natural analogue study has concentrated on processes that may affect radionuclide migration and retardation in the type of fractured, metamorphic rocks which will contain a Finnish and a Swedish repository, and which may be similar to repository host rocks in some other countries.

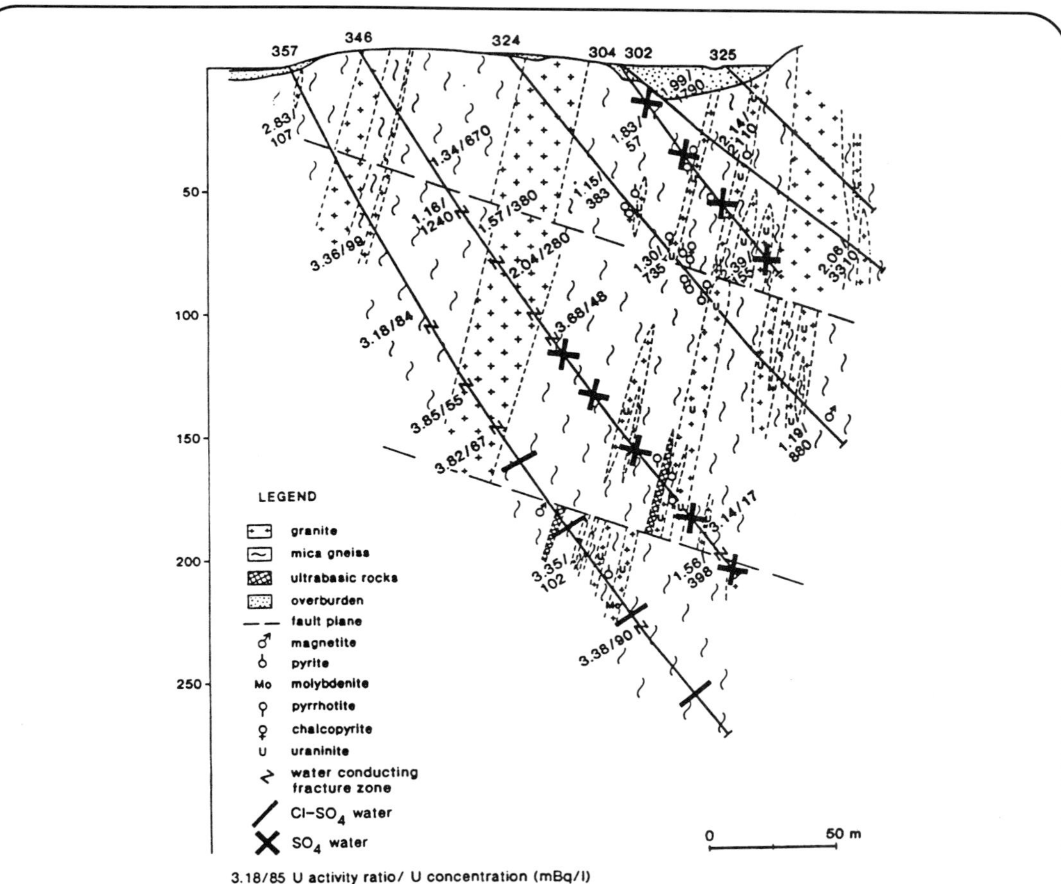

Figure B16.2: Cross-section of the Palmottu site showing the boreholes on which much of the study has been focussed. The variation in groundwater types with depth is evident from the location of Cl-SO_4 waters and the changing uranium content. From Blomqvist et al. (1995).

The natural analogue study at Palmottu has particularly examined:

- the stability and longevity of uranium minerals (see Section 4.2),
- blind predictive geochemical modelling studies (see Section 5.1),
- radionuclide transport by colloids (see Section 5.6),
- redox processes (see Section 5.5), and
- radionuclide retardation by matrix diffusion (see Section 5.3).

These natural analogue studies have been reported in a number of publications. The most recent being papers in the last EC Natural Analogue Working Group proceedings (von Maravic and Alexander, 2000) which include an overview paper describing results from the latest stage of investigations (Blomqvist et al., 2000) and a paper summarising the implications of the analogue data for repository performance assessment (Grundfelt et al., 2000).

Interpretation of the geochemical and micro-structural data from El Berrocal core samples (see Box 17) indicated that, not only was matrix diffusion limited to the first few tens of millimetres of rock adjacent to the fracture surface but that, within the rock matrix, the mobilised uranium was associated with secondary phases and is located in thin microfissures and along grain boundaries (Heath, 1995). There was a very good correlation between the distribution of the mobilised uranium, the redox conditions and the isotopic disequilibrium. The $^{234}U/^{238}U$ data suggested that uranium mobilisation at El Berrocal was a geologically recent event. Combining all of the data, Heath (1995) concluded that matrix diffusion alone had not taken place but, instead, a complex combination of matrix diffusion and chemical interaction had occurred between the rock and the mobile phases. This conclusion is important because it means that a measured concentration profile in a rock adjacent to a fracture may not represent matrix diffusion alone and, thus, their use in performance assessment must be redefined. The data from El Berrocal provide good evidence that once radionuclides migrate into the rock matrix from the flowing fracture, they are effectively immobilised irrespective of the process involved.

Smellie et al. (1993) presented natural radio-nuclide decay series, Mössbauer and stable isotope analyses of a section of rock extending from a single, water-conducting fracture in an oxidising groundwater environment into a granite from the Kamlunge test site in northern Sweden. Hydrothermal alteration affected the rock in a zone some 20 to 30 mm adjacent to the fracture. Long-term loss of uranium due to rock-water interactions on a timescale in the order of 10^5 years, with preferential loss of ^{234}U (probably under reducing conditions), affected the complete 70 mm section of rock, implying matrix diffusion on this scale. More recent (within the last 10^5 years or less), rapid removal of uranium from the rock marginal to the fracture by oxidising groundwater is restricted to a narrow zone of 30 to 40 mm into the rock from the fracture. Porosity and diffusivity measurements from altered and 'fresh' rock portions showed no significant differences. Uranium retardation via scavenging by secondary minerals on the fracture surface has been observed in other granitic rocks but does not appear to be effective at Kamlunge.

Matrix diffusion processes were also investigated at the Marysvale analogue site in Utah, USA. This site is located in a mining area with abundant mineralisations in the form of a networking of hydrothermal veins containing uraninite, pyrite and fluorite. These veins cross-cut both volcanic rocks (ash flows, breccias and tuffs) and intrusive rocks (quartz monazite and granite). Mineralisation took place around 19 million years ago, at a depth of approximately 450 m. The hydrothermal fluids were acidic (pH 2 to 4), with a temperature of around 200°C, and the duration of the hydrothermal event was around 10 000 years.

The objective of the analogue study was to investigate the mobilisation of uranium and other elements into the host rocks from the mineralised veins, see Figure 5.15. To this end, a number of elemental and isotopic profiles were determined in the rocks adjacent to the mineralised veins (Shea, 1984, 1999). It was found that sodium, magnesium, potassium, calcium, rubidium, barium, rare-earth elements and uranium all exhibited concentration gradients which could be modelled for coupled advective-diffusive transport. The depth into the rock through which transport could be identified on the basis of the profiles was 10 to 20 mm for the rare-earth elements, and 20 to 50 mm for uranium.

The data also showed that uranium concentrations in the mineralised veins exceeded 50 000 mg/kg but dropped to below 50 mg/kg in smaller veins and fracture coatings. The distance away from the veins at which uranium concentrations (due to mobilisation into the rock) are half of the concentration in the vein (the source) were determined and, on this basis, the mass of

Figure 5.15: Photograph of one of the mineralised veins in the Central Mining Area of Utah, and investigated in the Marysvale natural analogue study. Uranium and rare-earth element profiles into the rock were modelled using coupled advective-diffusive transport assumptions. Photograph courtesy of Mike Shea.

uranium mobilised into the bulk rock was determined to be 5.7×10^5 kg around the major veins and 6.8×10^2 kg around the smaller veins. This highlights the very large potential for the rock matrix to retard radionuclides released from a repository. However, care needs to be exercised in extrapolating these data directly to typical repository host rocks because these are unlikely to have been substantially affected by large-scale hydrothermal processes. The geochemical data from the Marysvale analogue study were also used to quantify diffusion parameters, as discussed below.

Valkiainen (1992) gives a compilation of laboratory and field data on matrix diffusion, including values for porosity as a function of depth (distance) from an active fracture in cores from various crystalline rocks. Typically, these data show that the rock in the first tens of millimetres has roughly 2 to 3 times higher porosity than the rest of the rock mass. This is the zone in which matrix diffusion would be most likely to occur and this agrees with the findings from the other studies mentioned above. However, problems arise when examining porosity, or elemental or isotopic data from profiles in cores adjacent to fractures because they are generally very erratic due to the heterogeneity of the rock in these locations.

Quantification of diffusivities in this zone is complicated, not only by the heterogeneity, but also because initial and boundary conditions, and retardation coefficients for sorbing species are not accurately known. Consequently, most quantitative information for matrix diffusion relates only to the depth of interconnected porosity, rather than for diffusivity values.

In general, it appears that more altered crystalline rocks have the greater potential for matrix diffusion than fresh rocks. It follows that these results confirm the theories of Glueckauf (1980) and Hadermann and Roesel (1985), and the experimental interpretation of Alexander et al. (1990a), in that matrix diffusion in crystalline rock is generally limited to only a small volume of rock close to fractures and does not extend throughout the entire rock volume.

However, it should be noted that even a small volume can make a significant difference to contaminant retardation. As an example, Alexander et al. (1990b) noted that, in one case using the RANCHMD transport code (Hadermann and Jakob, 1987), varying the effective diffusion

Box 17: The El Berrocal

El Berrocal is an area approximately 100 km south west of Madrid close to Toledo. The site and the analogue study take their name from the El Berrocal granite which forms a large hill. The granite is in the southwestern part of the Spanish Central Massif, close to the southeastern extent of the Sierra de Gredos and the Tertiary Basin of the Tajo River. The granite contains a number of small, vein-hosted uranium orebodies which have been exploited but are now abandoned. One of these orebodies was the focus of the analogue investigations.

Figure B17.1: Photograph of the El Berrocal site, showing a couple of mobile geochemical laboratories which were used for the sampling and analysis of groundwaters from boreholes.

This orebody had previously been mined an a number of undergrouned tunnels had been closed. However, one horizontal adit was still open and this allowed the granite and some veins to be characterised at depth.

The El Berrocal granite has a uranium content that averages 16 mg/kg, with primary uranium occurring as accessory uraninite dispersed in the granite matrix. Post emplacement hydrothermal alteration mobilised uranium, thorium and rare-earth elements from the granite and redeposited a proportion of the them in a large, 2 m wide, steeply dipping quartz vein. Uranium-series disequilibrium data indicate that this hydrothermal mobilisation event occurred more than one million years ago.

Erosion and weathering is exposing the vein-hosted mineralisation, causing further elemental mobilisation and transport. The El Berrocal project had the objective of investigating these present-day, low-temperature mobilisation processes as well as the processes responsible for elemental retardation in the granite.

In particular, the study investigated:

- uranium mineral stability, degradation and dissolution (see Section 4.2);
- uranium solubility and speciation, including blind predictive testing of geochemical codes and databases (see Section 5.1);
- matrix diffusion in the rock adjacent to fractures (see Section 5.3); and
- colloid associated radionuclide transport (see Section 5.6).

In addition to these analogue studies, a great deal of other work was undertaken to develop techniques and methodologies to allow the site to be characterised in detail in terms of geology, mineralogy, geochemistry and hydrogeology. This work including diverse studies such as detailed structural and mineralogical studies, hydrogeological measurements, tracer tests and microbiological investigations.

The results from the project indicated that mobilised uranium and thorium tended to be associated by sorption and coprecipitation with certain fracture coating minerals, notably iron oxyhydroxides and calcite. Enrichment of uranium by a factor of up to 6 was observed, and up to 3 for thorium, relative to the fresh granite. Considerable effort was made to understand these coprecipitation processes in order to improve thermodynamic models and databases for performance assessment (Section 5.1).

Figure B17.2: Photograph of the inside the adit at El Berrocal site which allowed some underground characterisation, showing the tops of the boreholes drilled into the adit floor. The combination of the adit and the boreholes allowed the site to be investigated in three dimensions, providing comprehensive characterisation of the mineralisation.

The El Berrocal project has been described widely in a number of publications. The most detailed are a four volume report series (ENRESA, 1996) and a final summary report (Rivas et al., 1997).

coefficient (of a non-sorbing radionuclide in the rock matrix) by three orders of magnitude produced no significant enhancement of calculated radionuclide retardation. In contrast, changing the assumed maximum depth of matrix diffusion by over an order of magnitude (from 1 mm in the code to 30 mm), increased the breakthrough time for a long-lived radionuclide pulse by an order of magnitude and decreased the peak activity by an order of magnitude. Further, for radionuclides where the transit time (through the fracture) is comparable to the radionuclide half-life, the effect of increased retardation becomes strongly non-linear, as the additional effects of radioactive decay also decrease releases.

A geological phenomenon that may provide further useful information on elemental migration through the bulk rock matrix is that of *reduction spots*, an example of which is given in Figure 5.16. These are discussed later in this volume with regard to redox fronts in Section 5.5. Although they have not been investigated with the specific intention of quantifying matrix diffusion, they may be used for this purpose due to the fact that they represent locations where net elemental migration has occurred, solely by diffusion, in the rock matrix.

Reduction spots do not form by matrix diffusion in the sense in which this term is used by most performance assessment models, because the elemental migration does not occur from a fracture into the bulk rock. This is, in fact, very useful because reduction spots represent diffusion in the rock where it has not noticeably been affected by chemical and physical alteration. Although reduction spots occur most commonly in slates and shales, they do occur infrequently in granites. In one investigation, reduction spots from a number of granites have been examined and their mode and time of formation calculated (Hofmann, 1990a). It was concluded that reductions spots with a total diameter of 50 mm, but with a core of about 10 mm diameter, took some 10^6 years to form. What is important here is not the calculation of apparent or effective diffusion coefficients, rather it is the clear indication that connected porosity and, therefore, the possibility of radionuclide retardation exists, even at depth in the rock matrix (i.e. at some distance from a fracture).

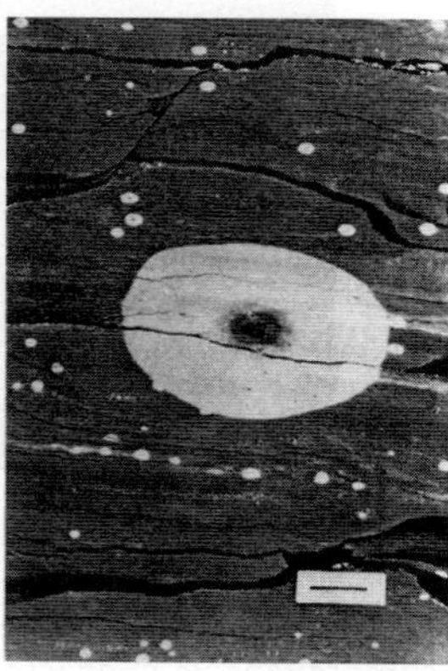

Figure 5.16: Top: Large reduction spot in a Permian red-bed clearly showing the morphology with a central core, core margin, pale and hematite zones and the sharp redox front: this spot is approximately 40 mm in diameter. From Hofmann et al. (1987). Bottom: Reduction spots in clay-rich sandstones: the scale bar is 1 cm. From Hofmann (1990a).

Bulk rock chemical buffering capacity

The solubility of many key radionuclides is substantially lower under reducing conditions than under oxidising conditions. It is clearly important, therefore, that the near-field geochemical environment of a repository should be capable of buffering the redox potential to maintain reducing conditions in the event of radiolytic oxidant production.

For a repository without large volumes of iron in the engineered barrier system (e.g. a spent fuel repository that uses copper canister), it is the ferrous iron in the host rock which must act as the largest buffer to redox conditions (Neretnieks, 1986b,c). Some ferrous iron will be present in fracture coating minerals or where the adjacent rock is hydrothermally altered, and will be directly accessible to groundwater by advection. However, in the case of fractured crystalline rock, a much larger amount of ferrous iron is held in minerals, such as biotite and the amphiboles, within the rock matrix. In order for the redox buffering capacity of this ferrous iron to be realised, these minerals must be accessible to groundwater by matrix diffusion. It follows that, in general, the greater the depth of connected microporosity, the higher the redox buffering capacity of the rock.

However, the form of the porosity is also important, for if only a few large pores are present then a smaller proportion of the total iron will be available than if a dense network of micropores or microfractures are present. Despite the obviously important link between matrix diffusion and redox buffering, this aspect of matrix diffusion has received scant attention to date. The work of Smellie et al. (1993), however, suggests that oxidising groundwaters are effectively buffered in the host rock within 30 to 40 mm from the conducting fracture. It was noted by Alexander et al. (1990a) that, not only is the buffering capacity of the rock (and thus matrix diffusion) important for reducing the solubility and mobility of radionuclides, but also that the 'redox front' thus created, if confined within the microporous, diffusion controlled rock mass makes the formation of colloids less problematic than if they had formed in an open fracture where advective transport occurs.

The redox buffering capacity of the host rock deserves more attention than it currently receives. In future studies of matrix diffusion, it would be informative to calculate the mass of ferrous iron that is accessible to groundwater through the connected microporosity, by matrix diffusion. This is, of course, a site-specific quantity, but no more so than the matrix diffusion capacity of the rock itself, which receives so much attention.

The extent of matrix diffusion in sedimentary formations

Only a few natural analogue studies have examined matrix diffusion in sedimentary formations. However, it is recommended that additional studies are performed because, despite the commonly held simplistic view that all sedimentary rocks are homogeneous porous media, in reality many are fractured and have a dual porosity in which matrix diffusion can occur.

Unambiguous results have been obtained from matrix diffusion studies in fractured sedimentary rocks in Switzerland. In one study (Mazurek et al., 1996) the Opalinus Clay formation was investigated and the results are of particular interest because this formation is a potential host rock for a Swiss HLW repository (see Box 2). These studies suggest that matrix diffusion occurs through the clay rocks to a depth of 80 mm from fracture surfaces. However, since the samples investigated have experienced glacial off-loading, the stress relaxation may have had an impact on the depth of interconnected porosity. Deeper rocks, at repository depth may not see matrix diffusion occurring in so much of the rock volume. Unfortunately, this has proved to impossible to check as no evidence of water conducting fractures have yet been found in deeper Opalinus Clay horizons.

In a second study (Alexander et al., 2000b) matrix diffusion in samples from the potential L/ILW repository site at Wellenburg, Switzerland, were examined. In this study, the samples were taken from repository relevant depths and the preliminary results indicate that matrix diffusion is operating to depths of a least 60 to 70 mm in these limestones, biomicrites and clay biomicrites. Of particular interest in this work is that even those water-conducting fracture surfaces which are coated with secondary minerals (calcite in this case) still display evidence of matrix diffusion in the rock.

The potential effect of fracture surface sealing on matrix diffusion was also investigated in the Maqarin study (see Box 11). Here, the surfaces of the water-conducting fractures are sealed with a wide range of secondary cement phases, such as calcite, tobermorite, ettringite, various zeolites etc. (Milodowski et al., 1998), produced by reaction between the hyperalkaline groundwaters leached from the local cements and the clay biomicrite host rock (Alexander and Smellie, 2000). Mineralogical examination of the fracture surface coating phases indicated a very low degree of interconnectivity between the fracture and the rock matrix porosity, suggesting that the secondary phases could very effectively seal the rock matrix porosity to radionuclides transported in the fractures.

A detailed study of the extent of rock matrix diffusion was carried out on four profiles taken perpendicular to water-conducting fractures from one of the adits at Maqarin (Smellie, 1998). Each

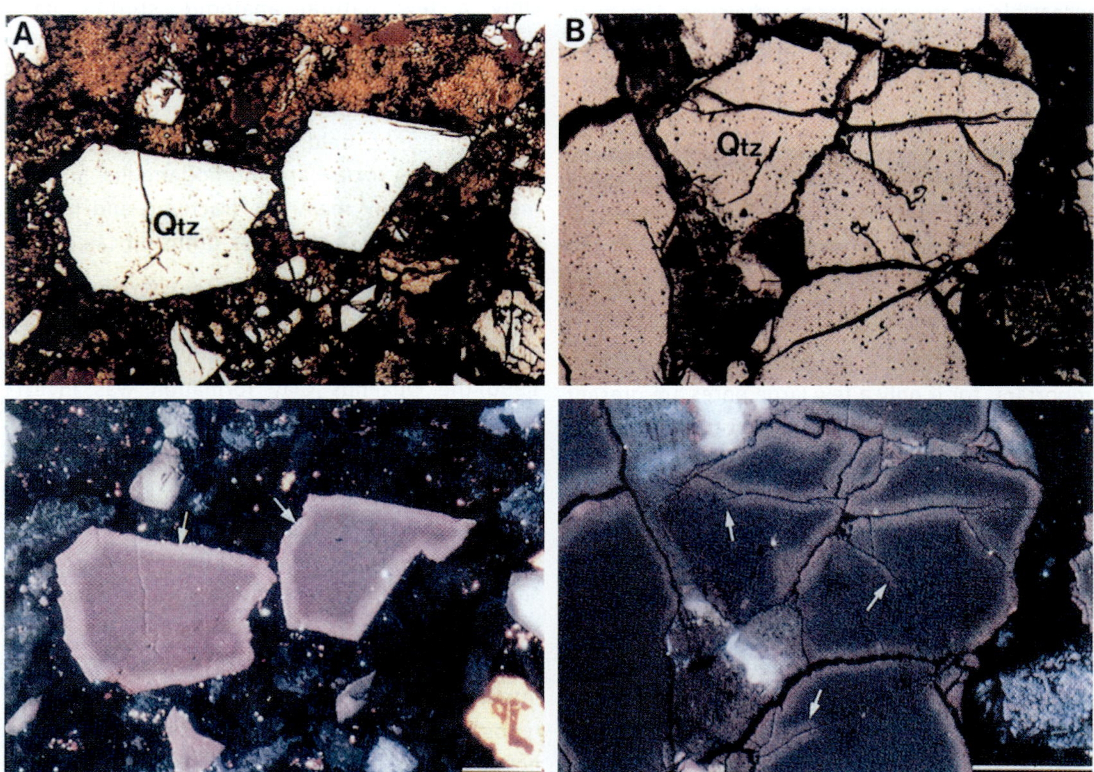

Figure 5.17: Photomicrographs and cathodoluminescence images of samples of the sedimentary rocks from Tono. Radiation haloes (the lighter zones in the lower images) around and within minerals indicate matrix diffusion of uranium in the microfractures within grains and along the grain boundaries. The arrows indicate the location of microfractures which have acted as migration pathways for uranium into grains. Qtz = quartz. From Yoshida (1994).

profile was analysed for a suite of elements and natural decay series radionuclides along with porosity variations but the results are highly ambiguous. Significant variations in the unaltered clay biomicrite signature have made it impossible to detect any potential perturbations due to hyperalkaline water/rock matrix interaction (i.e. the background noise is too great). In only one case is there some suggestion of a clear signal: the $^{226}Ra/^{238}U$ ratios in two of the profiles suggest relatively recent rock/water interaction at up to 40 to 70 mm into the rock. However, these depths should be treated cautiously considering that all four samples are heavily influenced by microfracture networks extending several tens of millimetres into the rock and by pre-existing lithological variations (bedding etc). Due to the importance of the implications of this work, new samples are currently being analysed to provide an unambiguous answer.

Matrix diffusion in the sedimentary rocks at the Tono site in Japan (see Box 18) have also been investigated (Yoshida, 1994). At Tono the uranium orebody is hosted by fluviolacustrine, lignite-bearing sediments which have relatively low porosity and permeability (hydraulic conductivity between 10^{-8} and 10^{-11} cm/s). Samples from the rock around the area of the orebody were investigated by optical, chemical and isotopic methods. Optical studies indicated that microfractures within grains and the grain boundaries have acted as pathways for uranium diffusion in association with the porewaters, as shown in Figure 5.17. A relationship was suggested between the microfabric of the sediments, particularly the connectivity of the porespaces, and the degree of uranium mobilisation.

Different minerals showed different uranium migration/retardation characteristics in their microfractures due to their textural characteristics. Although the extent of matrix diffusion was not quantitatively described, it was clear that a very large proportion of the rock was available for matrix diffusion and that the total surface area in microfractures and grain boundaries on which radionuclides could sorb is very large. As a consequence, these rocks have both a high dilution capacity from the porosity distribution and a high sorption capacity from the mineralogy.

Estimation of diffusion coefficients

The most common approach to estimating matrix diffusion coefficients using data from natural systems is to back-calculate from an elemental profile assumed to be produced by diffusion into or out of the bulk rock. In addition to the concentration gradient, information is also required on the duration of the process.

Ideally, only well-constrained systems are analysed, for example the flushing of saline groundwater into freshwater filled fractures as a result of a change in sea-level (Olin and Valkiainen, 1990), or systems which can be dated directly using an appropriate isotopic system (e.g. Latham and Schwarcz, 1989). The problem in many of the former examples is that the constraints on the systems are often model dependent (e.g. rate of postglacial land uplift), so uncertainty increases. In the latter examples, the radioisotopes do not behave as ideal, non-interacting tracers, thus causing underestimation of actual diffusion rates.

The major advantage of examples from natural systems is that they once again avoid the significant perturbations associated with laboratory experiments and it is, therefore, worthwhile comparing a small selection of the available diffusion data (Table 5.3) with the larger volume of data available elsewhere on laboratory based values (e.g. Neretnieks, 1990; Brandberg and Skagius, 1991). Perhaps the most striking feature of the table is the large range in calculated apparent diffusivity (D_a) values from 10^{-21} m^2/s to 10^{-9} m^2/s. The first value can be rejected, as it is in the range of solid state diffusion coefficients, as can the last, as it is greater than the related value for molecular diffusivity in pure water

Box 18: The Tono uranium orebody

The Tono region, located some 350 km south west of Tokyo, is the site of Japan's most extensive uranium deposits. The largest of these deposits, the Tsukiyoshi uranium orebody, has been the focus of most of the natural analogue studies in the area. This orebody has not been commercially exploited and only one gallery at a depth of 130 m below the ground surface has been constructed so the orebody can be examined in a relatively undisturbed state.

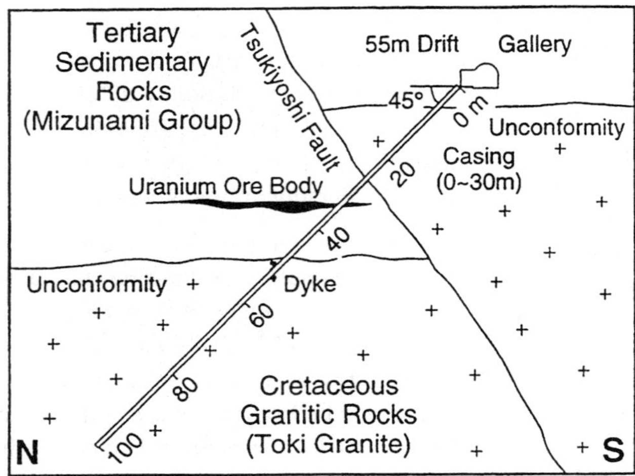

Figure B18.1: Cross-section through the Tsukiyoshi orebody at Tono showing its position at the unconformity and its faulted nature. Despite the large fault, no significant radionuclide transport along the fault has been observed. Illustration courtesy of JNC.

The Tsukiyoshi orebody is approximately 3.4 km long, between 300 to 700 m wide and is a few metres thick. However, this orebody was split into two sections, between 10 and 5 million years ago, by the Tsukiyoshi fault which has a vertical displacement of about 30 m and which does not continue upwards into the youngest sedimentary rocks in the region (the Seto Group).

The orebodies at Tono lie in palaeo-channels in the unconformity between Cretaceous granitic basement rocks (the Toki Granite) and overlying Miocene fluvio-lacustrine sediments, which themselves form the lowest unit in a pile of Miocene and Pliocene marine and lacustrine sediments. These sediments generally are around 200 m thick, although they are up to about 370 m thick in places.

The basement granitic rocks contain about 6 ppm uranium and are considered to be the source of the Tono uranium mineralisation. The sediments at the unconformity (the Toki Lignite-Bearing Formation) contain significant quantities of carbonaceous material and pyrite, in a clearly highly reducing environment. The uranium mineralisation itself occurs in conglomerate, sandstone and the lignite-bearing formations.

The primary, unoxidised uranium ore in the Tono deposit appears grey or black and comprises accumulations of coffinite and pitchblende, closely associated with pyrite, altered biotite or coaly plant materials in or around the porosity of the sediments. The secondary, oxidised uranium mineralisation appears yellowish and is comprised of a variety of uranium-bearing minerals including autunite, zippeite and uranocircite, and is accompanied by montmorillonite, limonite and other minerals in the oxidised zones.

The orebody is thought to have formed when oxidising groundwaters leached the uranium from the Toki Granite and transported it upwards to the lignite-bearing rocks, where the uranium was precipitated or adsorbed, or both, under the more reducing conditions that prevailed there. This initial uranium concentration process occurred around 10 million years ago. No substantial remobilisation of the uranium has occurred since this time, despite the later uplift, erosion and faulting history of the area.

Figure B18.2: Core sample from Tono showing the Tsukiyoshi fault which is indicated on Figure B18.1. Photograph courtesy of JNC.

Figure B18.3: Photograph underground at Tono in one of the tunnels showing equipment used for monitoring the groundwater chemistry. Photograph courtesy of JNC.

The hydraulic conductivity of the sedimentary rocks that host the uranium ore is low, between 10^{-8} and 10^{-11} cm/s. The groundwater in the region of the uranium ore is of the Na^{+}-HCO_3^{-} type, is strongly reducing and slightly alkaline (ph 8.7 to 9.5). The uranium content of the groundwater is generally low, about 0.05 to 0.2 ppb, although geochemical calculations suggest that the groundwater is saturated with respect to uraninite and coffinite. These thermodynamic solubility and speciation calculations also show that the redox environment is controlled by siderite or pyrite. Hydrogen and oxygen stable isotope studies show that the deep groundwater has a meteoric origin and preliminary ^{14}C measurements suggest an age between a few thousand and ten thousand years old.

The natural analogue studies at Tono are important because they demonstrate that the uranium ore has remained largely unaffected by the continued tectonic activity in the area over the last 10 million years since it formed. Although one fault cuts through the orebody itself, and other large faults lie close to the orebody, there is no evidence that significant uranium transport has occurred along these faults.

This is a particularly important finding because, in most repository concepts in fractured hard rock, large faults are considered to be the only transport route for radionuclides to return to the surface. The fact that significant fracture-based transport has not occurred at Tono suggests that, provided the chemical environment in a repository near-field remains stable, tectonic activity will not necessarily cause radionuclide releases to the surface.

Other investigations at Tono have focussed on uranium mobilisation at the mineral grain scale in the ore and surrounding rocks. To this end, hundreds of samples collected from the mine and from boreholes were analysed in uranium-series disequilibrium studies. Results indicate that reducing conditions have been maintained for, at least, the last million years and the uranium mobilisation has been limited to very slow diffusion in the rock matrix and in microfractures in mineral grains.

The Tono natural analogue studies have been reported in a number of conference proceedings and journal papers. The most recent being Seo and Yoshida (1993), Yoshida et al. (1996), Yoshida (1997) and Tsubota et al. (2000).

(see comments in Valkiainen, 1992, on the use of an inappropriate model for these particular calculations).

In the early Swiss Project Gewähr assessment, a pore diffusivity coefficient (D_p) of 1.5×10^{-10} m^2/s was taken as the base case, and this translates to a D_a of 1.5×10^{-15} m^2/s. Of the data presented in the table, the radium based D_a of Alexander et al. (1990b) and the recalculated uranium and rare-earth based D_a values of Shea (1998) are in this general area. However, it is of note that the uranium data (from the Böttstein and Grimsel samples) for Swiss crystalline rocks are significantly higher, at 10^{-10} m^2/s, than the Gewähr base case. Similarly, the data of Mazurek (1998) for aplitic and pegmatitic fracture zones produce Da values in the 10^{-10} m^2/s range. While this represents only a limited data set, it does suggest that the Project Gewähr base case D_a value of 10^{-15} m^2/s is over-conservative. Further comparison with laboratory derived data is also advisable.

Conclusions

Matrix diffusion in fractured crystalline rocks has been extensively investigated and further, generic work seems unnecessary. Site-specific studies are probably required and, indeed, several are currently underway (e.g. Alexander et al., 2000b). In general, matrix diffusion in fractured crystalline rocks appears to operate to depths of 10 to 20 mm in fresh crystalline rock. However, evidence for greater depths of matrix diffusion has been provided in some natural analogue (and site characterisation) studies where the rock is strongly altered. However, even if a conservative value of between 10 and 20 mm depth is assumed in performance assessment, this will provide a significant degree of retardation to radionuclides migrating through fractured rock.

Comprehensive geochemical and petrophysical characterisation reveals that the region of enhanced trace element mobility in the vicinity of a fracture often corresponds to zones of physical or hydrothermal alteration.

Matrix diffusion in sedimentary rocks has also been investigated, although far fewer studies have been performed in comparison to crystalline rocks. Initial data suggest that matrix diffusion depths in sediments may be deeper than in crystalline rocks (70 to 80 mm compared to 10 to 20 mm). However, until further data are available, it remains unclear how representative are these values. Further studies are to be encouraged.

*Table 5.3: Estimated diffusion coefficients in crystalline rock as compiled from a number of studies on natural samples using a variety of experimental methods. *For method of calculation, see Alexander et al. (1990b).*

D_a (m^2/s)	Method
10^{-21} - 10^{-18}	Fitting uranium data from chlorite grains around hydrothermal veins (Shea, 1984).
10^{-19} - 10^{-16}	Uranium concentration gradient in the bulk rock around hydrothermal veins (Shea, 1984).
1.1×10^{-17} - 1.6×10^{-15}	Uranium gradients in bulk rock around hydrothermal veins using coupled advection-diffusion calculations (Shea, 1998).
3.3×10^{-16}	Recalculation of Shea (1984) bulk rock uranium profiles using coupled advection-diffusion calculations (Shea, 1998).
2.0×10^{-18} - 2.5×10^{-16}	Rare-earth element gradients in bulk rock around hydrothermal veins using coupled advection-diffusion calculations (Shea, 1998).
0.3×10^{-14} - 1.8×10^{-14}	Radium concentration gradient in bulk rock around a fracture at Grimsel (Alexander et al., 1990b).
1.1×10^{-10} - 5.8×10^{-10}	Calculated* from uranium distribution in bulk rock around pegmatitic vein from Bottstein (Smellie et al., 1986).
1.0×10^{-10} - 5.1×10^{-10}	Calculated* from uranium distribution in bulk rock around a fracture at Grimsel (Alexander et al., 1990a).
2×10^{-9}	Back diffusion of chlorine from a core sample into distilled water (Lehikoinen et al., 1992).

The extent of rock available for matrix diffusion also controls the ability the host rock to buffer the redox conditions by restricting access of groundwater to ferrous iron-rich minerals. Little attention has been paid to this aspect of the matrix diffusion issue and more work along these lines would be worthwhile.

Any colloids formed within the rock matrix as a result of redox reactions will be physically trapped and, as a consequence, can have no part in advective transport of radionuclides. This area has not been studied to date and may be worth further investigation.

5.4 Radiolysis

Significant radiation fields may occur in the near-fields of repositories for spent fuel, HLW and certain forms of high-activity ILW. This radiation may cause radiolysis of a number of materials in the near-field (e.g. cellulose and bitumen), but it is radiolysis of groundwater which is most significant. Radiolysis of water is essentially the splitting of the water molecule into component charged radicals (e.g. $H^{\cdot}$) by the action of radiation. These radicals may recombine or react with other dissolved species to form new molecular products.

During radiolysis of water, equal amounts of oxidants and reductants are produced. The radiolytic yields of radicals such as e^-_{aq}, $H^{\cdot}$, $OH^{\cdot}$, $HO_2^{\cdot}$ and the molecular products H_2O_2, H_2 and O_2 are strongly dependent on the linear energy transfer (LET) of the radiation involved (Eriksen and Ndalamba, 1988). The most powerful reducing species are e^-_{aq}, $H^{\cdot}$ and H_2, whilst the most powerful oxidising species are $OH^{\cdot}$ and H_2O_2 (Vovk, 1987). These primary products may react with each other, and with other species dissolved in the near-field, to yield secondary products, the nature of which is very dependent on the chemistry of the engineered barrier system.

Most theoretical analyses assume that a net oxidant build up would occur in the near-field because the principal molecular reductant, H_2, is relatively chemically inert and would rapidly migrate out of the system due to its high diffusivity. The oxidants, in contrast, would be more reactive and have a lower diffusivity. The result could be to create a slowly moving redox front (oxidising) that migrates outwards from the waste into the other engineered barriers and the host rock (Neretnieks, 1982; Neretnieks and Åslund, 1983a,b). The extent of redox front migration would be controlled by the balance between oxidant production and oxidant consumption (buffering) by iron-rich materials in the engineered barriers and the rock (McKinley, 1985; Smith and Curti, 1995).

There is a considerable volume of laboratory and theoretical work whose aim has been to replicate and understand the processes involved in the production of radiolytic oxidants. The majority of this work on radiolysis has focussed on spent fuel because the potential for radiolysis is greatest for this waste type, although the same principles apply to other types of high activity waste.

Much of the work done has considered the effect of radiolytic oxidants on canister corrosion, wasteform dissolution, and subsequent radionuclide solubility and speciation (e.g. Christensen and Bjergbakke, 1982; Grenthe et al., 1983; Johnson et al., 1983; Christensen and Bjergbakke, 1984a,b; Forsyth et al., 1985; Vovk, 1987; Sunder et al., 1989; Werme et al., 1990; Neretnieks and Faghihi, 1991; Smith and Curti, 1995). In the repository environment, the bentonite porewater experiences a radiation dose that changes with time and with distance from the wasteform. Initially, once the buffer material surrounding the intact waste canisters has become saturated, the porewater will be exposed to low LET, high intensity gamma and neutron fluxes. The greatest radiolytic effects will occur at the canister wall.

Calculation of the net increase in oxidant is problematic. In a HLW repository with many steel canisters, the oxidants will enhance the corrosion of the steel canisters but, in the process, will be consumed, thus maintaining reducing conditions, at least until all the steel has been oxidised. In comparison, if canisters are made from a non-ferrous metal (e.g. copper), canister corrosion will be less affected by oxidant production. In this case the canister may maintain its integrity but the near-field will become progressively more oxidising. In both cases, however, the net radiolysis of bentonite porewater is negligible at this stage due to shielding by the canister while it remains intact.

Eventually, however, all canisters will become perforated, exposing the waste to direct contact with bentonite porewater. At this stage, in addition to the gamma and neutron fluxes, the porewater is exposed to high LET, alpha and beta radiation from which it was previously shielded by the canister. The effect of alpha radiation is much more important than that of the beta, gamma or neutron radiation (Neretnieks and Faghihi, 1991). The alpha and beta radiation doses to the porewater will depend on many factors which include the exposed wasteform surface area and the duration of canister containment. Once the canisters have been perforated, any steel present will still consume oxidants, helping to maintain a reducing near-field. Otherwise, an increasingly oxidising near-field may cause more rapid dissolution of the wasteform, or enhance solubilities of some important radionuclides released from the waste. If non-ferrous metal canisters are used, the near-field may already be oxidising when the canisters are perforated, causing the wasteform dissolution rate to be faster than it would otherwise be.

When oxidants come into contact with the wasteform, the strongest reductants in the system will be oxidised first (Grenthe et al., 1983). This generally leaves the less reactive actinides in the oxidation state they exhibited initially in the wasteform. After some time a protective layer of metal oxides will form on the strongest reductants, decreasing their redox buffering capacity. This may result in an increase in the rate of oxidation of the actinides. As the near-field becomes progressively more oxidising, the solubilities of many key radionuclides will increase. If the redox buffering capacity of the canisters can be overcome, a redox front may migrate out towards the host rock. This is unlikely to happen in a repository with steel canisters (see discussions in Smith and Curti, 1995; Alexander and McKinley, 1999) but it could potentially occur in other repository designs with less steel present (e.g. one with copper canisters). If it did occur, in the oxidising environment behind the redox front, the soluble radionuclides could migrate, by diffusion, through the buffer into the far-field.

However, this discussion assumes that the other barrier materials and the host rock have no redox buffering capacity. This is not strictly true (e.g. bentonite typically contains several percent iron) but the extent of the redox buffering capacity provided by other materials will be highly design and site specific. The effect of the redox front migrating out into the rock, and the consequences this has for radionuclide transport, are discussed in detail in Section 5.5.

In a modelling exercise for a spent fuel repository, Neretnieks and Faghihi (1991) proposed two mechanisms which would accompany radiolysis at a perforated canister and which would act to limit the production of oxidants by several orders of magnitude. These are shown diagrammatically in Figure 5.18. Once a canister has been perforated, groundwater will flow through the hole and fill the gap between the canister and the waste. In the early theoretical descriptions of radiolysis in the near-field, it was assumed that once a canister was perforated it had no further role to play. This is an overly conservative assumption, and Smith and Curti (1995) show that even the corrosion of only a minor amount of iron in a steel canister can act as a significant redox buffer. Neretnieks and Faghihi (1991) suggest that, once decomposition of the waste commences, the corrosion products will rapidly accumulate in this gap between the waste and the canister. The corrosion products have a higher density than the water and will adsorb the alpha radiation faster. This, together with the reduction in the volume of water adjacent to the waste, causes less water to be radiolysed.

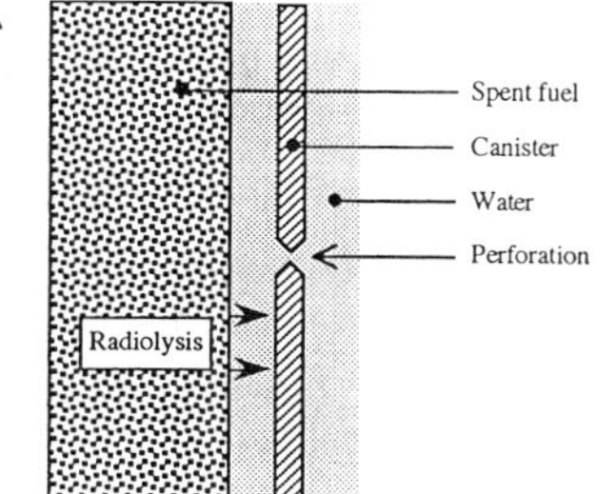

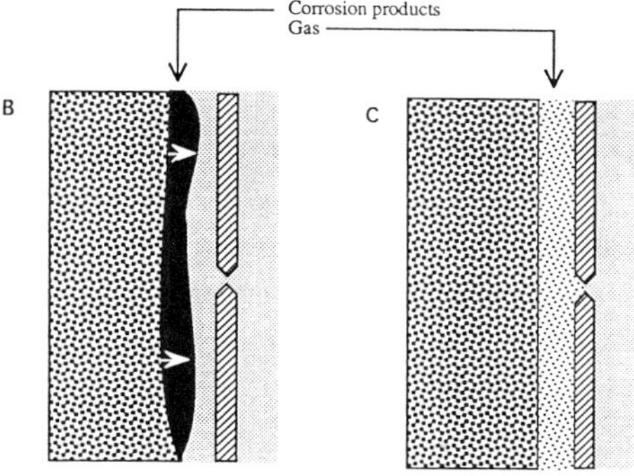

Figure 5.18: A: when the canister is perforated, water fills the gap between the spent fuel and the canister wall and is radiolysed. However, Neretnieks and Faghihi (1991) propose two independent processes that may limit radiolysis of the water. B: a layer of corrosion products form which reduces the volume of water in the canister and adsorbs the alpha-radiation. C: radiolytic hydrogen accumulates in the canister and an overpressure builds-up preventing further water from entering the canister.

A second, unrelated mechanism may also operate to limit radiolysis. The radiolytic hydrogen produced from the first stages of radiolysis may form a discrete gas phase in the gap between the waste and the canister. Some of this gas may dissolve and diffuse out of the system but, because of the high capillarity of the bentonite, a gas overpressure may build up. This overpressure will prevent more water from flowing into the gap to replenish that lost by radiolysis.

The effect from both of these processes is dependent on the size of the perforation in the canister and is not easily quantified. However, it is likely that initial perforations will be small, and the reduction in radiolysis greatest, at the time when radiolytic effects would be most critical (i.e. when the activity is high). Although this work was done for a spent fuel design, the same basic mechanisms could apply for other wastes and canister designs.

The majority of the investigations into radiolysis have been undertaken in the laboratory or as modelling studies but there have been a few natural analogue studies which have examined radiolysis. The relatively small number of analogue studies that have looked at this issue probably reflects that fact that radiolysis has not been considered a high priority in most repository development programmes (apart from in Sweden and Finland) and also because few suitable analogue sites have been identified.

The issues of most relevance to radiolysis that have been (or potentially could be) addressed in natural analogue studies are:

- the processes involved in radiolysis of groundwater;
- how common is radiolysis in nature; and
- the potential buffering capacity of reduced iron corrosion phases from corroding engineered barriers.

These issues are discussed in the following sections.

The processes involved in radiolysis of groundwater

The first detailed natural analogue investigation of radiolysis is that performed by Curtis and Gancarz (1983) at the Oklo natural fission reactors (see Box 4). That investigation indicated that radiolysis did occur within the reactor zones, that a redox front had formed and that it had migrated out into the host rocks. At Oklo, beta radiation predominated over alpha radiation during the period when the fission reactors were operating but the potential significance of this fact for interpretation of the analogue data from Oklo is unclear.

An inventory of radiogenic elements in the reactor zones at Oklo showed that approximately 80 % of molybdenum, 35 % technetium and 25 % ruthenium were missing (see Section 4.2). Curtis and Gancarz (1983) suggested that these elements had been converted to soluble oxyanions by reaction with radiolytic oxidants which had subsequently migrated out of the reactor zones with the redox front. These elements were then reduced and precipitated, in depositional haloes, which mark the limit of the redox front migration.

A number of important conclusions were reached in this study. First, radiolytic hydrogen appears not to have behaved in an entirely inert manner and did not escape from the system. This is indicated by the presence of reduced iron in the reactor zones, assumed to have been reduced by radiolytic hydrogen. This observation is apparently contradictory to the theoretical model proposed by various workers and outlined earlier. Instead it was suggested that the hydrogen diffused into the surrounding clay and reduced the iron present there. This has clear implications for a repository with an iron-rich bentonite buffer. Second, the redox conditions were much more complicated than the simple models suggest. Although the Fe^{2+}/Fe^{3+} ratios in the reactor zones indicate general reducing conditions, the actual redox environment is unknown. The oxidising conditions required to mobilise some of the radionuclides must have been very localised. For example, whilst oxidising conditions were maintained in the vicinity of uraninite grains, elsewhere the radiolytic hydrogen was concurrently reducing the iron in the clays.

Radiolysis of water in reactor zones 7, 8 and 9 is directly suggested by the presence of H_2-bearing fluid inclusions observed in quartz using Raman spectroscopy techniques (Dubessy et al., 1988; Savary et al., 1993). However, there is some doubt over the interpretation of these results because hydrogen may be able to leak from the inclusions at high temperature (Mavrogenes and Bodnar, 1994). No oxygen was detected in these inclusions suggesting that conditions were reducing, supporting the conclusions of Curtis and Gancarz (1983). The most likely explanation for the reducing environment is the abundance of organic material in the orebodies which buffered the redox conditions (Gauthier-Lafaye and Weber, 1993; Nagy, 1993; Nagy et al., 1991, 1993), probably aided by organic matter radiolysis.

The maximum temperatures at Oklo (about 600°C) were much higher than those that will be reached in a repository and these high temperatures could have had a significant effect on both the rate of hydrogen production and the behaviour of the hydrogen once it evolved. In addition, at Oklo the radiation dose to the groundwater was estimated to be 100 to 500 times lower than that predicted for a repository containing a similar quantity of spent fuel with a comparable burn-up history (Christensen and Bjerbakke, 1982). Given these differences, it is not possible to take the Oklo results wholesale and apply them directly to the repository situation. Nonetheless, the clay layer at Oklo is somewhat analogous to the bentonite buffer in a HLW repository and the fact that this clay apparently impeded the migration of hydrogen at Oklo may suggest that this processes will occur in the repository, although it might require comparable iron contents in the bentonite.

Radiolysis has also been the focus of investigation at the Cigar Lake analogue study (see Box 5). Early studies at Cigar Lake were focussed on establishing whether radiolysis products could affect the oxidation and degradation of the uraninite, and the role of radiolysis in the formation of the aureole of ferric iron oxide at the ore/clay interface (Karlsson et al., 1994; Liu et al., 1994; Christensen, 1994, Cramer and Smellie, 1994b). At Cigar Lake, radiolysis of water in contact with the ore generates oxidants (for example, H_2O_2, O_2 and OH-radicals) and hydrogen. In contrast to Oklo, criticality was not achieved at Cigar Lake and thus it is alpha radiation that dominates over beta radiation and this fact may explain some of the differences in the observations between the two sites.

As hydrogen is not very reactive at ambient temperatures, it is believed to escape from the ore zone by diffusion through the water-filled matrices of the ore and the surrounding clay halo. The net chemical effect of radiolysis is thus oxidation of components in the groundwater, rock and ore minerals. In principle, the following reductants present at Cigar Lake can react with the oxidants produced by radiolytically generated reactions:

- Fe^{2+}, HS^- and dissolved organic carbon in the groundwater, and
- Fe(II) (e.g. siderite), sulphide (e.g. FeS_2, PbS), solid organic carbon and U(IV) (uraninite) in solid mineral form.

Mineralogical and geochemical observations at Cigar Lake showed that radiolysis products were present in groundwaters and minerals. In groundwaters, the end products of radiolysis were identified as H_2 and SO_4^{2-}, the latter resulting from sulphide oxidation. The release rates which, assuming equilibrium, equate to radiolytic production rates were calculated to be 2.3×10^{-12} and 1.7×10^{-12} equivalents/m^3/s, respectively for these radiolytic products (Liu et al., 1994). However, given analytical difficulties and the small number of samples investigated, these results should be treated with caution. Applying a radiolytic model (Hofmann, 1992) for homogeneous systems to the Cigar Lake orebody gives an H_2 production rate of

2.3×10^{-12} equivalents/m^3/s assuming a radiolysis efficiency of 1 % (Hofmann, 1996).

Fluid inclusions in minerals from Cluff Lake and Rabbit Lake (both associated with Cigar Lake) contained both free O_2 and H_2 (Dubessy et al., 1988). These data are interpreted as due to primary trapping of a radiolysed fluid in the ore. However, these findings are not easily directly consistent with the present-day hydrogen levels in the groundwaters (Liu et al., 1994). The presence of both radiolytic oxygen and hydrogen in the same fluid inclusions is very interesting because it demonstrates that these species do not necessarily back-react at low temperatures.

Early studies at Cigar Lake (Sunder et al., 1988) revealed that surface layers on the uranium ore had been oxidised to higher mixed oxides (U_4O_9 to U_3O_7) but that the process appeared to have been limited because the threshold value for corrosion was not exceeded (Christensen, 1994). Understanding the oxidation of UO_2 is important because, for spent fuel, if oxidation proceeds beyond U_3O_7 to U_3O_8, the crystal lattice structure becomes significantly altered, causing radionuclides to be expelled from the matrix during the oxidative conversion process. Therefore, the potential for uraninite oxidation as a consequence of radiolysis has been examined in some detail at Cigar Lake.

According to some early radiolysis models, the Cigar Lake ore should have been totally oxidised within 200 million years after formation. Clearly, this is not the case and, thus, these radiolysis models are grossly overconservative (Karlsson et al., 1994). In an attempt to improve radiolysis models used in the performance assessment of spent fuel repositories, recent studies at Cigar Lake aimed to establish a more theoretical basis for radiolytic models to understand the radiation fields around uraninite grains and to establish what fraction of the radiation reaches the groundwater to initiate radiolysis. These modelling studies have found that only a very small fraction of the total radiation impacts on the groundwater to cause radiolysis and revised models more closely replicate the extent of oxidation observed in the Cigar Lake ore (Smellie and Karlsson, 1996). This is a good example of the use of natural analogue data to develop and improve performance assessment models.

The earlier studies at Cigar Lake concluded that the large accumulations of ferric iron oxide in the aureole at the ore/clay interface were evidence of an outward propagating redox front driven by a continuous supply of oxidants produced by radiolysis. However, as the revised radiolysis models now predict much lesser oxidant production, this explanation for the iron rich aureole has been discarded. The groundwater composition in the vicinity of the ore would also mean that iron would be insoluble and could not migrate out to the clay, which also counts against the redox front theory.

The current explanation for the iron rich halo is that it represents a fossil reaction front dating back to the time of orebody formation. As the temperature in the newly formed ore decreased, hematite became stable and formed pervasively throughout the orebody and into the surrounding clay. It may be that a similar process was also responsible for the iron rich halo at Oklo.

In uranium orebodies, not only is radiolysis of groundwater observable but so is radiolytic alteration of organic materials, including aromatisation, dehydrogenisation, polymerisation and oxidation (Hofmann, 1996). These processes have been observed in a number of different orebody types, including roll-front deposits, uraninite placer deposits and black shales (e.g. Leventhal et al., 1986; Lewen and Buchardt, 1989; Meyer et al., 1991; Landais, 1993; Nagy et al., 1993). Most of these were investigated as academic investigations unrelated to analogue studies. The observed radiolytic alteration processes result in an immobilisation of liquid or soluble organic matter, a destruction of biomarkers and shifts in the carbon isotopic composition. However, little is known about possible mobile, low molecular weight radiolysis products in natural systems. The formation of surface films consisting of aromatic organic polymers on radioactive minerals from Australia was described by Rasmussen et al. (1989, 1993).

Similar processes observed at Oklo have been used by Nagy et al. (1991) to suggest that these systems are good analogues for the bitumen wasteform. However, these suggestions were criticised by Alexander and Miller (1994) because of key differences with the bitumen immobilisation matrix. In particular, at Oklo uraninite exists as discrete crystals or aggregates of crystals in the bitumen, leading to highly localised radiation effects, whereas in the bitumen immobilisation matrix, waste is homogeneously dispersed throughout the bitumen leading to low, non-localised doses. Further, the natural bitumen (and kerogen) at Oklo is unlike the technical bitumen used in waste encapsulation and, indeed, at several instances in Oklo, the material is more akin to graphite, as discussed in Section 4.7.

How common is radiolysis in nature ?

In a review of uranium minerals at various ore deposits as possible natural analogues for the alteration of spent fuel, Finch and Ewing (1991) discussed radiolysis but also concluded that no known studies, at that time, adequately address the significance of this process. These authors, however, did note that the natural occurrence of some uranium peroxides, specifically studtite ($UO_4{\cdot}4H_2O$) and metastudtite ($UO_4{\cdot}2H_2O$), indicate the existence of highly oxidising conditions which may result from radiolysis in the geological environment.

In the same study, Finch and Ewing (1991) report that these minerals occur at the Shinkolobwe mine in Zaire, although this line of inquiry has not been taken any further. A detailed description of the Shinkolobwe mine is given in Finch and Ewing (1989). Hofmann (1996) reports that these minerals also occur at Menzenschwand in Germany. In both locations, studtite is associated with primary pitchblende and secondary uranyl silicates in near-surface weathering environments. Radiolysis may only be responsible for peroxide formation, while uranium oxidation is most likely the result of interaction with the atmosphere.

The identification of radiolysis at Oklo, Cigar Lake, Cluff Lake, Rabbit Lake, Shinkolobwe and Menzenschwand which, collectively, represent a range of geological environments and chemical conditions, suggest that radiolysis is a common feature in nature in systems where natural high-radiation fields occur. It is probable, therefore, that the radiolytic processes that would occur in a repository would be identical in mechanism to those observable in natural systems. Only the rates of radiolysis might be different due to the different radiation fields in a HLW or spent fuel repository compared to a uranium orebody.

However, there are indications that radiolysis may occur in nature at locations without particularly high radiation fields. Hofmann (1992) discusses the common occurrence of reduction spots (local reduction phenomena) in red bed sediments without any evident source of reductants. These have been interpreted as the result of porewater radiolysis followed by catalysed reduction of trace elements by H_2. These reduction spot features are relevant to the behaviour of redox fronts and are discussed in more detail in Section 5.5.

Conclusions

Radiolysis of groundwater appears readily to occur in nature in a range of geological environments where radioactive minerals can be found. This suggests that similar radiolysis processes will occur in a repository near-field, but at a rate controlled by the activity of the waste.

Models of radiolysis in the near-field environment assume that hydrogen will escape rapidly from the repository near-field and the resultant redox front will migrate outwards in all directions from the canister in a simple manner. This model does not explain the observed redox environment at the Oklo natural reactors where, although radiolysis did occur, hydrogen did not completely escape the system but apparently stayed and maintained 'pockets' of reducing conditions. This has important implications for performance assessment because, if hydrogen behaved in the same way in a repository, then the models could be overly conservative. However, it is not clear if the redox conditions at Oklo result from specific physico-chemical conditions (e.g. the extreme temperatures) which are not representative of the repository environment. If this is true then the theoretical models may still accurately describe the radiolytic development of the near-field. Experimental studies on material from the Cigar Lake site confirm that radiolytic oxidation of UO_2 can occur.

Due to the uncertain relating to the likelihood and impact of radiolysis in the repository and in nature, further natural analogue studies in relevant geological environments should be encouraged.

5.5 Redox fronts

Redox fronts are created at the boundary between two rock/groundwater systems with different oxidation environments. The development of redox fronts in the near and far-fields of repositories of all designs is usually unavoidable. The normal condition of rocks and groundwaters at depth is reducing and, as a consequence, the introduction of air and oxidising waters into a repository during its excavation and construction will cause oxidation of the exposed rock surfaces, and a redox front will be established.

If the excavated repository is left open for any length of time, the continuous supply of oxidants will cause the redox front to migrate from the excavation walls into the rock. This might occur if a repository is left open for an extended period of time for monitoring and retrievability considerations, before a decision is made to close and seal the facility. A good example of this process was observed after the excavation of the Grimsel Test Site (Baertschi et al., 1991), where uranium and iron rich exudations coated the tunnel walls after aerated porewaters migrated into the rock, causing remobilisation and outward diffusion of uranium held in the rock matrix. After final repository closure, air held in porespaces and dissolved in the local groundwaters will begin to be consumed in redox reactions and, over some decades, the near-field will become reducing. At the same time, in a cementitious repository, the pH will begin to rise, as shown in Figure 5.19.

Other redox fronts may occur in a repository far-field where recharging groundwaters, which will be oxidising, infiltrate downwards and create a redox front at the interface with the deeper reducing waters. However, the most significant cause of redox fronts in HLW and spent fuel repositories will be the waste-induced radiolysis discussed in Section 5.4.

The engineered barriers of most repository designs will include large volumes of steel which will act as redox buffers, scavenging the oxidising species to maintain reducing conditions until all the iron has been oxidised. However, in the case of a repository without much steel in the near-field (such as a spent fuel repository design with copper canisters), the redox buffering capacity may be limited by the accessibility of iron in the barrier materials and in the rock. However, if bentonite is present, this can have a potentially large redox buffering capacity. For example, in the current design for the proposed Swiss HLW repository, there will be some 50 m^3 of bentonite per canister. According to the analysis of Müller-Vonmoos and Kahr (1983), the iron content of the bentonite MX-80 is, taking pyrite in the bentonite into account, 2.4 mg/g of Fe^{2+} and 25 mg/g of Fe^{3+}. However, experimental work by Eriksen and Ndalamba (1988) shows that not all of the Fe^{3+} is available for reaction with radiolytically produced reductants (mainly hydrogen) and only a fraction

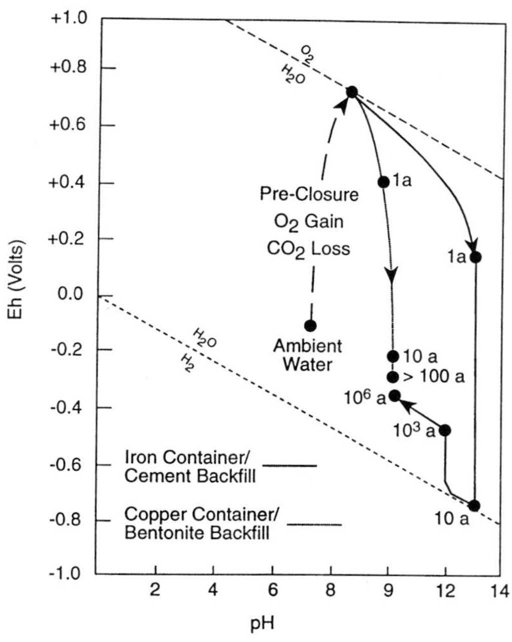

Figure 5.19: The evolution of the near-field chemistry for two repository designs. In all cases, the near-field will become oxidising during the operational phase, creating a redox front in the near-field rock, but will return to reducing conditions after closure.

of the Fe^{2+} is accessible for reaction with radiolytic oxidants (mainly hydrogen peroxide).

If the accessibility of iron in the near-field is limited, the near-field may become oxidising and a redox front will be established. In this case, the distance the redox front migrates into the far-field is controlled by the oxidant production rate, and the content and accessibility of ferrous iron in the rock and barrier materials available for redox reactions. However, it is unlikely that a redox front may move great distances and, as a guideline, it has been calculated that in a rock with 0.2 % Fe^{2+}, a redox front may move approximately 50 m in 10^6 years (Neretnieks and Åslund, 1983a,b).

As the solubility and speciation of many radionuclides is strongly influenced by the redox environment, their potential for transport changes significantly across a redox front. Simplistically, if the near-field of a repository is mildly oxidising due to groundwater radiolysis, radionuclides migrating outwards would precipitate and accumulate once they reached more reducing conditions at the redox front. This would, possibly, be favourable for radionuclide retention but could be problematic if such precipitates were in the form of mobile colloids. In the very unlikely event of a redox front 'breaking through' into oxidising waters in a deep fracture zone, a high concentration pulse of radionuclides might be released.

Redox fronts, and elemental accumulations at them, occur naturally in rock formations where ever a groundwater passes from reducing to oxidising conditions, or vice versa. Some economic ore deposits (and uneconomic trace element accumulations) form in these situations when mineral-rich fluids precipitate dissolved species on encountering a change in the physico-chemical environment, i.e. at a geochemical discontinuity where either temperature, pressure, pH or Eh changes significantly. As some of these ore deposits are formed at redox fronts they may, at first sight, be considered analogous to the process of radionuclide accumulation at redox fronts in the repository. However, in many cases, ore deposits form from precipitation from hydrothermal or pneumatolitic fluids, whose temperatures (which range up to 500 or 600°C) are significantly above the temperatures expected in the repository environment and, as such, they are not directly relevant as natural analogues. Furthermore, high levels of organic material are often involved in immobilising uranium from the fluids, such as the uranium mineralisation at Oklo, and this further limits their analogue potential.

However, there are certain types of orebody that are relevant and one type that has been proposed as a suitable natural analogue (Chapman et al., 1984) is the secondary uranium ore or *roll-front* deposits, because some of these may have formed at temperatures similar to those expected in a repository. Their action as redox traps and their uraniferous nature means that they could potentially be useful as natural analogues of a repository redox front. However, microbial activity and organic material have been implicated in the formation of some of these roll front deposits, so care must be taken to establish the extent of any analogue because microbes and organic material would be expected to be present in a deep repository in lower concentrations than occur at roll-fronts.

Figure 5.20: Photograph of a redox front in a recently exposed face at the Osamu Utsumi mine. The oxidised rock is to the right and the reduced rock to the left. The sharp redox front boundary is well defined.

Although some studies of the implications of redox fronts have been undertaken in the laboratory or in modelling studies, this is an issue which is particularly suited to field investigation. A summary of the significance of redox fronts in a repository and in nature is given by Hofmann (1999).

The issues of most relevance to redox fronts which have been (or potentially could be) addressed in natural analogue studies are:

- redox front formation and behaviour in crystalline rocks;
- redox front formation and behaviour in argillaceous rocks; and
- modelling radionuclide migration at a redox front.

These issues are discussed in the following sections.

Redox front formation and behaviour in crystalline rocks

One of the most detailed investigations of redox fronts performed so far was undertaken at the Osamu Utsumi uranium mine, Brazil, as part of the Poços de Caldas natural analogue project (see Box 14). The rocks at the Osamu Utsumi mine are predominantly phonolites: the overlying rocks are weathered and oxidised and are a brownish-red colour due to the presence of iron oxyhydroxides, while the underlying rock is fresh and reduced and is blue-grey in colour, containing disseminated pyrite (Figure 5.20).

The redox front between the two rock types is generally very sharp, but irregular in profile due to variations in the physical properties of the rock and the influence of hydraulically active fractures and faults along which oxidising surface waters have penetrated deeper into the rock mass, as can be seen in Figure 5.21. Secondary uranium mineralisation occurs at the redox front itself, and the mine has, consequently, been excavated down to this level.

The redox fronts at the Osamu Utsumi mine were the focus of much study as part of the Poços de Caldas project and detailed mineralogical, hydrochemical and chemical and isotopic investigations of the redox fronts were undertaken (Waber, 1990; Nordstrom et al., 1990b; MacKenzie et al., 1990a). These included uranium-series disequilibrium studies, which allowed the rate of front movement to be approximated. These investigations revealed

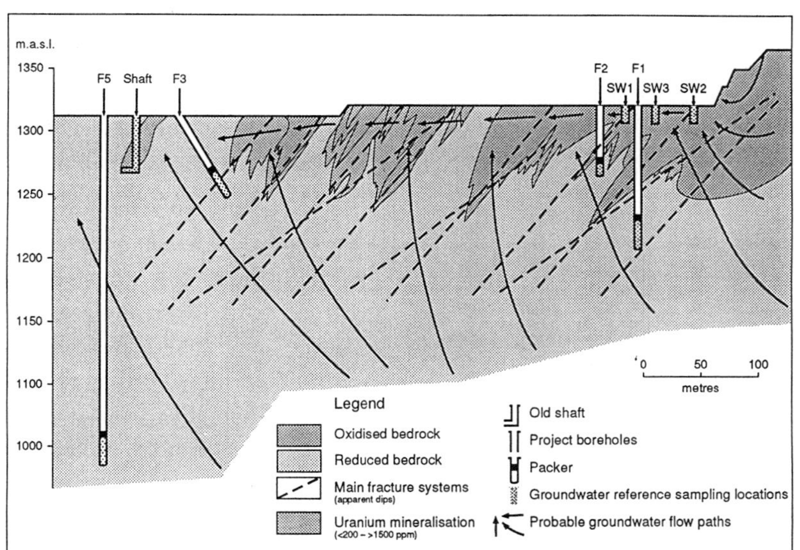

Figure 5.21: Cross-section through the Osamu Utsumi mine marking the position of the redox fronts. It is clear that the positions of the redox fronts are controlled by the fractures as 'fingers' of oxidation extend downwards into the deeper rock along them. From Cross et al. (1990).

that, over a period of some 10^6 years, advective groundwater flow has occurred along the fractures and parallel to the redox front. Diffusive transport of dissolved species occurred perpendicular to the fractures.

There is unambiguous evidence at the Osamu Utsumi mine for uranium precipitation and enrichment on the reducing side of redox fronts within the last 3×10^5 years. This uranium exists as a thin precipitate coating on other minerals and as nodules. A lesser degree of uranium enrichment has also occurred on the oxidising side of the redox front over the past 7×10^5 years, as sorption in association with iron oxides. It was concluded that the redox fronts migrate at rates of 1 to 20 m in 10^6 years, which is about the same as the rate of erosion at the site. As the redox front migrates, uraniferous nodules which formed on the reducing side become overtaken by oxidising conditions and slowly dissolve, as can be seen in Figure 5.22. Some of this uranium is held by the iron oxides on the oxidising side and sufficient uranium may be held by this process to cause a net uranium enrichment of the oxidised rock. However, the majority diffuses through the redox front, back to the reducing side. As the redox front migrates further, the uranium in the rock through which it passes becomes entrained in the process and it follows that the concentration of uranium at the reducing side of the redox front gradually increases.

In general, all those elements subject to oxidative redistribution migrate and accumulate with uranium. Another very important observation was that many redox insensitive trace elements were also concentrated, to greater or lesser degrees, at the redox front due to co-precipitation, solid solution or sorption processes involving iron oxyhydroxides that formed on the oxidised side of the redox front. In addition, some redox insensitive trace elements were concentrated at the reducing side of the front, possibly as a result of incorporation in minerals such as secondary UO_2 or FeS_2.

The primary difference between the redox fronts at the Osamu Utsumi mine and those which might occur in a repository near-field is that those in the mine are connected to and driven by the Earth's atmospheric system and near-surface groundwater flow, while those in the repository are not.

However, there are natural examples of deep redox fronts which also are disconnected from the atmosphere and near-surface waters. These include red-beds which were originally deposited as iron-rich sediments in redox equilibrium with the atmosphere (Hofmann, 1999). Some of these red-beds can now be found deeply buried beneath reduced sedimentary rocks and, thus, generate an

Figure 5.22: Hand specimen from the Osamu Utsumi mine clearly showing the sharp redox front separating oxidised rock (reddish-brown) from reduced rock (bluish-grey). This specimen is highly mineralised with black pitchblende nodules on the reduced side while the circular white areas in the oxidised rock represent relicts of dissolved pitchblende nodules. From Waber et al. (1990).

oxidising geochemical environment at depth and a redox front between the red-beds and the overlying reducing sediments. Such systems provide an opportunity to investigate redox front processes in an environment more closely related to a repository host rock than near-surface redox fronts. However, as far as is known, these deeply buried red-beds have not yet been investigated in detail as a natural analogue.

Much smaller natural systems which provide redox fronts at depth, also unconnected to the atmosphere and near-surface waters, are *reduction spots*. These are small, centimetre sized features found in some rocks which are comprised of roughly spherical accumulations of redox sensitive elements surrounded by a hematite dissolution halo. An example can be seen in Figure 5.16. Reduction spots may form in a variety of rock types, both sedimentary and crystalline, but are common in both continental and marine red-bed sequences (Hofmann, 1999).

Examples include altered igneous rocks, such as the Permian volcanics in the Saar-Nahe trough, Germany; lamprophyre, gneiss and granite in Northern Switzerland; and felsic volcanics and granite in the Black Forest, Germany (Hofmann, 1990b).

The actual mechanism of formation of reduction spots is still unclear but involves a strong initial oxidation reaction that may be microbially controlled. Regardless of the mode of formation, a redox zone migrates outwards from the centre causing redistribution of a number of elements in the rock.

A number of the elements involved have relevance for radioactive waste disposal and include uranium, palladium and rare-earth elements (as chemical analogues, as discussed in Section 3.1).

Very high concentration gradients of uranium and other elements may persist close to the redox front in the reduction spots for up to 10^8 years, demonstrating the immobility of the elements once reducing conditions are established. It is possible to model the growth of the reduction spots and hence the rate of redox front migration, although this may bear no relation to the rate of redox front migration in a repository because it must be controlled, at least partially, by the supply of reductants.

Redox front formation and behaviour in argillaceous rocks

Redox fronts form part of a number of systems which have been investigated as natural analogues of radionuclide transport and retardation in sediments and sedimentary rocks. Perhaps the best example of redox fronts in sediments are those investigated at the Madeira Abyssal Plain

(Colley and Thompson, 1991) which was discussed in Section 5.2.

At the Madeira Abyssal Plain, redox fronts migrated downwards as oxygenated seawater penetrated marine sediments. The downward migration of the redox front stopped when a new layer of sediment was laid down (a turbidite sequence) which effectively cut off the supply of oxygenated seawater to the front making it immobile (Figure 5.8). The aim of this study was to investigate the transport of radionuclides away from the inactive redox fronts, i.e. after a new turbidite sequence was deposited. Unfortunately, it follows that little information was produced regarding the behaviour of the active redox fronts.

Such redox fronts are still active in the upper layers of marine sediments today, and these may prove amenable to natural analogue investigation. In particular, as there is often a distinct spatial separation between the various types of chemical front found (e.g. NO_2/NO_3, iron, manganese, SO_4 etc.), it may be possible to examine the specific associations of trace metals of interest with the various chemical fronts.

Alternatively, similar redox fronts also form in lakes but these may be more difficult to study due to the generally faster sedimentation rate in enclosed lakes as opposed to deep sea areas. The sedimentation rate will need to be accounted for when measuring redox front penetration depths.

Reduction spots may also be used to examine redox front behaviour in argillaceous rocks in the same manner previously discussed for crystalline rock. Hofmann (1990a) examined many reduction spots from hematitic rocks from northern Switzerland, including argillaceous samples. The mechanism of reduction spot formation is independent of rock type although the kinetics of the process may be not be.

Modelling radionuclide migration at a redox front

Since there is little doubt that redox conditions in a repository potentially can influence radionuclide migration, there has been much interest in developing geochemical codes capable of modelling redox front processes. However, because the physico-chemical environment at a redox front is complicated, with such large concentration gradients involved, the ability of standard thermodynamic solubility and speciation codes to cope adequately with such a perturbed system is questionable.

One of the component activities of the Poços de Caldas study involved the development and testing of various modelling approaches to describe the reactions occurring in the redox fronts at the Osamu Utsumi mine (Cross et al., 1990). The main conclusions of these tests were that, although the codes could simulate the main pyrite oxidation to iron oxyhydroxide and pitchblende redox reactions, they could not simulate the behaviour of trace elements or demonstrate any true predictive capabilities at the redox front.

Mass balance calculations tended to overpredict the rate of redox front movement, whilst coupled thermodynamic/transport codes could predict major mineralogical changes but poorly simulated the important pH/Eh buffering reactions. Kinetic models were able to provide a detailed representation of the redox front but only after much 'fitting' to the system and hence have a poor 'blind' predictive capability.

It follows that much more code and thermodynamic database development is required before such codes could be used with confidence in performance assessment. As part of code development it would be necessary to test new codes on other well-characterised redox fronts in the same manner as standard geochemical codes are tested using the blind predictive modelling method discussed in Section 5.1. This would require analogue studies at a number of locations and, as a consequence, additional suitable natural analogue redox front environments must be identified. Similarly, other types of naturally occurring chemical fronts would also be useful for testing these geochemical codes.

Conclusions

Conceptual understanding of redox front formation and behaviour in fractured crystalline rock has progressed as a result of the Poços de Caldas natural analogue study. It is clear from this investigation that redox front locations are controlled by a system of hydraulically-active fractures but that elemental transfer over the redox front is predominantly diffusive. This will also be the case for radiolytically induced redox fronts migrating out of the near-fields of HLW, spent fuel and high activity ILW repositories.

The redox front controls strongly the mobility of redox sensitive elements with progressively higher concentrations forming at the leading edge of the redox front. Some redox insensitive trace elements are also concentrated at the redox front due to co-precipitation, solid solution or sorption processes involving secondary iron minerals formed around the front.

In the short-term, redox fronts may inhibit radionuclide migration by acting as a 'barrier' beyond which mobile radionuclides cannot migrate. In the longer-term, however, if a redox front continued to progress into the far-field, driven by radiolytic oxidants, then it potentially could transport radionuclides into a zone where they could be more readily mobilised by oxidising groundwaters, causing a pulse release. Although this scale of movement of a redox front seems highly unlikely, given the very slow rates of movement found even at sites with a very high rate of supply of oxidants (e.g. Poços de Caldas), the concept may require consideration in scoping calculations.

Redox front systems in argillaceous rocks have been poorly studied. Some investigations have been performed in marine and lake sediments, where the principal objective has been to examine the diffusion of radionuclides away from the redox front after it has become inactive. This is of direct relevance to the type of modelling which a performance assessment would need to include, and very slow rates of movement have been determined. It is possible that examination of the active redox fronts in sediments, at the top of the sediment pile, may provide some additional useful information in the absence of other useful redox systems in sedimentary rocks.

The Poços de Caldas natural analogue study has provided the only test so far of the ability of thermodynamic solubility and speciation codes to replicate a natural redox front system. In general, it was found that none of the codes tested was able accurately to model the mineralogy observed at the redox front and the rate of front movement. More code and database development is necessary before these codes could be used with confidence in a performance assessment to model an active radiolysis driven redox front in a repository near-field.

5.6 Colloids

Standard radionuclide transport models are based on the solution chemistry of radionuclides leached from the wasteform and components of the natural rock-groundwater system. An implicit assumption in these models is that radionuclides are transported only as dissolved species in the groundwater, as discussed in Section 5.1. As a result, these models may be overly simplistic and possibly non-conservative due to the fact that they neglect the potential effect of advective transport of radionuclides bound to colloids (e.g. McCarthy and Degueldre, 1991; Moulin and Ouzounian, 1992).

Colloids, in the sense used here, comprise suspended material in the size range 1 μm to 1 nm

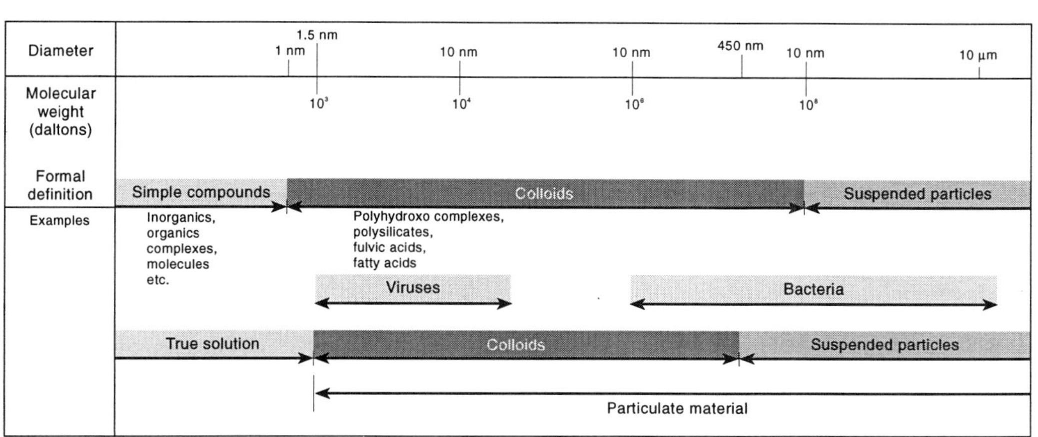

Figure 5.23: Diagram showing the relative sizes of operationally defined true solution, colloids and suspended particles, together with viruses and bacteria. Examples of compounds that lie in these size ranges are given.

which are dispersed in groundwater (Stumm and Morgan, 1981). In reality, it is difficult to measure the sizes of fine material in solution and, often, more pragmatic size limits are defined on the basis of filter cut-off sizes, as shown diagrammatically in Figure 5.23.

It is possible to define two types of colloids on the basis of their mode of radionuclide uptake (Avogadro and de Marsily, 1984; Kim et al., 1984). The first group are colloids produced by the processes of nucleation and growth of radionuclide-bearing solid species from solution and the second group comprise preexisting groundwater colloids (such as silica colloids and clay materials) containing adsorbed radionuclides. This second group includes colloids formed from the breakdown of the wasteform or the engineered barriers and may also include organic macromolecules, such as humic and fulvic acids, which complex strongly with some radionuclide ions. These two groups are sometimes referred to in the radioactive waste literature as *true-colloids* and *pseudo-colloids* respectively. This distinction is not, however, considered helpful since all colloids are potentially capable of affecting radionuclide transport and, thus, for the remainder of this discussion, all stable suspended material in the colloid size range, regardless of mode of radionuclide uptake, will simply be termed colloids.

The very small size of colloids means that a given mass will possess an extremely high specific surface area; up to 10^3 m^2/g for the smallest. This means that the physico-chemical behaviour of colloids is strongly controlled by surface reactions and interactions; for example they have a substantial capacity to adsorb molecular or ionic species in solution. It follows that a large colloidal population in groundwaters may have important consequences for the behaviour of radionuclides. As an example, the potential interactions between colloids, radionuclides and mineral surfaces in a fractured rock system are indicated on Figure 5.24. Firstly, the presence of colloids would mean that the traditional two component (rock-water) system would be invalid for modelling solute transport if radionuclides were substantially taken up by the colloids. This would have particular importance for the determination of in situ K_ds because the radionuclide concentration measured from solution would then include a significant component of radionuclides present as colloids. In fact, numerous laboratory studies have demonstrated that radionuclide uptake by colloids does occur, at least under these controlled conditions. Field studies have also identified radionuclides in

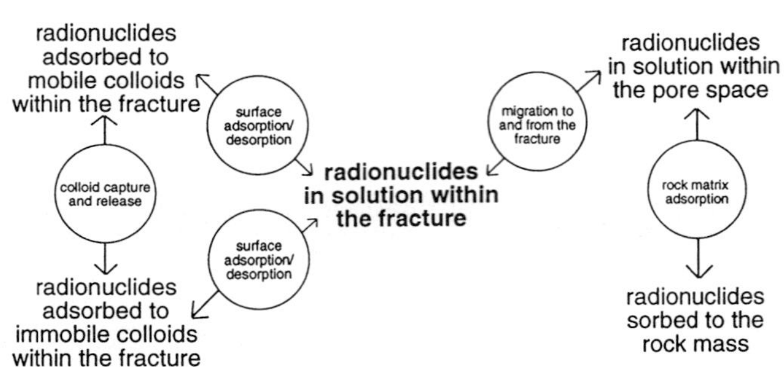

Figure 5.24: Diagram showing the potential interactions between colloids, radionuclides and mineral surfaces in a fractured rock system. The significance of these colloidal processes for repository performance is difficult to assess.

association with colloids. Secondly, radionuclides sorbed onto, or precipitated as, colloids potentially could be advectively transported at the groundwater flow rate. Such transport has been shown to occur in studies performed on shallow groundwater systems (McCarthy and Zacchara, 1989; Ryan and Gschwend, 1990; Gschwend et al., 1990). Nonetheless, before colloids could lead to significant radionuclide transport from a repository to the biosphere, a number of obstacles would need to be overcome. The circumstances under which colloids would become significant for repository safety are illustrated in Figure 5.25.

The combination of laboratory studies, which show radionuclide uptake on colloids, and field experiments, which show colloid transport in shallow systems, have highlighted the need to try and assess colloidal transport of radionuclides in deep groundwaters, i.e. under repository-relevant conditions. Several field studies have measured sizeable colloidal populations in deep groundwaters, but this information alone is not sufficient. What is required is an unambiguous demonstration of colloid transport in an appropriate geological formation over repository relevant distances, and this can only be obtained by natural analogue studies. The ideal location for such a demonstration would be a deep ore body (a point source) from which any mobile colloids could be followed downstream. This is basically the approach adopted in the Cigar Lake and Poços de Caldas natural analogue studies, which are discussed later.

Within the repository environment, colloidal transport is considered potentially most significant in the far-field. In the near-field of a HLW or spent fuel repository, a particular safety role of the compacted bentonite buffer is to act as a colloidal and macromolecule filter. It is generally assumed that the pore spaces within the compacted bentonite are too small to allow advection of colloids to occur. Laboratory studies indicate that this is either true or that mobility is extremely slow (Torstenfelt et al., 1982a,b; Eriksen and Jacobsson, 1982). The cementitious material in the near-field of a L/ILW should also act as a colloidal filter but degradation of concrete may produce colloids which could be mobile at the outer edges of the cement mass.

The growing appreciation of the potential importance of colloids has meant that a number of natural analogue studies have attempted to characterise colloids in deep groundwaters and to quantify the effects they might have for repository safety. It is worth noting at this stage that few of these natural analogue studies have realised their objectives. This is partly due to the operational difficulties inherent in measuring and characterising these particles. Indeed, the very act of sampling colloids may influence the nature of the colloids; for example, drilling and sampling is very likely to cause colloids to form. Also, cation uptake on filters may alter colloid:solution ratios (Degueldre et al., 1990). Therefore, it is always difficult to know if measured colloid populations and characteristics reflect the true nature of colloids at depth.

Two analytical methods are commonly used to characterise the particle size distribution in groundwater samples: *ultracentrifugation* and *ultrafiltration*. Both of these techniques have their drawbacks; for example, ultracentrifugation requires equipment too large to be used in the field, requiring samples to be returned to the laboratory. The resulting delay may mean that the nature of the colloids changes prior to measurement, e.g. due to dissolution or precipitation reactions, or by colloid coagulation. Furthermore, this technique cannot deal with low concentrations of colloids because, as the centrifuge stops spinning, substantial proportions of colloids may rediffuse back into the groundwater.

The alternative technique, ultrafiltration, may be used in the field with care but suffers from the physical action of filtration through fine membranes which can break down colloids, causing a change in their nature and size distribution. A detailed description of colloid sampling and measurement techniques is given in McCarthy and Degueldre (1991). Despite the analytical problems, the potential importance of colloids for repository safety requires that they be studied. However, it is important that a distinction is drawn between colloid studies that are effectively a component of a site characterisation exercise and those that are true natural analogue studies.

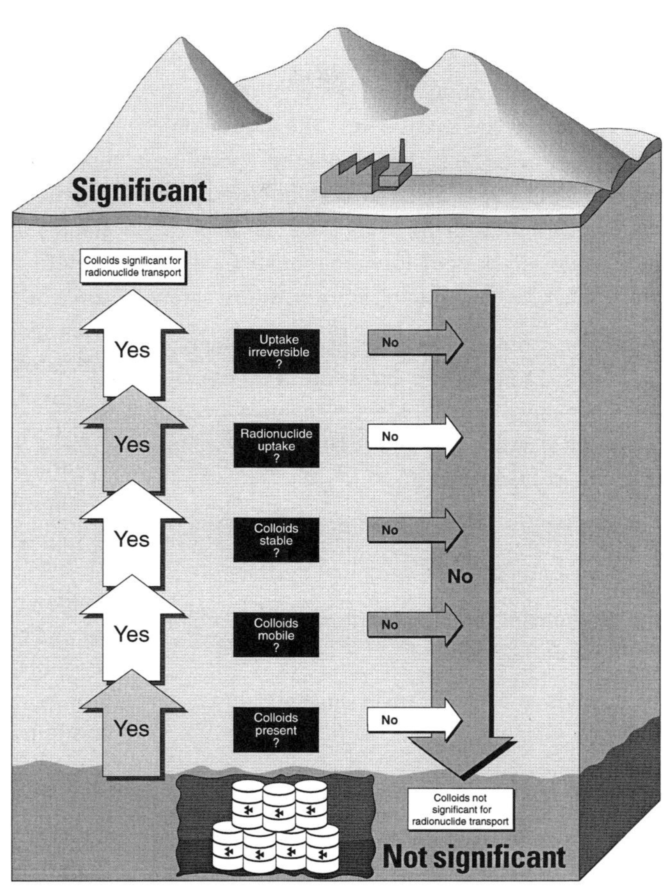

Figure 5.25: Diagram showing when colloid behaviour becomes significant for repository safety. Generally, if colloids are either absent, or immobile, unstable and do not sorb radionuclides they are not significant.

Whilst some of the colloid studies performed to date are interesting in their own right, they do not say much about the likely effects colloid populations might have on radionuclide behaviour at possible repository locations, i.e. the information obtained is not transferable. This is principally due to the fact that there is little to be gained from a performance assessment standpoint from studying colloids as an isolated phenomenon. Instead, colloids need to be studied within the context of the groundwater flow and radionuclide transport system.

As mentioned earlier, there have been a number of natural analogue and field studies that have attempted to investigate colloids in a range of geological environments, including deep crystalline rocks (e.g. Degueldre et al. 1989; Billion et al. 1991; Laaksoharju and Degueldre, 1994), sedimentary sequences (e.g. Kim et al., 1984), unsaturated volcanic tuffs (e.g. Kingston, 1989) and host rocks around orebodies (Miekeley et al., 1990b, 1992; Short et al., 1988; Ivanovich and Harmon, 1992; Vilks et al., 1993). However, few of these studies have been notable successes. The others were less successful for a number of reasons, not least being the fact that in some cases the colloid populations were extremely low, for example, at the Grimsel Test Site (Degueldre et al., 1990). This fact should not be ignored or misinterpreted; it suggests that high populations of colloids in locations analogous to that of a radioactive waste repository are probably unusual, which must be encouraging for repository safety.

Given the potential significance of colloids for enhancing radionuclide transport, the issues of most relevance to colloids which have been (or potentially could be) addressed in natural analogue studies are:

- populations of colloids in natural systems;
- stability of colloids in natural systems;
- radionuclide uptake and transport by colloids in natural systems;
- colloids in anthropogenic systems; and
- effect of biocolloids.

These issues are discussed in the following sections.

Populations of colloids in natural systems

Colloid populations have been examined in several subsurface environments since the potential significance of colloids has been highlighted and more reliable methods for colloid sampling and characterisation have been devised. The results and data obtained from many of these studies has been compiled by McCarthy and Degueldre (1993).

Unfortunately, most of the studies mentioned in this compilation are field investigations involving injection experiments and are not true analogue studies. In fact, only a few analogue studies have investigated colloids in detail: these are Cigar Lake, Poços de Caldas, Oklo, Palmottu and Alligator Rivers. The results from some of these studies are discussed below.

As an integral part of the Cigar Lake natural analogue study (see Box 5), colloids have been examined in order to evaluate their impact on radionuclide transport within, and away from, the orebody (Vilks et al., 1991a, 1993). The physico-chemical conditions at Cigar Lake are sufficiently similar to a repository environment to make it a useful site for the purpose of predicting the likely consequences of colloids within a repository. Analysis of the deep groundwaters at Cigar Lake indicates that the colloid population is approximately 8 mg/l, and it has been calculated that less than 0.01 % of radionuclides, in a given volume of rock, would be likely to sorb onto these colloids at any one time (Vilks et al., 1991). Whilst this is apparently a low figure, if it were to apply to the repository environment, then over the lifetime of the repository it may still be significant for radionuclide transport. The average uranium concentration of the colloids in the ore zone is 0.06 % uranium, compared to an average uranium concentration of the ore of 12 % uranium. These low concentrations suggest that the majority of the colloids at Cigar Lake have been formed from the clay and not from the ore. However, colloids from the sandstone (separated from the ore by the clay) have even lower uranium concentrations. The isotopic evidence indicates that the majority of the sorbed uranium and radium is derived from the groundwater and, furthermore, that some of the colloids have retained uranium for up to 8000 years.

The colloid population is too low significantly to affect radionuclide migration if sorption is reversible. However, the fact that some uranium may have been sorbed for thousands of years indicates that some sorption, at least, is long-lasting and that the colloids are stable. In view of this, it cannot be concluded that the low colloid population is insignificant with respect to radionuclide transport over geological timescales. The low uranium concentration of colloids from the sandstone suggests that the clay formation acts as an efficient barrier to radionuclide migration, as discussed in Section 4.5. This is consistent with other information which suggests that the clay is plastic and forms effective hydraulic seals between the ore and nearby water-bearing fractures.

One of the principal aims of the Poços de Caldas study (see Box 14) was to quantify the importance of colloids with regard to elemental transport. Two studied sites presented different rock-groundwater systems in which to investigate colloid behaviour and these are discussed separately. Full details of the sampling methodology and experimental results are given by Miekeley et al. (1990b, 1992). At the Osamu Utsumi uranium mine, abrupt changes in the groundwater chemistry were measured. There was a clear distinction between the nature of colloids identified from the shallow and deep waters. In the shallow zone, there was a colloid population of 0.8 mg/l, which consisted predominantly of clay particles. In the deeper zone, the colloid population was lower, 0.05 to 0.5 mg/l, consisting of mostly amorphous iron oxyhydroxides. The $^{234}U/^{238}U$ ratios of the colloids were similar to those of the groundwater indicating equilibrium between the particles and the groundwater. However, the $^{230}Th/^{234}U$ ratios for the particles are much higher than for the groundwater, indicating that either thorium is preferentially taken up by, or uranium is preferentially lost from, the colloids. This is confirmed by comparative analysis of the true solution and the colloid fractions which showed that uranium exists mostly in true solution while more than 70 % of thorium is partitioned on the colloids (Figure 5.26).

At the Morro do Ferro thorium orebody, there was an opportunity to sample groundwaters and associated colloids along an entire flow system, from the recharge zone along the flow lines to the discharge zone. The system was well defined, with the recharge zone located at the ore body thus providing an identifiable point-source of thorium and rare-earth elements from which particulate transport could be traced. The characteristics of the colloids identified at Morro do Ferro are similar to those from Osamu Utsumi, being clay particles in the shallow zones and iron oxyhydroxides at depth. In contrast to Osamu Utsumi, however, the colloidal populations at Morro do Ferro were generally higher, from 0.1 to 3.1 mg/l. The uranium and thorium distributions were also similar for the two sites, with high fractions of the thorium and rare-earth elements partitioned on the colloids. Comparative analyses of colloidal fractions along the flow paths, from the ore body to the discharge zone, did not indicate that any significant colloid transport took place. The same conclusion was reached for the larger suspended particles. However, this may be due to the fact that the source of the thorium and rare-earths actually lies in the unsaturated zone. It may well be that any colloids and particles washed out of this zone are efficiently filtered before reaching the saturated zone. As a consequence, the results of the Morro do Ferro study are ambiguous in this regard.

A number of groundwater sampling campaigns were performed at the Alligator Rivers natural analogue study site (see Box 15) in order to determine the extent to which colloids influence radionuclide transport (Edghill and Davey, 1988; Seo, 1991; Seo and Payne 1994; Seo et al., 1994). However, early attempts were hampered by low colloidal populations, but high populations of larger particles which clogged filters and caused general operational difficulties. Nevertheless, some general conclusions can be drawn from the results obtained from these studies. First, the actual numbers of colloids in the groundwaters are very low, usually in the range 10^6 colloids/litre, or less (c.f. Grimsel Test Site, with around 10^{11} colloids/litre). To put this into perspective, 10^6 colloids, each 1 μm in diameter and with a density of 2.65 g/cm, would weigh 1.4 μg. Whilst the number of colloids appears large, their mass approaches insignificance when compared to the mass of some of the radionuclides dissolved in the groundwater

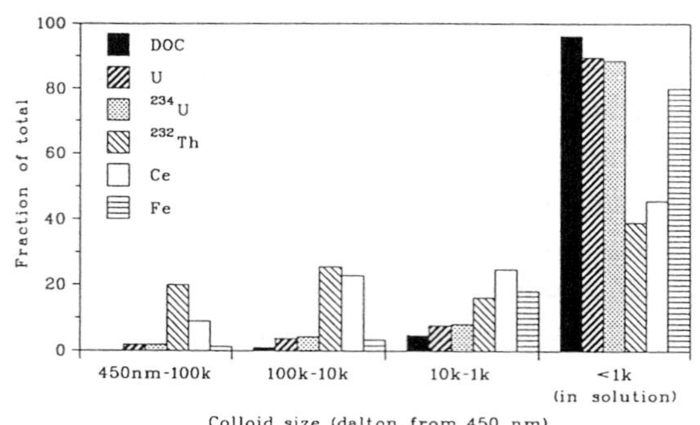

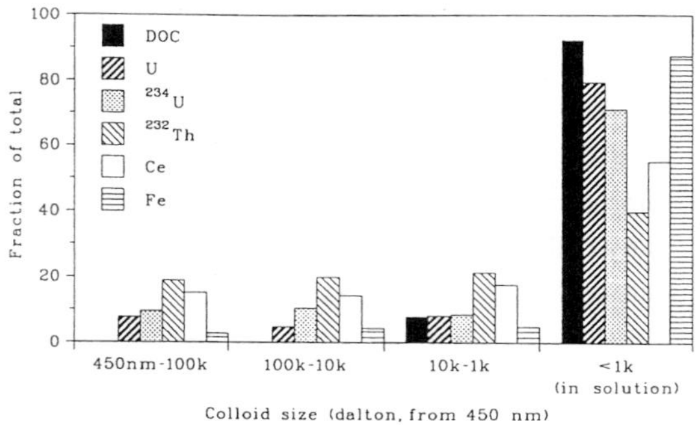

Figure 5.26: Distribution of organic carbon, iron and trace elements on colloids separated into various size ranges. Top: from shallow groundwaters from the Osamu Utsumi mine. Bottom: from deep groundwaters from the Osamu Utsumi mine. It is clear that in both groundwaters the predominant elemental load is in true solution, rather than sorbed onto colloids. From Miekeley et al. (1990b).

(Edghill and Davey, 1988). However, it is important to bear in mind that even a small mass of colloids has a very large surface area available for sorption. Second, measurable radionuclide concentrations associated with colloids are only recorded from groundwater samples from the centre of the orebody, where radionuclide concentrations are at their highest. Analysis of this groundwater makes it evident that almost all the uranium is present in dissolved form, i.e. it passes through the ultrafiltration equipment. In contrast, almost all the thorium in these groundwaters is associated with particles larger than 1 μm.

The dissolved uranium concentrations in the groundwater at Cigar Lake are far below saturation for any U^{6+} species, even though the pH/Eh data suggest that the groundwater is oxidising with respect to U^{4+}. It follows that uranium-bearing colloids could form by precipitation only in pockets of supersaturation, which may only occur where groundwater is in contact with uranium-rich rock for extended periods. Colloids formed by such a process would be short-lived because they would become unstable as soon as the water moved away from the rock and mixed with the generally more oxidising groundwaters. However, radionuclide sorption may also occur onto preexisting colloids. Thus, transport of radionuclides by colloids at the Alligator Rivers study site appears to be negligible. It is, however, difficult to extrapolate data from such a near-surface, oxidising environment to deep disposal systems.

The behaviour of colloids in cementitious L/ILW repositories may be different to that in a HLW repository due to the presence of the cement and possible changes in colloid stability and radionuclide speciation in the high pH environment. In particular, degradation of the cement minerals may provide an abundant source of colloids. These issues were addressed as part of the Maqarin natural analogue study (see Box 11). Evidence from Maqarin is that colloid populations in natural high pH groundwaters are very low. In the first phase of work at the site, a preliminary assessment of the colloidal population was carried out. The results indicated maximum colloidal populations were around 1 ppm. These colloids consisted mostly of $Ca(OH)_2$ and $Fe(OH)_3$ and contained insignificant quantities of uranium (West et al., 1992). Further colloid studies were undertaken in a later phase of the project (Wetton et al., 1998), which indicated populations of 10^7 colloids/litre. Both these results indicate that the colloidal populations at Maqarin are low in comparison to other near-surface waters (Smellie, 1998). Furthermore, no uranium was found to be associated with the colloids, suggesting that radionuclide transport on colloids in high pH environments would be insignificant. However, it is unclear if these results can be applied directly to a L/ILW repository because the groundwaters at Maqarin are oxidising, compared to the reducing waters in a repository. Both the stability of colloids derived from cement minerals and the association of radionuclides with them may be greater in reducing conditions. If this were to be the case, applying the colloid data from Maqarin directly to performance assessment would be non-conservative. Therefore, further analogue investigations into this issue are recommended.

Even after combining the results from all of the analogue and field studies on colloids, it is not possible to predict the populations of colloids as a function of the chemical, hydrological and mineralogical properties of host rocks. However, some general relationships can be drawn.

Colloids appear to be ubiquitous in all groundwater systems and no study has ever revealed a zero population (McCarthy, 1996). However, in some systems, particularly in deep crystalline rocks with a stable geochemical system, the colloid populations are low, typically in the region of around 25 µg/l. In contrast, shallow aquifer systems generally appear to have the largest colloid populations, even in the absence of geochemical instabilities.

Enhanced colloidal populations occur in all rock types when there is some hydrogeochemical perturbation to the system. For example, in fractured granitic systems, colloid populations are 20 to 1000 higher in groundwaters affected by inputs of surface water or in hydrothermal zones with large temperature and pressure gradients, compared to stable hydrogeochemical systems (Degueldre, 1994).

In most systems studied, the composition of inorganic colloids relates directly to the mineralogy of the parent rock and its alteration products. Thus, in fractured granitic systems, the observed silica, clay and mica colloids are consistent with production of primary and secondary minerals from geochemical alteration of the parent rock. Similar associations are observed in other rock types.

Stability of colloids in natural systems

The stability of colloids is important because it is partly responsible for the capacity of colloids to transport radionuclides over long distances. Laboratory investigations of colloids recovered from natural groundwaters tend to suggest that most colloids carry a negative charge and are relatively stable in the groundwater (e.g. Longworth and Ivanovich, 1989; Liang et al., 1993). This is also true for colloids of iron oxide which, in most groundwater pH systems, would be expected to have a positive charge. This charge reversal probably relates to an association with organic matter in the groundwater system (Miekeley et al., 1992; Liang et al., 1993) which may also be able to stabilise colloids which may otherwise be considered unstable in the hydrochemical environment.

The long-term stability of colloids is difficult to assess directly. Colloids formed by the alteration of host rock or fracture coating minerals may be expected to form (precipitate) or disappear (dissolve) along a flow path as geochemical conditions change. Such a process was observed at the Osamu Utsumi mine as part of the Poços de Caldas study. At Osamu Utsumi, iron oxide particles recovered from groundwater were considered to have been generated at depth by the upflowing, reducing waters rather than by the downward transport of material from the surface during rainfall (Miekeley et al., 1992). According to this mechanism, colloids would precipitate in the groundwater if conditions exceeded the solubility for a given solid phase in a zone of chemical saturation. This is thought to explain the presence of uranium colloids at the Alligator Rivers analogue site when the general groundwater chemistry was below saturation for any U(VI) species, as discussed earlier. Here, colloids formed when 'pockets' of groundwater were in contact with regions of uranium-enriched rock for extended periods but became unstable when the water flowed away.

Radionuclide uptake and transport by colloids in natural systems

The uptake and transport of radionuclides by colloids has been investigated in several analogue studies of natural systems with enhanced radionuclide content e.g. at Cigar River (Vilks et al., 1993), Whiteshell (Vilks et al., 1991), Alligator Rivers (Seo and Payne, 1994), Grimsel Test Site (Longworth et al., 1989b), El Berrocal (Rivas et al., 1997) and Poços de Caldas (Miekeley et al., 1992).

In most of these studies it was found that some proportion of the total uranium, thorium and rare-earth elements in the groundwater was associated with the colloids. This proportion was higher for thorium than for uranium because of thorium's lower solubility in most groundwaters. The rare-earths generally show an affinity for colloids that is intermediate between uranium and thorium (Miekeley et al., 1990a, 1992).

This observation was well supported by the analogue studies undertaken at the Steenkampskraal mine in South Africa. This mine is the richest monazite mine in the world, containing 45 % rare-earth element oxides, 4 % ThO_2 and 600 ppm UO_2. Geochemical evidence indicates that some monazite is dissolving in regions where the silicate matrix is exposed to oxidising groundwaters. Given the very high source term concentrations this mine was chosen as a good site to examine colloid transport (Jarvis et al., 1997).

Measured inorganic colloid concentrations in the Steenkampskraal groundwaters ranged between 3×10^{10} to 3×10^{11} colloids/litre. The elemental partitioning between the colloids and the groundwater (true solution) favoured the groundwater, with only 1 % of uranium and thorium, and less than 3 % of the total rare-earth elements being associated with the colloids. Measured solution concentrations of uranium and thorium are very high, with uranium up to 300 ppb and thorium up to 3 ppb. These thorium concentrations are the highest in the world, exceeding even those from Morro do Ferro thorium deposit at Poços de Caldas (see Box 14). However, sampling was performed in the oxidising waters and, thus, it is not surprising that elemental solubilities are high and associations with colloids low. The analogue studies at Steenkampskraal are in an early stage and it is planned to extend the colloid investigations to cover a wider range of water compositions.

Evidence from natural analogue studies for the reversible uptake of radionuclides on colloids is based largely on data from uranium-series disequilibrium studies. However, such measurements are unable to distinguish easily between adsorption and precipitation on colloids in the same way as it cannot on rock and fracture mineral surfaces (as discussed in Section 5.2). Furthermore, it is not clear to what extent the composition of the core of a colloid, which may not interact with the water, affects adsorption at the colloid-groundwater interface. Nonetheless, analogue studies can help to evaluate the extent of radionuclide exchange between colloids and aqueous phases in groundwater. The $^{234}U/^{238}U$ activity ratios for groundwater and inorganic colloidal phases are generally very similar, indicating that uranium in the two phases is in equilibrium (Longworth and Ivanovich, 1989) and that uranium uptake on colloids is readily reversible. In contrast, colloidal and groundwater $^{230}Th/^{234}U$ ratios are often quite different to each other but below unity, indicating that the two phases are not in equilibrium, and that the uptake of more strongly adsorbed elements (such as thorium and, perhaps, plutonium) is not readily reversible.

The reversibility of radionuclide sorption onto colloids is an issue that has not been resolved. If sorption is reversible, then the influence of colloids on the net transport of radionuclides will be limited for the concentrations and sorption properties of colloids measured in most natural deep water systems (Allard et al., 1991). However, if sorption is irreversible, then colloid transport becomes a potentially more significant process, if the colloids are both stable and mobile. Of course, truly irreversible sorption is a thermodynamic impossibility; irreversibility in practical geochemical terms refers to a duration of sorption that is long relative to the colloid transport rate in the deep groundwater system.

Unambiguous evidence from natural systems indicating colloidal transport over long distances is quite rare. Only one example is known that suggests transport on the kilometre scale. This is at Menzenschwand in the Black Forest of Germany, where the chemical signature of the natural colloids (magnesium:titanium ratio and rare-earth element contents) was significantly different to that of the host granite but similar to that of a neighbouring gneiss several kilometres distant and upgradient (Alexander et al., 1990c). These results should not, however, be over-emphasised because the flow system at Menzenschwand is not particularly representative of a repository host rock in that it is perturbed by the presence of a mine with associated fast groundwater flow. However, it does prove the general point that, if conditions are suitable, colloidal transport can occur over long distances.

However, supporting data from natural analogue studies to indicate colloid transport are quite limited and generally only from systems with dimensions of a few metres. In contrast, there are many analogue studies which suggest that colloid transport in natural systems can be significantly restricted. At the Cigar Lake site, the uranium and radium contents of colloids in the ore and the surrounding clay zones are significantly higher than in colloids from the sandstone, indicating that the clay is an effective barrier to colloid migration (Vilks et al., 1993). Similar results at Alligator Rivers and Morro de Ferro also suggest that colloids have a limited capacity for migration because the concentrations of colloid-bound radionuclides outside the orebodies are relatively low (Miekeley et al., 1992; Short et al., 1988).

Colloids in anthropogenic systems

Much of the data suggesting that colloids are mobile derive from observations of movement of radionuclides downgradient of an anthropogenic source of radionuclides, such as underground nuclear bomb test-sites or shallow LLW disposal facilities.

Radionuclide migration associated with underground nuclear tests at the Nevada Test Site in the US has been extensively studied in field investigations (e.g. Thompson, 1984). The host rocks at the test site are hydraulically-unsaturated rhyolitic lavas and are essentially the same as the rocks at the proposed Yucca Mountain HLW repository (see Section 2.3.1), which is located on the outer edge of the test site restricted area. Thus, analogue results from the test site have potential significance for performance assessments for the Yucca Mountain repository but are less relevant to other repository concepts in saturated rocks.

Studies at the test site began in the 1950s, although these early projects were not perceived as 'analogue' studies. Some of these studies have found radionuclide transport at the test site to be much faster than predicted on the basis of laboratory K_d values. In once study, (Coles and Ramspott, 1982) a borehole 91 m away from a detonation cavity was pumped to induce flow. No radioactivity was observed for the first 2 years but, with continued pumping, the concentration of both ^{3}H (tritium) and ^{106}Ru increased at the same rates, suggesting that both travelled at the same velocity from the detonation cavity. This is surprising because tritium is non-sorbing and, thus, moves at the same speed as the groundwater in contrast to ruthenium which is considered to be highly sorbing and, thus, was expected to travel some 30 000 times slower than tritium. It is commonly accepted that laboratory K_d values are poor predictors of field behaviour (see Section 5.2) but the very large discrepancy between predicted and measured ruthenium break-through times was so great, it was suggested that ruthenium transport must have involved colloids.

In another well documented study at the test site (Buddemeier and Hunt, 1988), field investigations focussed on the Cheshire explosion which took place on 14th February 1976 in a cavity at a depth of 1167 m on Pahute Mesa, Silent Canyon Caldera. Water samples were analysed in 1983, 1984 and 1985 from within the explosion cavity and some 300 m outside. In both sample locations, inorganic colloidal particles (3 to 50 nm) were associated with sorbed transition metals (manganese and cobalt) and lanthanide (cerium, europium) radionuclides. Colloid populations were an order of magnitude higher in the cavity than outside. It was concluded that radionuclide transport from the cavity took place by advection of colloids. More recently, plutonium bearing colloids have been found 1.3 km away from the site of the 'Benham' test (Kersting et al., 1999).

Waste water from the Los Alamos National Laboratory has routinely been disposed of in the tuffs of Mortandad Canyon since 1963 (Raloff, 1990). Original transport calculations indicated that movement would be restricted to only a few metres before being retained by the soils. However, Penrose et al. (1990) have detected americium and plutonium in wells some 3390 m away from the disposal point. Over that distance plutonium concentrations decrease exponentially, whereas americium concentrations showed no systematic variation. The radionuclide mobility was ascribed to colloid formation. Colloids responsible for americium mobility were smaller than those responsible for plutonium migration, a finding which corresponds to the apparent greater mobility of americium. There is no indication given of the nature of the colloids involved in the transport.

Whilst these types of study are useful and indicate that colloid transport is potentially significant, the analogues are not particularly close to most repository systems, with the possible exception of Yucca Mountain. In particular, the nuclear detonations at the test site will have severely affected the structural characteristics of the rock and there remains the possibility that some radionuclide transport may be associated with the blast rather than subsequent colloidal transport. As such, no direct application of these field data can be made to performance assessment. Nonetheless, these observations do lend support to the concept that radionuclide transport in the far-field can be facilitated by colloids. Unfortunately, no natural analogue studies have, to date, been able to quantify the importance of this process.

Biocolloids

Microbes will occur in all repositories and can have significant effects on material degradation and the formation of gas. These issues are discussed in Section 5.7. However, microbes can also act as biological colloids, *biocolloids*, and they potentially can enhance radionuclide migration if they are mobile.

It is well understood that microbes can biosorb radionuclides and metals, partly due to the presence of polysaccharides in cells walls that undergo ion exchange with metals in solution. Different microbe species exhibit wide variations in biosorption capacity. As an indication of typical biosorption capacity, thorium association with microbes in soils is comparable in magnitude to thorium sorption to groundwater colloids (Tsezos and Volesky, 1981).

The significance of biocolloids for radionuclide transport is dependent on the relative populations of mobile inorganic colloids and microbes, and their capacity to associate with radionuclides. Clearly, these are site specific issues. However, once associated with microbes, subsequent transport behaviour may be different from inorganic colloids because some microbes can be self-propelled and because most microbes can change certain transport properties (size or surface characteristics) in response to environmental conditions, further complicating predictions of the rate and extent of migration.

No comprehensive natural analogue studies are known to have addressed this issue in detail, although some field-based experiments suggest that transport of biocolloids can occur very rapidly and over long distances. For example, microbes injected into groundwater have been reported to travel at rates of up to 300 m/day (Wood and Ehrlich 1978; Keswick et al. 1982; Harvey et al., 1989). However, most of these field studies have been undertaken in permeable near-surface aquifers and, thus, these data have little relevance to transport in the far-field of a repository.

Conclusions

Several natural analogue studies have investigated colloid populations and their effect on radionuclide transport in repository-relevant conditions but performance assessment relevant conclusions from these studies are conspicuously few. This is partly because colloid populations in deep groundwaters are often too low to be measured effectively, meaning that studies are often inconclusive. However, this general observation does not imply that colloids would not affect radionuclide transport at specific repository sites.

Any assessment of colloid populations in performance assessment cannot be limited only to consideration of the natural colloid population occurring prior to development of the repository, as indicated by analogue studies. Instead, the performance assessment must consider the potential for generating new types or increased populations of colloids due to geochemical disturbances arising from the near-field environment, such as the alkaline plume expected to migrate away from cementitious L/ILW repositories.

It is apparent that, where colloid populations are low but measurable, they may still have the potential to affect radionuclide transport if they are stable and if sorption onto them is not instantaneously reversible. If this is the case, then low colloid populations could account for large fluxes of transported radionuclides over very long time periods if the colloids are mobile.

Although the Cigar Lake and Poços de Caldas environments are dissimilar to each other, similar conclusions were reached at both. The Cigar Lake study illustrates that, under repository-relevant conditions, this combination of stable colloids and long-term sorption can occur. Nonetheless, it would appear that colloids are not very mobile in deep groundwaters. At Cigar Lake, colloid immobility was assumed to be due to effective filtration in a clay formation with no open fractures and small void spaces. At Poços de Caldas, colloids were also immobile, again presumably due to effective filtration.

Further colloid studies should be encouraged if they investigate repository relevant conditions and transport distances. Although Cigar Lake and Poços de Caldas are both very comprehensive natural analogue studies, it would be unwise to conclude from only two detailed studies that colloids are not a significant problem for repository safety. In addition, the reason why higher populations of colloids were found at these two sites than at other analogue study sites should be investigated, particularly in view of the close approximation of some features of Cigar Lake to the repository environment.

Any new study needs to be properly representative of a repository far-field. For example, in a crystalline basement different groundwaters are likely to exist, ranging from deep, reducing waters to shallower oxidising waters. It may not be possible for colloids to migrate across the boundaries of these very different waters without becoming unstable. An ideal natural analogue study of colloid transport would be large scale and would encompass the probable suite of groundwaters that would occur in a repository far-field and examine colloid stability at all the groundwater interfaces. Of particular importance to crystalline rock environments would be the study of colloid behaviour in a saturated fractured rock mass. This has not been adequately addressed to date.

5.7 Microbial activity

Microbes exist naturally in deep rock/groundwater systems, and diverse and active populations of microbes have been found in all underground and sub-seafloor environments investigated. However, their populations are generally not considered to be large compared to those found in near-surface and above-surface environments but, nonetheless, their effect on groundwater chemistry can be significant (Stumm and Morgan, 1981).

In addition to the natural microbial populations, larger numbers and different species of microbes will be introduced to deep rock/water systems during construction and operation of the repository. These microbes will be introduced in the air, on construction and engineering materials, on equipment and on the human workers. Considerable populations may also be associated with the waste, particularly the organic waste component of LLW and some ILW. Thus, at repository closure, significant populations of different microbial species will be present in the repository near-field. The activity of these microbes may influence many chemical reactions in the near-field, especially waste degradation, metal corrosion and those involving redox equilibria, as shown diagrammatically in Figure 5.27.

However, the significance of these microbial reactions depends, in part, on the viability of the microbial populations in the closed repository. Many of the introduced species may not be viable in the repository once it has been closed, the near-field becomes water saturated and the barrier materials buffer the near-field chemistry. The actual extent to which microbial populations will be able colonise various repository environments, and survive and enhance reaction rates, is thus clearly important in performance assessment (see, for example, McKinley et al., 1985; Grogan and McKinley, 1990).

In a HLW or spent fuel repository, the radiation field is likely to affect the viability of some microbial species, although the extent to which this will occur is difficult to establish. Furthermore, in the near-field, microbes will have to compete with the bentonite for water, since the bentonite adsorbs water when it swells. Microbes in the buffer may be killed-off by dehydration as their water is adsorbed into the bentonite (Pedersen et al., 1995; Motamedi et al., 1996).

In general, most assessments of microbes in HLW repositories have found that they are unlikely significantly to affect repository safety. In contrast, in LLW repositories, microbial activity is accepted as one of the principal driving forces for waste degradation (see, for example, Kidby and Billington, 1992), but it has generally not been felt necessary to be able to model the processes in detail for the purposes of performance assessment. However, for ILW repositories, it may well be important to be able to evaluate microbial effects quantitatively. In all cases, though, it is necessary to be able qualitatively to describe the overall impacts of microbial activity in presenting a safety case.

Observations from natural systems on the existence and behaviour of microbes in environments analogous to repositories, particularly with regards to controls on their activity and on the net results of microbial mediation in hydrochemical reactions can provide useful support to

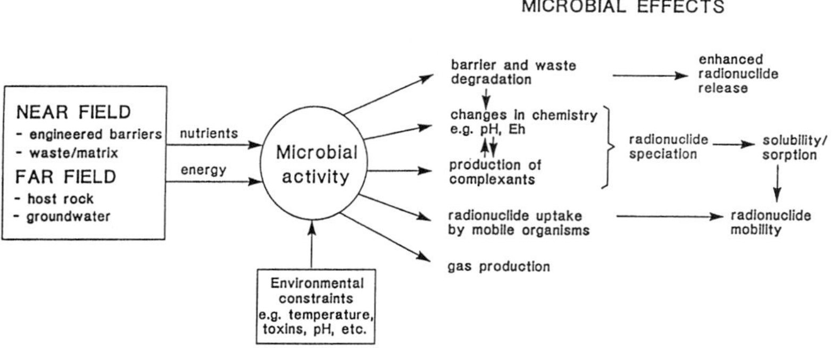

Figure 5.27: Diagram indicating the effects of microbial activity in the near-field of a L/ILW repository. This shows that microbes may affect virtually all processes that occur in the near and far-fields, although the effect may not be significant. From Grogan and McKinley (1990).

performance assessment, and can act as test-beds for models of microbial behaviour.

Natural analogue investigations of microbes have been performed in a limited number of analogue studies, both to reconcile the obvious presence of microbes in the systems with the geochemical reactions being observed, and to assess the potential role of microbes in the repository environment. Since microbes are most significant for L/ILW repository behaviour, much of the analogue work has focussed on alkaline environments.

The issues of most relevance to microbial activity that have been (or potentially could be) addressed in natural analogue studies are:

- microbial populations at depth in natural systems;
- tolerance to hyperalkaline conditions; and
- nutrient and energy availability.

These issues are discussed in the following sections.

Microbial populations at depth in natural systems

That microbes are recorded at depth in natural rock/water systems is not in doubt and they certainly have been observed in potential repository host rocks at planned repository depths (e.g. in fractured, cryst-alline rocks: Pedersen and Karlsson, 1995; Haveman et al., 1998).

There has been some discussion as to whether these populations are indigenous to the deep rock/water system and some researchers have argued that the observed populations are simply artefacts of drilling and sampling (see discussion in Pedersen, 1993). However, with more recent and better controlled groundwater sampling, it is now fairly well agreed that microbial populations do indeed inhabit continental crustal rocks.

What is uncertain is how large these microbial populations are, how active they are and how they might react during repository construction and after repository closure. In an attempt to understand the populations in granitic rocks, a microbial study was undertaken as part of the Äspö Hard Rock Laboratory investigations in Sweden. This study involved very careful gene sequencing and culturing methods to compare microbial populations in sampled groundwaters and on drilling equipment in order to assess whether microbial contamination at depth occurs. The results of this study confirmed that most microbes sampled at depth were intrinsic and active at low but significant levels. These results are consistent with other studies (e.g. Ekendahl and Pedersen, 1994). Furthermore, it was concluded that microbes introduced by sampling contamination did not become established. The reason given for this was that the 'foreign' microbes from the surface were unable to adapt to the prevailing oligotrophic, anaerobic and low temperature groundwaters (Pedersen, 1997). This does not, however, mean that microbial populations introduced to a repository would not be viable in the near-field environment.

Given that microbial populations exist at depth in all groundwater systems, it raises the question as to how microbes in very old (10 000 year or more) groundwaters at depth obtain the necessary energy to support their existence. Investigations into this issue suggest that the microbial populations use hydrogen as an energy source (Pedersen and Albinsson, 1992; Stevens and McKinley, 1995) and carbon dioxide as a carbon source. Other nutrients, such as phosphorous are available from minerals such as apatite. Photomicrographs of some microbe species from natural waters are shown in Figure 5.28.

Figure 5.28: Photomicrographs of microbes isolated from the deep groundwaters in Sweden. From Pedersen (1997).

Tolerance to hyperalkaline conditions

The tolerance of microbes to a hyperalkaline environment has been studied in Oman (Bath et al., 1987a) and Maqarin (Alexander, 1992; Smellie, 1998) at groundwater springs which are naturally hyperalkaline (pH 11.2 to 12.9). The absolute populations of microbes in these waters were low, although a wide range of microbes had, apparently, adapted to the conditions.

In the Oman system, some microbes require aerobic conditions to survive and, as a consequence, would not be viable in a deep repository environment, although they could be relevant for a shallow LLW disposal system. Among the other microbes identified at Oman were sulphate reducing bacteria which are important due to their involvement in steel corrosion and concrete degradation. They are known to tolerate extreme pressures, temperatures and radiation doses.

Cultures of these microbes were successfully grown at pH 10.2 in the laboratory but this does not necessarily imply that they would prove viable in the cement pore waters of a L/ILW repository. It was concluded in the Oman study that, although the hyperalkaline conditions in the natural springs contributed to the low microbial populations, the controlling factor was lack of nutrients in the system.

In the Maqarin study (see Box 11), microbes were found to be present in all of the hyperalkaline waters investigated, at populations of around 10^5 microbes/ml. However, it was not possible to determine conclusively whether these microbes were actually living (viable) in the high pH spring waters or whether they had only recently been transported to the sampling site by pH neutral recharge groundwaters. This latter suggestion has some credence because the types of microbes identified in the spring waters were similar to those found in other neutral pH, deep groundwater types. At the same time, none of the types were typical of microbe species known to be tolerant to alkaline conditions. If this is the case, then the conclusion may be that the very high pH of the spring waters at Maqarin (pH 12 to 13) is too extreme for active life. However, this has yet to be demonstrated.

Nonetheless, a further suggestion for the low microbe populations at Maqarin may be that they were poisoned by the high selenium concentrations, rather than affected by the high pH. In earlier work at Maqarin, Alexander (1992) noted that sulphate reducing bacteria from Maqarin could not be cultivated in the laboratory and it was suggested that they utilise selenium in preference to sulphur in this system and effectively intoxicate themselves, leading to cell death or at least metabolic shutdown. The effect of this microbial activity on selenium redox chemistry is, as yet, unknown but is potentially significant.

Given these uncertainties, it is not possible to draw firm conclusions from microbe studies at either the Oman or Maqarin sites which are applicable to performance assessment for cementitious repositories.

Nutrient and energy availability

Microbes are only significant to near-field repository evolution if the supply of nutrients and energy to them is sufficient to support their activity. The principal nutrients required to support any microbial activity are carbon, nitrogen, phosphorous and sulphur. As the material placed in any repository system will be carefully characterised, the availability of these nutrients can be quantified and the maximum microbial populations which these nutrients are able to support calculated. The total availability of nutrients is clearly repository-specific. The principal source of energy in a deep repository environment would come from the oxidation of steel, with some energy possibly supplied by methane and hydrogen from depth.

This nutrient and energy supply approach has been used to assess the consequences of microbial activity within various repository concepts (West and McKinley, 1984; McKinley et al., 1985). It was concluded in these studies that the nutrient and energy supplies would be limited for most deep HLW and spent fuel repositories. In this regard, it is worth noting that lack of nutrients was considered to be responsible for the low microbial populations in the naturally hyperalkaline groundwaters at Oman.

In LLW and some ILW repositories containing significant volumes of organic waste, the nutrient and energy supplies will be high. Thus, it is likely that large, viable microbial populations will be established in LLW repositories. However, due to the low radioactivity of this type of waste, the radiological significance of microbial activity in these repositories will be limited.

From the work on nutrient and energy supply, a modelling methodology has been established which aims to predict maximum microbial activity levels in a repository system (Grogan and McKinley, 1990). Some aspects of this modelling approach were tested as part of the Poços de Caldas natural analogue study (West et al., 1990). In particular, a comparison was made between model predictions of microbial activity and field observations. In addition, an evaluation was made of the assumption that microbes utilise chemical energy produced at a redox front and the consequences of microbial activity on the geochemistry at the redox front were determined. It was discovered that the maximum populations predicted by the modelling approach were in general agreement with the measured populations. It was further shown that microbes may catalyse specific redox reactions and that they may be important in defining the chemistry and mineralogy at a redox front, in particular by catalysing pyrite oxidation and influencing aqueous sulphur speciation.

Conclusions

Investigations of the microbial populations in hyperalkaline springs indicate that a high pH environment does limit activity but that the principal control is the lack of available nutrients.

In a deep repository environment, the viability of microbial populations will be restricted by the radiation field in the near-field and also the limited nutrient supply. Models of microbial populations in HLW environments suggest that their activity will be minimal. Tests of the models at Poços de Caldas produced results which were compatible with field measurements. Further tests of the models are in hyperalkaline systems are required.

In a shallow LLW repository, the energy and nutrient supply will be high and viable microbial populations should be established. However, the low radioactivity of these wastes means that the radiological significance of microbial activity will be limited.

5.8 Gas generation and migration

Gases may be generated in a repository shortly after closure, when conditions remain oxidising, and may continue to be generated in the longer term when conditions are chemically reducing. Shortly after closure of a deep repository, when free oxygen is available (oxidising environment), this free oxygen will be consumed through the aerobic corrosion of steel and aerobic microbial action on any organic materials present. The latter process will generate volumes of CO_2, equivalent to the volume of oxygen consumed, together with some volumes of inert gases, mainly H_2 and methane (CH_4).

These processes will rapidly consume free oxygen in the near-field system and anaerobic conditions will be established, and should be maintained due to the low oxygen content of deep groundwaters. After anaerobic conditions have been established, gases potentially can be produced by three principal processes:

1) anaerobic corrosion of steel;
2) microbially-induced anaerobic degradation of organic materials; and
3) radiolysis of water and organic material.

The significance of these three processes will clearly depend on the type of repository and the waste it contains. However, L/ILW repositories will generate the largest volumes of gas due to the presence of organic materials in the waste. For example, in the case of the proposed UK L/ILW repository, it has been calculated that 50 times more hydrogen will be produced than all the other gases combined (Rees and Rodwell, 1988). As an indication, the calculated volumes of gas likely to be produced in the proposed Swiss L/ILW repository are shown in Figure 5.29.

The most easily biodegradable material in the LLW is cellulose which may occur in the form of wood and paper etc., although materials such as these are may be incinerated rather than being emplaced 'raw' in the repository. Microbial degradation of cellulose creates CO_2, CH_4 and H_2 gases. Gas production from other organic materials is not considered to be as significant as from cellulose. The majority of the other organic materials in L/ILW waste (such as ion-exchange resins and bitumen) are generally more resistant to chemical degradation, meaning that gas generation is less significant. Anaerobic corrosion of steel found in all repository designs will produce hydrogen under anaerobic conditions:

$$3Fe + 4H_2O \rightarrow Fe_3O_4 + 4H_2$$

The rate of this reaction will be controlled by the accessibility of the metal surfaces and the chemical conditions. In a L/ILW repository, the alkaline conditions may slow the corrosion rate compared to the pH neutral conditions in a HLW or spent fuel repository.

Gas production rates due to radiolysis are likely to be much less than those due to anaerobic iron corrosion or organic material degradation and are generally not considered significant for performance assessment.

Initially, all gas evolved would dissolve in the groundwater. However, as gas production continues, free gas may form in the near-field if the gas production rate exceeds the rate at which gas can escape to the near-field. The formation of a free gas phase would be of concern for repository safety because the formation of a gas overpressure could cause physical damage to the engineered barriers or to the near-field rock. The potential for gas overpressure causing physical damage to the rock is unclear. It has been suggested that crack propagation around igneous intrusions might provide a form of natural analogue investigation could look at this mechanism (Daemen, 1994). However, the significant differences in terms of temperature, pressure and geological environment, between an active volcanic systems and a repository probably mean that no useful information could be obtained that would be applicable to performance assessment.

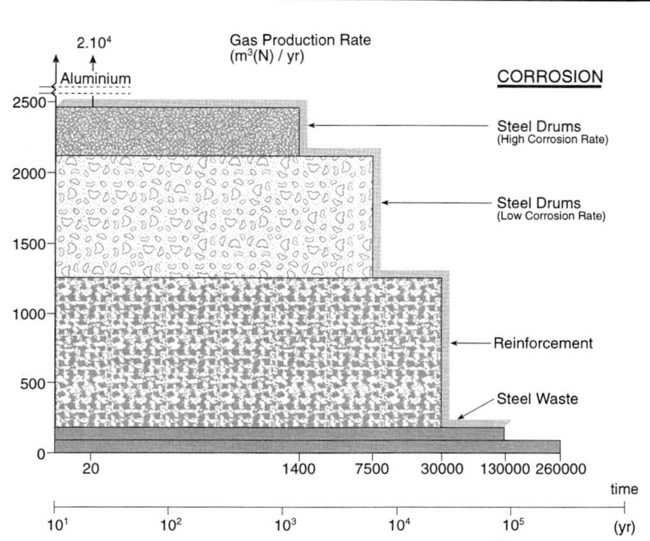

Figure 5.29: Diagram showing the comparative gas (mostly hydrogen) evolution rate due to the corrosion of steel objects in the proposed Swiss L/ILW repository. (N) implies normal cubic metres at standard temperature and pressure. After Wiborgh et al. (1986).

Furthermore, a mobile gas phase leaving the near-field, may also impact on radionuclide transport by expelling groundwater containing dissolved radionuclides. In a repository containing a compacted bentonite buffer, if the gas pressure rises, then a critical point may be reached allowing the gas to breakthrough the bentonite and flow (Horseman and Harrington, 1997). This may cause some localised increase to the hydraulic conductivity of the buffer.

Gas evolved in the repository will eventually escape to the far-field and may even reach the surface. If large volumes were able to escape along a particular fracture pathway and accumulated near the surface, this gas may conceivably pose an explosive or flammable hazard, as indicated in Figure 5.30. In addition, repository-derived gas at the surface would also pose a radiological hazard as the radioactive isotopes ^{3}H or ^{14}C may have substituted for the non-radioactive equivalents in hydrogen, methane or other hydrocarbons. It is very unlikely that any ^{222}Rn (the gaseous daughter of ^{238}U) generated in the repository would also reach the surface, because its half-life is very short.

The production of large amounts of methane in a repository by microbial degradation of organics could also lead to the methylation of various metallic radionuclides. Some methylated metals are volatile, and are rapidly assimilated by biota (Ridley et al., 1977). The possibility of their production and transport in a gas flux may require evaluation.

Gas production was not generally recognised to be a potential problem until the mid-1980s. To date, no natural analogue studies are known which specifically address the issue of gas transport or interactions in the geosphere, although a few field studies have examined the issue. Much of the

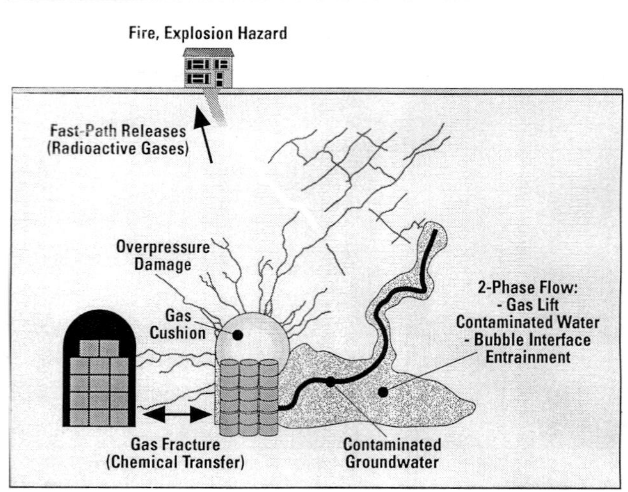

Figure 5.30: Gas formation and the consequences of gas in the near-field of a hypothetical L/ILW. If a suitable fracture system was available to the gas, then gas arriving at the surface may create an explosion hazard.

following discussion is thus based on geological investigations of the sources of hydrogen and methane in crustal rocks, and their possible reactions with the geosphere and biosphere.

Given the potential significance of gas on repository evolution, the important issues which might need to be addressed in any future analogue studies are:

- gas production rates;
- gas migration and reaction in the geosphere; and
- gas migration effects on solute transport.

These issues are discussed in the following sections.

Gas production rates

The rates of gas production in the near-fields of repositories are uncertain, especially for steel corrosion under hyperalkaline conditions, but also for microbial degradation of organic materials.

Natural analogue studies may have potential for providing quantitative information on gas production due to metal corrosion, by determining the corrosion rate of reinforcing bars in old concrete structures such as dams etc., where the environment may be water-saturated and reducing. For HLW and spent fuel disposal, a similar approach for steel objects buried in bentonites or other clays may be appropriate if relevant analogue systems could be found. The composition of the steel and porewaters and the degree of corrosion could be measured and, possibly, gas production calculated. This information may then be extrapolated to the repository environment.

Much natural decomposition of organic materials by microbial degradation occurs under neutral to slightly acid pH and oxidising conditions in the soil zone. There is a considerable body of literature on microbial decomposition of organic detritus in wetlands, and reducing lake and marine sediments. Methanogenesis in freshwater sediments occurs largely as a result of the breakdown of acetate. Much of the methane produced may be oxidised in the upper parts of the sediment pile before it can be released. In marine sediments, methanogenesis occurs largely by reduction of CO_2. Some information also exists on the rate of release of methane from reducing sediments, particularly in the wetland environments. The various processes proposed for methane production in the geosphere have been reviewed by Schoell (1988) and Schlesinger (1991). As a result of the large amount of work carried out in this area, the mechanisms of gas production are well understood. It may not, however, be a trivial task to transfer information on production rates to the repository environment, although a mass-balance approach may be feasible.

Information on gas production rates from known amounts of organic carbon in a particular sedimentary environments might be combined with information on amounts of carbon present in an environment analogous to a L/ILW repository near-field (e.g. the Maqarin hyperalkaline springs). If evidence of similar modes of microbial activity were present in both situations, then it may be possible to extrapolate gas production rates to the analogue environment. Given the amount of information already available, this area may warrant a much more detailed analysis than is possible within the scope of this book.

One simple calculation of this type was undertaken by Rennerfelt and Meijer (1986) who estimated gas production rates for the Swedish L/ILW repository (see Box 3) from comparison with activity in peat bogs and sediments generated by the release of cellulose fibres from pulp and paper mills. Both the repository and sediment environments are reducing, although the sediment environment does not approximate to the high pH conditions that will occur in the cementitious repository. The calculated gas production rate from this study was around 1 m^3/tonne of waste.

Gas migration and reaction with the geosphere

Considerable amounts of information on gas migration in the geosphere has been obtained by the oil and gas industries but is not always readily available. Other than in hydrocarbon reservoirs, gas production occurs at natural sources, mostly in hydrothermal areas, which are not particularly relevant to possible repository sites. Nonetheless, their study may provide useful information on the behaviour of gas in the geosphere. Large volumes of hydrogen are found in groundwaters from ophiolites, such as at the Semail Ophiolite in Oman (Neal and Stanger, 1983), and are associated with the hyperalkaline springs which have been studied as natural analogues of cementitious repositories (see Section 5.1). Hydrogen can exist in concentrations of a few percent in equilibrium with CO_2 and CH_4 in the crust even at depths of a couple of kilometres (Takach et al., 1987). This has been demonstrated by occurrences of methane and hydrogen at depth in the Canadian Shield (Sherwood et al., 1988) and in Finland (e.g. Pitkänen and Luukkonen, 1998).

Migration of gas in the geosphere will occur either in solution at depth or as a separate gas phase nearer the surface. Gas migration is controlled by the fracture network in most hard rocks and by the geometry of low permeability formations. Hence, a gas seepage at the surface does not necessarily indicate the presence of a gas source vertically below; lateral gas migration over tens to hundreds of kilometres has been documented (Jenden et al., 1988). Nonetheless, surface gas anomalies have been widely exploited for the discovery of hydrocarbon reserves and considerable experience has been acquired in the petroleum industry in interpreting these in terms of the local geology and structure (e.g. Philp and Crisp, 1982). Surface gas anomalies are also associated with trace element enhancements and ore prospecting has taken advantage of these anomalous enrichments. In addition, much effort has gone into understanding the mechanisms involved in the migration of gas through the geosphere (e.g. Leythaeuser et al., 1982; Goth, 1985; Bell 1989).

As gas migrates through the rock it may react with the minerals present (Stenhouse and Grogan, 1991). Many laboratory experiments have been performed to investigate such reactions, but not with a radioactive waste perspective and at temperatures and pressures very dissimilar to those expected in a repository. It does, however, seem likely that some gas will be consumed in redox reactions with species such as sulphate, nitrate, Mn(IV) and Fe(III) from the rock. These reactions are thermodynamically possible but may be inhibited by slow kinetics. This will be the case particularly where gas is migrating rapidly in a separate gas phase. If gas transport occurs in solution, such reactions may be more likely. There is a need to acquire more kinetic data for these reactions, to determine how efficient they can be in consuming the volumes of gas evolved in a repository.

Gas migration effects on solute transport

Gas migration through the geosphere may perturb groundwater flow and, hence, affect solute transport by a number of processes, such as:

1) bubble formation may change the apparent viscosity of the groundwater and induce or accelerate movement;
2) bubbles may push groundwaters in front of them along preferred pathways, i.e. causing *fingering*;
3) immobile bubbles may change the direction of groundwater flow by effectively sealing certain pathways;
4) bubbles may scavenge radionuclides and colloids from the groundwater which then attach to the gas/water interface and move with the bubble by the processes of *gas flotation*; and
5) gas bubbles coming out of solution due to a drop in confining pressures as groundwater rises may change the water chemistry, notably altering the pH as gaseous carbon dioxide is evolved.

Whilst some of these processes would appear potentially significant, no field or analogue evaluation of their likely impact on repository performance has been carried out. As with gas production rates, a substantial amount of information exists on some of these mechanisms. Data and experience reside in the hydrocarbon industry, in marine sciences, and in process engineering practices. It is recommended that scoping calculations are first attempted to evaluate the circumstances under which any of these mechanisms might be of significance in a repository and that a thorough review of the industrial and engineering literature is performed to ensure that the mechanisms addressed are comprehensive.

If any mechanisms are identified as important, then it may be appropriate to seek analogue evidence of their behaviour in the geosphere. However, the applicability of natural analogues to this issue seems tenuous.

Conclusions

Rates of gas production from corrosion of metal and microbial degradation of organic material in environments similar to a repository near-field are poorly quantified. It is possible that examination of reinforcing bars in old cement structures or clays may help quantify gas production rates due to metal corrosion. In addition, a thorough evaluation of the extensive pool of biogeochemical data on gas production and consumption rates in sediments may lead to the identification of suitable analogue studies. However, gas production due to microbial activity is probably best studied in the laboratory.

The physical processes involved in gas migration through the crust have been well-studied by the petroleum industry. However, geochemical information on the interaction of gas with the rock during migration has not usually been addressed. It is possible that the gases will be consumed during migration through the geosphere by redox reactions with minerals but, although these reactions are thermodynamically feasible, they may be kinetically inhibited.

A number of gas sources occur naturally other than in hydrocarbon reservoirs, but most are in geological environments grossly dissimilar to possible repository locations. Nonetheless, natural analogue study of these natural gas sources could prove useful in examining gas-rock interactions. No natural analogue studies have specifically addressed this issue. Gas migration may affect solute transport by a number of mechanisms. Whilst these processes may be potentially significant, their likely impacts have not been evaluated to the level where the value of analogue study can be demonstrated. Indeed, more useful information may come from the hydrocarbon, groundwater and process engineering industries.

Chapter 6: The application of analogue information

As discussed in Section 1.4, natural analogue studies originally developed out of a need to obtain information on the long-term natural processes which could affect the repository. This information is needed to help develop and support performance assessment. It is important that the models and supporting databases which underpin performance assessment are based on correct conceptual understanding of actual systems and processes, and can simulate repository evolution in an adequate manner. Natural analogue studies were seen as a way of circumventing the inherent limitations of short-term laboratory experiments, which generally are unable to replicate either the complexity or timescales of natural processes.

In the twenty years or so since natural analogue studies were first proposed, their use and application has evolved. Their true supporting role in performance assessment is now increasingly being acknowledged, after a long period of uncertainty, and they are now also being accepted as a main provider of illustrative information for use in non-technical demonstrations of safety. In the last few years, the application of natural analogue studies has also expanded beyond issues related to radioactive waste disposal and there is a growing awareness of their potential application to other environmental issues, notably toxic waste disposal, as was discussed in Section 1.6.

6.1 Natural analogues in the support of performance assessment

When considering how natural analogues can be used to support performance assessment, it is important to realise that they are actually only one component of the supporting research arsenal, alongside laboratory and field-based experimental work, and modelling studies. Unfortunately, there has sometimes been a tendency to consider natural analogues as a replacement for laboratory experiment. As shown earlier in Table 1.1, both natural analogues and laboratory studies have their advantages and disadvantages. The best means of supporting performance assessment is to draw on the strengths from both types of study. The need for both types comes about because the development of a working assessment code requires the following two types of supporting data:

i) soft, qualitative information, and

ii) hard, quantitative information.

The recognition of these two types of data, and of the types of studies that provide them, is now more widely recognised (Chapman and Miller, 1994). The distinction between the two is not a consequence of the environment or system being investigated but is a result of the methodology employed and its limitations or, rather, what the researcher is hoping to find. This is best explained by example.

Consider researchers who wish to develop a new performance assessment model (code) to simulate the impact of colloids on radionuclide transport in a particular rock and groundwater environment. They may begin by sampling groundwater in the vicinity of a uranium orebody as part of a natural analogue study, with the objective of learning about which types of colloid associate with radionuclides, and which are the primary processes controlling radionuclide uptake and which are secondary. This is an example of a qualitative study (no numerical data may be obtained) and the qualitative information it provides may be used in the construction of a conceptual model which describes colloid transport.

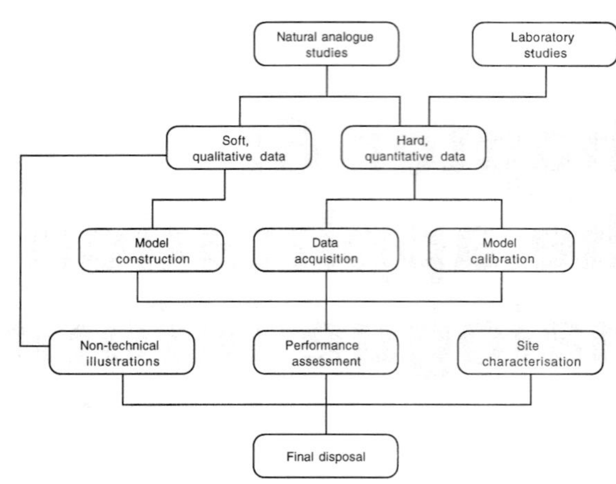

Figure 6.1: A model of the relationships and interdependencies between the various components necessary for a full safety case for a repository. Natural analogues provide key information for both the performance assessment and illustrative aspects. From Miller (1996a).

Once the important mechanisms and processes have been identified, another study may be undertaken to measure colloid populations and the rates of the primary processes controlling radionuclide uptake. This second study is an example of a quantitative study (numerical data is obtained) and the quantitative information may be used as input parameters to a computer model (code) developed from the conceptual model. Depending on how well the natural analogue system can be characterised and its boundary conditions defined, this quantitative study may be undertaken either at the analogue site or in the laboratory.

A little thought will indicate that the soft, qualitative form of investigation must precede the hard, quantitative form in the development of every performance assessment code because a process can not be quantified and modelled mathematically until it has been identified and its importance assessed.

Combining the concepts of hard and soft data with the need for both natural analogue and laboratory studies leads to a model for how various information types can be used to support performance assessment, and this is shown in Figure 6.1 (Miller, 1996a). This figure is a simple representation of the process of developing a safety case but it demonstrates unequivocally that natural analogue studies are essential for performance assessment support. In Section 1.3, the following stages in the development of a performance assessment code were identified:

1) construction of a conceptual model which describes the system and includes all of the important processes and their couplings;
2) translation of the conceptual model into a mathematical model and coding in the form of a computer program;
3) acquisition of quantitative input data for all the variable and constant parameter values included in the code;
4) verification of the numerical 'correctness' of the computer code; and

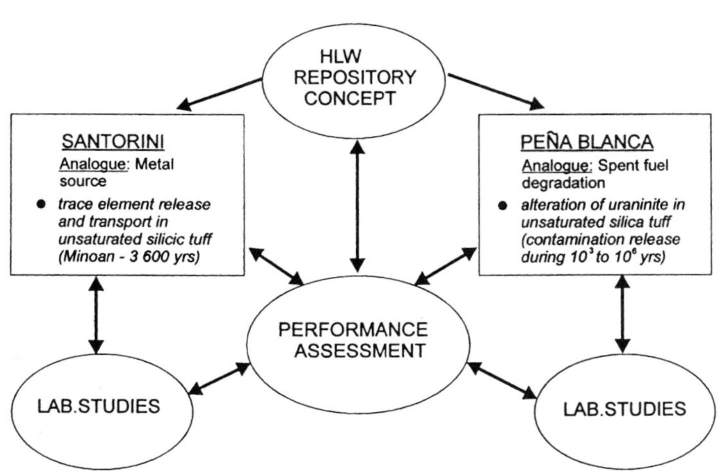

Figure 6.2: The complementary nature of natural analogue and laboratory studies in their support to performance assessment and model development. The example given refers to the assessment for the proposed US repository at Yucca Mountain. From Smellie et al. (1997).

5) validation of the code's 'applicability' to the repository system to assess its predictive capabilities.

Stages 1, 3 and 5 are shown on Figure 6.1 as *'conceptual model development'*, *'data acquisition'* and *'model testing and validation'* and, as mentioned in Section 1.3, represent the specific aspects of performance assessment code development which require direct research support in the form of natural analogue studies.

The manner in which natural analogues can provide this support varies between these stages: the conceptual model development stage requires mostly qualitative information, as indicated above, while the data acquisition and model testing and validation stages require more quantitative information. The varied supporting role of analogues to these stages is discussed in more detail below.

Figure 6.1 also highlights the fact that natural analogue and laboratory studies are complementary in their support of performance assessment because they can provide different information and opportunities to develop and test assessment models. This complementary nature was demonstrated by Smellie et al. (1997) by reference to analogue studies to investigate uraninite degradation at Peña Blanca (see Section 4.2) and elemental migration in volcanic ash at Santorini (see Section 5.2). Both these studies were undertaken to support performance assessments for the proposed US Yucca Mountain repository which is located in unsaturated volcanic tuffs (as described in Section 2.3.1) and both required complementary laboratory studies in order to gain maximum advantage from the natural analogue information. The links between the analogue and laboratory studies, and performance assessment for the Yucca Mountain repository are shown in Figure 6.2.

Model construction

The construction of a conceptual model involves deciding which processes need to be incorporated and, if a process is to be incorporated, how that process should be modelled. It is clear that the development of many conceptual models of repository processes may proceed only on the back of a good basic understanding of how the natural environment operates. The development of a conceptual model ought to involve natural analogue studies, as it is only after close observation of natural systems that it is possible to make the necessary decisions regarding which processes to model. For example, the transport component of an assessment may apply a simple, reversible sorption model when there is, in reality, a number of alternative retardation models whose inclusion may either be more appropriate or essential even for describing properly the processes active at the site under study.

Examples of this include the significance of precipitation at geochemical discontinuities, such as redox fronts, which may be present naturally in the rock or be produced by the presence of the repository, and the significance of amorphous secondary phases in retarding radionuclides being transported in solution. Also, natural analogue studies indicate that many processes do not operate in the linear fashion assumed by some models and, as a consequence, it is not necessarily correct to extrapolate short-term laboratory results to the long time-frames required for performance assessment. Given this potential complication, at the very least, natural analogue studies may demonstrate that a certain process is linear and that extrapolation is a legitimate procedure.

Thus, natural analogues should be used to tell the performance assessment model builder:

- which processes to include;
- which processes are likely to be dominant and which of secondary importance;
- which process interactions to model;
- the spatial and temporal scales with which the model should cope; and
- whether the basic premises of a model are appropriate (e.g. linear extrapolation to long time-frames).

It is clear, therefore, that conceptual model construction is the performance assessment end-point for the soft, qualitative form of natural analogue investigation. This is a necessary procedure because it is the foundations upon which the entire performance assessment will be built. It is unfortunate then, that most published performance assessments give little mention of this vital role of analogues, as will be discussed later.

Data acquisition

Natural analogue studies should be examined as one potential way (and, in some cases, the only way) in which to obtain certain quantitative information. This is the area where, historically, performance assessment modellers have had high expectations of natural analogues, to provide the 'missing' data they require for input to their models. Unfortunately, the inherent complexity of natural systems sometimes makes it difficult, if not impossible, to obtain such precise results on demand for modelling purposes. In fact, the number of cases in which analogue derived data have been used directly in performance assessment is quite limited. The few known examples include matrix diffusion depths and metal pitting factors.

Although it has become clear, from analogue and field studies, that certain laboratory derived data may be inappropriate, it has often not been possible for analogues to fill the consequent gaps. The very nature of analogues makes it difficult to extract precise parameter values from them, as would be required in a performance assessment model. The difficulty in extracting unambiguous K_d or diffusivity values, as described in Section 5.2, is typical of the overall problem.

Information derived from natural analogue studies is very often semi-quantitative, which is why it cannot often directly satisfy the parameter value requirements of a performance assessment code. However, laboratory data, though precisely measured, have uncertainty related to the lack of similarity between the laboratory and repository systems. Therefore, a role for analogues in data acquisition is to provide a form of validation for the laboratory data. Even though the analogue data may be imprecise, they can be used to give confidence in the reliability of the laboratory data if the two sets of data agree, within reasonable limits.

Thus, the following conclusions can be drawn on data acquisition from analogue studies:

- extraction of well-controlled parameter values for assessments is fundamentally very difficult when compared with laboratory data derivation, and laboratory studies will remain the principal means of data acquisition;
- analogues should be used to identify when laboratory derived data are appropriate and when not and, as such, are an adjunct to laboratory tests; and
- analogues are also invaluable in providing bounding values (maxima and minima) for processes which are too slow to be observed in the laboratory or where large laboratory-induced artefacts exist or are suspected, such as in the case of matrix diffusion depths.

Model testing and validation

Natural analogues provide an excellent means for testing and validating performance assessment codes and databases to ensure that they are applicable to the repository system. This is an area which has grown considerably in importance over the last several years.

Thermodynamic solubility and speciation codes and databases have been extensively tested in a number of natural analogue studies (see Section 5.1). The issue at stake with solubility and speciation models is whether the solubility-controlling mineral species and complexes in solution, which are generally specified from generic hydrochemical databases or from theoretical or laboratory information, are appropriate to the site-specific conditions being modelled, whether equilibrium actually occurs, and whether the system is controlled by kinetics.

At the present level of development, the modeller must make use of the available generic conceptual models and databases. These will need to be validated for the conditions expected at any potential repository site, in terms of the unique mineralogical and hydrochemical conditions found. That this will be necessary, and that 'off-the-shelf' thermodynamic codes and databases cannot be used with total confidence to make important predictions, has been well illustrated by analogue projects such as those in El Berrocal, Maqarin, Poços de Caldas, Cigar Lake and Palmottu.

In all of these studies, a rigorous blind predictive modelling approach was used to test the suitability of thermodynamic codes and databases for defining elemental behaviour in solution, as discussed in Section 5.1. The approach adopted in each project was to pose a problem in essentially the same way that it will be put to a performance assessment modeller of a repository site. The basic geochemical properties of the rocks and waters are provided, as they would be derived during any type of thorough site characterisation exercise, and the modeller is asked to use these data to predict how specific trace elements will behave. In the analogue case, these trace elements are naturally present in the waters, and their actual speciation and concentrations can be measured separately and compared with the modellers' predictions.

In the case of a repository performance assessment, the trace elements will be introduced into the system from the waste, and the modeller needs to be able to predict how they will behave under the measured conditions at the site, and also over long periods into the future during which hydrochemical conditions may change. The main lesson learned from these analogue studies is that, while the basic codes used appear adequate (provided the conceptual model of rock-water interactions in the system is appropriate), the databases used and the assumptions made by the modeller can cause errors in prediction.

There appears to be some scope for extending this model testing and validation approach to other types of model, particularly of system dynamics, rather than just static chemical equilibria. A limited attempt has been made on models of redox front movement, at Poços de Caldas for example. Perhaps the most obvious target would be solute transport models, where validation is currently based on short-term artificial tracer migration tests. The design of a predictive model testing exercise on this issue in an analogue study would require a simple and very well characterised system and extensive planning, but is clearly worth the effort. Thus, of all the possible applications of analogues in the performance assessment model building process, model testing is seen as being the area where most potential now lies.

6.1.1 The reality of analogue application to performance assessment

Having examined throughout this book the wealth of information which has been generated from natural analogue studies, and discussed how it potentially can be used to support performance assessment, it is instructive to examine the extent to which actual, published performance assessments have really addressed and used analogues.

Two recent reviews have examined the level of direct and acknowledged use of natural analogues in published performance assessment documents (McKinley and Alexander, 1996; IAEA, 1999). The results of these reviews are surprising. They show that very few performance assessment documents provided any detailed discussion of how analogues were used to support specific aspects of the assessments. Some did not discuss how natural analogues support geological disposal in even a general way, and a few did not mention natural analogues at all. It is particularly interesting that several national disposal programmes which are actively involved in natural analogue studies did not make natural analogues a central theme in their performance assessment discussions.

It appears to be the case that, within the three stages of performance assessment model development where analogues have a supporting role, only those analogues used for data acquisition or model validation usually get mentioned in the final performance assessment reports. This misrepresents the actual importance of analogues to a disposal programme because it does not pay due regard to the essential role of analogues for providing a general conceptual basis for geological disposal, and the understanding of specific processes which occur in nature that are explicitly modelled in some performance assessment codes.

McKinley and Alexander (1996) noted that references to natural analogues in the top-level documents for the US Yucca Mountain and Canadian performance assessments are very few. They suggested that this may be due to the fact that these performance assessments employ probabilistic modelling approaches and that analogues seem to be less well developed, or inherently less applicable to these modelling methods. If this is the case, then there is a requirement for the modellers and analogue researchers work together to identify what additional analogue information may be required to support this new breed of performance assessment codes.

Smellie et al. (1997) discussed the hidden or indirect use of analogues in performance assessment and highlighted the use of analogues in the systematic development of scenarios, where a large number of features, events and processes (FEPs) of importance to repository evolution require to be identified and described (e.g. Eng et al., 1994; Chapman et al., 1995). The role of analogue information here is to support (together with other sources of information) the inclusion or exclusion of different FEPs in scenarios to be analysed in performance assessment. The example given was of a scenario case for criticality in a spent fuel repository which was tested by comparison with known understanding of conditions and processes at the Oklo natural fission reactors (see Box 4). Smellie et al. (1997) noted that this indirect use of analogue information is not generally acknowledged in performance assessment scenario reports but can only come about because of the increasing availability of analogue information in the open literature. The current trend is for performance assessments to consider FEPs and scenarios in greater detail than was previously done. As a result, this is an area where future use of analogue information may increase.

The use of analogues for model development, data provision and model validation in a number of recent performance assessments is summarised in Table 5.1. This table indicates the actual use of analogues within these performance assessments as a whole. However, very few of these uses were explicitly mentioned in the top-level documents describing these assessments. In general, then, examination of recent assessments and consideration of the reasons given for the presence or absence of detailed analogue discussion in the top-level performance assessment reports leads to a number of conclusions:

- the role of natural analogues in providing a general conceptual basis for the geological disposal of radioactive wastes and for specific mechanisms is largely unacknowledged;
- there are only a few clear examples of parameter values being provided by natural analogues which may be directly input to an assessment model, such as matrix diffusion depths and metal corrosion rates;
- a semi-quantitative use of natural analogues has been to provide bounding limits to the ranges of parameter values obtained from laboratory studies, in the sense of checking their likelihood of being correct; and
- the most valuable quantitative role of natural analogues (which cannot be replicated in laboratory studies) is to provide test-beds for the validation of performance assessment models; so far, this has only been seriously attempted for static geochemical modelling but the methodology can be applied to other types of codes.

In essence, what is required from every analogue study is a separate performance assessment implications report (or at least a separate chapter within a report) which presents a distillation of the project and spells out, in simple terms, what has been learnt from the study that is relevant to performance assessment (Miller, 2000).

There is a very real desire in many disposal organisations to use the support of natural analogues in making a safety case. In some countries, this is actually backed-up by a regulatory requirement to use analogues. The real problem is that of translating this desire into something real and useful. Most of the problems expressed above are slowly disappearing as there is increasing interaction between the analogue researchers and the performance assessment teams. In some small national programmes, the issue may be simpler because the same people often work in both the performance assessment and analogue researcher groups. Smellie et al. (1997) note that several of the recent or ongoing analogue studies (e.g. Tono, Palmottu and Peña Blanca) involved performance assessment modellers in their planning, execution and evaluation.

A further advantage of greater cooperation is that existing data from previous analogue studies is now being re-evaluated by combined performance assessment modeller and analogue researcher teams, enabling greater benefit to be derived from the data. Smellie et al. (1997) further highlight the case of the investigations of radiolysis performed at Cigar Lake (see Section 5.4) which have recently been reevaluated (Smellie and Karlsson, 1996) allowing a new radiolysis model to be developed which more closely matches predicted and measured oxidant production.

Table 6.1: The use of natural analogues for model development, data provision and model validation in support of some recent performance assessments. Based on IAEA (1999).

Safety case	Conceptual model development	Data provision	Model validation
KBS-3 (Sweden, 1983)	▪ Radiolytic oxidation of spent fuel against observations from Oklo	▪ Maximum pitting corrosion factor for copper ▪ Bentonite stability at temperature <100°C	
Projekt Gewähr (Switzerland, 1985)	▪ Stability of borosilicate glasses ▪ Stability and instability of concretes and mortars ▪ Stability of bitumen ▪ Radionuclide release concepts against Oklo observations	▪ Long-term steel corrosion rates ▪ Constrain illitisation of bentonite	
SKB-91 (Sweden, 1991)	▪ Support of bentonite stability from observations in Sweden ▪ Redox front model supported by Poços de Caldas observations ▪ Inclusion of matrix diffusion	▪ Limit relevance of colloid transport by using data from Poços de Caldas ▪ Demonstrate conservatism in estimating radiolytic oxidation by using information from Cigar Lake	▪ Radionuclide solubility model testing and comparison with observed solubilities at Poços de Caldas and Cigar Lake
TVO (Finland, 1991)	▪ Use of palaeohydro-geological data in the development of ice-age scenarios ▪ Observations from copper deposits and Kronan canon to support corrosion estimates ▪ Use of colloidal and microbial information from Poços de Caldas and Palmottu to develop models	▪ Matrix diffusion profiles surveyed from various natural analogues	▪ Testing of UO_2 spent fuel dissolution models using information from Cigar Lake
AECL EIS (Canada, 1994)	▪ Support development of conceptual models for: - fuel dissolution - copper corrosion - clay buffer; and - radionuclide retardation, particularly the role of colloids and organics	▪ Geochemical processes and parameter values for: - Redox control on UO_2 stability (with radiolysis bounding values) - copper corrosion - bentonite to illite conversion; and - radionuclide retardation and matrix diffusion bounding values	▪ Testing of models and databases for: - radionuclide solubility - colloid formation and organic complexation; and - copper corrosion, using observations from Cigar Lake, the Canadian Shield and Kronan cannon
Kristallin-I (Switzerland, 1993)	▪ Back-up in scenario development	▪ Bounding conditions on redox front development using information from Poços de Caldas ▪ Depths of matrix diffusion penetration	▪ Radionuclide solubility model testing and comparison with observed solubilities at Poços de Caldas, Oman and Maqarin ▪ Testing models for redox front development

Safety case	Conceptual model development	Data provision	Model validation
NRC IPA (USA, 1995)	▪ Disruptive scenario development (volcanism) ▪ Back-up source term conceptual model from Peña Blanca ▪ Relative importance of microfractures and matrix transport at Peña Blanca ▪ Back-up for vapour phase transport from Valles Caldera ▪ Back-up conceptual model for transport in fractures	▪ Identification of secondary phases for long-term release at Peña Blanca	▪ Model testing for elemental transport in unsaturated media at Akrotiri
TILA-99 (Finland, 1999)		▪ Support for conservatism in assumptions regarding: - spent fuel dissolution rate using observations from Cigar Lake - occurrence of matrix diffusion; and - canister life time with reference to the Hyrkkölä native copper occurrence	
SR-97 (SKB, 1999)	▪ Use of permafrost data in development of ice-age scenarios ▪ Use of post-glacial tectonic data in development of ice-age scenarios	▪ Bentonite stability related to: - temperature effects - availability of potassium ▪ Clay as a barrier to microbial activity (i.e. Dunarobba) ▪ Gas transport in shales ▪ Insignificant colloid concentrations at repository depths ▪ Bounding calculations supporting reducing conditions at repository depths: - incursion of oxidising meteoric waters - lack of mineralogical evidence for Fe(II) oxidation	▪ Justification of model for radiolytic oxidation of UO_2 ▪ Reference to matrix diffusion data for model testing (Palmottu and Cigar Lake) ▪ Testing models of redox front propagation using observations from Poços de Caldas ▪ Development and testing of groundwater mixing model
SFR (SKB, 1999)	▪ Support for long-term durability of concrete barrier system using observations from Northern Ireland, Maqarin and ancient/aging concrete structures ▪ Hyperalkaline plume scenario using observations from Maqarin	▪ Hydrogeochemical processes and parameter values for: - released hydroxides due to leaching - CSH and CASH phases - zeolite phases - pH reduction due to reaction with silicate minerals - colloids, microbes and organics	▪ Blind modelling and testing of thermodynamic databases at Oman and Maqarin

6.2 Natural analogues in non-technical demonstrations of safety

Natural analogues have a very important role, beyond their application to performance assessment, as providers of illustrative and sometimes non-technical information to a broad range of audiences. This aspect of their use is indicated on Figure 6.1 which indicates that both a successful performance assessment and valid illustrations of safety are required to support a repository development programme.

From outside the confines of the performance assessment community, this aspect is sometimes perceived as the main reason why natural analogues are studied. Natural analogues, or comparisons with natural systems, are frequently mentioned as important components of the process of evaluating and accepting disposal concepts (e.g. IAEA, 1999). Among all levels of reviewer, from technical peer review panels, to non-technical audiences, there is a clear belief that performance assessments are only credible if shown to have strong natural parallels. For many audiences, the nature of predictive assessments themselves is difficult to understand and accept. Take, for example, the following quotation from from Blowers et al. (1991):

"Enormous scientific effort has been expended in Europe and North America researching and demonstrating the proposition that such repositories will be safe, for all practical purposes, for ever. Yet clearly the assertion is preposterous. The safety of an untried method cannot be proven until repositories have been constructed and monitored over many generations and the radionuclides have decayed to safe levels. Running such an empirical experiment is inconceivable. Sophisticated geological analysis, risk assessment or modelling of repository behaviour must rest upon heroic assumptions and are no substitute for empirical knowledge. Scientific predictions for periods of 10000 years or more lie in the realm of fantasy, not rationality. In conditions of such uncertainty it must be concluded that there is no technical solution to the problem of radioactive waste."

Although the authors of this quote come from a non-technical background, the strength of disbelief in long-term predictions voiced by all kinds of audiences cannot readily be dismissed by scientists. In the field of prediction, everyone has a more or less valid view, based on generations of widely publicised experience. No amount of scientific argument or proof is going to convince many people of the truth of a safety case that is inherently very complex and extends predictions far into the future. Doubts are not the prerogative of the non-technical audience and are pervasive throughout the concerned community.

It is thus necessary that appropriate demonstrations of safety are made to all the stakeholder groups involved in radioactive waste disposal. One of these groups is the public at large, but other groups can also be identified, such as politicians, decision-makers, academic peers and supporters of environmental pressure groups. What these groups have in common is that they generally are not familiar with the conceptual and technical aspects of geological disposal, even though they may have (in the case of academic peers) a high level of scientific training. For all of these interested groups, it is necessary to make demonstrations of safety that are appropriate to their level of understanding and concern. Quite rightly, many groups may find the concepts and conclusions from a performance assessment credible only if provided with natural parallels for comparison.

However, we should not forget that even the performance assessment modellers require illustrations from natural analogues that indicate that the underlying theory is correct (Chapman and Miller, 1994). The conceptual model development role of analogues is essentially illustrative and represents the soft, qualitative form of study. It is the identification of processes, the broad evaluation of material stability, and process rates and interaction that the qualitative study provides which are used in all forms of illustrative literature.

In some cases, this illustrative application takes the form of providing information to a wide public audience, with analogue studies being described in advertisements and promotional literature as simple comparisons between nature and the repository.

This form of use of analogues in material for wide public dissemination has value and should be encouraged, provided that the material used is relevant, informative and, above all, honest. The types of message that analogues can provide include:

- simple illustrations of the overall disposal concept although when applying analogues to performance assessment, we do not consider any analogue site to be a complete (global) analogue of a repository system;
- demonstrations of the similarity in materials, contaminants and radiation between the repository and nature are important because we can demonstrate that essentially none of the components of a repository system are beyond our experience;
- recognition of the important processes and events that control the repository behaviour can be demonstrated from geological analogues - for example, never has an analogue site revealed any transport or retardation processes that has not previously been recorded or predicted and, hence, no such unique process should take place in the repository; and
- our understanding of the relevant timescales is an important message because the extended time periods of interest to disposal are generally beyond those usually considered by the public.

The last point, understanding the relevant timescales, is a difficult issue to communicate. Comparison of future times with past history is one way to put repository times in context, as is illustrated in Figure 6.3.

Unfortunately, analogue information or illustrative cases can be misrepresented. The nuclear industry itself has not always represented natural analogues honestly, sometimes tending to oversimplify and overstate the facts. No doubt it may be argued that some simplification is necessary to make the scientific data readily digestible but there is a fine line between necessary simplification and misrepresentation of the facts, as discussed by Lindqvist (1996).

The natural analogue that is most frequently oversimplified is Oklo, often to the point of being misleading. To the scientist, Oklo does not provide unequivocal proof that a repository will be safe because Oklo is not a complete analogue for a repository and the situation at Oklo is simply too complex for that conclusion to be reached. Yet, Oklo is often presented in nuclear industry literature as *proof*. An example of going too far is provided by the British Nuclear Forum (1991), who said:

"The Oklo reactors ran gently at the kilowatt-power level for millions of years. They never blew up. The radiation and waste from them did not deter surrounding life forms. Over immense timescales, the waste has barely moved away from the reactor site. As a result, scientists today are confident that waste in man-made stores and repositories is likely to move even less..."

Such statements challenge belief at all levels, apart from leaving themselves open to scientific ridicule if ever they have to be defended.

On the positive side, there are a number of good examples of the use of natural analogues illustrations for public communications and education. In particular, there is the *'Traces of the Future'* video (Güntensperger, 1993) which was co-funded by many national and international bodies involved with radioactive waste disposal, as well as many brochures, leaflets and advertisements. A few examples are shown in Figure 6.4..

However, good as these illustrative uses of analogues have been, there is a need to develop new and better illustrative materials. While it is hard to define exactly what form these materials should take, it is important that the analogues they discuss should be presented as simply and as unambiguously as possible, but without overstating the interpretation of the analogue.

When choosing analogues to use in illustrative material, there is a wide choice from nature and from archaeology. Archaeological analogues can be particularly suitable for illustrations because they can relate to sites or systems with which the general public may have some familiarity or interest, as exemplified by the Kronan cannon analogue in Sweden and the Hadrian's Wall analogue in the UK.

6.3 Natural analogues applied to other environmental issues

As discussed in Section 1.5, there has recently been growing interest in applying the natural analogue methodology outside radioactive waste disposal to other environmental issues. In particular, there has been some work on analogues for toxic waste disposal.

Although, in principle, analogues could be used to support safety assessments for toxic waste disposal in the very same way they are used to support radioactive waste performance assessment (e.g. Bengtsson, 1989), in practice this is

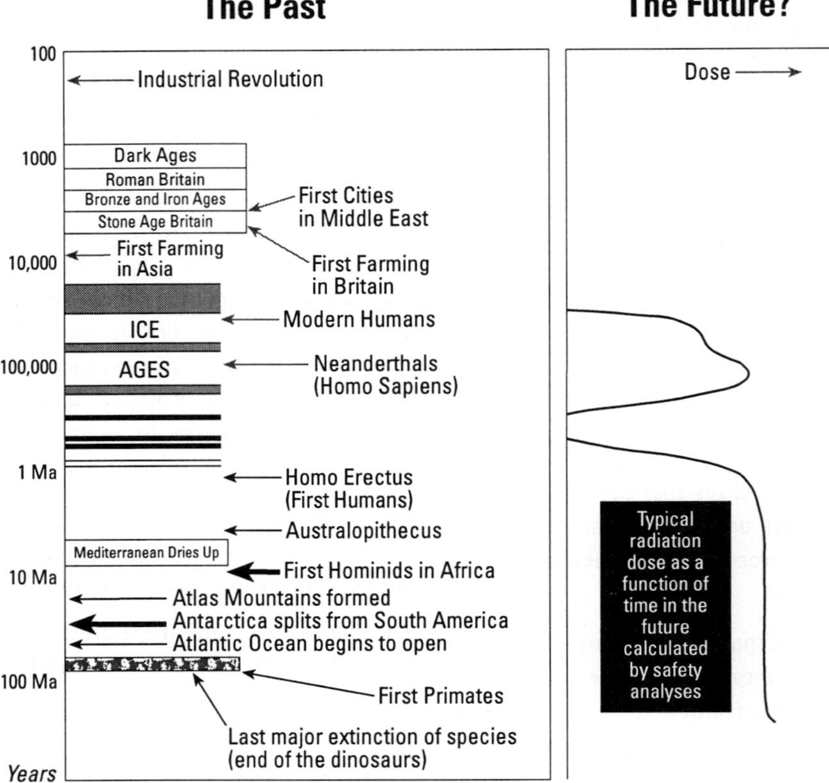

Figure 6.3: Comparison of future time showing predictions of typical releases calculated in a performance assessment against past history. Comparisons such as these help to place the repository assessment results in a meaningful context. From Chapman (1994).

Figure 6.4: Examples of the illustrative use of natural analogues. Left, from Sweden, this advert compares the Cigar Lake analogue to the Swedish spent fuel repository concept. Right, from the UK, this advert presents Roman (1700 year old) concrete from Hadrian's Wall and makes the simple point that cement and concrete can have a very long life-time in certain conditions. Illustrations courtesy of SKB and UK Nirex Ltd.

unlikely to happen. The reason for this is that the regulatory requirements for predicting the future behaviour and safety of a toxic waste disposal facility do not require the same level of attention as that required for radioactive waste performance assessment. Thus, it is not likely that comprehensive, predictive assessment codes for toxic wastes will be developed, at least not until there is a legislative requirement to do so.

Therefore, for the present time, it is most probable that analogues for toxic waste disposal will remain qualitative or, at best, semi-quantitative and be focussed on issues such as providing illustrations for the durability of solid waste matrices (e.g. Côme et al., 1997) and general contaminant transport processes in various near-surface materials (e.g. Bowell et al., 1997). In essence, this corresponds to the 'conceptual model development' stage of radioactive waste performance assessment development whereby the analogues are being used to ensure that the basic processes are understood.

If regulatory requirements are strengthened, then more detailed use of analogues may be required to support toxic waste disposal safety assessments. However, if this were to occur, it is unlikely that an entire new set of codes and assessment methodologies specific to toxic waste would be developed. Instead, it would be sensible first to test the validity of applying radioactive waste assessment tools to toxic wastes, with some necessary modifications. In the first instance, all that may be required would be expansion of the databases to cover additional non-radioactive elements and compounds. However, validating the applicability of the modified codes to toxic wastes and their disposal environments would require the identification of new analogue sites. Clearly, the type of site would be both waste and disposal system specific but it could be imagined that a wider range of metallic orebodies would be appropriate, as might an increased number of sites of anthropogenic contamination and industrial analogues.

Since history tends to suggest that environmental legislation always becomes stricter over time, with new laws and lower admissible contaminant concentrations continually being defined, it is to be expected that the scope for natural analogue studies will expand in the future as the overlap between radioactive waste and toxic wastes becomes more apparent.

Chapter 7: Summary, conclusions and recommendations

The natural analogue studies described in this report have contributed greatly to our confidence in the safety of radioactive waste repositories, through the provision of quantitative and qualitative information, and simple illustrations. This chapter summarises the results of these studies, the uses made of the analogue information and suggests future developments of analogue studies.

7.1 Summary of analogue results

The following section examines each of the materials and processes considered in this report in turn and briefly summarises those issues that have been successfully investigated and those that would benefit from another look.

7.1.1 Analogues of repository materials

Borosilicate glass

On the whole, the mechanisms by which glass alters and dissolves are well-understood from laboratory studies, even though their long-term kinetics are not well characterised. The most important process missing from current models is probably the incorporation of radionuclides in secondary alteration phases, as these will limit effective radionuclide solubility.

This is a prime area for further natural analogue study. Sorption of radionuclides onto secondary alteration phases is probably less important than direct incorporation but may, nevertheless, benefit from further natural analogue study.

Spent fuel

It is believed that, in the repository environment, dissolution of spent fuel will proceed at a sufficiently slow rate to ensure adequate margins of repository safety but, nevertheless, further natural analogue studies to investigate dissolution mechanisms and rates may be warranted. The nature of the secondary phases formed as a result of spent fuel dissolution is not well-defined and neither is their potential for retarding radionuclide transport.

Further natural analogue investigations of the nature of minerals formed during the dissolution of uraninite in relevant conditions would be worthwhile. It may also be possible to use some uranium ore bodies to provide quantitative test-beds to help formally to validate different mechanistic models for spent fuel corrosion. Appropriate natural analogue studies do not exist for mixed oxide fuel (MOX).

Ceramic and mineral waste forms

Natural zirconolite and pyrochlores are good analogue minerals for synthetic component minerals in SYNROC. However, these natural minerals are very rare and generally are acquired as detrital grains rather than from their place of formation. The limited natural analogue information on these minerals suggests that they are very stable and long-lived, and are suitable solidification and immobilisation matrices for liquid HLW produced from spent fuel reprocessing operations.

However, little quantitative information can be gained from the analogue studies that would be appropriate for input to performance assessment code development. Laboratory studies will probably remain the best means of investigating the stability of these mineral phases.

Metals

Geological evidence points to the generally reactive nature of iron in oxidising conditions. Long-lived native iron occurrences are limited to examples which have been isolated from air and water by impermeable host rocks or by an excess of iron to buffer the local redox conditions. Rates of iron corrosion seem to be adequately quantified and further study may not be justified. A potential problem concerning steel corrosion is the large volumes of hydrogen evolved, which may affect the near-field of a L/ILW repository if the engineered barriers have a low gas permeability. Industrial analogues may prove useful to assess such impacts.

The steel in old, reinforced concrete pier foundations should be in a reducing, alkaline environment analogous to that in a cementitious repository. Careful examination of samples of the concrete foundations may reveal if the cement has suffered any damage due to hydrogen gas build up. Iron and steel corrosion products may also sorb many key radionuclides, but this process is not normally included in performance assessment and may be an area where further analogue support would be useful.

Natural analogue studies of copper archaeological artefacts and native copper have clearly demonstrated the stable behaviour of copper in the repository environment. No further natural analogue studies are thought to be necessary on copper stability and longevity. Natural analogue investigation into the nature of copper corrosion products likely to form in a clay-rich near-field is restricted to the Kronan cannon. Further natural analogue studies of the nature and reactivity of such corrosion products may be useful, although the extremely slow copper corrosion rate probably makes such an investigation of very low priority.

Bentonite

The rate of illitisation of bentonite has been demonstrated in natural analogue studies to be slow, and probably unimportant for repository safety. No further natural analogue studies of this issue are required. Further work is, however, required to understand better the nature of the interactions between bentonite and other repository materials, in particular cement.

Cementation of the bentonite is a potential problem that may result in fracturing and allow advective transport through the near-field. Cementation could occur as the result of steam generation in the near-field. It would be useful if suitable natural systems could be found in which to study this problem.

A prime role of the bentonite is to act as a colloid filter, and a useful natural analogue study would thus be one where groundwaters carrying colloids flowed from rock to bentonite and the filtration of colloids could be quantified. The sinking of a dense waste package in bentonite may also be problematic, though it is difficult to imagine an appropriate analogue for this. The production of colloids by erosion of bentonite is another potential problem area where analogue input could be envisaged.

Cement and concrete

Natural analogue and laboratory studies have demonstrated that the calcium silicate hydrate (CSH) compounds that bind cement together are stable over historical and geological time periods. Theoretical studies of cement evolution are well established. The problems arise when attempts are made to validate the models. Laboratory studies are of some use but the long timescales of interest (up to several thousands of years) mean that the laboratory experimental data must be extrapolated beyond reasonable limits. Obviously, natural analogue studies of appropriate natural cements (and some archaeological cements) are useful here.

The issue of the impact of the hyperalkaline plume on on host rock mineralogy, has not yet been adequately investigated and natural analogue studies to investigate this would be welcome. The interaction of cement and cement pore waters with other repository materials, particularly clay, also warrants further study. Investigations of porewater compositions in old cements may be a useful test of geochemical models of their evolution.

Bitumen

There appear to be few natural analogues for bitumen that are sufficiently similar in composition to technological bitumens to allow quantitative conclusions to be reached. If appropriate analogues can be identified, the stability of bitumen in hyperalkaline conditions may be a useful issue to investigate. The effects of relatively high radiation doses on bitumen over long timescales might also be profitably examined. The natural degradation products of bitumen in the repository environment are poorly characterised, as is their potential as organic complexing agents which will influence radionuclide solubility and speciation and, thus, transport. This issue could be investigated in natural analogue studies, possibly in connection with investigations of bitumen degradation in the presence of microorganisms. In the absence of better analogues, only illustrative uses for analogues on bitumen can be envisaged.

Organic materials

The only useful natural analogue investigation of cellulose decomposition is that performed on the fossil forest at Dunarobba. However, this is not a close analogy to a cementitious L/ILW repository. It is important that further studies examine the mechanisms and rate of cellulose degradation in a hyperalkaline environment but it is not clear if many (or any) natural systems exist in which to investigate this issue. Old wood from the hyperalkaline springs at Jordan may give some indications as to its stability in such an environment. A potential problem associated with cellulose degradation is the evolution of large volumes of carbon dioxide and methane, but this issue is probably best investigated in the laboratory.

Natural resins exist in the form of ambers and rubbers but their chemical compositions are dissimilar to technological organic materials found in a repository. In the absence of any other information, qualitative observations on natural resins surviving for geological time periods may be useful, if only for illustrative purposes, especially if found under hyperalkaline conditions. The degradation products of polymers and resins may act as efficient organic complexing agents, which could enhance radionuclide transport. This issue needs to be investigated, but natural analogue studies are not a suitable technique for addressing this problem, owing to the lack of appropriate resins, noted above.

7.1.2 Analogues of transport and retardation

Solubility and speciation

Natural analogue studies are especially useful with regard to the issue of testing models of radionuclide solubility and speciation. First, natural analogue studies provide direct evidence to help define the physico-chemical stability fields in which radionuclides are significantly soluble and may be transported, plus an indication of which organic and inorganic ligands are abundant in repository-relevant conditions and their effect on radionuclide speciation.

Second, natural analogues have proved invaluable for directly testing thermodynamic solubility and speciation codes and their corresponding thermodynamic databases. To a large degree, the second of these applications is now the most important in the performance assessment process. Code development is by nature iterative, and the need for many more natural analogue studies is foreseen for this issue. The wider application of available techniques for measuring speciation in situ would greatly increase the value of such studies.

Sorption and retardation

Further information on the sorption behaviour of near-field degradation products would contribute usefully to performance assessment. Information on radionuclide interactions with alteration and corrosion products of cement, steel, copper and other materials from natural environments would be valuable .

The mechanisms of transport and retardation in crystalline rock are reasonably well understood, but not well quantified. Matrix diffusion apart, the principal retardation process is generally sorption onto secondary minerals in fractures. In situ K_d studies have met with limited success, but substantial development of the techniques is required to improve confidence in the results and interpretation. In addition, further thought has to be expended on defining meaningful methods of establishing in situ K_ds as, to date, most models and associated techniques are of highly questionable quality. Few natural analogue studies have examined transport and retardation in sedimentary rocks, but a number have examined unconsolidated sediments and useful results (e.g. diffusion coefficients) have been obtained. More studies in sedimentary rocks are required to investigate transport and retardation, on both micro and macroscopic scales.

Likewise, few natural analogue studies hav successfully addressed the issue of radionuclid transport in fractured crystalline rock. The whole issue of irreversible sorption remains to be resolved. At present, performance assessment makes the conservative assumption that sorption is reversible, as no useable data exist to the contrary. If analogues were able to demonstrate convincingly the circumstances under which sorption could be irreversible, they would contribute greatly to the reality of models and reduce the pessimism of assessments.

Radionuclide behaviour at the geosphere-biosphere interface can be investigated by natural analogue studies on the migration of radionuclides released from spills, leaks, underground bomb tests and accidents such as Chernobyl. An ideal study might be one of an archaeological site containing metal smelting wastes disposed of below the water table.

Matrix diffusion

A number of natural analogue studies clearly show that in fractured, crystalline rock the volume of rock likely to be available for matrix diffusion may be restricted to a zone a few millimetres to a few centimetres wide, adjacent to hydraulically active fractures. No further generic studies are required, but site or rock-specific data would be very useful in a particular repository performance assessment. Only a small number of analogue studies have reported on matrix diffusion in sedimentary rocks and these suggest that matrix diffusion will also be limited to a small proportion of the rock, but at greater depths than in crystalline rock. More studies are needed in this area.

Natural analogue studies of matrix diffusion have, so far, paid scant attention to the controlling influence of rock-matrix diffusion on the redox buffering capacity of the rock. The ability of the host rock to maintain a reducing near-field is particularly important in the case of a repository with limited steel to buffer redox conditions.

Colloids

Further natural analogue studies are required to quantify the real potential for long-distance radionuclide transport occurring on colloidal material. Analogue studies suggest that colloids provide an inefficient mechanism for transport due to low populations, limited radionuclide uptake and filtration by the rock. However, to conclude definitely that colloids are an unimportant factor for repository safety will require information from larger scale natural studies in relevant geo-chemical environments. Only if it is demonstrated that colloids can participate quantitatively in radionuclide transport in relevant environments should it then become necessary to define more clearly those influences which control the colloidal populations in groundwaters.

Radiolysis

Radiolysis might be significant in HLW and spent fuel repositories where the radiation flux is high. Whilst natural analogue studies may be able to define better the actual importance of radiolysis in the repository environment, there have been no convincing studies so far. The most useful data come from experimental studies of material from analogue sites, rather than from observations of natural processes in-situ The difficulties of attributing any oxidation in the vicinity of uranium ores specifically and uniquely to radiolysis of groundwaters lies at the root of the problem.

Redox fronts

Understanding of redox front formation and behaviour in crystalline rock has progressed as a result of the Poços de Caldas natural analogue study and it has become clear that redox fronts are potential traps for a wide range of relevant elements. If the redox front is confined to the near-field, this mechanism could greatly strengthen the barrier role of the near-field.

Applying thermodynamic solubility and speciation codes to the redox fronts at Poços de Caldas has shown that they can simulate major mineral changes but only poorly predict trace mineralogy or trace element behaviour. More code and database development is necessary before such codes could be used with confidence in performance assessment. This code development will best be done in parallel with other natural analogue studies of redox fronts and other coupled processes in repository-relevant conditions.

Microbiological populations

It is clear that some microorganisms are very tolerant of extreme environmental conditions to the extent that they will exist in the near-field of repositories or will colonise a repository near-field soon after peak temperatures and radioactivity have occurred. Both natural analogue and laboratory studies have indicated that, in a repository environment, it is nutrient and energy availability and not tolerance to environmental conditions that will limit a microbial population and, hence, activity.

Certain key nutrients will occur in restricted quantities in a repository near-field, particularly nitrogen and phosphorus, and, as a result, microbiological activity is not likely to be an important factor for repository safety. The models used to assert this conclusion should be further tested in natural analogue studies, and this is currently ongoing in Jordan. There are indications that microbes may play a key role in processes occurring at redox fronts, and this feature could be further investigated in future analogue studies.

Gas generation and migration

It is not yet clear if hydrogen and methane gases produced in a repository could cause safety problems, or whether they would simply dissolve or be consumed in redox reactions with the rock and soil zone during migration. In addition, information on the rates of gas production and escape from the region of formation would provide a useful control on the theoretical models available. The role of volatile methylated radionuclide species may also require evaluation.

As the consequences of gas production are potentially significant for some repository concepts, natural analogue studies of this issue may be warranted, although their potential contribution is considered marginal. It may be possible to derive useful data from old iron or steel objects encased in concretes or clay, but it is difficult to define the parameters of such a study.

A thorough review of information on petroleum reservoir leakage from the hydrocarbons industry may shed more light on the mobility of gas in the shallow geosphere. A parallel evaluation of information from the hydrocarbon, groundwater and process engineering industries is also considered to be more profitable than analogue studies for assessing the likely effects of gas movement on solute transport.

7.2 Conclusions

Since natural analogue studies were first developed in the 1970s, the subject has grown and matured to become one of the most important supporting activities to evaluating the safety of radioactive waste disposal. Almost all national disposal programmes are involved in natural analogue studies to some extent, either actively involving their own staff in direct research or by funding projects. National and international licensing agencies have also recognised the importance of analogues and many countries demand that analogue investigations are undertaken to back-up modelling and laboratory studies.

The performance assessment and illustrative roles of analogues are now clearly defined, and quite different in nature. They are equally important in presenting a safety case in a digestible manner.

It has now better appreciated that natural analogue studies are complementary with laboratory studies, and that both are necessary in the support of performance assessment (as was shown in Figure 6.1). Indeed, it is to be expected that future research programmes will combine laboratory and analogue investigations more closely to take full advantage of the different opportunities these investigative approaches offer to develop and test performance assessment models.

Three distinct applications of natural analogues to performance assessment modelling have been suggested: conceptual model building, provision and verification of data and model testing and validation. The first, qualitative function is apparent but is rarely acknowledged in performance assessment reports. The second function is quantitative but it now appears that analogues have only a limited role in data provision but a useful role in adding credence to data obtained elsewhere. The application is extremely powerful but has not yet been used to full advantage, still being limited largely to the testing of geochemical codes.

It is strongly recommended that model testing and validation activities should be developed and extended to other areas of performance assessment modelling; transport modelling is the key area, although several other models could, feasibly, be tested in the blind predictive mode developed for thermodynamic model testing.

The illustrative role of analogues should not be seen just as providing visually attractive material for brochures and exhibitions. Apart from giving examples which are easy to understand, analogues provide the vital natural context within which to evaluate assessments.

This is the second leg that should support a safety case; that of providing a temporal and environmental framework for the reader of an assessment to use as a backdrop against which to gauge the quantitative predictions of the assessment. This means showing evidence of what will be going on in the natural environment at the same time, in the same place, and at the same rates as the repository processes.

Some of the larger analogue studies have provided very useful dry-runs of many of the approaches to be used in repository site characterisation. Experience from several of the large analogues has highlighted the complexities involved in gathering relevant and adequate data but has also brought the limitations of site characterisation to the modellers' attention.

Although future analogue studies are likely to be increasingly focussed on specific, well-defined requirements, there is still scope for some large-scale, multi-objective studies. Certain issues, notably the still hazy appreciation of whether colloids are significant or not, would bear frequent overlapping studies to build up progressively a convincing database that would allow colloids to be treated with confidence in performance assessment. A number of suggestions are given below for the types of ideal analogue study which would help with some of the most central issues in assessments.

7.2.1 Suggested areas for future analogue investigation

Given the requirements for further investigations highlighted in this review, it is likely that most future natural analogue projects will be small-scale, issue or process-specific studies aimed at clarifying particular problems, and will need to be closely allied to laboratory investigations. Many of these natural analogue studies will probably be linked to individual national disposal programmes, rather than being multi-national research efforts.

New large-scale natural analogue studies, similar in scope to those at Alligator Rivers or Oklo, may still be undertaken but these should be focussed on geological environments more clearly similar to those of proposed repository sites (probably centred on ore bodies or other geochemical anomalies rich in relevant elements) and aimed at investigating geochemical processes of direct relevance to performance assessment.

It is not possible to produce a definitive list of the most critical issues for performance assessment which should be examined in future natural analogue studies. The reason for this is that there are many issues whose significance will be dependent on specific aspects of individual repository designs or proposed disposal environments.

However, it is possible to identify a small number of topics which are likely to be of relevance to several generic disposal concepts and which are more suited to analogue study in the first instance than to other investigative approaches. These relate to processes operating in repositories which:

1) have not yet been possible to resolve completely or with an acceptable level of uncertainty, or
2) are currently dealt with by making very conservative assumptions, or

3) are so central to the safety case that any additional supporting information would be very valuable.

The reason for selecting each item on the following list is given in parentheses.

- longevity of the near-field chemical environment, e.g. efficiency of buffering reactions for Eh and pH, (3)
- interactions of high pH plume with the host rock (1)
- rates of cellulose breakdown and nature of degradation products (1)
- solubility and speciation of radionuclides in various groundwater environments (2,3)
- sorption properties of engineered barrier corrosion products (1,3)
- irreversible sorption processes in the near and far-fields (2)
- site-specific matrix diffusion quantification (3)
- evidence for long-distance (hundreds of metres) colloid transport in relevant rock formations (1,2)

With this list in mind it is possible to define the properties of a number of ideal future analogue studies.

Archaeological studies of cementitious, wooden and metal artefacts that have been buried in saturated, reducing environments for known periods of time, where it is possible to examine the undisturbed surrounding soil in detail. This requires participation in the excavation of such materials, rather than simply access to artefacts after excavation.

Archaeological studies of metal mining or smelter wastes, preferably disposed of below the water table in a well or pit in clays or fractured rocks where the groundwater flow field can be reconstructed with confidence.

Any natural environment where groundwater chemistry is closely analogous to porewaters or groundwaters in some part of the repository system, and contains elevated concentrations of radionuclides and other relevant trace elements from a known source. For the purposes of equilibrium geochemical model testing, a very slowly flowing to stagnant environment would be preferable.

Discrete zones, rich in relevant trace elements (e.g. uranium) in fractured rocks intersected by identifiable preferential groundwater flow paths to allow the definition of a source region and transport pathway and a study of sorption processes on fracture surface minerals.

A natural or anthropogenic trace element source in a geochemical environment favouring colloid production and allowing subsequent migration along a definable groundwater flow path in fractured crystalline or sedimentary rocks. The source might be a well which has been used for metallic and chemical waste disposal.

References

Abelin H, Birgersson L, Gidland J, Moreno L, Neretnieks I and Tunbrant S (1986) Flow and tracer experiments in crystalline rocks; results from several Swedish in situ experiments. Materials Research Society Symposium Proceedings, 50, (Scientific Basis for Nuclear Waste Management, IX), 627-639.

Abraham H (1960) Asphalts and allied substances. van Nostrand, Princeton.

Adler M, Mäder U and Waber HN (1999) High-pH alteration of argillaceaous rocks: an experimental study. Scweiz Mineral Petrograph Mitt, 79, 445-454.

Ahlbom K and Tiren S (1991) Overview of geologic and hydrogeologic conditions at the Finnsjön site and its surroundings. SKB Technical Report, TR 91-08, SKB, Stockholm, Sweden.

Ahlbom K and Smellie JAT (1991, editors) Underground nuclear repository investigations at Finnsjön, Sweden. Journal of Hydrology, Special Issue, 126.

Airey PL (1984) Radionuclide migration around uranium ore bodies in the Alligator Rivers region of the Northern Territories of Australia. Analogue of radioactive waste repositories. In: Smellie JAT (editor) Natural analogues to the conditions around a final repository for high level radioactive waste. Proceedings of the natural analogue workshop held at Lake Geneva, Wisconsin, USA. SKB Technical Report, TR 84-18, SKB, Stockholm, Sweden.

Airey PL (1987) Application of natural analogue studies to the long-term prediction of far field migration at repository sites. In: Côme B and Chapman NA (editors) Natural analogues in radioactive waste disposal. CEC Radioactive Waste Management Series, EUR 11037, 32-41, CEC, Luxembourg.

Airey PL and Ivanovich M (1986) Geochemical analogues of high-level radioactive waste repositories. In: Côme B and Chapman NA (editors) Natural analogue working group, first meeting, Brussels, November 1985. CEC Nuclear Science and Technology Report, EUR 10315, 57-75, CEC, Luxembourg.

Alexander DH and van Luik AE (1991) Natural analogue studies useful in validating regulatory compliance analyses. Validation of Geosphere Flow and Transport Models GEOVAL-1990.

Alexander WR (1992, editor) A natural analogue study of cement buffered hyperalkaline groundwaters and their interaction with a sedimentary host rock, I. Source term description and geochemical database validation. Nagra Technical Report, NTB 91-10, Nagra, Wettingen, Switzerland.

Alexander WR (1995) Natural cements: how can they help us safely dispose of radioactive waste? Radwaste Magazine, September 1995, 62-69.

Alexander WR and McKinley IG (1992) A review of the application of natural analogues in performance assessment: improving models of radionuclide transport in groundwaters. Journal of Geochemical Exploration, 46, 83-116.

Alexander WR and Miller WM (1994) Natural analogues of bituminous waste - are there any? Proceedings of the Fourth International Conference on the Chemistry and Migration

Behaviour of Actinides and Fission Products (Migration '93), 559-563.

Alexander WR and Smellie JAT (1998) The Maqarin natural analogue project: synthesis report on Phases I, II and III. Unpublished Nagra Internal Report, Nagra, Wettingen, Switzerland.

Alexander WR and Smellie JAT (2000, in press) The Maqarin natural analogue project: an overview. In: von Maravic H and Alexander WR (editors) Natural analogue working group, eighth meeting, Strasbourg, March 1999. EC Nuclear Science and Technology Report, EC, Luxembourg.

Alexander WR, Scott RD, MacKenzie AB and McKinley IG (1988) A natural analogue study of radionuclide migration in a water conducting fracture in crystalline rock. Radiochimica Acta, 44/45, 283-289.

Alexander WR, MacKenzie AB, Scott RD and McKinley IG (1990a) Natural analogue studies in crystalline rock: the influence of water-bearing fractures on radionuclide immobilization in a granitic rock repository. Nagra Technical Report, NTB 87-08, Nagra, Wettingen, Switzerland.

Alexander WR, McKinley IG, MacKenzie AB and Scott RD (1990b) Verification of matrix diffusion by means of natural decay series disequilibria in a profile across a water conducting fracture in granitic rock. Materials Research Society Symposium Proceedings, 261, (Scientific Basis for Nuclear Waste Management, XIII), 567-576.

Alexander WR, Brutsch R, Degueldre C and Hofmann B (1990c) Evaluation of long distance transport of natural colloids in a crystalline groundwater. Paul Scherrer Institute, TM-43-90-20, Switzerland.

Alexander WR, Dayal R, Eagleson K, Eikenberg J, Hamilton E, Linklater CM, McKinley IG and Tweed CJ (1992a) A natural analogue of high pH cement pore waters from the Maqarin area of northern Jordan II: results of predictive geochemical calculations. Journal of Geochemical Exploration, 46, 133-146.

Alexander WR, Bradbury MH, McKinley IG, Heer W, Eikenberg J and Frick U (1992b) The current status of the radionuclide migration experiment at the Grimsel underground rock laboratory. Materials Research Society Symposium Proceedings, 257, (Scientific Basis for Nuclear Waste Management, XV), 721-728.

Alexander WR, Gautschi A and Zuidema P (1998a) Thorough testing of performance assessment models: the necessary integration of in situ experiments, natural analogues and laboratory work. Materials Research Society Symposium Proceedings, 506, (Scientific Basis for Nuclear Waste Management, XXI), 1013-1014.

Alexander WR, McKinley IG, Linklater CM, Tweed CJ, Casas S, Börjesson S and Sellin P (1998b) Testing the limits of the applicability of thermodynamic databases. In: Linklater CM (editor) A natural analogue study of cement buffered, hyperalkaline groundwaters and their interaction with a repository host rock II. Nirex Science Report, S/98/003.

Allard B, Eliasson L, Höglund S and Andersson K (1984) Sorption of Cs, I and actinides in concrete systems. SKB Technical Report, TR 84-15, SKB, Stockholm, Sweden.

Allard B, Persson G and Torstenfelt B (1985a) Actinide solubilities and speciation in a repository environment. Nagra Technical Report, NTB 85-18, Nagra, Wettingen, Switzerland.

Allard B, Persson G and Torstenfelt B (1985b) Radionuclide sorption on concrete. Nagra Technical Report NTB 85-21, Nagra, Wettingen, Switzerland.

Allard B, Karlsson F and Neretnieks I (1991) Concentration of particulate matter and humic substances in deep groundwaters and estimated effects on the adsorption and transport of radionuclides. SKB Technical Report, TR 91-50, SKB, Stockholm, Sweden.

Ambrosetti P, Basilici G, Gentili S, Biondi E, Cerquaglia Z and Girotti O (1992) La Foresta Fossile di Dunarobba. Ediart, Todi, Italy.

Amter S (1989) Natural analogues. Engineering Geology, 26, 431-440.

Anderson DM (1983, editor) Smectite alteration. SKB Technical Report, TR 83-03, SKB, Stockholm, Sweden.

Andersson DM and Fontain J (1981) Investigation of the chemical stability of clays employed as buffer materials in the storage of nuclear waste materials Interim report to the SKBF Project. Unpublished report, SKB, Stockholm, Sweden. Sweden.

Andersson K, Torstenfelt B and Allard B (1981) Diffusion of cesium in concrete. Scientific Basis for Nuclear Waste Management, III, 235-242.

Andersson K, Torstenfelt B and Allard B (1983) Sorption and diffusion of Cs and I in concrete. SKB Technical Report, TR 83-13, SKB, Stockholm, Sweden.

Andrews RW and Pearson FJ (1984) Transport of ^{14}C and uranium in the Carrizo aquifer of South Texas, a natural analogue of radionuclide migration. Materials Research Society Symposium Proceedings, 26, (Scientific Basis for Nuclear Waste Management, VII), 1085-1092.

Angeli F, Faucon P, Charpentier T, Petit JC and Virlet J (1998) Comparative structural study and dissolution of simplified glasses: a radioactive waste glass (R7T7) and a basaltic glass. Materials Research Society Symposium Proceedings, 506, (Scientific Basis for Nuclear Waste Management, XXI), 71-78.

Angus NS, Brown GT and Cleere HF (1962) The iron nails from the Roman legionary fortress at Inchtuthil, Perthshire. Journal of Iron and Steel Institute, 200, 956-968.

Apted MJ (1992) Natural analogues for predicting the reliability of the engineered barrier system for high-level waste. Journal of Geochemical Exploration, 46, 35-62.

Arai T, Yusa Y, Sasaki N, Tsunoda N and Takano H (1989) Natural analogue study of volcanic glass. A case study of basaltic glasses in pyroclastic fall deposits of Fuji volcano Japan. Materials Research Society Symposium Proceedings, 127, (Scientific Basis for Nuclear Waste Management, XII), 73-80.

Askarieh MM, Chambers AV, Daniel FBD, FitzGerald PL, Holtom GJ, Pilkington NJ and Rees JH (2000) The chemical and microbial degradation of cellulose in the near field of a repository for radioactive wastes. Waste Management, 20, 93-106.

Atkinson A (1985) The time-dependence of pH within a repository for radioactive waste disposal. UKAEA Technical Report, AERE-R-11777, Harwell, England.

Atkinson A, Boult DJ and Hearne JA (1986) An assessment of the long-term durability of concrete in radioactive waste environments. Materials Research Society Symposium Proceedings, 50, (Scientific Basis for Nuclear Waste Management, IX), 239-246.

Atkinson A, Ewart FT, Pugh SYR, Rees JH, Sharland SM, Tasker PW and Wilkins JD (1988a) Experimental and modelling studies of the near-field chemistry for Nirex repository concepts. Nirex Radioactive Waste Disposal: Safety Studies, NSS/R104, United Kingdom Nirex Ltd, Harwell, England.

Atkinson A, Ewart FT, Pugh SYR, Rees JH, Sharland SM, Tasker PW and Wilkins JD (1988b) Experimental and modelling studies of the near-field chemistry for Nirex repository concepts. NEA Workshop on Near-Field Assessment of Repositories for Low and Medium Level Radioactive Waste, Baden 1987, 143-157.

Australian Atomic Energy Commission (1987) Radionuclide migration around uranium ore bodies: analogue of radioactive waste repositories: Annual report for 1984-1985. US

Nuclear Regulatory Commission, NUREG/CR-5040, Washington DC, USA.

Avogadro A and de Marsily G (1984) The role of colloids in nuclear waste disposal. Materials Research Society Symposium Proceedings, 26, (Scientific Basis for Nuclear Waste Management, VII), 495-505.

Azam F (1984) The radioecological role of marine bacterioplankton. In: IUR/CEC workshop, role of microorganisms on the behaviour of radionuclides in aquatic and terrestrial systems and their transfer to man, 2-7.

Bachofen R and Luescher D (1984) Moegliche mikrobiologische Vorgaenge in unterirdischen Kavernen im Hinblick auf die Endlagerung radioaktiver Abfaelle (Literaturstudie). Nagra Technical Report, NTB 84-07, Nagra, Wettingen, Switzerland.

Bachofen R, Dubach AC, Tesch AW and Luescher D (1984) Literaturstudie ueber den Abbau von Bitumen durch Mikroorganismen. Nagra Technical Report, NTB 83-18, Nagra, Wettingen, Switzerland.

Baertschi P, Alexander WR and Dollinger H (1991) Uranium migration in crystalline rock: capillary solution transport in the granite of the Grimsel test site, Switzerland. Nagra Technical Report, NTB 90-15, Nagra, Wettingen, Switzerland.

Baeyens B and Bradbury MH (1991) A physico-chemical characteristion technique for determining the pore water chemistry in argillaceous rocks. Nagra Technical Report, NTB 90-40, Nagra, Wettingen, Switzerland.

Bailey MG, Johnson LH and Shoesmith DW (1985) The effects of alpha radiolysis on the corrosion of UO2.. Corrosion Science, 25, 233.

Bailey NJL, Jobson AM and Rogers MA (1973) Bacterial degradation of crude oil: comparison of field and experimental data. Chemical Geology, 11, 203-221.

Ball TK and Milodowski AE (1991) The geological geochemical topographical and hydrogeological characteristics of the Broubster natural analogue site, Caithness. CEC Nuclear Science and Technology Report, EUR 13275, CEC, Luxembourg.

Barenblatt GI, Zheltov IP and Kochina IN (1960) Basic concepts in the theory of seepage of homogeneous liquids in fissured rock. Journal of Applied Mathematical Mechanics, 24, 1286.

Baross JA and Deming JW (1983) Growth of 'black smoker' bacteria at temperatures of at least 250°C. Nature, 303, 423-426.

Basham IR, Milodowski AE, Hyslop EK and Pearce JM (1991) The location of uranium in source rocks and sites of secondary deposition at the Needle's Eye natural analogue site, Dumfries and Galloway. CEC Nuclear Science and Technology Report, EUR 13279, CEC, Luxembourg.

Bateman K, Entwistle DC, Kemp S and Savage D (1991) Bentonite-groundwater interactions: results of compression experiments. BGS Technical Report, WE/91/21C.

Bateman K, Coombs P, Noy DJ, Pearce JM and Wetton P (1995) Nagra/Nirex/SKB column experiments I: results of experiments and modelling. Unpublished Nagra Internal Report, Nagra, Wettingen, Switzerland.

Bateman K, Coombs P, Pearce JM, Noy DJ and Wetton P (2000) Nagra/Nirex/SKB column experiments II: fluid/rock interactions in the disturbed zone. Unpublished Nagra Internal Report, Nagra, Wettingen, Switzerland.

Bates JK, Ellison AJG, Emery JW and Hoh JC (1996) Glass as a waste form for the immobilisation of plutonium. Materials Research Society Symposium Proceedings, 412, (Scientific Basis for Nuclear Waste Management, XIX), 57-64.

Bath AH, Christofi N, Neal C, Philp JC, Cave MR, McKinley IG and Berner U (1987a) Trace element and microbiological studies of alkaline groundwaters in Oman Arabian Gulf: a natural analogue for cement pore waters. Nagra Technical Report, NTB 87-16, Nagra, Wettingen, Switzerland and BGS Technical Report, FLPU 87-2.

Bath AH, Berner U, Cave MR, McKinley IG and Neal C (1987b) Testing geochemical models in a hyperalkaline environment. In: Côme B and Chapman NA (editors) Natural analogues in radioactive waste disposal. CEC Radioactive Waste Management Series, EUR 11037, 167-178, CEC, Luxembourg.

Beavers JA and Durr CL (1991) Immersion studies on candidate container alloys for the tuff repository US Nuclear Regulatory Commission, NUREG/CR-5598, Washington DC, USA.

Beck CW, Greenlie J, Diamond MP, Macchiarulo AM, Hannenberg AA and Hauck MS (1978) The chemical identification of Baltic amber at the Celtic Oppidum Staré Hradisko in Moravia. Journal of Archaeological Science, 5, 343-354.

Behrensmeyer AK (1980) Fossils in the making. Chicago University Press.

Bell JS (1989) Case studies in Canadian petroleum geology: vertical migration of hydrocarbons at Alma offshore eastern Canada. Bulletin of the Canadian Association of Petroleum Geology, 37, 358-364.

Bell KG (1956) Uranium in precipitates and evaporites. USGS Professional Paper, 300, 381-386.

Bengtsson G (1989) Can the same principles be used for the mangement of radioactive and non-radioactive wastes? In: Safety assessment of radioactive waste repositories, 59-70. NEA-OECD, Paris.

Benvegnú, F, Brondi A and Polizzano C (1988) Natural analogues and evidence of long-term isolation capacity of clays occurring in Italy: contribution to the demonstration of geological disposal reliability of long-lived wastes in clay. CEC Nuclear Science and Technology Report, EUR 11896, CEC, Luxembourg.

Berner U (1986) Radionuclide speciation in the porewater of hydrated cement - I. The hydration model. EIR Technical Report, TM-45-86-28.

Berner U (1990) A thermodynamic description of the evolution of porewater chemistry and uranium speciation during the degradation of cement. Nagra Technical Report, NTB 90-12, Nagra, Wettingen, Switzerland.

Bibler NE, Ramsey WG, Meaker TF and Pareizs JM (1996) Durabilities and microstructures of radioactive glasses for the immobilisation of excess actinides at the Savannah River site. Materials Research Society Symposium Proceedings, 412, (Scientific Basis for Nuclear Waste Management, XIX), 65-72.

Billion A, Caceci M, Della Mea G, Dellis T, Dran JC, Moulin V, Nicholson S, Petit JC, Ramsay JDF, Russell PJ and Theyssier M (1991) The role of colloids in the transport of radionuclides in geological formations. CEC Nuclear Science and Technology Report, EUR 13506, CEC, Luxembourg.

BIOMOVS (1996a) Development of a reference biospheres methodology for radioactive waste disposal. BIOMOVS II Project Technical Report, 6.

BIOMOVS (1996b) Biosphere modelling for dose assessments of radioactive waste repositories. Final report of the Complementary Studies Working Group. BIOMOVS II Project Technical Report, 12.

Birchard GF and Alexander DH (1983) Natural analogues: a way to increase confidence in predictions of long-term performance of radioactive waste disposal. Materials Research Society Symposium Proceedings, 15, (Scientific Basis for Nuclear Waste Management, VI), 323-329.

Birgersson L and Neretnieks I (1982) Diffusion in the matrix of granitic rock. Field test in the Stripa mine, Part I. SKB Technical Report, TR 82-08, SKB, Stockholm, Sweden.

Birgersson L and Neretnieks I (1983) Diffusion in the matrix of granitic rock. Field test in the Stripa mine, Part II. SKB Technical Report, TR 83-39, SKB, Stockholm, Sweden.

Birgersson L and Neretnieks I (1988) Diffusion in the matrix of granitic rock. Field test in the Stripa mine, Final Report. SKB Technical Report, TR 88-08, SKB, Stockholm, Sweden.

Bischoff K, Wolf K and Heimgartner B (1987) Hydraulische leitfähigkeit porosität und uranrück-haltung von kristallin und mergel: bohrkern-infiltrationsversuche. Nagra Technical Report, NTB 85-42, Nagra, Wettingen, Switzerland.

Blanc P-L (1996) Oklo - natural analogue for a radioactive waste repository (Phase 1). Volume 1: acquirements of the project. EC Nuclear Science and Technology Report, EUR 16857, EC, Luxembourg.

Blomqvist R, Lindberg A, Räisänen E, Suutarinen R, Jaakkola T and Suksi J (1987) The occurrence and migration of natural radionuclides in groundwater, I. Preliminary results of investigations in the Palmottu U-Th deposit, Nummi-Pusula, SW Finland. Geological Survey of Finland, Nuclear Waste Disposal Research Report, YST-60.

Blomqvist R, Jaakkola T, Niini H and Ahonen L (1991) The Palmottu analogue project: progress report 1990. Geological Survey of Finland, Nuclear Waste Disposal Research Report, YST-73.

Blomqvist R, Marcos N, Ahonen L and Ruskeeniemi T (1997) Natural analogue studies supporting the disposal concept for high level nuclear waste in Finland. In: von Maravic H and Smellie J (editors) Natural analogue working group, seventh meeting, Stein am Rhein, October 1996. CEC Nuclear Science and Technology Report, EUR 15176, 213-221, CEC, Luxembourg.

Blomqvist R, Smellie JAT, Korkealaakso J, Jakobsson K and Grundfelt B (2000, in press) Overview and summary of main results at Palmottu. In: von Maravic H and Alexander WR (editors) Natural analogue working group, eighth meeting, Strasbourg, March 1999. EC Nuclear Science and Technology Report, EC, Luxembourg.

Blowers A, Lowry D and Solomon BD (1991) The international politics of nuclear waste. Macmillan Academic and Professional Ltd.

Boatner LA and Sales BC (1988) Monazite. In: Lutze W and Ewing RC (editors) Radioactive waste forms for the future. North Holland, 495-564.

Börgesson L and Pusch R (1989) Interim report on the settlement test in Stripa. SKB Technical Report, TR 89-29, SKB, Stockholm, Sweden.

Börgesson L, Pusch R, Fredriksson A, Hökmark H, Karnland O and Sandén T (1992) Final report of the rock sealing project - sealing of zones disturbed by blasting and stress release. Stripa Project Technical Report, 92-21.

Bors J, Martens R and Kuehn W (1984) Investigations on the influence of microorganisms on the translocation of radio-iodine in soil. In: IUR/CEC workshop, role of microorganisms on the behaviour of radionuclides in aquatic and terrestrial systems and their transfer to man, 219-227.

Bossart P and Mazurek M (1991) Structural geology and water flow paths in the migration shear zone. Nagra Technical Report, NTB 91-12, Nagra, Wettingen, Switzerland.

Bouchet A, Boisson JY, Kemp SJ, Parneix JC, Pellegrini R and Rochelle C (2000, in press) Mineralogical and chemical effects of volcanic intrusion on three clay formations. In: von Maravic H and Alexander WR (editors) Natural analogue working group, eighth meeting, Strasbourg, March 1999. EC Nuclear Science and Technology Report, EC, Luxembourg.

Bowell RJ, Chapman J, Connelly RJ, Cowan J, Dodds JE, Hockley D, Sadler PJK and Wood A (1997) Natural analogues for toxic waste disposal: a mining perspective. In: von Maravic H and Smellie J (editors) Natural analogue working group, seventh meeting, Stein am Rhein, October 1996. CEC Nuclear Science and Technology Report, EUR 15176, 225-247, CEC, Luxembourg.

Brace WF, Walsh JB and Frangos WT (1968) Permeability of granite under high pressure. Journal of Geophysical Research, 2225-2236.

Bradbury MH and Stephen IG (1986) Diffusion and permeability based sorption measurements in intact rock samples. Materials Research Society Symposium Proceedings, 50, (Scientific Basis for Nuclear Waste Management, IX), 81-90.

Bradbury MH and Baeyens B (1997) Far-field sorption data bases for PA of a L/ILW repository in a disturbed/altered Palfris Marl host rock. Nagra Technical Report NTB 96-06, Nagra, Wettingen, Switzerland.

Bradbury MH and van Loon LR (1997) Cementitious near-field sorption data bases for PA of a L/ILW repository in a Palfris Marl host rock. Nagra Technical Report, NTB 96-04, Nagra, Wettingen, Switzerland.

Bradbury MH, Baeyens B and Alexander WR (1990) Experimental proposals for procedures to investigate the water chemistry sorption and transport properties of marls. Nagra Technical Report, NTB 90-16, Nagra, Wettingen, Switzerland.

Brandberg F and Skagius K (1991) Porosity sorption and diffusivity data compiled for the SKB 91 study. SKB Technical Report, TR 91-16, SKB, Stockholm, Sweden.

Brandberg F, Grundfelt B, Höglund L-O, Karlsson F, Skagius K and Smellie JAT (1993) Studies of natural analogues and geological systems - their importance to performance assessment. YJT Technical Report, YJT-93-07, YJT, Helsinki, Finland.

Bremer PJ and Geesey GG (1991) Laboratory based model of microbiologically induced corrosion of copper. Applied and Environmental Microbiology, 57, 1956-1962.

Bresle A, Saers J and Arrhenius B (1983) Studies in pitting corrosion on archaeological bronzes. SKB Technical Report, TR 83-05, SKB, Stockholm, Sweden.

Brill RH (1975) Crizzling: a problem in glass conservation. In: Conservation in Archaeology and the Applied Arts. International Institute for Conservation of Historic and Artistic Works, 121-131.

British Nuclear Forum (1991) Mother earth's natural reactors. Nuclear Forum, September 1991. London, England.

Brookins DG (1984) Geochemical aspects of radioactive waste disposal. Springer-Verlag, New York.

Brookins DG (1986) Natural analogues for radwaste disposal: elemental migration in igneous contact zones. Chemical Geology, 55, 337-344.

Brookins DG (1987a) Natural and archaeological analogues: a review. In: Côme B and Chapman NA (editors) Natural analogues in radioactive waste disposal. CEC Radioactive Waste Management Series, EUR 11037, 42-56, CEC, Luxembourg.

Brookins DG (1987b) Sandstone uranium deposits: analogues for SURF disposal in some sedimentary rocks. In: Côme B and Chapman NA (editors) Natural analogues in radioactive waste disposal. CEC Radioactive Waste Management Series, EUR 11037, 73-81, CEC, Luxembourg.

Brookins DG (1990) Radionuclide behaviour at the Oklo nuclear reactor, Gabon. Waste Management, 10, 285-296.

Bruno J and Casas I (1991) Spent fuel dissolution modelling. AECL/SKB Cigar Lake Natural Analogue Project. Workshop Proceedings on Modelling Status, Forsmark, Sweden.

Bruno J and Casas I (1994) Spent fuel dissolution modelling. In: Cramer J and Smellie JAT (editors) Final report of the AECL/SKB Cigar Lake Analog Study. AECL Technical Report, AECL-10851, Whiteshell, Canada and SKB Technical Report, TR 94-04, Stockholm, Sweden.

Bruno J, Cross JE, Eikenberg J, McKinley IG, Read D, Sandino A and Sellin P (1990) Testing of geochemical models in the Poços de Caldas analogue study. SKB Technical Report, TR 90-20, SKB, Stockholm, Sweden; Nagra Technical Report, NTB 90-29, Nagra, Wettingen, Switzerland; UK DoE Technical Report, WR 90-051.

Bruno J, Duro L, De Pablo J, Casas I, Ayora C, Delgado J, Gimeno MJ, Peña J, Linklater C, Pérez del Villar L and Gómez P (1998) Estimation of the concentrations of trace metals in natural systems. The application of codissolution and coprecipitation approaches to El Berrocal (Spain) and Poços de Caldas (Brazil). Chemical Geology, 151, 277-291.

Bucher F, Kahr G, Madsen FT and Mayer PA (1993) Wechselwirkungen von abgebundenum Zement mit verdichtetem Bentonit: Quelldruckversuche mit anschliessenden mineralogischen Untersuchungen. Nagra Technical Report, NTB 93-25, Nagra, Wettingen, Switzerland.

Buddemeier RW and Hunt JR (1988) Transport of colloidal contaminants in groundwater: radionuclide migration at the Nevada test site. Applied Geochemistry, 3, 535-548.

Burakov BE, Anderson EB, Rovsha VS, Ushakov SV, Ewing RC, Lutze W and Weber WJ (1996) Synthesis of zircon for immobilisation of actinides. Materials Research Society Symposium Proceedings, 412, (Scientific Basis for Nuclear Waste Management, XIX), 33-39.

Burnay SG (1987) Comparative evaluation of alpha and gamma radiation effects in bitumenisate. Nuclear and Chemical Waste Management, 7, 107-127.

Byers CD, Jercinovic MJ and Ewing RC (1987) A study of natural glass analogues as applied to alteration of nuclear waste glass. US Nuclear Regulatory Commission, NUREG/CR-4842, Washington DC, USA.

Cadelli N (1987) Natural analogues and performance assessments: a point of view based on the PAGIS experience. In: Côme B and Chapman NA (editors) Natural analogues in radioactive waste disposal. CEC Radioactive Waste Management Series, EUR 11037, 3-11, CEC, Luxembourg.

Cadelli N, Cottone G, Orlowski S, Bertozzi G, Girardi F and Saltelli A (1988) Performance assessment of geological isolation systems for radioactive waste - summary. CEC Radioactive Waste Management Series, EUR 11775, CEC, Luxembourg.

Carlsson J (1988) The Swedish final repository for reactor waste (SFR): a summary of the SFR project with special emphasis on the near-field assessments. NEA Workshop on Near-Field Assessment of Repositories for Low and Medium Level Radioactive Waste, Baden 1987, 71-82.

Casas I and Bruno J (1994) Testing of solubility and speciation codes. In: Cramer J and Smellie JAT (editors) Final report of the AECL/SKB Cigar Lake Analog Study. AECL Technical Report, AECL-10851, Whiteshell, Canada and SKB Technical Report, TR 94-04, Stockholm, Sweden.

Cathelineau M and Vergneaud M (1989) U-Th-REE mobility and diffusion in granitic environments during alteration of accessory minerals and U-ores: a geochemical analogue to radwaste disposal. Materials Research Society Symposium Proceedings, 127, (Scientific Basis for Nuclear Waste Management, XII), 941-948.

Cathles LM and Shea ME (1990) Near-field high temperature transport: evidence from the genesis of the Osamu Utsumi uranium mine analogue site, Poços de Caldas, Brazil. SKB Technical Report, TR 90-22, SKB, Stockholm, Sweden; Nagra Technical Report, NTB 90-31, Nagra, Wettingen, Switzerland; UK DoE Technical Report, WR 90-053.

Cathles LM and Shea ME (1992) Near-field high temperature transport: evidence from the genesis of the Osamu Utsumi uranium mine, Poços de Caldas alkaline complex. Journal of Geochemical Exploration, 45, 565-603.

Chao TT (1984) Use of partial dissolution techniques in geochemical exploration. Journal of Geochemical Exploration, 20, 101-109.

Chapman NA (1986) Highly alkaline groundwaters in Oman: geomicrobiological and chemical speciation studies of a natural analogue of cement pore waters. In: Côme B and Chapman NA (editors)

Natural analogue working group, second meeting, Interlaken, June 1986. CEC Nuclear Science and Technology Report, EUR 10671, 135-137, CEC, Luxembourg.

Chapman NA (1988) Can natural analogues provide quantitative model validation? Validation of Geosphere Flow and Transport Models, GEOVAL-1988.

Chapman NA (1990) Natural analogues. In: Côme B (editor) CEC project Mirage-Second phase on migration of radionuclides in the geosphere. Third (and final) summary progress report (work period 1989). CEC Nuclear Science and Technology Report, EUR 12858, CEC, Luxembourg.

Chapman NA (1992) Natural analogues: the state of play in 1992. Proceedings of the Third International High-Level Radioactive Waste Management Conference, Las Vegas, 1695-1700.

Chapman NA (1994) The geologist's dilemma: predicting the future behaviour of buried radioactive wastes. Terra Nova, 6, 5-19.

Chapman NA and Smellie JAT (1986, editors) Natural analogues to the conditions around a final repository for high-level radioactive wastes. Chemical Geology: Special Issue, 55.

Chapman NA and McKinley IG (1987) The geological disposal of nuclear waste. John Wiley and Sons.

Chapman NA and McKinley IG (1990) Radioactive wastes: back to the future? New Scientist, 1715, 54-58.

Chapman NA and McEwen TJ (1992) The application of palaeohydrogeological information to repository performance assessment. In: Palaeohydrogeological Methods and their Applications for Radioactive Waste Disposal. NEA-OECD, Paris.

Chapman NA and Miller WM (1994) Using information from natural systems to build confidence in performance assessment. In: von Maravic, H and Smellie J (editors) Natural analogue working group, fifth meeting, Toledo, October 1992. CEC Nuclear Science and Technology Report, EUR 15176, 15-23, CEC, Luxembourg.

Chapman NA, McKinley IG and Smellie JAT (1984) The potential of natural analogues in assessing systems for deep disposal of high-level radioactive waste. SKB Technical Report, TR 84-16, SKB, Stockholm, Sweden; Nagra Technical Report, NTB 84-41, Nagra, Wettingen, Switzerland; EIR Technical Report, Nr 545.

Chapman NA, McKinley IG, Shea ME and Smellie JAT (1990) The Poços de Caldas Project: summary and implications for radioactive waste management. SKB Technical Report, TR 90-24, SKB, Stockholm, Sweden; Nagra Technical Report, NTB 90-33, Nagra, Wettingen, Switzerland; UK DoE Technical Report, WR 90-055.

Chapman NA, McKinley IG, Penna Franca E, Shea MJ and Smellie JAT (1992) The Poços de Caldas project: an introduction and summary of its implications for radioactive waste disposal. Journal of Geochemical Exploration, 45, 1-24.

Chapman NA, Andersson J, Robinson P, Skagius K, Wene C-O, Wiborgh M and Wingefors S. (1995) Systems analysis, scenario construction and consequence analysis definition for SITE-94. SKI Technical Report, 95:26, SKI, Stockholm, Sweden

Chapuis AM and Blanc, PL (1993) Oklo: natural analogue for transfer processes in a geological repository - present status of the programme. In: von Maravic H and Moreno J (editors) Migration of radionuclides in the geosphere. Proceedings of a Progress Meeting. CEC Nuclear Science and Technology Report, EUR 14690, CEC, Luxembourg.

Chermak J (1992) Low temperature experimental investigation of the effect of high pH NaOH solutions on the Opalinus shale, Switzerland. Clays and Clay Minerals, 40, 650-658.

Chermak J (1993) Low temperature experimental investigation of the effect of high pH KOH solutions on the Opalinus shale, Switzerland. Clays and Clay Minerals, 41, 365-372

Chermak J (1996) Interactions of the Palfris and Opalinus shales with high pH NaOH and KOH solutions at 75°C. Unpublished Nagra Internal Report, Nagra, Wettingen, Switzerland.

Chernis PJ (1981) Scanning electron microscope study of the microcrack structure of a granite sample from Pinawa, Manitoba. AECL Technical Report, TR-173, AECL, Whiteshell, Canada.

Chernis PJ (1983) Notes on the pore-microfracture structure of shallow and deep samples of the Lac du Bonnet granite. AECL Technical Report, TR-223, AECL, Whiteshell, Canada.

Chernis PJ (1984) Comparison of the pore-microcrack structure of some granitic samples from the Whiteshell nuclear research establishment. AECL Technical Report, TR-226, AECL, Whiteshell, Canada.

Chet I and Mitchell R (1976) Ecological aspects of microbial chemotactic behaviour. Annual Review in Microbiology, 30, 221-239.

Christensen H (1984) Formation of nitric and organic acids by the irradiation of ground water in a spent fuel repository. SKB Technical Report, TR 84-12, SKB, Stockholm, Sweden.

Christensen H (1990) Calculation of the effect of alpha-radiolysis on UO_2 oxidation. Studsvik Technical Report, NS-89/117, Studsvik Material AB, Sweden.

Christensen H (1994) Oxidation by water radiolysis products. In: Cramer J and Smellie JAT (editors) Final Report of the AECL/SKB Cigar Lake Analogue Study. AECL Research Report, AECL-10851; SKB Technical Report, TR 94-05.

Christensen H and Bjergbakke E (1982) Radiolysis of ground water from spent fuel. SKB Technical Report, TR 82-18, SKB, Stockholm, Sweden.

Christensen H and Bjergbakke E (1984a) Radiolysis of concrete. SKB Technical Report, TR 84-02, SKB, Stockholm, Sweden.

Christensen H and Bjergbakke E (1984b) Effect of beta-radiolysis on the products from alpha-radiolysis of groundwater. SKB Technical Report, TR 84-03, SKB, Stockholm, Sweden.

Christensen H, Sunder S and Shoesmith DW (1992a) Calculation of radiation induced dissolution of UO_2. Adjustment of the model based on alpha-radiolysis experiments Studvisk Material Report, M-92/20, Studsvik Laboratory, Sweden.

Christensen H, Sunder S and Shoesmith DW (1992b) Radiolysis modelling for the Cigar Lake uranium deposit. Studvisk Material Report, M-92/56, Studsvik Laboratory, Sweden.

Coles DG and Ramspott LD (1982) Migration of ruthenium-106 in a Nevada test Site aquifer: discrepancy between field and laboratory results. Science, 215, 1235-1237.

Collepardi M, Marcialis A and Turrizani R (1972) Penetration of chloride ions into cement pastes and concrete. Journal of the American Ceramic Society, 55, 534-535.

Colley S and Thomson J (1985) Recurrent uranium relocations in distal turbidites emplaced in pelagic conditions. Geochimica et Cosmochimica Acta, 49, 2339-2348.

Colley S and Thomson J (1991) Migration of uranium daughter radionuclides in natural sediments. CEC Nuclear Science and Technology Report, EUR 13182, CEC, Luxembourg.

Colley S, Thomson J, Wilson TRS and Higgs NC (1984) Post-depositional migration of elements during diagenesis in brown clay and turbidite sequences in the North East Atlantic. Geochimica et Cosmochimica Acta, 48, 1223-1234.

Côme B and Chapman NA (1986a, editors) Natural analogue working group, first meeting, Brussels, November 1985. CEC Nuclear Science and Technology Report, EUR 10315, CEC, Luxembourg.

Côme B and Chapman NA (1986b, editors) Natural analogue working group, second meeting, Interlaken, June 1986. CEC Nuclear Science and Technology Report, EUR 10671, CEC, Luxembourg.

Côme B and Chapman NA (1987, editors) Natural analogues in radioactive waste disposal. CEC Radioactive Waste Management Series, EUR 11037, CEC, Luxembourg.

Côme B and Chapman NA (1989, editors) Natural analogue working group, third meeting, Snowbird, June 1988. CEC Nuclear Science and Technology Report, EUR 11725, CEC, Luxembourg.

Côme B and Chapman NA (1991, editors) Natural analogue working group, fourth meeting and Poços de Caldas project final meeting, Pitlochry, June 1990. CEC Nuclear Science and Technology Report, EUR 13014, CEC, Luxembourg.

Côme B, Piantone P and Revin P (1997) Natural analogues in toxic waste disposal: an example in the French context. In: von Maravic H and Smellie J (editors) Natural analogue working group, seventh meeting, Stein am Rhein, October 1996. CEC Nuclear Science and Technology Report, EUR 15176, 257-266, CEC, Luxembourg.

Coons W, Bergström A, Gnirk P, Gray M, Knecht B, Pusch R, Steadman J, Stillborg B, Tokonami M and Vaajasaari M (1987) State-of-the-art report on potentially useful materials for sealing nuclear waste repositories. Nagra Technical Report, NTB 87-33, Nagra, Wettingen, Switzerland.

Coughtrey PJ and Thorne MJ (1982) Radionuclide distribution and transport in terrrestrial and aquatic ecosystems (Three volumes). AA Balkema.

Couture RA (1985) Steam rapidly reduces the swelling capacity of bentonite. Nature, 318, 50-52.

Cowan R and Ewing RC (1989) Freshwater alteration of basaltic glass, Hnauma Bay Oahu, Hawaii: a natural analogue for alteration of basaltic glass in freshwater. Materials Research Society Symposium Proceedings, 127, (Scientific Basis for Nuclear Waste Management, XII), 49-56.

Cramer JJ (1984) Sandstone-hosted uranium deposits in northern Saskatchewan as natural analogues to nuclear fuel waste disposal vaults. In: Smellie JAT (editor) Natural analogues to the conditions around a final repository for high level radioactive waste Proceedings of the natural analogue workshop held at Lake Geneva, Wisconsin, USA. SKB Technical Report, TR 84-18, SKB, Stockholm, Sweden.

Cramer JJ (1989) Natural analogue studies on the Cigar Lake uranium deposit: an update. In: Côme B and Chapman NA (editors) Natural analogue working group, third meeting, Snowbird, June 1988. CEC Nuclear Science and Technology Report, EUR 11725, 50-56, CEC, Luxembourg.

Cramer JJ (1991) Cigar Lake project: progress report for the period May-October 1991. AECL Cigar Lake Report, CLR-91-5, AECL, Whiteshell, Canada.

Cramer JJ and Sargent FP (1986) Cigar Lake project: a U-deposit natural analogue. In: Côme B and Chapman NA (editors) Natural analogue working group, first meeting, Brussels, November 1985. CEC Nuclear Science and Technology Report, EUR 10315, 216-223, CEC, Luxembourg.

Cramer JJ and Smellie JAT (1994a) The AECL/SKB Cigar Lake analog study: some implications for performance assessment. In: von Maravic, H and Smellie J (editors) Natural analogue working group, fifth meeting, Toledo, October 1992. CEC Nuclear Science and Technology Report, EUR 15176, CEC, Luxembourg.

Cramer JJ and Smellie JAT (1994b, editors) Final report of the AECL/SKB Cigar Lake Analog Study. AECL Technical Report, AECL-10851, Whiteshell, Canada and SKB Technical Report, TR 94-04, Stockholm, Sweden.

Cramer JJ, Vilks P and Larocque JPA (1987) Near-field analogue features from the Cigar Lake uranium deposit. In: Côme B and Chapman NA (editors) Natural analogues in radioactive waste disposal. CEC Radioactive Waste Management Series, EUR 11037, 59-72, CEC, Luxembourg.

Crissman D and Jacobs G (1982) Native copper deposits of the Portage Lake Volcanics, Michigan: their implications with respect to canister stability for nuclear waste isolation in Columbia River basalts beneath the Hanford site, Washington. Rockwell Hanford Operations Technical Report, RHO-BW-ST-26P.

Cross JE, Haworth A, Lichtner PC, MacKenzie AB, Moreno L, Neretnieks I, Nordstrom DK, Read D, Romero L, Scott RD, Sharland SM and Tweed CJ (1990) Testing models of redox front migration and geochemistry at the Osamu Utsumi mine and Morro do Ferro analogue study sites, Poços de Caldas, Brazil. SKB Technical Report, TR 90-21, SKB, Stockholm, Sweden; Nagra Technical Report, NTB 90-30, Nagra, Wettingen, Switzerland; UK DoE Technical Report, WR 90-052.

Crovisier J-L, Honnorez J, Fritz B, and Petit J-C (1992) Dissolution of subglacial volcanic glasses from Iceland: laboratory study and modelling. Applied Geochemistry, Supplementary Issue, 1, 55-82.

Crummy P (1997) Colchester: the Stanway burials. Current Archaeology, 153, 337-342.

Cullen TL and Penna Franca E (1977, editors) International symposium on areas of high natural radioactivity, Poços de Caldas, Brazil, June 1986.

Curtis DB (1985) The chemical coherence of natural spent fuel at the Oklo nuclear reactors. SKB Technical Report, TR 85-04, SKB, Stockholm, Sweden.

Curtis DB (1996) Radionuclide release rates from spent fuel for performance assessment modelling. In: von Maravic H and Smellie J (editors) Natural analogue working group, sixth meeting, Santa Fe, September 1994. CEC Nuclear Science and Technology Report, EUR 16761, 145-153, CEC, Luxembourg.

Curtis DB and Gancarz AJ (1983) Radiolysis in nature: evidence from the Oklo natural reactors. KBS Technical Report, TR 83-10.

Curtis DB, Benjamin TM and Gancarz AJ (1982) Transport of fission products at the Oklo natural reactor. Los Alamos National Laboratory Report, LA-UR-82-1276.

Curtis DB, Benjamin TM and Gancarz AJ (1983) The Oklo reactors: natural analogues to nuclear waste repositories. Los Alamos National Laboratory Report, LA-UR-81-3/83.

Curtis DB, Benjamin TM, Gancarz AJ, Loss R, Rosman JKR, DeLaeter JR, Delmore JE and Maeck WJ (1989) Fission product retention in the Oklo natural fission reactors. Journal of Applied Geochemistry, 4, 49-62.

Daemen JJK (1994) Gas pressure build-up and host rock mechanics - possible implications for the performance and repository assessment of a high level nuclear waste repository. In: von Maravic H and Smellie J (editors) Natural analogue working group, fifth meeting, Toledo, October 1992. CEC Nuclear Science and Technology Report, EUR 15176, 267-272, CEC, Luxembourg.

Daux V, Crovisier JL and Petit JC (1991) Rare-earth element behaviour during alteration of basaltic glasses: case of the weathering of Icelandic hyaloclastites. Materials Research Society Symposium Proceedings, 212, (Scientific Basis for Nuclear Waste Management, XIV), 107-114.

Dearlove JPL (1989) Analogue studies in natural rock systems: uranium series radionuclide and REE distribution and transport. Unpublished PhD Thesis, Cambridgeshire College of Arts and Technology, UK.

Dearlove JPL, Ivanovich M and Green DC (1989) Partition coefficients for the U series radionuclides in ion exchange sites and amorphous Fe phase in illite and granite samples. In: Miles DL (editor) Water Rock Interaction WRI-6. Proceedings of the 6th International Symposium on Water-Rock Interaction, Malvern, August 1989, 191-195.

Degueldre C (1994) Colloid properties in groundwater from crystalline formations. Nagra Technical Report, NTB 92-05, Nagra, Wettingen, Switzerland.

Degueldre C, Baeyens B, Görlich W, Riga J, Verbist J and Stadelmann P (1989) Colloids in water from a subsurface fracture in granitic rock, Grimsel Test

Site, Switzerland. Geochimica et Cosmochimica Acta, 53, 603-610.

Degueldre C, Longworth G, Moulin V and Vilks P (1990) Grimsel Test Site: Grimsel colloid exercise - an international intercomparison exercise on the sampling and characterisation of groundwater colloids. Nagra Technical Report, NTB 90-01, Nagra, Wettingen, Switzerland.

DeLaeter JR, Rosman JKR and Smith CL (1980) The Oklo natural reactor: cumulative fission yields and retentivity of the symmetric mass region fission products. Earth and Planetary Science Letters, 50, 238-246.

Del Nero M, Salah S, Miura T, Clément A and Gauthier-Lafaye F (2000, in press) Retention processes of uranium and REE in the Bangombé natural reactor zone, Gabon. In: von Maravic H and Alexander WR (editors) Natural analogue working group, eighth meeting, Strasbourg, March 1999. EC Nuclear Science and Technology Report, EC, Luxembourg.

Dexter SC (1986, editor) Biologically induced corrosion. Proceedings of the International Conference on Biologically Induced Corrosion, Houston.

Dresselaers J, Casteels F and Tas H (1983) Corrosion of construction materials in clay environments. Materials Research Society Symposium Proceedings, 6, (Scientific Basis for Nuclear Waste Management, IV), 311-319.

Drew D J and Vandergraaf, TT (1989) Construction and operation of a high pressure radionuclide migration apparatus. AECL Technical Report, TR-476, AECL, Whiteshell, Canada.

Dubessy J, Pagel M, Beny J-M, Christensen H, Hickel B, Zosztolanyi C and Poty B (1988) Radiolysis evidenced by H_2O_2 and H_2 bearing fluid inclusions in three uranium deposits. Geochimica et Cosmochimica Acta, 52, 1155-1167.

Duerden P, Golian C, Hardy CJ, Nightingale T and Payne T (1987) Alligator Rivers analogue project: review of research and its implications for model validation. In: Côme B and Chapman NA (editors) Natural analogues in radioactive waste disposal. CEC Radioactive Waste Management Series, EUR 11037, 82-91, CEC, Luxembourg.

Duerden P (1990, editor) Alligator Rivers Analogue Project: annual report 1989-1990. Australian Nuclear Science and Technology Organisation.

Eberl DD and Hower J (1976) Kinetics of illite formation. Bulletin of the Geological Society of America, 87, 1326-1330.

Edgehill R and Davey BG (1988) Colloids in Koongarra groundwater. In: Duerden P (editor) Alligator Rivers Analogue Project: progress report May 1988-August 1988. Australian Nuclear Science and Technology Organisation.

Eisenberg N Neretnieks I, Lever D Bruno J and MacKinley IG (1994) Why do we not see more recognition of natural analogues in performance assessment? Panel session notes. In: von Maravic H and Smellie J (editors) Natural analogue working group, fifth meeting, Toledo, October 1992. CEC Nuclear Science and Technology Report, EUR 15176, CEC, Luxembourg.

Eisenbud M, Lei W, Ballard R, Penna Franca E, Miekeley N, Cullen T and Krauskopf K (1982) Studies of the mobilisation of thorium from Morro do Ferro. Materials Research Society Symposium Proceedings, 11, (Scientific Basis for Nuclear Waste Management, V), 735-744.

Eisenbud M, Krauskopf K, Penna Franca E, Lei W, Ballard R, Linsalata P and Fujimori K (1984) Natural analogues for the transuranic actinide elements: an investigation in Minas Gerais Brazil. In: Smellie JAT (editor) Natural analogues to the conditions around a final repository for high level radioactive waste. Proceedings of the natural analogue workshop held at Lake Geneva, Wisconsin, USA. SKB Technical Report, TR 84-18, SKB, Stockholm, Sweden.

Ekendahl S and Pedersen K (1994) Carbon transformations by attached bacterial populations in granitic groundwater from deep crystalline bed rock of the Stripa research mine. Mincrobiology, 140, 1565-1573.

Eng T, Hudson J, Stephansson O, Skagius K and Wiborgh M (1994) Scenario development methodologies. SKB Technical Report, TR 94-28, SKB, Stockholm, Sweden.

ENRESA (1996) El Berrocal Project Topical Report Series. Four Volumes. ENRESA, Madrid, Spain.

Eriksen TE and Jacobsson A (1982) Diffusion of hydrogen hydrogen sulphide and large molecular weight anions in bentonite. SKB Technical Report, TR 82-17, SKB, Stockholm, Sweden.

Eriksen TE and Ndalamba P (1988) On the formation of a moving redox front by alpha-radiolysis of compacted water saturated bentonite. SKB Technical Report, TR 88-27, SKB, Stockholm, Sweden.

EWI (1983) Bitumen ein Verfestigungsmaterial für radioaktive Abfälle und seine historischen Analoga. Nagra Technical Report, NTB 85-25, Nagra, Wettingen, Switzerland.

Ewing RC (1979) Natural analogues: analogues for radioactive waste forms. (Scientific Basis for Nuclear Waste Management, I), 57-68.

Ewing RC and Haaker RF (1979) Natural glasses: analogues for radioactive waste forms. Battelle PNL Report, 2776/UC-70.

Ewing RC and Jercinovic MJ (1987) Natural analogues: their application to the prediction of the long-term behaviour of waste forms. Materials Research Society Symposium Proceedings, 84, (Scientific Basis for Nuclear Waste Management, X), 67-83.

Ewing RC, Chakoumakos BC, Lumpkin GR, Murakimi T, Greegor RB and Lytle FW (1988) Metamict minerals: natural analogues for radiation damage effects in ceramic nuclear waste forms. Nuclear Instruments and Methods in Physics Research B32, 487-497.

Fabryka-Martin J and Cutis D (1994) Uranium orebodies as source terms of radionuclides - measurements and models. In: von Maravic H and Smellie J (editors) Natural analogue working group, fifth meeting, Toledo, October 1992. CEC Nuclear Science and Technology Report, EUR 15176, 83-94, CEC, Luxembourg.

Falck WE and Hooker PJ (1990) Quantitative interpretation of Cl, Br and I porewater concentration profiles in lake sediments of Loch Lomond, Scotland. BGS Technical Report, WE/90/3.

Finch RJ and Ewing RC (1989) Alteration of natural UO_2 under oxidising conditions from Shinkolobwe, Katanga, Zaire: a natural analogue for the corrosion of spent fuel. SKB Technical Report, TR 89-37, SKB, Stockholm, Sweden.

Finch RJ and Ewing RC (1991) Uraninite alteration in an oxidising environment and its relevance to the disposal of spent nuclear fuel. SKB Technical Report, TR 91-15, SKB, Stockholm, Sweden.

Finch RJ and Ewing RC (1992) The corrosion of uraninite under oxidising conditions. Journal Nuclear Materials, 190, 133-156.

Fisher NS (1984) Concentration of radionuclides by marine phytoplankton. In: IUR/CEC workshop, role of microorganisms on the behaviour of radionuclides in aquatic and terrestrial systems and their transfer to man, 8-19.

Forbes RJ (1934) Aus der ältesten Geschichte des Bitumens. Bitumen, 6-11.

Forbes RJ (1938) Neues zur ältesten Geschichte des Bitumens. Bitumen, 128-134.

Forsman NF (1984) Durability and alteration of some Cretaceous and Palaeocene pyroclastic glasses in North Dakota. In: Pye LD O'Keefe JA and Fréchette VD (editors) Natural glasses, 449-461, North-Holland Publishers.

Forsyth RS, Werme LO and Bruno J (1985) The corrosion of spent UO_2 fuel in synthetic groundwater. SKB Technical Report, TR 85-16, SKB, Stockholm, Sweden.

Foster SSD (1975) The Chalk groundwater tritium anomaly - a possible explanation. Journal of Hydrology, 25, 159-165.

Francis AJ (1977) The cement industry 1796-1914: a history. David and Charles.

Frick U, Alexander WR, Baeyens B, Bossart P, Bradbury MH, Buhler C, Eikenberg J, Fierz T, Heer W, Hoehn E, McKinley IG and Smith PA (1992) Grimsel Test Site: The radionuclide migration experiment - overview of investigations 1985-1990. Nagra Technical Report, NTB 91-04, Nagra, Wettingen, Switzerland.

Fritz B, Kam M and Tardy Y (1984) Geochemical simulation of the evolution of granitic rocks and clay minerals submitted to a temperature increase in the vicinity of a repository for spent nuclear fuel. SKB Technical Report, TR 84-10, SKB, Stockholm, Sweden.

Fritz B, Kam M and Tardy Y (1985) Geochemical modelling of the alteration of a granitic rock around a repository for spent nuclear fuel. Materials Research Society Symposium Proceedings, 50, (Scientific Basis for Nuclear Waste Management, IX), 517-524.

Fritz B, Madé, B and Tardy Y (1988) Geochemical modelling of the evolution of a granite-concrete-water system around a repository for spent nuclear fuel. SKB Technical Report, TR 88-18, SKB, Stockholm, Sweden.

Gancarz, AJ, Cowan G, Curtis D and Maeck W (1980) ^{99}Tc, Pb and Ru migration around the Oklo natural fission reactors. (Scientific Basis for Nuclear Waste Management, II), 601-610.

Gani MSJ (1997) Cement and concrete. Chapman and Hall, London, UK.

Gardner MP, Holtom GJ and Swanton SW (1999, in press) Influence of colloids, microbes and other perturbations on the near-field source term. The Analyst.

Garrels RM, Dreyer RM and Howland AL (1949) Diffusion of ions through intergranular spaces in water saturated rocks. Bulletin of the Geological Society of America, 60, 1809-1924.

Garwin RL (1996) Options and choices in the disposition of excess weapon plutonium. Materials Research Society Symposium Proceedings, 412, (Scientific Basis for Nuclear Waste Management, XIX), 3-13.

Gauthier-Lafaye F (2000, in press) Oklo natural analogue phase II: general context and project objectives. In: von Maravic H and Alexander WR (editors) Natural analogue working group, eighth meeting, Strasbourg, March 1999. EC Nuclear Science and Technology Report, EC, Luxembourg.

Gauthier-Lafaye F and Weber F (1993) Uranium hydrocarbon association in Fancevillian uranium ore deposits, Lower Proterozoic of Gabon. In: Parnell J, Kucha H and Landais P (editors) Bitumens in ore deposits. Springer-Verlag.

Gera F, Andretta D, Bocala W, Chiantore V and Schneider A (1992) State of the art report: disposal of radioactive wastes in deep argillaceous formations. ENRESA Technical Report, 01/92. ENRESA, Madrid.

Gera F, Hueckel T and Pellegrini R (1994) Magmatic intrusions in clays as geomechanical natural analogues. In: von Maravic H and Smellie J (editors) Natural analogue working group, fifth meeting, Toledo, October 1992. CEC Nuclear Science and Technology Report, EUR 15176, 273-279, CEC, Luxembourg.

Gibb FGF (1999) High-temperature, very deep, geological disposal: a safer alternative for high-level radioactive waste? Waste Management, 19, 207-211.

Gieré R, Williams CT and Lumpkin GR (1998) Crystal chemistry of natural zirconolite: implications for high-level waste incorporation in SYNROC. Materials Research Society Symposium Proceedings, 506, (Scientific Basis for Nuclear Waste Management, XXI), 1031-1032.

Gilbert TL (1984) A natural analogue approach for estimating the health risks from release and migration of radionuclides from radioactive waste. Materials Research Society Symposium Proceedings, 26, (Scientific Basis for Nuclear Waste Management, VII), 935-942.

Gjrv OE and Vennesland O (1979) Diffusion of chloride ions from seawater into concrete. Cement and Concrete Research, 9, 229-238.

Glueckauf E (1980) The movement of solutes through aqueous fissures in porous rock. UKAEA Technical Report, AERE-R-9823, Harwell, England.

Golian C and Lever DA (1994) Transport modelling in the Koongara weathered zone. In: von Maravic H and Smellie J (editors) Natural analogue working group, fifth meeting, Toledo, October 1992. CEC Nuclear Science and Technology Report, EUR 15176, 115-120, CEC, Luxembourg.

Goodwin BW, Cramer JJ and McConnell DB (1989) The Cigar Lake uranium deposit: an analogue for nuclear fuel waste disposal. In: Natural analogues in performance assessments for the disposal of radioactive wastes. IAEA Technical Report, 304, International Atomic Energy Agency, Vienna, Austria.

Goth M (1985) Indication of methane movement from petroleum reservoir to surface, Löningen Oilfield, Germany. Journal of Geochemical Exploration, 23, 81-97.

Grambow B, Jercinovic MJ, Ewing RC and Byers CD (1986) Weathered basalt glass: a natural analogue for the effects of reaction progress on nuclear waste glass alteration. Materials Research Society Symposium Proceedings, 50, (Scientific Basis for Nuclear Waste Management, IX), 263-272.

Grandstaff DE (1976) A kinetic study of the dissolution of uraninite. Economic Geology, 71, 1493-1506.

Granger HC and Warren CG (1969) Unstable sulfur compounds and the origin of roll-type uranium deposits. Economic Geology, 64, 160-171.

Grauer R (1986) Bentonite as a backfill material in the high-level waste repository: chemical aspects. Nagra Technical Report, NTB 86-12E, Nagra, Wettingen, Switzerland.

Grauer R (1988) The corrosion behaviour of carbon steel in Portland cement. Nagra Technical Report, NTB 88-02E, Nagra, Wettingen, Switzerland.

Grauer R (1990) The chemical behaviour of montmorillonite in a repository backfill: selected aspects. Nagra Technical Report, NTB 88-24E, Nagra, Wettingen, Switzerland.

Grenthe I (1989). Chemical thermodynamics of uranium. NEA-OECD, Paris.

Grenthe I, Puigdoménech I and Bruno J (1983) The possible effects of alpha and beta radiolysis on the matrix dissolution of spent nuclear fuel. SKB Technical Report, TR 83-02, SKB, Stockholm, Sweden.

Grisak GE and Pickens JF (1980) Solute transport through fractured media, Part I: the effect of matrix diffusion. Water Resources Research, 16, 719-730.

Grogan HA and McKinley IG (1990) An approach to microbiological modelling: application to the near-field of a Swiss low and intermediate-level waste repository. Nagra Technical Report, NTB 89-06, Nagra, Wettingen, Switzerland.

Gschwend PM, Backhus D, MacFarlane JK and Page AL (1990) Mobilisation of colloids in groundwater due to infiltration of water at a coal-ash disposal site. Journal of Contaminant Hydrology, 6, 307-320.

Güntensperger M (1993) International video project on natural analogues European Nuclear Society PIME '93 Meeting, Karlovy Vary.

Gustafsson E, Skålberg M, Sundbland B, Karlberg O, Tullborg E-L, Ittner T, Carbol P, Eriksson N and Lampe S (1987) Radionuclide deposition and migration within the Gideå and Finnsjön study sites Sweden: A study of the fallout after the Chernobyl accident Phase I, initial survey. SKB Technical Report, TR 87-28, SKB, Stockholm, Sweden.

Guthrie VA (1989) Fission-track analysis of uranium distribution in granitic rocks. Chemical Geology, 77, 87-103.

Haaker RF and Ewing RC (1980) Uranium and thorium minerals: natural analogues for radioactive waste forms. (Scientific Basis for Nuclear Waste Management, III), 281-288.

Hack H (1988, editor) Galvanic corrosion. American Society for Testing and Manufacture, ASTM-STP 572, Philadelphia.

Hadermann J and Roesel F (1984) Matrix diffusion. Swiss Federal Institute for Reactor Research Technical Report, EIR/TM-45-84-26.

Hadermann J and Roesel F (1985) Radionuclide chain transport in homogeneous crystalline rocks: limited matrix diffusion and effective surface sorption. Nagra Technical Report, NTB 85-40, Nagra, Wettingen, Switzerland.

Hadermann J and Jakob, A (1987) Modelling small scale infiltration experiments into bore cores of crystalline rock and break-through curves. EIR-Bericht Nr 622, Eidgenossisches Institut fur Reaktorforschung, Wurenlingen, Switzerland.

Hallberg RO, Östlund P and Wadsten T (1987) A 17th century cannon as analogue for radioactive waste disposal. In: Côme B and Chapman NA (editors) Natural analogues in radioactive waste disposal. CEC Radioactive Waste Management Series, EUR 11037, 135-139, CEC, Luxembourg.

Hardy CJ and Duerden P (1989) Progress in the Alligator Rivers analogue project. In: Côme B and Chapman NA (editors) Natural analogue working group, third meeting, Snowbird, June 1988. CEC Nuclear Science and Technology Report, EUR 11725, 43-49, CEC, Luxembourg.

Hart KP, Lumpkin GR, Gieré R, Williams CT, McGlinn PJ and Payne TE (1996) Naturally-occurring zirconolites - analogues for the long-term encapsulation of actinides in SYNROC. Radiochimica Acta, 74, 309-312.

Hart KP, Lumpkin GR, Ellis DJ, Allen CM, Gieré R and Williams CT (1997) Further analysis of the applicability of naturally-occurring zirconolites as analogues for HLW waste matrices. In: von Maravic H and Smellie J (editors) Natural analogue working group, seventh meeting, Stein am Rhein, October 1996. CEC Nuclear Science and Technology Report, EUR 15176, 3-8, CEC, Luxembourg.

Harries KA (1995) Concrete construction in early Rome. Concrete International, 17, 58-62.

Harvey RW, George LH, Smith RL and LeBlanc DR (1989) Transport of fluorescent microsphere and indigenous bacteria through a sandy aquifer: results of natural and forced gradient tracer experiments. Environmental Science and technology, 23, 51-56.

Haveman S, Pedersen K and Ruotsalainen P (1998) Geomicrobiological investigations of groundwaters from Olkiluoto, Hästholmen, Kivetty and Romuvaara, Finland. Posiva Technical Report, 98-09, Posiva, Helsinki, Finland.

Haworth A, Sharland SM, Tasker PW and Tweed CJ (1987) Evolution of the groundwater chemistry around a nuclear waste repository. Materials Research Society Symposium Proceedings, 112, (Scientific Basis for Nuclear Waste Management, XI), 425-434.

Heath MJ (1995) Rock matrix diffusion as a mechanism for radionuclide retardation: natural radioelement migration in relation to the microfractography and petrophysics of fractured crystalline rock. CEC Nuclear Science and Technology Report, EUR 15977, CEC, Luxembourg.

Hellmuth K-H (1989a) Natural analogues of bitumen and bitumenized waste. Finnish Centre for Radiation and Nuclear Safety, STUK-B-VALO 58, Helsinki, Finland.

Hellmuth K-H (1989b) The long-term stability of natural bitumen. A case study at the bitumen-impregnated limestone deposit near Holzen, Lower Saxony, FRG. Finnish Centre for Radiation and Nuclear Safety, STUK-B-VALO 59, Helsinki, Finland.

Hellmuth K-H (1991a) The existence of native iron - implications for nuclear waste management, Part I: evidence from existing knowledge. Finnish Centre for Radiation and Nuclear Safety, STUK-B-VALO 67, Helsinki, Finland.

Hellmuth K-H (1991b) The existence of native iron - implications for nuclear waste management, Part II: evidence from investigation of samples of native iron. Finnish Centre for Radiation and Nuclear Safety, STUK-B-VALO 68, Helsinki, Finland.

Hellmuth K-H (1994) Natural analogue study on native iron. In: von Maravic H and Smellie J (editors) Natural analogue working group, fifth meeting, Toledo, October 1992. CEC Nuclear Science and Technology Report, EUR 15176, 333-342, CEC, Luxembourg.

Hellmuth K-H, Lindberg A and Tullborg EL (1994) High FeO olivine rock: a potential redox-active backfill working in a nature analogue way. In: von Maravic H and Smellie J (editors) Natural analogue working group, fifth meeting, Toledo, October 1992. CEC Nuclear Science and Technology Report, EUR 15176, 343-352, CEC, Luxembourg.

Hernán P and Astudillo J (2000, in press) Projects BARRA, MATRIX and ARCHEO: the ENRESA programme of natural analogues. In: von Maravic H and Alexander WR (editors) Natural analogue working group, eighth meeting, Strasbourg, March 1999. EC Nuclear Science and Technology Report, EC, Luxembourg.

Hietanen R, Kämäräinen EL and Alaluusua M (1984) Sorption of Sr, Cs, Ni, I and C in concrete. Nuclear Waste Commission of Finnish Power Companies Technical Report, YJT-84-04.

Higgo JJW (1989) Sorption studies of uranium in sediment-groundwater systems from the natural analogue sites of Needle's Eye and Broubster. BGS Technical Report, WE/89/40.

HMSO (1995) Making waste work - a strategy for sustainable waste management in England and Wales. Cm 3040, Her Majesty's Stationery Office, London.

Hodgkinson DP and Robinson PC (1987) Nirex near-surface repository project. Preliminary radiological assessment: Summary. Nirex Technical Report, NSS/RA100, Harwell, England.

Hodgkinson DP and McEwen TJ (1991) Spatial variability: a repository performance perspective. In: NEA/OECD Workshop on Heterogeneities in Hydrogeological Systems. NEA-OECD.

Hoek J (1998) Deep well injection of radioactive waste in Russia. In: Stenhouse MJ and Kirko VI (editors) Defense nuclear waste disposal in Russia: international perspective, 219-230. NATO Advanced Science Institute Series 1: Disarmament Technologies, Volume 18. Kluwer Academic Publishers.

Hofmann BA (1989) Geochemical analogue study in the Krunkelbach Mine, Menzenschwand, southern Germany: geology and water-rock interaction. Materials Research Society Symposium Proceedings, 127, (Scientific Basis for Nuclear Waste Management, XII), 921-926.

Hofmann BA (1990a) Reduction spheres in hematitic rocks from northern Switzerland: implications for the mobility of some rare elements. Nagra Technical Report, NTB 89-17, Nagra, Wettingen, Switzerland.

Hofmann BA (1990b) Reduction spheroids from northern Switzerland: mineralogy geochemistry and genetic models. Chemical Geology, 81, 55-81.

Hofmann BA (1992) Isolated reduction phenomena in red beds: a result of pore water radiolysis? In: Kharaka YK and Maest AS (editors) Proceedings of the 7th International Symposium on Water-Rock Interaction, 503-506.

Hofmann BA (1996) Natural analogues of radiolytic processes. In: von Maravic H and Smellie J (editors) Natural analogue working group, sixth meeting, Santa Fe, September 1994. CEC Nuclear Science and Technology Report, EUR 16761, 175-184, CEC, Luxembourg.

Hofmann BA (1999) Geochemistry of natural redox fronts - a review. Nagra Technical Report, NTB 99-05, Nagra, Wettingen, Switzerland.

Hofmann B, Dearlove JPL, Ivanovich M, Lever DA, Green DC, Baertschi P and Peters Tj (1987) Evidence of fossil and recent diffusive element migration in reduction haloes from Permian red-beds of northern Switzerland. In: Côme B and Chapman NA (editors) Natural analogues in radioactive waste disposal. CEC Radioactive Waste Management Series, EUR 11037, 217-238, CEC, Luxembourg.

Höglund LO (1987) Degradation of concrete in a LLW/ILW repository. SKB Technical Report, TR 86-15, SKB, Stockholm, Sweden.

Höglund LO and Bengtsson A (1991) Some chemical and physical processes related to the long term performance of the SFR repository. SKB Progress Report, 91-06, SKB, Stockholm, Sweden.

Holmes DC, Pitty AE and Noy DJ (1990) Geomorphological and hydrogeological features of the Poços de Caldas caldera and the Osamu Utsumi mine and Morro do Ferro analogue study sites. SKB Technical Report, TR 90-14, SKB, Stockholm, Sweden; Nagra Technical Report, NTB 90-23, Nagra, Wettingen, Switzerland; UK DoE Technical Report, WR 90-045.

Hollinger P (1992) Geochemical and isotopic characterisation of the reactor zones. In: von Maravic, H (editor) Second Oklo Working Group Meeting. CEC Radioactive Waste Management Series, CEC, Luxembourg.

Hooker PJ, Ivanovich M, Milodowski AE, Ball TK, Dawes A and Read D (1989) Uranium migration at the South Terras mine, Cornwall. BGS Technical Report, WE/89/13.

Hooker PJ (1990) The geology hydrogeology and geochemistry of the Needle's Eye natural analogue site. BGS Technical Report, WE/90/5.

Hooker PJ (1991) UK Natural Analogue Studies. In: Read D and Hooker PJ (editors) UK Natural Analogue Co-ordinating Group: Fourth Annual Report. UK DoE Report, DoE/HMIP/RR/92/009.

Hooker PJ, MacKenzie AB, Scott RD, Ridgway I, McKinley IG and West JM (1985) A study of natural and long term (10^3 - 10^4 year) elemental migration in saturated clays and sediments, part III. BGS Technical Report, FLPU 85-9; CEC Radioactive Waste Management Series, EUR 10788/2, CEC, Luxembourg.

Horseman ST and Harrington JF (1997) Study of gas migration in MX80 buffer bentonite. BGS Technical Report, WE/97/7.

Hubbard N, Laul JC and Perkins RW (1984) The use of natural radionuclides to predict the behaviour of radwaste radionuclides in far-field aquifers. Materials Research Society Symposium Proceedings, 26, (Scientific Basis for Nuclear Waste Management, VII), 891-897.

Husain L and Schaeffer OA (1973) Lunar volcanism: age of the glass in the Apollo 17 orange soil. Science, 180, 1358-1360.

IAEA (1975) Proceedings of a symposium on the Oklo phenomena. IAEA Technical Report, STI/PUB/405, International Atomic Energy Agency, Vienna, Austria.

IAEA (1978) Proceedings of the technical committee meeting on natural fission reactors. IAEA Technical Report, STI/PUB/475, International Atomic Energy Agency, Vienna, Austria.

IAEA (1989) Natural analogues in performance assessments for the disposal of radioactive wastes. IAEA Technical Report, 304, International Atomic Energy Agency, Vienna, Austria.

IAEA (1992) Storage of radioactive waste. IAEA Technical Report, 653, International Atomic Energy Agency, Vienna, Austria.

IAEA (1993) Bitumenization processes to condition radioactive wastes. IAEA Technical Report, 352, International Atomic Energy Agency, Vienna, Austria.

IAEA (1994) Safety indicators in different time frames for the safety assessment of underground radioactive waste repositories: first report of the INMAC subgroup on principles and criteria for radioactive waste disposal. IAEA Technical Report, 767, International Atomic Energy Agency, Vienna, Austria.

IAEA (1999) Use of natural analogues to support radionuclide transport models for deep geological repositories for long lived radioactive wastes. IAEA Technical Report 1109, International Atomic Energy Agency, Vienna, Austria.

Idorn GM and Thaulow N (1983) Examination of 136 years old Portland cement concrete. Cement and Concrete Research, 13, 739-743.

Isobe H, Murakami T and Ewing RC (1992) Alteration of uranium minerals in the Koongarra deposit, Australia: unweathered zone. Journal of Nuclear Materials, 190, 174-190.

Ittner T, Torstenfelt B and Allard B (1988) Diffusion of Np, Pu and Am in granitic rock. Radiochimica Acta, 44/45, 171-177.

Ittner T, Gustafsson E and Nordqvist R (1991) Radionuclide content in surface and groundwater transformed into breakthrough curves. A Chernobyl fallout study in a forested area in northern Sweden. SKB Technical Report, TR 91-28, SKB, Stockholm, Sweden.

Ivanovich M and Hardy CJ (1986) Identification and measurement of colliods in groundwater. In: Côme B and Chapman NA (editors) Natural analogue working group, second meeting, Interlaken, June 1986. CEC Nuclear Science and Technology Report, EUR 10671, 227-260, CEC, Luxembourg.

Ivanovich M and Harmon RS (1992, editors) Uranium series disequilibrium: applications to earth, marine and environmental sciences, second edition. Claredon Press, Oxford, UK.

Ivanovich M, Duerden P, Payne T, Nightingale T, Longworth G, Wilkins MA, Edghill RB, Cockayne DJ and Davey BG (1987) Natural analogue study of the distribution of uranium series radionuclides between the colloid and solute phases in the hydrological system of the Koongarra uranium deposit, Australia. In: Côme B and Chapman NA (editors) Natural analogues in radioactive waste disposal. CEC Radioactive Waste Management Series, EUR 11037, 300-313, CEC, Luxembourg.

Ivanovich M, Duerden P, Payne T, Nightingale T, Longworth G, Wilkins MA, Hasler SE, Edghill RB, Cockayne DJ and Davey BG (1988) Natural analogue study of the distribution of uranium series radionuclides between the colloid and solute phases in the hydrological system of the Koongarra uranium deposit, Australia. UKAEA Technical Report, AERE-R-12975, Harwell, England.

Iverson WP (1987) Microbial corrosion of metals. Advances in Applied Microbiology, 32, 5.

Izett GA (1991) Tektites in Cretaceous-Tertiary boundary rocks on Haiti and their bearing on the Alvarez impact extinction hypothesis. Journal of Geophysical Research 96, 20879-20905.

Izzo G (1986) Bacteria in ancient sediments. In: Côme B and Chapman NA (editors) Natural analogue working group, second meeting, Interlaken, June 1986. CEC Nuclear Science and Technology Report, EUR 10671, 138-152, CEC, Luxembourg.

Jaakkola T, Suksi J, Suutarinen R, Niini H, Ruskeeniemi T, Söderholm B, Vesterinen M, Blomqvist R, Halonen S and Lindberg A (1989) The behaviour of radionuclides in and around uranium deposits 2: Results of investigations at the Palmottu analogue study site, SW Finland. Geological Survey of Finland, Nuclear Waste Disposal Research Report, YST-64.

Jacobs G (1984) Geochemical modelling. In: Kelmers AD (editor) Progress in evaluation of radionuclide geochemical information developed by DOE high-level nuclear waste repository projects. Oak Ridge National Laboratory Technical Report, NUREG/CR-4381, Oak Ridge, Tennessee, USA.

Jakobsson SP and Moore JG (1986) Hydrothermal minerals and alteration rates at Surtsey volcano. Bulletin of the Geological Society of America, 97, 648-659.

Jakubick AT and Church W (1986) Oklo natural reactors: geological and geochemical conditions - A review. Atomic Energy Board of Canada Research Report, INFO-0179, Ottawa, Canada.

Janeczek J and Ewing RC (1992) Dissolution and alteration of uraninite under reducing conditions. Journal of Nuclear Materials, 190, 157-173.

Jarvis NV, Andreoli MAG and Read D (1997) The Steenkampskraal natural analogue study and nuclear waste disposal in South Africa. In: von Maravic H and Smellie J (editors) Natural analogue working group, seventh meeting, Stein am Rhein, October 1996. CEC Nuclear Science and Technology Report, EUR 15176, 9-26, CEC, Luxembourg.

Jebrak M, Lemiere B, Piantone P, Sureau, JF and Griffault L (1987) Hydrothermal alteration systems as analogues of nuclear waste repositories in granitic rocks. In: Côme B and Chapman NA (editors) Natural analogues in radioactive waste disposal. CEC Radioactive Waste Management Series, EUR 11037, 191-204, CEC, Luxembourg.

Jefferies NL (1987) Long-term solute diffusion in a granite block immersed in sea water. In: Côme B and Chapman NA (editors) Natural analogues in radioactive waste disposal. CEC Radioactive Waste Management Series, EUR 11037, 249-260, CEC, Luxembourg.

Jenden PD, Newell KD, Kaplan IR and Watney WL (1988) Composition and stable-isotope geochemistry of natural gases from Kansas, Midcontinent USA. Chemical Geology, 71, 117-147.

Jenkins PM (1936). Reports of the Percy Sloden expedition to some rift valley lakes in Kenya in 1929. VII: summary of the ecological results, with special reference to the alkaline lakes. Annual Magazine of Natural History, 18, 133-181.

Jercinovic MJ and Ewing RC (1988) Basaltic glasses from Iceland and the deep sea: natural analogues to borosilicate nuclear waste-form glass. SKB Technical Report, TR 88-01, SKB, Stockholm, Sweden.

Johnson AB (1989) Lessons in metal durability from the ancient metals. In: Côme B and Chapman NA (editors) Natural Analogue Working group, third meeting, Snowbird, June 1988. CEC Nuclear Science and Technology Report, EUR 11725, 228-231, CEC, Luxembourg.

Johnson AB and Francis B (1980) Durability of metals from archaeological objects metal meteorites and native metals. Battelle Pacific Northwest Laboratory, PNL-3198.

Johnson LH and Shoesmith DW (1988) Spent fuel. In: Lutze W and Ewing RC (editors) Radioactive waste forms for the future, 635-698. North Holland

Johnson LH, Stroes-Gascoyne S, Shoesmith DW, Bailey MG and Sellinger DM (1983) Leaching and radiolysis studies on UO_2 fuel. In: Werme L (editor) Proceedings of the Third Spent Fuel Workshop. SKB Technical Report, TR 83-76, SKB, Stockholm, Sweden.

Jull SP and Lees TP (1990) Studies of historic concrete. CEC Nuclear Science and Technology Report EUR 12972, CEC, Luxembourg.

Kamineni DC (1986) Distribution of uranium thorium and rare-earth elements in the Eye-Dashwa Lakes pluton - a study of some analogue elements. Chemical Geology, 55, 361-373.

Kaplan MF (1979) Ancient materials data as a basis for waste form integrity projections. US DoE Report, TR-1746-1.

Kaplan MF (1980a) Characterisation of weathered glass by analysing ancient artefacts (Scientific Basis for Nuclear Waste Management, II), 85-92.

Kaplan MF (1980b) An archaeological perspective on a modern issue: nuclear waste disposal. Journal of Field Archaeology, 7, 265-267.

Kaplan MF (1982) Marking a nuclear waste repository: an archaeologist's perspective. Transactions of the American Nuclear Society, 41, 96-97.

Kaplan MF (1986) Mankind's future: using the past to protect the future. Archaeology and the disposal of highly radioactive wastes. Interdisciplinary Science Reviews, 11, 257-268.

Karlsson F, Smellie JAT and Höglund L-O (1994) The application of natural analogues to the Swedish SKB 91 performance assessment. In: von Maravic H and Smellie J (editors) Natural analogue working group, fifth meeting, Toledo, October 1992. CEC Nuclear Science and Technology Report, EUR 15176, 307-319, CEC, Luxembourg.

Katz JJ, Seaborg GT and Morse LR (1986, editors) The chemistry of the actinide elements. Chapman and Hall, London.

KBS (1983) Final storage of spent nuclear fuel (Five Volumes). SKBF/KBS Technical Report.

Kersting AB, Efurd DW, Finnegan DL, Rokop DJ, Smith DK and Thompson JL (1999) Migration of plutonium in ground water at the Nevada Test Site. Nature, 397, 56-59.

Keswick BH, Wang DS and Gerba CP (1982) The use of microorganisms as groundwater tracers: a review. Ground Water, 20, 142-149.

Khoury HN and Salameh E (1986) The origin of high temperature minerals from Suweilah area, Jordan. Dirasat, 13, 261-269.

Khoury HN, Salameh E and Abdul-Jaber Q (1985) Characteristics of an unusual highly alkaline water from the Maqarin area, northern Jordan. Journal of Hydrology, 81, 79-91.

Khoury HN, Salameh E, Clark IP, Fritz P, Milodowski AE, Cave MR, Bajjali W and Alexander WR (1992) A natural analogue of high pH cement pore waters from the Maqarin area of northern Jordan II: introduction to the site. Journal of Geochemical Exploration 46, 117-132.

Kickmaier W, Vomvoris S, Alexander WR, Marschall P, Frieg B, Wanner W and Huertas F (2000, in press) GTS - Nagra's underground research laboratory in crystalline rock: status of the scientific programme 1997 - 2002. Hydrogeology.

Kidby DW and Billington RS (1992) Microbial aspects of gas generation from low level radioactive waste simulant. UK Department of the Environment Report, DOE/HMIP/RR/92/087, London, England.

Kim JI, Buckau, G, Baumgartner F, Moon HC and Lux D (1984) Colloid generation and the actinide migration in Gorleben groundwaters. Materials Research Society Symposium Proceedings, 26, (Scientific Basis for Nuclear Waste Management, VII), 31-40.

Kim JI, Buckau G and Klenze R (1987) Natural colloids and generation of actinide pseudocolloids in groundwater. In: Côme B and Chapman NA (editors) Natural analogues in radioactive waste disposal. CEC Radioactive Waste Management Series, EUR 11037, 289-299, CEC, Luxembourg.

Kindness A, Lachowski E, Minocha A and Glasser F (1994) Immobilisation and fixation of molybdenum (VI) by Portland cement. Waste Management, 14, 97-102

Kingston WL (1989) Characterisation of colloids found in various groundwater environments in central and southern Nevada. Unpublished MSc Thesis, University of Nevada, Reno, USA.

Klinkenberg LJ (1951) Analogy between diffusion and electrical conductivity in porous rocks. Bulletin of the Geological Society of America, 62, 559-568.

Kopajtic Z, Laske D, Linder HP, Mohos M, Nellen M and Zwicky HU (1989) Characterisation of bituminous intermediate-level waste products. Materials Research Society Symposium Proceedings, 127, (Scientific Basis for Nuclear Waste Management, XII), 527-534.

Krauskopf KB (1986) Thorium and rare-earth metals as analogues for actinide elements. Chemical Geology, 55, 323-325.

Krishnaswami S, Graustein WC and Turekian KK (1982) Radium thorium and radioactive lead isotopes in groundwaters: Application to the in situ determination of adsorption-desorption rate constants and retardation factors. Water Resources Research, 18, 1633-1675.

Kumpulainen H, Melamed A, Pitkänen P, Valkiainen M and Manninen P (1992) Elemental mobility in crystalline rock around open fractures at Palmottu. Palmottu Project Progress Report. Geological Survey of Finland Technical Report, YST-78, Espoo, Finland.

Laaksohaiju M and Degueldre C (1994) Colloids from the swedish granitic groundwater. SKB Technical Report, TR 94-07, SKB, Stockholm, Sweden.

Landais P (1993) Bitumens in uranium deposits. In: Parnell J, Kucha H and Landais P (editors) Bitumens in ore deposits. Springer-Verlag.

Landström O and Sundbland B (1986) Migration of thorium, uranium, radium and 137-caesium in till soils and their uptake in organic matter and peat. SKB Technical Report, TR 86-24, SKB, Stockholm, Sweden.

Landström O and Tullborg E (1990) The influence of fracture mineral/groundwater interaction on the mobility of U, Th, REE and other trace elements. SKB Technical Report, TR 90-37, SKB, Stockholm, Sweden.

Landström O and Tullborg EL (1995) Interactions of trace elements with fracture filling minerals from the Äspö Hard Rock Laboratory. SKB Technical Report, TR 95-13, SKB, Stockholm, Sweden.

Langmuir D (1997) Aqueous environmental geochemistry. Prentice Hall, New Jersey.

Langmuir D and Herman JS (1980) The mobility of thorium in natural waters at low temperatures. Geochimica et Cosmochimica Acta, 44, 1753-1766

Latham AG and Schwarcz HP (1989) Review of the modelling of radionuclide transport from U-series disequilibria and of its use in assessing the safe disposal of nuclear waste in crystalline rock. Applied Geochemistry, 527-537.

Laul JC and Papike JJ (1982) Chemical migration by contact metamorphism between granite and silt-carbonate systems. In: Environmental migration of the long-lived radionuclides. IAEA Technical Report, STI/PUB/597, International Atomic Energy Agency, Vienna, Austria.

Laul JC and Smith MR (1988) Disequilibrium study of natural radionuclides of uranium and thorium series in cores and briny groundwaters from the Palo Duro Basin, Texas. Radioactive Waste Management and the Nuclear Fuel Cycle, 11, 169-225.

Laul JC, Smith MR and Hubbard N (1985) Behaviour of natural U, Th and Ra isotopes in the Wolfcamp brine aquifers Palo Duro Basin, Texas. Scientific Basis for Nuclear Waste Management, VIII, 475-482.

Laul JC, Smith MR and Hubbard N (1986) $^{234}U/^{230}Th$ ratio as an indicator of redox state and U, Th and Ra behaviour in briny aquifers. Materials Research Society Symposium Proceedings, 50, (Scientific Basis for Nuclear Waste Management, IX), 475-482.

Laul JC, Walker RJ, Shearer CK, Papike JJ and Simon SB (1984). Chemical migration by contact metamorphism between pegmatite and country rocks: natural analogues for radionuclide migration. Materials Research Society Symposium Proceedings, 26, (Scientific Basis for Nuclear Waste Management, VII), 951-958.

Lea FM (1970) The chemistry of cement and concrete: third edition. Edward Arnold, London.

Lee CF (1986) A case history on long-term effectiveness of clay sealant. In: Côme B and

Chapman NA (editors) Natural analogue working group, second meeting, Interlaken, June 1986. CEC Nuclear Science and Technology Report, EUR 10671, 172-190, CEC, Luxembourg.

Lehikoinen J, Muurinen A, Olin M, Uusheimo K and Valkiainen M (1992) Diffusivity and porosity studies in rock matrix - the effect of salinity. VTT Research Notes VTT, Espoo, Finland.

Lei W, Linsalata P, Penna Franca E and Eisenbud M (1986) Distribution and mobilization of cerium lanthanum and neodymium in the Morro do Ferro basin Brazil. Chemical Geology, 55, 313-321.

Leventahal JS, Daws TA and Frye JS (1986) Organic geochemical analysis of sedimentary organic matter associated with uranium. Applied Geochemistry, 1, 241-247.

Lever DA (1987) Natural analogues and radionuclide transport model validation. In: Côme B and Chapman NA (editors) Natural analogues in radioactive waste disposal. CEC Radioactive Waste Management Series, EUR 11037, 23-31, CEC, Luxembourg.

Lever DA and Bradbury MH (1985) Rock matrix diffusion and its implication for radionuclide migration. Mineralogical Magazine, 49, 245-254.

Lewen MD and Buchardt B (1989) Irradiation of organic matter by uranium decay in Alum Shale, Sweden. Geochimica et Cosmochimica Acta, 53, 1307-1322.

Leythaeuser D, Schaefer RG and Yukler A (1982) Role of diffusion in primary migration of hydrocarbons. Bulletin of the American Association of Petroleum Geologists, 67, 932-952.

Liang L, McCarthy JF, Jolley LW, McNabb JA and Mehlhorn TW (1993) Iron dynamics: transformation of Fe(II)/Fe(III) during injection of natural organic matter in a sandy aquifer. Geochimica et Cosmochimica Acta, 57, 1987-1999.

Lindberg RD and Runnels DD (1984) Groundwater redox reactions: an analysis of the equilibrium state applied to Eh measurements and geochemical modelling. Science, 225, 925-927.

Lindqvist J (1996) The use of natural analogues in public relations. In: von Maravic H and Smellie J (editors) Natural analogue working group, sixth meeting, Santa Fe, September 1994. CEC Nuclear Science and Technology Report, EUR 16761, 283-286, CEC, Luxembourg.

Lindqvist K and Laitakari I (1980) Glass and amygdales in Precambrian diabases from Orivesi, southern Finland. Geological Society of Finland Bulletin, 52, 221-229.

Linklater CM (1998, editor) A natural analogue study of cement-buffered, hyperalkaline groundwaters and their interaction with a repository host rock: phase II. Nirex Science Report, S/98/003, Nirex, Harwell, UK.

Linsley GS and Sjoeblom KL (1994) The international Arctic seas assessment project. Radwaste Magazine, 1, 64-68.

Lippard SJ, Shelton AW and Gass IG (1986) The ophiolite of northern Oman. Geological Society of London, Memoir 11.

Liu J, Ji-Wei Y and Neretnieks I (1994) Transport modelling and model validation in the natural analogue study of the Cigar Lake uranium deposit. Proceedings of the Fourth International Conference on the Chemistry and Migration Behaviour of Actinides and Fission Products (Migration '93), 787-795.

Lombardi S and Valentini G (1996) The Dunarobba forest as natural analogue: analysis of the geoenvironmental factors controlling the wood preservation. In: von Maravic H and Smellie J (editors) Natural analogue working group, sixth meeting, Santa Fe, September 1994. CEC Nuclear Science and Technology Report, EUR 16761, 127-133, CEC, Luxembourg.

Longworth G and Ivanovich M (1989) The sampling and characterisation of natural groundwater colloids: studies in aquifers in slate, granite and glacial sand. Nirex Radioactive Waste

Disposal: Safety Studies, NSS/R1003, United Kingdom Nirex Ltd, Harwell, England.

Longworth G, Ivanovich M and Wilkins MA (1989a) Uranium series disequilibrium studies at the Broubster analogue site. UKAEA Technical Report, AERE-R-13609, Harwell, England.

Longworth G, Ross CAM, Degueldre C and Ivanovich M (1989b) Interlaboratory study of sampling and characterisation techniques for groundwater colloids. UKAEA Technical Report, AERE-R-13393, Harwell, England.

Loss RD, Rosman JKR and DeLaeter JR (1984) Transport of symmetric mass region fission products at the Oklo natural reactor. Earth and Planetary Science Letters, 68, 240-248.

Loss RD, Rosman KJR, DeLaeter JR, Curtis DB, Benjamin TM, Gancarz AJ, Maeck WJ and Delmore JE (1989) Fission product retentivity in peripheral rocks at the Oklo natural fission reactors. Chemical Geology, 76, 71-84.

Lowson RT, Short SA, Davey BG and Gray D (1986) $^{234}U/^{238}U$ and $^{230}Th/^{234}U$ activity ratios in mineral phases of a lateritic weathered zone. Geochimica et Cosmochimica Acta, 50, 1697-1702.

Lumpkin GR and Ewing RC (1989) Alpha-decay damage and annealing effects in natural pyrochlore: analogues for long-term radiation damage effects in actinide pyrochlore structure types. Materials Research Society Symposium Proceedings, 127, (Scientific Basis for Nuclear Waste Management, XII), 253-260.

Lumpkin GR and Mariano AN (1996) Natural occurrence and stability of pyrochlore in carbonatites, related hydrothermal systems and weathering environments. Materials Research Society Symposium Proceedings, 412, (Scientific Basis for Nuclear Waste Management, XIX), 831-838.

Lumpkin GR, Hart KP, McGlinn PJ and Payne TE (1994) Retention of actinides in natural pyrochlores and zirconolites. Radiochimica Acta, 66/67, 469-474.

Lumpkin GR, Colella M, Smith KL, Mitchell RH and Olav Larsen A (1998) Chemical composition, geochemical alteration and radiation damage effects in natural perovskite. Materials Research Society Symposium Proceedings, 506, (Scientific Basis for Nuclear Waste Management, XXI), 207-214.

Lutze W (1988) Silicate glasses. In: Lutze W and Ewing RC (editors) Radioactive waste forms for the future, 1-160. North Holland.

Lutze W and Ewing RC (1988, editors) Radioactive waste forms for the future. North Holland, Amsterdam.

Lutze W, Malow G, Ewing RC, Jercinovic MJ and Keil K (1985) Alteration of basalt glasses: implications for modelling the long-term stability of nuclear glasses. Nature, 314, 252-255.

Lutze W, Grambow B, Ewing RC and Jercinovic MJ (1987) The use of natural analogues in the long-term extrapolation of glass corrosion processes. In: Côme B and Chapman NA (editors) Natural analogues in radioactive waste disposal. CEC Radioactive Waste Management Series, EUR 11037, 142-152, CEC, Luxembourg.

McCarthy JF (1996) Natural analogue studies of the role of colloids, natural organics and microorganisms on radionuclide transport. In: von Maravic H and Smellie J (editors) Natural analogue working group, sixth meeting, Santa Fe, September 1994. CEC Nuclear Science and Technology Report, EUR 16761, 195-210, CEC, Luxembourg.

McCarthy JF and Zacchara JM (1989) Subsurface transport of contaminants. Environmental Science and Technology, 23, 496-502.

McCarthy J and Degueldre C (1991) Sampling and characterisation of colloids and particles in groundwater for studying their role in contaminant transport. In: van Leeuwen HP and Buffle J (editors)

Environmental particles. IUPAC Environmental Analytical and Physical Chemistry, Series II.

McCombie C (1991) Critical uncertainties in safety assessments and how to address them In: Côme B and Chapman NA (editors) Natural analogue working group, fourth meeting and Poços de Caldas project final meeting, Pitlochry, June 1990. CEC Nuclear Science and Technology Report, EUR 13014, 19-28, CEC, Luxembourg.

McCombie C, McKinley IG and Zuidema P (1990) Sufficient validation: the value of robustness in performance assessment and system design. GEOVAL-1990.

McConnell JDC (1954) The hydrated calcium silicates riversideite, tobermoreite and plombierite. Mineralogical Magazine, 30, 293-305.

McConnell JDC (1955) The hydration of larnite and bredigite and the properties of the resulting gelatinous mineral plombierite. Mineralogical Magazine, 31, 672-680.

McKinley IG (1985) The geochemistry of the near-field. Nagra Technical Report, NTB 84-48, Nagra, Wettingen, Switzerland.

McKinley IG (1989) Applying natural analogues in predictive performance assessment. Unpublished Nagra Internal Report, Nagra, Wettingen, Switzerland.

McKinley IG and Alexander WR (1992a) A review of the use of natural analogues to test performance assessment models of a cementitious near field. Waste Management, 12, 253-259.

McKinley IG and Alexander WR (1992b) Constraints on the applicability of in situ distribution coefficient values. Journal of Environmental Radioactivity, 15, 19-34.

McKinley IG and Alexander WR (1993a) Assessment of radionuclide retardation: uses and abuses of natural analogue studies. Journal of Contaminant Hydrology, 13, 249-259.

McKinley IG and Alexander WR (1993b) Reply to comments on: McKinley IG and Alexander WR (1993a) Assessment of radionuclide retardation: uses and abuses of natural analogue studies. Journal of Contaminant Hydrology, 13, 271-275.

McKinley IG and Alexander WR (1996) The uses of natural analogue input in repository performance assessment: an overview. In: von Maravic H and Smellie J (editors) Natural analogue working group, sixth meeting, Santa Fe, September 1994. CEC Nuclear Science and Technology Report, EUR 16761, 273-2282, CEC, Luxembourg.

McKinley IG and Grogan HA (1986) Natural analogue support for geomicrobiological modelling. In: Côme B and Chapman NA (editors) Natural analogue working group, second meeting, Interlaken, June 1986. CEC Nuclear Science and Technology Report, EUR 10671, 153-159, CEC, Luxembourg.

McKinley IG and Hadermann J (1984) Radionuclide sorption database for Swiss safety assessment. Nagra Technical Report, NTB 84-40, Nagra, Wettingen, Switzerland.

McKinley IG and Savage D (1996) Comparison of solubility databases used for HLW performance assessment. Journal of Contaminant Hydrology, 21, 335-350.

McKinley IG, MacKenzie AB, West JM and Scott RD (1984) A natural analogue study of radionuclide migration in clays. Materials Research Society Symposium Proceedings, 26, (Scientific Basis for Nuclear Waste Management, VII), 851-857.

McKinley IG, West JM and Grogan HA (1985) An analytical overview of the consequences of microbial activity in a Swiss HLW repository. Nagra Technical Report, NTB 85-43, Nagra, Wettingen, Switzerland.

McKinley IG, Berner U and Wanner H (1987) Predictions of radionuclide chemistry in a highly alkaline environment. In: Chemie und Migrationsverhalten der Aktinide und Spaltprodukte in naturlichen aquatischen Systemen. PTB-SE-14, 77-89.

McKinley IG, Bath AH, Berner U, Cave M and Neal C (1988) Results of the Oman analogue study. Radiochimica Acta, 44/45, 311-316.

McKinley IG, Alexander WR, McCombie C and Zuidema P (1992a) Application of results from the Poços de Caldas project in the Kristallin-I HLW performance assessment. Proceedings of the Third International High-Level Radioactive Waste Management Conference, Las Vegas, 357-361.

McKinley IG, Smith PA, and Curti E (1992b) Can the Kristallin-I near-field model be considered robust? Proceedings of the Third International High-Level Radioactive Waste Management Conference, Las Vegas, 1770-1776.

McKinley IG, Alexander WR, Gautschi A and Waber HN (1998) An approach to validation of solubility databases for performance assessment. Radiochim Acta, 82, 407-412.

MacKenzie AB, Scott RD, McKinley IG and West JM (1983) A study of long term (10^3 - 10^4 year) elemental migration in saturated clays and sediments. Institute of Geological Sciences Technical Report, FLPU 83-6.

MacKenzie AB, Scott RD, Ridgway IM, McKinley IG and West JM (1984) A study of long term (10^3 - 10^4 year) elemental migration in saturated clays and sediments, part II. BGS Technical Report, FLPU 84-11; CEC Radioactive Waste Management Series EUR 10788/1, CEC, Luxembourg.

MacKenzie AB, Scott RD, Houston CM and Hooker PJ (1989a) Natural decay series radionuclide studies at the Needle's Eye natural analogue site, 1986-1989. BGS Technical Report, WE/90/4.

MacKenzie AB, Shimmield TM, Scott RD and Houston CM (1989b) Development of an analytical method for the analysis of I and Br concentrations in lacustrine sediment interstitial water. BGS Technical Report, WE/89/65.

MacKenzie AB, Scott RD, Linsalata P, Miekeley N, Osmond JK and Curtis DB (1990a) Natural radionuclide and stable element studies of rock samples from the Osamu Utsumi mine and Morro do Ferro analogue study sites, Poços de Caldas, Brazil. SKB Technical Report, TR 90-16, SKB, Stockholm, Sweden; Nagra Technical Report, NTB 90-25, Nagra, Wettingen, Switzerland; UK DoE Technical Report, WR 90-047.

MacKenzie AB, Shimmield TM, Scott RD, Davidson CM and Hooker PJ (1990b) Chloride bromide and iodide distributions in Loch Lomond sediment interstitial water. BGS Technical Report, WE/90/2.

MacKenzie AB, Whitton AN, Shimmield TM, Jemielita RA, Scott RD and Hooker PJ (1991) Natural decay series radionuclide studies at the Needle's Eye natural analogue site II, 1986-1989. BGS Technical Report, WE/91/37.

MacKenzie AB, Scott RD, Linsalata P and Miekeley N (1992) Natural decay series studies of the redox front system in the Poços de Caldas uranium mineralisation. Journal of Geochemical Exploration, 45, 289-322.

Magonthier M-C, Petit J-C and Dran J-C (1992) Rhyolitic glasses as natural analogues of nuclear glasses: behaviour of an Icelandic glass upon aqueous corrosion. Applied Geochemistry, Supplementary Issue, 1, 83-94.

Malinowski R and Garfinkel Y (1991) Prehistory of concrete. Concrete International, 13, 62-68.

Mallinson LG and Davies (1987) A historical examination of concrete. CEC Nuclear Science and Technology Report, EUR 10937, CEC, Luxembourg.

Marcos N (1989) Native copper as a natural analogue for copper canisters. Nuclear Waste Commission of Finnish Power Companies Technical Report, YJT-89-18.

Marcos N (1996) The Hyrkkölä native copper mineralisation as a natural analogue for copper canisters. Posiva Technical Report, 96:15, Posiva, Helsinki, Finland.

Marcos N and Ahonen L (1999) New data on the Hyrkkölä U-Cu mineralisation: the behaviour of native copper in a natural environment. Posiva Technical Report, 99-23, Posiva, Helsinki, Finland.

Marshall RR (1961) Devitrification of natural glass. Bulletin of the Geological Society of America, 72, 1493-1520.

de Marsily G, Berhendt V, Ensminger D, Flebus C, Hutchinson B, Kane P, Karpf A, Klett R, Mobbs S, Poulin M and Stanner D (1988) Feasibility of disposal of high-level radioactive waste into the seabed. Volume 2: Radiological protection. NEA-OECD, Paris.

Martin JM, Nirel P and Thomas AJ (1987) Sequential extraction techniques: promises and problems. Marine Chemistry, 22, 313-341.

Marx G, Amarantos S, Lyon CE, Nominé JC, Stammose D and Vejmelka P (1991) Radionuclide release from low and intermediate level waste forms. In: Cecille L (editor) Proceedings of the 3rd Conference on Radioactive Waste Management and Disposal.

Mattson E (1983) Corrosion resistance of a copper canister for spent nuclear fuel. KBS Technical Report, 83-24, KBS, Stockholm, Sweden

Mavrogenes JA and Bodnar RJ (1994) Hydrogen movement into and out of fluid inclusions in quartz: experimental evidence and geologic implications. Geochimica et Cosmochimica Acta, 58, 141-148.

Mazer JJ (1994) The role of natural glasses as analogues in projecting the long-term alteration of high-level nuclear waste glasses, part I. Materials Research Society Symposium Proceedings, 333, (Scientific Basis for Nuclear Waste Management, XVII), 159-165.

Mazer JJ, Bates JK, Bradley CR and Stevenson CM (1992) Water diffusion in tektites: an example of the use of natural analogues in evaluating the long-term reaction of glass with water. Journal of Nuclear Materials, 190, 277-287.

Mazurek M (1990) OBS: the interface between marl and concrete lining in Seelisberg tunnel. Unpublished Nagra Internal Report, Nagra, Wettingen, Switzerland.

Mazurek M (1998) Geology of the crystalline basement of Northern Switzerland and derivation of geological input data for safety assessment models. Nagra Technical Report, NTB 93-12, Nagra, Wettingen, Switzerland.

Mazurek M, Smith PA and Gautschi A (1992a) Application of a realistic geological database to safety assessment calculations: an exercise in interdisciplinary communication. In: Kharaka YK and Maest AS (editors) Proceedings of the 7th International Symposium on Water-Rock Interaction.

Mazurek M, Gautschi A and Vomvoris S (1992b) Deriving input data for discrete transport models from deep borehole investigations: an approach for crystalline rocks. In: IAEA/CEC International Symposium on Geological Disposal of Spent Fuel High Level and Alpha-Bearing Waste. International Atomic Energy Agency, Vienna, Austria.

Ménager M-T, Menet C and Petit J-C (1989) U, Th, REE mobilization during water-rock interactions in a U-mineralized granite. In: Miles DL (editor) Water Rock Interaction, WRI-6. Proceedings of the 6th International Symposium on Water-Rock Interaction, Malvern, August 1989.

Ménager M-T, Petit J-C and Brocandel M (1992a) The migration of radionuclides in granite: a review based on natural analogues. Applied Geochemistry, Supplementary Issue, 1, 217-238.

Ménager M-T, Menet C and Petit J-C, Cathelineau M and Côme BC (1992b) Dispersion of U, Th and REE by water-rock interaction around an intergranitic U-vein, Jalerys Mine, Morvan, France. Applied Geochemistry, Supplementary Issue, 1, 239-252.

Meyer FM, Drennan GR, Robb LJ, Cathelineau M, Dubessy J and Landais P (1991) Conditions of Au-U mineralisation in Witwatersrand reef. In Pagel M and Leroy JL (editors) Source, transport and deposition of metals, 681-684. Balkema, Rotterdam.

Michel J (1984) Redistribution of uranium and thorium series isotopes during isovolumetric weathering of granite. Geochimica et Cosmochimica Acta, 48, 1249-1255.

Miekeley N, Vale MGR, Taveres TM and Lei W (1982) Some aspects of the influence of surface and groundwater chemistry on the mobility of thorium in the Morro do Ferro environment. Materials Research Society Symposium Proceedings, 11, (Scientific Basis for Nuclear Waste Management, V), 725-733.

Miekeley N, Coutinho de Jesus H, Porto da Silveira CL, Morse R and Osmond K (1990a) Natural series radionuclide and rare-earth element geochemistry of waters from the Osamu Utsumi and Morro do Ferro analogue study sites, Poços de Caldas, Brazil. SKB Technical Report, TR 90-17, SKB, Stockholm, Sweden; Nagra Technical Report, NTB 90-26, Nagra, Wettingen, Switzerland; UK DoE Technical Report, WR 90-048.

Miekeley N, Coutinho de Jesus H, Porto da Silveira CL and Degueldre C (1990b). Chemical and physical characterisation of suspended particles and colloids in waters from the Osamu Utsumi and Morro do Ferro analogue study sites, Poços de Caldas, Brazil. SKB Technical Report, TR 90-18, SKB, Stockholm, Sweden; Nagra Technical Report, NTB 90-27, Nagra, Wettingen, Switzerland; UK DoE Technical Report, WR 90-049.

Miekeley N, Coutinho de Jesus H, Porto da Silveira CL and Degueldre C (1992). Chemical and physical characterisation of suspended particles and colloids in waters from the Osamu Utsumi mine and Morro do Ferro analogue study sites, Poços de Caldas, Brazil. Journal of Geochemical Exploration, 45, 409-438.

Mill JS (1874) A system of logic, ratiocinative and inductive: being a connected view of the principles of evidence and the methods of scientific investigation. Harper and Brothers.

Miller JDA (1981) Metals. In: Rose AH (editor) Economic Microbiology. Academic Press, New York.

Miller WM (1996a) The value of natural analogues. In: von Maravic H and Smellie J (editors) Natural analogue working group, sixth meeting, Santa Fe, September 1994. CEC Nuclear Science and Technology Report, EUR 16761, 23-31, CEC, Luxembourg.

Miller WM (1996b) Archaeological materials as natural analogues for the geological disposal of radioactive waste. In: Corfield M, Hinton P, Nixon T and Pollard M (editors) Preserving archaeological remains in situ - conference proceedings. Museum of London.

Miller WM (2000, in press) Communication with natural analogues. In: von Maravic H and Alexander WR (editors) Natural analogue working group, eighth meeting, Strasbourg, March 1999. EC Nuclear Science and Technology Report, EC, Luxembourg.

Miller WM and Chapman (1995) Postcards from the past: archaeological and industrial analogues for deep repository materials. Radwaste Magazine, 2, 32-42.

Miller WM, Smith GM, Savage D, Towler P and Wingefors S (1996) Natural radionuclide fluxes and their contribution to defining licensing criteria for deep geological repositories for radioactive wastes. Radiochimica Acta, 74, 289-295.

Milnes AG (1985) Geology and radwaste. Academic Press, Geology Series.

Milodowski AE, Basham IR, Hyslop EK and Pearce JM (1989a) The uranium source-term mineralogy and geochemistry at the Broubster natural analogue site, Caithness. BGS Technical Report, WE/89/50; CEC Nuclear Science and Technology Report EUR 13280, CEC, Luxembourg.

Milodowski AE, Nancarrow PHA and Spiro B (1989b) A mineralogical and stable isotope study of natural analogues of ordinary portland cement (OPC) and CaO-Si_2-H_2O (CSH) compounds. Nirex Radioactive Waste Disposal: Safety Studies, NSS/R2404, United Kingdom Nirex Ltd, Harwell, England.

Milodowski AE, Kemp SJ, Pearce JM and Hughes CR (1990) Characterisation of bentonite alteration in reacted cement-bentonite blocks from swelling test experiments. Unpublished Nagra Internal Report, Nagra, Wettingen, Switzerland.

Milodowski AE, Hyslop EK, Pearce JM, Wetton PD, Kemp SJ, Longworth G, Hodginson E and Hughes CR (1998) Mineralogy, petrology and geochemistry. In: Smellie JAT (editor) Maqarin natural analogue study: phase III. SKB Technical Report, TR 98-04, SKB, Stockholm, Sweden.

Mobbs SF, Charles D, Delow CE and McColl NP (1988) Performance assessment of geological isolation systems for radioactive waste (PAGIS): disposal into the sub-seabed. CEC Nuclear Science and Technology Report EUR 11779, CEC, Luxembourg.

Montoto M, Rodriguez Rey A, Ruiz de Argandoña VG, Calleja L, Menendez, B and Heath MJ (1992) Matrix diffusion for radionuclide retardation in relation to crystalline rock. In: ENRESA Ponencias/Informes, 1988-1991, ENRESA, Madrid.

Montoto M, Heath MJ, Rodriguez Rey A, Ruiz de Argandoña VG, Calleja L and Menendez, B (1991a) Matrix diffusion for radionuclide retardation in relation to crystalline rock. EURATOM-AECL.

Montoto M, Rodriguez Rey A, Ruiz de Argandoña VG, Menendez, B and Heath MJ (1991b) Natural analogues and microstructural studies in relation to radionuclide retardation by rock matrix diffusion in granites: final report CEC Radioactive Waste Management Series.

Mossman DJ and Nagy B (1996) Solid bitumens: an assessment of their characteristics, genesis and role in geological processes. Terra Nova, 8, 114-128.

Motamedi M, Karnland O and Pedersen K (1996) Survival of sulphate reducing bacteria at different water activities in compacted bentonite. Microbiological Letters, 141, 83-87.

Moulin V and Ouzounian G (1992) Role of colloids and humic substances in the transport of radio-elements through the geosphere. Applied Geochemistry, Supplementary Issue, 1, 179-186.

Müller-Vonmoos M and Kahr G (1983) Mineralogische Untersuchungen von Wyoming Bentonit MX-80 und Montigel. Nagra Technical Report, NTB 83-12, Nagra, Wettingen, Switzerland.

Murphy WM (1992) Natural analog studies for geologic disposal of nuclear waste. Technology Today, June 1992.

Murphy WM (1996) In situ distribution coefficients derived from uranium and thorium decay series isotopes in water-rock systems: promise and practice. In: von Maravic H and Smellie J (editors) Natural analogue working group, sixth meeting, Santa Fe, September 1994. CEC Nuclear Science and Technology Report, EUR 16761, 233-241, CEC, Luxembourg.

Murphy WM and Pearcy EC (1994) Performance assessment significance of natural analog studies at Peña Blanca. Mexico and at Santorini, Greece. In: von Maravic H and Smellie J (editors) Natural analogue working group, fifth meeting, Toledo, October 1992. CEC Nuclear Science and Technology Report, 219-224, EUR 15176, CEC, Luxembourg.

Murphy WM and Pearcy EC (1996) Natural analog support for unsaturated transport modelling using data from the Akrotiri archaeological site. Materials Research Society Symposium Proceedings, 412, (Scientific Basis for Nuclear Waste Management, XIX), 817-822.

Murphy WM, Pearcy EC and Goodell PC (1991) Possible analogue research sites for the proposed high-level nuclear waste repository in hydrologically unsaturated tuff at Yucca Mountain, Nevada. In: Côme B and Chapman NA (editors) Natural analogue working group, fourth meeting and Poços de Caldas project final meeting, Pitlochry, June 1990. CEC Nuclear Science and Technology Report, EUR 13014, 267-276, CEC, Luxembourg.

Murphy WM, Pearcy EC, Green RT, Prikryl JD, Mohanty S, Leslie BW and Nedungadi A (1997) A test of the long term, predictive, geochemical transport modeling at the Akrotiri archaeological site. Journal of Contaminant Hydrology, 29, 245-279.

Murphy WM, Pickett DA, Pearcy EC and Turner DR (2000, in press) Peña Blanca data in source term and performance assessment models for Yucca Mountain. In: von Maravic H and Alexander WR (editors) Natural analogue working group, eighth meeting, Strasbourg, March 1999. EC Nuclear Science and Technology Report, EC, Luxembourg.

Myers GE and McCready RGL (1966) Bacteria can penetrate rock. Canadian Journal of Microbiology, 12, 477-484.

Nagra (1985) Nukleare Entsorgung Schweiz: Konzept und übersicht über das Projekt Gewähr 1985. Nagra Gewähr Reports NGB 85-01 to 85-08, Nagra, Wettingen, Switzerland. (English Volume is NTB 85-09).

Nagra (1988) Sediment study - Disposal options for long-lived radioactive waste in Swiss sedimentary formations: executive summary. Nagra Technical Report, NTB 88-25E , Nagra, Wettingen, Switzerland.

Nagra (1992) Nukleare Entsorgung Schweiz - Konzept und Realisierungsplan. Nagra Technical Report, NTB 92-02, Nagra, Wettingen, Switzerland.

Nagra (1993) Geologie und Hydrogeologie des Kristallins der Nordschweiz. Nagra Technical Report, NTB 93-01, Nagra, Wettingen, Switzerland.

Nagra (1994) Kristallin-I: safety assessment report. Nagra Technical Report, NTB 93-22E, Nagra, Wettingen, Switzerland.

Nagy B (1993) Kerogens and bitumens in Precambrian uraniferous ore deposits: Witwatersand, South Africa; Elliot Lake, Canada and the natural fission reactors, Oklo, Gabon. In: Parnell J, Kucha H and Landais P (editors) Bitumens in ore deposits. Springer-Verlag.

Nagy B, Gauthier-Lafaye F, Holliger P, Davis DW, Mossman DJ, Leventhal JS, Rigali MJ and Parnell J (1991) Organic matter and containment of uranium and fissiogenic isotopes at the Oklo natural reactors. Nature, 354, 472-475.

Nagy B, Gauthier-Lafaye F, Holliger P, Mossman DJ, Leventhal JS and Rigali MJ (1993) Role of organic matter in the Proterozoic Oklo natural fission reactors, Gabon, Africa. Geology, 21, 655-658.

Nakashima S and Nakamura H (1987) Mechanisms and quantitative evaluations of radionuclide fixation in geosphere. In: Côme B and Chapman NA (editors) Natural analogues in radioactive waste disposal. CEC Radioactive Waste Management Series, EUR 11037, 386-396, CEC, Luxembourg.

NEA (1988) Feasibility of disposal of high-level radioactive waste into the seabed. Volume 1: Overview of research and conclusions. NEA-OECD, Paris.

NEA (1989) Plutonium fuel: an assessment. NEA-OECD, Paris.

NEA (1991) Radiation protection and safety criteria, disposal of high level radioactive waste. Proceedings of NEA workshop. NEA-OECD, Paris.

NEA (1993) The International INTRAVAL Project Phase 1. Summary Report. NEA-OECD, Paris.

NEA (1997) Management of separated plutonium - the technical options. NEA/OECD.

Neal C and Stanger G (1983) Hydrogen generation from mantle source rocks in Oman. Earth and Planetary Science Letters, 6, 315-320.

Neall FB (1994). Modelling of the near-field chemistry of the SMA repository at the Wellenberg site. PSI Bericht 94-18, Paul Scherrer Institute, Wuerenlingen, Switzerland.

Neall F, Smith P, Sumerling T and Umeki H (1995) Putting HLW performance results in perspective. Nagra Bulletin, 25, 47-55.

Neretnieks I (1980) Diffusion in the rock matrix: an important factor in radionuclide migration? Journal of Geophysical Research, 85, 4379-4397.

Neretnieks I (1982) Diffusivities of some dissolved constistuents in compacted wet bentonite clay-MX80 and the impact on radionuclide migration in the buffer. SKB Technical Report, TR 82-27, SKB, Stockholm, Sweden.

Neretnieks I (1986a) Investigations of old bronze cannons. In: Côme B and Chapman NA (editors) Natural analogue working group, second meeting, Interlaken, June 1986. CEC Nuclear Science and Technology Report, EUR 10671, 191-197, CEC, Luxembourg.

Neretnieks I (1986b) The need for geologic evidence for radionuclide migration in the geosphere. In: Côme B and Chapman NA (editors) Natural analogue working group, first meeting, Brussels, November 1985. CEC Nuclear Science and Technology Report, EUR 10351, 32-36, CEC, Luxembourg.

Neretnieks I (1986c) Some uses for natural analogues in assessing the function of a HLW repository. Chemical Geology, 55, 175-188.

Neretnieks I (1990) Solute transport in fractured rock: applications to radioactive waste repositories. SKB Technical Report, TR 90-38, SKB, Stockholm, Sweden.

Neretnieks I (1996) Matrix diffusion - how confident are we? In: von Maravic H and Smellie J (editors) Natural analogue working group, sixth meeting, Santa Fe, September 1994. CEC Nuclear Science and Technology Report, EUR 16761, 211-232, CEC, Luxembourg.

Neretnieks I and Aslund B (1983a) The movement of radionuclides past a redox front KBS Technical Report, TR 83-66, KBS, Stockholm, Sweden.

Neretnieks I and Aslund B (1983b) Two dimensional movements of a redox front downstream from a repository for nuclear waste KBS Technical Report, TR 83-68, KBS, Stockholm, Sweden.

Neretnieks I and Faghihi M (1991) Some mechanisms which may reduce radiolysis. SKB Technical Report, TR 91-46, SKB, Stockholm, Sweden.

Nightingale TJ (1988) Mobilisation and redistribution of radionuclides during weathering of a uranium ore body. Unpublished MSc Thesis, University of Sydney, Australia.

Nirex (1989) Deep repository project. Nirex Report 71, United Kingdom Nirex Ltd, Harwell, England.

Nirex (1991) The repository project: an engineering progress report describing the preferred design concept. United Kingdom Nirex Ltd, Harwell, England.

Nohara T, Ochiai Y, Seo T and Yoshida H (1992) Uranium-series disequilibrium studies in the Tono uranium deposit, Japan. Radiochimica Acta.

Nordic Radiation Protection and Nuclear Safety Authorities (1993) Disposal of high-level radioactive waste, consideration of some basic criteria. (The so-called Nordic Flag Book: Edition Two).

Nordstrom DK (1996) Geochemical transport modelling. In: von Maravic H and Smellie J (editors) Natural analogue working group, sixth meeting, Santa Fe, September 1994. CEC Nuclear Science and Technology Report, EUR 16761, 243-254, CEC, Luxembourg.

Nordstrom DK, Smellie JAT and Wolf M (1990a). Chemical and isotopic compositions of groundwaters and their seasonal variability at the Osamu Utsumi and Morro do Ferro analogue study sites, Poços de Caldas, Brazil. SKB Technical Report, TR 90-15, SKB, Stockholm, Sweden; Nagra Technical Report, NTB 90-24, Nagra, Wettingen, Switzerland; UK DoE Technical Report, WR 90-046.

Nordstrom DK, Puigdoménech I and McNutt RH (1990b) Geochemical modelling of water-rock interactions at the Osamu Utsumi mine and Morro do Ferro analogue study sites, Poços de Caldas, Brazil. SKB Technical Report, TR 90-23, SKB, Stockholm, Sweden; Nagra Technical Report, NTB

90-32, Nagra, Wettingen, Switzerland; UK DoE Technical Report, WR 90-054.

Nordstrom DK, McNutt RH, Puigdoménech I, Smellie JAT and Wolf M (1992) Groundwater chemistry and geochemical modelling of water-rock interactions at the Osamu Utsumi mine and Morro do Ferro analogue study sites, Poços de Caldas, Minas Gerais, Brazil. Journal of Geochemical Exploration, 45, 249-288.

Norman S (1991) Verification of HYDRASTAR - a code for stochastic continuum simulation of groundwater flow. SKB Technical Report, TR 91-27, SKB, Stockholm, Sweden.

Ohlsson Y and Neretnieks I (1995) Literature survey of matrix diffusion theory and of experiments and data including natural analogues. SKB Technical Report, TR 95-12, SKB, Stockholm, Sweden.

Olin M and Valkiainen M (1990) Concentration profiles of anions in granite bedrock caused by postglacial land uplift and matrix diffusion. IVO Technical Report, TR 90-1.

Oversby VM (2000, in press) Oklo natural analogue phase II - project summary and PA applications. In: von Maravic H and Alexander WR (editors) Natural analogue working group, eighth meeting, Strasbourg, March 1999. EC Nuclear Science and Technology Report, EC, Luxembourg.

Palmer HC, Tazaki K, Fyfe WS and Zhou, Z (1988) Precambrian glass. Geology, 16, 221-224.

Papp T (1987) The role of natural analogues in safety assessment and acceptability. In: Côme B and Chapman NA (editors) Natural analogues in radioactive waste disposal. CEC Radioactive Waste Management Series, EUR 11037, 12-22, CEC, Luxembourg.

Parks GA and Pohl DC (1985) Hydrothermal solubility of uraninite. US Department of Energy, DOE/ER12016-1.

Parneix JC, Menager M, Trotignon L and Petit JC (1987) Hydrothrmal alteration in the Auriat granite, Massif Central. In: Côme B and Chapman NA (editors) Natural analogues in radioactive waste disposal. CEC Radioactive Waste Management Series, EUR 11037, 449-461, CEC, Luxembourg.

Parneix JC (1992) Effects of hydrothermal alteration on radioelement migration from a hypothetical disposal site for high level radioactive waste: example from the Auriat granite France. Applied Geochemistry, Supplementary Issue, 1, 253-268.

Parnell J, Kucha H and Landais P (1993, editors) Bitumens in ore deposits. Springer-Verlag.

Pate SM, Alexander WR and McKinley IG (1994) Use of natural analogue test cases to evaluate a new performance assessment TDB. In: von Maravic H and Smellie J (editors) Natural analogue working group, fifth meeting, Toledo, October 1992. CEC Nuclear Science and Technology Report, EUR 15176, CEC, Luxembourg.

Pearcy EC and Murphy WM (1991) Geochemical natural analogues: literature review. Nuclear Regulatory Commission, NRC-02-88-005, Washington DC, USA.

Pearcy EC, Prikryl JD, Murphy WM and Leslie BW (1994) Alteration of uraninite from the Nopal I deposit, Peña Blanca District, Chihuahua, Mexico, compared to degradation of spent nuclear fuel in the proposed US high-level nuclear waste repository at Yucca Mountain, Nevada. Applied Geochemistry, 9, 713-732.

Pearson FJ, Noronna CJ and Andrews RW (1983) Mathematical modelling of the distribution of natural ^{14}C, ^{234}U and ^{238}U in a regional groundwater system. Radiocarbon, 25, 291-300.

Pearson FJ and Berner U (1991) Nagra thermochemical database I: Core data. Nagra Technical Report, NTB 91-17, Nagra, Wettingen, Switzerland.

Pedersen K (1993) The deep subterranean biosphere. Earth Science review, 34, 243-260.

Pedersen K (1997) Investigations of subterranean microorganisms and their importance for performance assessment of radioactive waste disposal: results and conclusions achieved during the period 1995 to 1997. SKB Technical Report, TR 97-22, SKB, Stockholm, Sweden.

Pedersen K and Albinsson Y (1992) Possible effects of bacteria on trace element migration in crystalline bed-rock. Radiochimica Acta, 58/59, 365-369.

Pedersen K and Karlsson F (1995) Investigations of subterranean microorganisms - their importance for performance assessment of radioactive waste disposal. SKB Technical Report, TR 95-10, SKB, Stockholm, Sweden.

Pedersen K, Motamedi M and Karnland O (1995) Survival of bacteria in nuclear waste buffer materials: the influence of nutrients, temperature and water activity. SKB Technical Report, TR 95-27, SKB, Stockholm, Sweden.

Pellegrini R, Horseman S, Kemp S, Rochelle C, Boisson J-Y, Lombardi S, Bouchet A and Parneix J-C (1999) Natural analogues of the thermo-hydro-chemical and thermo-hydro-mechanical response. EC Nuclear Science and Technology Report, EUR 19114, EC, Luxembourg.

Pellegrini R, Lombardi S, Rochelle C, Boisson J-Y and Parneix J-C (2000, in press) Thermal effects of clay barrier materials: stress related effects. In: von Maravic H and Alexander WR (editors) Natural analogue working group, eighth meeting, Strasbourg, March 1999. EC Nuclear Science and Technology Report, EC, Luxembourg.

Peltonen E (1985, editor) Safety analysis of disposal of spent nuclear fuel - normal and disturbed evolution scenarios. Nuclear Waste Commission of Finnish Power Companies Technical Report, YJT-85-22.

Penna Franca E, Campos MT, Lobao N, Trindade H, Sachett I and Eisenbud M (1984) Radium mobilisation and transport at a large thorium ore deposit in Brazil. In: Smellie JAT (editor) Natural analogues to the conditions around a final repository for high level radioactive waste Proceedings of the natural analogue workshop held at Lake Geneva, Wisconsin, USA. SKB Technical Report, TR 84-18, SKB, Stockholm, Sweden.

Penrose WR, Polzer WL, Essington EH, Nelon DM and Orlandini KA (1990) Mobility of plutonium and americium through a shallow aquifer in a semiarid region. Environmental Science and Technology, 24, 228-234.

Percival JB and Kodama H (1989) Sudoite from Cigar Lake, Saskatchewan. Canadian Mineralogist, 27, 633-641.

Pérez del Villar L, Pelayo M, Cózar JS, de la Cruz B, Pardillo J, Reyes E, Cabellero E, Delgado R, Nuñez R, Ivanovich M and Hasler S (1997) Mineralogical and geochemical evidence of the migration and retention processes of U and Th in fracture fillings from the El Berrocal granitic site (Spain). Journal of Contaminant Hydrology, 26, 45-60.

Pescatore C (1995) Validation: the eluding definition. Radioactive Waste Management and Environmental Restoration, 20, 13-22.

Petit JC (1991a) Migration of radionuclides in the geosphere: what can we learn from natural analogues. Radiochimica Acta, 51, 181-188.

Petit JC (1991b) Design and performance assessment of radioactive waste forms; what can we learn from natural analogues. In: Côme B and Chapman NA (editors) Natural analogue working group, fourth meeting and Poços de Caldas project final meeting, Pitlochry, June 1990. CEC Nuclear Science and Technology Report, EUR 13014, 31-72, CEC, Luxembourg.

Petit JC (1992a) Natural analogues for the design and performance assessment of radioactive waste forms: a review. Journal of Geochemical Exploration, 46, 1-34.

Petit JC (1992b) Reasoning by analogy: rational foundation of natural analogue studies. Applied Geochemistry, Supplementary Issue, 1, 9-12.

Petit JC, Dran J-C, Trotignon L, Casabonne JM, Paccagnella A and Della Mea G (1989) Mechanism of heavy element retention in hydrated layers formed on leached silicate glasses. Materials Research Society Symposium Proceedings, 127, (Scientific Basis for Nuclear Waste Management, XII), 33-40.

Petts J and Eduljee G (1994) Environmental impact assessment for waste treatment and disposal facilities. Johh Wiley and Sons.

Philp RP and Crisp PT (1982) Surface geochemical methods used for oil and gas prospecting: a review. Journal of Geochemical Exploration, 17, 1-34.

Philpotts AR and Miller JA (1963) A Precambrian glass from St Alexis-des-Montes, Quebec. Geological Magazine, 100, 337-344.

Piantone P (1989) Analogue naturel de la migration des radioéléments en formation granitique par l'étude des paléoaltérations hydrothermales. CEC Nuclear Science and Technology Report, EUR 12297, CEC, Luxembourg.

Pickett DA and Murphy WM (2000, in press) Uranium chemistry and isotopy in waters and rocks at Peña Blanca. In: von Maravic H and Alexander WR (editors) Natural analogue working group, eighth meeting, Strasbourg, March 1999. EC Nuclear Science and Technology Report, EC, Luxembourg.

Pinto Coelho P (1987) Element distribution across veins in the East Bull Lake gabbro anorthosite layered intrusion, Algoma District, Ontario - an evaluation of matrix diffusion. In: Côme B and Chapman NA (editors) Natural analogues in radioactive waste disposal. CEC Radioactive Waste Management Series, EUR 11037, 261-274, CEC, Luxembourg.

Pitkänen P and Luukkonen A (1998). Geochemical modelling of groundwater evolution and residence time at the Olkiluoto site. Posiva Technical Report, 98-10, Posiva, Helsinki, Finland.

Pitts L and St Joseph A (1985) Inchtuthil Roman legionary fortress excavation 1952-1965. Allan Sutton.

Pusch R (1982a) Copper-bentonite interactions. SKB Technical Report, TR 82-07, SKB, Stockholm, Sweden.

Pusch R (1982b). Chemical interaction of clay buffer materials and concrete. SKB Technical Report, TR 82-01, SKB, Stockholm, Sweden.

Pusch R (1983) Stability of deep-sited smectite minerals in crystalline rock: chemical aspects. SKB Technical Report, TR 83-16, SKB, Stockholm, Sweden.

Pusch R (1986) Settlement of canisters with smectite clay envelopes in deposition holes. SKB Technical Report, TR 86-23, SKB, Stockholm, Sweden.

Pusch R and Karnland O (1988) Geological evidence of smectite longevity. The Sardinian and Götland cases. SKB Technical Report, TR 88-26, SKB, Stockholm, Sweden.

Pusch R, Börgesson L and Erlström M (1987) Alteration of isolating properties of dense smectite clay in repository environment as exemplified by seven pre-quaternary clays. SKB Technical Report, TR 87-29, SKB, Stockholm, Sweden.

Rainey TP and Rosenbaum MS (1989) The adverse influence of geology and groundwater on the behaviour of London Underground railway tunnels near Old Street Station. Proceedings of the Geological Association, 100, 123-134.

Raloff J (1990) The colloid threat. Science News, 17, 169-170.

Ramsey JDF (1985) The role of colloids in the release of radionuclides from nuclear waste. UKAEA Technical Report, AERE-R-11823, Harwell, England.

Rasmuson A and Neretnieks I (1981) Migration of radionuclides in fissured rock: the influence of micropore diffusion and longditudinal dispersion. Journal of Geophysical Research, 86, 3749-3758.

Rasmussen B, Glover JE and Alexander R (1989) Hydrocarbon rims on monazite in Permian-Triassic arenites, northern Perth basin, Australia - pointers to the former presence of oil. Geology, 17, 115-118.

Rasmussen B, Glover JE and Foster CB (1993) Polymerisation of hydrocarbons by radioactive minerals in sedimentary rock: diagenetic and economic significance. In: Parnell J, Kucha H and Landais P (editors) Bitumens in ore deposits. Springer-Verlag.

Rassineux F, Petit J-C and Meunier A (1989) Ancient analogues of modern cement: calcium hydrosilicates in mortars and concretes from Gallo-Roman thermal baths of western France. Journal of the American Ceramic Society, 72, 1026-1032.

Rautenschlein M, Jenner GA, Hertogen J, Hofmann AW, Kerrich R, Schmincke H-U and White WM (1985) Isotopic and trace element composition of volcanic glasses from the Akaki Canyon Cyprus: implications for the origin of the Troodos ophiolite. Earth and Planetary Science Letters, 75, 369-383.

RAWMAC (1990) Radioactive waste management advisory committee; eleventh annual report. HMSO, London.

Rayment DL and Pettifer K (1987) Examination of durable mortar from Hadrian's Wall. Materials Science and Technology, 3, 997-1004.

Read D (1988) Geochemical modelling of the Broubster natural analogue site, Caithness, Scotland. BGS Technical Report, WE/88/43.

Read D (1991) Equilibrium speciation and chemical transport modelling. In: Read D and Hooker PJ (editors) UK Natural Analogue Co-ordinating Group: Fourth Annual Report. UK DoE Report, DoE/HMIP/RR/92/009.

Read D (1992) Geochemical modelling of uranium redistribution in the Osamu Utsumi mine, Poços de Caldas. Journal of Geochemical Exploration, 45, 503-520.

Read D and Hooker P (1989) The speciation of uranium and thorium at the Broubster natural analogue site, Caithness, Scotland. Materials Research Society Symposium Proceedings, 127, (Scientific Basis for Nuclear Waste Management, XII), 715-722.

Read D and Hooker P (1991) Using hydrogeochemical data from natural environments to improve models of radionuclide speciation in groundwaters. In: Côme B and Chapman NA (editors) Natural analogue working group, fourth meeting and Poços de Caldas project final meeting, Pitlochry, June 1990. CEC Nuclear Science and Technology Report, EUR 13014, 95-117, CEC, Luxembourg.

Rees JH and Rodwell WR (1988) Gas evolution and migration in repositories - current status. Nirex Radioactive Waste Disposal: Safety Studies, NSS/G104, United Kingdom Nirex Ltd, Harwell, England.

Rennerfelt J and Meijer JE (1986) Gasproduktion vid anaerob nedbrytning av cellulosa etc. Unpublished SKB Report, SKB, Stockholm, Sweden

Rice EE and Priest CC (1981) An overview of nuclear waste disposal in space. In Hofmann PL (editor) The technology of high-level nuclear waste disposal. US Department of Energy Technical Report, DOE/TIC-4621, 370-386, USDOE, Washington, USA.

Ridley WP, Dizikes LJ and Wood JM (1977) Biomethylation of toxic elements in the environment. Science, 197, 329-332.

Ringwood AE, Kesson SE, Ware NG, Hibberson W and Major A (1979) Immobilisation of high-level nuclear reactor wastes in SYNROC. Nature, 278, 219-223.

Ringwood AE, Kesson SE, Reeve KD, Levins D and Ramm EJ (1988) Synroc. In: Lutze W and Ewing RC (editors) Radioactive waste forms for the future. North Holland, 495-564.

Rivas P, Hernán P, Bruno J, Carrera J, Gómez P, Guimerà J, Marín C and Pérez del Villar L (1997) El

Berrocal project. Characterisation and validation of natural radionuclide migration processes under real conditions on the fissured granitic environment. CEC Nuclear Science and Technology Report, EUR 17478, CEC, Luxembourg.

Roberson HE and Lahann RW (1981) Smectite to illite conversion rates: effects of solution chemistry. Clays and Clay Minerals, 29, 129-135.

Roberts PD, Ball TK, Hooker PJ and Milodowski AE (1989) A uranium geochemical study at the natural analogue site of Needle's Eye, SW Scotland. Materials Research Society Symposium Proceedings, 127, (Scientific Basis for Nuclear Waste Management, XII), 933-940.

Robinson PT, Melson WG, O'Hearn T and Schmincke H-U (1983) Volcanic glass compositions from the Troodos ophiolite, Cyprus. Geology, 11, 400-404.

Rolfe WDI and Brett DW (1969) Fossilization process. In: Eglinton G and Murphy MTJ (editors) Organic Geochemistry. Springer Verlag, Berlin.

Romero L, Neretnieks I, and Moreno L (1992) Movement of the redox front at the Osamu Utsumi mine, Poços de Caldas, Brazil. Journal of Geochemical Exploration, 45, 471-502.

Root-Bernstein RS (1988) Setting the stage for discovery: breakthroughs depend on more than luck. The Sciences, 28, 26-34.

Roxburgh IS (1987) Geology of high-level nuclear waste disposal. Chapman and Hall.

Ruskeeniemi T, Niini H, Söderholm B, Vesterinen M, Blomqvist R, Halonen S, Lindberg A, Jaakkola T, Suksi J and Suutarinen R (1989) The Palmottu U-Th deposit in SW Finland as a natural analogue to the behaviour of spent nuclear fuel in bedrock: a preliminary report. In: Miles DL (editor) Water Rock Interaction, WRI-6. Proceedings of the 6th International Symposium on Water-Rock Interaction, Malvern, August 1989, 601-604.

Ryan JN and Gschwend PM (1990) Colloid mobilisation in two Atlantic coastal plain aquifers. Water Resources Research, 20, 307-22.

Ryan J, McPhail D, Rogers P and Oakley V (1993) Glass deterioration in the museum environment. Chemistry and Industry, July, 498-501.

Rybalchenko A (1998) Deep well injection of liquid radioactive waste in Russia: present situation. In: Stenhouse MJ and Kirko VI (editors) Defense nuclear waste disposal in Russia: international perspective, 199-217. NATO Advanced Science Institute Series 1: Disarmament Technologies, Volume 18. Kluwer Academic Publishers.

Sano Y, Urabe A, Wakita H and Wushiki H (1993) Origin of hydrogen-nitrogen gas seeps Oman. Applied Geochemistry, 8, 1-8.

Savage D (1995, editor) The scientific and regulatory basis for the geological disposal of radioactive waste. John Wiley and Sons.

Savage D (1998). A review of zeolite occurrence and behaviour. In: Smellie JAT (editor) Maqarin natural analogue study: phase III. SKB Technical Report, TR 98-04, SKB, Stockholm, Sweden.

Savage D and Rochelle C (1993) Modelling reactions between cement pore fluids and rock: implications for porosity change. Journal of Contaminant Hydrology, 13, 365-378.

Savage D, Bateman K, Hill P, Hughes C, Milodowski A, Pearce J, Rae E and Rochelle C (1992) Rate and mechanism of the reaction of silicates with cement pore fluids. Applied Clay Science, 7, 33-45.

Savary V, Pagel M, Pironon J, Holliger P and Gauthier-Lafaye F (1993) Redox conditions in the Oklo nuclear reactor zones. Terra Nova, 5, 510-511.

Schlesinger WH (1991) Biogeochemistry. An Analysis of Global Change. Academic Press.

Schoell M (1988) Multiple origins of methane in the Earth. Chemical Geology, 71, 1-10.

Schorscher HD and Shea ME (1990) Outline of regional geology mineralogy and geochemistry, Poços de Caldas, Minas Gerais, Brazil. SKB Technical Report, TR 90-10, SKB, Stockholm, Sweden; Nagra Technical Report, NTB 90-19, Nagra, Wettingen, Switzerland; UK DoE Technical Report, WR 90-041.

Schwarcz HP, Gascoyne M and Ford D C (1982) U-series disequilibrium studies of granitic rocks. Chemical Geology, 36, 87-102.

Schweingruber M (1983) Actinide solubility in deep groundwater; estimates for upper limits based on chemical disequilibrium calculations. Nagra Technical Report, NTB 83-24, Nagra, Wettingen, Switzerland.

Seo T (1991) Uranium distribution in the colloidal and solute phases at the Koongara uranium deposit. In: Duerden P (editor) Alligator Rivers Analogue Project: second annual report 1989-1990. Australian Nuclear Science and Technology Organisation, Sydney, Australia.

Seo T and Payne TE (1994) A study of colloids in groundwaters at the Konngara uranium deposit. CEC Nuclear Science and Technology Report, EUR 11725, CEC, Luxembourg.

Seo T and Yoshida H (1994) Natural analogue studies of the Tono uranium deposit in Japan. In: von Maravic H and Smellie J (editors) Natural analogue working group, fifth meeting, Toledo, October 1992. CEC Nuclear Science and Technology Report, EUR 15176, CEC, Luxembourg.

Seo T, Edis R and Payne TE (1994) A study of colloids in groundwaters at the Koongara uranium deposit. In: von Maravic H and Smellie J (editors) Natural analogue working group, fifth meeting, Toledo, October 1992. CEC Nuclear Science and Technology Report, EUR 15176, 71-76, CEC, Luxembourg.

Shea M (1984) Uranium migration at some hydrothermal veins near Marysvale, Utah. A natural analogue for radwaste isolation. Material Research Society Symposium Proceedings, 26, (Scientific Basis for Nuclear Waste Management, VII), 227-238.

Shea M (1987) Marysvale natural analogue study: feasibility phase analytical results. In: Côme B and Chapman NA (editors) Natural analogues in radioactive waste disposal. CEC Radioactive Waste Management Series, EUR 11037, 275-286, CEC, Luxembourg.

Shea M (1990) Isotopic geochemical characterisation of selected nepheline syenites and phonolites from the Poços de Caldas alkaline complex, Minas Gerais, Brazil. SKB Technical Report, TR 90-13, SKB, Stockholm, Sweden; Nagra Technical Report, NTB 90-22, Nagra, Wettingen, Switzerland; UK DoE Technical Report, WR 90-044.

Shea M (1998) Hydrologic, thermal and chemical processes related to fracture controlled hydrothermal water-rock interaction. Unpublished PhD Thesis, University of Chicago, USA.

Sherwood B, Fritz P, Frape SK, Macko SA, Weise SM and Welhan JA (1988) Methane occurrences in the Canadian Shield. Chemical Geology, 71, 223-236.

Shin WS (1982) The long-term stability of cement and concrete for nuclear waste disposal under normal geologic conditions. Nagra Technical Report, NTB 82-03, Nagra, Wettingen, Switzerland.

Shoesmith DW and Sunder S (1991) An electrochemistry-based model for the dissolution of UO_2. SKB Technical Report, TR 91-63, SKB, Stockholm, Sweden.

Shoesmith DW and Sunder S (1993) The prediction of nuclear fuel (UO_2) dissolution rates under waste disposal conditions. Journal of Nuclear Materials.

Sholkovitz ER (1989) Artefacts associated with the chemical leaching of sediments for the REEs. Chemical Geology, 77, 47-51.

Short SA, Lowson RT and Ellis J (1988) $^{234}U/^{238}U$ and $^{230}Th/^{234}U$ activity ratios in the colloidal phases of aquifers in lateritic weathered zones. Geochimica et Cosmochimica Acta, 52, 2555-2563.

Sibley TH and Myttenaere C (1986) Application of distribution coefficients to radiological assessment models. Elsevier, Amsterdam.

Simpson HJ, Trier RM, Li YH and Anderson RF (1984) Field experiment determinations of distribution coefficients of actinide elements in alkaline lake environments. In: Alexander DH and Birchard GF (editors) NRC Nuclear Waste Geochemistry '83, NUREG / CP-0052.

Simpson JP (1983) Experiments on container materials for Swiss high-level waste disposal projects: part I. Nagra Technical Report, NTB 83-05, Nagra, Wettingen, Switzerland.

Simpson JP (1984) Experiments on container materials for Swiss high-level waste disposal projects: part II. Nagra Technical Report, NTB 84-01, Nagra, Wettingen, Switzerland.

Simpson JP (1989) Experiments on container materials for Swiss high-level waste disposal projects: part IV. Nagra Technical Report, NTB 89-19, Nagra, Wettingen, Switzerland.

Simpson JP and Vallotton PH (1986) Experiments on container materials for Swiss high-level waste disposal projects: part III. Nagra Technical Report, NTB 86-25, Nagra, Wettingen, Switzerland.

Sjoeblom KL and Linsley GS (1994) Sea disposal of radioactive wastes: the London Dumping Convention 1972. IAEA Builletin, 2/1994. International Atomic Energy Agency, Vienna, Austria.

Sinnock S, Lin YT and Brannen JP (1987) Preliminary bounds on the expected postclosure performance of the Yucca Mountain repository site southern Nevada. Journal of Geophysical Research, 92, 7820-7842.

Skagius K (1986) Diffusion of dissolved species in the matrix of some Swedish crystalline rocks. Unpublished PhD Thesis, Royal Institute of Technology, Stockholm, Sweden.

Skagius K and Neretnieks I (1982) Diffusion in crystalline rocks of some sorbing and non-sorbing species. SKB Technical Report, TR 82-12, SKB, Stockholm, Sweden.

Skagius K and Neretnieks I (1983) Diffusion measurements in crystalline rocks. SKB Technical Report, TR 83-15, SKB, Stockholm, Sweden.

Skagius K and Neretnieks I (1985a) Diffusivities in crystalline rock materials. Materials Research Society Symposium Proceedings, 50, (Scientific Basis for Nuclear Waste Management, IX), 73-80.

Skagius K and Neretnieks I (1985b) Porosities and diffusivities of some non-sorbing species in crystalline rock. SKB Technical Report, TR 85-03, SKB, Stockholm, Sweden.

Skagius K and Neretnieks I (1986) Porosities and diffusivities of some nonsorbing species in crystalline rocks. Water Resources Research, 22, 389-397.

SKB (1989) WP-CAVE - assessment of feasibility safety and development potential. SKB Technical Report, TR 89-20, SKB, Stockholm, Sweden.

SKB (1990) SKB annual report 1990. SKB Technical Report, TR 90-46, SKB, Stockholm, Sweden.

SKB (1992) Final disposal of spent nuclear fuel. Importance of the bedrock for safety. SKB Technical Report, TR 92-20, SKB, Stockholm, Sweden.

SKB (1999) Deep repository for spent nuclear fuel SR 97 – Post closure safety. SKB Technical Report, TR-99-06. SKB, Stockholm, Sweden.

SKI (1991) SKI Project-90. SKI Technical Report, TR 91-23, SKI, Stockholm.

SKI (1996) SKI Site-94. Deep Repository Performance Assessment Project. SKI report 96:36. SKI, Stockholm. (2 volumes).

Smellie JAT (1984, editor) Natural analogues to the conditions around a final repository for high level radioactive waste Proceedings of the natural analogue workshop held at Lake Geneva, Wisconsin, USA. SKB Technical Report, TR 84-18, SKB, Stockholm, Sweden.

Smellie JAT (1994) The Cigar Lake natural analogue study. In: von Maravic H and Smellie J (editors) Natural analogue working group, fifth meeting, Toledo, October 1992. CEC Nuclear Science and Technology Report, EUR 15176, CEC, Luxembourg.

Smellie JAT (1998, editor) Maqarin natural analogue study: phase III. SKB Technical Report, TR 98-04, SKB, Stockholm, Sweden.

Smellie JAT and Rosholt JN (1984) Radioactive disequilibria in mineralised fracture samples from two uranium occurrences in northern Sweden. Lithos, 17, 215-225.

Smellie JAT and Karlsson F (1996, editors) The Cigar Lake analogue project: a reappraisal of some key issues and their relevance to repository performance assessment. SKB Technical Report, TR 96-08, SKB, Stockholm, Sweden.

Smellie JAT and Karlsson F (1999) The use of natural analogues to assess radionuclide transport. Engineering Geology, 52, 193-220.

Smellie JAT, MacKenzie AB and Scott RD (1986a) An analogue validation study of natural radionuclide migration in crystalline rocks using uranium-series disequilibrium studies. In: Côme B and Chapman NA (editors) Natural analogue working group, first meeting, Brussels, November 1985. CEC Nuclear Science and Technology Report, EUR 10315, 93-100, CEC, Luxembourg.

Smellie JAT, MacKenzie AB and Scott RD (1986b) An analogue validation study of natural radionuclide migration in crystalline rocks using uranium-series disequilibrium studies. Chemical Geology, 55, 233-254.

Smellie JAT, MacKenzie AB and Scott RD (1986c) An analogue validation study of natural radionuclide migration in crystalline rock using uranium-series disequilibrium studies II: A comparison of neutron activation and alpha spectroscopy analyses of thorium in crystalline rocks. SKB Technical Report, TR 86-01, SKB, Stockholm, Sweden.

Smellie JAT, Karlsson F and Alexander WR (1997) Natural analogue studies: present status and performance assessment implications. Journal of Contaminant Hydrology, 26, 3-17.

Smellie JAT, Chapman NA, McKinley IG, Penna Franca E and Shea M (1989) Testing safety assessment models using natural analogues in high natural series groundwaters. Materials Research Society Symposium Proceedings, 127, (Scientific Basis for Nuclear Waste Management, XII), 863-870.

Smellie JAT, Fallick AE, Hållenius U, MacKenzie AB, Scott RD and Tullborg E-L (1993) Isotopic and Mössbauer studies of a single conducting fracture in crystalline rock

Smith DK (1991) Mineralogical textural and compositional data on the alteration of basaltic glass from Kilauea Hawaii to 300°C: insights to the corrosion of a borosilicate glass waste form. Materials Research Society Symposium Proceedings, 212, (Scientific Basis for Nuclear Waste Management, XIV), 115-121.

Smith PA and Curti E (1995) Some variations of the Kristallin-1 near-field model. Nagra Technical Report, NTB 95-09. Nagra, Wettingen, Switzerland.

Smith KL, Lumpkin GR, Blackford MG, Day RA, and Hart KP (1992) The durability of SYNROC. Journal of Nuclear Materials, 190, 287-294.

Snelling A (1980) A geochemical study of the Koongarra uranium deposit, Northern Territory, Australia. Unpublished PhD Thesis, University of Sydney, Australia.

Snellman M, Uotila H and Rantanen J (1987) Laboratory and modelling studies of sodium bentonite groundwater interaction. Scientific Basis for Nuclear Waste Management, X, 781-790.

Stanley CC (1979) Highlights in the history of concrete. Cement and Concrete Association.

Steadman JA (1986) Archaeological concretes as analogues. In: Côme B and Chapman NA (editors) Natural analogue working group, second meeting, Interlaken, June 1986. CEC Nuclear Science and

Technology Report, EUR 10671, 165-171, CEC, Luxembourg.

Steefel C and Lichtner P (1994) Diffusion and reaction in rock matrix bordering a hyperalkaline fluid-filled fracture. Geochimica et Cosmochimica Acta, 58, 3595-3612.

Stenhouse MJ and Grogan H (1991) Review of reactions of hydrogen and methane in the geosphere and biosphere. Nirex Radioactive Waste Disposal: Safety Studies, NSS/R262, United Kingdom Nirex Ltd, Harwell, England.

Stevens T and McKinley JP (1995) Lithoautotrophic microbial ecosystem in deep basaltic aquifers. Science, 270, 450-453.

Strack S and Mueller A (1984) Studies of the microbiological influences on the behaviour of iodine-125 in humus soil. In: IUR/CEC workshop, role of microorganisms on the behaviour of radionuclides in aquatic and terrestrial systems and their transfer to man, 207-218.

Stumm W and Morgan J (1981) Aquatic Chemistry. Wiley-Interscience, New York.

Sudicky EA and Frind EO (1981) 14-C dating of groundwater in confined aquifers: implications of aquitard diffusion. Water Resources Research, 17, 1060-1064.

Suksi J and Ruskeeniemi T (1992) Matrix diffusion - evidence from drill cores at Palmottu. Palmottu Project Progress Report. Geological Survey of Finland Technical Report, YST-78, Espoo, Finland.

Suksi J, Ruskeeniemi T, Linberg A and Jaakkola T (1991) The distribution of natural radionuclides on fracture surfaces in Palmottu analogue study site in SW Finland. Radiochimica Acta, 52/53, 367-372.

Sunder S, Taylor P and Cramer JJ (1988) XPS and XRD studies of uranium rich minerals from Cigar Lake, Saskatchewan. Material Research Society Symposium Proceedings, 112, (Scientific Basis for Nuclear Waste Management, XI), 465-472.

Sunder S, Shoesmith DW, Christensen H, Bailey MG and Miller NH (1989) Electrochemical and x-ray photoelectron spectroscopic studies of UO_2 fuel oxidation by specific radicals formed during radiolysis of groundwater. Materials Research Society Symposium Proceedings, 127, (Scientific Basis for Nuclear Waste Management, XII), 317-324.

Suutarinen R, Blomqvist R, Halonen S and Jaakkola T (1991) Uranium in groundwater in Palmottu analogue study site in Finland. Radiochimica Acta, 52/53, 373-380.

Sverjensky DA (1991) Geochemical investigations of uranium mobility in the Koongarra ore deposit: a natural analogue for the migration of radionuclides from a nuclear waste repository. In: Duerden P (editor) Alligator Rivers Analogue Project: annual report 1989-1990. Australian Nuclear Science and Technology Organisation, Sydney, Australia.

Swedish Corrosion Research Institute (1983) Corrosion resistance of a copper canister for spent nuclear fuel. SKB Technical Report, TR 83-24, SKB, Stockholm, Sweden.

Takach NE, Barker C and Kemp MK (1987) Stability of natural gas in the deep subsurface: thermodynamic calculation of equilibrium compositions. Bulletin of the American Association of Petroleum Geologists, 71, 322-333.

Takahashi M (1987, editor) Properties of bentonite clay as a buffer material in high-level waste geological disposal Part I: chemical species contained in bentonite. Nuclear Technology 76, 221-229.

Taylor RP and Fryer BJ (1982) REE geochemistry as an aid to interpreting hydrothermal ore deposits. In: Evans AM (editor) Metallization associated with acid magmatism, 357-365. J Wiley and Sons, London.

Thomassin JH and Rassineux, F (1992) Ancient analogues of cement-base materials: stability of calcium silicate hydrates. Applied Geochemistry, Supplementary Issue, 1, 137-142.

Thompson JL (1984) Laboratory and field studies related to the radionuclide migration project. Los Alamos National Laboratory Technical Report, LA-10372-PR, Los Alamos, New Mexico USA.

Thomson J, Colley S, Higgs NC, Hydes DJ, Wilson TRS and Sorensen J (1987) Geochemical oxidation fronts in North East Atlantic distal turbidites and their effects in the sedimentary records. In: Weaver PPE and Thomson J (editors) Geology and geochemistry of abyssal plains. Geological Society of London Special Publication, 31.

Thurber D (1965) The concentrations of some natural radioelements in the waters of the Great Basin. Bulletin of Volcanology, 28, 195-201.

Tomabechi K (1995) Uranium glass. Iwanami Book Service Centre, Tokyo.

Torstenfelt B, Ittner T, Allard B, Andersson K and Olofsson U (1982a) Mobilities of radionuclides in fresh and fractured crystalline rock. SKB Technical Report, TR 82-26, SKB, Stockholm, Sweden.

Torstenfelt B, Andersson K, Allard B and Olofsson U (1982b) Diffusion measurements in compacted bentonite. Materials Research Society Symposium Proceedings, 6, (Scientific Basis for Nuclear Waste Management, IV), 295-302.

Trägårdh J and Lagerblad B (1998) Leaching of 90-year old concrete mortar in contact with stagnant water. SKB Technical Report, TR 98-11, SKB, Stockholm, Sweden.

Tsezos M and Volesky B (1981) Biosorption of uranium and thorium. Biotechnology and Bioengineering, 23, 583-604.

Tullborg EL, Landström O and Wallin B (1997) Low temperature trace element mobility influenced by hydraulic activity - indications from isotopic and trace element analyses of fracture calcite and pyrite. Unpublished PhD Thesis, Göteborg University, Sweden.

TVO (1985) Käytetyn ydinpolttoaineen loppus-ijoitus Suomen kalloiperään TVO Technical Report, YJT-85-30, TVO, Helsinki, Finland.

Tweed CJ and Milodowski AE (1994) An overview of the Maqarin natural analogue project - a natural analogue study of a hyperalkaline cement groundwater system. In: von Maravic H and Smellie J (editors) Natural analogue working group, fifth meeting, Toledo, October 1992. CEC Nuclear Science and Technology Report, EUR 15176, CEC, Luxembourg.

Tylecote RF (1977) Durable materials for seawater: the archaeological avidence. The International Journal of Nautical Archaeology and Underwater Exploration, 6, 269-283.

Tylecote RF (1979) The effect of soil conditions on the long-term corrosion of buried tin-bronzes and copper. Journal of Archaeological Science, 6, 345-368.

Tylecote RF (1983) The behaviour of lead as a corrosion resistant medium undersea and in soils. Journal of Archaeological Science, 10, 397-409.

Ulff-Möller F (1990) Formation of native iron in sediment-contaminated magma: a case study of the Hanekammen Complex on Disko Island, West Greenland. Geochimica et Cosmochimica Acta, 54, 57-70.

USDOE (1980) Final environmental impact statement. manegemnt of commercially-generated radioactive waste. US Department of Energy Technical Report, DOE/EIS-0046F, USDOE, Washington, USA.

US National Research Council (1990) Rethinking high-level radioactive waste disposal. Commission on Geosciences, Environment and Resources. US National Research Council. National Academy Press.

Valkiainen M (1992) Diffusion in the rock matrix - a review of laboratory tests and field studies. Nuclear Waste Commission of Finnish Power Companies, Report 92-04, YJT, Helsinki, Finland.

Vance ER, Jostsons A, Day RA, Ball CJ, Begg BD and Angel PJ (1996) Excess Pu disposition in zirconolite-rich SYNROC. Materials Research

Society Symposium Proceedings, 412, (Scientific Basis for Nuclear Waste Management, XIX), 41-47.

Vandecasteele CM, Delmotte A, Henrot J, Van Hove D and Cogreau, M (1984) Interaction between technetium and nitrogen fixing organisms. In: IUR/CEC workshop, role of microorganisms on the behaviour of radionuclides in aquatic and terrestrial systems and their transfer to man, 158-172.

van Loon LR and Kopajtic Z (1990) Complexation of Cu^{2+}, Ni^{2+} and UO_2 by radiolytic degradation products of bitumen. Nagra Technical Report, NTB 90-18, Nagra, Wettingen, Switzerland.

van Luik AE (1987) Uranium in selected endorheic basins as partial analogue for spent nuclear fuel behaviour in salt. In: Côme B and Chapman NA (editors) Natural analogues in radioactive waste disposal. CEC Radioactive Waste Management Series, EUR 11037, 92-103, CEC, Luxembourg.

van Orden AC (1989) Corrosion mechanisms relevant to high-level waste repositories. Engineering Geology, 26, 331-349.

Vandiver PB (1994) Corrosion of synthesised glasses and glazes as analogs for nuclear waste glass degradation. Materials Research Society Symposium Proceedings, 333, (Scientific Basis for Nuclear Waste Management, XVII), 969-982.

Varoufakis G and Stathis EC (1971) A contribution to the study of the corrosion of ancient bronzes. Metallurgia, 5, 141-144.

Vieno T and Nordman H (1999) Safety Assessment of Spent Fuel Disposal in Hästholmen, Kivetty, Olkiluoto and Romuvaara TILA-99. Posiva Oy, Helsinki, Finland.

Vilks P and Bachinski DB (1994) Colloid studies. In: Cramer JJ and Smellie JAT (editors) Final report of the AECL/SKB Cigar Lake Analog Study. AECL Technical Report, AECL-10851, Whiteshell, Canada and SKB Technical Report, TR 94-04, Stockholm, Sweden.

Vilks P, Miller HG and Doern DC (1991) Natural colloids and suspended particles in the Whiteshell research area and potential effects on radiocolloid formation. Applied Geochemistry, 6, 565-574.

Vilks P, Cramer J, Bachinski DB, Doern DC and Miller HG (1993) Studies of colloids and suspended particles, Cigar Lake uranium deposit, Saskatchewan, Canada. Applied Geochemistry, 8, 605-616.

Vira J (1996) Natural analogues for canister performance. In: von Maravic H and Smellie J (editors) Natural analogue working group, sixth meeting, Santa Fe, September 1994. CEC Nuclear Science and Technology Report, EUR 16761, 163-174, CEC, Luxembourg.

Vitruvius (27 BC - 14 AD) De Architectura. Translated as: 'The Ten Books on Architecture', Morgan HH (1960) Dover Publications, New York.

von Maravic H and Smellie JAT (1994, editors) Natural Analogue Working Group, fifth meeting, Toledo, October 1992. CEC Nuclear Science and Technology Report, EUR 15176, CEC, Luxembourg.

von Maravic H and Smellie JAT (1996, editors) Natural Analogue Working Group, sixth meeting, Santa Fe, September 1994. CEC Nuclear Science and Technology Report, EUR 15176, CEC, Luxembourg.

von Maravic H and Smellie JAT (1997, editors) Natural Analogue Working Group, seventh meeting, Stein am Rhein, October 1996. CEC Nuclear Science and Technology Report, EUR 17851, CEC, Luxembourg.

von Maravic H and Alexander WR (2000, editors, in press) Natural analogue working group, eighth meeting, Strasbourg, March 1999. EC Nuclear Science and Technology Report, EC, Luxembourg.

Vovk IF (1987) Some geochemical and mineralogical peculiarities of deposits of radioactive materials as evidence for radiolysis in nature. In: Côme B and Chapman NA (editors) Natural analogues in radioactive waste disposal.

CEC Radioactive Waste Management Series, EUR 11037, 205-216, CEC, Luxembourg.

Vovk IF (1988) The IAEA report on the role of natural analogues in performance assessment. In: Côme B and Chapman NA (editors) Natural analogue working group, third meeting, Snowbird, June 1988. CEC Nuclear Science and Technology Report, EUR 11725, 141-145, CEC, Luxembourg.

Waber N (1990) Mineralogy petrology and geochemistry of the Poços de Caldas analogue study sites, Minas Gerais, Brazil II: Morro do Ferro. SKB Technical Report, TR 90-12, SKB, Stockholm; Nagra Technical Report, NTB 90-21, Nagra, Wettingen, Switzerland; UK DoE Technical Report, WR 90-043.

Waber N, Schorscher HD, MacKenzie AB and Peters T (1990) Mineralogy petrology and geochemistry of the Poços de Caldas analogue study sites, Minas Gerais, Brazil I: Osamu Utsumi uranium mine. SKB Technical Report, TR 90-11, SKB, Stockholm; Nagra Technical Report, NTB 90-20, Nagra, Wettingen, Switzerland; UK DoE Technical Report, WR 90-042.

Waber N, Schorscher HD and Peters T (1992) Hydrothermal and supergene uranium mineralisation at the Osamu Utsumi mine, Poços de Caldas, Minas Gerais, Brazil. Journal of Geochemical Exploration, 45, 53-112.

Werme L (1990) Near-field performance of the advanced cold process canister. SKB Technical Report, TR 90-31, SKB, Stockholm, Sweden.

Werme L, Sellin P and Forsyth R (1990) Radiolytically induced oxidative dissolution of spent nuclear fuel. SKB Technical Report, TR 90-08, SKB, Stockholm, Sweden.

Werme L, Sellin P and Kjellbert N (1992) Copper canisters for nuclear high level waste disposal Corrosion aspects. SKB Technical Report, TR 92-26, SKB, Stockholm, Sweden.

West JM and McKinley IG (1984) The geomicrobiology of nuclear waste disposal. Materials Research Society Symposium Proceedings, 26, (Scientific Basis for Nuclear Waste Management, VII), 487-494.

West JM, McKinley IG and Vialta A (1989) The influence of microbial activity on the movement of uranium at Osamu Utsumi, Poços de Caldas, Brazil. Materials Research Society Symposium Proceedings, 127, (Scientific Basis for Nuclear Waste Management, XII), 771-777.

West JM, Christofi N, Philp JC and Arme SC (1986) Investigations on the populations of introduced and resident micro-organisms in deep repositories and their effects on containment of radioactive wastes. CEC Nuclear Science and Technology Report, EUR 10405, CEC, Luxembourg.

West JM, Vialta A and McKinley IG (1990) Microbiological analysis at the Osamu Utsumi and Morro do Ferro analogue study sites, Poços de Caldas, Brazil. SKB Technical Report, TR 90-19, SKB, Stockholm, Sweden; Nagra Technical Report, NTB 90-28, Nagra, Wettingen, Switzerland; UK DoE Technical Report, WR 90-050.

West JM, Degueldre C, Allen M, Brütsch R, Gardner SJ, Ince S and Milodowski AE (1992) Microbial and colloidal populations in the Maqarin groundwaters. In: Alexander WR (editor) A natural analogue study of cement-buffered hyperalkaline groundwaters and their interaction with a sedimentary host rock - I: Source-term description and geochemical code database validation. NAGRA Technical Report Series (NTB 91-10), Nagra, Wettingen, Switzerland.

Wetton PD, Pearce JM, Alexander WR, Milodowski AE, Reeder S, Wragg J and Salameh E (1998) Production of colloids at the cement/host rock interface. In: Smellie JAT (editor) Maqarin Natural Analogue Study: Phase III. SKB Technical Report. TR 98-04. SKB, Stockholm, Sweden.

White WS (1968) The native-copper deposits of northern Michigan. In: Ridge JD (editor) Ore deposits of the United States. American Institute of Mining and Metallurgy and Petroleum Engineers, 303-325.

Wiborgh M, Höglund LO and Pers K (1986) Gas formation in a L/ILW repository and gas transport in the host rock. Nagra Technical Report, NTB 85-17, Nagra, Wettingen, Switzerland.

Wieland E (1997) Colloid concentrations in cementitious backfill: monokorn morter and quartz in contact with hyperalkaline cement pore water. Unpublished Internal Technical Note, Paul Scherrer Institute, Villigen, Switzerland.

Winberg A and Stevenson D (1994) Hydrogeological modelling. In: Cramer JJ and Smellie JAT (editors) Final report of the AECL/SKB Cigar Lake Analog Study. AECL Technical Report, AECL-10851, Whiteshell, Canada and SKB Technical Report, TR 94-04, Stockholm, Sweden.

Wilson TRS, Thomson J, Colley S, Hydes DJ, Higgs NC and Sorensen J (1985) Early organic diagenesis: the significance of progressive subsurface oxidation fronts in pelagic sediments. Geochimica et Cosmochimica Acta, 49, 811-822.

Wilson TRS, Thomson J, Hydes DJ, Colley S, Culkin F and Sorensen J (1986) Oxidation fronts in pelagic sediments: diagenetic formation of metal-rich layers. Science, 232, 972-975.

Winograd IJ (1986) Archaeology and public perception of a trans-scientific problem: disposal of toxic wastes in the unsaturated zone. US Geological Survey Open File Report, 86-136.

Wollenberg HA and Flexser S (1984) Contact zones and hydrothermal systems as analogues to repository conditions. In: Smellie JAT (editor) Natural analogues to the conditions around a final repository for high level radioactive waste. Proceedings of the natural analogue workshop held at Lake Geneva, Wisconsin, USA. SKB Technical Report, TR 84-18, SKB, Stockholm, Sweden.

Wollenberg HA, Brookins DG, Cohen LH, Flexser S, Abashain M, Murphy M and Williams AE (1984) Uranium thorium and trace elements in geologic occurrences as analogues of nuclear waste repository conditions. In: Alexander DH and Birchard GF (editors) NRC Nuclear Waste Geochemistry '83, NUREG / CP-0052, Washington DC, USA.

Wood WW and Ehrlich GG (1978) Use of baker's yeast to trace microbial movement in groundwater. Ground Water, 16, 398-403.

Yamakawa M (1991) Geochemical behaviour of natural uranium-series nuclides in geological formation. In: Advanced Nuclear Energy Research, Global Environment and Nuclear Energy. 3rd International Symposium Ibaraki, Japan.

Yanase N and Isobe H (1991) Uranium distribution in mineral phases of rock extracted by a sequential procedure: effect of reagents on the minerals. In: Duerden P (editor) Alligator Rivers Analogue Project: Progress report December 1989 to February 1990. Australian Nuclear Science and Technology Organisation, Sydney, Australia.

Yanase N and Isobe H (1991) Uranium distribution in mineral species of rock by a sequential extraction procedure. In: Duerden P (editor) Alligator Rivers Analogue Project: Second Annual Report 1989 to 1990. Australian Nuclear Science and Technology Organisation, Sydney, Australia.

Yoshida H (1994) Relation between U-series nuclide migration and microstructural properties of sedimentary rocks. Applied Geochemistry, 9, 479-490.

Young JF (1995) Engineering advanced cement-based materials for new applications. In: Aguado A, Gettu R and Shah SP (editors) Concrete technology: new trends, industrial applications. E & FN Spon, London, UK.

Yusa Y, Kamei G and Arai T (1991) Some aspects of natural analogue studies for assessment of long-term durability of engineered barrier materials - recent activities at PNC Tokai Japan. In: Côme B and Chapman NA (editors) Natural analogue working group, fourth meeting and Poços de Caldas project final meeting, Pitlochry, June 1990. CEC Nuclear Science and Technology Report, EUR 13014, 215-232, CEC, Luxembourg.

Zhou ZH, Fyfe WS and Tazaki K (1987) Glass stability in the marine environment. In: Côme B and Chapman NA (editors) Natural analogues in radioactive waste disposal. CEC Radioactive Waste Management Series, EUR 11037, 153-164, CEC, Luxembourg.

Zobell CE and Molecke MA (1978) Survey of microbial degradation of asphalts with notes on relationship to nuclear waste management. Sandia Laboratories Technical Report, SAND-78-1371, Albuquerque, USA.

Index

D

E

F

G

H

I

J

K

L

M

N

O

P

Contents

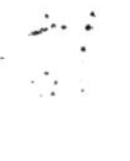

Preface

Xining, Qinghai Province, People's Republic of China (PRC), was the site of the International Conference on Continental Dynamics and Environmental Change of the Tibetan Plateau, 18–21 June 2006, followed by a field trip from Xining to Dunhuang from 22 to 24 June. The meeting was sponsored by The Chinese Academy of Sciences, the Province of Qinghai, and the Geological Society of America (GSA). This marked a milestone because it was the first meeting cosponsored by GSA in China. The meeting was opened by short introductions of welcome by Hu Xian Lai, secretary general of Qinghai Province; Jack Hess, executive director of GSA; B.C. Burchfiel, co-chair of the scientific committee; and Erchie Wang, conference secretary general. The meeting attracted 96 participants, 31 from outside of China. The diversity of papers and backgrounds of the participants led to an extraordinary experience centered around sharing data and ideas and discussing a broad range of interpretations; topics incorporated the many subdisciplines of the Earth Sciences focused on understanding the processes of Tibetan Plateau evolution. Meeting presentations covered a wide range of subjects, such as tectonics, sedimentation, petrology, geodesy, geophysics, dynamic modeling, magmatism, and climatic change. Two of the important aspects of the meeting were the large number of young Chinese scientist attendees and discussions and friendships fostered between Chinese and foreign scientists.

Xining lies within one of the many nonmarine Cenozoic basins bounded by mountain ranges of pre-Cenozoic rocks that form part of a Cenozoic deformational province that is currently being incorporated into the NE part of the Tibetan Plateau. Thus, Xining was a perfect venue for the conference, located within an area that displays the great variety of features that are forming the plateau today. It also lies close to a transition between the large Qsidam Basin and linear Qilian Shan ranges in the west and the smaller arcuate mountain ranges enclosing rhomb-shaped basins to the east. Global positioning system (GPS) and longer-term geological data suggest that this pattern of basins and ranges is the result of general NE-SW shortening in the west and progressively more E-W shortening and strike slip in a transpressional environment to the east. A number of the tectonic features of the NE Tibetan Plateau were visited during field trips both during and after the meeting.

In this volume, Meng and Fang present an overview of the sedimentary and tectonic evolution for the Qsidam Basin and its adjacent mountain ranges in the western part of the NE Tibetan Plateau, which reveal that the northern Tibetan Plateau expanded by means of stepwise propagation of separate fold-and-thrust belts and simultaneous development of intervening basins. The Qsidam Basin originated by fragmentation from the Tarim Basin, whereas other smaller basins behaved as piggyback basins. These basins were filled with sediments at rates that kept pace with uplift of adjacent deforming structural belts. Ongoing deformation of sediments should make the basins evolve into folded domains that will be integrated into the Tibetan Plateau, perhaps similar to Paleogene basins present in the central part of the plateau to the south.

The Altyn Tagh fault zone, one of the longest and most active strike-slip fault zones in the world, bounds the northern margin of the Tibetan Plateau and has a tectonic development that varies along strike. Wang et al. describe one such tectonic variation, interpreting the Xorkuli Basin as an extensional basin formed by vertical-axis bending of the adjacent Altyn Tagh mountain range to the north. The basin formed within a dilatent nose of a large fold in the left-lateral Altyn Tagh shear system. The amount of left-lateral shear absorbed by this fold and basin formation is about 60 km. On the south side of the basin, across a branch of the Altyn Tagh fault zone, shortening formed folds and thrust faults that shortened around a horizontal axis and are very discordant to the structures on the north side of the fault zone. The contrast of deformational style on the two sides of the Altyn Tagh fault zone is perhaps a unique feature of this segment of the fault zone.

The Ailao Shan shear zone in SE China is another of the great fault zones related to formation of the Tibetan Plateau. This shear zone bounds the north side of the Indochina lithospheric fragment extruded to the SE from SE Tibet during early Cenozoic time. Burchfiel et al. show that the Ailao Shan shear zone has a very complex geometry and cannot be considered to be a simple through-going shear zone from China into the Day Nui Con Voi shear zone in Vietnam. The two shear zones are en echelon, and the Day Nui Con Voi shear zone forms an antiform where it enters China that plunges beneath Mesozoic rocks belonging to the South China Yangtze platform. The contact between the Mesozoic rocks and the mylonitic rocks of the Day Nui Con Voi is interpreted to be a low-angle, north-dipping, left-lateral transpressional shear zone, bounded at the top by a low-angle contact that has a component of normal down-to-the-north movement as the hot weak lower crust extrudes upward. Late Cenozoic brittle right-lateral displacement along the Red River fault zone has disrupted the antiform, folding and displacing the early Cenozoic shear zone.

GPS studies have shown the dramatic pattern of present-day crustal movements across the Tibetan Plateau and around the Eastern Himalayan syntaxis. Zhang and Gan present a summary of evidence that compares short-term GPS data and long-term data from neotectonic studies; results indicate that crustal deformation differs significantly from plate-like models that require large slip rates of 20–30 mm/yr along major block-bounding strike-slip faults. They further suggest that present-day tectonics in the Tibetan Plateau are best described as a result of deformation on a network of faults that resulted from deformation within a continuous medium at depth. Regions of high rigidity behave with rigid block-like movements, and those of low rigidity are dominated by continuous deformation. Rheological flow in the lower crust and upper mantle plays an important role in controlling deformation of the upper crust, and present-day tectonic deformation of continental China can be described in terms of a combined model of rigid block movement and continuous deformation.

The GPS studies also show large areas of active rotation of modern crustal movements in the Tibetan Plateau, and Dupont-Nivet et al. present data that show that rotations have also been important at various times during the Cenozoic. Paleomagnetic results throughout Paleogene and Neogene strata in the Xining Basin, recently dated using magnetostratigraphy between 52 and 17 Ma, show that some 25° of clockwise rotation with respect to the stable Eurasian continent occurred ca. 41 Ma. In view of the regional compilation of existing paleomagnetic data from the northeastern Tibetan Plateau, these results suggest that this region experienced phases of clockwise rotations that took place in the regional Paleocene-Miocene basin system, including rotation in the Xining Basin at ca. 41 Ma, thus establishing the existence of widespread deformation at this time. During a mid-Miocene phase, between 17 and 11 Ma, clockwise rotations were restricted to the Miocene-Quaternary basin system, implying that the Laji Shan thrust belt, separating two basin systems, was active during this time interval.

A focus of many recent Tibetan studies has been the nature of the crust and lithosphere below Tibet and its relation to the shallow crust and Tibetan landscape. Jin and Wang review the variations in lateral compensation of two-dimensional (2-D) transects from the lowlands to the highlands around the plateau, the lateral compensation and the weakening of the Indian plate subducting beneath the Tibetan Plateau in the south, the possible support of the Tarim lithosphere underplating the northern edge of the Tibetan Plateau, and the possible lateral compensation style beneath the eastern plateau associated with the western foredeep of the Sichuan Basin. The lateral variation of lithosphere strength is consistent with the irregular geometry of the tectonic units over the Tibetan Plateau and its vicinity based on three-dimensional (3-D) flexural modeling in a spatial domain. Rheological modeling with a simple thermal structure at the various locations over the Tibetan Plateau reveals that the base of the strong upper crust of Tibet is at some 30–35 km and that it tends to move on a flowing weak, ductile lower crust (or middle and lower crust), and depends on the thickness of the weak lower crust. The thicker the weak lower crust is, the easier the upper crust flows relative to the strong upper mantle lithosphere; thus, the Moho geometry plays a key role in the tectonics of Tibet.

Extension within obliquely convergent zones is a feature common in many mountain ranges, and it takes a variety of different forms. Studies have shown the Tibetan Plateau has many different types of extensional features, often related to complex relations between shortening and strike-slip movement, and the study by Nadimi and Nadimi describes late Cenozoic to active extensional structures that expose very deep parts of the crustal section within the Zagros Mountains of Iran. They show an evolutionary history that begins with major shortening by thrusting that exposed basement rocks from below a 6–15 km sedimentary section, followed by extensional faulting after middle Miocene time that led to more extensive exposures of deep crustal rocks within horsts and grabens parallel to the shortening direction. These events were followed by strike-slip movements active from Pliocene to present day. The sequence of events is similar to that found in parts of Tibet.

It is unfortunate that more papers were not submitted to this volume from this conference as there was a wide range of many very important papers presented. Many of the Chinese scientists did not submit papers as they have been requested not to submit papers for publications not cited by the Citation Index. Since most of the papers were presented by Chinese scientists, this greatly reduced the number of papers submitted. Nevertheless, the meeting was a major success in bringing an international audience to China under the auspices of the Geological Society of America and the Chinese Academy of Sciences to exchange data and views with their Chinese colleagues on Tibetan geology We hope that this represents the beginning of closer relations between American and Chinese geoscientists.

B. Clark Burchfiel (Co-Chair of the Scientific Committee)
Department of Earth, Atmospheric and Planetary Sciences
Massachusetts Institute of Technology
Cambridge, Massachusetts, USA

Erchie Wang (Conference Secretary General)
Institute of Geology and Geophysics
Chinese Academy of Sciences
Beijing, 100029, PR China

The Geological Society of America
Special Paper 444
2008

Cenozoic tectonic development of the Qaidam Basin in the northeastern Tibetan Plateau

Qing-Ren Meng
Institute of Geology and Geophysics, Chinese Academy of Sciences, Beijing 100029, China

Xiang Fang
Institute of Petroleum Exploration and Development, China National Petroleum Company, Beijing 100083, China

ABSTRACT

The Qaidam Basin constitutes a major portion of the northeastern Tibetan Plateau, and an understanding of its tectonic development will help decipher how the Tibetan Plateau was formed. This study presents key subsurface data in conjunction with observations and analysis of the stratigraphic and sedimentary evolution to reconstruct Cenozoic tectonic history of the Qaidam Basin. We show that Late Cretaceous–Paleocene deposits of the southwestern Qaidam Basin can be well correlated with their counterparts of the southwestern Tarim Basin, implying that the two regions were originally connected or were in the same depositional basin during that period of time. The Qaidam Basin commenced subsiding due to crustal shortening in the Eocene, and it has subsequently evolved into an independent basin since the Miocene. The main depocenter was noticeably persistent in the middle of the western Qaidam Basin from Eocene to Miocene time, and then it shifted to the east. On the basis of spatial stratigraphic correlation and restoration of sedimentary processes, we surmise that there existed a proto–Qaidam Basin during the Paleogene, where the Suhai and Kumukol Basins represent its northern and southern margins, respectively. The Suhai and Kumukol Basins were subsequently isolated from the Qaidam Basin as a result of basinward thrusting in basin-margin areas.

We suggest that the Qaidam Basin was generated as a result of crustal buckling or folding, manifesting itself as a synclinal depression. The crustal folding model can account for a number of observations, including localization of the depocenter in the middle of the basin, nearly concomitant deformation on the southern and northern sides of the Qaidam Basin, occurrence of major high-angle reverse faults at basin margins, and generation of adjacent intermontane Suhai and Kumukol Basins. It is further shown that the western Qaidam Basin experienced three distinct stages: the first stage was characterized by a simple synclinal depression; the second stage was marked by occurrence of reverse faults at inflection points of the megafold and continuous subsidence in the middle of the basin; and the third stage featured intrabasinal deformation and uplift. The eastern Qaidam Basin underwent a diverse evolution and became the main depositional area in the Quaternary. A tectonic model is accordingly advanced to illustrate Cenozoic tectonics of the Qaidam Basin.

Keywords: Tibetan Plateau, Qaidam Basin, Cenozoic, tectonics, crustal folding.

Meng, Q.-R., and Fang, X., 2008, Cenozoic tectonic development of the Qaidam Basin in the northeastern Tibetan Plateau, *in* Burchfiel, B.C., and Wang, E., eds., Investigations into the Tectonics of the Tibetan Plateau: Geological Society of America Special Paper 444, p. 1–24, doi: 10.1130/2008.2444(01). For permission to copy, contact editing@geosociety.org.

INTRODUCTION

The Tibetan Plateau is generally accepted to have formed as a result of collision of the Eurasian and Indian continents around 50 Ma (Patriat and Achache, 1984), but the way(s) in which it grew vertically and expanded laterally has been a matter of debate (Tapponnier et al., 1986; Houseman and England, 1993). Cenozoic tectonics of the northeastern Tibetan Plateau are considered to be the consequence of the far-field effect of continued Eurasia–India convergence (Tapponnier et al., 2001). Previous studies have centered primarily on crustal and lithospheric deformations of mountain belts, such as the Karakorum (Searle, 1996; Lacassin et al., 2004), the eastern Kunlun (Burchfiel et al., 1989b; Van der Woerd et al., 1998; Mock et al., 1999; Jolivet et al., 2003; Fu and Atwata, 2007), the Altyn Tagh fault (Peltzer et al., 1989; CSBS, 1992; Meyer et al., 1998; Sobel et al., 2001; Yin et al., 2002; Cowgill et al., 2003; Liu et al., 2007), and the Qilian (Tapponnier et al., 1990; Ding et al., 2004).

The Qaidam Basin constitutes a considerable portion of the northeastern Tibetan Plateau (Fig. 1), and it is bounded on all sides by active faults and structural belts (Fig. 2). The Qaidam Basin is expressed as a huge intermontane basin and is regarded to have originated from flexural subsidence due to basinward displacement of the eastern Kunlun on the south and the southern Qilian on the north (Gu and Di, 1989; Wang and Coward, 1990; Tapponnier et al., 1990; Meyer et al., 1998). Hsü (1988) postulated that the Qaidam Basin evolved in a back-arc setting and was floored with oceanic crust. A number of studies were conducted in an attempt to unravel the tectono-sedimentary development of the Qaidam Basin (Hsü, 1988; Wang and Coward, 1990; Zhu et al., 1994; Huang et al., 1996; Métivier et al., 1998; Xia et al., 2001) but arrived at different conclusions. Xia et al. (2001) proposed that the Qaidam Basin experienced a

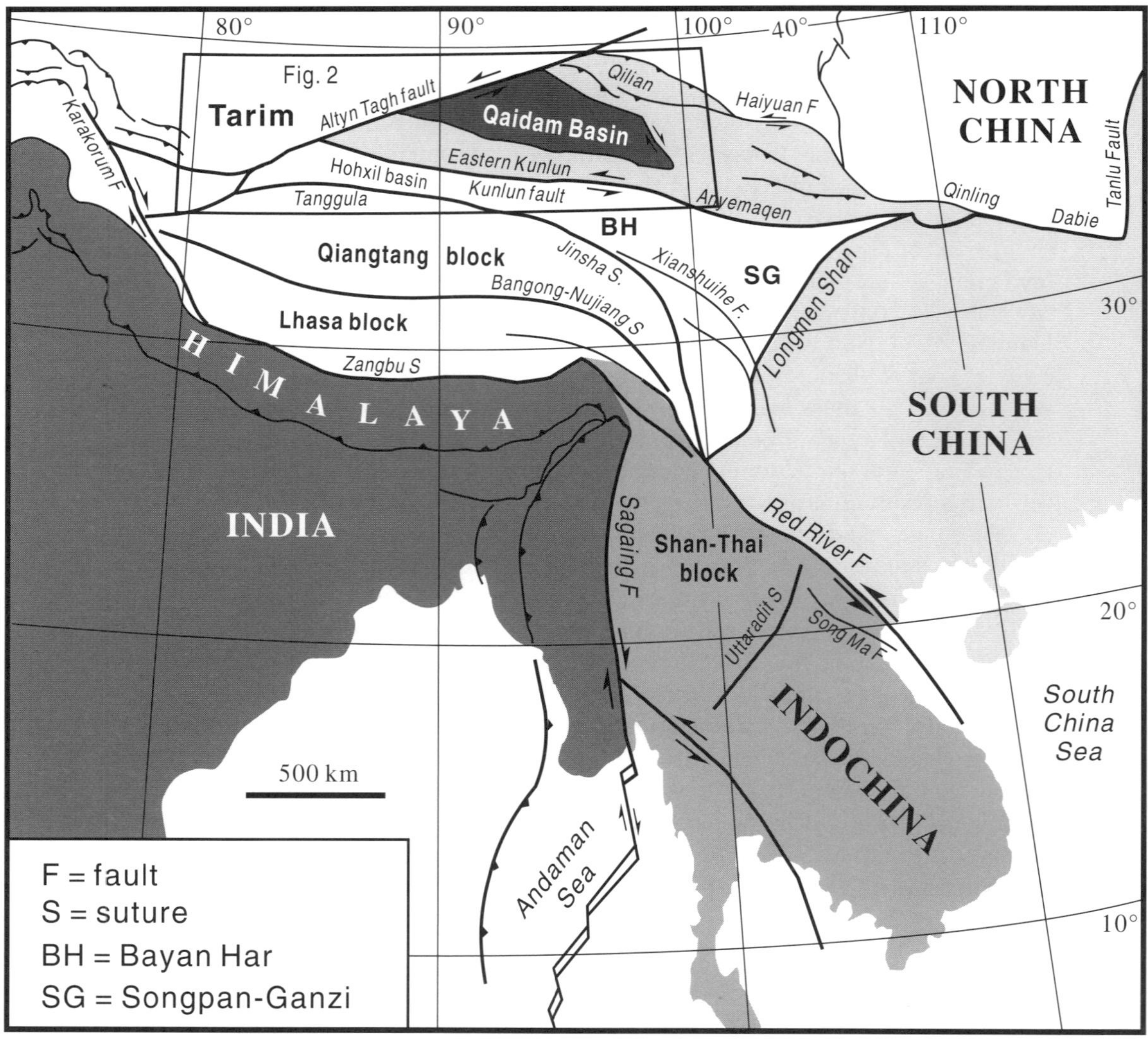

Figure 1. Simplified tectonic map of the Tibetan Plateau and its adjacent regions, showing the position of the Qaidam Basin and its present tectonic setting.

two-stage evolution, a first-stage in an extensional setting in the early Tertiary and a second stage of basin inversion. Other studies, however, showed that the Qaidam Basin evolved in a contractional setting throughout the Cenozoic that was controlled predominantly by basinward displacement of fold-and-thrust belts on both the northern and southern sides (Gu and Di, 1989; Song and Wang, 1993; Chen et al., 1999; Yin et al., 2002). Cenozoic sedimentary successions can be up to 17 km thick in the middle of the basin (Gu and Di, 1989), but dispersal processes of sediments are debated (Métivier et al., 1998; Wang et al., 2006). Métivier et al. (1998) claimed that sediments mainly originated from adjacent rising structural belts, and deposition in the Qaidam Basin was described as a "bathtub infilling" process. Wang et al. (2006), however, conjectured that most of the sediments were transported into the basin a great distance from the western Kunlun by a paleoriver, which was subsequently eliminated as a result of the left-lateral slip and shortening along the Altyn Tagh fault. Studies by Rieser et al. (2005, 2006) using $^{40}Ar/^{39}Ar$ dating of detrital muscovites were recently carried out to constrain the provenance of basin fill, and they showed that the Altyn Tagh, the Qiman Tagh, and the southern Qilian were all possible sources for the sedimentary fill.

Although the Cenozoic Qaidam Basin has been extensively studied, numerous questions remain unresolved. One of the prominent uncertainties is the driving mechanism of subsidence of the basin. This study presents some key subsurface data, including basin-scale geological profiles, isopach maps of different stratigraphic units, and local seismic sections, which address the tectonic development of the Qaidam Basin. Our study suggests that the Qaidam Basin was created as a consequence of crustal buckling in horizontal compression, and the buckling model can well account for both the tectonic subsidence and depositional processes within the basin.

REGIONAL GEOLOGY

The Qaidam Basin is one of the major Cenozoic oil-bearing basins of China (Gu and Di, 1989; Huo, 1990; Wang and Coward, 1990; Tang et al., 2000), and petroleum explorations of the basin have yielded a wealth of information about its internal structures, stratigraphy, and depositional history. The present-day Qaidam Basin covers a rhomb-shaped area of ~120,000 km^2, and it has an elevation of ~3000 m. The original framework and position of the basin, however, are not well understood due to strong crustal shortening in northeastern Tibet during the late Cenozoic (Burchfiel et al., 1989a; Tapponnier et al., 1990) and its large-magnitude NE-directed lateral displacement (Ritts and Biffi, 2000; Meng et al., 2001; Wang et al., 2006). The nature of the basement of the basin remains uncertain because few drilling holes have reached the basement rocks in the middle of the Qaidam Basin. Proterozoic crystalline rocks crop out in the adjoining mountains, such as the Dakendaban Group in the southern Qilian (or sometimes called the Nan Shan), which might represent the metamorphic basement of the basin. Early Paleozoic basement of the Qaidam Basin has either been regarded as an accretionary complex (Hsü et al., 1995) or interpreted as a magmatic arc system (Gehrels et al., 2003). Early Paleozoic suture zones have been recognized in the mountain belts around the Qaidam Basin, such as in the northern Qilian (Xu et al., 2000; Xia et al., 2003), at the northern edge of the Qaidam Basin (Yang et al., 2002), along the Altyn Tagh (Sobel and Arnaud, 1999), and in the eastern Kunlun (Yang et al., 1996; Bian et al., 2004), indicating that the older rocks in the northeastern Tibetan Plateau are made up of mosaic terranes that were amalgamated at the end of the early Paleozoic. Late Paleozoic time saw a period of widespread marine transgression, as characterized by Carboniferous shallow-marine limestone. The Upper Permian and Triassic strata are mostly missing in the

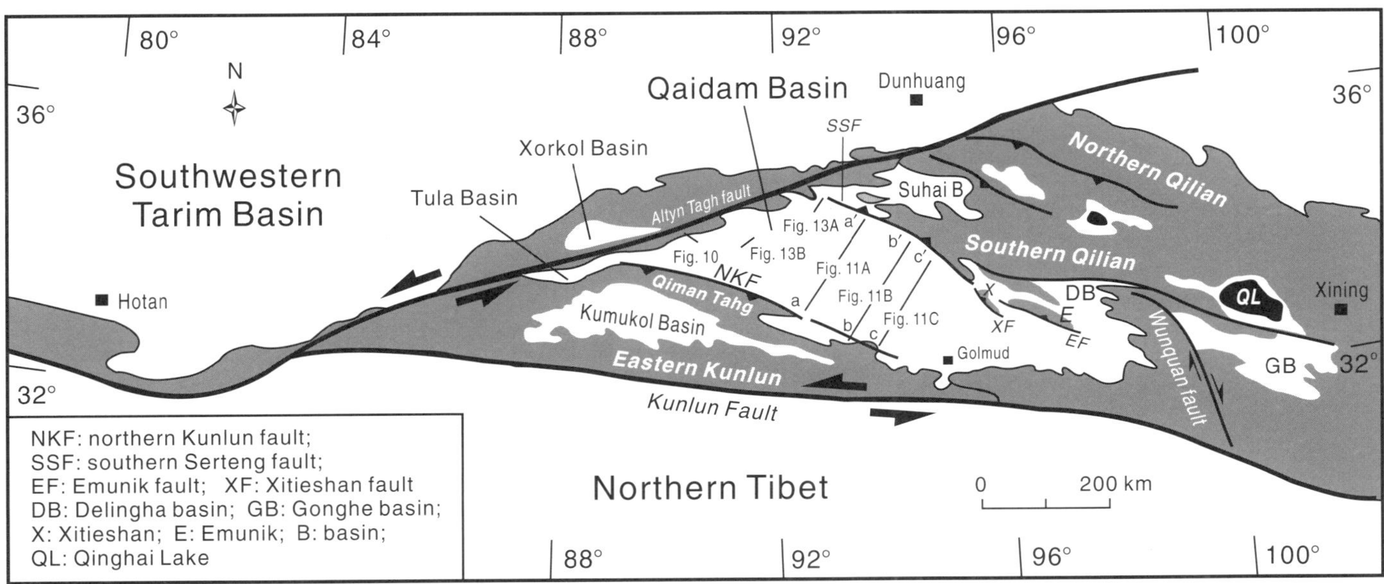

Figure 2. Tectonic sketch showing the Qaidam Basin and its relationship with adjacent intermontane basins.

Qaidam Basin, and their absence is ascribed to the Indosinian crustal shortening event (Tang et al., 2000), which has been usually regarded to have resulted from the closure of paleo-Tethyan Ocean in Triassic time (Roger et al., 2003).

The Mesozoic successions are composed of Jurassic and Cretaceous continental strata in the Qaidam Basin, but their tectonic settings and the basin evolution at that time have been controversial. The Mesozoic Qaidam Basin was considered to have formed in response to contractional deformation of the Qilian to the north (Ritts and Biffi, 2001), and therefore developed as a moderate flexural basin (Zhou et al., 2003). Other studies, however, have argued that the Qaidam Basin evolved in an extensional setting in the Mesozoic (Tang et al., 2000; Chen et al., 2003). Some seismic profiles across the northern Qaidam Basin provide some evidence in support of an extensional setting, showing that Lower–Middle Jurassic deposition was controlled by listric normal faults (Liu et al., 2004). The Upper Jurassic is mostly missing, whereas Cretaceous sedimentation occurred not only in the north but also in the western Qaidam Basin. The absence of Upper Jurassic strata might imply a crustal shortening event that led to the inversion of Early–Middle Jurassic extensional basins and regional uplift of the Qaidam Basin. Crustal extension resumed in Cretaceous time (Xia et al., 2001), and a marine transgression might have occurred in the southwestern part of the Qaidam Basin in Late Cretaceous time (Meng et al., 2001). Paleozoic and Mesozoic tectonics of the Qaidam Basin are still poorly known, and much additional work is needed.

Present-day framework and topographic expressions of the Qaidam Basin and adjacent areas are primarily the consequence of late Cenozoic deformation. The basin is now bounded on the north by a number of N-dipping reverse faults and on the south by S-dipping reverse faults (Fig. 2). These reverse faults have placed the Proterozoic and Mesozoic rocks over Cenozoic rocks as young as the Pleistocene, implying that the faults are still active. The left-lateral Altyn Tagh fault and right-lateral Wenquan fault developed on the northwestern and eastern sides of the Qaidam Basin, respectively (Fig. 2), and they are thought to have accommodated the northward indentation of the Qaidam Basin (Dupont-Nivet et al., 2004; Wang and Burchfiel, 2004). Cenozoic strata of the western Qaidam Basin have been folded, as displayed by surface traces of fold layers (Fig. 3) and seismic profiles. In contrast, the eastern Qaidam Basin has experienced little internal deformation. It is worth noting that there exist two subordinate basins, the Suhai Basin to the north and the Kumukol Basin (also called the Ayakkum Basin) to the south of the Qaidam Basin, respectively (Fig. 2). Evolution of these two basins is shown to have a bearing on the deformation of the Qaidam Basin, and it is addressed in detail in the following sections.

STRATIGRAPHY

The Cenozoic lithostratigraphy of the Qaidam Basin has long been established (Fig. 4). The Tertiary succession is divided into six formations, which are, from the base upward, the Lulehe, Xiaganchaigou (also termed as Xia Ganchaigou or Lower Ganchaigou in literature), Shangganchaigou (also termed as Shang Ganchaigou or Upper Ganchaigou), Xiayoushashan (also termed as Xia Youshashan or Lower Youshashan), Shangyoushashan (also termed as Shang Youshashan or Upper Youshashan), and Shizigou Formations, respectively. The Quaternary sequence is divided from bottom to top into the Qigequan, Chaidam, Sanhu, and Budaxun Formations, respectively (Shen et al., 1993). The age of individual units, however, has not been well established, particularly for the lower Tertiary units (Xia et al., 2001; Qiu et al., 2003). Previous studies used to simply correlate lithostratigraphic units with chronostratigraphic epochs because of the lack of absolute age constraints. The currently used Cenozoic stratigraphy is mainly based upon the study of lithofacies sequences in the western Qaidam Basin, where the sedimentary sequences are dominated by fine-gained fluvial-lacustrine deposits that contain abundant fossils, such as ostracod, pollen, and spores (Huo, 1990). Pollen and ostracod assemblages have been widely used to subdivide and correlate lithostratigraphic units of the Qaidam Basin by Huang et al. (1996) and Yang et al. (2000).

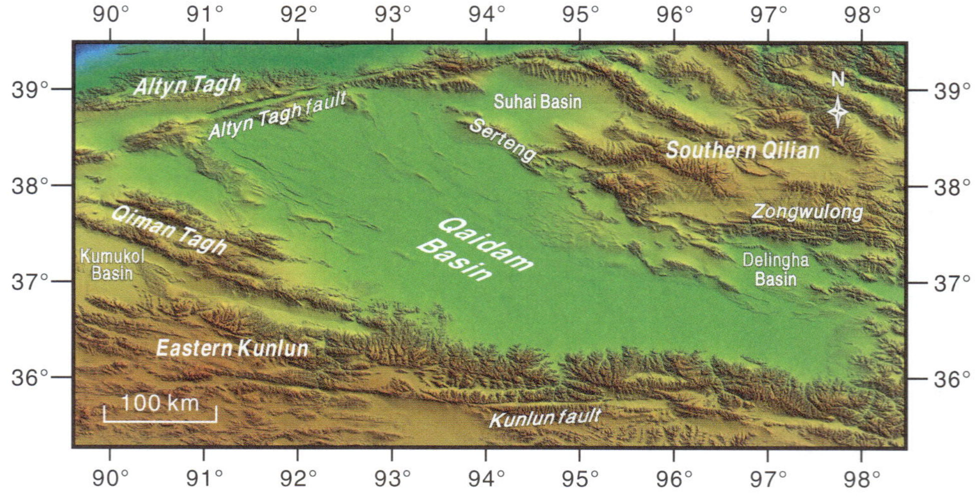

Figure 3. Digital elevation model (DEM) image showing topography of the Qaidam Basin and its surrounding mountains. Note that the western Qaidam has been folded, whereas the eastern Qaidam exhibits little deformation. The Altyn Tagh fault and the Kunlun fault are expressed by their strikingly linear traces.

Figure 5 displays the typical ostracod assemblages that are used to subdivide the lithostratigraphic units of the Qaidam Basin.

Liu et al. (1996) made an attempt to assign absolute ages to the lower Tertiary units by dating gypsum by using the fission-track method and by correlating stratigraphic units of the Qaidam Basin with the deep-sea oxygen isotope record. They claimed that the Lulehe Formation might have been deposited from the Late Cretaceous into Paleocene time, whereas the Xiaganchaigou Formation began since ca. 58 Ma and was followed by the Shangganchaigou Formation at ca. 41 Ma. The Shangganchaigou Formation was considered to span from the Bartonian of the Eocene to the end of Oligocene (Liu et al., 1996). In order to place tighter constraints on ages of the lithostratigraphic units, magnetostratigraphic studies have been carried out since the early 1990s. Yang et al. (1992) established the first Tertiary magnetostratigraphic column for the Qaidam Basin. Their preliminary study, however, provided no detailed database, and the validity of the age assignments to individual lithostratigraphic units cannot be accurately judged. Other magnetostratigraphic investigations were focused upon subdivision of Quaternary lacustrine sequences in the central and eastern Qaidam Basin to understand climatic changes (Liu et al., 1996, 1998; Zhu et al., 1994). Recently, Sun et al. (2005) carried out a detailed study of magnetostratigraphy of lower Tertiary sedimentary sequence in the western Qaidam Basin, showing that the Xiaganchaigou Formation corresponds to the subchrons C16n–C18, which range in age from older than 40 Ma to 35.5 Ma, and the Shangganchaigou Formation corresponds to subchrons C8.2n–C15, which range in age from 35.5 Ma to 26.5 Ma. The age of the Lulehe Formation is, unfortunately, not constrained because it has not been studied. Fang et al. (2007) reported magnetostratigraphic results for an Upper Tertiary to Pleistocene succession in the northeastern Qaidam Basin, and they concluded that the ages of the Xiayoushashan, Shangyoushashan, Shizigou, and Qigequan Formations are >15.3 Ma, 15.3–8.1 Ma, 8.1–2.5 Ma, and <2.5 Ma, respectively. The chronology of the Cenozoic stratigraphy in this study is thus primarily based upon the recent magnetostratigraphic studies of Sun et al. (2005) and Fang et al. (2007).

Subdivision and correlation of Cenozoic strata of the Qaidam Basin were mainly based upon pollen and ostracod

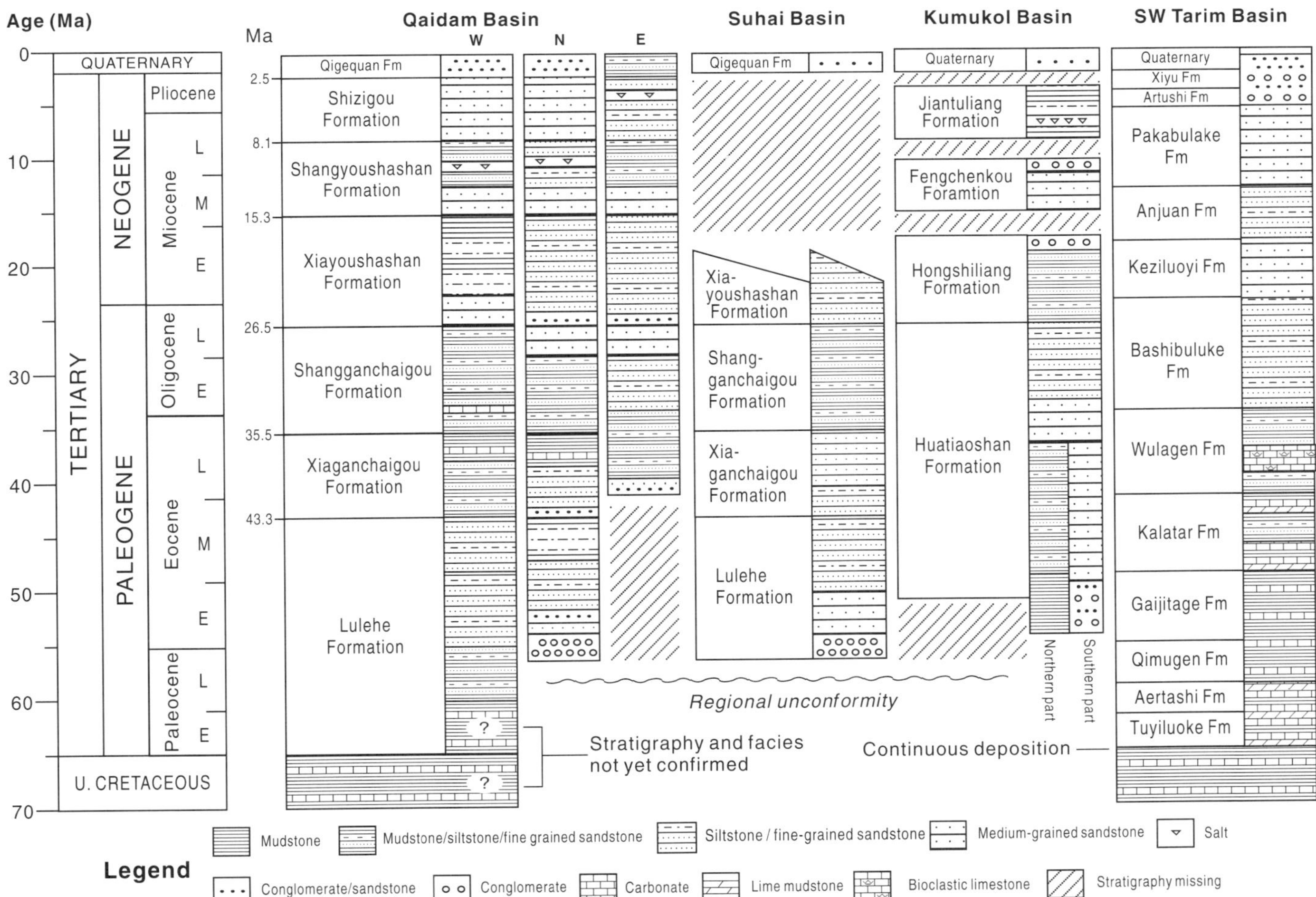

Figure 4. Stratigraphy of the Qaidam Basin and its correlation with the Suhai, Kumukol, and southwestern Tarim Basins, compiled from Tang et al. (1989), Wang et al. (1999), Hao et al. (2001), Sun et al. (2005), Xiao et al. (2005), and Fang et al. (2007).

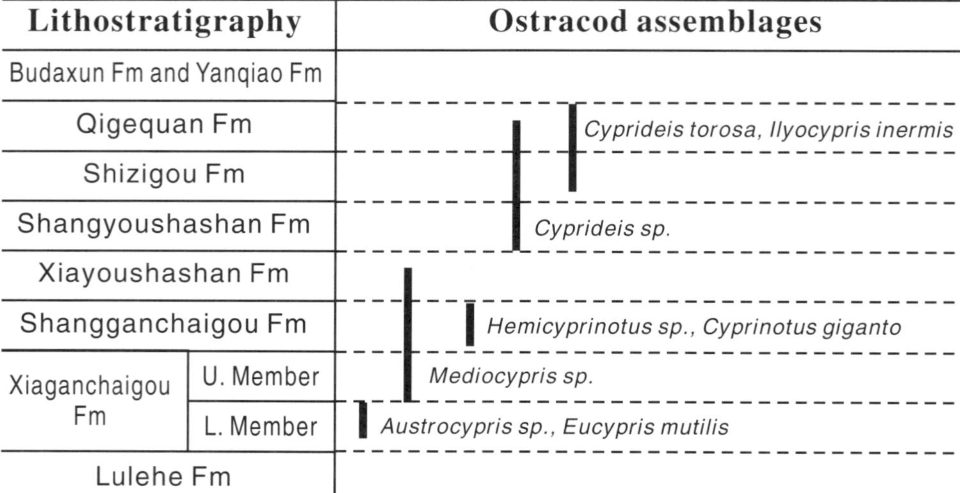

Figure 5. Subdivision of lithostratigraphic succession based on the ostracod biostratigraphy, compiled mainly from Huang et al. (1996), Yang et al. (2000), and Sun et al. (2005).

assemblages (Fig. 5). Basinwide correlations of Cenozoic strata are achieved by tracing seismic-reflection horizons of stratigraphic units in conjunction with borehole data (Zhou et al., 2006). The main boundaries of individual lithostratigraphic units have been fairly well defined by characteristic reflection horizons, and they are labeled from bottom to top as T_R, T_5, T_4, T_3, T_2, T_2', T_1, and T_0, respectively (Zhou et al., 2006). T_R represents a regional unconformity between the Cenozoic strata and the underlying rocks of different ages, and it is characterized by a high-amplitude and continuous reflection. T_5 marks the boundary between the Lulehe and Xiaganchaigou Formations in the western Qaidam Basin, but it merges with the T_R to the east, indicating the pinch-out of the Lulehe Formation toward the eastern Qaidam Basin. T_4 divides the Xiaganchaigou Formation into the lower and upper members, which possess distinct lithofacies in some localities.

Figure 4 represents a generalized stratigraphic chart that displays a correlation of stratigraphic and sedimentological evolution of the Qaidam Basin, adjacent Suhai and Kumukol Basins, as well as the southwestern Tarim Basin. Cenozoic stratigraphy is relatively complete in the western Qaidam Basin, but the lower Tertiary stratigraphy is not well developed in the east. In addition, the Paleocene–Eocene sequences of the northern Kumukol Basin show sedimentary facies associations similar to those of the western Qaidam Basin, and the Suhai Basin displays the same lithology for the Lulehe Formation as in the northern Qaidam Basin. It is worthy to note that the Lower Tertiary sequences are also quite similar in stratigraphy and lithology between the western Qaidam and the southwestern Tarim Basins (Fig. 4). A comparison of Neoproterozoic–Cenozoic stratigraphic sequences between the two basins also demonstrates their identical stratigraphic evolution (Fig. 6). Another striking fact is the occurrence of marine deposits in both the western Qaidam and the southwestern Tarim Basins in Late Cretaceous to Paleocene time (Tang et al., 1989; Hao et al., 2001; Meng et al., 2001), indicating that they must have been connected at that time.

SEDIMENTOLOGY

Cenozoic successions in the Qaidam Basin are composed predominantly of alluvial, fluvial, and lacustrine deposits, and they display marked spatial and temporal facies variations (Wang and Coward, 1990). A brief description is given here of sedimentary facies and facies associations of each lithostratigraphic unit, followed by interpretations of the corresponding depositional processes and environments.

Lulehe Formation

The Lulehe Formation represents the earliest deposits of Cenozoic successions of the Qaidam Basin, resting unconformably over underlying units of differing ages. Conglomeratic lithofacies mainly occur in the northern margin of the Qaidam Basin and mostly in the basal part of the formation. Conglomerate is usually thick-bedded matrix- and clast-supported, and it consists primarily of metamorphic gravels (Fig. 7A). Paleocurrent directions can be restored using pebble imbrication and cross-bedding, and they show that gravels originated from the Qilian in the north (Figs. 8A and 8B). Conglomeratic lithofacies pass upward into finer facies, such as red-colored and cross-bedded sandstone and siltstone. The Lulehe Formation in other parts of the Qaidam Basin, however, is dominated by parallel- and cross-stratified sandstone and siltstone (Fig. 7B). Dark-colored massive and horizontally laminated mudstones have been reached in a number of boreholes in the western Qaidam Basin, and these are considered as potential source rocks (Gu and Di, 1989). The conglomeratic facies are interpreted as debris-flow deposits on alluvial fans because of their massive structure, whereas the cross-stratified sandstone and siltstone facies were deposited in a fluvial environment. The dark-colored and horizontally laminated fine-grained facies in the western Qaidam Basin have usually been considered as products of lacustrine sedimentation, but Meng et al. (2001) reinterpreted them as marine deposits.

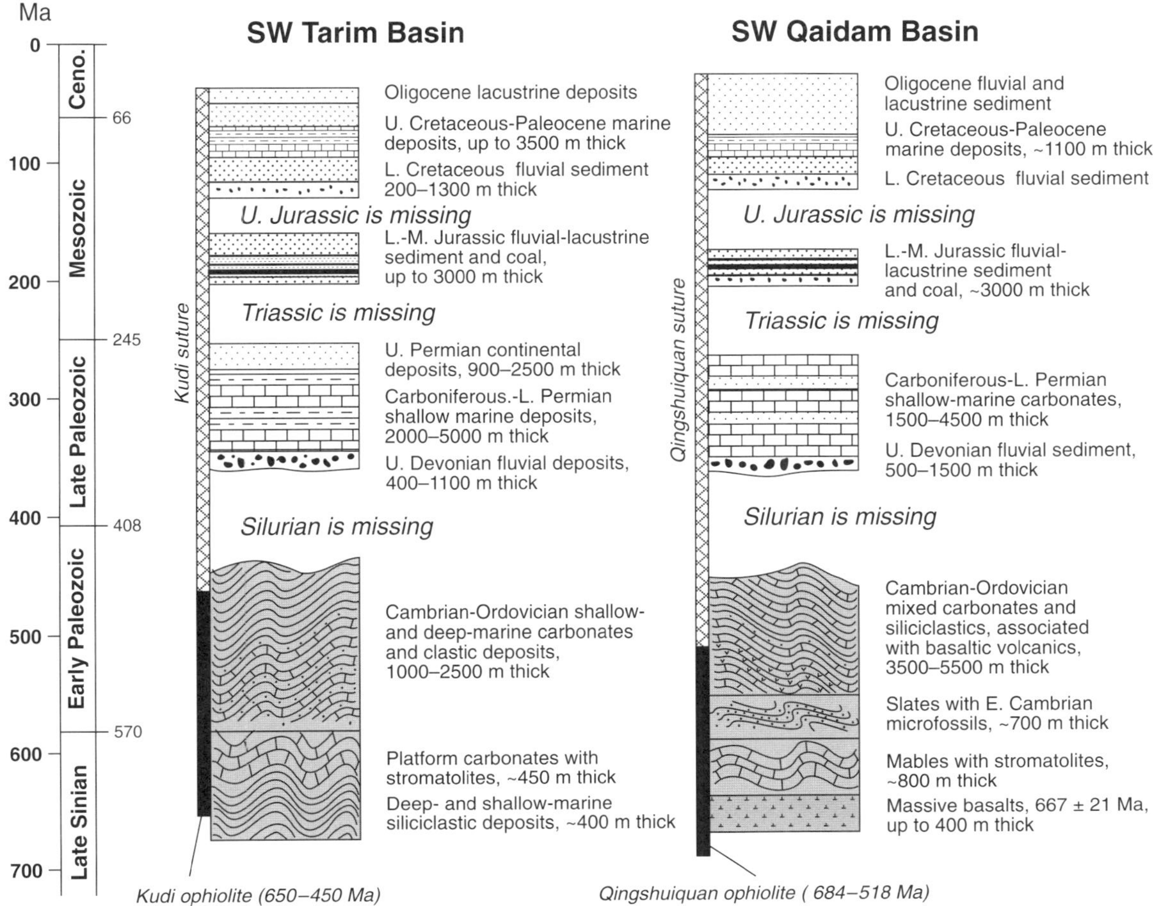

Figure 6. Composite stratigraphic columns and facies interpretation for the southwestern margin of the Tarim Basin and southwestern portion of the Qaidam Basin. Note that the two regions display remarkably similar stratigraphic and depositional sequences (from late Neoproterozoic to early Paleogene). (Modified from Meng et al., 2001.)

The Lulehe Formation only occurs in the western portion of the Qaidam Basin, and it pinches out toward the east (Fig. 9A). Although several depocenters can be identified by the spatial variations in thickness of the Lulehe Formation, the main depocenter, called the Yiliping depression, is located in the middle of depositional area (Fig. 9A). The Lulehe Formation also occurs in the Suhai Basin to the north (Fig. 2), where it is characterized by coarser proximal lithofacies similar to those in the northern Qaidam Basin (Fig. 4). The similarity in lithofacies implies that the Suhai Basin was presumably an integrated part of the proto–Qaidam Basin during Eocene time. In addition, the thickness of the Lulehe Formation, ~500 m, shows little change across the Serteng, as determined from both seismic profiles and borehole data, supporting the idea that the two basins were originally connected during deposition of the Lulehe Formation. The finer facies of the Lulehe Formation in the western Qaidam Basin is quite similar to its equivalents in both the northern Kumukol and the southwestern Tarim Basins (Fig. 4), implying that these three areas were probably not mutually separated in the early part of Paleogene time (Meng et al., 2001).

Xiaganchaigou Formation

The Xiaganchaigou Formation is usually divided into two members. The lower member is dominated by coarse-grained facies of conglomerate and coarse-grained sandstone, whereas the upper member is characterized by fine-grained facies such as siltstone, mudstone, and muddy limestone. The lower and upper members of the Xiaganchaigou Formation are fairly distinct in the northern Qaidam Basin but become less distinguishable in the west. Conglomerate in the lower member appears thick-bedded and matrix- and clast-supported, occasionally displaying low-angle cross-bedding, and it is mostly present at the base of the formation. Conglomeratic beds pass upward into cross-bedded sandstone and thin-bedded siltstone/mudstone, thereby exhibiting several stacked fining-upward sequences of different thickness. The fine-grained facies in the upper member are characterized by red thin-bedded ripple-laminated siltstone and mudstone, and they contain thick- and medium-bedded sandstone interlayers with limited lateral extent. The coarse-grained facies of the lower member is interpreted as fluvial channel deposits, as indicated by

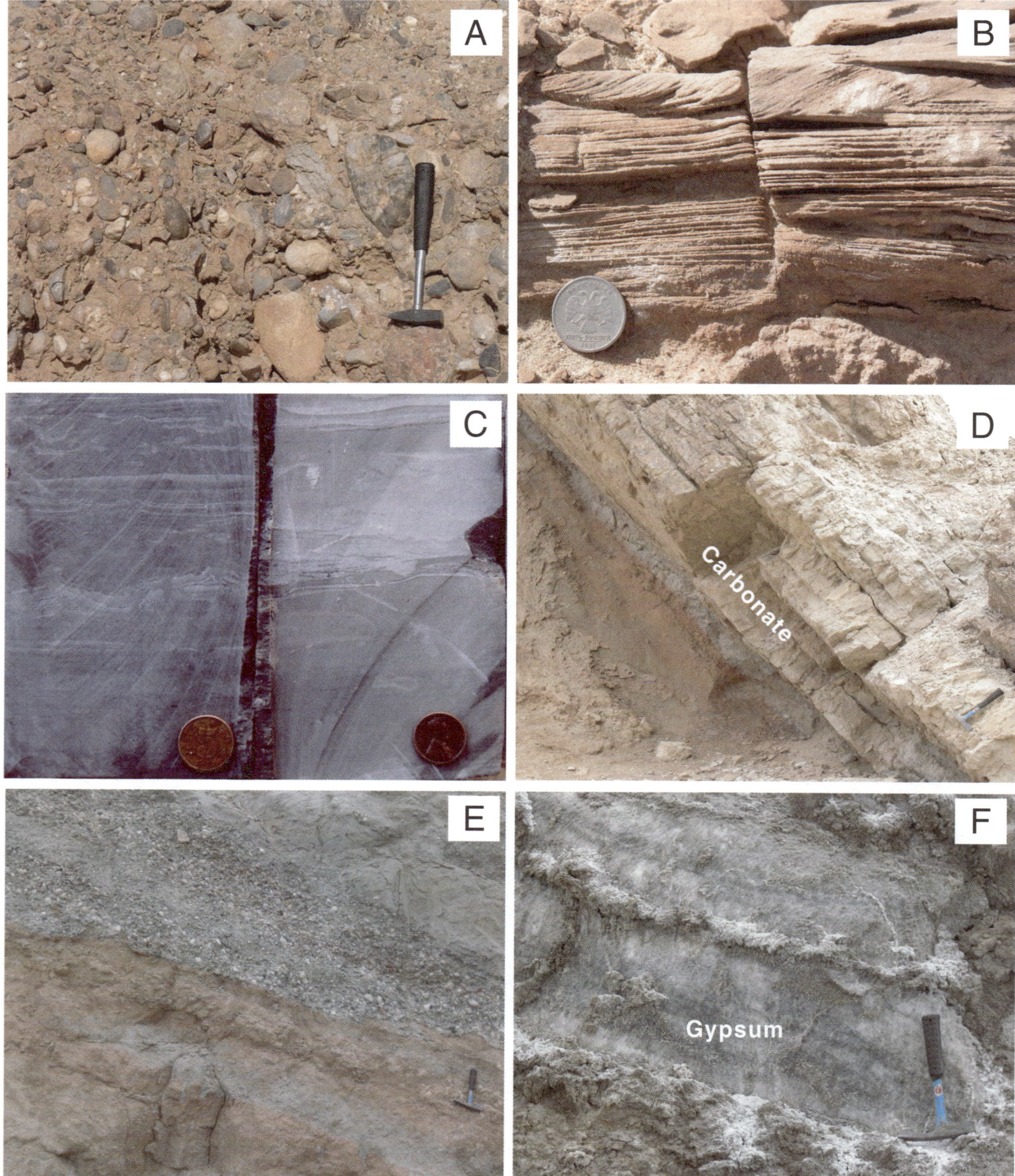

Figure 7. Field photos of some Tertiary sedimentary facies: (A) Conglomerate in basal part of the Lulehe Formation with gravels containing mostly metamorphic rocks, in the northern margin of the western Qaidam Basin. Hammer is 35 cm long. (B) Parallel- and cross-stratified sandstone, the Lulehe Formation east of the Xitieshan. Coin is 2 cm across. (C) Cores of dark-colored siltstone and silty mudstone of the Xiaganchaigou Formation from the southwest of the Qaidam Basin. Coins are 1.5 cm across. (D) Carbonate layer in the Shangganchaigou Formation in the southwestern part of the Qaidam Basin. Hammer is 25 cm long. (E) Low-angle cross-bedded conglomerate and sandstone of the Xiayoushashan Formation from the northern Qaidam Basin. Hammer is 25 cm long. (F) Gypsum layer in the Shizigou Formation from the western Qaidam Basin. Hammer is 25 cm long.

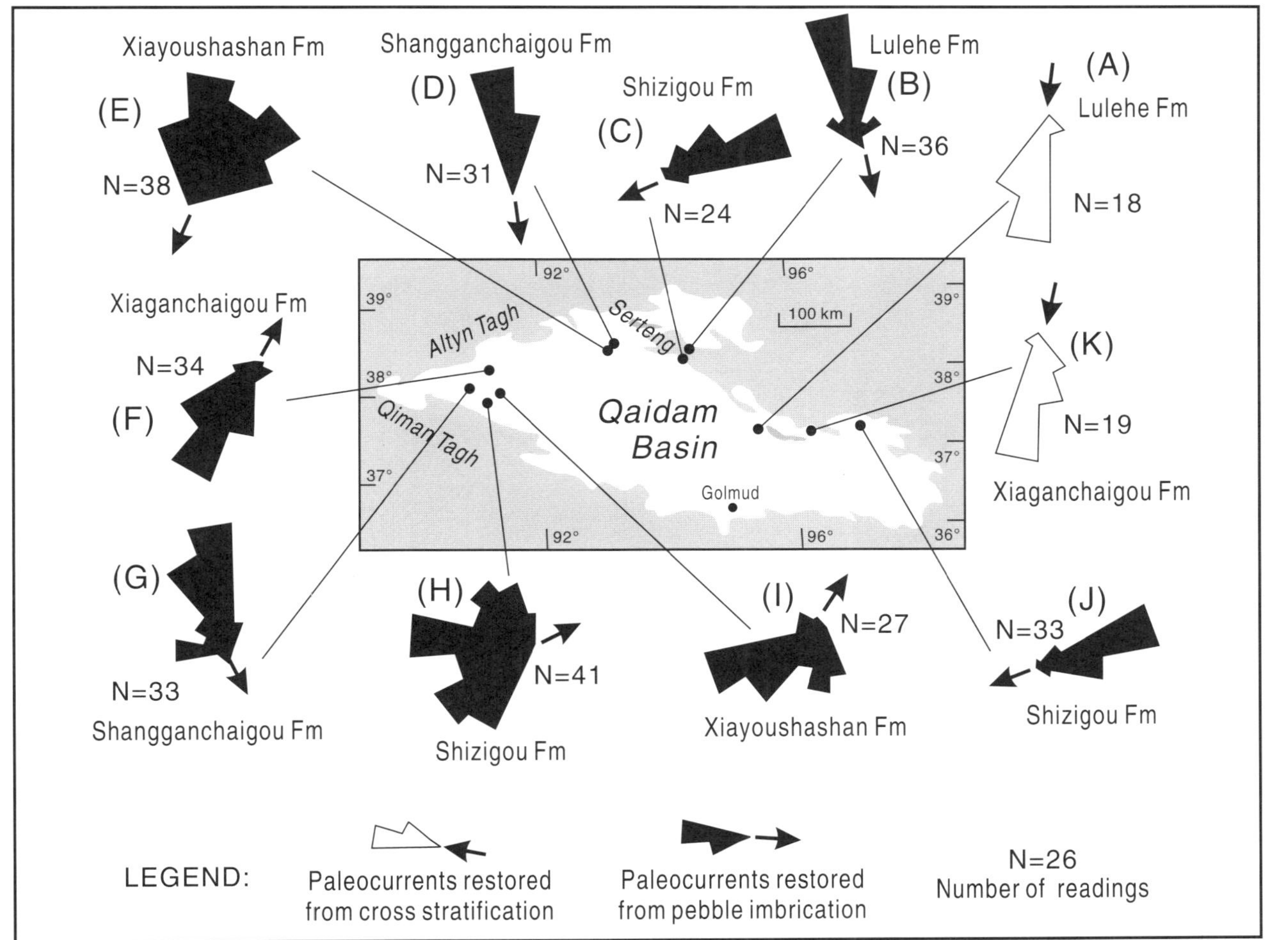

Figure 8. Paleoflows restored from pebble imbrication and cross-stratification in different stratigraphic units of the Qaidam Basin.

occurrence of variable-scale cross-bedding and channelization within conglomeratic and sandstone beds. The fine-grained facies are thought to have formed in a floodplain setting, with the sandwiched sandstone layers as crevasse-splay deposits. Paleocurrent restorations show that the sediments in the northern part of the basin were derived from the southern Qilian (Fig. 8K), but in the western basin, they were derived from the southwest (Fig. 8F).

It is worthwhile to note the occurrence of dark-colored mudstone (Fig. 7C), lime mudstone, and limestone in the Xiaganchaigou Formation in the western Qaidam Basin, which can be up to 2000 m thick in the Mangnai depression (Figs. 9B and 9C) because they are considered to be the main source rocks of the Qaidam Basin (Huo, 1990). In addition, the accumulated thickness of limestone and lime mudstone is >300 m, as revealed by boreholes. It has been generally regarded in literature that these dark-colored mudstone and associated limestones were formed in a lacustrine setting (Gu and Di, 1989), but this claim is not well substantiated. An alternative interpretation is that the dark-colored mudstone and limestone association might have been developed in a marine environment that was connected with the epeiric sea in the southwestern Tarim Basin during early Paleogene time (Meng et al., 2001).

The lower member of the Xiaganchaigou Formation displays similar spatial distribution and facies zonation to the strata in the underlying Lulehe Formation, and its main depocenter was basically in the same place as the underlying formation (Fig. 9B). The upper member, however, covers a broader area and occurs almost in the whole present-day Qaidam Basin (Fig. 9C). The Xiaganchaigou Formation was also developed in the Suhai Basin, where its thickness ranges from 250 to 500 m comparable to that in the northern margin of the Qaidam Basin. The middle part of the Huatiaoshan Formation in the Kumukol Basin is stratigraphically equivalent to the Xiaganchaigou Formation, and its dark-colored fine-grained facies in the north is similar to the Xiaganchaigou Formation in the western Qaidam Basin.

Shangganchaigou Formation

The Shangganchaigou Formation is distributed throughout the Qaidam Basin, and it is composed of finer facies, such as fine-grained sandstone, siltstone/mudstone, and muddy limestone (Fig. 7D). Coarse-grained facies, such as gravelly sandstone and conglomerate, are only developed at the basin margins.

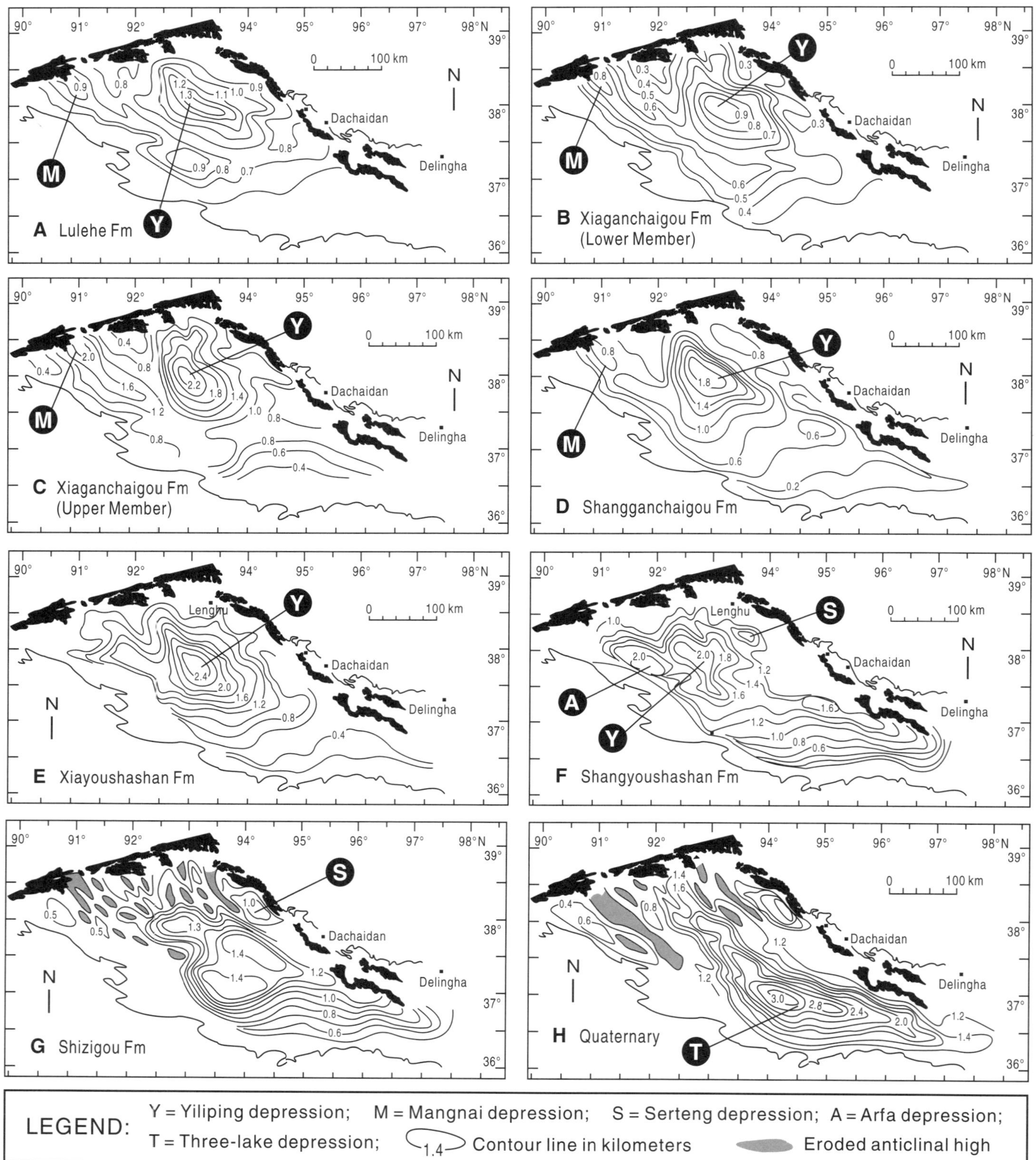

Figure 9. Isopach maps of stratigraphic units of the Qaidam Basin. Note that the main depocenter is located in the middle of the western Qaidam Basin during the periods of deposition of the Lulehe Formation (A), the Xiaganchaigou Formation (B and C), Shangganchaigou Formation (D), and Xiayoushashan Formation (E). Thickness variation of the Shangyoushashan Formation implies development of several depocenters in the western Qaidam Basin (F). Sedimentation of the Shizigou Formation (G) and the Quaternary (H) appears synchronous with folding in the western Qaidam Basin, with main depocenter shifting to the east. The data are from Zhu et al. (1994), Huang et al. (1996), and available reports of the Qinghai Oil Company.

Siltstone and mudstone are thin-bedded, internally rippled, or horizontally laminated and contain abundant ostracods. Thin- and medium-bedded sandstone occurs as interlayers of differing thickness. The finer facies are dark-gray or gray in color in the lower and middle portions of the formation, in particular, in the western Qaidam Basin, and become red upward. Muddy limestone primarily develops in the lower part of the sequence in association with dark-gray siltstone/mudstone. In contrast, the upper red fine-grained facies frequently contains thick and cross-bedded sandstone beds with channelized bases. Conglomerate and gravelly sandstones are thick-bedded, showing flat bedding and cross-stratification. Gravels are mostly well-rounded pebbles and cobbles, exhibiting marked imbrication. The finer facies are interpreted as lacustrine products by virtue of their dark color, fine grain size, and horizontal lamination, as well as by the occurrence of freshwater fossils and limestone. In contrast, the finer facies with sandwiched sandstone beds of varying thickness in the upper portion of the Shangganchaigou Formation are interpreted to have been deposited by fluvial processes. Flat-bedded or low-angle cross-bedded conglomerate and gravelly sandstone facies are considered to have formed by sheetflooding, whereas the gravelly sandstone was deposited as fluvial channel deposits. Therefore, the coarse-grained facies are thought to have formed in a braided fluvial environment. From the pebble imbrication, paleocurrent restoration indicates that the gravels came from the north in front of the Serteng and from the northwest in from the Altyn Tagh (Figs. 8D and 8G).

Notably, the Shangganchaigou Formation exhibits a marked upward-coarsening trend at the basin margins, varying from interlayered fine-grained sandstone, siltstone, and mudstone upward into gravelly sandstone and pebbly conglomerate. It is also interesting to note the upward change in color within the Shangganchaigou Formation from dark brown and gray to red. The isopach map of the Shangganchaigou Formation clearly shows that the Yiliping depression was persistent in the middle of the western Qaidam Basin, with a thickness up to 2000 m (Fig. 9D). Though present in the Suhai Basin, the Shangganchaigou Formation is much thinner, ~300 m thick.

Xiayoushashan Formation

The Xiayoushashan Formation is similar to the underlying Shangganchaigou Formation in both spatial facies variations and localization of the main depocenter (Fig. 9E). Coarse-grained facies, however, are more common compared with the Shangganchaigou Formation. Conglomeratic layers become more pronounced along the margins of the present-day Qaidam Basin (Fig. 7E), and gravels most likely came from adjacent mountain belts, such as the Serteng (Fig. 8E) and the Qiman Tagh (Fig. 8I). Fine-grained facies are still dominant in the middle part of the basin and consist of thin-bedded siltstone, mudstone, and limestone, as revealed in both borehole cores and outcrops. The coarse- and fine-grained facies can be grouped into several distinct facies associations. Facies association 1 (FA1) consists primarily of thick-bedded pebbly conglomerate and coarse-grained sandstones with few internal sedimentary structures. Facies association 2 (FA2) is composed of medium-bedded cross- and horizontal-stratified sandstone and thin-bedded siltstone. Massive sandstone and soft-sediment deformations also occur in this association. Facies association 3 (FA3) is characterized by interlayered thin-bedded siltstone, mudstone, and limestone. FA1 is thought to have been deposited from alluvial-fluvial sedimentation and represent proximal deposits. FA2 is interpreted as a fluvial facies, but it might also have formed in a deltaic setting, since it possesses many intervals of soft-sediment deformation and massive sandstone. FA3 developed in the middle part of the basin and is considered to be a lacustrine sediment, indicative of a distal suspended sedimentation.

The present-day Qaidam Basin appears to have already been established during the deposition of the Xiayoushashan Formation, and its margins were defined by spatial distribution of the FA1. The main depocenter continued to form in the middle of basin, with the deposition of up to 2500 m of Xiayoushashan Formation (Fig. 9E). The Xiayoushashan Formation is absent in the Suhai Basin, whereas coarse-grained sediments like conglomerate represent equivalent strata in the Kumukol Basin.

Shangyoushashan Formation

The Shangyoushashan Formation shares the same spatial facies zonation as the Xiayoushashan Formation and is characterized by occurrence of coarser facies along the basin margins and finer facies in the middle of the basin. Differences, however, do exist. Coarse-grained facies became more and more pronounced and propagate significantly into the basin interior. Dark mudstone becomes less significant compared with the Xiayoushashan Formation in the middle of the basin. Instead, purple fine-grained sandstone and siltstone predominate. Conglomerate and coarse-grained sandstone in the proximal regions are interpreted as braided-river deposits, because they contain minor fine-grained facies, and they are internally cross-bedded and parallel stratified. Dark-colored horizontal laminated fine-grained strata in the middle part of the basin are interpreted as lacustrine facies, whereas the red fine-grained strata associated with cross-bedded sandstone in distal areas represent floodplain sedimentation.

As indicated by the isopach map (Fig. 9F), the Yiliping depression continued to develop in the middle of the basin, with thickness up to 2000 m, but some subordinate depocenters came into being along the northern and southern margins of the basin. A reconstruction of the paleoenvironments of the Shangyoushashan Formation reveals facies zonations, indicating spatial variations from a braided-river system along basin margins through a meandering river system in a transitional zone to a deltaic-lacustrine system in the basin interior (Gu and Di, 1989). In addition, the Shangyoushashan Formation is missing in the Suhai Basin. The equivalent unit, the Fengchenkou Formation, in the Kumukol Basin is characterized by coarse-grained facies, such as conglomerates and gravelly sandstone (Fig. 4).

Shizigou Formation

The Shizigou Formation is dominated by alluvial conglomeratic deposits along the basin margins that clearly originated from the adjacent highlands (Figs. 8H and 8J), but the main part of the formation consists of evaporate deposits (Fig. 7F), carbonaceous mudstone, and yellowish siltstone-mudstone. The main depocenter became less well defined, and a number of depocenters developed in the western Qaidam Basin (Fig. 9G). The main depositional areas appear to have begun to migrate toward the east, where they were filled with fluvial and lacustrine deposits (Zhu et al., 1994). The sedimentation of the Shizigou Formation was coeval with contractional deformation in the western Qaidam Basin, as evidenced by its cross-sectional geometry, which is characterized by typical growth strata (Zhou et al., 2006).

Quaternary

Quaternary lithostratigraphy, magnetostratigraphy, and sedimentation have been well documented by Shen et al. (1993), Gu (1993), and Guo (2006). The Qigequan Formation is considered to form the lowest part of the Quaternary strata, and it rests conformably above the Shizigou Formation. In general, Quaternary lithologies are dominated by conglomerate and sandstone in the western and marginal parts of the basin, and interpreted to have resulted from alluvial-fluvial deposition at basin margins. Fine-grained facies, such as thin-bedded siltstone, mudstone, and muddy limestone, constitute the bulk of the Quaternary fill in the eastern Qaidam Basin (Shen et al., 1993; Guo, 2006). The fine-grained facies, usually dark-gray or greenish, are thought to have formed in a freshwater lake, and halite is mainly present in the uppermost part of the Quaternary succession (Shen et al., 1993). The Quaternary depocenter, also termed as the three-lake (Sanhu) sag in Chinese literature, is clearly located in the eastern Qaidam Basin, and it has a stratigraphic thickness of up to 3000 m (Fig. 9H). A number of isolated smaller-scale salt lakes also developed in the western Qaidam Basin from the early Pleistocene and were bounded by the NW-SE–trending highlands resulting from folding.

BASIN ANALYSIS

Provenance

Sources for the Qaidam Basin sediments have been a matter of debate. Métivier et al. (1998) suggested that the Qaidam Basin was an internally drained basin filled mainly with sediments from the eastern Kunlun on the south. Wang et al. (2006) recently advanced a model postulating that the Qaidam was a displaced basin that received sediments mostly from western Kunlun rather than from adjacent mountain ranges during Paleogene time. Rieser et al. (2005) conducted a study of sandstone provenance for the Qaidam Basin, and they concluded that no dramatic changes took place in the composition of Tertiary succession. Previous $^{40}Ar/^{39}Ar$ dating of detrital white mica places further constraints on Tertiary provenance and suggests that the strata are mostly derived from the bordering mountains (Rieser et al., 2006). Although uncertainties exist about source regions for Paleogene sequence, the Neogene deposits in the Qaidam Basin were evidently derived from the adjacent mountain belts, as indicated by paleocurrent restorations (Fig. 8). The eastern Kunlun had been exhumed by ca. 30 Ma (Mock et al., 1999), and the southern Qilian is demonstrated to have undergone rapid denudation around the same period of time (Wang et al., 2004). In addition, the Altyn Tagh was also substantially uplifted in the early–middle Oligocene (Jolivet et al., 2001; Chen et al., 2001; Yin et al., 2002; Cowgill et al., 2003), thereby serving as another major source for the Qaidam Basin sediments on the northwest.

It is interesting to note that contour lines on isopach maps are obviously truncated by the Altyn Tagh fault for the Lulehe and Xiaganchaigou Formations (Figs. 9A–9C), whereas the contour lines go around the edges of topographic highs in front of the Altyn Tagh fault on isopach maps of the Shangganchaigou, Xiayoushashan, and Shangyoushashan Formations (Figs. 9D–9F). These relations suggest that the Altyn Tagh fault and adjacent highlands should not have served as the northwestern border of the Qaidam Basin until the beginning of sedimentation of the Shangganchaigou Formation. In other words, the Altyn Tagh was initiated in the Oligocene and then became a sediment source shedding detritus into the western Qaidam Basin. This inference is consistent with other studies about denudation history of the Altyn Tagh (Jolivet et al., 2001; Chen et al., 2001; Yin et al., 2002; Cowgill et al., 2003).

A seismic section shows that the Lulehe and Xiaganchaigou Formations are in fault contact with basement rocks of the Altyn Tagh, whereas the overlying units gradually overlap the Altyn Tagh to the northwest (Fig. 10). The continuous and parallel patterns of seismic reflections of the Lulehe and Xiaganchaigou Formations suggest that they are composed of laterally continuous fine-grained facies like thin-bedded siltstone and mudstone, and they do not represent proximal coarser deposits. In contrast, the overlying units rest unconformably above the Altyn Tagh and become thinner to the northwest, indicating proximal sedimentation.

Fault Systems

The present-day Qaidam Basin is bounded on the north and south by basin-directed reverse fault systems, in particular, in the western Qaidam Basin (Fig. 2). The major border faults on the south are termed the northern Kunlun fault system (NKF), which manifests itself as high-angle S-dipping reverse fault (Fig. 2). Interestingly, the Lulehe, Xiaganchaigou, and Shangganchaigou Formations show little thickness changes from the basin interior toward the northern Kunlun fault, whereas the overlying strata apparently become thinner (Fig. 11, b–b′). This observation suggests that the northern Kunlun fault was not initiated until deposition of the Xiayoushashan Formation, after which it began

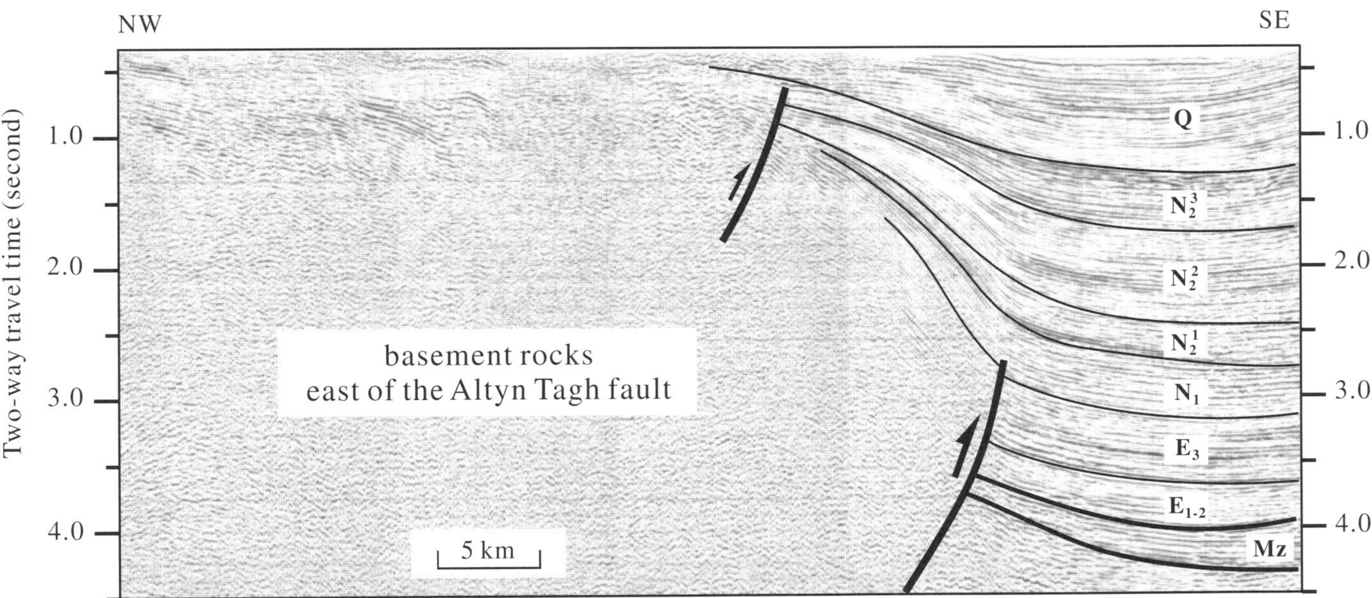

Figure 10. Seismic section showing that the Lulehe and Xiaganchaigou Formations are truncated by a high-angle transpressional reverse fault, whereas the overlying stratigraphic units become thinner and pinch out northwestward over basement rocks. Refer to Figure 2 for locality. Abbreviations: Mz—Mesozoic; E_{1-2}—Lulehe Formation; E_3—Xiaganchaigou Formation; N_1—Shangganchaigou Formation; N_2^1—Xiayoushashan Formation; N_2^2—Shangyoushashan Formation; N_2^3—Shizigou Formation; Q—Quaternary.

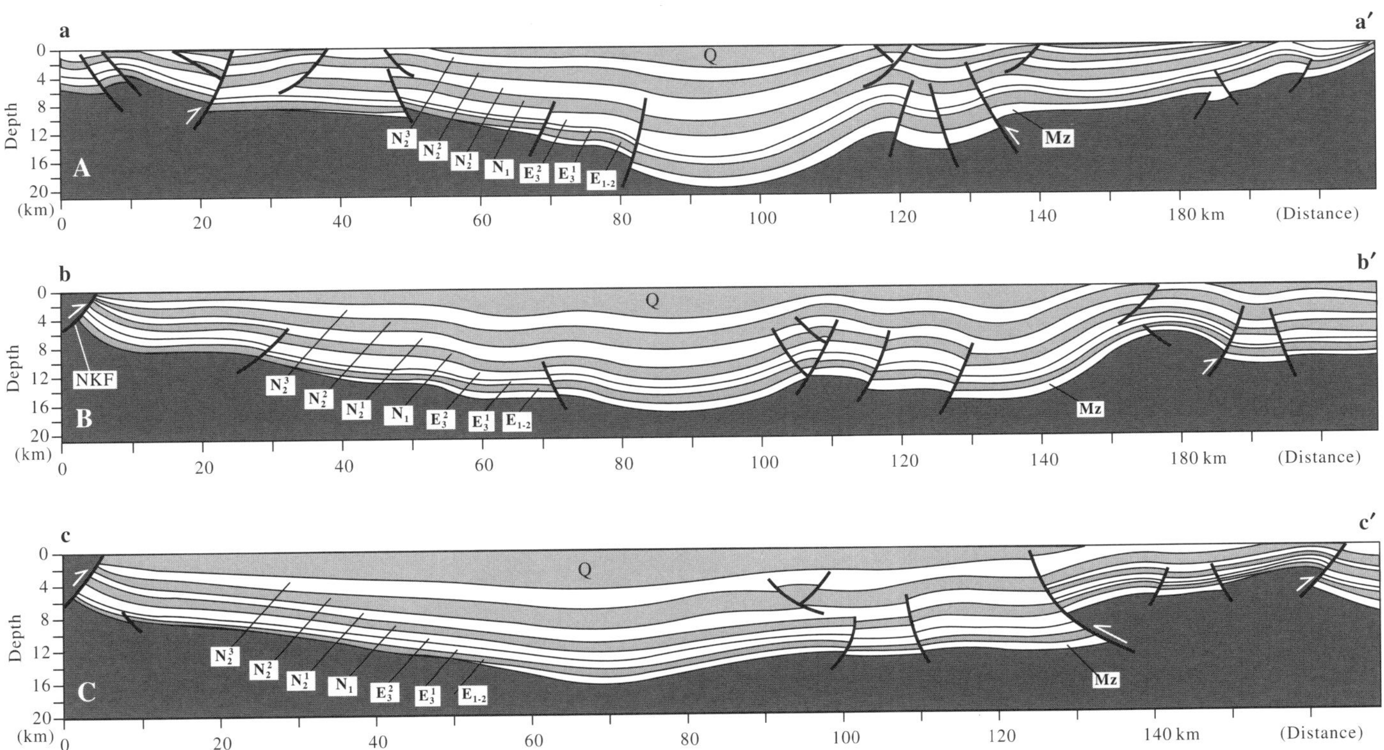

Figure 11. Basin-scale geologic profiles across Qaidam Basin, showing that Cenozoic depocenters are located in the middle of the basin, and cumulative thickness of Cenozoic successions in the western Qaidam Basin can be up to 17 km (A). Also noticeable is that the Lulehe (E_{1-2}), Xiaganchaigou (E_3^1 and E_3^2 are Lower and Upper Members), and Shangganchaigou (N_1) Formations exhibit no obvious thickness variation toward the northern Kunlun fault (NKF), whereas the Xiayoushashan Formation (N_2^1) gets thinner toward the northern Kunlun fault system (B). Basement-rooted high-angle reverse faults occur in the middle of the basin, particularly in the western Qaidam Basin, but they are less developed toward the east (C). Refer to Figure 2 for locations of profiles and Figure 10 for meanings of abbreviations of other stratigraphic units.

exerting control on sedimentation of subsequent stratigraphic units. The northern border fault system can be divided into several segments—the northwestern segment is called the southern Serteng fault (Fig. 2). Similar to the northern Kunlun fault system, the southern Serteng fault did not control deposition of the Lulehe and Xiaganchaigou formations because the two formations show no thickness variation across the southern Serteng fault. Therefore, the southern Serteng fault had not formed until the Oligocene, when the Shangganchaigou Formation began to be deposited. Along the southern Serteng fault, Proterozoic basement rocks are thrust over Mesozoic and Cenozoic units (Fig. 12A), and Mesozoic rocks are thrust over Cenozoic strata (Fig. 12B). The footwall Cenozoic rocks can be as young as Pliocene (the Shizigou Formation), suggesting that the southern Serteng fault must have been active in late Cenozoic time. Thrust faults are also present in the northern margin of the Qaidam Basin where Paleozoic metamorphic rocks are thrust above Cenozoic strata (Fig. 12C).

The Altyn Tagh fault on the northwest side of the Qaidam Basin is topographically expressed as a straight well-defined lineament of E-NE–trending valleys or, along the margins, similar-trending ranges (Fig. 3). The fault zone is ~100–1000 m wide, and it is characterized by brittle to semiductile left-lateral strike-slip deformation (Fig. 12D). Structural features and tectonic history of the Altyn Tagh fault have been intensively investigated (cf. Yin et al., 2002, and references cited therein), but controversies remain over a number of critical issues, such as timing of the fault inception. Our present study places some independent constraints on the Altyn Tagh fault evolution.

Another conspicuous feature is the occurrence of reverse faults in the middle of the Qaidam Basin, which are mostly basement-rooted and penetrate upward into the Cenozoic cover to different levels (Fig. 11). These reverse faults are usually high-angle, and they dip both to the north or south. Except for a few faults that extend upward to the Upper Miocene–Pliocene, or even reach the surface, most of the reverse faults die out upward into older strata (Fig. 11). The intrabasinal reverse faults occur primarily in the western Qaidam Basin, and they become less well developed and eventually absent toward the east. There are also a number of reverse faults within the Tertiary sedimentary succession (Fig. 11), and these faults are closely associated with folds, or they occur as conjugate faults at inflection points of fold limbs (Figs. 11 and 13B).

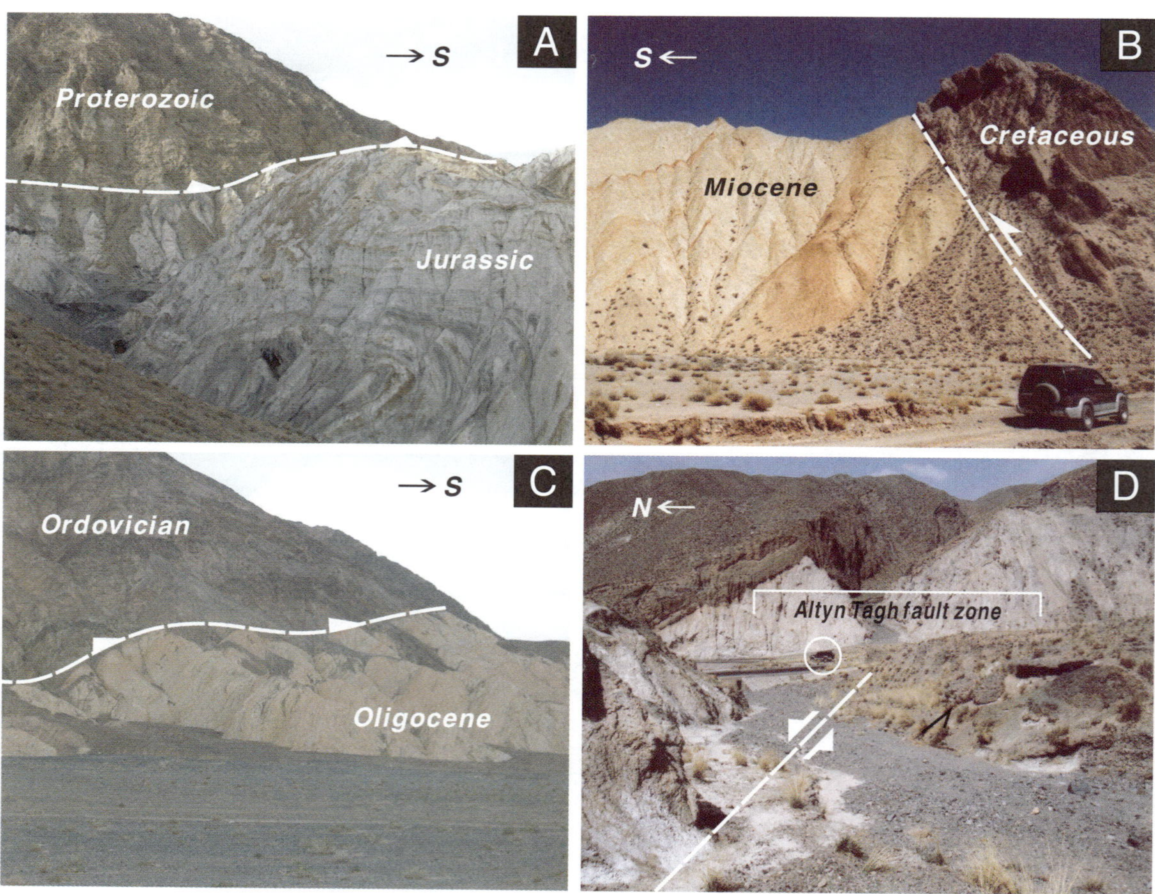

Figure 12. (A) Thrust fault south of the Zongwulong. Width of view is 150 m. (B) Reverse fault south of the Zongwulong. Car is 2 m high. (C) Thrust fault south of the Xitieshan. Width of view is 600 m. (D) The Altyn Tagh fault zone, the Dangjinshan Pass. Car is for scale.

Syndeformational Sedimentation

Tertiary sedimentary strata are folded in the western Qaidam Basin (Figs. 3 and 11). As clearly shown on seismic profiles (Fig. 13), deposition of some stratigraphic units was synchronous either with folding in the basin interior or with basement uplifting at basin margins. These syndeformational sedimentary successions, referred to as growth strata by Suppe et al. (1992), can be used to infer the timing, geometry, and kinematics of fold-and-thrust systems (Burbank and Vergés, 1994), as well as the interaction of sedimentation with coeval deformation (Burbank et al., 1996). Individual stratigraphic units of the Qaidam Basin might have been deposited as pregrowth strata in one region, but as growth strata in other areas. The Lulehe and Xiaganchaigou Formations on the northern fringe of the Qaidam Basin display constant thickness toward the Serteng (Fig. 13A), and their thickness is also similar to that of their counterparts in the Suhai Basin north of the Serteng. The Lulehe and Xiaganchaigou Formations are therefore interpreted to have formed prior to the uplift of the Serteng, and they represent the pregrowth strata. The overlying stratigraphic units, in contrast, thin toward the edge of the Serteng (Fig. 13A), thereby suggesting that they are growth strata. This observation indicates that the Serteng started to rise in the Oligocene, when the Shangganchaigou Formation was deposited. Successions from the Lulehe Formation up to the Xiayoushashan Formation in the basin interior, however, show

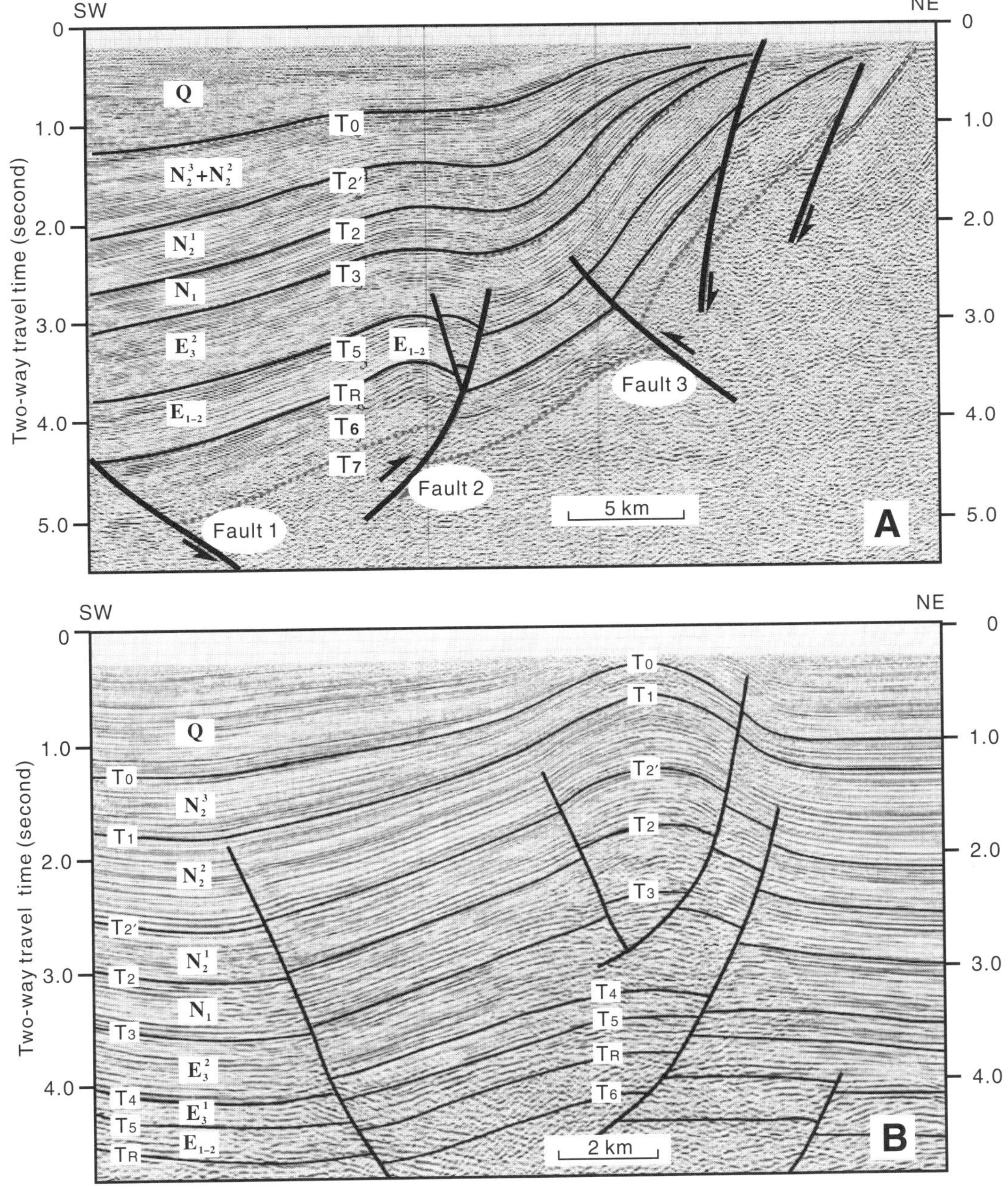

Figure 13. Seismic sections illustrating the initiation of growth strata in both the margin and interior of Qaidam Basin. (A) Section in the northeast of Qaidam Basin, with the Shangganchaigou Formation (N_1) as growth strata. (B) Section in basin interior, showing that growth strata had not developed until deposition of the Shangyoushashan Formation (N_2^2), which becomes thinner toward fold crest (modified from Zhou et al., 2006). Refer to Figure 2 for localities of sections. Abbreviations: E_{1-2}—Lulehe Formation; E_3—Xiaganchaigou Formation; N_1—Shangganchaigou Formation; N_2^1—Xiayoushashan Formation; N_2^2—Shangyoushashan Formation; N_2^3—Shizigou Formation; Q—Quaternary. T_0–T_6—seismic reflections or boundaries that separate different stratigraphic units.

no striking thickness variations and thus express themselves as pregrowth strata (Fig. 13B). Thinning from the limb to crest of folds is evident in the Shangyoushashan Formation and becomes more pronounced in the Shizigou Formation and in the Quaternary strata (Fig. 13B). This indicates that deformations had not taken place in the basin interior until the middle Miocene. Accordingly, Cenozoic deformation occurred earlier at the margin than in the interior of the western Qaidam Basin.

Depocenter

It has been recognized that the main depositional area was in the western Qaidam Basin during the Tertiary, and then shifted to the east during the Quaternary (Gu and Di, 1989; Wang and Coward, 1990). The eastward migration of the depocenter was interpreted to be the consequence of strong deformation and uplift of the western Qaidam Basin toward the end of the Tertiary (Wang and Coward, 1990). It is important, however, to note that the main depocenter had been persistently in the middle of the Qaidam Basin from Eocene to Miocene time (Fig. 9).

The NW-SE–trending Mangnai depression represents a prominent subordinate depocenter that lies in the southwestern part of the Qaidam Basin (Figs. 9A–9D). Notably, the Mangnai depression was filled dominantly with Paleogene fine-grained strata, which might have been deposited continuously from Upper Cretaceous marine facies (Meng et al., 2001). Truncation of Paleogene isopach contour lines in the Mangnai depression by the Altyn Tagh fault suggests that the persevered depression is only part of the original depocenter (Figs. 9B–9D).

Subordinate Basins

The Qaidam Basin is bounded on all sides by mountain ranges, and some subordinate basins are present within the surrounding mountains, such as the Suhai Basin in the southern Qilian, the Kumukol Basin in the eastern Kunlun, and the Tula and Xorkol Basins within the Altyn Tagh (Fig. 2). An understanding of Cenozoic stratigraphy and sedimentation of these basins provides insight into the tectonic development of the Qaidam Basin. We chose to analyze the Suhai and Kumukol Basins to reveal possible early-stage linkages of these subordinate basins with the Qaidam Basin.

The Suhai Basin occurs to the north of the Qaidam Basin and is separated from it by the Serteng (Figs. 2 and 3). According to available seismic profiles and borehole data, only the Lulehe, Xiaganchaigou, and Shangganchaigou Formations were deposited in the basin, and most other overlying stratigraphic units are absent (Fig. 4). The Lulehe Formation is composed of conglomerate and coarse-grained sandstone in the lower part and passes upward into siltstone and mudstone with interlayered sandstone (Fig. 4). It is ~550 m thick in the middle of the basin, and it shows no important thickness variations toward its southern margin on seismic profiles. The Xiaganchaigou Formation is dominated by sandstone and sandy siltstone, with minor conglomerate. It rests conformably above the Lulehe Formation, and its thickness ranges from 400 to 600 m. The lower portion of the Shangganchaigou Formation consists of sedimentary facies similar to those of the Xiaganchaigou Formation, but they become finer upward (Fig. 4). The Xiayoushashan Formation is only locally distributed and is overlain by Quaternary strata.

Importantly, sedimentary facies of the Lulehe and Xiaganchaigou Formations in the Suhai Basin are quite similar to coeval strata on the northwestern margin of the Qaidam Basin, and their thickness is also generally comparable. This suggests that the Suhai Basin probably was the northern continuation of the Qaidam Basin, and there were no topographic barriers separating the two at least before deposition of the Shangganchaigou Formation. Thinning of the Shangganchaigou Formation in both the Suhai and northwestern Qaidam Basins toward the Serteng indicates that the Serteng began to rise in the Oligocene. The Suhai Basin is thought to have been elevated since the late Oligocene because of the absence of most Neogene strata, and it manifested itself as an intermontane basin during the Quaternary.

The Kumukol Basin is located south of the Qiman Tagh, and it has been uplifted to an elevation of ~4500 m (Figs. 2 and 3). The basin contains an Eocene to Pliocene sequence that has been moderately deformed into a syncline (Fig. 14A). The Tertiary sequences can be divided into four stratigraphic units, from bottom to top, the Huatiaoshan, Hongshiliang, Fengchenkou, and Jiantuliang Formations (Fig. 4). The Huatiaoshan Formation, up to 3000 m thick, is dominated by coarse-grained facies, conglomerate and coarse-grained sandstone in the lower part in the southern Kumukol Basin that passes upward into fine-grained facies, such as interlayered fine-grained sandstone and siltstone/mudstone (Fig. 4). In contrast, the lower part of the Huatiaoshan Formation is dominated by fine-grained facies in the northern Kumukol Basin (Fig. 4). The Huatiaoshan Formation is assigned an Oligocene age based on fossils in fine-grained layers (Zhang et al., 1996). Given that the fossils occur in the upper portion of the Huatiaoshan Formation and some fossils are also common in Eocene strata, we thus consider the Huatiaoshan Formation to range from Eocene to Oligocene in age, and it can be correlated with the Xiaganchaigou and Shangganchaigou Formations of the southwestern Qaidam Basin. The Hongshiliang Formation is composed of thin-bedded sandstone and siltstone in its lower part and changes upward into thick- and medium-bedded coarse-grained sandstone and some conglomerate (Fig. 4). It is distributed throughout the basin, up to 1800 m thick, and rests conformably over the Huatiaoshan Formation. The Hongshiliang Formation is assigned an early Miocene age (Zhang et al., 1996). Coarse-grained facies, such as sandstone and conglomerate, make up the bulk of the Fengchenkou Formation (Fig. 4), which lies unconformably over the Hongshiliang Formation. This unit is assigned a middle Miocene age based on the presence of the estheria assemblages (Zhang et al., 1996). The Jiantuliang Formation is characterized by massive gypsum and associated calcareous mudstone in its lower part, which pass upward into red sandstone and thin-bedded siltstone and mudstone. Angular

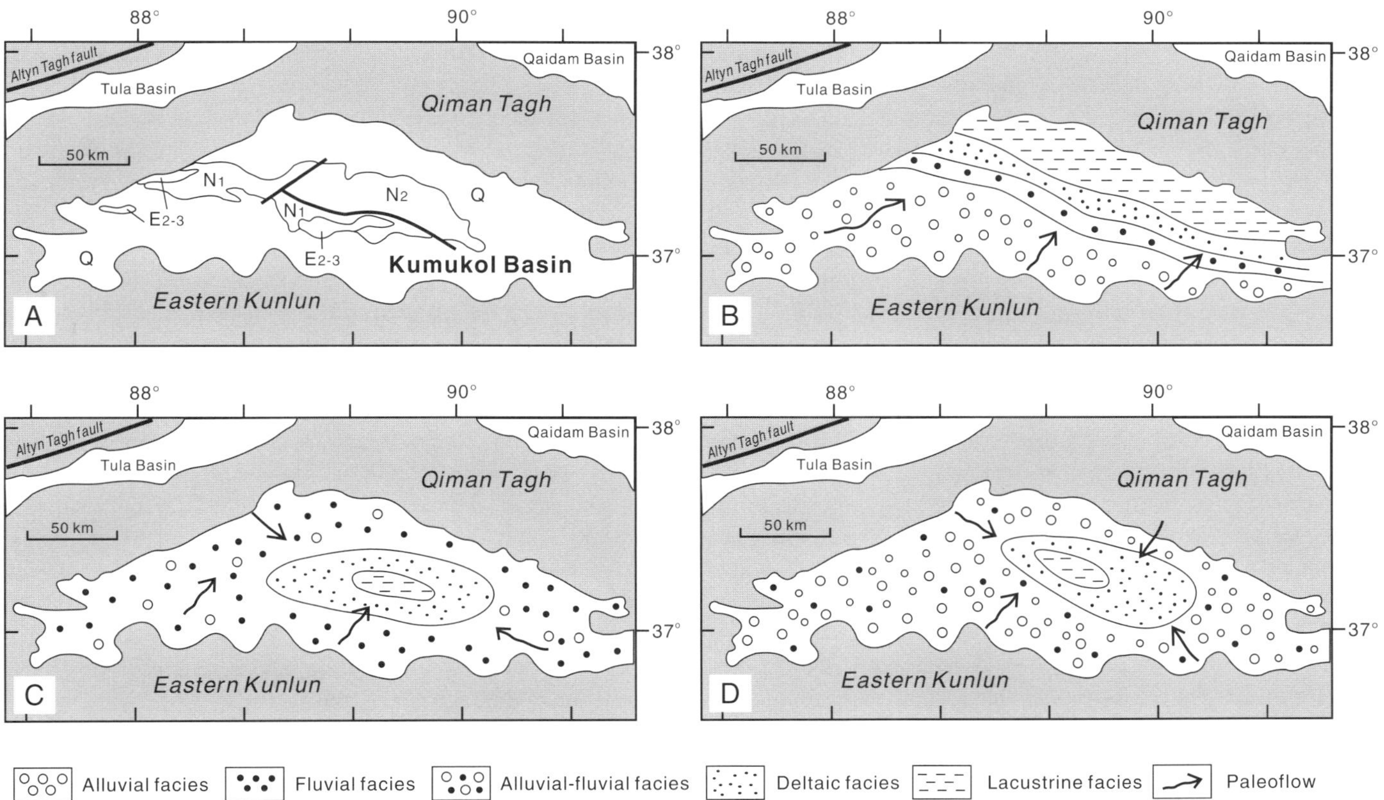

Figure 14. (A) Simplified geologic map. (B) Alluvial facies occur in the south of the basin, changing into fluvial, deltaic, and lacustrine facies toward the north during Huatiaoshan Formation deposition. (C) Basin is characterized by a concentric pattern of facies distribution during the period of Hongshiliang Formation deposition, with sediments debouching into basin from all sides. (D) The Fengchenkou and Jiantuliang Formations also exhibit a concentric distribution of sedimentary facies, receiving clastic sediments from surrounding mountains. Paleoenvironmental reconstruction and paleoflows are mainly after Zhang et al. (1996) and Xiao et al. (2005). Refer to text for detailed discussion.

unconformities are observed in places between the Jiantuliang Formation and the underlying units. Abundant fossils, such as bivalves and gastropods, are present in this unit and date the Jiantuliang Formation as Pliocene–Pleistocene (Zhang et al., 1996).

It is interesting to note that alluvial conglomeratic facies of the Huatiaoshan Formation were developed in the south of the Kumukol Basin, in contrast to the coeval dark-colored fine-grained facies in the north (Fig. 14B). The paleocurrent indicators from cross-stratification and pebble imbrication show a uniform northward flow (Fig. 14B), indicating that sediments of the Huatiaoshan Formation came from the south (Xiao et al., 2005). Also noticeable is paleoflow, which became multidirectional during the period of deposition of the Hongshiliang Formation (Fig. 14C). Sediments began debouching to the basin from the north when the Fengchenkou and Jiantuliang Formations were deposited, as indicated by the S-directed paleocurrent directions in the northern Kumukol Basin (Fig. 14D).

We infer that the southern border of the Kumukol Basin had already been established in Eocene and Oligocene time, but the Qiman Tagh, which forms the present northern border of the basin, did not exist at that time. This inference is supported by (1) proximal facies, which only occur in the south of the basin, in contrast with finer facies in the north (Fig. 14B); (2) paleocurrent data, which show a consistent northward flow (Fig. 14B); and (3) the finer facies in the northern Kumukol basin, which can be correlated with the fine-grained deposits of the Lulehe and Xiaganchaigou Formations in the southwest Qaidam Basin. Therefore, it is plausible that the Kumukol Basin was part of the southern portion of the proto–Qaidam Basin during Eocene and Oligocene time. The Kumukol basin evolved into an isolated and internally drained intermontane basin in the Miocene, when it began to develop its own depocenter surrounded by proximal alluvial-fluvial coarse-grained facies, as shown by the spatial distribution of facies in the Hongshiliang Formation (Fig. 14C). The Kumukol Basin continued to develop throughout the Neogene and was eventually uplifted to a high elevation with the eastern Kunlun.

IMPLICATIONS FOR BASIN TECTONICS

Several models have been advanced to account for tectonic evolution of the Qaidam Basin, and all of them share the idea that the subsidence of the Qaidam Basin resulted from basinward emplacement of thrust sheets along the northern and southern edges of the basin (Wang and Coward, 1990; Métivier et al., 1998; Meyer et al., 1998; Cowgill et al., 2003). It has been postulated that the N-directed thrust system on the southern margin of the basin could be attributed to hidden S-directed subduction of the mantle lithosphere that had been decoupled from the upper plate (the Qaidam Basin) through a weak zone in the lower crust (Tapponnier et al., 2001). It has further been suggested that a middle-crustal detachment might exist beneath the Qaidam Basin, which branches upward to form the Qilian thrust system in the north (cf. Tapponnier et al., 1990). The N-dipping reverse faults at the northern edge of the Qaidam Basin have been considered to be the consequence of the back thrusting of the Qilian thrust system (Meyer et al., 1998). Métivier et al. (1998) also claimed that a series of thrust-related basins was generated due to northward propagation of S-dipping thrusts/reverse faults that became successively younger to the north. The stepwise development of the basins has also been considered to be related to northward prorogation of left-lateral displacement on the Altyn Tagh fault during the Tertiary (Yin et al., 2002; Cowgill et al., 2003).

However, there are some flaws with these previous models because they are in conflict with some crucial observations. For instance, Tertiary depocenters of the Qaidam Basin were located in the middle of the basin, rather than close to basin margins. The cumulated Tertiary thickness in the Yiliping depression is up to 17 km, in striking contrast to that of basin margins (Figs. 9 and 11). If the basin were generated by crustal flexure due to tectonic loads imposed by thrust sheets, depositional loci would have been near thrust fronts (Beaumont, 1981; Flemings and Jordan, 1989). In addition, the thrusting and uplifting has been shown to be roughly synchronous in the eastern Kunlun and the southern Qilian Shan around ca. 30 Ma (Mock et al., 1999; Wang et al., 2004). Exhumation of the northern Qilian also appears to have occurred as early as the Early Miocene (George et al., 2001), much earlier than previous estimates of 6–2 Ma (e.g., Tapponnier et al., 1990; Meyer et al., 1998). The contemporaneity of deformation in the northeastern Tibetan Plateau thus throws doubt on the stepwise growth model.

In this study, we propose that the Qaidam Basin originated from crustal buckling or folding because the buckling mechanism can account for a number of key observations and geologic evolution of the basin, such as localization of a main depocenter, stratigraphic and sedimentary development, initiation and nature of reverse fault systems, and interrelationship between the Qaidam and adjacent intermontane basins (Fig. 15). Continental lithospheric-scale folding is considered to be one of primary responses to regional horizontal compression (Cloetingh et al., 1999), and it has been applied to explain active intracontinental deformation in some regions, such as the Tibetan Plateau (Burg and Podladchikov, 1999), Central Asia (Nikishin et al., 1993; Burov et al., 1993; Cobbold et al., 1993), European continent (Cloetingh et al., 2002), and central Australia (Stephenson and Lambeck, 1985). Many analogue and numerical models have been conducted to investigate periodic instability and folding of the lithosphere or the crust in horizontal compression (Martinod and Davy, 1994; Burg et al., 1994; Cloetingh et al., 1999; Sokoutis et al., 2005). They show that continental lithosphere or crust will experience folding at initial and intermediate stages during horizontal shortening, and that later faulting occurs at inflection points of folds. Further shortening will eventually lead to the closing of synclinal depressions (cf. Sokoutis et al., 2005). The modeling also indicates that fold wavelengths are primarily controlled by lithospheric thickness and thermal structure, where long wavelength corresponds to thick and cold lithosphere (Cloetingh et al., 1999). Lithospheric-scale folding usually displays long wavelengths of 250–360 km (Nikishin et al., 1993). In practice, periodicity of lithospheric- and crustal-scale folds can be quite irregular in wavelength and amplification in view of crust-mantle coupling and uncoupling, strain localization, and intensity of erosion (Cloetingh et al., 1999). Another interesting result of analogue and numerical experiments is that prefold tectonic fabrics, such as old sutures or faults, exert little influence on lithospheric folding (Martinod and Davy, 1994; Gerbault et al., 1999), and continuous folding behavior of faulted lithosphere is possibly due to fault locking as a result of gravity and friction (Cloetingh et al., 1999).

The modeling of continental lithospheric and crustal folding provides insight into the origin and tectonic evolution of the Qaidam Basin. We surmise that the Qaidam Basin underwent three stages of evolution in the Cenozoic (Fig. 15). The first stage was characterized by relative slow subsidence, possibly due to crustal folding in the Paleocene–early Oligocene interval, when the Lulehe and Xiaganchaigou Formations were Formed (Fig. 15B). A proto–Qaidam Basin manifested itself as a synclinal depression that consisted of both the Qaidam Basin and the Kumukol Basin on the south and the Suhai Basin on the north. There also developed simultaneously two antiformal uplifts, the eastern Kunlun on the south and the southern Qilian on the north. The depocenter was located in the middle of the basin, corresponding to the trough of the crust-scale megafold, the wavelength of which was ~300 km in the western Qaidam Basin. The raised Qilian and eastern Kunlun provided sediments to the proto–Qaidam Basin (Fig. 15B).

During the second stage, there was further development of the megafold during the late Oligocene to early Middle Miocene, when the Shangganchaigou and Xiayoushashan Formations were deposited. Reverse faults commenced at the basin margins, and the depocenter kept developing in the middle of the basin (Fig. 15C). The S-dipping northern Kunlun fault system on the south and N-dipping southern Serteng fault on the north represent the basin-margin reverse faults. With development of the reverse faults, proximal regions of the proto–Qaidam Basin were gradually isolated by fault-related uplifts, such as the Qiman Tagh and the Serteng, thereby forming the Suhai and Kumukol Basins on the north and south, respectively. Oligocene uplift of the Serteng is recorded by the inception of growth strata in the Shangganchaigou Formation on the southern side of the Serteng, coeval with the exhumation of the Qaidam Shan east of the Serteng

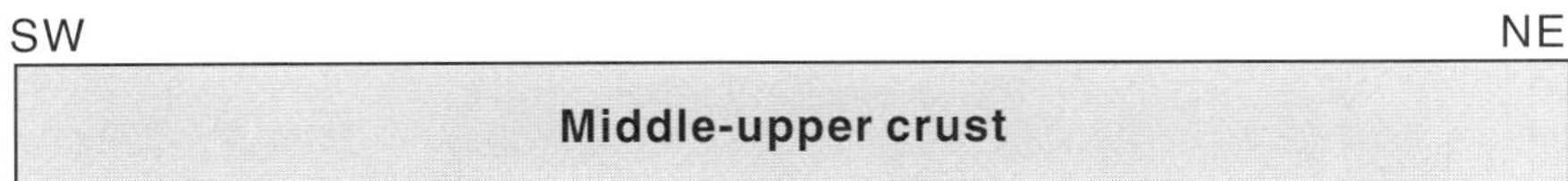

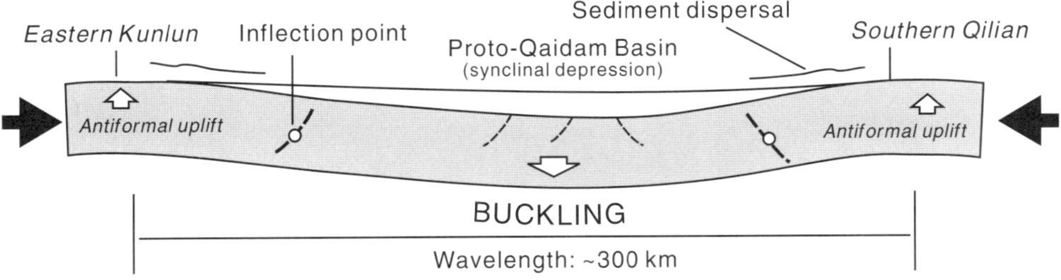

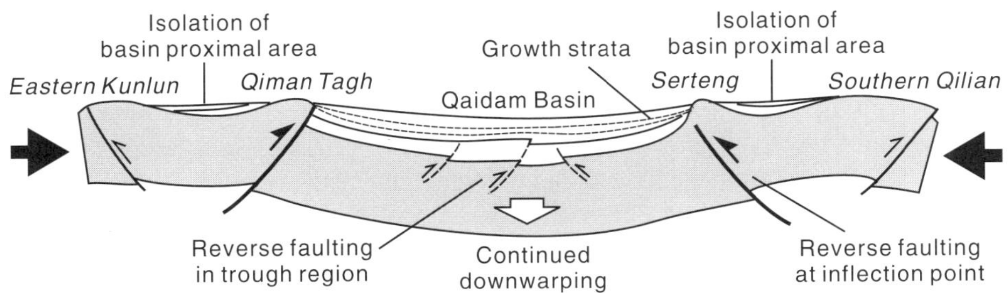

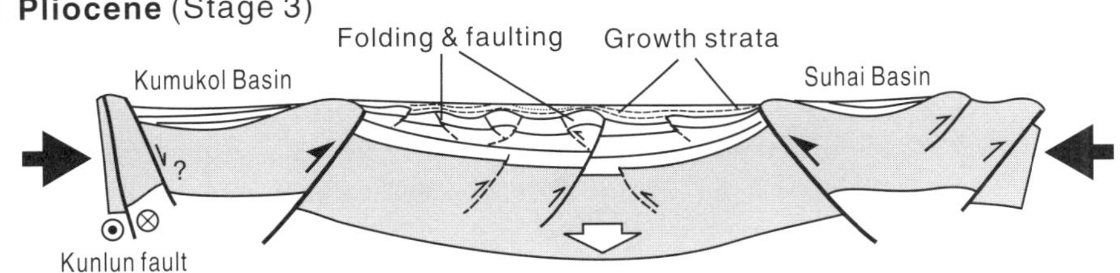

Figure 15. Two-dimensional (2-D) crustal-buckling model for Cenozoic three-stage development of the western Qaidam Basin: (A) Pre-buckling stage. (B) Northeastern Tibetan Plateau commenced buckling from the Eocene, with its wavelength at ~300 km. Proto–Qaidam Basin was initiated as a synclinal depression, and antiformal highs developed synchronously, creating the southern Qilian on the north and the eastern Kunlun on the south. Depocenter developed in the middle of the basin and was filled with sediments from adjacent antiformal highs. (C) Major reverse faults became active at inflection points of the megafold, leading to formation of the Qiman Tagh and the Serteng, as well as isolation of Suhai and Kumukol Basins at basin margins. (D) The western Qaidam Basin underwent intrabasinal folding and faulting due to continued horizontal compression. The Suhai Basin was raised, but the Kumukol Basin continued subsiding presumably in an extension setting. Refer to text for a full discussion.

(Wang et al., 2004). Tertiary stratigraphic development further suggests that the Suhai Basin had been gradually elevated since the Late Oligocene (Fig. 4) because the Shangganchaigou and Xiayoushashan Formations are relatively thin and locally distributed. Furthermore, Neogene strata are almost missing in the Suhai Basin. The Qiman Tagh was also exhumed ca. 30 Ma and has become an area of provenance for the Kumukol Basin since Early Miocene time. Occurrence of the Fengchenkou and Jiantuliang Formations indicates that the Kumukol Basin continued subsiding in Middle Miocene–Pliocene times, contrasting with the synchronous uplift of the Suhai Basin. Middle Miocene and Pliocene strata in the Kumukol Basin, up to 3500 m thick (Zhang et al., 1996; Xiao et al., 2005), presumably developed in an extensional setting, which has been shown to be prevalent in the eastern Kunlun in that time interval (Mock et al., 1999; Jolivet et al., 2003; Fu and Awata, 2007). In the context of crustal-scale folding, we suggest that the reverse faults on the basin margins were generated at inflection points of the resulting megafold with continued horizontal shortening (Fig. 15C). Therefore, the occurrence of reverse faults at basin margins can be envisioned as a corollary of the ongoing crustal folding. Accordingly, the N-dipping reverse faults on the northern border of the Qaidam Basin were not necessarily the result of back thrusting related to the northward-prograding Qilian or Nanshan thrust system (Meyer et al., 1998).

The Qaidam Basin entered the third stage when it underwent strong contractional deformations characterized by intrabasinal folding and faulting. Most of the intrabasinal reverse faults are basement-rooted, and some of them can penetrate upward to the surface (Fig. 11). Various-scale folds and faults are clearly related and have striking surface expressions in the western Qaidam Basin (Fig. 3). Neogene successions, such as the Shangyoushashan and Shizigou Formations, form typical growth strata (Fig. 15D), and as a consequence, the older main depocenter in the middle of the western Qaidam Basin was gradually disrupted and evolved into several subordinate depocenters separated by growing anticlinal highs (Fig. 9G). With increased horizontal shortening, the Suhai Basin was raised, but the Kumukol Basin continued subsiding in an extensional setting. Neogene extension of the Kumukol Basin is also implied by coeval occurrence of volcanism and extensional basins in adjacent regions (Jolivet et al., 2003; Q. Wang et al., 2005). Also noticeable is the eastern Kunlun fault, which was initiated in the late Miocene (around 10 Ma) and which acted as one of the major left-lateral strike-slip faults in northern Tibet (Fu and Awata, 2007). The left-lateral displacement or strain partitioning along the eastern Kunlun fault might have contributed in part to the Neogene extension in the eastern Kunlun (Jolivet et al., 2003).

It is assumed that the hypothesized crustal folding was decoupled from the mantle along a lower-crustal weak zone. There is ample evidence that indicates the existence of a low-density or weak lower crust in northern Tibet (Zhu and Helmberger, 1998), which is as thick as 20–30 km beneath the eastern Kunlun (Zhao et al., 2006). The weak lower crust might be able to flow and thus could accommodate the upper-crustal folding processes in compression. The lithospheric mantle beneath the Qaidam Basin might be subducted to the south, as suggested by previous studies (Meyer et al., 1998; Tapponnier et al., 2001). If this is the case, the overlying detached crust would be readily buckled during continued horizontal shortening.

Although the three-stage model can account for the tectono-sedimentary evolution of the western Qaidam Basin, the eastern Qaidam Basin has a different history. As shown by Figure 9, the eastern Qaidam Basin gradually evolved into a part of the Qaidam Basin since Oligocene time, and it became the main locus of deposition in Quaternary time, simultaneous with active intrabasinal faulting and folding in the western Qaidam (Fig. 9H). Interestingly, the depocenter was also localized in the middle of the eastern Qaidam, implying that the subsidence could be also attributed to crustal buckling. There exists another subordinate Cenozoic basin, the Delingha Basin, that is separated from the eastern Qaidam Basin by the Emunik high (Fig. 2). It is assumed that the Delingha Basin was originally the northern portion of the eastern Qaidam Basin because they share a similar stratigraphic succession. Separation of the Delingha Basin from the eastern Qaidam Basin is considered to have taken place at the end of the Tertiary. This inference is based upon two observations: (1) thickness of Tertiary strata is compatible across the Emunik ranges, as indicated by available seismic profiles, and (2) Tertiary successions are strongly harmonically folded, with little Quaternary sediment in the Delingha Basin. Compared with the three-stage evolution of the western Qaidam Basin, the eastern Qaidam Basin appears to have only experienced the first two stages of development. There are no major reverse faults breaking the southern margin of the eastern Qaidam Basin. The absence of major reverse faults is presumably due to reduction of orthogonal compression due to strain partitioning by left-lateral displacement of the Kunlun fault since the late Miocene (Fu and Awata, 2007).

The discrepancy of tectono-sedimentary histories between the western and eastern Qaidam Basin is thought to originate from the spatial variation of the stress field in the Cenozoic. The western Qaidam is bounded on the northwest by the Altyn Tagh fault, and the large-magnitude left-lateral displacement of the Altyn Tagh fault must have exerted a strong influence on the stress regime in the western Qaidam Basin. Recent activity of the Altyn Tagh fault system has been well documented (Peltzer et al., 1989; Bendick et al., 2000), but its geological evolution is controversial. It is generally thought that the Altyn Tagh fault did not initiate until the Oligocene (Ritts and Biffi, 2000; Meng et al., 2001; Yue et al., 2001; Chen et al., 2004), although other studies argue that inception of the Altyn Tagh fault could have been as early as late Mesozoic (Y. Wang et al., 2005; Liu et al., 2007). Seismic tomographic imaging has revealed present-day deep structure of the Altyn Tagh fault, showing that it cuts through the crust and penetrates downward as deep as ~140 km (Wittlinger et al., 1998). Accordingly, the Altyn Tagh fault would have been acting as an accommodation zone to adjust differing strain regimes between the Tarim and Qaidam Basins. It is suggested that progressive left-lateral slip of the Altyn Tagh fault promoted the crustal folding of the Qaidam Basin in the Cenozoic.

Admittedly, the crustal buckling model presented here is quite preliminary, and it cannot satisfactorily explain all the features in the study of the Qaidam Basin. For instance, we are not sure if the buckling could result in so strong a tectonic subsidence that the resultant basin received more than 15 km of sediment. The buckling-induced subsidence can be probably coupled with other thermo-mechanic processes, such as sublithospheric loading or phase changes in the deep crust. It is, however, not the purpose of this study to discuss mechanical processes of the crustal buckling and its possible linkage with other thermo-tectonic processes.

Figure 16 presents cartoons to illustrate sequential tectonic evolution of the Qaidam Basin during the Cenozoic. It is assumed that the Qaidam Basin was connected with the southwestern Tarim Basin in Late Cretaceous–Paleocene times, and a marine transgression reached the southwestern part of the Qaidam Basin (Fig. 16A). The remaining parts of the Qaidam Basin were highlands with scattered late Mesozoic relict extensional basins. The crust began to fold in the northeastern Tibetan Plateau in the Eocene, and the Qaidam Basin was initiated as a synclinal depression (Fig. 16B). Crustal folding in the western Qaidam was accommodated by left-lateral slip of the Altyn Tagh fault, and sediments were shed from both the northern and southern antiformal highs. Another possible provenance at this stage was the western Kunlun. With continued horizontal compression, reverse faulting became active at the basin margins or at inflection points of the crustal megafold (Fig. 16C). The resulting fault-related highs gradually divided the basin-margin areas, leading to the formation of isolated intermontane basins, such as the Suhai Basin. The Qaidam Basin was completely separated from the Tarim Basin with continued displacement on the Altyn Tagh fault and simultaneous formation of the Altyn Tagh range. The Qaidam depocenter was confined to the middle of the basin with the folding of the crust. The western Qaidam was folded and uplifted in the Quaternary as a result of the continued shortening, and the depositional area was forced to migrate eastward. The younger depocenter was located in

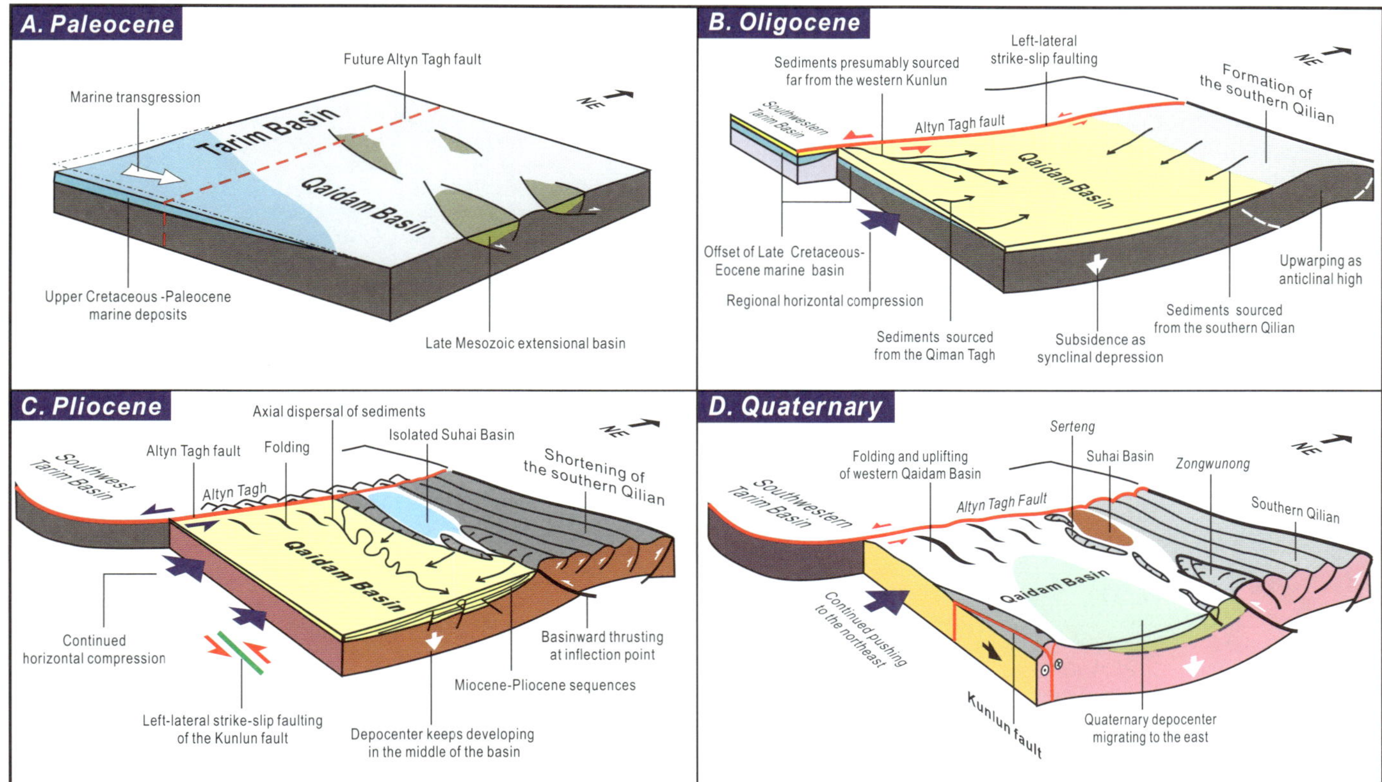

Figure 16. Cartoons illustrating tectonic scenario of the Qaidam Basin and its possible linkage with other tectonic processes in adjoining regions: (A) The Qaidam Basin was connected with the southwestern Tarim Basin and received Paleocene marine deposition. (B) The crust began folding in response to horizontal compression, and the Altyn Tagh fault was initiated to accommodate different stress fields between the Qaidam and Tarim Basins. The Qaidam Basin manifested itself as a synclinal depression with depocenter in the middle. (C) Continued compression generated major reverse faults at inflection points in basin-margin areas, and subordinate intermontane basins came into being, such as the Suhai Basin. (D) The western Qaidam Basin experienced intrabasinal folding and inversion, and the main depositional area was forced to migrate to the east.

the middle of the eastern Qaidam (Fig. 16D). Active left-lateral slip on the eastern Kunlun fault occurred from ca. 10 Ma and resulted in partitioning of the orthogonal stress on the southern margin of the eastern Qaidam Basin.

There exists another Cenozoic basin to the south of the Qaidam Basin, the Hohxil Basin, which is bounded on the north by the eastern Kunlun and on the south by the Tanggula (Fig. 1). That basin was initiated in the early Eocene and filled with fluvial-lacustrine sediments, as represented by Eocene Fenghuoshan Group and Lower Oligocene Yaxicuo Group (Liu et al., 1998; Wang et al., 2002). The basin then experienced contractional inversion at the end of the Oligocene, leading to strong deformation of Paleogene strata. The Lower Miocene Wudaoliang Group unconformably overlies the folded Paleogene succession, and it is mostly composed of lacustrine limestone up to 800 km thick (Liu and Wang, 2001). Of great interest is the depocenters, which were also persistently located in the middle of the basin, and a Paleogene cumulative thickness up to 5 km (Liu et al., 2001). The Hohxil Basin might also have arisen from the crustal folding, with an original wavelength of ~300 km, similar to that of the Qaidam Basin.

CONCLUSION

The Qaidam Basin experienced a three-stage evolution during the Cenozoic, and its tectonic subsidence is interpreted to have resulted from crustal folding or buckling in response to regional horizontal compression. The basin was initiated as a synclinal depression, ~300 km wide in the first stage during Eocene to Oligocene time, flanked by antiformal highs, the southern Qilian on the north and the eastern Kunlun on the south. Sediments were shed from both the south and north, whereas the Altyn Tagh did not exist at this time. The second stage was characterized by the development of reverse faults at the basin margins in the Miocene, which then evolved into the northern and southern border faults of the Qaidam Basin. The original basin-margin areas were isolated by these border reverse faults and related uplifts, and they became intermontane basins, forming as the Suhai Basin on the north and the Kumukol Basin on the south. During Pliocene and Quaternary time, the third stage began when the western Qaidam Basin was deformed and uplifted and when the main depocenter shifted to the east. The crustal folding model can account well for some key observations, such as the persistent localization of the depocenter in the middle of the basin, synchronous deformation on both sides of the basin, and the occurrence and basinward displacements of reverse faults at basin margins. It is suggested that the Qaidam Basin was presumably attached to the southwestern Tarim Basin in the Late Cretaceous and early Paleocene, after which it was displaced to the northeast. The Altyn Tagh fault is believed to have played a major role in accommodating the crustal folding in northeastern Tibet since the Oligocene.

ACKNOWLEDGMENTS

This study is supported by grants from the Knowledge Innovation Project of the Chinese Academy of Sciences (80567300), Chinese National Major Fundamental Research Developing Project (2006CB202305), and Natural Science Foundation of China (40721003). Paul Heller and Eric Kirby are thanked for their thoughtful comments and helpful suggestions.

REFERENCES CITED

Beaumont, C., 1981, Foreland basin: Geophysical Journal of the Royal Astronomical Society, v. 65, p. 291–329.

Bendick, R., Bilham, R., Freymueller, J., Larson, K., and Yin, G., 2000, Geodetic evidence for a low slip rate in the Altyn Tagh fault system: Nature, v. 404, p. 69–72, doi: 10.1038/35003555.

Bian, Q., Li, D., Pospelov, I., Yin, L., Li, H., Zhao, D., Chang, C., Luo, X., Gao, S., Astrakhantsev, O., and Chamov, N., 2004, Age, geochemistry and tectonic setting of Buqingshan ophiolites, North Qinghai-Tibet Plateau, China: Journal of Asian Earth Sciences, v. 23, p. 577–596, doi: 10.1016/j.jseaes.2003.09.003.

Burbank, D.W., and Vergés, J., 1994, Reconstruction of topography and related depositional systems during active thrusting: Journal of Geophysical Research, v. 99, p. 20,281–20,297, doi: 10.1029/94JB00463.

Burbank, D.W., Meigs, A., and Brozovic, N., 1996, Interactions of growing folds and coeval depositional systems: Basin Research, v. 8, p. 199–224, doi: 10.1046/j.1365-2117.1996.00181.x.

Burchfiel, B.C., Deng, Q., Molnar, P., Royden, L.H., Wang, Y., Zhang, P., and Zhang, W., 1989a, Intracrustal detachment within zones of continental deformation: Geology, v. 17, p. 748–752, doi: 10.1130/0091-7613(1989)017<0448:IDWZOC>2.3.CO;2.

Burchfiel, B.C., Molnar, P., Zhao, Z., K'uangyi, L., Wang, S., Huang, M., and Sutter, J., 1989b, Geology of the Ulugh Muztagh area, northern Tibet: Earth and Planetary Science Letters, v. 94, p. 57–70, doi: 10.1016/0012-821X(89)90083-6.

Burg, J.P., and Podladchikov, Y., 1999, Lithospheric scale folding: Numerical modelling and application to the Himalayan syntaxes: International Journal of Earth Sciences, v. 88, p. 190–200, doi: 10.1007/s005310050259.

Burg, J.P., Davy, P., and Martinod, J., 1994, Shortening of analogue models of the continental lithosphere: New hypothesis for the formation of the Tibetan Plateau: Tectonics, v. 13, p. 475–483, doi: 10.1029/93TC02738.

Burov, E.B., Lobkovsky, L.I., Cloetingh, S., and Nikishin, A.M., 1993, Continental lithosphere folding in Central Asia (II): Constraints from gravity and topography: Tectonophysics, v. 226, p. 73–87, doi: 10.1016/0040-1951(93)90111-V.

Chen, W., Chen, C., and Nabelek, J.L., 1999, Present-day deformation of the Qaidam Basin with implications for intra-continental tectonics: Tectonophysics, v. 305, p. 165–181, doi: 10.1016/S0040-1951(99)00006-2.

Chen, X., Yin, A., Gehrels, G.E., Cowgill, E., Grove, M., Harrison, M., and Wang, X.-F., 2003, Two phases of Mesozoic north-south extension in the eastern Altyn Tagh range, northern Tibetan Plateau: Tectonics, v. 22, p. 1053, doi: 10.1029/2001TC001336.

Chen, Z., Zhang, Y., Wang, X., Chen, X., Washburn, Z., and Arrowsmith, J.R., 2001, Fission track dating of apatite constraints on the Cenozoic uplift of the Altyn Tagh Mountain: Acta Geoscientia Sinica, v. 22, p. 413–418.

Chen, Z.L., Wang, X.F., Yin, A., Chen, B.L., and Chen, X.H., 2004, Cenozoic left-slip motion along the central Altyn Tagh fault as inferred from the sedimentary record: International Geology Review, v. 46, p. 839–856, doi: 10.2747/0020-6814.46.9.839.

Cloetingh, S., Burov, E., and Poliakov, A., 1999, Lithosphere folding: Primary response to compression? (from central Asia to Paris basin): Tectonics, v. 18, p. 1064–1083, doi: 10.1029/1999TC900040.

Cloetingh, S., Burov, E., Beekman, F., Andeweg, B., Andriessen, P.A.M., Garcia-Castellanos, D., de Vicente, G., and Vegas, R., 2002, Lithospheric folding in Iberia: Tectonics, v. 21, p. 1041, doi: 10.1029/2001TC901031.

Cobbold, P.R., Davy, P., Gapais, D., Rossello, E.A., Sabybakasov, J., Thomas, J.C., Tondji Biyo, J.J., and de Urreiztieta, M., 1993, Sedimentary basins and crustal thickening: Sedimentary Geology, v. 86, p. 77–89, doi: 10.1016/0037-0738(93)90134-Q.

Cowgill, E., Yin, A., Harrison, T.M., and Wang, X.-F., 2003, Reconstruction of the Altyn Tagh fault based on U-Pb geochronology: Role of back thrusts, mantle sutures, and heterogeneous crustal strength in forming the Tibetan Plateau: Journal of Geophysical Research, v. 108, p. 2346, doi: 10.1029/2002JB002080.

CSBS (Chinese State Bureau of Seismology), 1992, The Aerjin Active Fault Belt: Beijing, Seismology Publishing House, 319 p. (in Chinese).

Ding, G., Chen, J., Tian, Q., Shen, X., Xing, C., and Wei, K., 2004, Active faults and magnitudes of left-lateral displacement along the northern margin of the Tibetan Plateau: Tectonophysics, v. 380, p. 243–260, doi: 10.1016/j.tecto.2003.09.022.

Dupont-Nivet, G., Robinson, D., Butler, R., Yin, A., and Melosh, H.J., 2004, Concentration of crustal displacement along a weak Altyn Tagh fault: Evidence from paleomagnetism of the northern Tibetan Plateau: Tectonics, v. 23, p. TC1020, doi: 10.1029/2002TC001397.

Fang, X., Zhang, W., Meng, Q., Gao, J., Wang, X., King, J., Song, C., Dai, S., and Miao, Y., 2007, High-resolution magnetostratigraphy of the Neogene Huaitoutala section in the eastern Qaidam Basin on the NE Tibetan Plateau, Qinghai Province, China, and its implication on tectonic uplift of the NE Tibetan Plateau: Earth and Planetary Science Letters, v. 258, p. 293–306, doi: 10.1016/j.epsl.2007.03.042.

Flemings, P.B., and Jordan, T.E., 1989, A synthetic stratigraphic model of foreland basin development: Journal of Geophysical Research, v. 94, p. 3851–3866, doi: 10.1029/JB094iB04p03851.

Fu, B., and Awata, Y., 2007, Displacement and timing of left-lateral faulting in the Kunlun fault zone, northern Tibet, inferred from geologic and geomorphic features: Journal of Asian Earth Sciences, v. 29, p. 253–265, doi: 10.1016/j.jseaes.2006.03.004.

Gehrels, G.E., Yin, A., and Wang, X.-F., 2003, Detrital zircon geochronology of the northeastern Tibetan Plateau: Geological Society of America Bulletin, v. 115, p. 881–896, doi: 10.1130/0016-7606(2003)115<0881:DGOTNT>2.0.CO;2.

George, A.D., Marshallsea, S.J., Wyrwoll, K., Chen, J., and Lu, Y., 2001, Miocene cooling in the northern Qilian Shan, northeastern margin of the Tibetan Plateau, revealed by apatite fission-track and vitrinite-reflectance analysis: Geology, v. 29, p. 939–942, doi: 10.1130/0091-7613(2001)029<0939:MCITNQ>2.0.CO;2.

Gerbault, M., Burov, E.B., Poliakov, A., and Daigniéres, M., 1999, Do faults trigger folding in the lithosphere?: Geophysical Research Letters, v. 26, p. 271–274, doi: 10.1029/1998GL900293.

Gu, S., 1993, Conditions for Quaternary Gas Fields in the Eastern Qaidam Basin and Exploration Practice: Beijing, Petroleum Industry Press, 149 p. (in Chinese).

Gu, S., and Di, H., 1989, Mechanism of formation of the Qaidam Basin and its control on petroleum, *in* Zhu, X., ed., Chinese Sedimentary Basins: Amsterdam, The Netherlands, Elsevier, p. 45–51.

Guo, S., 2006, Application Sequence Stratigraphy of Quaternary of Sanhu Sag, Qaidam Basin: Beijing: Geological Publishing House, 109 p. (in Chinese).

Hao, Y., Guo, X., and Ye, L., 2001, The Boundary between the Marine Cretaceous and Tertiary in the Southwest Tarim Basin: Beijing, Science Press, 108 p. (in Chinese).

Houseman, G., and England, P., 1993, Crustal thickening versus lateral expulsion in the Indian Asian continental collision: Journal of Geophysical Research, v. 98, p. 12,233–12,249, doi: 10.1029/93JB00443.

Hsü, K.J., 1988, Relict back-arc basins: Principles of recognition and possible new examples from China, *in* Kunen, L.K., ed., New Perspective in Basin Analysis: New York, Springer-Verlag, p. 254–264.

Hsü, K.J., Pan, G., and Şengör, A.M.C., 1995, Tectonic evolution of Tibet and Himalayas: International Geology Review, v. 16, p. 418–421.

Huang, H., Huang, Q., and Ma, Y., 1996, Geology of Qaidam Basin and its Petroleum Prediction: Beijing, Geological Publishing House, 257 p. (in Chinese).

Huo, G.M., 1990, Petroleum Geology of China: Oil fields in Qinghai and Xizang: Beijing, Chinese Petroleum Industry Press, 483 p. (in Chinese).

Jolivet, M., Brunel, M., Seward, D., Xu, Z., Yang, J., Roger, F., Tapponnier, P., Malavieille, J., Arnaud, N., and Wu, C., 2001, Mesozoic and Cenozoic tectonics of the northern edge of the Tibetan Plateau: Fission-track constraints: Tectonophysics, v. 343, p. 111–134, doi: 10.1016/S0040-1951(01)00196-2.

Jolivet, M., Brunel, M., Seward, D., Xu, Z., Yang, J., Malavieille, J., Roger, F., Leyreloup, A., Arnaud, N., and Wu, C., 2003, Neogene extension and volcanism in the Kunlun fault zone, northern Tibet: New constraints on the age of the Kunlun fault: Tectonics, v. 22, p. 1052, doi: 10.1029/2002TC001428.

Lacassin, R., Valli, F., Arnaud, N., Leloup, P.H., Paquette, J.L., Li, H., Tapponnier, P., Chevalier, M., Guillot, S., Maheo, G., and Xu, Z., 2004, Large-scale geometry, offset and kinematic evolution of the Karakorum fault, Tibet: Earth and Planetary Science Letters, v. 219, p. 255–269, doi: 10.1016/S0012-821X(04)00006-8.

Liu, Y., Neubauer, F., Genser, J., Ge, X., Takasu, A., Yuan, S., Chang, L., and Li, W., 2007, Geochronology of the initiation and displacement of the Altyn strike-slip fault, western China: Journal of Asian Earth Sciences, v. 29, p. 243–252, doi: 10.1016/j.jseaes.2006.03.002.

Liu, Z., and Wang, C., 2001, Facies analysis and depositional systems of Cenozoic sediments in the Hoh Xil Basin, northern Tibet: Sedimentary Geology, v. 140, p. 251–270, doi: 10.1016/S0037-0738(00)00188-3.

Liu, Z., Wang, J., Wang, Y., Sun, S., Chen, Y., Zhang, J., Jiang, W., Fang, L., Li, J., Yang, F., Qu, P., and Chen, H., 1996, On Lower Tertiary chronostratigraphy and climatostratigraphy of Mangnai depression in western Qaidam Basin: Journal of Stratigraphy, v. 20, p. 104–113.

Liu, Z., Wang, Y., Chen, Y., Li, X., and Li, Q., 1998, Magnetostratigraphy and sedimentologically derived geochronology of the Quaternary lacustrine deposits of a 3000 m thick sequence in the central Qaidam Basin, western China: Palaeogeography, Palaeoclimatology, Palaeoecology, v. 140, p. 459–473, doi: 10.1016/S0031-0182(98)00048-0.

Liu, Z., Wang, C., and Yi, H., 2001, Evolution of mass accumulation of the Cenozoic Hoh Xil Basin, northern Tibet: Journal of Sedimentary Research, v. 71, p. 971–984, doi: 10.1306/030901710971.

Liu, Z., Yang, J., Wan, C., Liu, Z., and Zhang, L., 2004, Nature of Mesozoic basins in the northern edge of Qaidam Basin: Oil and Gas Geology, v. 25, p. 620–624.

Martinod, J., and Davy, P., 1994, Periodic instabilities during compression of the lithosphere: 2. Analogue experiments: Journal of Geophysical Research, v. 99, p. 12,057–12,069, doi: 10.1029/93JB03599.

Meng, Q.R., Hu, J.M., and Yang, F.Z., 2001, Timing and magnitude of displacement on the Altyn Tagh fault: Constraints from stratigraphic correlation of adjoining Tarim and Qaidam Basins: NW China: Terra Nova, v. 13, p. 86–91, doi: 10.1046/j.1365-3121.2001.00320.x.

Métivier, F., Gaudemer, Y., Tapponnier, P., and Meyer, B., 1998, Northeastward growth of the Tibet plateau deduced from balanced reconstruction of two depositional areas: The Qaidam and Hexi Corridor basins, China: Tectonics, v. 17, p. 823–842, doi: 10.1029/98TC02764.

Meyer, B., Tapponnier, P., Bourjot, L., Métivier, F., Gaudemer, Y., Peltzer, G., Shunmin, G., and Zhitai, C., 1998, Crustal thickening in Gansu-Qinghai, lithospheric mantle subduction, and oblique, strike-slip controlled growth of the Tibet Plateau: Geophysical Journal International, v. 135, p. 1–47, doi: 10.1046/j.1365-246X.1998.00567.x.

Mock, C., Arnaud, N., and Cantagrel, J.-M., 1999, An early unroofing in northeastern Tibet? Constraints from $^{40}Ar/^{39}Ar$ thermochronology on granitoids from the eastern Kunlun range (Qinghai, NW China): Earth and Planetary Science Letters, v. 171, p. 107–122, doi: 10.1016/S0012-821X(99)00133-8.

Nikishin, A.M., Cloetingh, S., Lobkovsky, L.I., Burov, E.B., and Lankreijer, A.L., 1993, Continental lithosphere folding in Central Asia (I): Constraints from rheological observations: Tectonophysics, v. 226, p. 59–72, doi: 10.1016/0040-1951(93)90110-6.

Patriat, P., and Achache, J., 1984, India–Eurasia collision chronology has implications for crustal shortening and driving mechanism of plates: Nature, v. 311, p. 615–621, doi: 10.1038/311615a0.

Peltzer, G., Tapponnier, P., and Amijo, R., 1989, Magnitude of late Quaternary left-lateral displacement along the north edge of Tibet: Science, v. 246, p. 1285–1289, doi: 10.1126/science.246.4935.1285.

Qiu, N., Kang, Y., and Jin, Z., 2003, Temperature and pressure field in the Tertiary succession of the western Qaidam Basin, northeast Qinghai-Tibet Plateau, China: Marine and Petroleum Geology, v. 20, p. 493–507, doi: 10.1016/S0264-8172(03)00080-1.

Rieser, A.B., Neubauer, F., Liu, Y., and Ge, X., 2005, Sandstone provenance of north-western sectors of the intracontinental Cenozoic Qaidam Basin, western China: Tectonic vs. climatic control: Sedimentary Geology, v. 177, p. 1–18, doi: 10.1016/j.sedgeo.2005.01.012.

Rieser, A.B., Liu, Y., Genser, J., Neubauer, F., Handler, R., Friedl, G., and Ge, X., 2006, $^{40}Ar/^{39}Ar$ ages of detrital white mica constrain the Cenozoic development of the intracontinental Qaidam Basin, China: Geological Society of America Bulletin, v. 118, p. 1522–1534, doi: 10.1130/B25962.1.

Ritts, B.D., and Biffi, U., 2000, Magnitude of post–Middle Jurassic (Bajocian) displacement on the central Altyn Tagh fault system, northwest China: Geological Society of America Bulletin, v. 112, p. 61–74, doi: 10.1130/0016-7606(2000)112<0061:MOPMJB>2.3.CO;2.

Ritts, B.D., and Biffi, U., 2001, Mesozoic northeast Qaidam Basin: Response to contractional reactivation of the Qilian Shan, and implications for the extent of Mesozoic intracontinental deformation in central Asia, *in* Hendrix, M.S., and Davis, G.A., eds., Paleozoic and Mesozoic Tectonic Evolution of Central and Eastern Asia: From Continental Assembly to Intracontinental Deformation: Geological Society of America Memoir 194, p. 293–316.

Roger, F., Arnaud, N., Gilder, S., Tapponnier, P., Jolivet, M., Brunel, M., Malavieille, J., Xu, Z., and Yang, J., 2003, Geochronological and geochemical constraints on Mesozoic suturing in east central Tibet: Tectonics, v. 22, p. 1037, doi: 10.1029/2002TC001466.

Searle, M.P., 1996, Geological evidence against large-scale pre-Holocene offsets along the Karakoram fault: Implications for the limited extrusion of the Tibetan Plateau: Tectonics, v. 15, p. 171–186, doi: 10.1029/95TC01693.

Shen, Z., Cheng, G., Le, C., and Liu, S., 1993, The Division and Sedimentary Environment of Quaternary Salt-Bearing Strata in Qaidam Basin: Beijing, Geological Publishing House, 162 p. (in Chinese).

Sobel, E.R., and Arnaud, N., 1999, A possible middle Paleozoic suture in the Altyn Tagh: NW China: Tectonics, v. 18, p. 64–74, doi: 10.1029/1998TC900023.

Sobel, E.R., Arnaud, N., Jolivet, M., Ritts, B.D., and Brunel, M., 2001, Jurassic to Cenozoic exhumation history of the Altyn Tagh range, NW China, constrained by $^{40}Ar/^{39}Ar$ and apatite fission track thermochronology, *in* Hendrix, M.S., and Davis, G.A., eds., Paleozoic and Mesozoic Tectonic Evolution of Central and Eastern Asia: From Continental Assembly to Intracontinental Deformation: Geological Society of America Memoir 194, p. 1–15.

Sokoutis, D., Burg, J.-P., Bonini, M., Corti, G., and Cloetingh, S., 2005, Lithospheric-scale structures from the perspective of analogue continental collision: Tectonophysics, v. 406, p. 1–15, doi: 10.1016/j.tecto.2005.05.025.

Song, T., and Wang, X., 1993, Structural styles and stratigraphic patterns of syndepositional faults in a contractional setting: Examples from Qaidam Basin, NW China: The American Association of Petroleum Geologists Bulletin, v. 77, p. 102–117.

Stephenson, R.A., and Lambeck, K., 1985, Isostatic response of the lithosphere with in-plane stress: Application to Central Australia: Journal of Geophysical Research, v. 90, p. 8581–8588, doi: 10.1029/JB090iB10p08581.

Sun, Z., Yang, Z., Pei, J., Ge, X., Wang, X., Yang, T., Li, W., and Yuan, S., 2005, Magnetostratigraphy of Paleogene sediments from northern Qaidam Basin, China: Implications for tectonic uplift and block rotation in northern Tibetan Plateau: Earth and Planetary Science Letters, v. 237, p. 635–646, doi: 10.1016/j.epsl.2005.07.007.

Suppe, J., Chou, G.T., and Hook, S.C., 1992, Rates of folding and faulting determined from growth strata, *in* McClay, K.R., ed., Thrust Tectonics: Suffolk, Chapman & Hall, p. 105–121.

Tang, L., Jin, Z., and Zhang, M., 2000, Tectonic development of the northern Qaidam Basin and reservoir phases of oil and gas: Petroleum Exploration and Development, v. 27, p. 36–39.

Tang, T., Yang, H., Land, X., Yu, C., Xue, Y., Zhang, Y., Wei, J., Hu, L., and Zhong, S., 1989, Cretaceous and Early Tertiary Marine-Facies Strata and Their Oil-Bearing Property, Western Tarim, Xinjiang: Beijing, Science Press, 155 p. (in Chinese).

Tapponnier, P., Peltzer, G., and Armijo, R., 1986, On the Mechanics of the Collision between India and Asia, *in* Coward, M.P., and Ries, A.C., eds., Collision Tectonics: Geological Society [London] Special Publication 19, p. 115–157.

Tapponnier, P., Meyer, B., Avouac, J.P., Peltzer, G., Gaudemer, Y., Shunmin, G., Hongfa, X., Kelun, Y., Zhitai, C., Shuahua, C., and Huagang, D., 1990, Active thrusting and folding in the Qilian Shan, and decoupling between upper crust and mantle in northeastern Tibet: Earth and Planetary Science Letters, v. 97, p. 382–403, doi: 10.1016/0012-821X(90)90053-Z.

Tapponnier, P., Xu, Z., Roger, F., Meyer, B., Arnaud, N., Wittlinger, G., and Yang, J., 2001, Oblique stepwise rise and growth of the Tibet Plateau: Science, v. 294, p. 1671–1677, doi: 10.1126/science.105978.

Van der Woerd, J., Ryerson, F.J., Tapponnier, P., Gaudemer, Y., Finkel, R.C., Meriaux, A.S., Caffee, M.W., Zhao, G., and He, Q., 1998, Holocene left-lateral slip-rate determined by cosmogenic surface dating on the Xidatan segment of the Kunlun fault (Qinghai, China): Geology, v. 26, p. 695–698, doi: 10.1130/0091-7613(1998)026<0695:HLSRDB>2.3.CO;2.

Wang, C., Liu, Z., and Yi, H., 2002, Tertiary crustal shortening and peneplanation in the Hoh Xil region: Implications for the tectonic history of the northern Tibetan Plateau: Journal of Asian Earth Sciences, v. 20, p. 221–223.

Wang, E., and Burchfiel, B.C., 2004, Late Cenozoic right-lateral movement along the Wenquan fault and its implications for the kinematics of the Qaidam Basin, the northeastern margin of the Tibetan Plateau: International Geology Review, v. 46, p. 861–879, doi: 10.2747/0020-6814.46.10.861.

Wang, E., Xu, F., Zhou, J., Wan, J., and Burchfiel, B.C., 2006, Eastward migration of the Qaidam Basin and its implications for Cenozoic evolution of the Altyn Tagh fault and associated river system: Geological Society of America Bulletin, v. 118, p. 349–365, doi: 10.1130/B25778.1.

Wang, F., Lo, C., Li, Q., Yeh, M., Wan, J., Zheng, D., and Wang, E., 2004, Onset timing of significant unroofing around Qaidam Basin, northern Tibet, China: Constraints from $^{40}Ar/^{39}Ar$ and FT thermochronology on granitoids: Journal of Asian Earth Sciences, v. 24, p. 59–69, doi: 10.1016/j.jseaes.2003.07.004.

Wang, J., Wang, Y., Liu, Z., Li, J., and Xi, P., 1999, Cenozoic environmental evolution of the Qaidam Basin and its implications for the uplift of the Tibetan Plateau and the drying of central Asia: Palaeogeography, Palaeoclimatology, Palaeoecology, v. 152, p. 37–47, doi: 10.1016/S0031-0182(99)00038-3.

Wang, Q., and Coward, M.P., 1990, The Chaidam Basin (NW China): Formation and hydrocarbon potential: Journal of Petroleum Geology, v. 13, p. 93–112, doi: 10.1111/j.1747-5457.1990.tb00255.x.

Wang, Q., McDermott, F., Xu, J.-F., Bellon, H., and Zhu, Y., 2005, Cenozoic K-rich adakitic volcanic rocks in the Hohxil area, northern Tibet: Lower-crustal melting in an intracontinental setting: Geology, v. 33, p. 465–468, doi: 10.1130/G21522.1.

Wang, Y., Zhang, X., Wang, E., Zhang, J., Li, Q., and Sun, G., 2005, $^{40}Ar/^{39}Ar$ thermochronological evidence for formation and Mesozoic evolution of the northern-central segment of the Altyn Tagh fault system in the northern Tibetan Plateau: Geological Society of America Bulletin, v. 117, p. 1336–1346, doi: 10.1130/B25685.1.

Wittlinger, G., Tapponnier, P., Oupinet, G., Jiang, M., Shi, D., Herquel, G., and Masson, F., 1998, Tomographic evidence for localized lithospheric shear along the Altyn Tagh fault: Science, v. 282, p. 74–76, doi: 10.1126/science.282.5386.74.

Xia, L., Xia, Z., and Xu, X., 2003, Magmagenesis in the Ordovician backarc basins of the Northern Qilian Mountains, China: Geological Society of America Bulletin, v. 115, p. 1510–1522, doi: 10.1130/B25269.1.

Xia, W., Zhang, N., Yuan, X., Fan, L., and Zhang, B., 2001, Cenozoic Qaidam Basin, China: A stronger tectonic inversed extensional rifted basin: The American Association of Petroleum Geologists Bulletin, v. 85, p. 715–736.

Xiao, A., Li, D., Li, X., Zhou, X., and Du, S., 2005, Evolution of the Kumukol Basin in Xinjiang: Geology of Shaanxi, v. 23, p. 59–69.

Xu, Z., Zhang, J., and Li, H., 2000, Architecture and orogeny of the northern Qilian orogenic belt, northwestern China: Geological Society of China Journal, v. 43, p. 125–141.

Yang, F., Ma, Z., Xu, T., and Ye, S., 1992, A Tertiary paleomagnetic stratigraphic profile in Qaidam Basin: Acta Petrolei Sinica, v. 13, p. 97–101.

Yang, J., Robinson, P.T., and Xu, Z., 1996, Ophiolites of the Kunlun Mountains, China and their tectonic implication: Tectonophysics, v. 258, p. 215–231, doi: 10.1016/0040-1951(95)00199-9.

Yang, J.S., Xu, Z.Q., Zhang, J.X., Song, S.G., Wu, C.L., Shi, R.D., Li, H.B., and Maurice, B., 2002, Early Paleozoic North Qaidam UHP metamorphic belt on the northeastern Tibetan Plateau and a paired subduction model: Terra Nova, v. 14, p. 397–404, doi: 10.1046/j.1365-3121.2002.00438.x.

Yang, P., Sun, Z., Li, D., Jing, M., Xu, F., and Liu, H., 2000, Ostracoda extinction and explosion events of the Mesozoic–Cenozoic Qaidam Basin, northwest China: Journal of Palaeogeography, v. 2, p. 69–74.

Yin, A., Rumelhart, P.E., Butler, R., Cowgill, E., Harrison, T.M., Foster, D.A., Ingersoll, R.V., Zhang, Q., Zhou, X.Q., Wang, X.F., Hanson, A., and Raza, A., 2002, Tectonic history of the Altyn Tagh fault system in northern Tibet inferred from Cenozoic sedimentation: Geological Society of America Bulletin, v. 114, p. 1257–1295, doi: 10.1130/0016-7606(2002)114<1257:THOTAT>2.0.CO;2.

Yue, Y., Ritts, B.D., and Graham, S.A., 2001, Initiation and long-term slip history of the Altyn Tagh fault: International Geology Review, v. 43, p. 1087–1093.

Zhang, Y., Che, Z., Liu, L., and Luo, J., 1996, Tertiary in the Kumukol Basin, Xinjiang: Regional Geology of China, v. 15, p. 311–317.

Zhao, J., Tang, W., Li, Y., Yao, C., Zhang, J., Wang, W., and Huang, Y., 2006, Lithospheric density and geomagnetic intensity in northeastern margin of the Tibetan Plateau and tectonic implications: Earth Sciences Frontier, v. 13, p. 391–400.

Zhou, J., Xu, F., and Hu, Y., 2003, Mesozoic and Cenozoic tectonism and its control on hydrocarbon accumulation in the northern Qaidam Basin of China: Acta Petrolei Sinica, v. 24, p. 19–24.

Zhou, J., Xu, F., Wang, T., Cao, A., and Yin, C., 2006, Cenozoic deformation history of the Qaidam Basin, NW China: Results from cross-section restoration and implications for Qinghai–Tibet Plateau tectonics: Earth and Planetary Science Letters, v. 243, p. 195–210, doi: 10.1016/j.epsl.2005.11.033.

Zhu, L., and Helmberger, D.V., 1998, Moho offset across the northern margin of the Tibetan Plateau: The Sciences, v. 281, p. 1170–1172.

Zhu, Y., Zhong, J., and Li, W., 1994, The Neotectonic Movement and the Evolution of Saline Lakes of the Qaidam Basin in Northeast China: Beijing, Geological Publishing House, 132 p. (in Chinese).

Manuscript Accepted by the Society 04 April 2008

Printed in the USA

The Geological Society of America
Special Paper 444
2008

Vertical-axis bending of the Altyn Tagh belt along the Altyn Tagh fault: Evidence from late Cenozoic deformation within and around the Xorkol Basin

Erchie Wang
State Key Laboratory of Lithospheric Evolution, Institute of Geology and Geophysics, Chinese Academy of Sciences, Beijing 100029, People's Republic of China

Feng-Yin Xu
Qinghai Petroleum Sub-corporation of PetroChina Company Ltd., Dunhuang 736202, People's Republic of China

Jian-Xun Zhou
Earth Science Department, University of Petroleum, Beijing 102249, People's Republic of China

Shifeng Wang
Chung Fan
Gang Wang
Institute of Geology and Geophysics, Chinese Academy of Sciences, Beijing 100029, People's Republic of China

ABSTRACT

The Altyn Tagh fault is one of the largest intracontinental strike-slip faults in the world, extending linearly ~1500 km along the northern edge of the Tibetan Plateau. All tectonic units bounded by the fault on the south, such as the Qaidam Basin and the Qilian Shan thrust belt, have experienced intensive shortening and uplift due to transfer motion from the Altyn Tagh. However, questions as to whether the tectonic units north of the Altyn Tagh fault have experienced associated deformation and whether the Altyn Tagh fault itself has experienced deformation remain unexplored. Our field study shows that the middle part of the Altyn Tagh fault separates the Altyn Tagh belt to the north from the Qaidam Basin to the south. The former was formed as a NE-SW–trending, lens-shaped block, consisting mainly of metamorphic rocks of Proterozoic age, and the latter is a basin that has been elongated in the E-W direction and filled with a thick succession of fluvial and lacustrine sediments of Cenozoic age. The Altyn Tagh fault has been considered to progressively propagate linearly to the northeast, accommodating the northeastward growth of the Tibetan Plateau by transfer of horizontal motion into intensive crustal shortening and uplift along the northeastern margin of the plateau. However, structural data from the main part of the fault show that this part of the fault was dominated by transtensional deformation in late Cenozoic time, which resulted in the subsidence of the large Xorkol Basin

Wang, E., Xu., F.-Y., Zhou, J.-X., Wang, S., Fan, C., and Wang, G., 2008, Vertical-axis bending of the Altyn Tagh belt along the Altyn Tagh fault: Evidence from late Cenozoic deformation within and around the Xorkol Basin, *in* Burchfiel, B.C., and Wang, E., eds., Investigations into the Tectonics of the Tibetan Plateau: Geological Society of America Special Paper 444, p. 25–44, doi: 10.1130/2008.2444(02). For permission to copy, contact editing@geosociety.org.

and deposition of a succession of fluvial and lacustrine sediments as old as Pliocene age. The foliation in the early Proterozoic metamorphic rocks within the Altyn Tagh belt generally strikes NE-SW, but the foliation along the main part of the belt is bent around the Xorkol Basin to strike NW-SE. Such deformation caused the increase in the width of the belt. These data led us to infer that the Altyn Tagh belt has experienced oroclinal bending along the main part of the Altyn Tagh fault around a vertical axis. From the difference in length along the northern and southern edges of the basin, as much as 60 km of late Cenozoic left-lateral motion along the main part of the Altyn Tagh fault was absorbed by the crustal bending and associated extension. The bending of the Altyn Tagh belt implies that the southwestern movement of the Altyn Tagh belt relative to the Qaidam Basin met with strong resistance, which is interpreted to have been generated by NE-SW compression along an E-W–trending segment of the Altyn Tagh fault to the southwest of the Xorkol Basin as part of a restraining bend.

Keywords: Tibetan plateau, Altyn Tagh belt, Altyn Tagh fault, bending, late Cenozoic time.

INTRODUCTION

Marginal areas of the Tibetan Plateau occur where the stresses generated by the India-Eurasia convergence have been released by different strains, characterized by compression, extension, and strike-slip deformation. The strain along the northern edge of the plateau is mainly dominated by left-lateral strike-slip motion on the Altyn Tagh fault, one of the largest intracontinental strike-slip faults in the world, which resulted in growth of the northern and northeastern margins of the plateau by associated shortening, uplift, extension, and rotation (Fig. 1). This fault is a prominent NE-SW–trending linear feature extending along the northern periphery of the plateau for ~1500 km. It separates the Tarim Basin and the Altyn Tagh belt to the northwest from the Qilian Shan belt, the Qaidam Basin, and the Kunlun belt to the southeast. Because of its central role in the intracontinental deformation of the Tibetan Plateau, its evolution has been the focus of a number of studies (Peltzer, and Tapponnier, 1988; Burchfiel et al., 1989; Peltzer et al., 1989; Avouac and Tapponnier, 1993; Matte et al., 1996; Wang, 1997; Wittlinger et al., 1998; Rumelhart et al., 1999; Ritts and Biffi, 2000; Tapponnier et al., 2001; Yin et al., 2002; Cowgill et al., 2004; Mériaux et al., 2004). These studies have focused mainly on the deformation, sedimentation, and thermochronologic events along the fault to determine timing and magnitude of the left-lateral movement. To do so, it was assumed that the fault remained as a linear feature during its

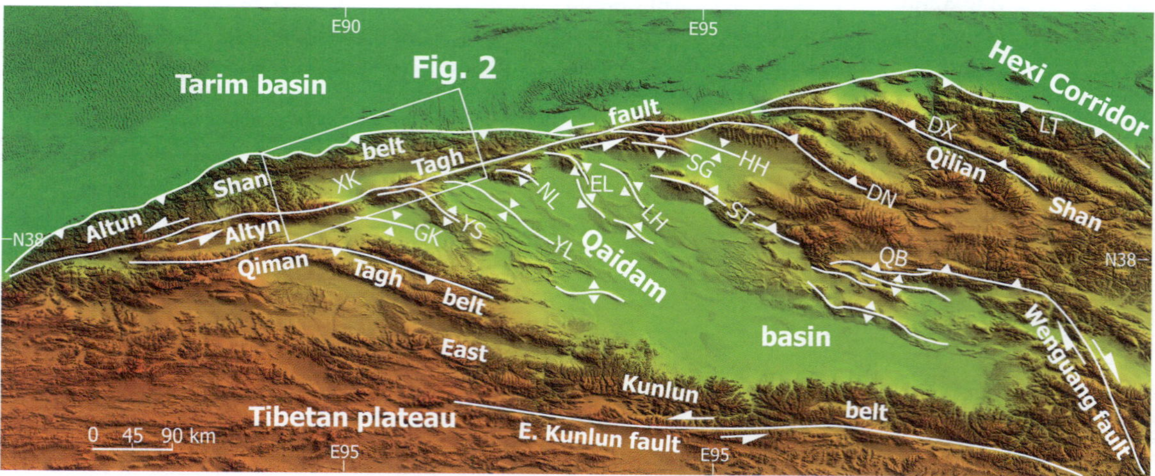

Figure 1. Shuttle Radar Topography Mission (SRTM) image showing the topographic features of the northern and northeastern margins of the Tibetan Plateau and some of the main active strike-slip faults on SRTM. Inset shows location of the region relative to major tectonic elements of the Tibetan Plateau. Location of Figure 2 is indicated. XK—Xorkol Basin; GK—Gasikule syncline; YS—Yousha Shan anticline; YL—Yiliping syncline; NL—Niubiliang anticline; EL—Eboliang anticline; LH—Lenghu fold-and-thrust belt; SG—Sugan syncline; ST—Saishiteng fold-and-thrust belt; QB—Qaibeiyuan fold-and-thrust belt; DN—Danghenan Shan fold-and-thrust belt; DX—Daxue Shan fold-and-thrust belt; LT—range-front thrust fault of the Qilian Shan belt.

propagation to the northeast. However, the Altyn Tagh fault has had a prolonged evolutionary history (Yin et al., 2002), and it marks the boundary of several major fault blocks with contrasting geological character. The blocks include the Kunlun belt, the Qaidam Basin, and the Qilian Shan belt to the southeast, and the Tarim Basin and the Altyn Tagh belt to the northwest. Due to transfer motion from the Altyn Tagh fault, the tectonic units south of the fault have been intensively shortened in a NE-SW direction, forming NW-SE–trending thrust faults and folds, which are often described as the horst-tail structures (Tapponnier et al., 2001). However, it is unknown whether the Tarim Basin and the Altyn Tagh belt north of the fault have experienced associated deformation. Moreover, the Altyn Tagh fault is one of the oldest Cenozoic structures within the Tibetan Plateau, and its left-lateral movement is considered to have been initiated in Eocene time (Tapponnier et al., 1981; Yin et al., 2002). Thus, the geometry of the fault as it propagated to the north through time is an unexplored question. In addressing these questions, we conducted a field study on the late Cenozoic deformation on the middle segment of the Altyn Tagh fault during 2004–2005. This part of the fault contains complex geometric features, in particular, a widening where a large Cenozoic basin, the Xorkol Basin, and several high mountains are present.

GEOLOGIC BACKGROUND

Although many aspects of the Altyn Tagh fault evolution remain either unclear or in dispute, there is general agreement that its left-lateral displacement decreases northeastward by transfer of displacement to a series of NW-SE–trending folds and thrust faults, known as horst-tail structures. These structures cause the deformation along the northern and the northeastern margins of the plateau by shortening, extension, uplift, and rotation. Here, the plateau margin contains four tectonic units, which include the Altyn Tagh belt north of the Altyn Tagh fault and the Qilian Shan, Qaidam Basin, and Kunlun Shan south of the fault (Yin et al., 2002; Wang et al., 2006) (Fig. 1). The Altyn Tagh belt (Altyn Tagh mean cypress mountain in Mongolian) marks the northern margin of the Tibetan Plateau, bounding the Tarim Basin on the southeast. It was formed as a lens-shaped block, ~650 km long and 80 km wide in its middle part, and it is underlain by metamorphic rocks that range from late Archean to Proterozoic in age (Xijiang BGMG, 1993) (Fig. 2). The Archean rocks are assigned to the Milan Group, which has been dated as 2462.5 Ma by U-Pb method. The Proterozoic unit is divided into two parts, the lower part is assigned to the Altyn Tagh Group, dated as early Proterozoic, and the upper part is assigned to the Bashikurgan Group, dated as middle to late Proterozoic age. Some Lower and Middle Ordovician strata crop out along the northeastern margin of the belt and consist mainly of basic volcanic rocks mixed with ultrabasic rocks. The Qaidam Basin lies between the Qilian Shan belt to the northeast and the Kunlun belt to the southwest and is bounded by the Altyn Tagh fault on the northwest. It is the largest Cenozoic intermountain basin within the Tibetan Plateau, it is filled with a thick sequence of Cenozoic terrestrial sediments, and it forms a NW-SE–trending synclinorium by transfer of left-lateral movement along the Altyn Tagh fault into shortening (Burchfiel et al., 1989; Yin et al., 2002; Wang et al., 2006). The basin also contains a series of smaller-scale folds and thrust faults that merge with the Altyn Tagh fault at their western ends.

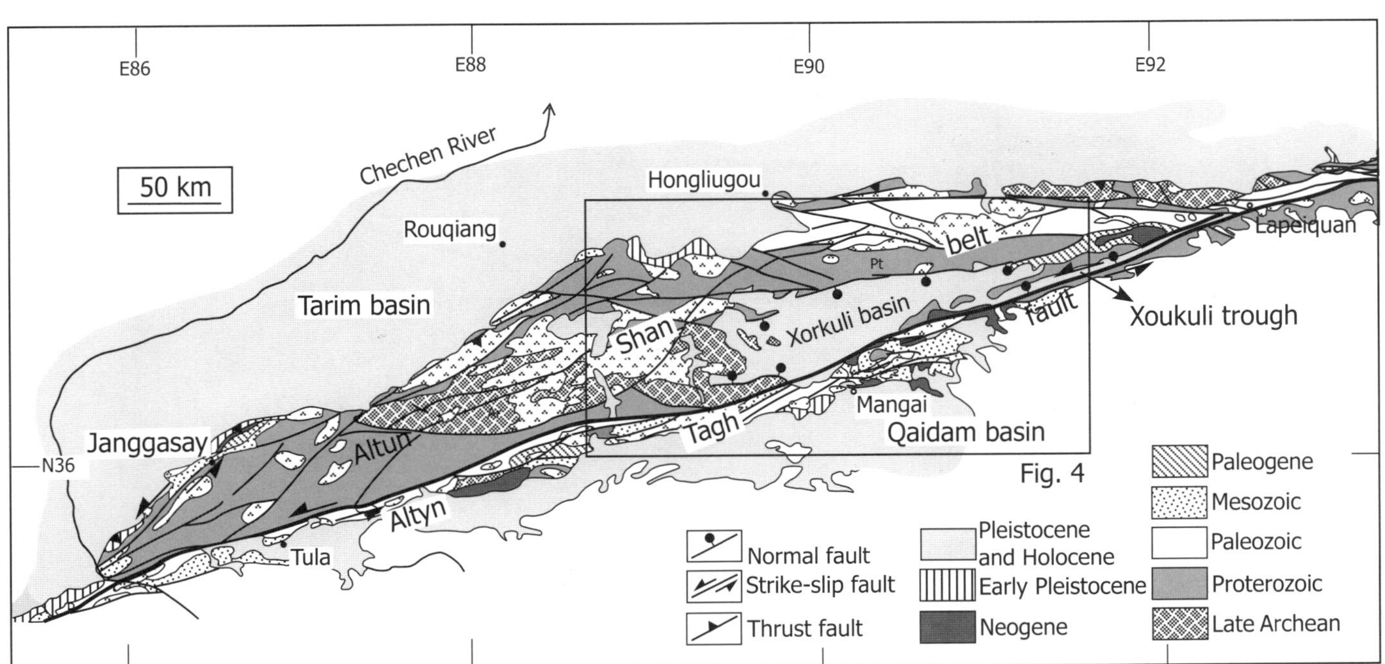

Figure 2. Simplified geological map of the Altyn Tagh belt, showing the first-order structural features and rock units.

The largest of these folds is the Yousha Shan anticline, which is cored by various pre-Cenozoic rocks flanked by Tertiary clastic rocks. The northern part of the E-W–trending anticline forms the Akatengneng Shan, which bounds the Xorkol Basin along the Altyn Tagh fault.

The Altyn Tagh fault has a clear surface expression in the field and on satellite images marked by linear troughs, steep fault scarps, stream courses, and elongated slivers along which series of streams are offset left-laterally including the Chechen River (Yin et al., 2002). Strike-slip motion on the Altyn Tagh fault merges into the Qilian Shan thrust belt along its northeast end and into the West Kunlun thrust belt along its southwest end. Based on such relationships, the Altyn Tagh fault is a major transfer fault that transfers strike-slip displacement to crustal shortening from one place to another and is interpreted to be limited within the crust (Burchfiel et al., 1989). This model is supported by some geophysical evidence (Zhao et al., 2006). Others have argued that this fault extends into the upper mantle (Wittlinger et al., 1998; Bedrosian et al., 2001). Based on the relationships with the tectonic units that are cut by it, we divide the Altyn Tagh fault into three segments: a northeastern segment that divides the Tarim Basin to the northwest from the Qilian Shan to the southeast, a middle segment between the Altyn Tagh belt and the Qaidam Basin, and a southwestern segment that offsets the Kunlun belt for 500–600 km (Peltzer and Tapponnier, 1988; Matte et al., 1996).

The Xorkol Basin developed along the middle segment of the Altyn Tagh fault, and it is 150 km long and 40 km wide in the west and 5 km wide in the east (Figs. 3 and 4; see Fig. 5 for index map of all figures). It pinches out to the northeast along the 100-km-long and 2-km-wide linear Xorkol trough between the Altyn Tagh range to the north and the marginal range of the Qaidam Basin to the south. The basin is largely covered by Quaternary alluvial fans shed from the Altyn Tagh belt to the north, the East Akato Tagh to the southwest, and the Akatengneng Mountains to the south. A seasonal river flows from northeast to southwest along the center of the basin, ending in Wushuxiao Lake to the southwest. The Altyn Tagh belt, which bounds the basin on the north, is mainly underlain by various types of metamorphic rocks of Archean and Proterozoic age, whereas the belt that bounds the basin on the northeast contains some Tertiary red beds, which have been assigned an Eocene and Miocene age.

The Xorkol Basin was suggested by Guo et al. (1998) to have formed as a pull-apart basin along a releasing bend along the Altyn Tagh fault, based on a mechanical modeling study. Chen et al (2004) proposed that this basin formed as an erosional basin that was previously connected with the Qaidam Basin. However, large parts of this basin and adjacent areas have not

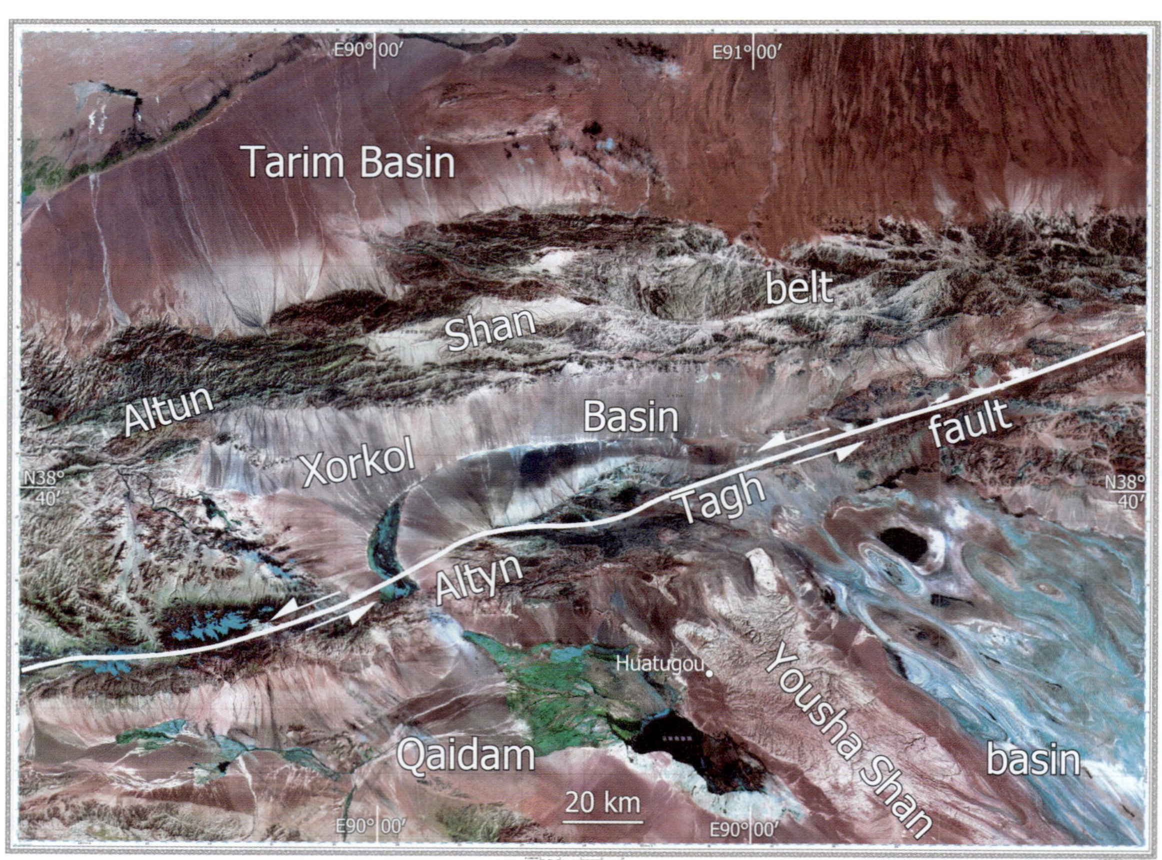

Figure 3. Digital elevation model (DEM) image of topographic features of the Xorkol Basin and adjacent areas.

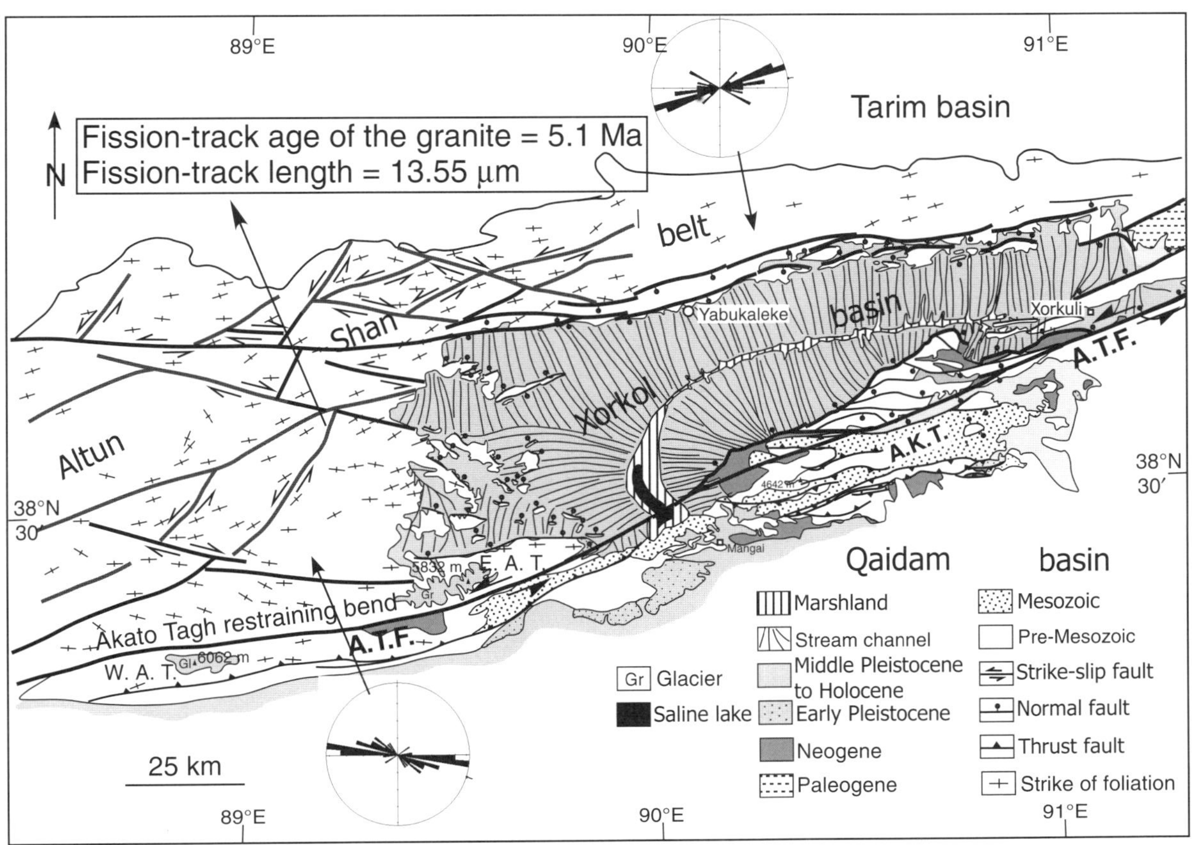

Figure 4. Simplified geological map of the Xorkol Basin and adjacent areas. A.T.F.—Altyn Tagh fault; A.K.T.—Akatengneng Shan; W.A.T.—West Akato Tagh (6062 m); E.A.T. —East Akato Tagh (5832 m).

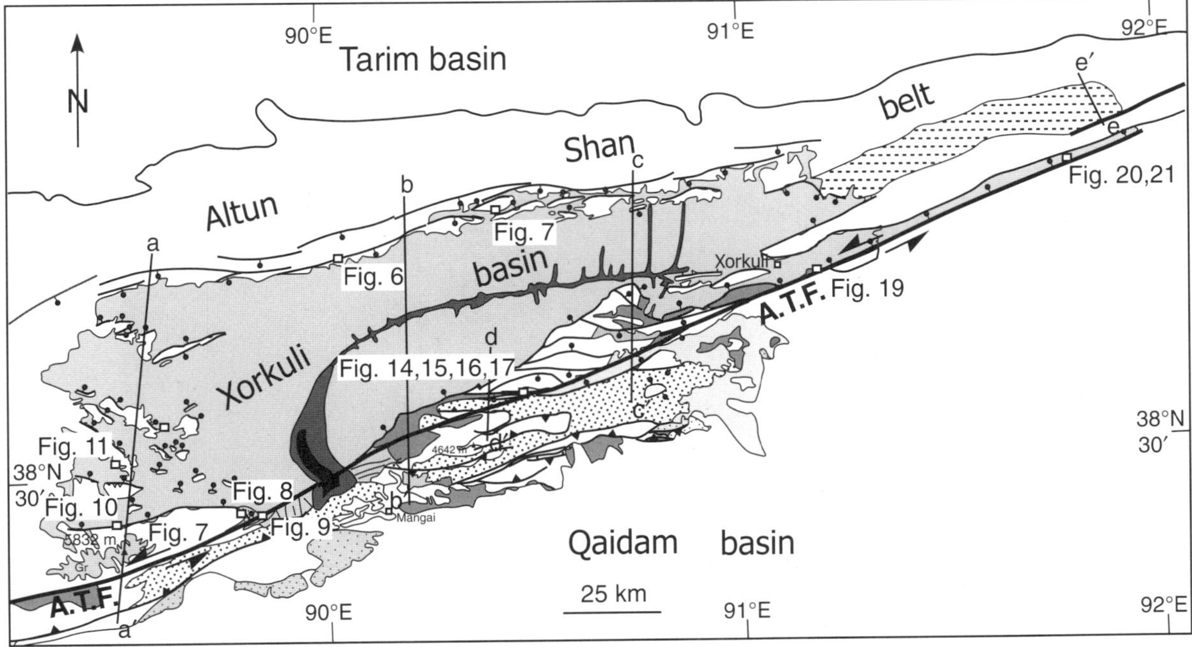

Figure 5. Index map of the figures. The location of all photographic figures is shown. Locations of cross sections a–a′, b–b′, c–c′, d–d′, and e–e′ shown in Figures 12, 13, 18, and 22 are indicated. A.T.F.—Altyn Tagh fault.

been investigated through systematic field study, so its origin and tectonic correlation with the middle segment of the Altyn Tagh fault remain somewhat unclear.

XORKOL BASIN

Northern Boundary Deformation of the Xorkol Basin

The northern margin of the Xorkol Basin is topographically steep and defined by a series of discontinuous normal faults that dip mostly south at ~30°–50°. The Proterozoic metamorphic rocks of the Altyn Tagh belt exposed along the faults, characterized by schist, quartzite, and marble, are deformed along a series of narrow grabens coupled with a horst (Figs. 3 and 4). In the Yabukaleke area, the northern boundary normal fault lies along a coupled graben and horst, ~1 km wide and 5 km long. The graben is filled with Quaternary sediments, and the horst is underlain by dark-gray, well-bedded marble, tilted to the north. Rocks within the normal fault zones are mostly brecciated and penetrated by a set of concentrated cleavage that dips to the south at high angle (Figs. 6 and 7). A main branch of the northern boundary normal fault system of the basin can be traced westward into the Altyn Tagh belt, where it juxtaposes Tertiary purple-colored clastic rocks in the hanging wall against the Proterozoic metamorphic rocks in the footwall. The clastic strata are dated as Neogene and tilted to the north 35°–70°, constraining the age of the fault.

Southern Boundary Deformation of the Xorkol Basin

The southern boundary of the Xorkol Basin is approximately defined by the middle segment of the Altyn Tagh fault, which bounds the East Akato Tagh and the Akatengneng Shan on the south and north, respectively (Figs. 3 and 4). The East Akato Tagh, with its highest peak at 5832 m, marks one of the highest topographic features in the northern margin of the plateau along the southern margin of the Altyn Tagh belt, and it is composed entirely of late Archean and Proterozoic metamorphic rocks (Fig. 8). The middle segment of the Altyn Tagh fault lies within a steep-sided valley, dips steeply north, and bounds the East Akato Tagh on the south. At the eastern end of the East Akato Tagh, the fault passes through an elongated hill, along a 20- to 100-m-wide notch (Fig. 9). The hill is about 20 m high and is underlain by late Cenozoic sediments separated into two parts by an unconformity. The lower part consists of rhythmically layered sediments assigned to the Pliocene, and the upper part consists of poorly sorted, unconsolidated, coarse-grained conglomerate, presumably glacial deposits of late Pleistocene in age. These sediments are consistently tilted to the south at 35° along the Altyn Tagh fault, which dips to the north at 40°. The relations here show that the Altyn Tagh fault has a normal component and was still active after late Pleistocene time. To the northeast, the fault ends within Wushuxiao Lake, where the dark-gray modern lacustrine deposits are strongly disturbed and squeezed to the surface. From there, northeastward to

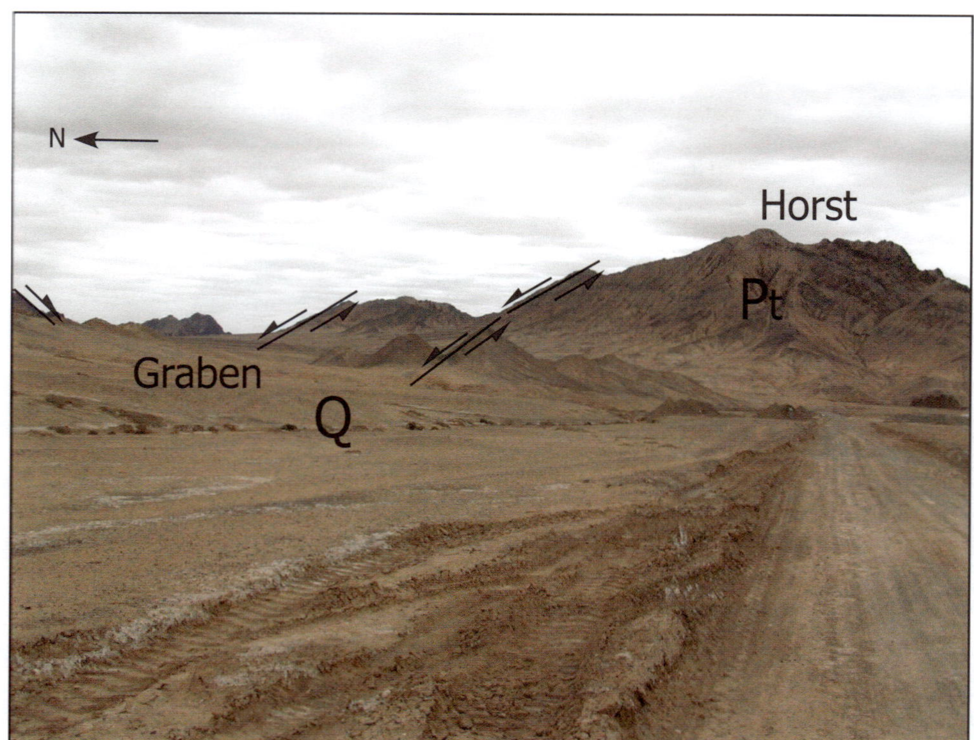

Figure 6. Looking east along a coupled graben and horst developed along the northern edge of the Xorkol Basin. Pt—Proterozoic rocks; Q—Quaternary.

the Akatengneng Mountains, the fault loses its clear identity within the alluvial-fan deposits.

The northern range front of the East Akato Tagh is characterized by a steep normal fault that strikes NW-SE and merges with the middle segment of the Altyn Tagh fault to the southeast and with the northern boundary fault of the Xorkol Basin to the northwest (Figs. 8 and 10). It separates the Archean metamorphic rocks in the footwall from the Quaternary sediments of the Xorkol Basin in the hanging wall. The metamorphic rocks are characterized by marble, schist, gneiss, and quartzite, the foliation of which strikes mostly NW-SE instead of NE-SW, as do the first-order structures of the belt. The normal fault scarps dip to the north at 40° and are up to ~1000 m high. The metamorphic rocks, mostly marble, distributed along the range front normal fault are sheared

Figure 7. Looking west at breccias penetrated by cleavages within the northern boundary normal fault of the Xorkol Basin. Arrows show the sense of normal shear. Pt—Proterozoic rocks.

Figure 8. View to the west at the eastern flank of the East Akato Tagh (5832 m), one of the highest mountains along the northern margin of the plateau along the trace of the Altyn Tagh fault. The mountain is shown as an asymmetric topographic feature, tilted to the south along its northern boundary fault, which is recognized as a N-dipping normal fault. The mountain is limited on the south by a NE-trending steep valley, along which rocks as young as Pliocene in age are cut by the Altyn Tagh fault. Note the hills in the piedmont of the mountain, which are formed as fault blocks, tilted to the south against the mountain along the range-front normal fault.

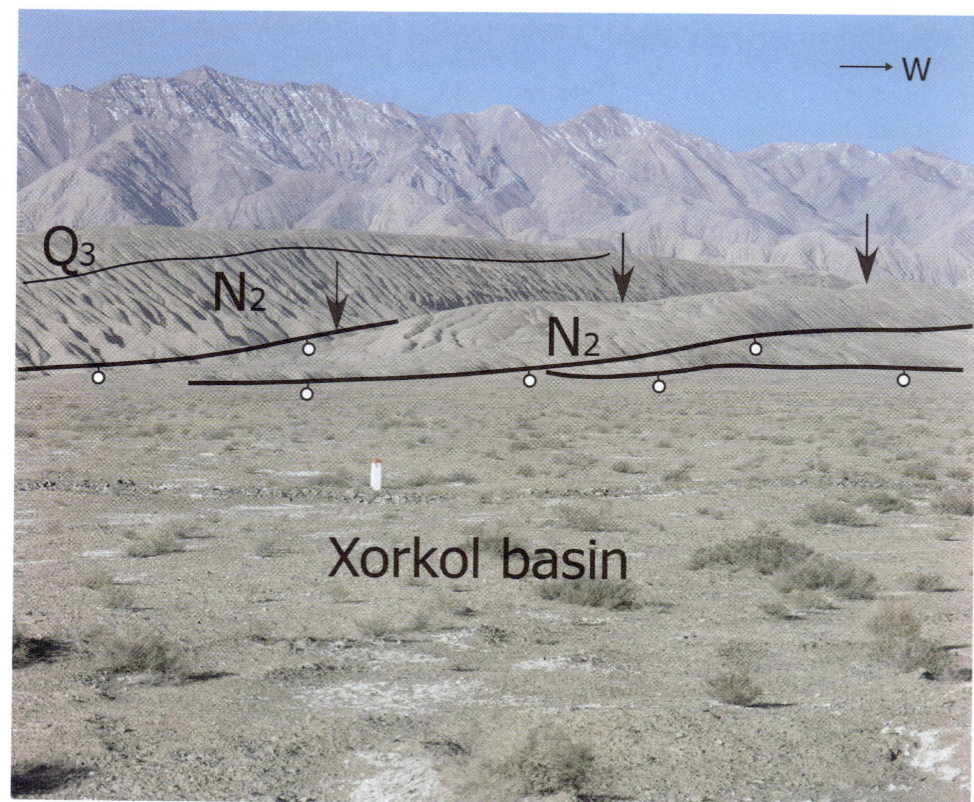

Figure 9. View to the southwest at the Altyn Tagh fault, which has offset the Pliocene fine-grained conglomerate by dip-slip movement. The Pliocene rocks in the hanging wall and footwall are consistently tilted to the south along the fault. Arrows point to the trace of the fault. Q_3—Late Pleistocene; N^2—Pleistocene.

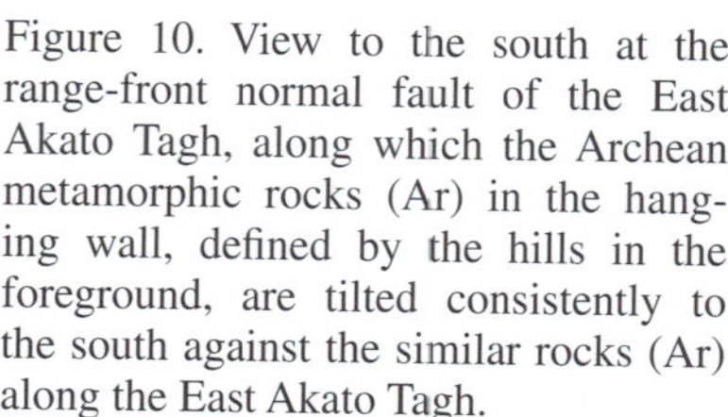
Figure 10. View to the south at the range-front normal fault of the East Akato Tagh, along which the Archean metamorphic rocks (Ar) in the hanging wall, defined by the hills in the foreground, are tilted consistently to the south against the similar rocks (Ar) along the East Akato Tagh.

into breccias and fault gouge intercalated with the metamorphic rocks. The fault scarp forms a series of triangle fault facets that have been incised by glacial activity along a series of U-shaped valleys, within which enormous coarse-grained moraine deposits were shed into the piedmont of the mountain from its summit. Strikingly, a number of elongated fault blocks with various sizes within the west part of the Xorkol Basin are underlain by a variety of metamorphic rocks of Archean and Proterozoic age, similar to those along the Altyn Tagh belt, rising 10–200 m above the piedmont surface (Fig. 11). The strike of the metamorphic rocks in these fault blocks is consistently oriented E-W, parallel to the northern and southern boundary normal faults of the basin. As demonstrated by the asymmetric topography of these fault blocks, the metamorphic rocks are tilted both to the south and north. Based on the geologic and geomorphic evidence, these fault blocks are interpreted to have formed as horsts resulting from N-S extension across the Xorkol Basin. In fact, the East Akato Tagh is also topographically asymmetric, indicating that it was also tilted south as a whole along its northern and southern boundary faults (Fig. 8). As shown in cross section a–a′ in Figure 12, a large fault block lying

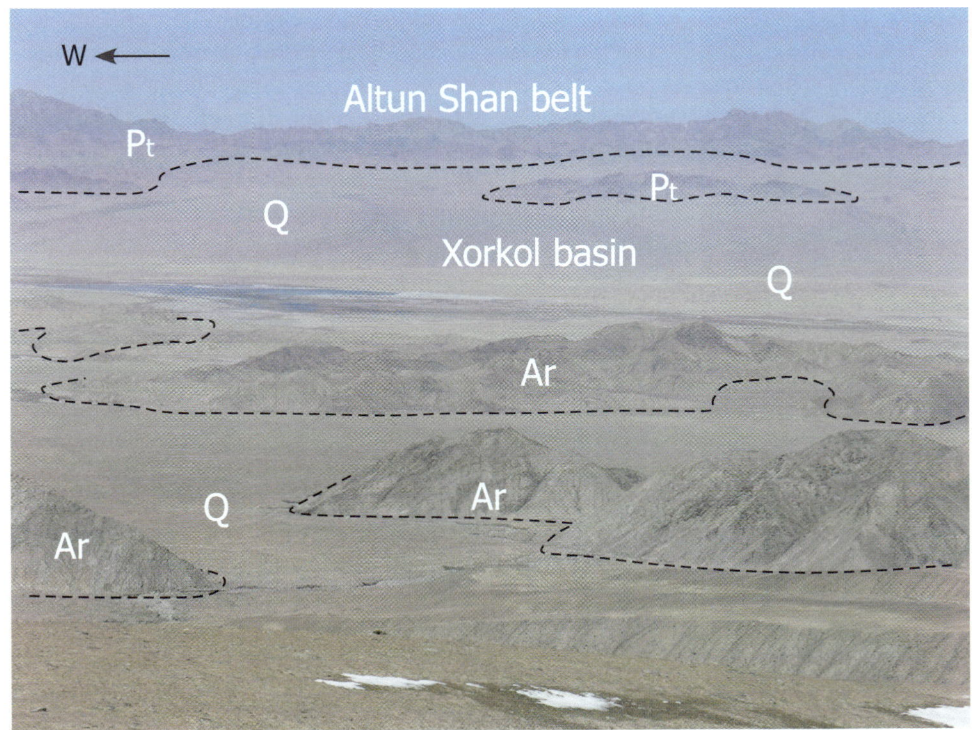

Figure 11. Looking down to the north into the western part of the Xorkol Basin and isolated hills within the basin, elongated in E-W direction, that mark horsts. Ar—Archean rocks, Pt—Proterozoic rocks; Q—Quaternary.

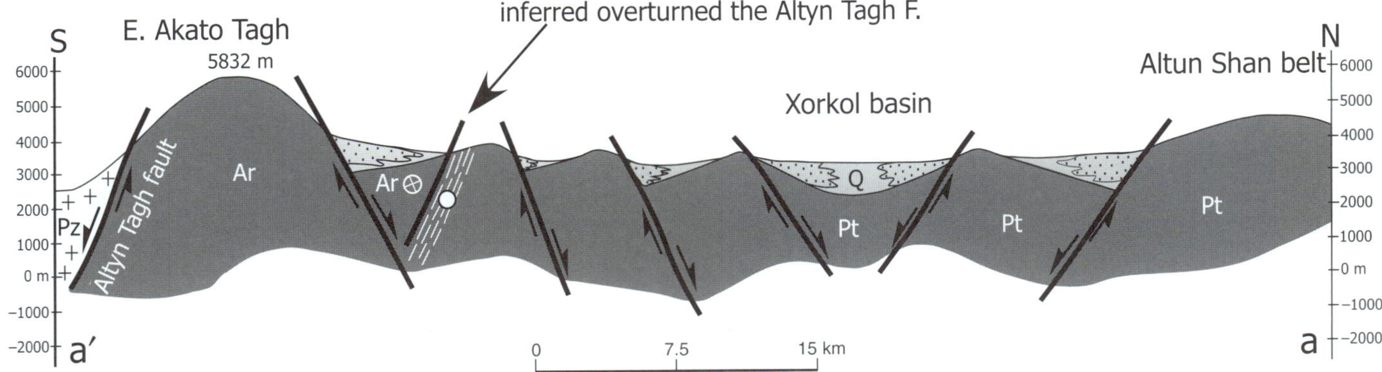

Figure 12. The cross sections of the western part of the Xorkol Basin in roughly N-S direction, constructed based on our surface observations. See Figures 2–4 for location.

to the north of the range-front normal fault of the East Akato Tagh is composed of Proterozoic granite. Rocks along northern margin of the block are sheared into mylonite, dipping to the south at 40°. The mylonite contains clear structural features, such as S-C fabric and rotated crystals, that indicate right-lateral faulting. This fault block can be inferred to have been derived from the top of the East Akato Tagh along the range-front normal fault, along which the mylonite was overturned, and originally they may have resulted from the left-lateral faulting along the Altyn Tagh fault. In contrast to the fault blocks along the southern margin of the basin, the fault blocks along the northern margin of the basin were tilted to the north, by normal faulting along the northern boundary fault of the basin. Cross section a–a′ in Figure 12 was constructed based on the surface observations discussed previously, which indicate that the west part of the Xorkol Basin is a complex graben formed by a series of normal faults that dip both to the north or south.

The Akatengneng Shan separates the Xorkol Basin from the Qaidam Basin in roughly an E-W direction and is underlain by four lithological units of Proterozoic, early Paleozoic, Jurassic, and Tertiary ages that were shortened to form a complex anticline, plunging to the southeast (Figs. 3–4). The core of the anticline is formed by gneiss and granite of Proterozoic age and metamorphosed basic volcanic rocks of early Paleozoic age intruded by granite of early Paleozoic age. They are overlain by a set of highly deformed Lower to Middle Jurassic clastic rocks on the north, east, and southeast. The Pliocene rocks, characterized by light-brown fine-grade clastic rocks, discontinuously flank the anticline on the north and south. To the southeast, the anticline joins obliquely with the Yousha Shan anticline, one of the largest en-echelon structures within the Qaidam Basin that exposes mainly a Tertiary succession (Wang et al., 2006).

The northern part of the Akatengneng Mountains is bounded by two primary branches of the Altyn Tagh fault and forms an isolated, rhombic-shaped block, culminating in the Akatengneng Shan at 4642 m (Fig. 13; see Fig. 5 for location). The fault flanking the block on the north is a normal fault that dips 40°–60°N and is marked by a series of triangle fault facets about 200–500 m high. The rocks truncated by the fault are mostly light-brown clastic rocks of Pliocene age that dip 25°–45°N. The Pliocene rocks in the hanging wall of the range-front normal fault of the northern part of the Akatengneng Shan are tilted to the south against the early Paleozoic volcanic rocks in the footwall to form a series of low-relief hills, along which the groundwater comes to the surface. A primary branch fault of the Altyn Tagh fault bounds the northern part of the Akatengneng Shan on the south and is marked by a magnificent south-sloping cliff, about 500 m high, that divides the early Paleozoic metamorphosed volcanic rocks in the footwall from Jurassic dark-gray sandstone and slate in the hanging wall (Fig. 14). The fault scarp developed along a 5-m-thick vein of white quartz that was sheared into breccia and fault gouge, which is marked by slickenlines that plunge to the south, suggestive of hanging-wall-down displacement (Fig. 15). About 200 m to the east, the multicolored Proterozoic volcanic rocks are offset by the Altyn Tagh fault with a normal separation of about 100 m (Fig. 16). Additionally, series of stream channels are deflected left-laterally along the foot of the fault scarps about 20 m (Fig. 17), indicating that the middle segment of the Altyn Tagh fault in this region was dominated by left-lateral transtensional deformation. From this main branch fault southwestward, much of the middle segment of the Altyn Tagh fault cuts through the Pliocene clastic rocks that dip south 40° along the western flank of the Akatengneng Shan. The fault demonstrates clear evidence for transtensional deformation, as shown by the left-lateral

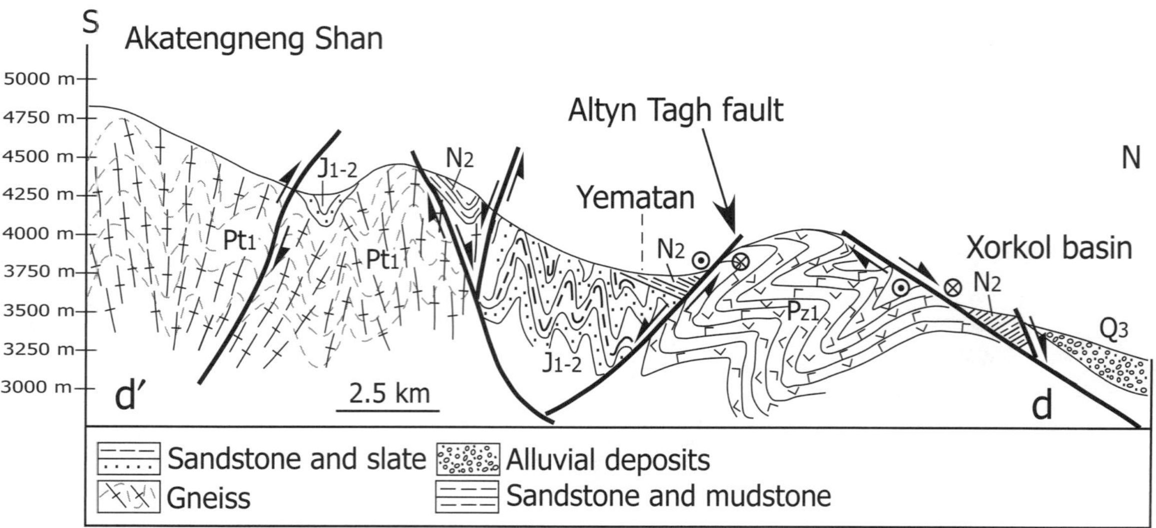

Figure 13. The cross section of the northern part of the Akatengneng Shan in roughly N-S direction. The transtensional deformation of the fault is shown by left-lateral offset of a series of tectonic units and by the juxtaposition of the tilted Pliocene rocks with the strongly deformed pre-Cenozoic rocks.

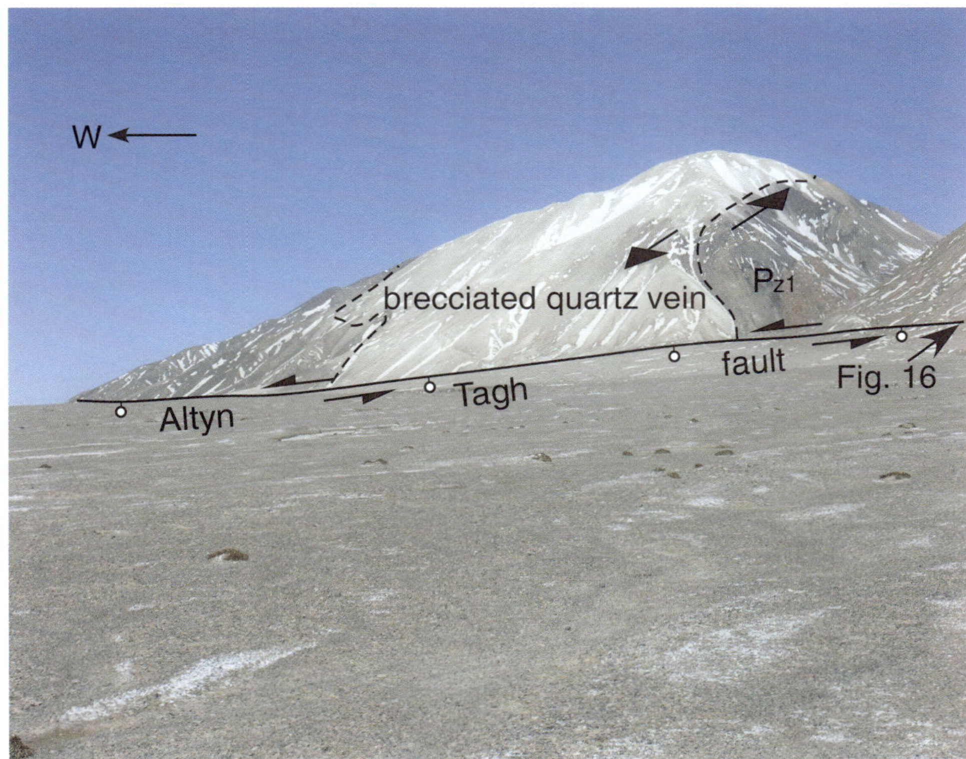

Figure 14. View to the northwest to a primary branch fault of the Altyn Tagh fault that limits the southern edge of the northern part of the Akatengneng Shan. The fault scarp is about 1000 m high, and it is defined by a 5- to 200-m-thick white, brecciated quartz vein. Pz1—early Paleozoic. Location of Figure 17 is indicated.

Figure 15. A close-up of the breccias in Figure 14 and fault features on the slickenside surface indicate hanging-wall-down displacement.

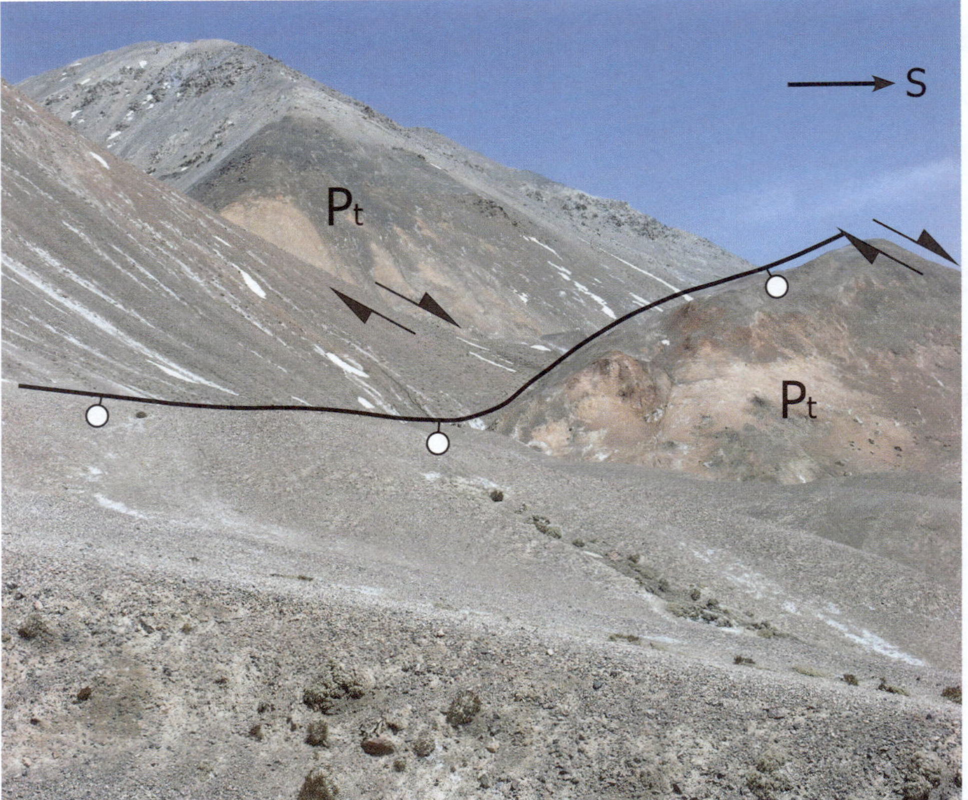

Figure 16. Looking northeast at the Altyn Tagh fault along the southern edge of the northern part of the Akatengneng Shan. The dip-slip movement of the fault is shown by the offset of the purple-colored volcanic rocks of Proterozoic age (Pt).

Figure 17. Looking south at the Altyn Tagh fault, flanking the northern part of the Akatengneng Shan on the south. The left-lateral movement of the fault is shown by consistent offset of two stream channels, with a displacement of 50 m.

offset of a series of tectonic units, including Paleozoic volcanic rocks, Jurassic slate, and Pliocene clastic rocks of 2000 m to 5000 m, respectively, and by northward tilting of the Pliocene clastic rocks in the hanging wall, juxtaposed with early Paleozoic volcanic rocks in the footwall.

The deep structure in the east part of the Xorkol Basin is shown on two N-S cross sections in Figure 18, which were constructed from electronic prospecting data provided by Qinghai Oil Company. It confirms that the northern and southern boundary faults of the Xorkol Basin formed as normal faults and demonstrates that the basin is a graben, in which the oldest Cenozoic sediments are late Neogene in age, with a maximum thickness of 1000 m, suggesting the subsidence of the basin initiated at that time.

Deformation along the Xorkol Trough

The Xorkol Basin pinches out northeastward along the linear Xorkol sedimentary trough, which is expressed as a prominent linear feature in the field and on satellite images (Figs. 3 and 4). It is about 150 km long and 5 km wide and bounded by the Altyn Tagh belt on the north and the marginal fold belt of the Qaidam Basin on the south along the middle segment of the Altyn Tagh fault. The former is composed of Proterozoic metamorphic rocks, whereas the latter is composed of the basement rocks of the Qaidam Basin, characterized by high-grade metamorphic rocks of Proterozoic age, volcanic rocks of early Paleozoic age, and gray clastic rocks of Jurassic age. The trough is dry and floored by alluvial and fluvial deposits, on which some short seasonal rivers flow to the southwest or northeast. The middle segment of the Altyn Tagh fault crops out along the northern or southern edges of the trough, marked by fault scarps, elongated fault slivers, and sag ponds, along which the stream channels are disturbed and modern sediments are ruptured. At the southwest end of the trough, the Altyn Tagh fault crops out along the southern edge of the trough where it is marked by a series of stepped normal faults with 200- to 500-m-high modified scarps, along which rocks as young as Neogene in age are dropped down to form terraces, tilted consistently 20°–45° to the south against the similar rocks in the footwall. Two kilometers south of Xorkol, the Neogene rocks truncated by the fault are divided into Miocene and Pliocene units (Lower Yousha Shan Formation and Shizigou Formation). In the area southeast of Xorkol, the Altyn Tagh fault crops out along the southern edge of the trough along a 5-m-thick, pure white quartz vein that dips north 35°. The quartz vein is sheared into breccia and gouge, suggestive of formation by normal faulting (Fig. 19). At the northeastern end of the trough, the Altyn Tagh fault is exposed along the northern edge of the trough, where it dips north 45° along the northern flank of an elongated fault sliver. The latter is about 1000 m long, 150 m wide, and 100 m high, and it is composed of light-brown clastic rocks of Pliocene age, titled to the south by a component of

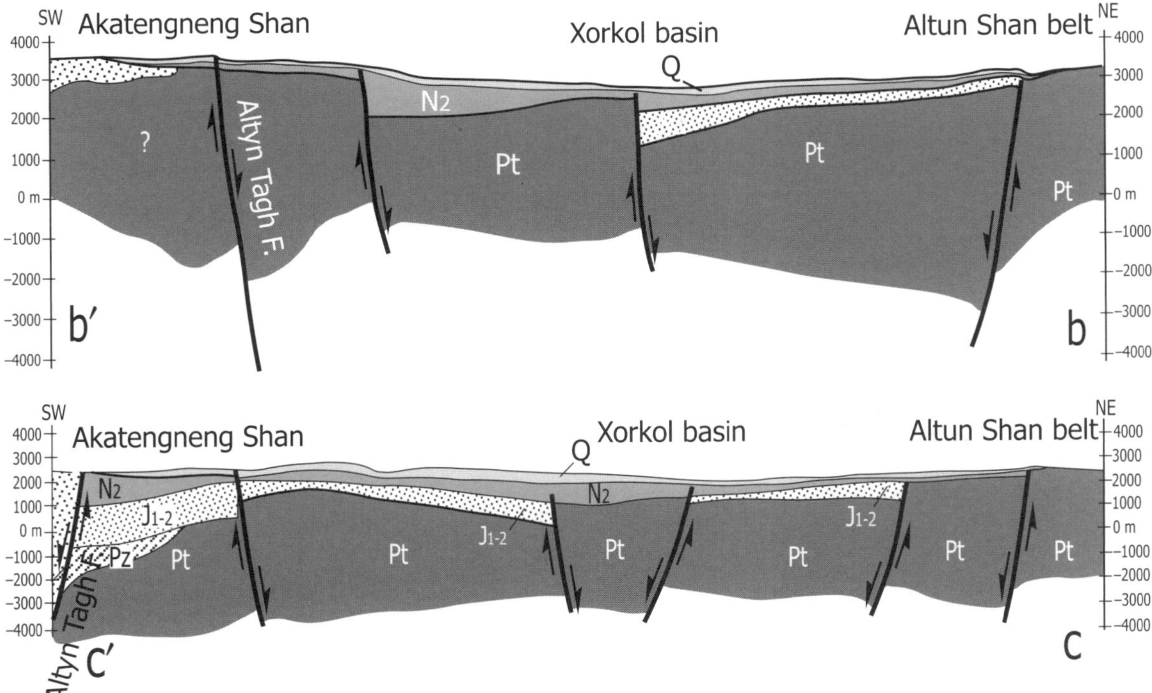

Figure 18. The cross sections of the middle and east parts of the Xorkol Basin and Akatengneng Shan, constructed based on the electronic prospecting data provided by Qinghai Petroleum Sub-corporation of PetroChina Company Limited Dunhuang. The basal sediments of the basin are Pliocene in age and have a maximum thickness of 1000 m. Q—Quaternary; N_2—Pliocene; J_{1-2}—Lower and middle Jurassic; Pz—Early Paleozoic; Pt—Proterozoic.

normal faulting on the Altyn Tagh. The Altyn Tagh fault along this sliver is also shows left-lateral movement, indicated by consistent offset of a series of stream channels from 10 to 100 m (Figs. 20 and 21). As shown already, characteristic features of the Xorkol trough are dominated by left-lateral transtensional deformation, but the left-lateral slip component seems to decrease southwestward toward the Xorkol Basin.

Internal Deformation of the Altyn Tagh Belt

The Precambrian rocks within the Altyn Tagh fault zone (Figs. 3 and 4) are metamorphosed to different grades. The late Archean Milan Group, composed of gneiss, granulite, granite, and migmatite, experienced middle- to high-pressure hornblende-granulite–facies metamorphism, whereas the Proterozoic Altyn Tagh and Bashikurgan Groups, composed of marble, schist, and quartzite, experienced greenschist-hornblende–facies metamorphism. They were intruded by Proterozoic and early Paleozoic granite. Associated with metamorphic and magmatic events, as shown above, the Altyn Tagh belt also experienced multiple deformations, and the first-order structural pattern is a series of NE-SW–trending folds and thrust faults, along which the Tertiary rocks are also deformed. Some of the main Cenozoic structures have been recognized by previous studies (Yin et al., 2002), such as the Jianglisai strike-slip fault that bounds the belt on the northwest, the Xorkol and Jinyan thrust faults within eastern part of the belt, and the Lapeiquan detachment fault that bounds the belt on the northeast. Along the northern margin of the belt in the Janggasay area, a 3000-m-thick sedimentary succession unconformably overlies the Precambrian metamorphic rocks and consists of Early Jurassic, Early Cretaceous, Paleogene, Neogene, and early and middle Pleistocene strata (Yin et al., 2002; Chen et al., 2005). These strata, except for the middle Pleistocene strata, were deformed into a NE-trending, asymmetric syncline, and beds along the northern flank of the syncline are nearly vertical and unconformably overlain by the subhorizontal middle Pleistocene conglomerate. The folded late Cenozoic strata contain evidence indicating that sedimentation and erosion rapidly increased in late Miocene to early Pleistocene time (Chen et al., 2005). The existing thermal evidence also shows that the Altyn Tagh belt experienced a rapid cooling event in late Miocene time (ca. 8 Ma; Chen et al., 2004; Jolivet et al., 2001). Moreover, to the east of the Xorkol Basin, a large outcrop of Tertiary red beds within the Altyn Tagh belt trends NE-SW (Fig. 4). Fossils form the basis for age determination of these red beds, which includes the Xaiganchaigou Formation of Oligocene age, Shangganchaigou Formation of early Miocene age, and Xiayoushan Shan Formation of late Miocene age (Yue et al., 2003). These rocks were considered to be a part of the basin fill of the Xorkol Basin (Yue et al., 2003). However, based our field observation: (1) these rocks have experienced strong deformation defining a syncline that is undergoing intensive erosion along numerous seasonal rivers, and (2) the southwestern end of the syncline is truncated by the northern boundary fault of the Xorkol Basin (Fig. 22). Thus, this outcrop of Tertiary rocks must have formed as a remnant of a middle Tertiary basin within the Altyn Tagh belt instead of being a part of the Xorkol Basin. Near the northeastern end of the syncline, the Miocene

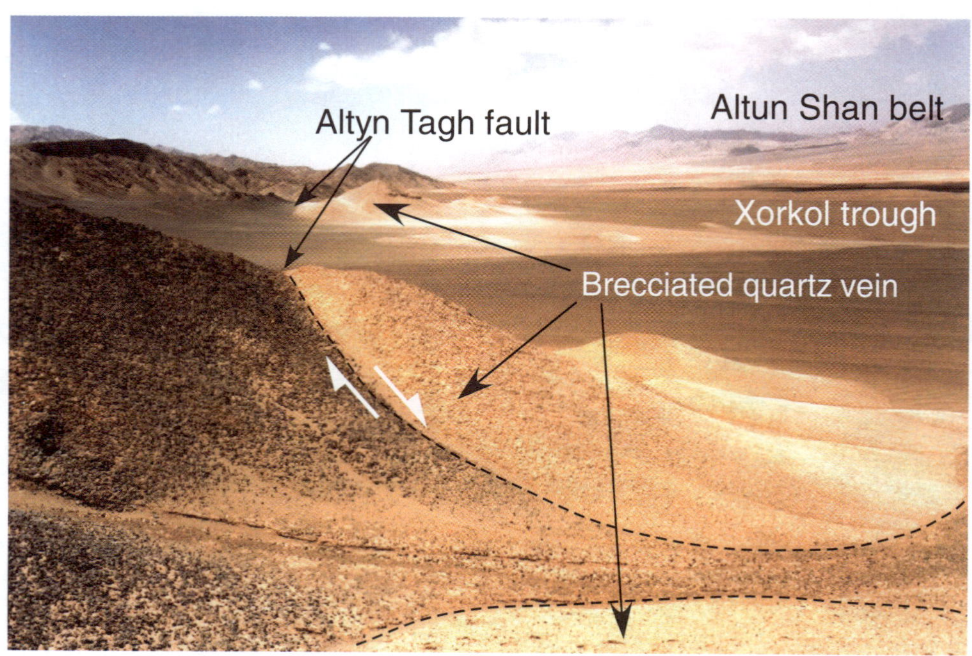

Figure 19. Looking southwest along the Altyn Tagh fault at the southern edge of the Xorkol trough. The trace of the fault is marked by a white, brecciated quartz vein dipping to the north.

Figure 20. Photo showing a large elongated fault sliver bounded by the Altyn Tagh fault along the northeastern end of the Xorkol trough. The sliver is composed of fine-grained conglomerate of Pliocene age (N_2).

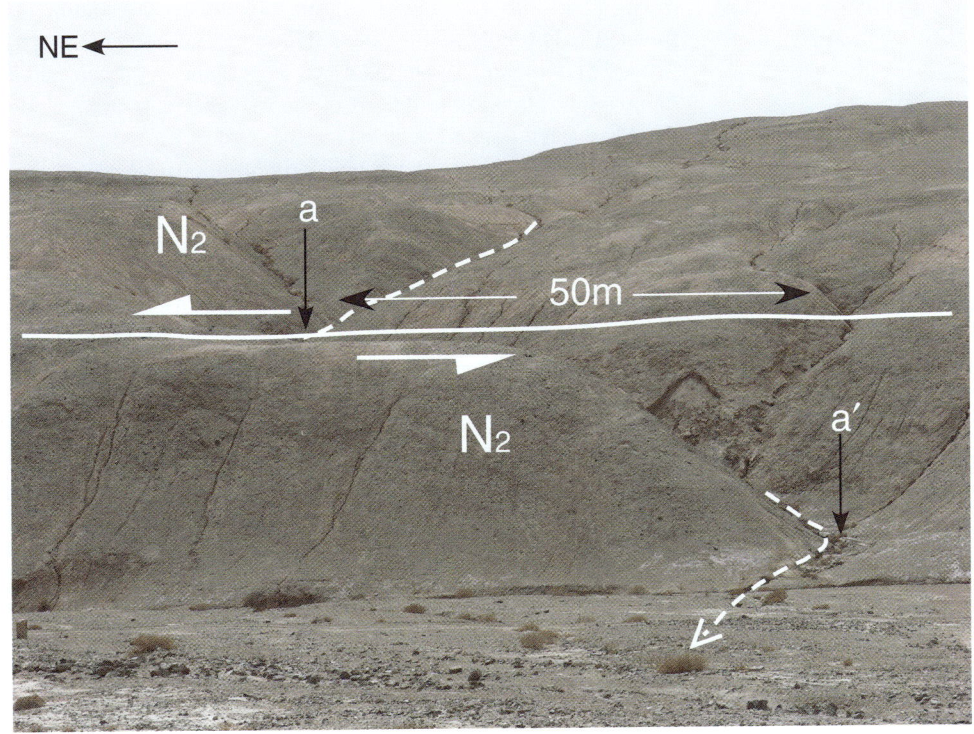

Figure 21. Looking south at the Altyn Tagh fault along the northern edge of the fault sliver shown in Figure 20. The left-lateral movement on the fault is demonstrated by the 50 m offset of a stream channel.

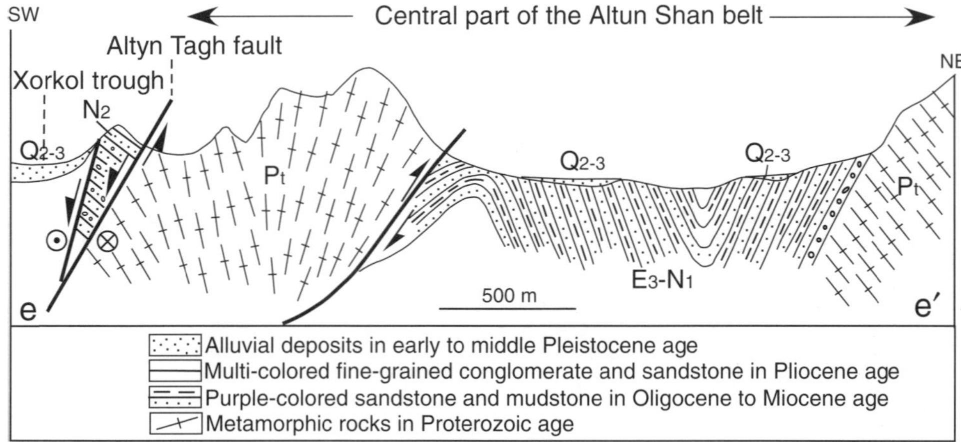

Figure 22. N-S cross section over the central part of the Altyn Tagh belt, 15 km north of the northeastern end of the Xorkol trough, showing the shortening deformation feature of the Miocene red beds. Their deformation, together with other deformations that occurred elsewhere in other places in the Altyn Tagh belt as presented herein, indicates that much of the present shape of the Altyn Tagh belt resulted from transpression.

red beds have been intensively deformed into a syncline and are juxtaposed against the Proterozoic metamorphic rocks by a series of NW-trending thrust faults that show no evidence for modern activity. In places, the folded red beds are unconformably overlain by poorly sorted conglomerate, probably glacial deposits of Pleistocene age, which places an upper time limit for the N-S shortening across the Altyn Tagh belt. At the northeastern tip of the Altyn Tagh belt, the Pliocene red beds are deformed into an E-trending anticline, cored by Proterozoic metamorphic rocks (Wang, 1997). All this evidence indicates that the whole Altyn Tagh belt underwent intensive strike-perpendicular shortening in late Tertiary time and the present shape of the belt largely resulted from the deformation and erosion.

Most of the metamorphic rocks along the Altyn Tagh belt extend in a NE-SW direction, parallel to the strike of the belt, but those along the middle part of the belt bend gradually into a NW-SE direction around the northwestern corner of the Xorkol Basin around a vertical axis (Figs. 2 and 4). Within the bend, the metamorphic rocks are cut by two groups of faults that strike NW-SE and NE-SW. The former are mostly right-lateral, and the latter are left-lateral. Based on their spatial correlation with the structural bend and their shear pattern, these two groups of faults can be interpreted to form a conjugate fault system that accommodated the bending of the Altyn Tagh belt.

To determine the timing of bending, we collected samples from a Proterozoic granite intruded into the metamorphic rocks for fission-track age dating, but, unfortunately, only one sample contained adequate fission-track lengths (13.55 μm), and it yielded a cooling age of 5.1 Ma (Fig. 4). Although the thermal data are limited, together with the sedimentary and erosion evidence mentioned previously, they support the interpretation that the Altyn Tagh belt underwent intensive deformation in late Tertiary time, coincident with the vertical-axis bending of the belt.

VERTICAL-AXIS BENDING MODEL

Although the Altyn Tagh fault is presently shown as a straight geological and geomorphic feature extending for 1500 km, it may not have functioned as a linear feature during all its history, because the fault has a prolonged evolution and forms the boundary of several crustal blocks with contrasting rock types and Cenozoic structures. The Altyn Tagh belt that bounds the fault on the northwest is different from the Qilian Shan belt, the Qaidam Basin, and the Kunlun belt that are bounded by the fault on the southeast. These tectonic elements to the south contrast not only lithologically but also in strike and trend parallel to instead of perpendicular to the Altyn Tagh fault. Our recent work demonstrates that the foliations of the metamorphic rocks along the Altyn Tagh belt are bent around the Xorkol Basin, indicating that the belt itself may have bent around a vertical axis forming the widened middle part of the belt (Fig. 23). Based on the limited fission-track age combined with the regional structural and sedimentary data within the Altyn Tagh belt, the timing of the initiation of the bending can be dated as late Miocene (8–5 Ma). The Xorkol Basin is bounded by the metamorphic rocks undergoing bending along the Altyn Tagh belt on the north, and its extension was coeval with the bending. Based on their special and temporal correlations, we interpret the extension across the Xorkol Basin to have been generated by the NW-SE divergent movement between the Altyn Tagh belt and the Qaidam Basin, resulting from the vertical-axis bending of the metamorphic rocks. Taking this interpretation further, it can be inferred that the northern boundary normal fault of the Xorkol Basin also experienced vertical-axis bending along with the Altyn Tagh belt, and prior to the bending, it used to be a branch of the Altyn Tagh fault in the hanging wall of the extensional fault that lay below the Xorkol Basin, whereas its counterpart, the Altyn Tagh fault

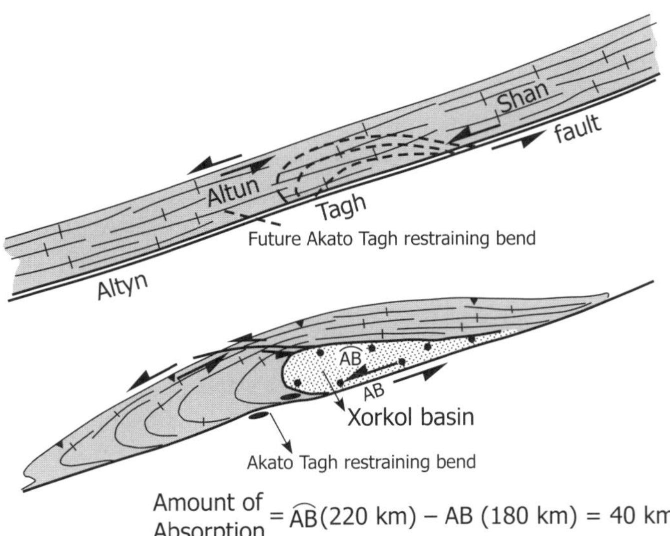

Figure 23. Tectonic model showing the bending of the Altyn Tagh belt along the Altyn Tagh fault and formation of the Xorkol Basin by extension within the inner arc of the bend. Note the contrasting change in geometry between the Altyn Tagh fault in the hanging wall and footwall. The bending of the Altyn Tagh belt also accommodated additional left-lateral movement along its northern boundary fault relative to the southern boundary fault. The lengths of the northern and southern boundaries of the Xorkol Basin are about 220 km and 160 km, respectively, from which it can be estimated that as much as 60 km of left-lateral movement along the middle segment of the Altyn Tagh fault has been absorbed by the N-S extension across the Xorkol Basin and vertical-axis bending of the Altyn Tagh belt.

in the footwall, remained straight through time, and its left-lateral displacement was partitioned into the normal and the left-lateral displacements, causing subsidence of the Xorkol Basin. The modern depocenter of the Xorkol Basin is a very prominent arcuate feature that consists of the Wushuxiao Lake and a seasonal river, limited between a series of alluvial fans. That area can be interpreted to mark the center of extension across the Xorkol Basin; otherwise, it would have been buried by the growing alluvial-fan deposits shed from the north and south.

The occurrence of the vertical-axis bending of the Altyn Tagh belt implies that the southwestward movement of the belt relative to the Qaidam Basin along the middle segment of the Altyn Tagh fault met with strong resistance. To the southwest of the Xorkol Basin, a 50-km-long segment of rocks within the Altyn Tagh fault zone strikes E-W instead of NE-SW and forms the Akato Tagh, which was recognized as a young restraining bend by Cowgill et al. (2004), with NE-SW compression, as manifested by the uplift of two high mountains built at its inside corners; the west corner has a peak at 6062 m elevation, and the east corner is the East Akato Tagh (5832 m) (Fig. 4). It is postulated that such resistance along the middle segment of the Altyn Tagh fault zone was generated by the compression along this restraining bend. Nevertheless, as shown previously, the East Akato Tagh also experienced N-S extension in late Cenozoic time, so its original compressional deformation as described by Cowgill et al. (2004) has been largely modified. The transpressional convergence between the Qaidam Basin and the Altyn Tagh belt is also demonstrated by the south-directed thrusting along the boundary fault between the Qaidam Basin and the Akatengneng Shan, which displaced Paleozoic rocks southward onto Cenozoic rocks as young as Pliocene (Fig. 4). As shown in Figure 23, the lengths of the northern and southern edges of the Xorkol Basin are 160 km and 220 km long, respectively, so that from their difference in length, we can estimate that as much as 60 km of the late Cenozoic left-lateral motion along the middle segment of the Altyn Tagh fault was absorbed by the crustal bending and extension during late Cenozoic time. Besides the left-lateral motion, much of the northern boundary fault of the Altyn Tagh belt also shows compressional deformation, which can be also attributed to the widening of the middle segment of the Altyn Tagh belt associated with the vertical-axis bending of the belt.

DISCUSSION AND CONCLUSIONS

Our study shows that: (1) left-lateral placement along the Altyn Tagh fault was partitioned into left-lateral and normal displacement along the edge of the Xorkol Basin, and (2) within the rhombic-shaped Altyn Tagh belt, the rocks were folded around a vertical axis that caused an increase in its width. Data from the deformation, sedimentation history, and limited fission-track data demonstrate that the folding was initiated in late Miocene time (8–5.1 Ma). The vertical-axis folding is interpreted to have been the result of a restraining bend along the Altyn Tagh fault. The Xorkol Basin was formed between the Altyn Tagh belt to the north and the Altyn Tagh fault to the south, and its northern boundary fault is also bent around a vertical axis, along with the foliation in the metamorphic rocks of the Altyn Tagh belt. The bending around a vertical axis also caused extension across the Xorkol Basin by divergent movement between the Altyn Tagh belt and the Qaidam Basin. In our interpretation, the arcuate northern boundary fault of the Xorkol Basin, presently dominated by normal faulting, used to be the hanging wall fault of the Altyn Tagh fault prior to the bending of the Altyn Tagh belt, whereas the footwall branch of the Altyn Tagh fault remained straight and is presently dominated by left-lateral transtensional motion bounding the Xorkol Basin on the south. The folding of the Altyn Tagh belt indicates that the southwestward movement of the belt relative to the Qaidam Basin along the left-lateral Altyn Tagh fault met resistance, as indicated by the formation of a restraining bend, and this caused NE-SW compression along the Akato Tagh lying to the southwest of the Xorkol Basin. The difference in length between the northern and southern boundary faults of the Xorkol Basin is as much as 60 km and indicates that 60 km of left-lateral movement along the Altyn Tagh fault has been absorbed by the extension across the Xorkol Basin and vertical-axis bending of the Altyn Tagh belt since late Miocene time (8–5 Ma). Active normal faulting across the Xorkol Basin indicates that the Altyn Tagh belt is still undergoing vertical-axis bending. The geometric model in Figure 24 illustrates different deformational styles of the crust on either side of the Altyn Tagh fault. The Altyn Tagh belt, shown as case 1, was subject to vertical-axis bending because it is mainly composed of the metamorphic rocks with a near-vertical foliation that was trending parallel to the strike of the Altyn Tagh fault. In contrast, the Qaidam Basin consists of Cenozoic sedimentary rocks that accumulated coeval with the subsidence of the basin, so that these strata along the Yousha Shan anticline had horizontal bedding, as shown as case 2, and were subject to horizontal axial bending.

The resistance to the southwestward movement of the Altyn Tagh belt relative to the Qaidam Basin may have resulted either from heterogeneity of the crust or from constantly changing stress along the fault. In particular, the sedimentary record within the Qaidam Basin shows that a huge increase in the accumulation rates after the beginning of the Pliocene and a southeastward shift of the depocenters (Metivier et al., 1998; Wang et al., 2006), reflecting an enhancement of the transpressional interaction between the Qaidam and Tarim Basins along the Altyn Tagh fault during that time, resulted in the occurrence of the strike-parallel deformation along the middle segment of the Altyn Tagh fault. The deformational study presented here only reveals a very short period of evolution history of the Altyn Tagh fault. If our model is correct, then it reflects the complex nature of the growth of the Tibetan Plateau along the Altyn Tagh fault, which cannot be explained by simple strike-slip movement as some studies have proposed. It is stressed that the occurrence of the vertical-axis bending of the Altyn Tagh belt was not coeval with the horizontal-axis bending of the Yousha Shan anticline, the latter of which occurred in the Oligocene time (Wang et al., 2006). Based on the fact that the Xorkol Basin is relatively small in size and shallow in depth, we interpret the Xorkol Basin to die out downward in the upper crust (Fig. 24). Thus, taking this interpretation further, the vertical-axis bending of the Altyn Tagh belt, as a tectonic block moving along the Altyn Tagh fault, must have occurred within the upper crust, and thus so does the Altyn Tagh fault, as suggested by Burchfiel et al. (1989).

Deformation of the Altyn Tagh fault, like that which occurs on the middle segment of the Altyn Tagh fault along the Altyn Tagh belt, is a common feature. For example, similar vertical-axis bending occurs in the southeastern margin of the Tibetan Plateau, where two large, active, strike-slip faults, the

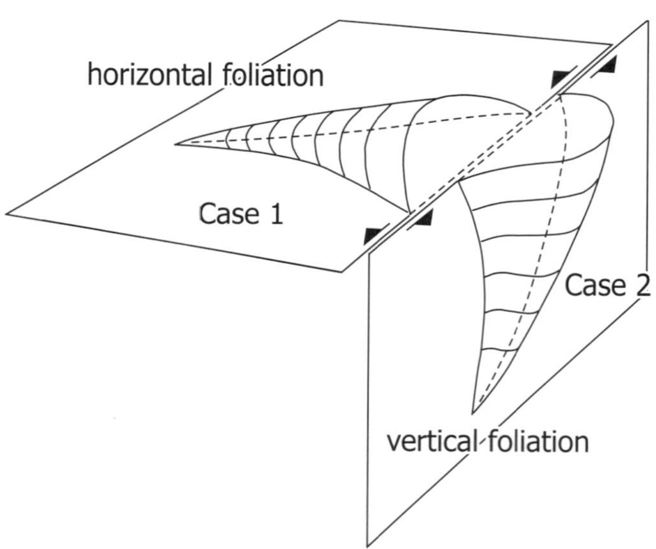

Figure 24. An illustrated three-dimensional geometric model showing the differential deformation of the crust along a strike-slip fault. Case 1: Rocks with foliation striking parallel to the strike-slip fault are subject to bending around a vertical axis, as occurred to the Altyn Tagh belt along the middle segment of the Altyn Tagh fault. Case 2: Rocks with horizontal foliation along the strike-slip fault are subject to horizontal-axis bending, as occurred in the Yousha Shan anticline and many other en echelon structures within the Qaidam Basin along and south of the Altyn Tagh fault.

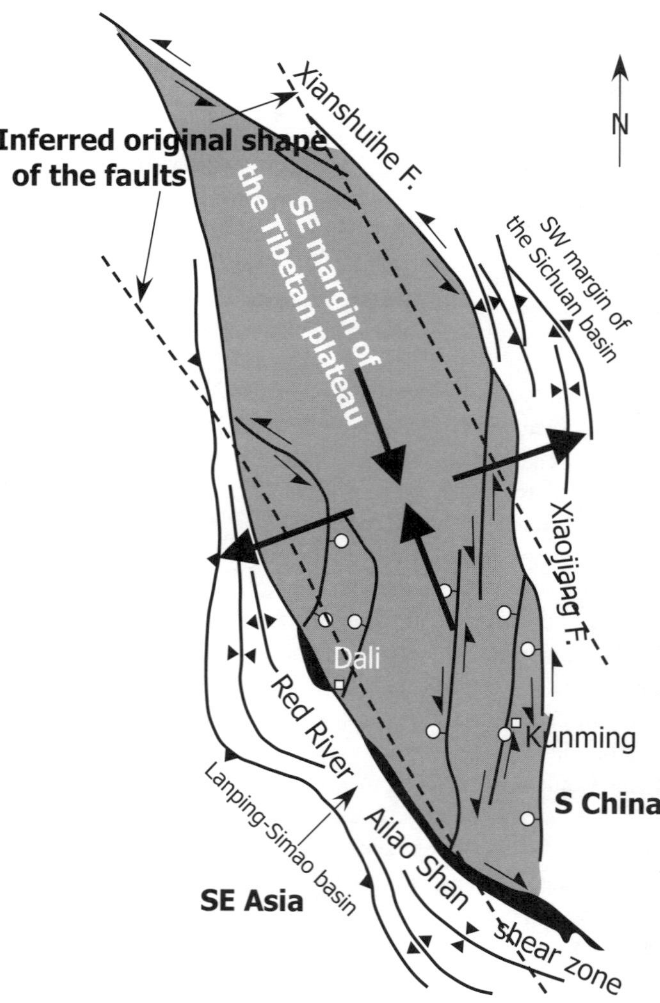

Figure 25. Tectonic map of the Chuan-Dian block on the southeastern margin of the Tibetan Plateau. As indicated by its shape, both the Xianshuihe-Xiaojiang and the Red River–Ailao Shan fault zones have experienced vertical-axis bending that resulted in the NE-SW shortening of the southwestern margin of the Sichuan Basin and the Lanping-Simao Basin.

Xianshuihe-Xiaojiang fault zone and the Red River–Ailao Shan shear zone, bound the large rhombic-shaped Chuan-Dian block on its northeast and southwest, respectively, suggesting that these two faults experienced a vertical-axis bending to the northeast and southwest. The Chuan-Dian block was interpreted to have undergone a clockwise rotation around the eastern Himalayan syntaxis in late Cenozoic time as part of the accommodation of N-S convergence between India and Eurasia (Fig. 25) (Wang et al., 1998). Intensive NE-SE shortening occurred along the southwestern margin of the Sichuan Basin and along the Lanping-Simao Basin, which can be interpreted to have been caused by the NE-SW compression generated by the vertical-axis bending of the Xianshuihe-Xiaojiang fault zone and the Red River–Ailao Shan shear zone, respectively. We can infer that as the oblique movement of the Chun-Dian block proceeded, these two fault zones would have been further bent to the northeast and southwest. Similarly, paleomagnetic data (Rumelhart et al., 1999) indicate that the Karakax fault, the southwestward prolongation of the Altyn Tagh fault in the NW part of the Tibetan Plateau, has undergone clockwise rotation of between 19° and 27° in response to the northward indentation of India into Eurasia.

We conclude that: (1) the late Cenozoic left-lateral movement along the Altyn Tagh fault took place not only by left-lateral horizontal movement, but also by local differential vertical-axis bending; (2) the subsidence of the Xorkol Basin was due to the N-S extension generated by the vertical-axis bending of both the Altyn Tagh belt and the Altyn Tagh fault, and the age of the oldest sediments within the basin constrain the bending to be of Pliocene age; and (3) the strike-slip faults may be subject to deformation by vertical-axis bending, such as that which occurred along the Altyn Tagh fault as well as on the Xianshuihe-Xiaojiang fault zone and the Red River–Ailao Shan shear zone.

ACKNOWLEDGMENTS

The authors would like to thank J. Ramon Arrowsmith and D. Bradley Ritts for critical and constructive comments and suggestions. We thank B. Clark Burchfiel for helpful discussions and suggestions. This study was jointly supported by grants from Knowledge Innovating Project of Chinese Academy of Sciences (kzcx2-yw-12), National Science Foundation of China (40672151; 40721003; 40672151; 407211003), and Chinese National Key Project (2002CB412601).

REFERENCES CITED

Avouac, J.P., and Tapponnier, P., 1993, Kinematic model of active deformation in central Asia: Geophysical Research Letters, v. 20, p. 895–898, doi: 10.1029/93GL00128.

Bedrosian, P.A., Unsworth, M.J., and Wang, F., 2001, Structure of the Altyn Tagh fault and Daxue Shan from magnetotelluric surveys: Implications for faulting associated with the rise of the Tibetan Plateau: Tectonics, v. 20, p. 474–486, doi: 10.1029/2000TC001215.

Burchfiel, B.C., Deng, Q., Molnar, P., Royden, L.H., Wang, Y., Zhang, P., and Zhang, W., 1989, Intracrustal detachment within zones of continental deformation: Geology, v. 17, p. 748–752, doi: 10.1130/0091-7613(1989) 017<0448:IDWZOC>2.3.CO;2.

Chen, X., Yin, A., Gehrels, G.E., Cowgill, E.S., Grove, M., Harrison, T.M., and Wang, X., 2003, Two phases of Mesozoic north-south extension in the eastern Altyn Tagh range, northern Tibetan Plateau: Tectonics, v. 22, no. 5, 1053, doi: 10.1029/2001TC001336.

Chen, Z., Wang, X., Yin, A., Chen, B., and Chen, X., 2004, Cenozoic left-slip motion along the central Altyn Tagh fault as inferred from the sedimentary record: International Geology Review, v. 46, p. 839–856, doi: 10.2747/0020-6814.46.9.839.

Chen, Z., Liu, J., Sun, Z., Wang, X., Pei, J., and Gong, H., 2005, Cenozoic erosional history of the Altyn Tagh Mountains inferred from the sedimentary record of the foreland basin: Geological Bulletin of China, v. 24, p. 302–308.

Cowgill, E., Yin, A., Arrowsmith, J.R., Wang, X., and Zhang, S. 2004, The Akato Tagh bend along the Altyn Tagh fault, northwest Tibet. 1: Smoothing by vertical-axis rotation and the effect of topographic stresses on bend-flanking faults: Geological Society of America Bulletin, v. 116, p. 1423–1442.

Guo, Z., Zhang, Z., and Wang, J., 1998, Formation and evolution of the Xorkol Basin and its relation to the Altyn Tagh fault: Geological Journal of China Universities, v. 4, p. 59–63.

Jolivet, M., Brunel, M., Seward, D., Xu, Z., Yang, J., Roger, F., Tapponnier, P., Malavieille, J., Arnaud, N., and Wu, C., 2001, Mesozoic and Cenozoic tectonics of the northern edge of the Tibetan Plateau: Fission-track constraints: Tectonophysics, v. 343, p. 111–134, doi: 10.1016/S0040-1951 (01)00196-2.

Matte, P., Tapponnier, P., Arnaud, N., Bourjot, L., and Avouac, J.P., 1996, Tectonics of western Tibet, between the Tarim and the Indus: Earth and Planetary Science Letters, v. 142, p. 311–330, doi: 10.1016/0012-821X (96)00086-6.

Mériaux, A.-S, Ryerson, F.J., Tapponnier, P., Van der Woerd, J., Finkel, R.C., Xu, X., Xu, Z., and Caffee, M.W., 2004, Rapid slip along the central Altyn Tagh fault: Morphochronologic evidence from Chechen He and Sulamu Tagh: Journal of Geophysical Research, v. 109, B06401, doi: 10.1029/2003JB002558.

Metivier, F., Gaudemer, Y., Tapponnier, P., and Meyer, B., 1998, Northeastward growth of the Tibetan Plateau deduced from balanced reconstruction of two depositional areas: The Qaidam and Hexi Corridor basins, China: Tectonics, v. 17, p. 823–842, doi: 10.1029/98TC02764.

Peltzer, G., and Tapponnier, P., 1988, Formation and evolution of strike-slip faults, rifts, and basins during the India-Asia collision: An experimental approach: Journal of Geophysical Research, v. 93, p. 15,085–15,117, doi: 10.1029/JB093iB12p15085.

Peltzer, G., Tapponnier, P., and Armijo, R., 1989, Magnitude of late Quaternary left-lateral displacements along the north edge of Tibet: Science, v. 246, p. 1285–1289, doi: 10.1126/science.246.4935.1285.

Ritts, B.D., and Biffi, U., 2000, Magnitude of post–Middle Jurassic (Bajocian) displacement of the Altyn Tagh fault, northwest China: Geological Society of America Bulletin, v. 112, p. 61–74, doi: 10.1130/0016-7606(2000) 112<0061:MOPMJB>2.3.CO;2.

Rumelhart, P.E., Yin, A., Cowgill, E., Butler, R., Zhang, Q., and Wang, X., 1999, Cenozoic vertical-axis rotation of the Altyn Tagh fault system: Geology, v. 27, p. 819–822, doi: 10.1130/0091-7613(1999)027 <0819:CVAROT>2.3.CO;2.

Tapponnier, P., Mattauer, M., Proust, F., and Cassaigneau, C., 1981, Mesozoic ophiolites, sutures, and large-scale tectonic movements in Afghanistan: Earth Planet. Sci. Lett., v. 52, p. 355–371.

Tapponnier, P., Xu, Z., Roger, F., Meyer, B., Arnaud, N., Wittlinger, G., and Yang, J., 2001, Oblique stepwise rise and growth of the Tibet Plateau: Science, v. 294, p. 1671–1677, doi: 10.1126/science.105978.

Wang, E., 1997, Displacement and timing along the northern strand of the Altyn Tagh fault zone, northern Tibet: Earth and Planetary Science Letters, v. 150, p. 55–64, doi: 10.1016/S0012-821X(97)00085-X.

Wang, E., Burchfiel, B.C., Royden, L.H., Chen, L., Chen, J., Li, W., and Chen, Z., 1998, Late Cenozoic Xianshuihe-Xiaojiang, Red River, and Dali fault systems of southwestern Sichuan and central Yunnan, China: Geological Society of America Special Paper 327, 108 p.

Wang, E., Xu, F., Zhou, J., Wan, J., and Burchfiel, B.C., 2006, Eastward migration of the Qaidam Basin and its implications for Cenozoic evolution of the Altyn Tagh fault and associated river systems: Geological Society of America Bulletin, v. 118, p. 349–365, doi: 10.1130/B25778.1.

Wittlinger, G., Tapponnier, P., Poupinet, G., Jiang, M., Shi, D., Herquel, G., and Masson, F., 1998, Tomographic evidence for localized lithospheric shear along the Altyn Tagh fault: Science, v. 282, p. 74–76, doi: 10.1126/science.282.5386.74.

Xinjiang BGMG, 1993, Beijing, Geological Publishing House.

Yin, A., Rumelhard, P.E., Butler, R., Cowgill, E., Harrison, T.M., Foster, D.A., Ingersoll, R.V., Zhang, Q., Zhou, X., Wang, X., Hanson, A., and Raza, A., 2002, Tectonic history of the Altyn Tagh fault system in northern Tibet inferred from Cenozoic sedimentation: Geological Society of America Bulletin, v. 114, p. 1257–1295, doi: 10.1130/0016-7606(2002)114 <1257:THOTAT>2.0.CO;2.

Yue, Y., Ritts, B., Graham, S.A., Wooden, J.L., Gehrels, G.E., and Zhang, Z., 2003, Slowing extrusion tectonics: Lowered estimate of post–early Miocene slip rate for the Altyn Tagh fault: Earth and Planetary Science Letters, v. 217, p. 111–122, doi: 10.1016/S0012-821X(03)00544-2.

Zhao, J., Mooney, W.D., Zhang, X., Li, Z., Jin, Z., and Okaya, N., 2006, Crustal structure across the Altyn Tagh range at the northern margin of the Tibetan Plateau and tectonic implications: Earth and Planetary Science Letters, v. 241, p. 804–814, doi: 10.1016/j.epsl.2005.11.003.

Manuscript Accepted by the Society 04 April 2008

Printed in the USA

The Geological Society of America
Special Paper 444
2008

Preliminary investigation into the complexities of the Ailao Shan and Day Nui Con Voi shear zones of SE Yunnan and Vietnam

B.C. Burchfiel
Department of Earth, Atmospheric and Planetary Sciences, Massachusetts Institute of Technology, Cambridge, Massachusetts 02139, USA

L. Chen
Yunnan Institute of Geological Sciences, Kunming, Yunnan 650000, China

E. Wang
Institute of Geology and Geophysics, Chinese Academy of Sciences, Beijing 100029, China

E. Swanson
Department of Earth, Atmospheric and Planetary Sciences, Massachusetts Institute of Technology, Cambridge, Massachusetts 02139, USA

ABSTRACT

The present-day distributions of mylonitic belts within Yunnan and north Vietnam, along the northern boundary of early Cenozoic extruded crustal fragments around the eastern Himalayan syntaxis, collectively called the Ailao Shan–Day Nui Con Voi shear zone, cannot be traced directly into one another. There are at least three belts of mylonitic rocks, the Ailao Shan, a middle belt, and the Day Nui Con Voi, and, in Yunnan, these are separated by zones of weakly to unmetamorphosed sedimentary rocks of Triassic and possibly early Paleozoic age, narrow fault-bounded deposits of Neogene age, and young or active faults of the Red River fault zone. These three belts of mylonitic rocks are interpreted to have been originally a single through-going zone of left-lateral shear that was disrupted by younger events in an evolving shear zone. The Day Nui Con Voi belt terminates to the NW in Yunnan in a NW-plunging antiform, where it is surrounded by Triassic rocks that belong to the Yangtze platform of South China. Within Yunnan, a carapace of metasedimentary Triassic rocks locally forms the cover for the Day Nui Con Voi mylonite, but the cover is interpreted to be a tectonic cover with an original low-angle normal fault at its base. Ordovician to Triassic strata near the Vietnamese border can be inferred to be the cover of the Ailao Shan based on geological relations between the sedimentary rocks and the metamorphic and mylonitic rocks, which appear to have been joined in Mesozoic time by plutons that intruded into both sequences. The nature of the cover, whether the sedimentary rocks are the normal stratigraphic cover or a tectonic cover, is unknown. Disruption of the mylonitic belts by late Cenozoic right-lateral faulting occurred along the Red River fault zone.

Burchfiel, B.C., Chen, L., Wang, E., and Swanson, E., 2008, Preliminary investigation into the complexities of the Ailao Shan and Day Nui Con Voi shear zones of SE Yunnan and Vietnam, *in* Burchfiel, B.C., and Wang, E., eds., Investigations into the Tectonics of the Tibetan Plateau: Geological Society of America Special Paper 444, p. 45–58, doi: 10.1130/2008.2444(03). For permission to copy, contact editing@geosociety.org.

The preliminary tectonic conclusions suggest: (1) Triassic metamorphism, plutonism, and tectonism may have been important in the Vietnamese part of the mylonitic belts, but the extent of this event into Yunnan cannot be determined at present. (2) The early Cenozoic SE extrusion of Indochina was bounded on the NE side by a narrow mylonitic belt formed from Precambrian rocks and Paleozoic to Cenozoic igneous rocks. (3) During early Cenozoic transpressional extrusion, rocks in the evolving NE-dipping mylonitic belt were decoupled from rocks in the footwall and hanging wall by a thrust below and normal fault above. (4) While left-lateral shear was the dominant mode of deformation, upward extrusion of the ductile lower- and middle-crustal mylonitic rocks occurred early in the deformational process. (5) Decoupling of the mylonitic belt from adjacent rocks in the crust and also from rocks in the lower crust and mantle indicates that all rock units have moved relative to one another and makes determination of the magnitude of displacement difficult. (6) The Paleozoic to Triassic cover of the Ailao Shan correlates best with rocks near Erhai Lake, suggesting ~250–300 km of displacement relative to South China rocks in the hanging wall of the normal fault at the top of the mylonite belt, but they are still separated by an unknown amount of displacement from the Lanping-Simao tectonic unit farther to the SW. (7) The magnitude of displacement during disruption of the original continuous mylonite zone later in evolving transpressional left-lateral shear is limited by the broad continuity of the South China Triassic cover north of the Day Nui Con Voi and may be no more than 100 km. (8) The magnitude of late Cenozoic disruption by right-lateral brittle faults of the Red River fault zone is likewise limited by the distribution of the South China rocks and is probably ~50 km.

Keywords: extrusion tectonics, Ailao Shan shear zone, Day Nui Con Voi shear zone, Yunnan, China, Vietnam.

INTRODUCTION

The Ailao Shan shear zone and the Red River fault zone have figured prominently in early and late Cenozoic tectonic scenarios for deformation in southeastern Tibet, possible extrusion of Indochina, and the tectonics around the eastern Himalayan syntaxis. During early Cenozoic time following the start of the collision of India and Eurasia, Tapponnier et al. (1982), Leloup et al. (1995, 2001), and Replumaz and Tapponnier (2003) all have proposed that there was a major displacement of continental lithosphere from the area north of the Eastern Himalayan syntaxis eastward into southeast Asia. Continental lithospheric fragments were bounded by major strike-slip fault zones, the left-lateral Ailao Shan shear zone on the north, and right-lateral faults such as the Gaolingong shear zone on west (Fig. 1). They proposed that deformation occurred mainly by movement on large faults with little internal deformation or rotation of the lithospheric fragments. Other hypotheses have suggested important clockwise rotation and internal deformation of the lithospheric fragments during extrusion (Wang and Burchfiel 1997; J.W. Geissman, 2008, personal commun.) or little or no eastward movement of rocks north and east of the syntaxis but varying degrees of clockwise rotation (England and Houseman, 1989; Houseman and England, 1993). The Ailao Shan shear zone, a belt of variably mylonitized metamorphic rocks, forms the northern boundary for extruding fragments in interpretations that favor major eastward extrusion. Younger late Cenozoic eastward extrusion has been proposed for continental lithospheric fragments north of the Ailao Shan shear zone and south of the Altyn Tagh fault zone and their postulated eastward continuations (Tapponnier et al., 1982). Such later movement reversed the sense of displacement on the Ailao Shan shear zone and was accommodated on shallow crustal right-lateral faults of the Red River fault zone. Early Cenozoic movement as preserved in the Ailao Shan shear zone was mainly through ductile shear, while later movement on the Red River fault zone was brittle. Even though the Red River fault zone and the Ailao Shan shear zone are parallel with and adjacent to one another, Wang et al. (1998) and Wang and Burchfiel (1997) suggested that the two features should be regarded as separate structures. This suggestion is adopted here for reasons that become apparent in the following discussions.

In all the interpretations of the mylonitic rocks of the Ailao Shan and Day Nui Con Voi shear zones, there is no direct evidence for the presence of cover rocks that overlie them. Although Wang and Burchfiel (1997) and Jolivet et al. (2001) suggested a Phanerozoic cover for the shear zone rocks, now present mainly to the north of the shear zone, such rocks cannot be shown to directly overlie the shear zone rocks. The lack of known cover rocks has greatly limited the ability to match rocks across the shear zone, to determine the possible protoliths for the shear zone rocks, and also to match the extruded fragments with rocks in other areas to determine their predeformational position.

Here, we present evidence that the Day Nui Con Voi shear zone extends from Vietnam, where it derives its name, into Yunnan, China, where it is overlain by Mesozoic strata and, by inference, by Paleozoic strata of South China. We also present evidence that the Ailao Shan shear zone protolith was overlain by a Paleozoic and Mesozoic stratigraphic succession in SE Yunnan that also correlates with South China stratigraphy, but not with rocks directly to the north of the shear zone. Mylonitic rocks in the Day Nui Con Voi also contain numerous rocks of granitic composition, often containing tourmaline, which may be Cenozoic in age. Furthermore, evidence indicates that the Day Nui Con Voi shear zone is presently not the direct continuation of the Ailao Shan shear zone, suggesting that these shear zones have a more complex relation than has been previously envisioned.

MYLONITIC BELTS IN SE YUNNAN AND ADJACENT VIETNAM

Early Cenozoic extrusion of rocks east of the eastern Himalayan syntaxis is commonly shown to be bounded on the north by a single through-going boundary, usually taken to be a continuous Ailao Shan shear zone in China and Day Nui Con Voi shear zone in Vietnam. These shear zones are characterized by mylonitic rocks with left-lateral shear kinematic indicators. Our preliminary study in SE Yunnan and adjacent Vietnam shows that there are at least three belts of mylonitic rocks that cannot be traced directly from one to the other because they are separated by young faults and zones of sedimentary or metasedimentary rocks of Paleozoic, Mesozoic, and Cenozoic age. The three belts of mylonitic rocks are the Ailao Shan, middle belt, and Day Nui Con Voi (Fig. 2; shown as mylonitic belts I, II, III, respectively); however, with additional work, there may be other belts that will be identified. Here, we show that while at one time there may have been a continuous zone of left-lateral mylonitic shear through these belts, they have been disrupted by younger deformation to become separated. Perhaps equally as important, evidence will be presented that the mylonitic belt probably formed beneath a cover of largely Paleozoic and Mesozoic rocks of South China affinity.

Identification of the Three Mylonitic Belts

The three mylonitic belts are identified by their separation by sedimentary and metasedimentary belts of Mesozoic and Paleozoic rocks, young postmylonitic faults mostly belonging to the Red River fault zone, and by narrow belts of postmylonitic Neogene strata that are present along the Red River faults. Because the general geology of the rocks surrounding and adjacent to the mylonitic belts is important in their identification, much of this paper focuses on the general geology of the area in SE Yunnan and adjacent Vietnam. Importantly, we discuss for the first time our interpretation that two of these belts have a cover of Paleozoic and Mesozoic rocks that provides evidence for the magnitude of displacement on the shear zones.

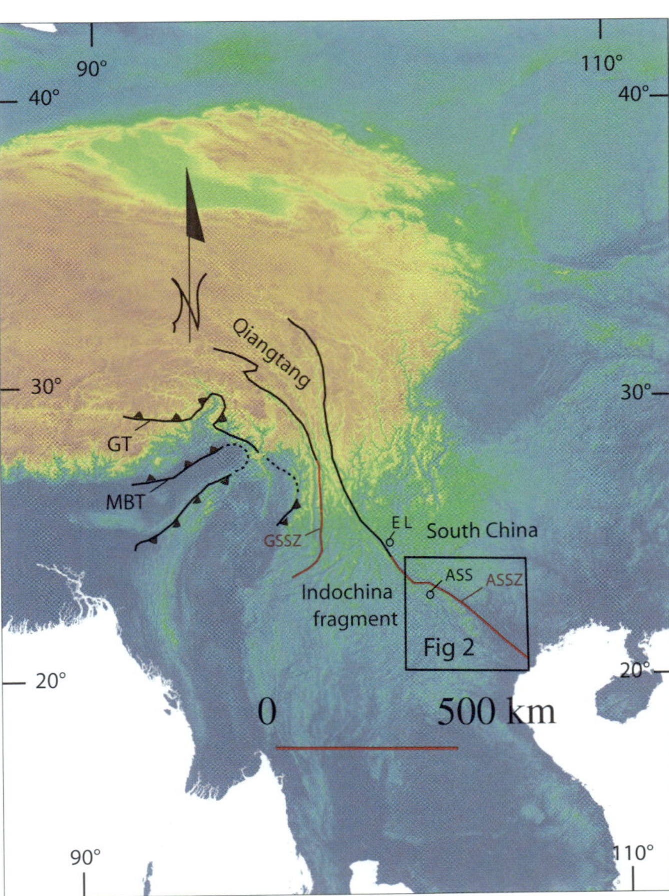

Figure 1. Location map of the Ailao Shan (China)–Day Nui Con Voi (Vietnam) region, shown in Figure 2, relative to the Eastern Himalayan syntaxis. The extruded Indochina fragment is labeled, along with its bounding shear zones, the Ailao Shan shear zone (ASSZ) on the north and the Gaolingong shear zone (GSSZ) on the west. EL—Erhai Lake; ASS—position of the Ailao Shan cover strata along the SE side of the Ailao Shan; GT—Gangdese thrust; MBT—Main Boundary thrust.

General Geology Around and Adjacent to the Three Mylonitic Belts

Day Nui Con Voi Belt (III, Fig. 2)

The Day Nui Con Voi shear zone in north Vietnam is characterized by generally steeply dipping mylonitic foliation with abundant left-lateral shear indicators (see Jolivet et al., 2001, and references therein). The shear zone rocks are bounded by two sharply defined faults, the Song Chay and Song Hong (Red River) faults to the north and south, respectively (Fig. 2). Along both of these faults, there are narrow, commonly steeply dipping fault-bounded belts of Miocene and younger strata dominated by thick boulder conglomerate. The southern belt can be traced discontinuously from Vietnam into Yunnan, where it forms a narrow belt, nearly 100 km long, along the north side of the Ailao Shan shear zone, which was recently studied by Schoenbohm et al. (2005) (Fig. 2). The southern belt of Neogene rocks and its bounding faults are part of the regional late Cenozoic Red River

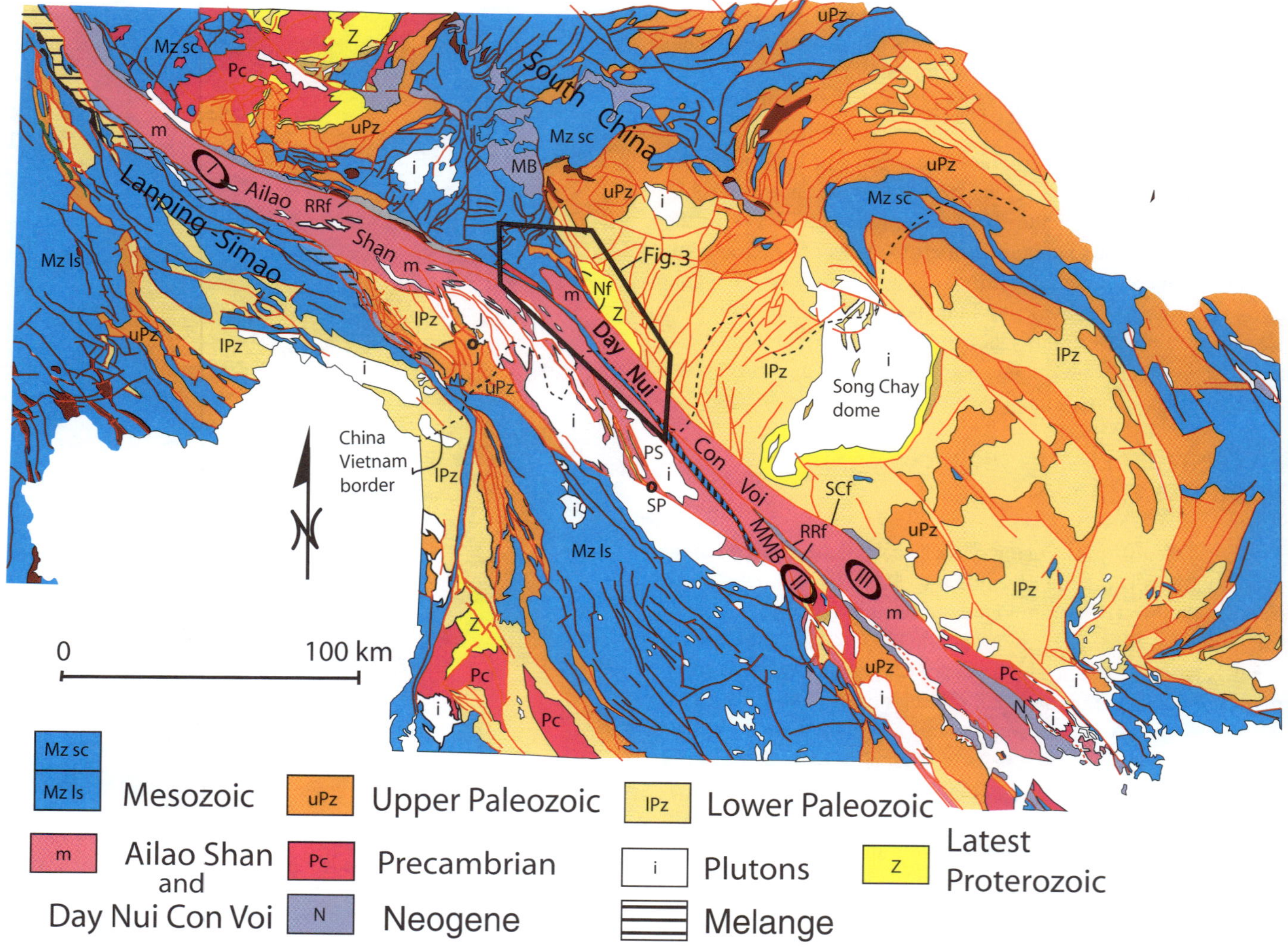

Figure 2. Generalized geological map of the Ailao Shan (I)–middle belt (II)–Day Nui Con Voi (III) area in Yunnan and Vietnam. Location of Figure 3 is indicated. J—Jinping; MB—Meng Zhi Basin; MMB—middle mylonite belt; Mz ls—Mesozoic rocks of the Lanping Simao tectonic unit; Mz sc—Mesozoic rocks of the South China tectonic unit; Nf—Nanxihe fault; PS—Po Sen orthogneiss; RRf—Red River fault; SCf—Song Chay fault; SP—Sa Pa; solid black—ophiolite; vertical black stripes—Cambrian (?) rocks in Vietnam.

fault zone. The northern belt of Neogene strata in Vietnam ends before entering Yunnan; however, the Song Chay fault continues into Yunnan, where it splays into several branches, one of which continues as the Nanxihe fault into the Meng Zhi Basin (Fig. 2). The Neogene conglomerates commonly contain cobbles and boulders of Paleozoic, Mesozoic, and, most importantly, mylonitic rocks similar to mylonitic rocks within the shear zones, although no detailed study of the mylonitic clasts has been done to correlate them with adjacent sources.

Along the north side of the Day Nui Con Voi shear zone and north of the Neogene conglomerates within Vietnam and southeasternmost Yunnan, there is a thick succession of lower Paleozoic and some upper Proterozoic strata (Fig. 2). They occur in the large Song Chay culmination cored by a granite, which yielded an U/Pb age of 428 ± 5 Ma, and which was later metamorphosed and sheared into orthogneiss with gently dipping foliation during Triassic time (Roger et al., 2000; Maluski et al., 2001; Fig. 2). To the north and west in Yunnan, the Paleozoic rocks are conformably overlain by Mesozoic, mostly Triassic strata, but on the SW side, they are bounded by the Song Chay fault and its Nanxihe branch. Near the village of Dimi, the Day Nui Con Voi ends in a NW-plunging antiform below a broad area north of the Ailao Shan. This area is underlain by mainly Triassic strata that are continuous with Triassic strata north of the Paleozoic culmination (Figs. 2 and 3). Farther west, these Triassic rocks are underlain by Proterozoic metamorphic and sedimentary rocks and lower and upper Paleozoic strata of typical South China facies (Fig. 2). These rocks show no evidence for the Silurian intrusive event recorded within the Song Chay dome in Vietnam.

Southeast from where the Day Nui Con Voi ends, it is separated from the Ailao Shan mylonitic rocks by two fault-bounded

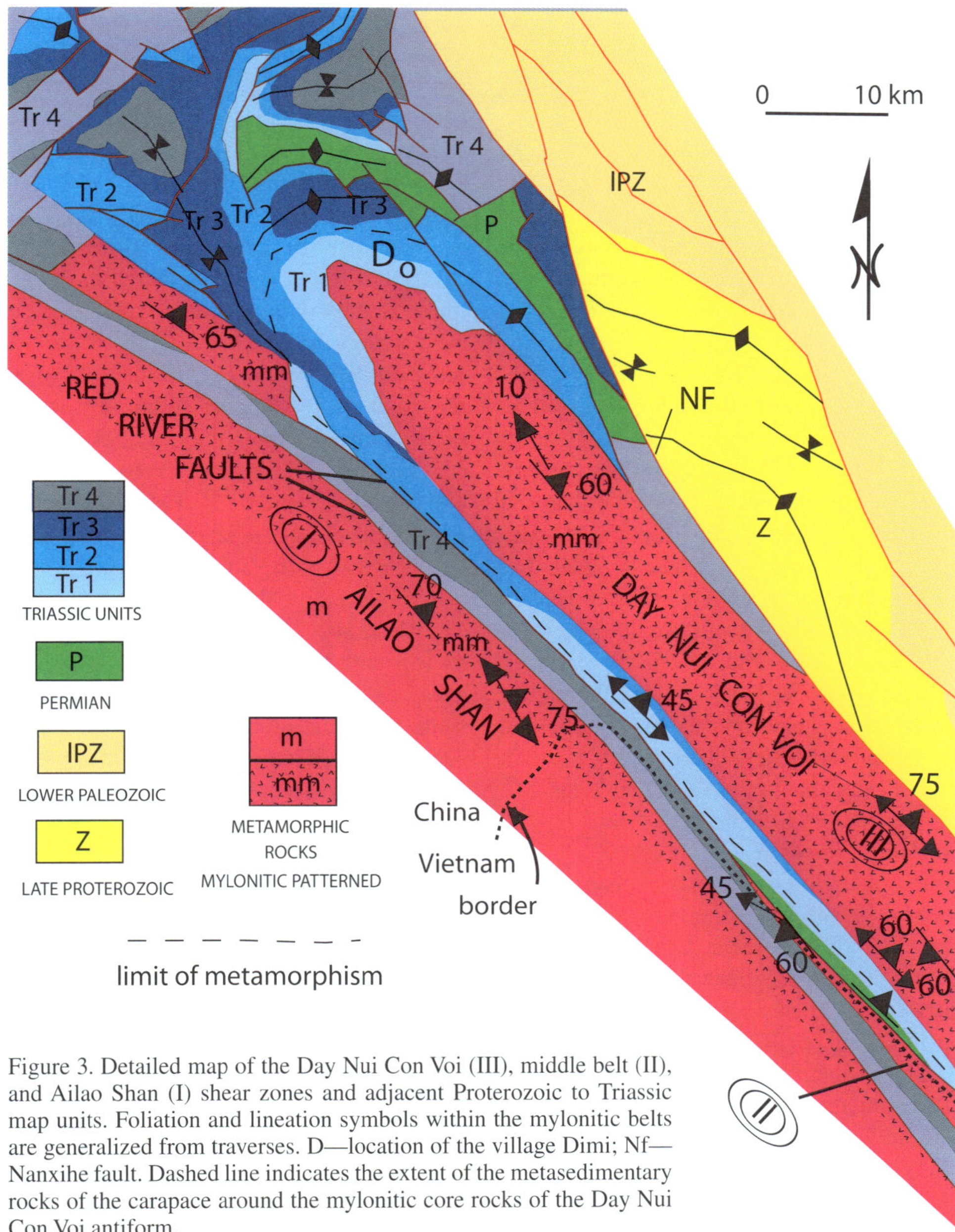

Figure 3. Detailed map of the Day Nui Con Voi (III), middle belt (II), and Ailao Shan (I) shear zones and adjacent Proterozoic to Triassic map units. Foliation and lineation symbols within the mylonitic belts are generalized from traverses. D—location of the village Dimi; Nf—Nanxihe fault. Dashed line indicates the extent of the metasedimentary rocks of the carapace around the mylonitic core rocks of the Day Nui Con Voi antiform.

belts of Triassic rocks, a belt of metamorphosed Lower Triassic schist to the northeast and a belt unmetamorphosed Middle and Upper Triassic strata to the southwest (Fig. 3). The metamorphosed strata are pelitic and sandy schist with rare marble. Their age is based entirely upon lithologic correlation with, and stratigraphic position relative to, unmetamorphosed strata in areas to the northwest. There are very few parts of the South China section other than the Lower Triassic that could be the protolith for these rocks. The belt of unmetamorphosed strata consists of shale, mudstone, sandstone, and limestone that characteristically contain coal beds that are easily followed by a string of black mine dumps, mainly along the south side of the Red River. They contain fossils and are Middle and Late Triassic in age. These two belts of Triassic rocks separating the two mylonitic shear zones can be traced to the Vietnam border, but they become complexly faulted near the border.

Middle Belt of Mylonitic Rocks (II, Figs. 2 and 3)

Just north of the Vietnam-Yunnan border, a thin zone of mylonitic granite associated with mafic igneous rocks wedges in and widens southeastward into Vietnam; we refer to this belt here as the middle belt (II, Figs. 2 and 3). At the border, Vietnamese maps continue the southern belt of Triassic rocks in Yunnan as a narrow fault-bounded belt of Cambrian strata that become metamorphosed to greenschist facies (shown with vertical lined pattern in Fig. 2). The reason for the change in age assignment is unknown, but the rocks in Yunnan are definitely Triassic. The rocks in Vietnam are metasediments, the ages of which need further study, but an Early Triassic age remains possible, perhaps correlative to the metasedimentary rocks assigned a Triassic age along the north side of the middle belt in Yunnan (see Fig. 2). These strata can be traced continuously for ~100 km into Vietnam, and they form the southern boundary of the middle belt of mylonitic rocks. The fault on the north side of the these metasediments is a brittle NW-striking fault of the Red River fault zone, although it lacks Neogene rocks along it, like the fault zone on the north side of the middle belt of mylonitic rocks.

On Vietnamese maps, the mylonitic rocks are shown as Precambrian, but the mylonitization is certainly Cenozoic based on ages presented by Jolivet et al. (2001) from Vietnam. We have not studied the middle belt in Vietnam, but maps show that it widens to the SE, where it becomes faulted and juxtaposed with abundant Paleozoic strata as old as Ordovician. All of the contacts between the Paleozoic and metamorphic rocks are shown as faults. The nature of these contacts and the stratigraphy of the sedimentary rocks need additional study because they may provide important evidence for possible cover rocks and predisplacement origin for the middle belt of mylonitic rocks. In the SE part of the middle belt, Vietnamese maps show additional NW-striking faults within what we designate as the middle belt, suggesting that this belt may be much more complicated than we presently envisage.

Ailao Shan Belt of Mylonitic Rocks (I, Figs. 2 and 3)

The Ailao Shan belt of mylonitic rocks has been studied and described by many workers (for example, see Leloup et al., 1995, 2001; Zhang et al., 2006). Its separation from the Day Nui Con Voi and the middle belt has been described already. The northern boundary of the Ailao Shan is one of the active major through-going faults of the Red River fault zone. Its southern boundary and internal structure are discussed in more detail here. Within Yunnan, the mylonitic and metamorphic rocks of the Ailao Shan are fault bounded. The belt widens from NW to SE, and near the China-Vietnam border, it is intruded by numerous plutons that are abundant in Vietnam (Fig. 2; see following). Within Vietnam, the Ailao Shan loses its character and becomes much more complexly faulted.

Structure of the Mylonitic Belts

At the international border, there are three fault-bounded belts of mylonitic granitic rocks. The southern one, the continuation of the Ailao Shan (I, Fig. 2), broadens to the southeast and is intruded by abundant plutons that range from diorite to granite. The Ailao Shan is separated from the middle belt, which widens to the SE in Vietnam, by definite Triassic strata in Yunnan and their southeastward continuation, assigned a Cambrian(?) age, in Vietnam (Fig. 2). The metamorphosed Triassic(?) strata north of the middle belt wedge out to the SE between faults that appear to merge into the Hong He (Red River) fault zone in Vietnam. The Hong He fault separates the middle belt of mylonitic rocks from the Day Nui Con Voi in Vietnam and is marked by fault-bounded lenses of Miocene and younger conglomerate, which become more abundant in Vietnam. None of the three fault-bounded belts of mylonitic rocks can be traced directly into one another.

Day Nui Con Voi (III, Fig. 2)

The mylonitic rocks of the Day Nui Con Voi shear zone have been the focus of numerous studies within Vietnam (see Jolivet et al., 2001, for review and references). Jolivet et al. (2001) have interpreted two mylonitic events, an older, high-temperature event followed by a lower-temperature event. The older event (pressure-temperature [*P-T*] conditions of 6–7 kbar and 690 °C) produced gently dipping foliation that was overprinted by a younger, steeper mylonitic foliation that formed at lower temperatures and pressures (Jolivet et al., 2001; *P-T* conditions at 480 °C and less than 3 kbar reported by Tran Ngoc Nam et al., 1998). Both events are interpreted to be Cenozoic in age and may be as old as 33 Ma (Gilley et al., 2003).

However, the timing of metamorphism within the Day Nui Con Voi remains somewhat controversial. Unlike thermochonological data from the Ailao Shan (see following), a range of ages has been reported for analyses of monazites from inclusions in garnet and from matrix monazites in the mylonitic rocks. Monazite ages for the inclusions range from ca. 220 to 44 Ma, where younger ages are reported for rim inclusions and older ages are from the more internal inclusions. Matrix monazite age ranges are younger, commonly 33–21 Ma, but ages as old as 107 Ma have also been also reported (Gilley et al., 2003). Just north of the Day Nui Con Voi shear zone, the Song Chay dome contains gently dipping foliation with top-to-the-north kinematic indicators within orthogneisses. Maluski et al. (2001) reported ^{40}Ar-^{39}Ar ages that range from 236 Ma in its southern part to 160 Ma in its core, and they regarded the metamorphism to be of Early Triassic age. The older ages for metamorphism in the Day Nui Con Voi may be interpreted as part of that deformation, but this would require a strong overprinted signature from the Cenozoic mylonitic foliation. Jolivet et al. (2001) interpreted the general Cenozoic structure of the Day Nui Con Voi shear zone to be an open NW-trending antiform based on local subhorizontal foliation within Vietnam. Within the foliation, there is a subhorizontal stretching lineation.

The mylonitic foliation within Yunnan is a direct continuation of foliations mapped within Vietnam, but virtually no work has been done on these rocks in Yunnan. Here, we divide the rock units into two sequences, a core sequence of mylonitic metaigneous and metasedimentary rocks surrounded by outer metasedimentary to unmetamorphosed sedimentary rocks largely of Triassic age (Fig. 3). Our preliminary study of the core rocks shows little evidence for the gently dipping foliation reported by Jolivet et al. (2001). The mylonitic foliation within the Day Nui Con Voi shear zone strikes consistently N45–60°W and mainly dips steeply NE (and locally SW); however, an antiformal structure within the mylonitic rocks cannot as yet be demonstrated. Near the town of Hekou at the Vietnam border, foliation attitudes are more variable with strikes locally N15–30°W. Lineation within the foliation, interpreted to be a stretching lineation, commonly plunges 10–30°NW but is locally subhorizontal or more rarely plunges gently east or southeast. However, the general features around the NW end of the Day Nui Con Voi shear zone make up a NW-trending, NW-plunging antiform defined by the Mesozoic rocks that surround the mylonite (Fig. 3, see following). We have no direct evidence that the mylonitic foliation is folded into the antiform defined by the Triassic rocks, largely because of the poor outcrop. Thus, even though the mylonitic foliation is consistent with it being axial planar to the Day Nui Con Voi antiform, this cannot be completely demonstrated at present.

In Yunnan, the core rocks within the Day Nui Con Voi shear zone in Yunnan contain granite, garnet mica schist, and mylonitized and unmylonitized leucogranite. The mylonitization occurred at greenschist grade (Fig. 4), but the subhorizontal foliation described by Jolivet et al. (2001) has not been observed. The degree of mylonitization ranges from weak in some of the granitic rocks to very strong in some metasedimentary gneiss. It is hard to generalize how the degree of mylonitization or the rock units vary across the belt because of poor exposure. However, on the road just north of Hekou (town on the Chinese side of the international border), a 200-m-long section across strike was exposed in the walls of a new road cut (Fig. 5), and similar rocks occur in recent exposures in building sites around the town (weathering is so rapid that within one year, the exposures are mostly gone). Beneath ~3–4 m of saprolite cover, different rock types, mylonitic granite, leucogranite, and different metasedimentary rock types occur as lenses embedded in a matrix of greenschist-grade mylonitic schist, phyllite, and gneiss (Fig. 5). All the rocks within the lenses are mylonitic, and their foliation is generally parallel with the foliation in the matrix mylonite, but they are folded and boudinaged. The grade of metamorphism in the lenses appears higher than that in the retrograded matrix. The rocks are garnet, muscovite schist with sillimanite, and leucogranite, whereas the matrix is lower grade and is devoid of leucogranite. This outcrop indicates that generalizations about the nature of the mylonitic belt from incomplete exposure are risky at best. There is the indication of two mylonitic events at two different grades of metamorphism, but whether they are part of a continuous event from high to low grade or are two separate events is unknown. It is clear that the rocks assigned to the core of the Day Nui Con Voi structure in Yunnan have protoliths that are unlike most of the adjacent surrounding metasedimentary rocks. They are tentatively assigned a Precambrian protolith age with strong deformational and magmatic modification by Phanerozoic, most likely Mesozoic(?) and Cenozoic, events.

The rocks that mantle the core mylonites consist of metasedimentary rocks that grade outward into unmetamorphosed strata and lack the abundant plutonic rocks of the core, but they do contain rare tourmaline-bearing leucogranite and pegamatite dikes that grade outward into unmetamorphosed strata. The metamorphosed strata form a carapace around the core and have sillimanite-bearing assemblages that indicate they were metamorphosed to the same grade as rocks in the core assemblage. West of the village of Dimi (Fig. 3), Chinese maps show that the core mylonitic rocks of the Day Nui Con Voi shear zone terminate in a NW-plunging antiform, and the metasedimentary rocks

Figure 4. Mylonitic orthogneiss and schist of the Day Nui Con Voi core rocks near the northern border of the belt shown in Figure 3. Stretching lineation is horizontal. These rocks are typical of the orthogneiss mylonitic rocks that make up the belt. The width of the image is 3 m.

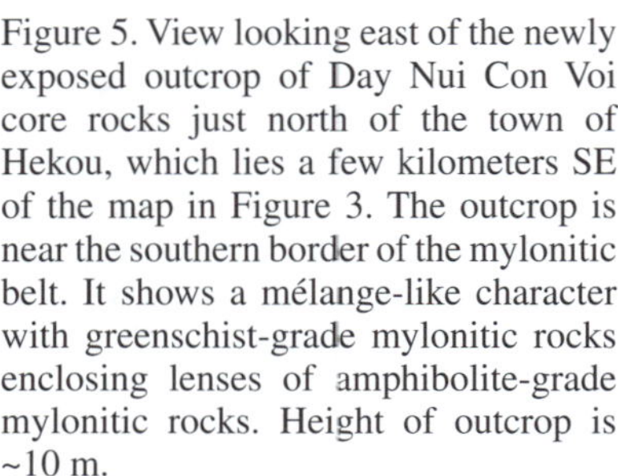

Figure 5. View looking east of the newly exposed outcrop of Day Nui Con Voi core rocks just north of the town of Hekou, which lies a few kilometers SE of the map in Figure 3. The outcrop is near the southern border of the mylonitic belt. It shows a mélange-like character with greenschist-grade mylonitic rocks enclosing lenses of amphibolite-grade mylonitic rocks. Height of outcrop is ~10 m.

of the carapace are assigned a Lower and Middle Triassic age. The metasedimentary strata grade upward into unmetamorphosed fossiliferous Middle and Upper Triassic strata that are part of the extensive outcrops of Triassic rocks of South China (see previous; Figs. 2 and 3). The contact between the Triassic strata and the mylonitic core rocks of the Day Nui Con Voi shear zone is shown on Chinese maps as an unconformity at the plunging nose of the antiform where two younger WNW-striking, subvertical faults that bound the Day Nui Con Voi core rocks diverge to the northwest (Figs. 2 and 3). We attempted to confirm the unconformable relationship in the field but were not successful in this heavily vegetated and deeply soil covered part of Yunnan. We tentatively propose that this contact is a folded fault (see following). If it is a fault, then the identification of rocks mapped by the Chinese as metamorphosed Lower and Middle Triassic becomes less certain and depends entirely upon lithological correlations of their protoliths, and continuity with overlying unmetamorphosed Triassic strata. The protoliths of the metasedimentary rocks were mainly shale, siltstone, and sandstone, with limestone interbeds becoming more common upward, grading into thick marble units. The metamorphosed sequence is lithologically very similar to that of the lower part of the Triassic succession of South China, but because of the complex structure in the area, this correlation must be considered tentative. At the nose of the antiform, they grade laterally into unmetamorphosed strata that contain Triassic fossils and have a stratigraphy that is typical for the broad surrounding area of Triassic rocks of South China (see Fig. 2). Units in both the metasedimentary carapace and the unmetamorphosed strata wrap around the plunging nose of the mylonitic rocks. The structure in these Triassic rocks at the nose of the antiform is complicated. The metasedimentary rocks contain abundant small-scale folds, and the axial traces of map-scale folds in the unmetamorphosed strata curve around the antiformal nose and are suggestive of folds refolded by the antiform. Regionally, the Triassic strata and their underlying Paleozoic rocks of South China lie within a wide fold belt that is regarded to be Triassic in age, but the actual age of earlier folding is uncertain (see discussion in Wang et al., 1998). If the Day Nui Con Voi antiform is a Cenozoic structure, as we argue here, the refolding of older folds is expected.

Mylonitic rocks at the nose of the antiform are very poorly exposed. Foliation strikes N45–60°W, and an antiformal closure is not exposed. Foliation in overlying strata is also rarely exposed and strikes N45–60°W but is often folded into small-scale tight folds and kinks that plunge gently to the NW. On the SW side of the antiform, the Lower Triassic schist generally dips ~30–45°NE, and the stratigraphy of the units suggests that they are overturned. These metasedimentary rocks are strongly folded on a small scale, with fold hinges plunging 10–30°NW; the folds suggest that the antiform also plunges NW, as does the mapped contact with Triassic units. Research is continuing to better to define the nature of the termination of the mylonites.

Even though the metasedimentary rocks at the nose of the antiform are poorly dated, structurally higher unmetamorphosed fossil-dated Middle and Upper Triassic rocks of South China lie stratigraphically above the termination of the core and metasedimentary carapace rocks of Day Nui Con Voi (Figs. 2 and 3). East of the Nanxihe fault, which forms the eastern boundary of the Day Nui Con Voi antiform, Triassic rocks lie conformably above a thick succession of strata of latest Proterozoic to Permian age (see Fig. 2), suggesting that the Nanxihe fault, or an older fault in this position, has major displacement at this latitude. The older rocks are dominated by a thick sequence of pelitic Upper Proterozoic and Lower Cambrian sedimentary rocks overlain by Middle and Upper Cambrian and Lower Ordovician limestone and dolomite. This part of the sequence is different from the thinner and incomplete section of strata that lies to the north, and which forms the characteristic lower Paleozoic succession typical of this part of South China. Devonian to Permian rocks east of the Nanxihe fault are dominated by carbonate strata and the Upper Permian Emei Shan basalt, a succession similar to South China. To the NW, along the Nanxihe fault, the separation of stratigraphic units on either side of the fault becomes less until the fault strikes into the Meng Zhi Basin, where Triassic strata are present on both sides. These Triassic strata form the broad area of Triassic strata of South China, indicating that the Nanxihe fault at this latitude has much less displacement (see Fig. 2).

The Proterozoic and Paleozoic sequences east of the Nanxihe fault continue into Vietnam north of the Day Nui Con Voi shear zone, where they form a large culmination cored by the Song Chay dome (Roger et al., 2000; Fig. 2). The paleogeographic positions of the lower Paleozoic and Proterozoic rocks east of the Nanxihe fault are poorly known, but they may represent a more offshore environment along the South China margin; the upper Paleozoic rocks, by contrast, are typical for South China. However, the Silurian magmatic event within the Song Chay dome is not recognized within Yunnan. Likewise, the Triassic metamorphic event within the dome has no obvious counterpart. The broad belt of Triassic strata within this succession continues west and forms the rocks that overlie the hinge of the Day Nui Con Voi antiform (Figs. 2 and 3). Because of the regional continuity of the Triassic and Upper Paleozoic rocks and the lack of a major unconformity below the Triassic rocks elsewhere in this part of South China, and the presence of the Upper Permian Emei Shan basalt just north of the antiformal core, we interpret the contact between the metasedimentary carapace rocks assigned a Lower and Middle Triassic age and the metamorphic rocks of the Day Nui Con Voi as a fault, probably a low-angle fault that cuts out the Paleozoic rocks of the South China succession. Whether this interpretation is correct or not, the geological relations place a cover, albeit a tectonic cover, of South China stratigraphic units above the rocks of the core mylonitic rocks of the Day Nui Con Voi antiform. The structural position of the uppermost Precambrian and Lower Paleozoic rocks east of the Nanxihe fault in relation to the rocks of the Day Nui Con Voi antiform remains unclear, but a similar argument to that given previously also places them as the tectonic cover.

Because the Triassic and Upper Paleozoic rocks above the mylonitic core of the Day Nui Con Voi shear zone are continuous across South China to the Red River fault zone, strike-slip displacement along the boundary faults of the Day Nui Con Voi cannot be large. This places important constraints on the tectonic history of Day Nui Con Voi shear zone (see following).

Middle Mylonitic Belt (II, Figs. 2 and 3)

Little is known of the middle belt. Within Yunnan, it forms a narrow wedge that appears at the fault-forming contact between the two belts of Triassic rocks that lie between the Day Nui Con Voi and Ailao Shan. Both sides of the middle belt are formed by WNW-striking active brittle faults (Figs. 2 and 3). The middle belt is composed of mylonitic granite with a well-developed foliation with a N30W, 60°S attitude and a lineation that plunges 45°NW. It is associated with fine-grained mafic rocks, of probable Permian age, that are unfoliated to weakly foliated. The relations between the mylonitic granite and mafic rocks are unclear, but it appears that the mafic rock intrudes the granite. The middle belt lies mainly in the Red River so that exposures are poor. The Red River forms the border between China and Vietnam for ~30 km, and we did not follow this belt into Vietnam, but from Vietnamese maps, the belt of granite widens to the SE.

Along the projection of the middle belt to the northwest in SE Yunnan, there is a narrow, 15-km-long, poorly exposed area of mylonitic granite, leucogranite, and garnet-grade metasedimentary rocks. These rocks are similar lithologically to the core rocks in the Day Nui Con Voi shear zone. Foliation in these mylonitic rocks dips steeply to the NE and contains a subhorizontal to gently NW-plunging lineation. While not continuous with the middle belt rocks, these rocks occupy a similar position bounded by WNW-striking faults of the Red River fault zone. At its east and west ends, Chinese maps show the contact with adjacent Triassic rocks to be an unconformity and a fault, respectively; however, we have not been able to confirm these relationships, and we interpret the boundaries as a fault, similar to, or the same as the fault at the plunging nose of the Day Nui Con Voi antiform (Fig. 3).

Ailao Shan Mylonite Belt (I, Figs. 2 and 3)

The Ailao Shan shear zone has been studied by several workers who report conflicting observations about the degree and timing of mylonitization across the Ailao Shan metamorphic belt. Leloup and Kienast (1993) suggested that the rocks are uniformly strongly mylonitized across the belt, whereas Zhang et al. (2006) and Schoenbohm et al. (2005) reported that the rocks are not uniformly mylonitized but are partitioned into several separate zones of differential mylonitic character, and that may be of different ages (Zhang et al., 2006). The age of mylonitization has been studied from Vietnam through Yunnan to the NW end of the Ailao Shan. This work has suggested that mylonitization began at ca. 27 Ma (possibly as old as 34 Ma; Gilley et al., 2003), and ended at ca. 17 Ma, but it was diachronous, becoming younger from SE to NW (Harrison et al., 1996). Two matrix monazite grains from one section (South Mosha) yielded ages of 74.1 and 52.9 Ma interpreted by Gilley et al. (2003) to be incompletely reset detrital grains. The most recent study by Zhang et al. (2006) suggested that partitioning of shear occurred within the Ailao Shan shear zone in Yunnan and that the zones of different left-lateral mylonitic character are of different ages; early high-grade L-tectonites formed at ca. 58–56 Ma, L-S tectonites formed near the brittle-ductile transition at ca. 27–22 Ma, and low-grade phyllitic deformation took place mainly west of the Ailao Shan metamorphic belt at ca. 13–12 Ma. The interpretations of the metamorphism and deformation within the mylonitic rocks from Vietnam by Jolivet et al. (2001; see previous), Gilley et al. (2003), and Zhang et al. (2006) begin to raise the question: how complicated are these shear zones?

Studies of the deformation and the fabric elements of the shear zone generally have been conducted where roads cross the Ailao Shan, and ours is no exception. In the southernmost section across the Ailao Shan in Yunnan from the Red River to Jinping (Figs. 2 and 3), well-developed mylonite is present only in the northeasternmost ~1–2 km, and across the remainder of the Ailao Shan, it is locally poorly developed, restricted to narrow zones, or is not present as the metamorphic rocks of the Ailao Shan widen from 15 km in its central part to more than 30 km in the Jinping cross section. Additionally, there is a rapid increase to the south in the abundance of plutonic rocks within the Ailao Shan, beginning ~20 km north of the Jinping cross section and continuing into Vietnam (Fig. 2).

In the Jinping area, the relations between the plutonic, metamorphic, and sedimentary rocks is important for understanding the regional structural position and original position of the Ailao Shan rocks. Along the southern part of the Ailao Shan, there is a sequence of Ordovician to Triassic, mainly sedimentary rocks, which end to the north but become progressively wider to the SE and are separated from the metamorphic and plutonic rocks by a north-dipping thrust fault (Fig. 2). The Paleozoic rocks are strongly folded into south-vergent often-overturned folds below the thrust. A section of Ordovician rocks forms the base of the exposed section, and they lie below the north-dipping thrust fault that carries the metamorphic rocks of the Ailao Shan in the hanging wall. The Paleozoic section has important differences from similar age rocks in the South China region. The Ordovician rocks are shown on Chinese maps as more than 3000 m thick (thicknesses are often suspect in this highly deformed and poorly exposed area of Yunnan) and consisting of graptolite-bearing shale and fine-grained clastic rocks, much thicker than similar age rocks east of the Nanxihe fault to the north. Thick Silurian limestones (1600 m) overlie the Ordovician strata but are missing in the section east of the Nanxihe fault. Devonian to Upper Permian rocks have similarities in both sections but are consistently thicker in the Ailao Shan area. Basal Devonian rocks are clastic, and Middle Devonian to Lower Permian rocks are dominantly limestone. In both sections, they are overlain by Emei Shan basalt that is much thicker in the Ailao Shan section (4000 + m) than in areas to the north (1200 m). The Ailao Shan section is overlain by ~500 m of Middle Triassic limestone, whereas Triassic rocks to the north range from a thick Lower Triassic unit (lower clastics and upper limestone) through a thick Middle and Upper Triassic limestone section at the top. The contrasting stratigraphy of these rocks must reflect their original paleogeographic relationships before tectonic rearrangement (by extrusion; see following); however, more detailed knowledge of the stratigraphy of these rocks is required before conclusions can be reached. Devonian rocks on the east limb of the synform are intruded by plutons that form part of the Ailao Shan plutonic complex. The age of the folding is unclear, but it is younger than the Upper Triassic rocks involved in the folds at the Vietnam border.

Separating the south-vergent folded rocks along the Ailao Shan from the Lanping-Simao tectonic unit to the south (which consists of largely Mesozoic nonmarine red beds; see Wang and Burchfiel, 1987), there is a discontinuous belt of phyllitic dark slates and, in some places, red beds that have yielded Middle Triassic fossils (Fig. 2). The slates have the character in places of mélange, and we interpret them to represent the continuation of a mélange belt that contains late Paleozoic fossils to the north. The age of the strata yields only an upper limit for mélange formation. The mélange zone in SE Yunnan is unconformably overlain by uncleaved Upper Triassic strata of the main Lanping-Simao tectonic unit. We interpret these relations in the following way: Eastward underthrusting of the Ailao Shan unit by the Lanping-Simao tectonic unit occurred in Triassic time and ended by early late Triassic time.

The age of the plutonism in the Jinping area of the Ailao Shan is important in understanding the folded Paleozoic rocks on their SE side. The age of the plutonic rocks is poorly established. Zhang and Scharer (1999) dated samples from an alkali granite by U-Pb analysis of titanite, which yielded ages of 35.0 ± 3 Ma to 38.6 ± 5.7 Ma. The alkali granite is not mylonitic and is shown on their inset location map as a single large pluton that extends into Vietnam. However, our observations and those shown on Chinese maps show a plutonic complex that ranges from diorite to granite. Chinese maps also show the ages of these plutons to range from Triassic to Cenozoic, although the dating is old and probably suspect. These plutons require additional detailed study. The plutons intrude the Devonian rocks within the folded Paleozoic rocks along the south side of the Ailao Shan. These relations indicate that the Paleozoic rocks were part of a metamorphic-plutonic complex tied to the Ailao Shan by the time of intrusion, and some of the metamorphism and folding may be of Triassic age. We interpret the Paleozoic rocks on the south side of the Ailao Shan to have been the cover rocks for the Ailao Shan rocks. The nature of the cover, whether it was stratigraphic or tectonic, remains unknown, but the plutons tie it to the Ailao Shan by the time of mylonitization and metamorphism of the Ailao Shan rocks. The protoliths for the Ailao Shan metamorphic and mylonitic rocks are most likely Precambrian; the protoliths for the metamorphic rocks in the Ailao Shan do not have any obvious counterparts in the Paleozoic section, which is dominated from the Silurian upward by carbonate rocks or Permian basalt. Some of the rocks in Vietnam have yielded Precambrian ages (see following).

While some mylonitization of the Ailao Shan rocks may have occurred in Triassic time, it was largely a Cenozoic event within a zone of left-lateral transpression (Wang and Burchfiel, 1997). Folding of the Paleozoic rocks on the south side of the

Ailao Shan may have been partly or dominantly Cenozoic because the folds are thrust over steeply dipping Cenozoic strata to the south, which are mapped as Neogene, but their color (maroon) and degree of induration suggest that they may be partly Paleogene. In summary, we interpret the southern boundary of the Ailao Shan to be marked by a mélange belt, which was formed by northeastward underthrusting or transpressional underthrusting in early Late Triassic time but modified by additional thrusting within a left-lateral transpressional tectonic setting during Cenozoic time (see Wang and Burchfiel, 1997). The intensity of mylonitization in the metamorphic rocks becomes less and the zone narrows to the SE, and it is represented by only a very narrow zone along the northern part of the Ailao Shan unit in this area.

The continuation of the Ailao Shan rocks into Vietnam is characterized by gneiss and schist of probable Precambrian age, the Po Sen orthogneiss, and abundant large plutons, the ages of which are poorly known; they are shown on Vietnamese maps as mainly Proterozoic, but they have yielded Mesozoic and Cenozoic isotopic ages. There are a few narrow belts of weakly metamorphosed to unmetamorphosed sedimentary rocks, shown as Cambrian, Ordovician, and Devonian, in fault contact with the metamorphic rocks. In one place, Cambrian strata are shown in depositional contact with the metamorphic rocks, and if true, this relation would support the Precambrian age for the protoliths of some of the metamorphic rocks. The gneissic part of the belt contains narrow zones of NW-striking, steeply dipping mylonitic rocks, with generally subhorizontal stretching lineations, a pattern not unlike the Ailao Shan belt in Yunnan (Jolivet et al., 2001). Tran et al. (1998) reported K-Ar hornblende ages of 1669 ± 24 Ma, 1940 ± 26 Ma, and 2032 ± 27 Ma for rocks farther to the SE, and Carter et. al. (2001) reported a U-Pb Permian-Triassic age of 245 ± 5 Ma for an orthogneiss within this belt. Near Sapa (Fig. 2), Jolivet et al. (2001) showed the Po Sen orthogneiss to be thrust south along a narrow mylonitic zone above a narrow belt of weakly metamorphosed sedimentary rocks shown on Vietnamese maps as Cambrian in age (Fig. 2). They report Ar/Ar ages of 40 and 28 Ma for rocks from the Po Sen orthogneiss and interpreted them to be the probable age of deformation within the Po Sen unit, related to south-vergent thrusting near Sapa. To the southeast, the Ailao Shan begins to lose the characteristics it has in China, and the metamorphic rocks and plutons pass southward into a large area dominated by volcanic rocks and small plutons shown on Vietnamese maps as Jurassic and Cretaceous in age.

The northeastern boundary of the Po Sen unit (continuation of the Ailao Shan in Vietnam) is a north-dipping fault of the Red River fault system that separates it from the belt of unmetamorphosed Triassic strata and metasedimentary rocks to the NE. These units are the ones that separate the Ailao Shan from the continuation of the middle belt of mylonitic rocks in Yunnan. This fault and the associated flanking units can be followed to the SE, where the Po Sen orthogneiss (Ailao Shan) loses its character (Fig. 2).

DISCUSSION AND INTERPRETATION

From the geological relations discussed here, we make the following tentative tectonic interpretation. The protolith age is Proterozoic for many of the metamorphic rocks of the three belts of mylonitic rocks (Ailao Shan, middle belt, and Day Nui Con Voi core rocks), but parts of these gneisses may be igneous protoliths of Triassic to Cenozoic age (Carter et al., 2001; Zhang and Scharer, 1999). The presence and extent of Triassic metamorphism within the rocks now forming the mylonitic belts remain uncertain. Triassic metamorphism can only be documented in the Song Chay dome north of the Day Nui Con Voi shear zone in Vietnam (Maluski et al., 2001; Fig. 2) and probably within the Day Nui Con Voi, although Carter et al. (2001) suggested that Triassic metamorphism may be widespread in the basement rocks of Vietnam. Triassic metamorphic rocks within the Song Chay dome come within a few kilometers of the Day Nui Con Voi mylonite belt in Vietnam, but it cannot be assumed these Triassic rocks continue directly into the Day Nui Con Voi because during Cenozoic time, left-lateral shear within the Day Nui Con Voi indicates that the Song Chay rocks probably would have originated much farther to the SE.

During Middle and possibly Late Triassic time, the Lanping-Simao tectonic unit was thrust NE below the rocks now in the Ailao Shan along a zone now marked by pre–Upper Triassic mélange, which may continue into Vietnam along either the Song Ma suture or farther west in western Vietnam and Laos (see Lan et al., 2000; Metcalfe, 2000). Carter et al. (2001) suggested that during Permian-Triassic time, there was an important deformational-thermal event in north Vietnam during the accretion of the Sibumasu unit to Indochina–South China. Triassic deformation within the protoliths of the Ailao Shan, middle belt, and Day Nui Con Voi mylonitic belts may be part of such an Indosinian event, but most of the evidence for this Triassic deformational, thermal, and magmatic event is in north Vietnam, and its spatial extent within Yunnan remains uncertain and needs considerably more study.

During early Cenozoic deformation, the metamorphic rocks of all three belts of mylonitic rocks in SE Yunnan and north Vietnam were part of one unit subject to left-lateral transpression as rocks of the Lanping-Simao thrust belt in Yunnan were thrust to the northeast below the mylonite belt (Fig. 6; Wang and Burchfiel, 1997). During early Cenozoic time, the rocks now in the three mylonitic belts were all within the middle crust, metamorphosed to amphibolite grade, and were extruded upward to the SW during left-lateral oblique-slip transpression. The upward extrusion of these ductile rocks was marked by an oblique thrust at their base and by oblique-slip normal faults with top-to-the-NE displacement at the top (Fig. 6; see Wang and Burchfiel, 1997), although left-lateral shear was the dominant mode of deformation. This is similar to the interpretation by Jolivet et al. (2001), who also considered the mylonitic rocks to be bounded by a thrust at the base and a normal fault at the top; however, they considered this geometry to be the result of early transpression followed by younger transtension.

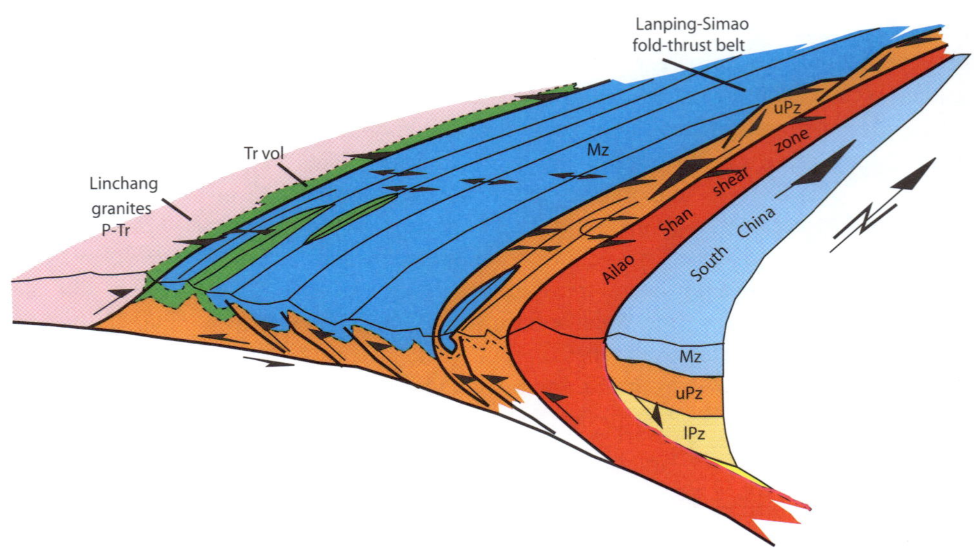

Figure 6. Tectonic interpretation of the early Cenozoic mylonitic belt. Its lower contact is at this time a thrust fault, along which the Lanping Simao tectonic unit is thrust beneath the mylonitic belt, and its upper contact at this time is a north-dipping normal fault (purple) with South China rocks in its hanging wall. The main deformation is left-lateral shear as the Lanping Simao tectonic unit is extruded to the SE, with the mylonitic rocks extruded upward in a transpressional setting. lPz—Lower Paleozoic sedimentary rocks; uPz—Upper Paleozoic mainly sedimentary rocks; P-Tr—Permian Triassic granites; Tr vol—Triassic volcanic rocks; Mz—Mesozoic mainly sedimentary rocks.

Mylonitization accompanied the dominantly left-oblique shear. The low-angle oblique normal fault on the northeast boundary of the mylonitic rocks juxtaposed the South China sedimentary cover in the hanging wall with the mylonitic rocks in the footwall (Fig. 6). The normal fault would have been a ductile normal fault that juxtaposed rocks of similar metamorphic grade (forming the core and carapace relations present in the Day Nui Con Voi belt in Yunnan). The distance that these two rock units were originally separated across this normal fault is not known. Continued left-lateral shear at lower metamorphic grade formed younger low-grade mylonites that overprinted the older higher-grade mylonites and disrupted the mid-crustal metamorphic rocks shearing them into (at least) three fault slivers that cut obliquely at a low-angle across an originally continuous and broad mylonitic zone, so that the older mylonitic rocks within the Ailao Shan mylonitic belt were separated into at least three discrete mylonitic belts in SE Yunnan and adjacent Vietnam (Fig. 7). The once-continuous belt forms a left-stepping en-echelon pattern from Ailao Shan mylonites to the middle mylonitic belt, to the Day Nui Con Voi mylonitic belt. During the younger left-oblique shearing, the South China cover rocks lay structurally above Day Nui Con Voi rocks, and probably the middle belt of mylonitic rocks in Yunnan, where they were disrupted to be become fault-bounded between the Ailao Shan and Middle mylonitic belt and folded around the Day Nui Con Voi mylonitic belt (Fig. 7).

During late Cenozoic time, these three mylonitic belts were further disrupted by brittle right-lateral strike-slip faults of the Red River fault system, which presently bound the northeast side of the Ailao Shan and both sides of the middle belt and Day Nui Con Voi core rocks (Fig. 7). The distribution of South China strata restricts the magnitude of the overall right-lateral displacement to ~50 km or less.

In our interpretation, the mylonitic rocks of the middle belt and Day Nui Con Voi core rocks in SE Yunnan and Vietnam were structurally overlain by sedimentary rocks of latest Proterozoic to Triassic age belonging to South China during their oblique extrusion from the lower crust. The magnitude of the displacement between the extruded rocks and their South China cover is unknown during the early Cenozoic oblique shearing because they are tectonically separated. Such deformation was confined to the middle and upper crust and decoupled from deeper lower-crust and mantle deformation.

Our interpretation places the Paleozoic and Triassic rocks along the southeast side of the Ailao Shan as its cover, but whether this relation was tectonic or stratigraphic remains unknown. Paleomagnetic data (J.W. Geissman, 2008, personal commun.) suggest that extrusion of rocks in the Lanping-Simao tectonic unit to the southwest of the Ailao Shan, relative to South China, was at least 500 km, but these rocks in our interpretation and that of Wang and Burchfiel (1997) were separated from the Ailao Shan along a major thrust fault zone. Thus, the mylonitic belt was bounded both at the base and top by faults, and its original position remains subject to interpretation. Because it participated in part of the left-lateral displacement within the extruded rocks, the magnitude of its movement should be less than that of the Lanping-Simao. If our interpretation is correct, that the Paleozoic rocks along the SW side of the Ailao Shan are its cover, then they must be compared to rocks to the northeast to find their original position and that of the Ailao Shan protolith. The Paleozoic rocks have some similarity but several major differences to the rocks of South China directly to their north. In particular, the very thick fine-grained clastic rocks of Ordovician age and the very thick Emei Shan basalt are markedly different. The best correlation that we suggest for these rocks is ~250–300 km to the NW near Erhai Lake (see Fig. 1). This would require the rocks in the Ailao Shan mylonite belt, and its other disrupted parts, to be separated from their South China cover by ~250–300 km along a transpressional mylonitic zone that flattens into the middle and lower crust below South

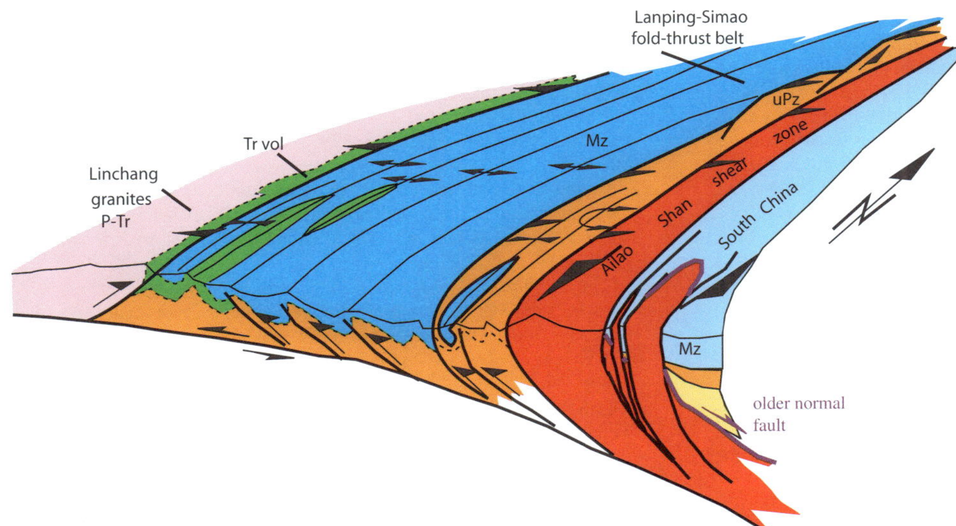

Figure 7. Tectonic interpretation of the later stages of the early Cenozoic left-lateral transpression when younger mylonitic rocks at greenschist grade disrupted the older, higher-grade mylonites of the shear zone. The South China tectonic cover in the hanging wall of the earlier normal fault (purple) was faulted between belts of mylonite, and it was folded around the NW-plunging nose of the Day Nui Con Voi antiform. Symbols and colors are the same as in Figure 6.

China. The probable presence of Triassic metamorphism within the Vietnamese part of the Day Nui Con Voi shear zone directly south of Triassic orthogneiss in the Song Chay dome may suggest that the major displacement within the mylonite belt lies within its southern part, and it may suggest a complex partitioning of displacement within the mylonitic belt. The magnitudes of displacement within the two mylonite-forming events remain unknown, but the second event that disrupted the original continuous mylonite belt, producing the present en echelon pattern, could be quite large. Because South China stratigraphic units can be followed across the NW projections of the Day Nui Con Voi and Middle belts, the major displacement may be between the middle belt and the Ailao Shan belt. From the present distribution of rocks, the displacement could be more than 100 km.

The younger late Cenozoic right-lateral displacement took place along several faults forming the Red River fault zone and is probably ~50 km or less, determined from the maximum horizontal overlap of the mylonitic belts and the continuity of the South China sedimentary cover NW of the end of the Day Nui Con Voi core rocks. This right-lateral displacement needs to be added back to the early Cenozoic left-lateral displacement to determine the total early Cenozoic left-lateral displacement. Since the paleomagnetic studies in the Lanping Simao by J.W. Geissman (2008, personal commun.) indicate at least 500 km of southward transport of the Lanping-Simao tectonic unit, part of this total displacement was absorbed along the early Cenozoic mylonite belt of the Ailao Shan and its disrupted parts to the SE.

The geological relations in SE Yunnan and Vietnam do not support the present-day continuity of the Ailao Shan shear zone from China through Vietnam to Hanoi. However, they do support an early Cenozoic continuity of the mylonite belt. The structural relations suggest a complex and evolving zone of early Cenozoic left-oblique transpressional shear within possible South China basement rocks, which is now decoupled from South China to the NE and the Lanping-Simao tectonic unit to the SW.

ACKNOWLEDGMENTS

This study was supported by National Science Foundation (NSF) grants EAR-0003571 and EAR-8904096 awarded to Massachusetts Institute of Technology (MIT). We thank Liangzhong Chen of the Yunnan Geological Institute and Nguyen Van Pho of the Institute of Geological Sciences of the Vietnamese Academy of Sciences and Technology, Vietnam, for their support and assistance during this project. We also thank David Rowley and William Kidd, whose reviews greatly improved the manuscript.

REFERENCES CITED

Carter, A., Roques, D., Bristow, C., and Kinny, P., 2001, Understanding Mesozoic accretion in southeast Asia: Significance of Triassic thermotectonism (Indosinian orogeny) in Vietnam: Geology, v. 29, p. 211–214, doi: 10.1130/0091-7613(2001)029<0211:UMAISA>2.0.CO;2.

England, P., and Houseman, G., 1989, Extension during continental convergence, with application to the Tibetan Plateau: Journal of Geophysical Research, v. 94, p. 17,561–17,579.

Gilley, L.D., Harrison, T.M., Leloup, P.H., Ryerson, F.J., Lovera, O., and Wang, J.-H., 2003, Direct dating of left-lateral deformation along the Red River shear zone, China and Vietnam: Journal of Geophysical Research, v. 108, doi: 10.1029/2001JB001726.

Harrison, T.M., Leloup, P.H., Ryerson, F.J., Tapponnier, P., Lacassin, R., and Wenji, C., 1996, Diachronous initiation of transtension along the Ailao Shan–Red River shear zone, Yunnan and Vietnam, *in* Yin, A., and Harrison, T.M., eds., The Tectonic Evolution of Asian: Cambridge, Cambridge University Press, p. 208–225.

Houseman, G., and England, P., 1993, Crustal thickening versus lateral expulsion in the Indian-Asian continental collision: Journal of Geophysical Research, v. 98, p. 12,233–12,249, doi: 10.1029/93JB00443.

Jolivet, L., Beyssac, O., Goffe, G., Avigad, D., Lepvrier, C., Maluske, H., and Thang, T.T., 2001, Oligo-Miocene midcrustal subhorizontal shear zone in Indochina: Tectonics, v. 20, p. 46–57, doi: 10.1029/2000TC900021.

Lan, C.-Y., Chung, S.-L., Shen, J.J.-S., Lo, C.-H., Wang, P.-L., Hoa, T.T., Thanh, H.H., and Mertzman, S.A., 2000, Geochemical and Sr-Nd isotopic characteristics of granitic rocks from northern Vietnam: Journal of Asian Earth Sciences, v. 18, p. 267–280, doi: 10.1016/S1367-9120(99)00063-2.

Leloup, P.H., and Kienast, J.-R., 1993, High-temperature metamorphism in a major strike-slip zone: The Ailaoshan–Red River, People's Republic of China: Earth and Planetary Science Letters, v. 118, p. 213–234.

Leloup, P.H., Lacassin, R., Tapponnier, P., Scharer, U., Zhong, D., Liu, X., Zhang, L., Ji, S., and Trinh, P.T., 1995, The Ailao Shan–Red River shear zone (Yunnan, China), Tertiary transform boundary of Indochina: Tectonophysics, v. 251, p. 3–84, doi: 10.1016/0040-1951(95)00070-4.

Leloup, P., Arnaud, N., Lacassin, R., Kienast, J., Harrison, T., Trong, T., Replumaz, A., and Tapponnier, P., 2001, New constraints on the structure, thermochronology, and timing of the Ailao Shan–Red River shear zone, SE Asia: Journal of Geophysical Research, v. 106, no. B4, p. 6683–6732, doi: 10.1029/2000JB900322.

Maluski, H., Lepaviri, C., Jolivet, L., Carter, A., Roques, D., Beyssac, O., Tang, T.T., Thang, N.D., and Avigad, D., 2001, Ar-Ar and fission track ages in the Song Chay Massif: Early Triassic and Cenozoic tectonics in northern Vietnam: Journal of Asian Earth Sciences, v. 19, p. 233–248, doi: 10.1016/S1367-9120(00)00038-9.

Metcalfe, I., 2000, The Bentong-Raub suture zone: Journal of Asian Earth Sciences, v. 18, p. 691–712, doi: 10.1016/S1367-9120(00)00043-2.

Replumaz, A., and Tapponnier, P., 2003, Reconstruction of the deformed collision zone between India and Asia by backward motion of lithospheric blocks: Journal of Geophysical Research, v. 108, no. B6, p. 2285, doi: 10.1029/2001JB000661.

Roger, F., Leloup, P.H., Jolivet, M., Lacassin, R., Trinh, P.T., Brunel, M., and Seward, D., 2000, Long and complex thermal history of the Song Chay metamorphic dome (northern Vietnam) by multi-system geochronology: Tectonophysics, v. 321, p. 449–466, doi: 10.1016/S0040-1951(00)00085-8.

Schoenbohm, L.M., Burchfiel, B.C., Chen, L., and Yin, J., 2005, Exhumation of the Ailao Shan shear zone recorded by Cenozoic sedimentary rocks, Yunnan, Province, China: Tectonics, v. 24, doi: 10.1029/2005TC001803.

Tapponnier, P., Peltzer, G., Le Dain, A.Y., Armijo, R., and Cobbold, P., 1982, Propagating extrusion tectonics in Asia: New insights from simple experiments with plasticine: Geology, v. 10, p. 611–616.

Tran, N.N., Mitsushiro, T., and Tetsumaru, I., 1998, *P-T-t* paths and post-metamorphic exhumation of the Day Nui Con Voi shear zone in Vietnam: Tectonophysics, v. 290, p. 299–318, doi: 10.1016/S0040-1951(98)00054-7.

Wang, E., and Burchfiel, B.C., 1997, Interpretation of Cenozoic tectonics in the right-lateral accommodation zone between the Ailao Shan shear zone and the eastern Himalayan syntaxis: International Geology Review, v. 39, p. 191–219.

Wang, E., Burchfiel, B.C., Royden, L.H., Chen Liangzhong, Chen Jishen, Li Wenxin, and Chen Zhiliang, 1998, Late Cenozoic Xianshuihe-Xiaojiang, Red River, and Dali fault systems of southwestern Sichuan and central Yunnan, China: Geological Society of America Special Paper 327, 108 p.

Zhang, L.-S., and Scharer, U., 1999, Age and origin of magmatism along the Cenozoic Red River shear belt, China: Contributions to Mineralogy and Petrology, v. 134, p. 67–85, doi: 10.1007/s004100050469.

Zhang, J., Zhong, D., Sang, H., and Zhou, Y., 2006, Structural and geochronological evidence for multiple episodes of Tertiary deformation along the Ailaoshan–Red River shear zone, southeastern Asia, since the Paleocene: Acta Geologica Sinica, v. 80, no. 1, p. 79–96.

Manuscript Accepted by the Society 04 April 2008

Printed in the USA

The Geological Society of America
Special Paper 444
2008

Combined model of rigid-block motion with continuous deformation: Patterns of present-day deformation in continental China

P.-Zh. Zhang
Weijun Gan
State Key Laboratory of Earthquake Dynamics, Institute of Geology, China Seismological Bureau, Beijing 100029, China

ABSTRACT

One of the critical questions of continental dynamics is whether the preliminary assumption of rigid plate rotation is applicable to a continental lithosphere that possesses a long history of geological evolution. Different theoretical models give diverse answers that are difficult to test because of the lack of coherent and reliable kinematical constraints. In this paper, we study kinematic patterns of present-day tectonic deformation (or crustal movement) of continental China on the basis of 1350 global positioning system (GPS) measurements in combination with active faults and seismic activity studies. The present-day tectonic deformation of continental China is characterized by a combination of rigid-block movement and continuous deformation. For example, the Tarim Basin, the Ordos, and South China behave as coherent blocks similar to oceanic rigid blocks without internal deformation, whereas the Tibetan Plateau and Tianshan seem to deform continuously with significant internal deformation. Mechanical and rheological properties of the lithosphere dictate the style of regional deformation. Regions of high rigidity behave with rigid block-like movements, and those of low rigidity are dominated by continuous deformation. Rheological flow in the lower crust and upper mantle plays an important role in controlling deformation of the upper crust. Present-day tectonic deformation of continental China can be described in terms of a combined model of rigid-block movement and continuous deformation.

Keywords: rigid block movement, continuous deformation, continental dynamics, GPS measurements.

INTRODUCTION

Following the recognition of plate tectonics, the essence of that idea, of rigid body movement of plates or blocks on the surface of the planet, was promptly applied to continental regions undergoing widespread deformation (McKenzie, 1972; Dewey et al., 1973). In interpretations applied to eastern Asia, deformation was thought to occur as large parts of the region acted as microplates that moved eastward out of the India's northward path (sometimes termed "continental escape") (Tapponnier et al., 2001; Avouac and Tapponnier, 1993; Replumaz and Tapponnier, 2003). However, it has been recognized for many years that the deformation within continents represents a significant departure from the principles of plate-tectonic rigid plates (Molnar and Tapponnier, 1975). For example, deformation of continental Asia is spread over a width of thousands of kilometers and cannot be adequately accounted for in terms of the relative motion of a few large plates, although some of the seismicity and deformation appear to occur in narrow linear zones. Continuous deformation of either the entire lithosphere (England and Houseman, 1986;

Zhang, P.-Zh., and Gan, W., 2008, Combined model of rigid-block motion with continuous deformation: Patterns of present-day deformation in continental China, *in* Burchfiel, B.C., and Wang, E., eds., Investigations into the Tectonics of the Tibetan Plateau: Geological Society of America Special Paper 444, p. 59–71, doi: 10.1130/2008.2444(04). For permission to copy, contact editing@geosociety.org.

Houseman and England, 1993; England and Molnar, 1997; Holt et al., 2000) or flow in the lower crust (Royden et al., 1997; Clark and Royden, 2000) has also been proposed to account for the pattern of distributed deformation.

High-precision phase measurements using the global positioning system (GPS), developed in last decade, provide a powerful means to directly measure the kinematic pattern of present-day crustal deformation in continental China and its vicinity (referred to as continental China hereafter). These new GPS data provide unprecedented information for the study of continental deformation and provide new insights into the dynamics of contemporary tectonic deformation (Chen et al., 2000; Wang et al., 2001; Wang et al., 2003; Calais et al., 2003; Zhang et al., 2003a, 2004). Using both geological slip rates and GPS velocities, England and Molnar (2005) calculated a velocity field for eastern Asia that shows plate- or block-like motion within deforming Asia only in the Tarim, South China, and the Amurian region, and, elsewhere, the kinematics of rigid blocks do not provide a useful description of the motion.

Meade (2007) used the velocity field of Zhang et al. (2004) and solved for the angular velocities of relative motion not only among Eurasia, India, South China, and Tarim, but also among 12 other smaller elastic blocks. Using much the same data, Thatcher (2007) solved for relative velocities of 11 "quasi-rigid blocks" in eastern Tibet, in a regions where Meade used 10 blocks (including Tarim and South China). Calais et al. (2006) subdivided eastern Asia into nine regions, including Eurasia, India, Tarim, and South China, and they used a combination of GPS observations from several networks to address the relative motion of the regions and internal deformation within them. They ignored southern Tibet and the Himalaya, and among the other eight regions, they reported, according to the data, that the velocity field of Tibet could not be described by the movements of two rigid blocks.

A description in terms of blocks may or may not be satisfactory, depending on how one imagines continental deformation to occur and on the degree to which existing GPS data can resolve such a question. Thatcher (2007) shows that the transition between block and continuum models is gradational rather than abrupt, and the two converge as the number of faults increase and the block size decreases. In this paper, we show that the present-day tectonic deformation of continental China is characterized by combined rigid-block movements and continuous deformation. The behavior of a region, either as a rigid block or a continuum with significant internal deformation, depends on the rheological properties of the lithosphere of the region.

DATA AND DATA PROCESSING

Significant advancement for the monitoring of crustal deformation in eastern Asia was accomplished in 1998 when the Crustal Motion Observation Network of China (CMONOC) was established. The principal data used for this study come from the CMONOC collected since 1998, including 25 continuous stations, 56 annually observed stations with an occupation of at least 7 d (~168 h data collection) in each survey, and 961 regional stations observed in 1999 and 2001 with an occupation of at least 3 d (~72 h data collection) in each survey.

The data were processed in four steps (Z.K. Shen et al., 2000, 2001). First, we combined the daily observations to solve for loosely constrained station coordinates and satellite orbits using the GAMIT software (King and Bock, 1995). Second, we combined the regional daily solution with the loosely constrained global solutions of ~80 IGS (International GPS Service) tracking stations produced at the Scripps Orbital and Position Analysis Center (Bock et al., 1997) using the GLOBK software (Herring, 1995). The merged daily solution included the loosely constrained station coordinates, polar motion, and satellite orbit parameters, and the variance covariance matrix. Third, we estimated station positions and velocities in the ITRF2000 (International Terrestrial Reference Frame) reference frame using the QOCA software (Dong et al., 1998). The QOCA modeling of the data was done through sequential Kalman filtering, allowing adjustment for global translation and rotation of each daily solution. In the last step, we transformed the velocity solution to a Eurasia-fixed reference frame using the angular velocity of Eurasia with respect to the ITRF deduced from 11 IGS stations (NYAL, ONSA, HERS, WSRT, KOSG, WTZR, VILL, GLSV, IRKT, TIXI) in the stable Eurasia plate (Z.K. Shen et al., 2000, 2001; M. Wang et al., 2003).

Besides the CMONOC data, we collected four additional data sets of station velocities from Calais et al. (2003), Paul et al. (2001), Wang et al. (2001), and Banerjee and Burgmann (2002) to increase the coverage and station density in the Mongolia, India, Himalayan, and central Tibetan regions. Velocity data from Paul et al. (2001) (13 stations in India and the Himalaya) are in an India fixed reference frame, whereas velocity data from Calais et al. (2003) (48 stations in Mongolia), Q. Wang et al. (2001) (41 stations distributed in India, the Himalaya, and central Tibet), and Banerjee and Burgmann (2002) (24 stations in the western Himalaya) are in a Eurasia-fixed reference frame, which differs slightly from the Eurasia-fixed reference frame we employed. Since each of the additional velocity data sets has some common stations with the CMONOC data set, we chose common stations for the three data sets and transformed them to the Eurasia-fixed reference frame of the CMONOC data set by minimizing the velocity differences of the common stations in the corresponding reference frames. After the transformation, the maximum difference of the velocities for each common station in different data sets was less than 2.9 mm/yr and 2.6 mm/yr for the east and north components, respectively, which are within two standard deviations of the velocity components. Thus, we calculated the weighted average of the velocity components for the common stations and estimated their standard deviations. We finally obtained velocities for 1350 stations in eastern Asia, which provided adequate spatial coverage to interpret the magnitude and style of deformation (Fig. 1). The data that Zhang et al. (2004) used to study present-day deformation in Tibetan Plateau is a subset of this data set.

CHARACTERISTICS AND KINEMATICS OF RIGID BLOCKS

The rigidity of a crustal block can be tested through rotation surround an Euler pole on a sphere. For a rigid block, the angular velocity of points within the block should increase as the radius increases away from the pole of rotation (Euler pole). The angular velocity should remain the same along a small circle. Relative motion or significant deformation should occur only along the boundary of two neighboring blocks, and there should be no significant deformation within the blocks. These features have been successfully used to describe the relative motions of rigid oceanic

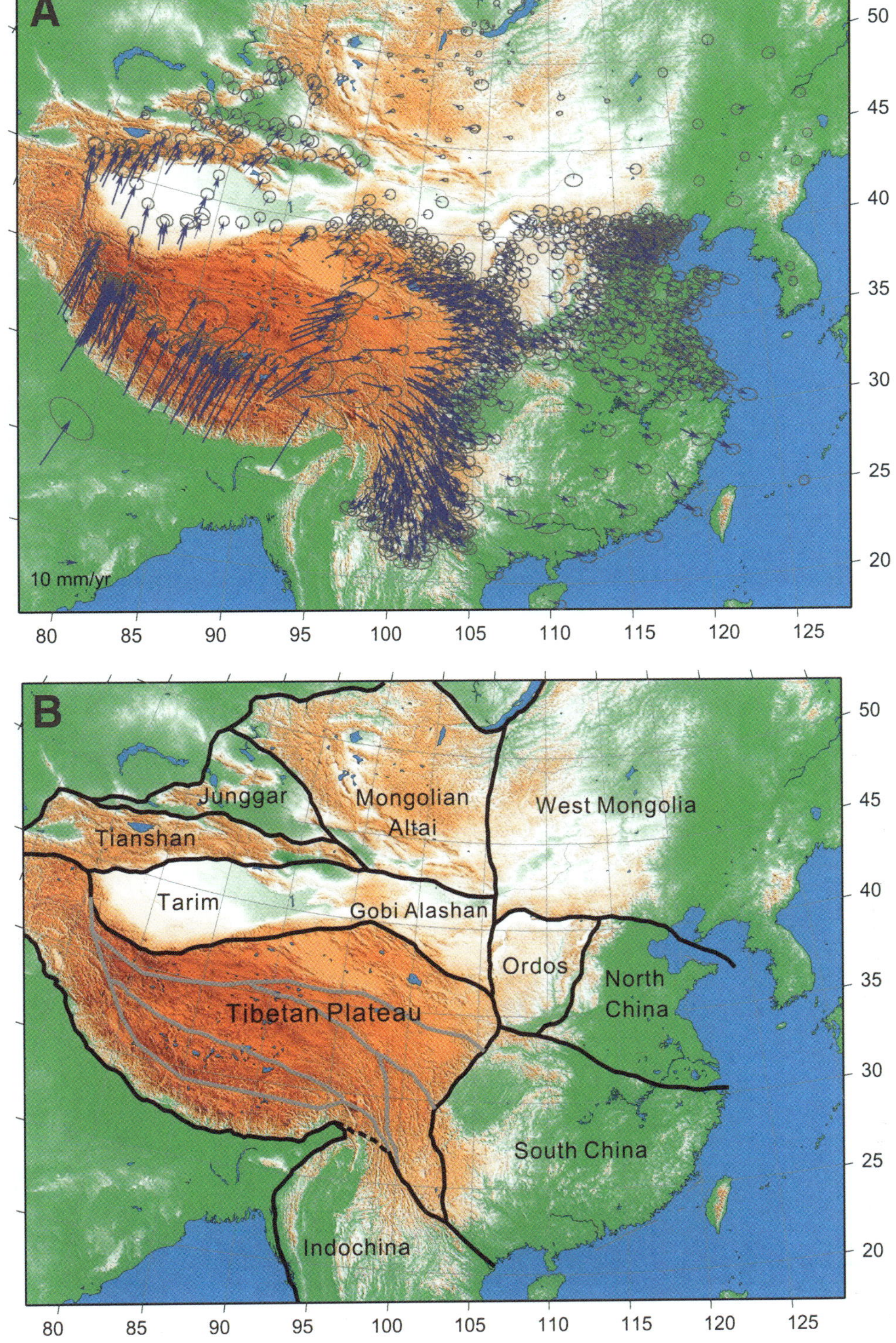

Figure 1. (A) Velocity field of present-day crustal movement in continental China constrained from global positioning system (GPS) measurements. Each arrow indicates a GPS velocity vector; scale of velocity vector is located at the lower left corner of the figure. (B) Crustal blocks studied in this paper. Blocks within the Tibetan Plateau outlined by gray lines do not behave as rigid blocks.

plates during the establishment of plate tectonics in the 1960s and 1970s of the last century. We can also use these features and the velocity field shown in Figure 1 to test the rigidities of crustal blocks in continental China.

The Tarim block has been regarded as a rigid or quasi-rigid block based on seismological and geological studies (Molnar and Tapponnier, 1975; Ma, 1989). It is bounded by the strongly deformed Tianshan in the north and the Tibetan Plateau in the south (Fig. 1). Tectonic activity in the Tarim block is very weak—no earthquake with magnitude over 6 has occurred during the history of instrumental earthquake recording. No active fault has been reported in the Tarim block (Deng et al., 2003). Using a subset of GPS data, Z.K. Shen et al. (2001) identified clockwise rotation of the Tarim as a rigid block around a pole at 37.5° ± 0.4°N and 96.2° ± 0.8°E. The blue arrows in Figure 2 are updated GPS data within the Tarim block. The direction of GPS velocity vectors is generally toward NS ± 10°, similar to the direction of relative motion between the Indian and Eurasian plates. In the vicinity of longitude 88°E, the velocity is in the range of 4–5 mm/yr, and it increases to 10–12 mm/yr near longitude 83°E. Further west to 77°E, the velocity increases to 18–19 mm/yr (Fig. 2). This westward increase supports the notion of clockwise rigid-block rotation (Avouac et al., 1993; F. Shen et al., 2001). We find that velocity field of the Tarim block can be modeled by a clockwise rigid-block rotation around an Euler pole located at 93.3° ± 0.9°E, 37.3° ± 0.3°N, at rate of –0.679 ± 0.059°/m.y., similar to that found by Z.K. Shen et al. (2001). Red arrows in Figure 2 are modeled velocities using rigid-block rotation, which fits the observed velocity shown by blue generally well except a few points that have large deviations (Fig. 2). Thus, the Tarim block probably does behave as a rigid block.

The Ordos block is located northeast of the Tibetan Plateau. It consists of relatively undeformed Triassic rocks with a Quaternary eolian loess cover a few hundred meters thick. The block is bounded by active normal faults and associated grabens around its southern, eastern, northern, and northwestern sides (Deng et al., 1984). Active thrust faulting and folding characterize late Cenozoic tectonic deformation along its southwestern side (Burchfiel et al., 1991; Zhang et al., 1991). There have been seven earthquakes with magnitude over 7 around the block during about two thousand years of documented history, but none has occurred in the interior of the block (Deng et al., 1984). Active tectonic studies indicate the absence of active faulting and folding in the interior of the block, corroborating the very weak seismic activity.

There are 25 GPS stations within the Ordos block that show coherent movement toward the southeast (N115°–125°E). The velocities of these stations are in the range of 5–7 mm/yr (Fig. 3). We modeled the motion of the Ordos block as rigid-block rotation around an Euler pole at 47.4° ± 1.1°N, 119.0° ± 1.4°E with an angular velocity of 0.2205° ± 0.0250°/m.y. in a counterclockwise fashion. Red arrows are modeled velocities that are generally in good agreement with the observed ones shown in blue (Fig. 3), suggesting that the Ordos block may indeed behave as a rigid block.

The South China block is regarded as a stable block due to the absence of active faults and strong earthquakes of magnitude over 7 (Deng et al., 2003; Zhang et al., 2003b). Previous studies using a subset of our GPS data (less than 21 stations) have indicated that the entire South China block moves coherently east-southeastward (110°–30°) at 8–10 mm/yr (King et al., 1997; Chen et al., 2000; Q. Wang et al., 2001; Shen et al., 2000; Calais et al., 2003). No velocity gradient has been identified within the South China block, which further suggests good rigidity of the block.

We used 102 GPS stations distributed within the South China block to show the coherent movement of the entire block (Fig. 4).

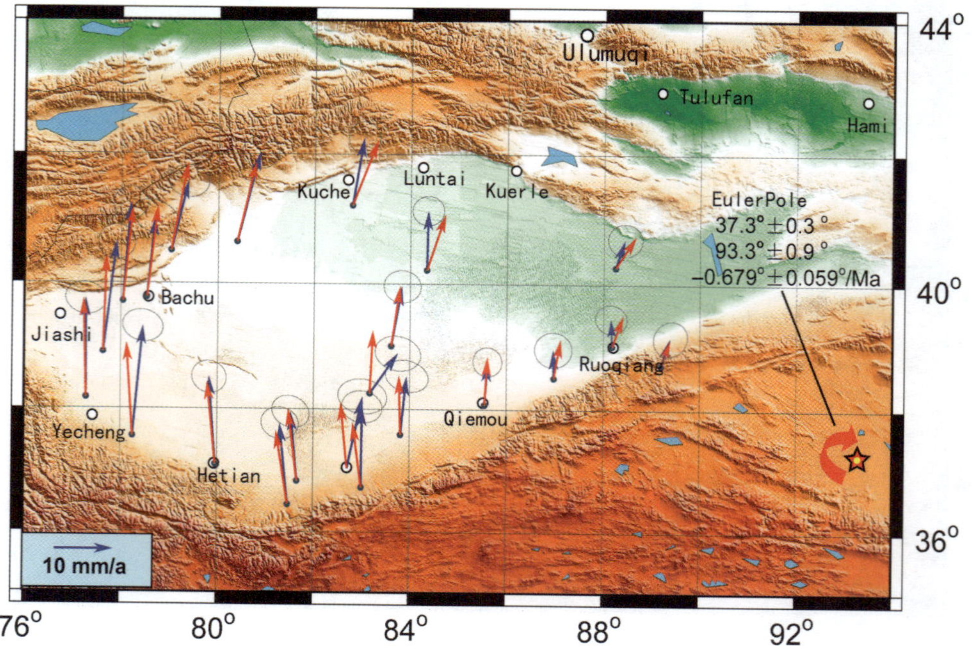

Figure 2. Global positioning system (GPS) velocity field in Tarim Basin observed from GPS measurements and predicted by rigid-block rotation. Blue arrows are observed GPS velocity. Red arrows are predicted velocity. The Euler rotation pole was obtained by fitting the observed GPS velocities in the Tarim Basin.

Although the number of stations that we used is much greater than the previous studies, our data support the previous findings obtained with a subset of data. By means of numerical modeling of rigid-block rotation, we found an Euler pole located at −50.5° ± 4.8°S, 60.9° ± 6.9°E, with an angular velocity of −0.0706° ± 0.0009°/m.y. The predicted velocities agree with the observed values at the sites within the entire South China block (Fig. 4). The good agreement suggests rigid-block rotation.

Because both the Ordos and South China blocks are likely to be rigid blocks, the velocity profile across them should show a rise or a step of certain width across the boundary due to interseismic strain accumulation. We can also use this property to test the rigidities of the blocks, and to calculate interseismic strain within the blocks and across the boundary. Figure 5 shows a velocity profile across the two blocks spanning longitudes from 106°E to 109°E. The N110°E velocity components within both the Ordos and South China blocks vary within a range of <3 mm/yr over a distance of more than 500 km, but the difference across the 100-km-wide boundary between the two blocks reaches ~5–6 mm/yr. The average strain rates within the South China and Ordos blocks are 2.9×10^{-9} and 5.8×10^{-9}, respectively, whereas the average strain rate across the block boundary is 2.7×10^{-8}, which is an order of magnitude greater than that within the block (Fig. 5). This test also suggests the rigid block property of the two blocks.

CONTINUOUS DEFORMATION IN THE TIBETAN PLATEAU AND TIANSHAN

The present tectonic deformation of the Tibetan Plateau and Tianshan cannot be described by rigid-block movements (Q. Wang et al., 2001; Zhang et al., 2004). Figure 6 shows distribution of GPS stations used in the study. Zhang et al. (2004) drew four profiles (A–A′, B–B′, C–C′, and D–D′) across the plateau along the N20°E direction, the inferred India-Eurasia convergence direction, to calculate the shortening across different parts of the plateau (Fig. 6). Taking 36–40 mm/yr as total relative motion

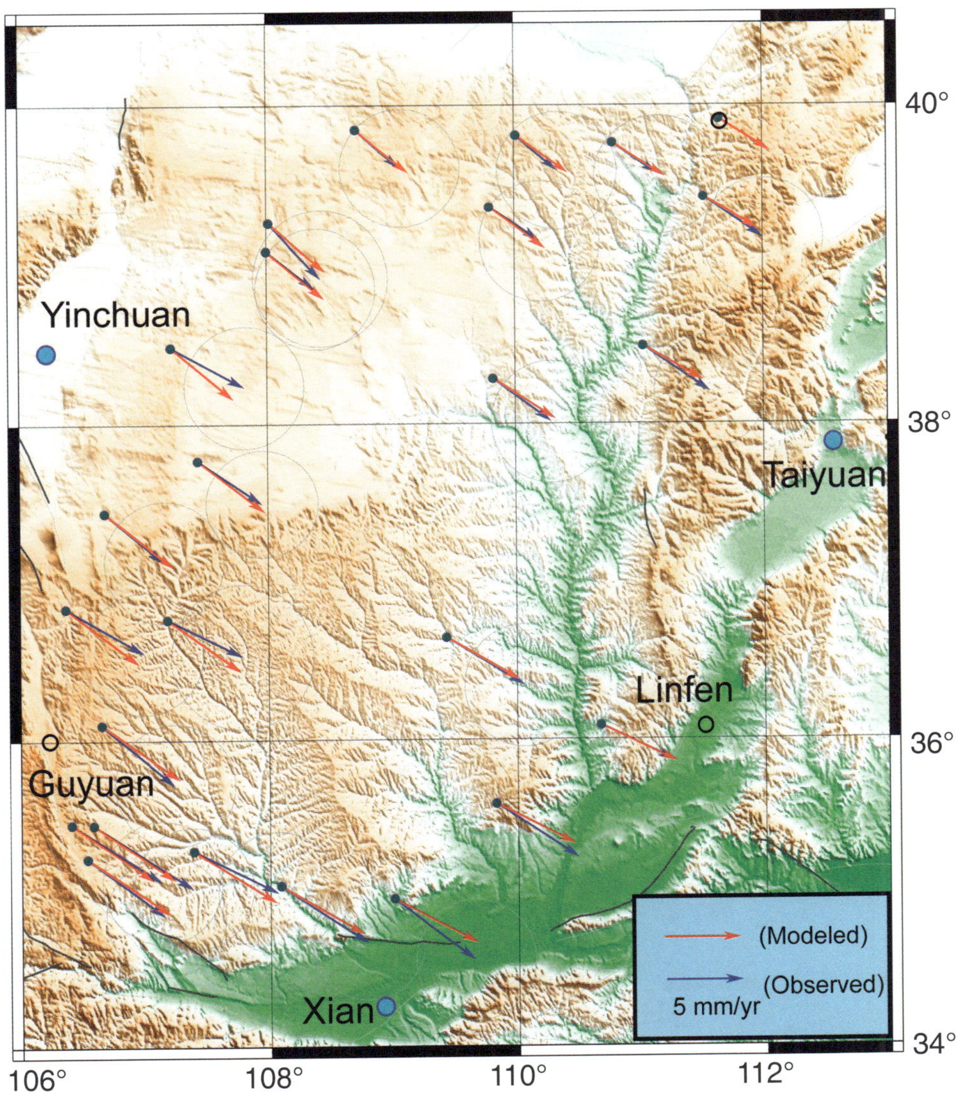

Figure 3. Global positioning system (GPS) velocities within the Ordos block shown by blue arrows in comparison with the modeled ones (see text).

between India and Eurasia, the eastern Tibetan Plateau and its margins (profiles A–A′ and B–B′) accommodates 85%–94% of the total motion, whereas western Tibet (profiles C–C′ and D–D′) absorbs 70%–91% of total convergence, and the rest is taken up by shortening across the Tianshan in the north (Abdrakhmatov et al., 1996; Reigber et al., 2001). Zhang et al. (2004) also reported linear gradients of velocity along all of the four profiles across the entire plateau, without steps where they cross major faults within the Tibetan Plateau, such as the Bangong-Nujiang suture and the Jinsha suture. The gradient in a specific velocity profile indicates internal strain within the region covered by the profile, and the linear gradient of the velocity profile suggests uniform strain rather than localized strain across a particular fault or tectonic zone. Even if the linear velocity gradient may be modeled by interseismic strain on faults, the number of faults must be many, and the slip on each fault must be minor, a few millimeters per year. This pattern of internal strain, slips on many faults, is significantly different from that within a rigid block such as in the Tarim or the Ordos block.

To show different interseismic strain accumulation in the Indian plate, Tibetan Plateau, Gobi Alashan, and South China, we drew another two long profiles across them in the direction N20°E and N110°E, respectively. Figure 7A shows velocity changes with distance across the Tibetan Plateau along the direction of N20°E. The velocity differences over 2000 km are 6.3 ± 2.0 and 6.7 ± 1.2 mm/yr across the Indian plate in the south and the stable Gobi Alashan block in the north, respectively. These velocity differences and distances correspond to strain rates of 2.3×10^{-9} and 4.4×10^{-9} yr^{-1}, respectively. The velocity difference over 2000 km reaches ~38 mm/yr, which takes up more than 90% of total relative motion between the Indian and Eurasian plates. The corresponding strain rate is 2.6×10^{-8} yr^{-1}, which is an order of magnitude

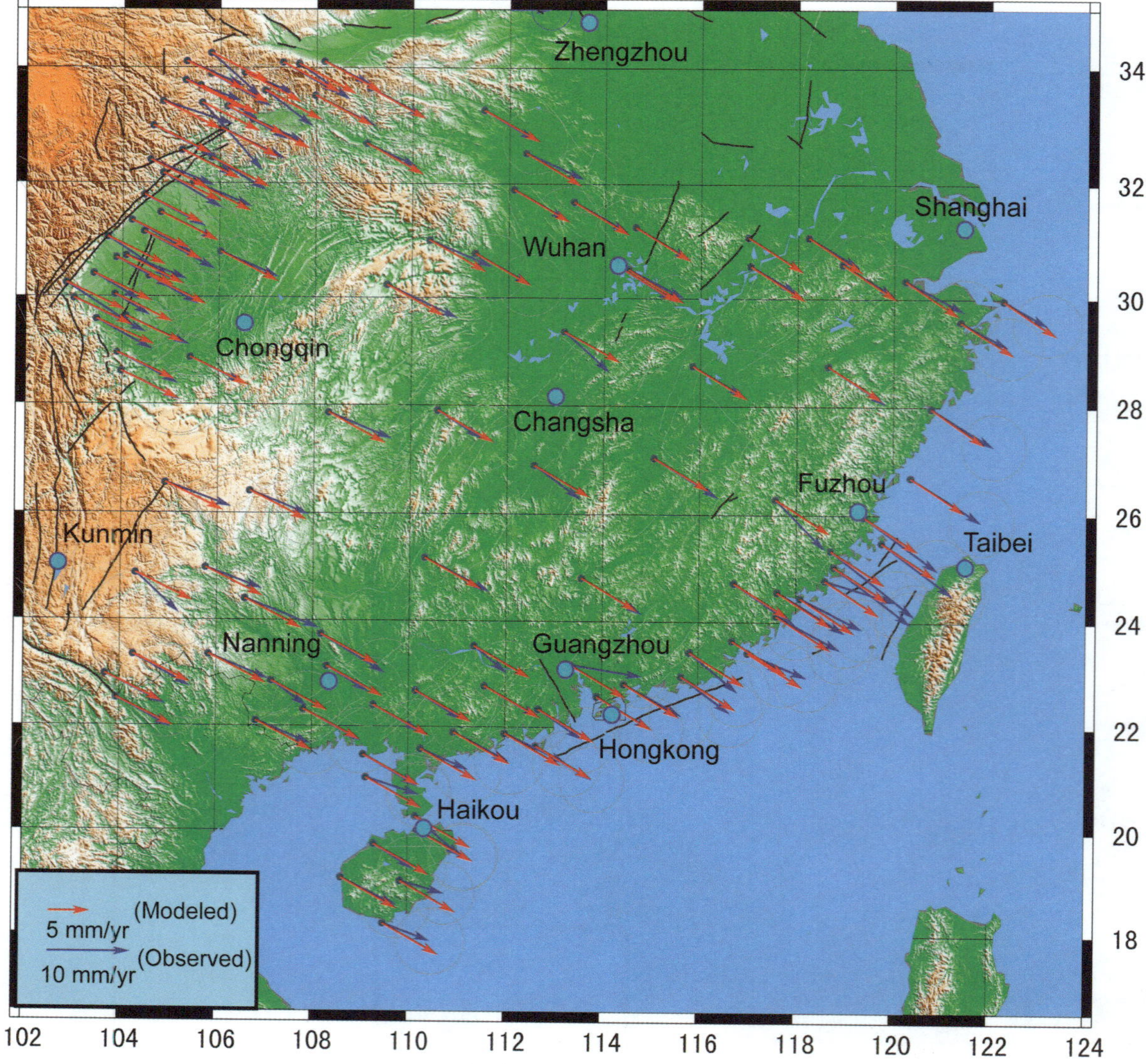

Figure 4. Global positioning system (GPS) velocities within the South China block shown by blue arrows in comparison with the modeled ones (see text).

greater than that across the rigid Indian plate and Gobi Alashan blocks. The velocity profile across the Tibetan Plateau in N110°E direction shows another pattern of strain. In the western part of the plateau, velocity increases eastward, and the velocity difference reaches ~25 ± 5.0 mm/yr. This corresponds to an extensional strain rate of 3.6×10^{-8} yr^{-1} (Fig. 7B). The eastern part of the Tibetan Plateau, however, is experiencing a compressional strain rate of 1.9×10^{-8} yr^{-1}. The GPS velocity profile across the South China block shows a strain rate of only 3.7×10^{-9} yr^{-1}, which is also an order of magnitude less than that across both the western and eastern parts of the Tibetan Plateau (Fig. 7B). Thus, the high strain rates across the Tibetan Plateau in both directions testify to the nonrigidity of the plateau in contrast to the low strain rates in the South China and the Gobi Alashan blocks, which manifest their property of relatively high rigidity.

How does deformation occur in the interior of the Tibetan Plateau? The NNE-SSW shortening of the plateau interior is accommodated by conjugate strike-slip faulting and orthogonal normal faulting, which do not require crustal thickening or thrust faulting to occur. Neither field investigations (Kidd and Molnar, 1988; Armijo et al., 1989) nor fault-plane solutions of earthquakes (Molnar and Lyon-Caen, 1989) reveals evidence for active thrust faulting. Instead, they, like recent field evidence (Taylor et al.,

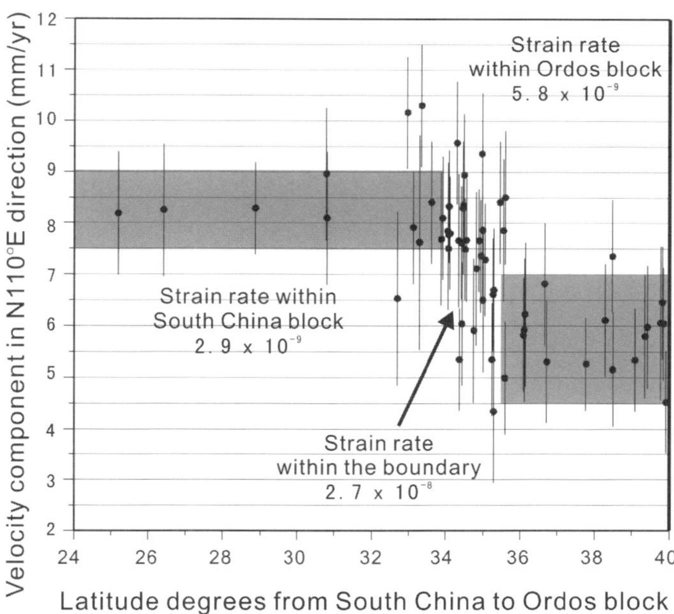

Figure 5. Relative motion between the South China and the Ordos blocks measured by global positioning system (GPS). The strain rate across the boundary between these two rigid blocks is an order of magnitude greater than that across the blocks.

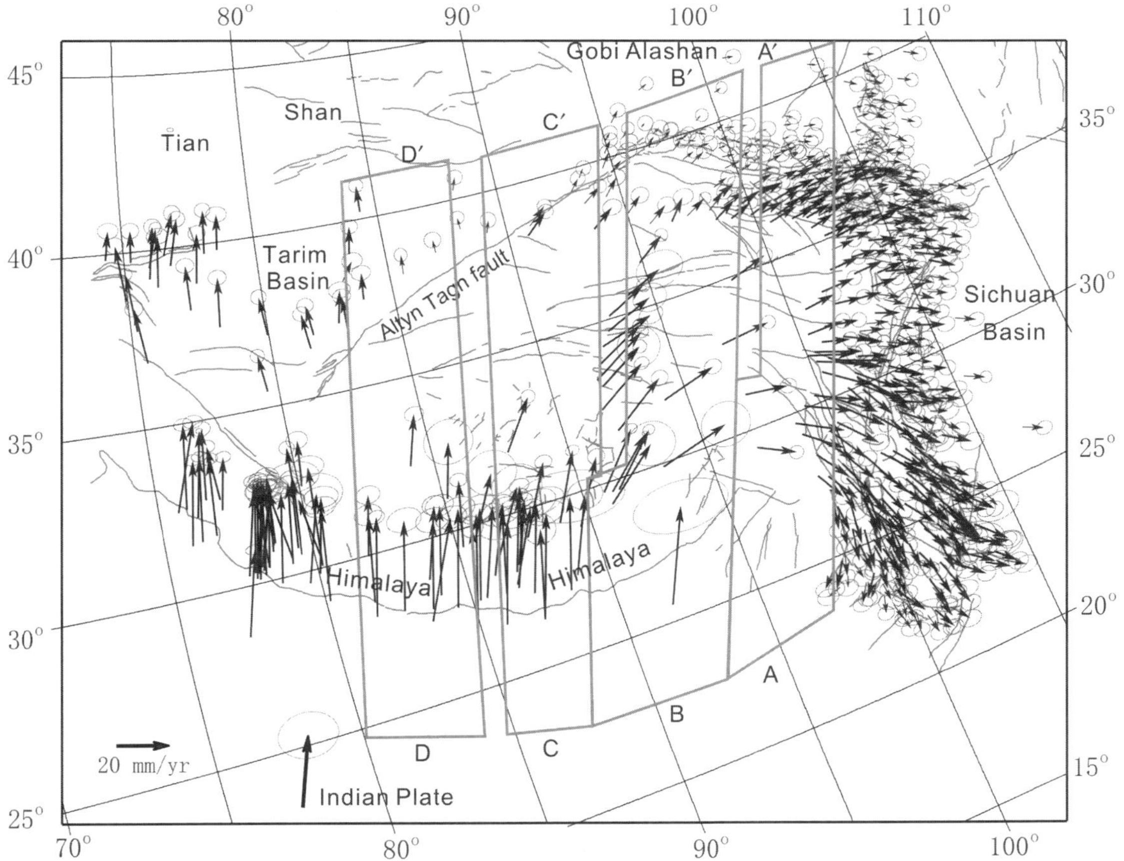

Figure 6. Global positioning system (GPS) velocities in and around the Tibetan Plateau used in this study (modified from Zhang et al., 2004).

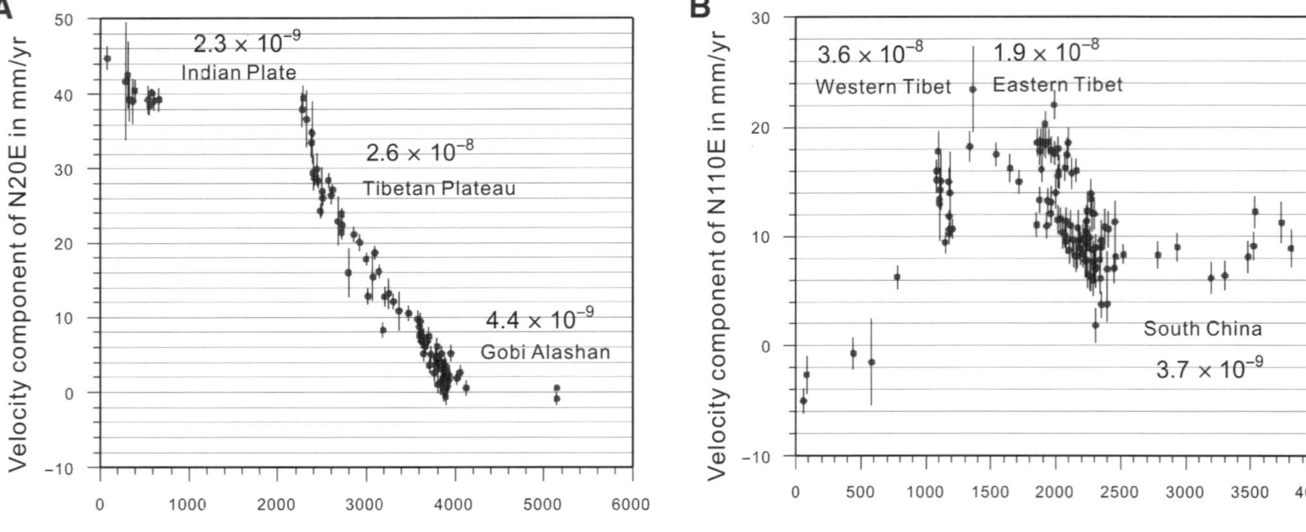

Figure 7. Global positioning system (GPS) velocity profile and average strain across the Tibetan Plateau and its adjacent blocks. (A) GPS velocity profile and average strain along the N20E direction across the Tibetan Plateau from the Indian plate to the Alashan block. (B) GPS velocity profile and average strain along the N110E direction across the Tibetan Plateau and the Huanan block.

2003), show ESE-WNW extension, which occurs by both normal faulting and conjugate strike-slip faulting. A sum of seismic moment tensors for earthquakes within northern and central Tibet suggests that ESE-WNW extensional strain dominates deformation of the plateau interior, where approximately half of that strain is accommodated by strike-slip faulting (NE-trending, left-lateral, and NW-trending, right-lateral) and half by normal faulting (Molnar and Lyon-Caen, 1989). GPS data concur with this pattern of strain partitioning. Components of velocity parallel to N110°E at stations in the interior of the plateau increase eastward to yield eastward stretching of 21.6 ± 2.5 mm/yr between longitudes 79°E and 93°E, which is roughly twice the N20°E convergence rate across the plateau interior of 10–14 mm/yr (Zhang et al., 2004).

The strain associated with the eastward transfer of crustal material can be illustrated by lateral movements orthogonal to the inferred India-Eurasia motion (Fig. 8). Lateral motions along profiles A–A′ and B–B′ in the eastern Tibet increase steadily northward from the Himalayan across the breadth of southern Tibet and then decrease further north across the broad northeastern Tibetan Plateau into the stable Gobi Alashan region (Fig. 8). The fastest east-southeastward motion occurs at latitudes of 31°N–33°N and 33°N–35°N along profiles A–A′ and B–B′, respectively. The core of rapid eastward flow of crustal material in the central plateau is bounded by two shear zones, several hundred kilometers wide: right-lateral in the southern Tibetan Plateau and left-lateral in central and northern Tibetan Plateau (Fig. 8). In profile A–A′, the right-lateral shear clearly exists but is difficult to calculate due to lack of stations. The left-lateral shear is 7.3 ± 1.5 mm/yr in the northern plateau interior. In profile B–B′, the right-lateral shear is 13 ± 2.0 mm/yr, and the left-lateral shear is 10 ± 1.5 mm/yr. These right-lateral and left-lateral shears are accommodated by distributed faults with right lateral strike-slip in the southern plateau (Armijo et al., 1989; Institute of Geology, 1993) and with left-lateral slip in the northern plateau, respectively (Kidd and Molnar, 1988; Zhang et al., 2007).

The flow of Tibetan crustal material rotates around the eastern Himalaya syntaxis, causing southeastward to southward and even southwestward velocities observed in southern and western Yunnan Province of China (Figs. 1 and 6). In addition to the rotation around the eastern Himalayan syntaxis reported previously (Le Dain et al., 1984; Molnar and Lyon-Caen, 1989; King et al., 1997; Royden et al., 1997; Q. Wang et al., 2001), our data show that clockwise rotation involves the entire eastern part of the Tibetan Plateau. This kind of rotation fundamentally differs from the rigid-block rotation that successfully describes plate motion, because for rigid-body movement, rates increase away from the rotation axis, and this seems to be opposite from the velocity field in Figures 1 and 6. Instead, it probably relates to the eastward flow of crustal material away from internal parts of the plateau in response to the northward movement of India and southern Tibet with respect to southeastern China (Le Dain et al., 1984; Molnar and Lyon- Caen, 1989; King et al., 1997; Royden et al., 1997; Q. Wang et al., 2001).

Although the Tibetan Plateau may be divided into blocks based on different data (Zhang et al., 2003b; Meade, 2007; Thatcher, 2007), the blocks have to be nonrigid or deformable in nature (Chen et al., 2004). In summary, while deformation in the shallow brittle crust does occur on a distributed network of faults governed by frictional sliding, the present-day tectonics in the Tibetan Plateau are best described as resulting from deformation of a continuous medium at depth.

The present tectonics of the Tianshan, a rejuvenated late Cenozoic intracontinental mountain building zone, also show

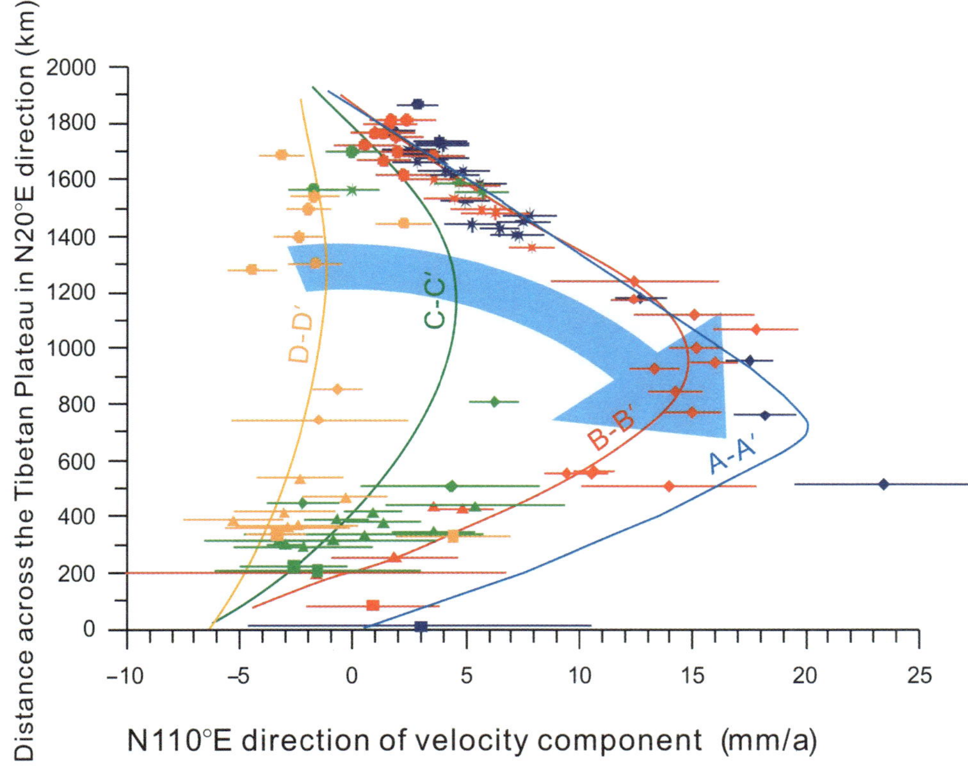

Figure 8. The N110°E components of velocities along the four profiles across the Tibetan Plateau. Different colors represent different profiles in order from west to east. Yellow color indicates profile D–D′, green color represents profile C–C′, red color shows profile B–B′, and blue color shows profile A–A′. Squares are global positioning system (GPS) stations located in the Indian plate, diamonds are stations inside the Himalayan Mountains, stars are stations within the interior of the plateau, and circles are the stations located in the Tarim and Alashan blocks. (Figure was modified from Zhang et al., 2004.)

uniform shortening and continuous deformation (Abdrakhmatov et al., 1996; Reigber et al., 2001) rather than concentrated deformation on range fronts on both sides of Tianshan. Our GPS data within Chinese territory support previous findings of present-day deformation. Figure 9 shows a velocity profile across the Chinese Tianshan near the longitude of 80.3°E. The shortening across the Tianshan cannot be totally accounted for by folding and thrust faulting across range-front fold-and-thrust belts on both sides of the mountain. Internal shortening of the Tianshan is required to take up a fraction of the total shortening. Thompson et al. (2003) reported remarkable agreement between geologically constrained shortening rates and those inferred from GPS measurements. However, the geologically constrained rates are distributed among at least six fault zones. If the geodetic profile were extended into the Tarim Basin across the entire Tianshan, there would be an additional three or four fault zones to take up the total geodetic shortening rate because of the existence of an active range-front fold-and-thrust fault belt in the southern Tianshan. If regions bounded by these active faults are regarded as blocks, the size of these blocks, with a width of several tens of kilometers perpendicular to the fault, would be one or two orders of magnitude less than the blocks discussed in this paper, such as the Tarim, the Ordos, or South China. We view the entire Tianshan as one block rather than a serious of small blocks because its size is comparable with other blocks discussed in this paper. Of course, the rigidity of a block made up of many small subblocks is not as good as one coherent block if subblock interactions are used to account for deformation in the Tianshan.

Similar to the geological observations and numerical modeling, GPS measurements also reveal eastward decrease of crustal shortening across the Tianshan (Avouac et al., 1993). Our results show that crustal shortening is ~18 mm/yr near longitude 75°E, ~13 mm/yr near longitude 76.8°E, ~6.2 mm/yr near 80.3°E, ~4 mm/yr near 83.4°E, <~2 mm/yr near longitude 87.5°E, and about zero near the eastern end of the Tianshan. This eastward decrease of crustal shortening suggests continuous deformation rather than movements of rigid blocks of large scale. This kinematic pattern can be interpreted as the mechanically "soft" Tianshan's crust being squished by clockwise rotation of the rigid and "strong" Tarim block (Chen et al., 1992; Avouac et al., 1993).

The North China Plain does not behave as a rigid block in the long-term view, as indicated by active faults beneath the Quaternary sedimentary cover and strong historical earthquakes. Present-day tectonic deformation obtained by GPS, however, reveals coherent movement, a characteristic of a rigid block. Since GPS measurements in the North China Plain only span ~10 yr, we are not sure if the rigid-block behavior is a transient phenomena or regional strain change (Shen et al., 2000; Q. Wang et al., 2001). Future measurement and study are needed to resolve the problem.

COMBINED MODEL OF RIGID-BLOCK MOTION WITH CONTINUOUS DEFORMATION

Our analysis based on GPS velocity field shows that present deformation of continental China involves both rigid-block movements and continuous deformation. Three kinds of crustal

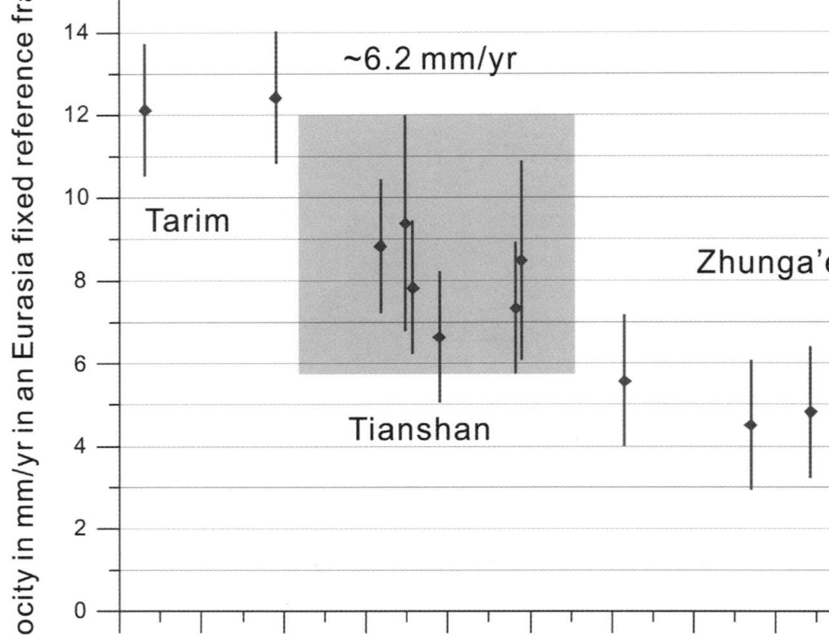

Figure 9. Global positioning system (GPS) velocity profile across the Tianshan near longitude 80.3°E. Horizontal dimension of the gray shadow shows the span of Tianshan region in this profile, and vertical dimension depicts velocity difference across the Tianshan that gives an estimation of crustal shortening rate of ~6.2 mm/yr. Both the Tarim and Zhunga'er blocks have less velocity gradient, suggesting less crustal shortening and less strain.

blocks constitute continental China (Zhang et al., 2003b). The first consists of rigid blocks such as the Tarim, Gobi Alashan, Ordos, and South China. In the interior of these blocks, adjacent GPS stations do not move significantly relative to each other, no major active faults are present, and strong historical earthquakes are absent. The second kind of crustal block has very poor rigidity such as the Tibetan Plateau and the Tianshan Mountains. Deformation occurs within the interior of this kind of block as manifested by widespread earthquakes and active faults, and GPS stations within this kind of block move relative to each other, depicting internal strain. The velocity field of the block interior cannot be described as rigid-block movement. The third kind of block has some internal deformation, but its intensity of deformation is far less than that along the boundaries of the block. We suspect that the North China block and the blocks within the Tibetan Plateau may belong to this kind.

Lithospheric property, strength, and structure dictate the rigidity of crustal blocks (Molnar, 1988; Jackson, 2002). Rigid crustal blocks often have relatively high seismic velocity and simple velocity structure. For example, P-wave velocities beneath the Tarim and Ordos blocks are 6.4 and 6.5 km/s (Zhang et al., 2003b), surface-wave tomography suggests ~3.6 km/s seismic shear wave velocity (Huang et al., 2003), and velocity structure is simple due to the absence of abnormal velocity zones and bodies (Zhang et al., 2003b). Blocks with poor rigidity possess low seismic velocity and complex velocity structure. Take the northeastern margin of Tibetan Plateau as an example: the average seismic P velocity is ~6.0–6.3 km/yr (Li et al., 2002), the seismic shear wave velocity is ~3.2 km/s (Huang et al., 2003), and numerous low-velocity zones and bodies make up complex seismic velocity structures.

Heat flow is also different from one kind of block to another. Hu et al. (2001) compiled the most complete and up to date heat-flow data for China. The average heat flow in Tibetan Plateau is commonly larger than 80 mW/m^2. The blocks with high rigidity, however, are characterized by low heat flow. Heat flow in the Tarim block is less than 70 W/m^2, less than 60 W/m^2 in the Ordos block, and less than 70 W/m^2 in the South China block.

To summarize, the high-rigidity blocks have relative low heat flow and high seismic velocity, and hence they are "cold" and "hard." These properties suggest strong lithosphere. The poor-rigidity blocks have relative high heat flow and low seismic velocity, and hence they are "hot" and "soft." These properties suggest weak lithosphere. Thus, the lithosphere of continental China consists of a number of "hard" and "soft" blocks, and interactions among these blocks constitute the tectonic deformation of continental China.

The GPS velocity field only represents the kinematic pattern of less than 10 yr. Historical earthquake history can be traced back from a few hundreds to more than one thousand years. Distributions of active faults represent tectonic activities of more than 10,000 yr. We compared patterns of tectonic activity depicted by the three different means for these three time intervals to test if GPS velocity field indeed represents deformation over an ~10,000 yr time period. Figure 10 is a seismotectonic map of continental

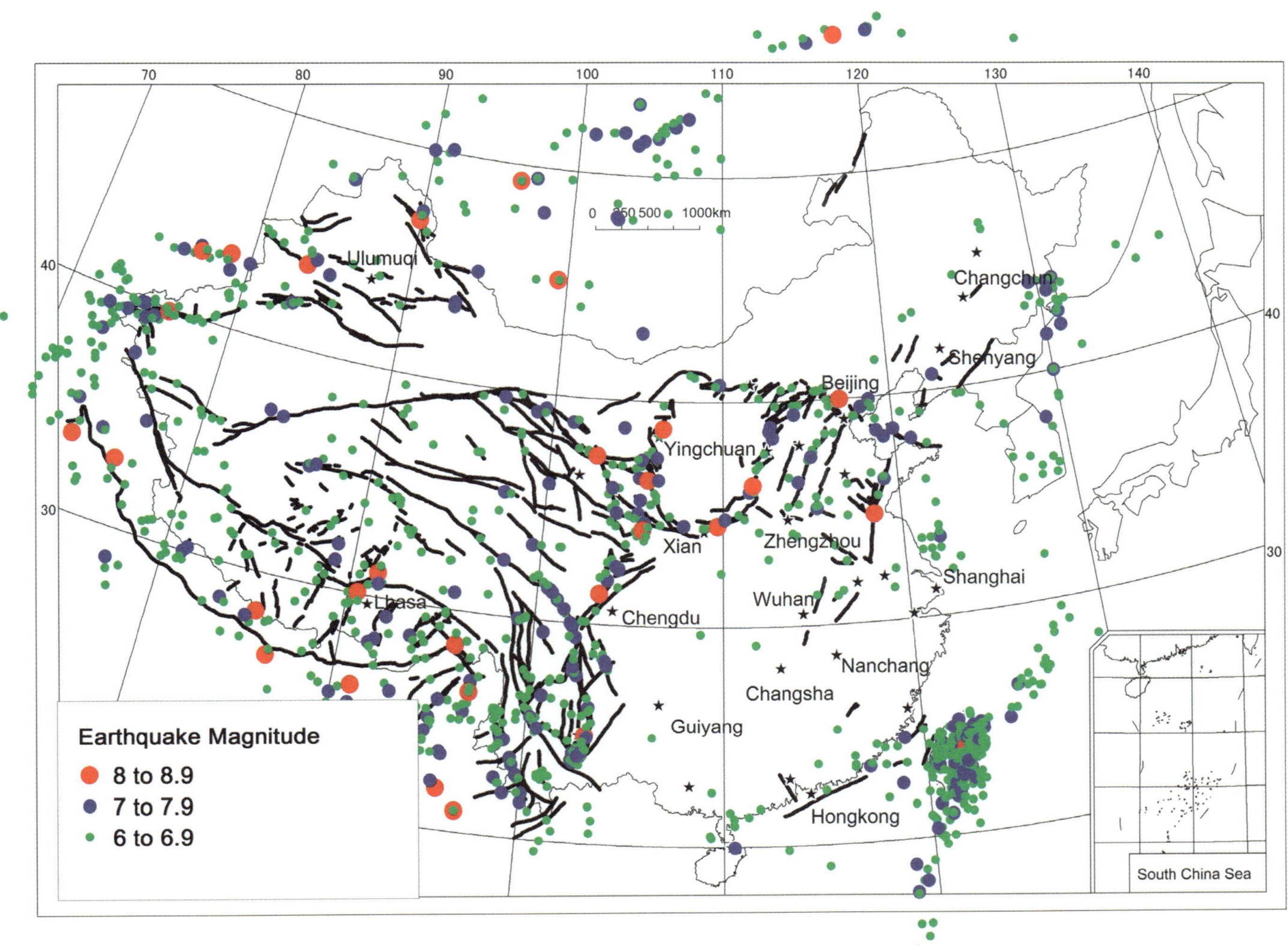

Figure 10. Distribution of Holocene active faults and major earthquakes in continental China. Thick black lines are Holocene active faults. Thin lines are Quaternary faults. Red dots are earthquakes of magnitude greater than 8, blue dots are earthquakes of magnitude between 7 and 7.9, and green dots are earthquakes of magnitude between 6 and 6.9.

China. It is obvious that almost no earthquake with magnitude over 7 and very few earthquakes with magnitudes between 6.0 and 6.9 occur in the interior of rigid blocks such as Tarim, Gibi Alashan, Ordos, and South China (Fig. 10). Major Holocene active faults form boundaries of these rigid blocks, which are the places of localized deformation and homes of strong earthquakes of magnitudes over 7. Blocks of poor rigidity, such as the Tibetan Plateau and Tianshan, show different patterns of internal deformation. Holocene active faults present all over the interior of those blocks are associated with earthquakes with magnitudes over 7 (Fig. 10). Strain rate calculations using GPS measurements indicate that the rigid blocks have low strain rates in the range of $\sim 10^{-8}$ yr^{-1}, which is an order of magnitude less than $\sim 10^{-9}$ yr^{-1} in the poor-rigidity blocks (Wang et al., 2003; Figure 7). Thus, the present patterns of tectonic activity obtained by GPS roughly agree with those from historical earthquakes in the past thousand years and those from active faulting in ~10,000 yr.

The lithosphere of continental China has been subjected to very complex tectonic processes during its long-term geological history. Each crustal block has its own mechanical and rheological property, which results in different strengths to resist tectonic deformation. The rheological strength of a particular crustal block is also different at different depths. The brittle upper crust is characterized by elastic deformation resulting in earthquakes. The lower crustal and upper mantle deform by ductile rheological flow, which drives movement of the overlying upper-crustal block (Molnar, 1988). During the processes of continental deformation, rheological flow in the lower crust and upper mantle drives movement of upper-crustal bocks. For the "hard" and "cold" lithospheric blocks, the deformation will be rigid block movement and rigid-block interactions. Deformation will be localized along boundaries among the blocks in a similar fashion to plate tectonics in this case. For the "soft" and "hot" lithospheric blocks, rigid-blocks movements and interaction do not work, and widespread deformation dominates tectonics in the interior of the blocks. For continental deformation, both rigid-block movements and continuous deformation are manifestations of rheological flow in the lower crust and upper mantle. Present-day tectonic deformation of continental China can be described in terms of a combined model of rigid-block movement and continuous deformation.

CONCLUSION

On the bases of 1350 GPS measurements in combination with active faults and seismic activity, we studied kinematical patterns of present-day tectonic deformation of continental China. Both rigid-block movements and continuous deformation are required to account for the velocity fields obtained by GPS. For example, the Tarim Basin, the Ordos, and the South China blocks behave as coherent blocks similar to oceanic rigid blocks without internal deformation, whereas the Tibetan Plateau and Tianshan seem to deform continuously with significant internal deformation. Mechanical and rheological properties of the lithosphere dictate the style of regional deformation. Thus, combined rigid-block movements with continuous deformation characterize the present-day tectonic deformation of continental China. The behavior of a region, whether it behaves as rigid block or a continuum with significant internal deformation, depends on the rheological properties of the lithosphere of the region.

ACKNOWLEDGMENTS

We thank the "Crustal Movement Observation Network of China" (CMONOC) for providing the global positioning system (GPS) data used in this study. We also thank Zheng-kang Shen and Min Wang for data processing. We are grateful for the encouragement of B.C. Burchfiel and Erchie Wang. We appreciate the comments and criticism by R. King and B. Meade that improved this paper. This work has been supported by the National Key Basic Research Program (2004CB418400), National Science Foundation of China (40234040), and National Major Scientific Infrastructure Program (CMONOC).

REFERENCES CITED

Abdrakhmatov, K.Y., Aldazhanov, S.A., Hager, B.H., Hamburger, M.W., Herring, T.A., Kalabaev, K.B., Makarov, V.I., Molnar, P., Panasyuk, S.V., Prilepin, M.T., Reilinger, R.E., Sadybakasov, I.S., Souter, B.J., Trapeznikov, Yu.A., Tsurkov, V.Ye., and Zubovich, A.V., 1996, Relatively recent construction of the Tien Shan inferred from GPS measurements of present-day crustal deformation rates: Nature, v. 384, p. 450–453, doi: 10.1038/384450a0.

Armijo, R., Tapponnier, P., and Tonglin, H., 1989, Late Cenozoic right-lateral strike-slip faulting in southern Tibet: Journal of Geophysical Research, v. 94, p. 2787–2838, doi: 10.1029/JB094iB03p02787.

Avouac, J.-P., and Tapponnier, P., 1993, Kinematic model of active deformation in central Asia: 7: Geophysical Research Letters, v. 20, p. 895–898, doi: 10.1029/93GL00128.

Avouac, J.-P., Tapponnier, P., Bai, M., You, H., and Wang, G., 1993, Active thrusting and folding along the northern Tien Shan and late Cenozoic rotation of the Tarim relative to Dzungeria and Kazakhstan: Journal of Geophysical Research, v. 98, p. 6755–6804, doi: 10.1029/92JB01963.

Banerjee, P., and Burgmann, R., 2002, Convergence across the northwest Himalaya from GPS measurement: Geophysical Research Letters, v. 29, no. 13, doi: 10.1029/2002GL015184.

Bock, Y., Wdowinski, S., Fang, P., Zhang, J., Williams, S., Johnson, H., Behr, J., Genrich, J., Dean, J., van Domselaar, M., Agnew, D., Wyatt, F., Stark, K., Oral, B., Hudnut, K., King, R., Herring, T., Dinardo, S., Young, W., Jackson, D., and Gurtner, W., 1997, Southern California Permanent GPS Geodetic Array: Continuous measurements of regional crustal deformation between the 1992 Landers and 1994 Northridge earthquakes: Journal of Geophysical Research, v. 102, p. 18,013–18,033, doi: 10.1029/97JB01379.

Burchfiel, B.C., Zhang, P., Wang, Y., Zhang, W., Jiao, D., Deng, Q., Song, F., Molnar, P., and Royden, L., 1991, Geology of the Haiyuan fault zone, Ningxia Autonomous Region, China, and its relation to the evolution of the northeastern margin of the Tibetan Plateau: Tectonics, v. 10, p. 1091–1110, doi: 10.1029/90TC02685.

Calais, E., Vergnolle, M., San'kov, V., Lukhnev, A., Miroshnitchenko, A., Amarjargal, S., and Déverchère, J., 2003, GPS measurements of crustal deformation in the Baikal-Mongolia area (1994–2002): Implications for current kinematics of Asia: Journal of Geophysical Research, v. 108, no. B10, p. 2501, doi: 10.1029/2002JB002373.

Calais, E., Dong, L., Wang, M., Shen, Z., and Vergnolle, M., 2006, Continental deformation in Asia from a combined GPS solution: Geophysical Research Letters, v. 33, p. L24319, doi: 10.1029/2006GL028433.

Chen, Q., Freymueller, J.T., Wang, Q., Yang, Z., Xu, C., and Liu, J., 2004, A deforming block model for the present day tectonics of Tibet: Journal of Geophysical Research, v. 109, no. B3, doi: 10.1029/2002JB002151.

Chen, Y., Cogné, J.P., and Courtillot, V., 1992, New Cretaceous paleomagnetic poles from the Tarim Basin, northwestern China: Earth and Planetary Science Letters, v. 114, p. 17–38, doi: 10.1016/0012-821X(92)90149-P.

Chen, Z., Burchfiel, B.C., Liu, Y., King, R.W., Royden, L.H., Tang, W., Wang, E., Zhao, J., and Zhang, X., 2000, Global positioning system measurements from eastern Tibet and their implications for India/Eurasia intercontinental deformation: Journal of Geophysical Research, v. 105, p. 16,215–16,227, doi: 10.1029/2000JB900092.

Clark, M.K., and Royden, L.H., 2000, Topographic ooze: Building the eastern margin of Tibet by lower crustal flow: Geology, v. 28, p. 703–706, doi: 10.1130/0091-7613(2000)28<703:TOBTEM>2.0.CO;2.

Deng, Q., Song, F., Zhu, Sh., Li, M., Wang, T., Zhang, W., Burchfiel, B.C., Molnar, P., and Zhang, P., 1984, Active faulting and tectonics of the Ningxia Hui Autonomous Region, China: Journal of Geophysical Research, v. 89, no. B6, p. 4427–4445, doi: 10.1029/JB089iB06p04427.

Deng, Q., Zhang, P., Ran, Y., Min, W., Yang, X., and Chu, Q., 2003, Basic characteristics of active tectonics of China: Science in China, v. 46, no. 4, p. 357–372.

Dewey, J.F., Pitman, W.C., III, Ryan, W.B.F., and Bonnin, J., 1973, Plate tectonics and the evolution of the Alpine system: Geological Society of America Bulletin, v. 84, p. 3137–3180, doi: 10.1130/0016-7606(1973)84<3137:PTATEO>2.0.CO;2.

Dong, D., Herring, T.A., and King, R.W., 1998, Estimating regional deformation from a combination of space and terrestrial geodetic data: Journal of Geodesy, v. 72, p. 200–214, doi: 10.1007/s001900050161.

England, P., and Houseman, G.A., 1986, Finite strain calculations of continental deformation 2. Comparison with the India-Asia collision: Journal of Geophysical Research, v. 91, p. 3664–3667.

England, P., and Molnar, P., 1997, The field of crustal velocity in Asia calculated from Quaternary rates of slip on faults: Geophysical Journal International, v. 130, p. 551–582.

England, P., and Molnar, P., 2005, Late Quaternary to decadal velocity fields in Asia: Journal of Geophysical Research, v. 110, no. B1, p. 2401, doi: 10.1029/2004JB003541.

Herring, T.A., 1995, GLOBK: Global Kalman Filter VLBI and GPS Analysis Program, Version 4.0: Cambridge, Massachusetts Institute of Technology.

Holt, W.E., Chamot-Rooke, N., Le Pichon, X., Haines, A.J., Shen-Tu, B., and Ren, J., 2000, Velocity field in Asia inferred from Quaternary fault slip rates and global positioning system observations: Journal of Geophysical Research, v. 105, p. 19,185–19,209, doi: 10.1029/2000JB900045.

Houseman, G., and England, P., 1993, Crustal thickening versus lateral expulsion in the Indian-Asian continental collision: Journal of Geophysical Research, v. 98, p. 12,233–12,249, doi: 10.1029/93JB00443.

Hu, Sh., He, L., and Wang, J., 2001, Heat flow database of continental China: Chinese Journal of Geophysics, v. 44, no. 5, p. 611–626.

Huang, Z., Su, W., Peng, Y., Zheng, Y., and Li, H., 2003, Rayleigh wave tomography of China and adjacent regions: Journal of Geophysical Research, v. 108, no. B2, p. 2073, doi: 10.1029/2001JB001696.

Jackson, J., 2002, Strength of the continental lithosphere: Time to abandon the jelly sandwich?: GSA Today, v. 12, no. 9, p. 4–10, doi: 10.1130/1052-5173(2002)012<0004:SOTCLT>2.0.CO;2.

Kidd, W.S.F., and Molnar, P., 1988, Quaternary and active faulting observed on the 1985 Academia Sinica-Royal Society geotraverse of Tibet: Royal Society of London Philosophical Transactions, ser. A, v. 327, p. 337–363.

King, R.W., and Bock, Y., 1995, Documentation for the MIT GPS Analysis Software: GAMIT, Version 9.3: Cambridge, Massachusetts Institute of Technology.

King, R.W., Shen, F., Burchfiel, B.C., Royden, L.H., Wang, E., Chen, Z., Liu, Y., Zhang, X.Y., Zhao, J.-X., and Li, Y., 1997, Geodetic measurement of crustal motion in southwest China: Geology, v. 25, p. 179–182, doi: 10.1130/0091–7613(1997)025<0179:GMOCMI>2.3.CO;2.

Le Dain, A.Y., Tapponnier, P., and Molnar, P., 1984, Active faulting and tectonics of Burma and surrounding regions: Journal of Geophysical Research, v. 89, p. 453–472, doi: 10.1029/JB089iB01p00453.

Li, S., Zhang, X., Zhao, J., Zhang, Ch., and Lai, X., 2002, Preliminary studies of crustal velocity structure along the Maqin-Lanzhou-Jinbian wide-angle seismic refraction profile: Chinese Journal of Geophysics, v. 45, no. 2, p. 210–217.

Ma, X., 1989, Atlas of Chinese Lithospheric Dynamics (1:400,000): Beijing, Map Publishing House.

McKenzie, D.P., 1972, Active tectonics of the Mediterranean system: Geophysical Journal of the Royal Astronomical Society, v. 30, p. 109–172.

Meade, B.J., 2007, Present-day kinematics at the India-Asia collision zone: Geology, v. 35, p. 81–84, doi: 10.1130/G22924A.1.

Molnar, P., and Tapponnier, P., 1975, Cenozoic tectonics of Asia: Effects of a continental collision: Science, v. 189, p. 419–426, doi: 10.1126/science.189.4201.419.

Molnar, P., 1988, Continental tectonics in the aftermath of plate tectonics: Nature, v. 335, p. 131–137, doi: 10.1038/335131a0.

Molnar, P., and Lyon-Caen, H., 1989, Fault plane solutions of earthquakes and active tectonics of the northern and eastern parts of the Tibetan Plateau: Geophysical Journal International, v. 99, p. 123–153, doi: 10.1111/j.1365-246X.1989.tb02020.x.

Paul, J., Bürgmann, R., Gaur, V., Bilham, R., Larson, K.M., Ananda, M.B., Jade, S., Mukal, M., Anupama, T.S., Satyal, G., and Kumar, D., 2001, The motion and active deformation of India: Geophysical Research Letters, v. 28, p. 647–650, doi: 10.1029/2000GL011832.

Reigber, Ch., Michel, G.W., Galas, R., Angermann, D., Klotz, J., Chen, J.Y., Papschev, A., Arslanov, R., Tzurkov, V.E., and Ishanov, M.C., 2001, New space geodetic constraints on the distribution of deformation in Central Asia: Earth and Planetary Science Letters, v. 191, p. 157–165, doi: 10.1016/S0012-821X(01)00414-9.

Replumaz, A., and Tapponnier, P., 2003, Reconstruction of the deformed collision zone between India and Asia by backward motion of lithospheric blocks: Journal of Geophysical Research, v. 108, no. B6, p. 2285, doi: 10.1029/2001JB000661.

Royden, L.H., Burchfiel, B.C., King, R.E., Wang, E., Chen, Z., Shen, F., and Liu, Y., 1997, Surface deformation and lower crustal flow in eastern Tibet: Science, v. 276, p. 788–790, doi: 10.1126/science.276.5313.788.

Shen, F., Royden, L.H., and Burchfiel, B.C., 2001, Large-scale crustal deformation of the Tibetan Plateau: Journal of Geophysical Research, v. 106, p. 6793–6816, doi: 10.1029/2000JB900389.

Shen, Z.K., Zhao, Ch., Yin, A., Li, Y., Jackson, D.D., Fang, P., and Dong, D., 2000, Contemporary crustal deformation in east Asia constrained by global positioning system measurement: Journal of Geophysical Research, v. 105, p. 5721–5734, doi: 10.1029/1999JB900391.

Shen, Z.K., Wang, M., Li, Y., Jackson, D.D., Yin, A., Dong, D., and Fang, P., 2001, Crustal deformation associated with the Altyn Tagh fault system, western China, from GPS: Journal of Geophysical Research, v. 106, p. 30,607–30,621, doi: 10.1029/2001JB000349.

Tapponnier, P., Xu, Zh., Roger, F., Meyer, B., Arnaud, N., Wittlinger, G., and Yang, J., 2001, Oblique stepwise rise and growth of the Tibetan Plateau: Science, v. 294, p. 1671–1677, doi: 10.1126/science.105978.

Taylor, M., Yin, A., Ryerson, F.J., Kapp, P., and Ding, L., 2003, Conjugate strike-slip faulting along the Bangong-Nujiang suture zone accommodates coeval east-west extension and north-south shortening in the interior of the Tibetan Plateau: Tectonics, v. 22, no. 4, p. 1044, doi: 10.1029/2002TC001361.

Thatcher, W., 2007, Microplate model for the present-day deformation of Tibet: Journal of Geophysical Research, v. 112, p. 1–13, doi: 10.1029/2005JB004244.

Thompson, S.C., Weldon, R.J., Rubin, C.M., Abdrakhmatov, K., Molnar, P., and Berger, G.W., 2002, Late Quaternary slip rates across the central Tien Shan: Kyrgyzstan, central Asia: Journal of Geophysical Research, v. 107, no. B9, doi: 10.1029/2001JB000596.

Wang, M., Shen, Zh., Niu, Zh., et al., 2003, Contemporary crustal deformation of Chinese continent and tectonic block model: Science in China, v. 33, supplement, p. 21–32.

Wang, Q., Zhang, P.-Zh., Freymueller, J., Bilham, R., Larson, K.M., Lai, X., You, X., Niu, Z., Wu, J., Li, Y., Liu, J., Yang, Z., and Chen, Q., 2001, Present-day crustal deformation in China constrained by global positioning system (GPS) measurements: Science, v. 294, p. 574–577, doi: 10.1126/science.1063647.

Zhang, P., Burchfiel, B.C., Molnar, P., Royden, L., Jiao, D., Zhang, W., Deng, Q., Wang, Y., and Song, F., 1991, Rate, amount, and style of late Cenozoic deformation of southern Ningxia, northeastern margin of Tibetan Plateau: Tectonics, v. 10, p. 1111–1129, doi: 10.1029/90TC02686.

Zhang, P., Wang, M., Gan, W., and Deng, Q., 2003a, Slip rates along major active faults from GPS measurements and constraints on contemporary continental tectonics: Earth Science Frontiers, v. 10, no. 3, p. 81–92.

Zhang, P., Deng, Q., Zhang, G., et al., 2003b, Active tectonic blocks and strong earthquakes in continental China: Science in China, v. 46, supplement, p. 13–24.

Zhang, P.-Zh., Shen, Z., Wang, M., Gan, W., Bürgmann, R., Molnar, P., Wang, Q., Niu, Z., Sun, J., Wu, J., Sun, H., and You, X., 2004, Continuous deformation of the Tibetan Plateau from GPS data: Geology, v. 32, no. 9, p. 809–812, doi: 10.1130/G20554.1.

Zhang, P.-Zh., Molnar, P., and Xu, X., 2007, Late Quaternary and present-day rates of slip along the Altyn Tagh fault, northern margin of the Tibetan Plateau: Tectonics, v. 26, 1–24, TC5010, doi:10.1029/2006TC002014.

Zhang, X., Li, S., Wang, F., Zhao, J., and Zhang, Ch., 2003, Differences of crustal structures from the northeastern margin of Tibetan Plateau, the Ordos Plateau and the Tangshan great earthquake regions—Evidence from deep seismic sounding results: Seismology and Geology, v. 25, no. 11, p. 78–86.

Manuscript Accepted by the Society 04 April 2008

Printed in the USA

The Geological Society of America
Special Paper 444
2008

Timing and distribution of tectonic rotations in the northeastern Tibetan Plateau

Guillaume Dupont-Nivet*
Paleomagnetic Laboratory "Fort Hoofddijk," Utrecht University, 3584 CD Utrecht, Netherlands

Shuang Dai
Key Laboratory of Western China's Environmental Systems, Ministry of Education of China & College of Resources and Environment, Lanzhou University, Gansu 730000, China

Xiaomin Fang
Institute of Tibetan Plateau Research, Chinese Academy of Science, P.O. Box 2871, Beijing 100085, China, and
Key Laboratory of Western China's Environmental Systems, Ministry of Education of China & College of Resources and Environment, Lanzhou University, Gansu 730000, China

Wout Krijgsman
Veronique Erens
Mariel Reitsma
Cor Langereis
Paleomagnetic Laboratory "Fort Hoofddijk," Utrecht University, 3584 CD Utrecht, Netherlands

ABSTRACT

We report paleomagnetic data from the northeastern margin of the Tibetan Plateau to help understand the timing and distribution of deformation (i.e., vertical-axis rotations) during the India-Asia collision. Paleomagnetic results throughout Xining Basin strata, recently dated using magnetostratigraphy to between 52 and 17 Ma, show that some 25° of clockwise rotation with respect to the stable Eurasian continent occurred at ca. 41 Ma. In view of a regional compilation of existing paleomagnetic data from the northeastern Tibetan Plateau, these results suggest that this region experienced clockwise rotations in the regional Paleocene-Miocene basin system, including rotation in the Xining Basin, ca. 41 Ma, thus establishing the existence of widespread deformation at this time. During a mid-Miocene phase, between 17 and 11 Ma, clockwise rotations were restricted to the Miocene-Quaternary basin system, implying that the Laji Shan thrust belt, which separates the two basin systems, was active during this time interval.

Keywords: tectonic rotations, paleomagnetism, Tibetan Plateau, Cenozoic.

*gdn@geo.uu.nl.

Dupont-Nivet, G., Dai, S., Fang, X., Krijgsman, W., Erens, V., Reitsma, M., and Langereis, C., 2008, Timing and distribution of tectonic rotations in the northeastern Tibetan Plateau, *in* Burchfiel, B.C., and Wang, E., eds., Investigations into the Tectonics of the Tibetan Plateau: Geological Society of America Special Paper 444, p. 73–87, doi: 10.1130/2008.2444(05). For permission to copy, contact editing@geosociety.org.

INTRODUCTION

The world's largest and highest orogenic system results from the collision and northward indentation of India into Asia since the Eocene (Argand, 1924; Yin and Harrison, 2000) (Fig. 1). The timing and mechanism of tectonic deformation have important implications for understanding lithospheric behavior in a continental collision setting, and for the possible impacts of uplift on regional and global climate (Molnar et al., 1993; Tapponnier et al., 2001; Ruddiman et al., 1997). Using paleomagnetism and quantitative determinations of tectonic rotations about a vertical axis we can test geodynamic models describing the kinematics of the India-Asia collision system (Avouac and Tapponnier, 1993; Holt et al., 2000; Dupont-Nivet et al., 2002a, 2003). A growing paleomagnetic data set provides first-order information showing that significant rotations are concentrated mainly along the margins of the deforming region in the western and eastern syntaxis (Dupont-Nivet et al., 2003). Unfortunately, precise age determinations of this component of deformation remain hindered by poor age control on the rocks studied, such that it is still difficult

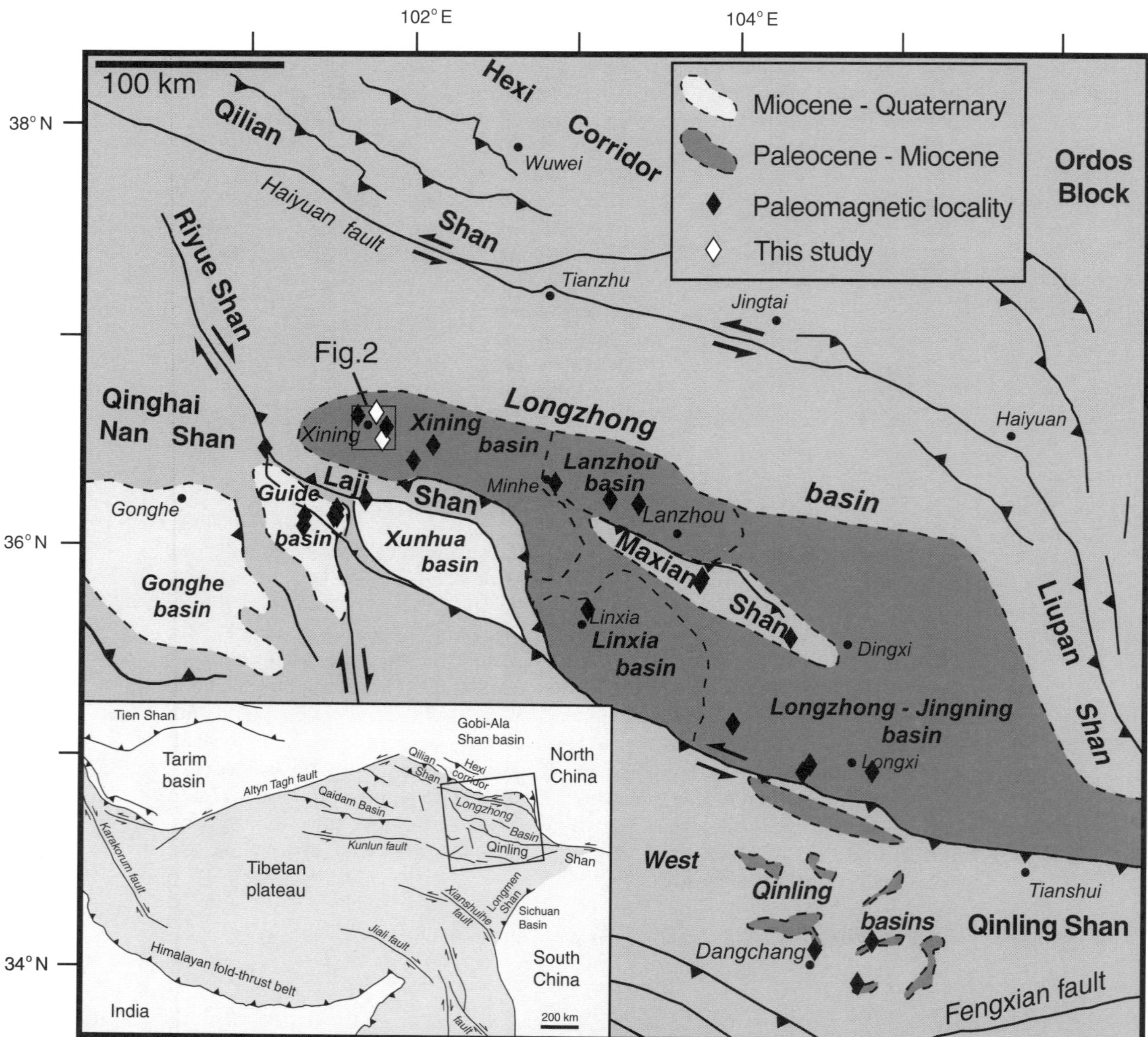

Figure 1. Simplified tectonic map of the northeastern Tibetan Plateau with approximate borders of Cenozoic basin systems (modified from Horton et al., 2004). Diamonds indicate locations of paleomagnetic sampling localities. Inset shows location in relation to the Tibetan Plateau and in the context of major tectonic features.

to relate tectonic rotations to regional geodynamics. Throughout the northeastern margin of the Tibetan Plateau, widespread clockwise tectonic rotations of ~25° have been reported (Dupont-Nivet et al., 2004; Yan et al., 2006), but the timing and geographic distribution of these rotations remain inconsistent. On the one hand, a mid-Miocene (17–11 Ma) age for the rotation is indicated in Miocene to Quaternary basins from rocks dated using magnetostratigraphy (Fang et al., 2005; Yan et al., 2006). On the other hand, an Eocene to Oligocene age has been assigned to the rotations based on previous work on rocks from Paleocene to Miocene basins (Dupont-Nivet et al., 2004). To accurately define this early phase of rotation, we provide new paleomagnetic data from the Xining Basin (Fig. 1) directly coupled to recent magnetostratigraphic dating (Dai et al., 2006). In addition, we evaluate the geographic distribution of rotations using a compilation of existing data over the northeastern Tibetan Plateau considered in a tectonic framework based on the current understanding of basin formation in the region (Horton et al., 2004).

GEOLOGIC SETTING

Two systems of Cenozoic basins are distinguished in the northeastern Tibetan Plateau (Fig. 1). The Paleocene-Miocene basin system is characterized by fining-upward sediments associated with a decreasing-upward accumulation rate (1–10 cm/k.y.) attributed to regional postrift thermal subsidence subsequent to Mesozoic extension (Horton et al., 2004). In contrast, the Miocene-Quaternary basin system is characterized by coarsening-upward sediment associated with an increasing-upward accumulation rate (20–200 cm/k.y.) in subbasins. This younger basin system is interpreted to result from progressive compartmentalization of the older larger basins by compressional and transpressional range uplift, such as the Laji Shan, which now separates the two basin systems (Horton et al., 2004; Fang et al., 2003, 2005; Pares et al., 2003).

This study was undertaken in the Xining Basin, a subbasin of the broad Longzhong Basin within the Paleocene-Miocene basin system (Figs. 1 and 2). The stratigraphic base is disconformable on the Upper Cretaceous Minhe Group. This basal discontinuity extends to the east into the adjacent Lanzhou Basin, where thicker conglomeratic series lie unconformably on folded Cretaceous rocks, suggesting post-Cretaceous tectonism east of the Xining Basin followed by regional subsidence and basin initiation or reactivation (Qiu et al., 2001; Qinghai Bureau of Geology and Mineral Resources, 1991; Zhai and Cai, 1984). Recent magnetostratigraphic dating of the particularly well-exposed, >1000-m-thick Cenozoic stratigraphy of the Xining Basin indicates subcontinuous deposition between ca. 52 Ma and 17 Ma (Dai et al., 2006; Fig. 3). The stratigraphy consists of basal sandy successions (Qijiachuan Formation, pre–52 Ma) overlain by red

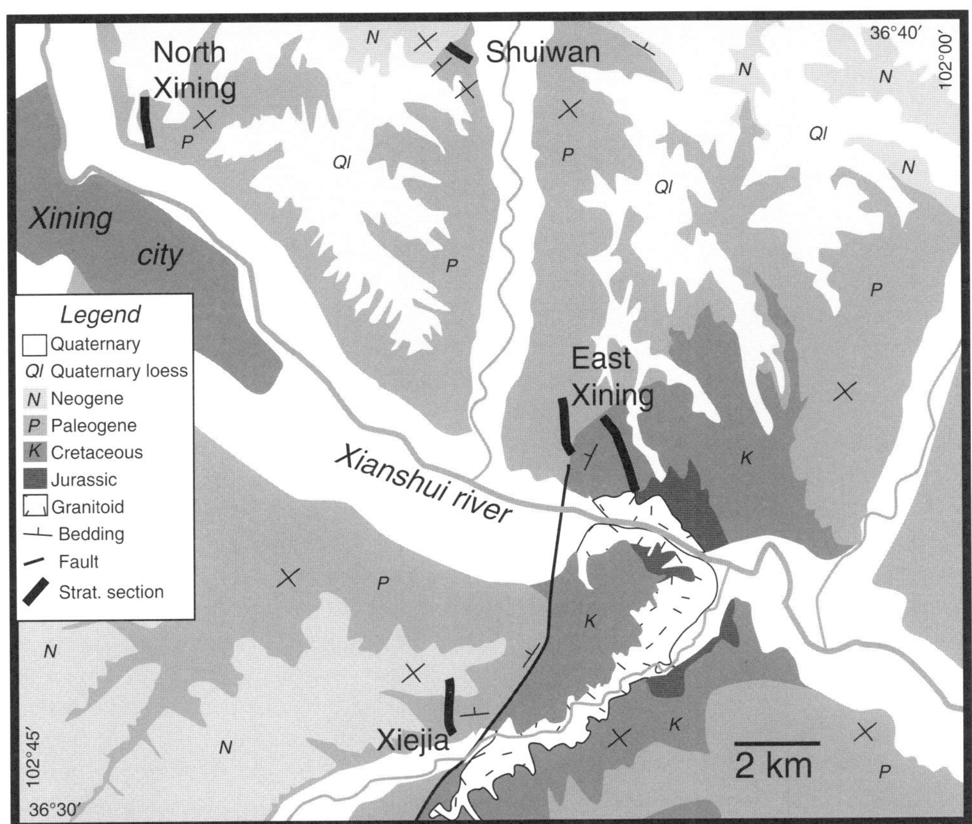

Figure 2. Geologic map of sampling locations of stratigraphic (Strat.) sections near Xining city.

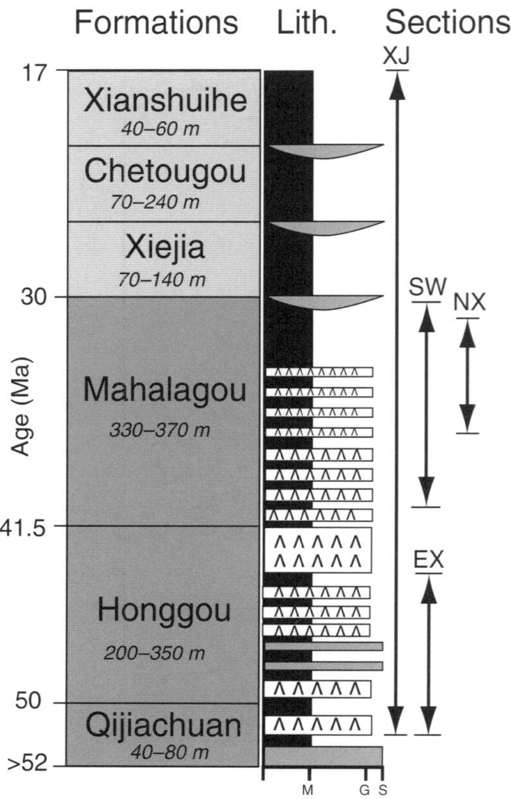

Figure 3. Generalized stratigraphic column for the Xining Basin showing formation names, ages, thickness ranges, and main lithologies (Lith.; M—mud, G—gypsum, S—sand). The sampled stratigraphic intervals are indicated within arrows for each section (XJ—Xiejia, SW—Shuiwan, NX—North Xining, EX—East Xining).

mudstones and distinctive gypsiferous intercalations (Honggou and Mahalagou Formations, 52–30 Ma) overlain by light brown to yellow mudstones with occasional sandy lenses (Xiejia, Chetougou and Xianshuihe Formations, 30–17 Ma). Apart from the disappearance of gypsum intercalations precisely correlated to the 34 Ma Eocene-Oligocene climate transition (Dupont-Nivet et al., 2007), deposition is virtually undisturbed, with slow accumulation (average 2.2 cm/k.y.) until 17 Ma, suggesting that major tectonic structures did not affect the Xining Basin until that time. Because these strata are now folded and faulted by a set of major regional E-W–trending (such as the major thrusts bounding the Laji Shan; Fig. 1) and local NW-SE–trending structures (such as the small fault and associated folding shown on Fig. 2), a post–17 Ma age is indicated for the observed deformation of the strata.

PALEOMAGNETIC SAMPLING AND ANALYSIS

Previous paleomagnetic results from the Xining Basin showed (1) ~20°–25° of clockwise rotation based on data from Cretaceous to Eocene strata up to the Honggou Formation at the East Xining section, and (2) no statistically significant rotation observed for the Oligocene Mahalagou Formation at the North Xining section (Figs. 2 and 3; Table 1; Dupont-Nivet et al., 2004). Based on these results, the timing of tectonic rotation has been estimated sometime between deposition of the Honggou Formation and the Mahalagou Formation; however, the age of these formations based on palynologic and mammalian paleontological content was only loosely constrained (Dupont-Nivet et al., 2004). Tighter age control on these formations was, however, recently provided by magnetostratigraphic studies in the Xining Basin (Dai et al., 2006). To better determine the timing of the tectonic rotations, we provide here new results from additional paleomagnetic sampling in the Xining Basin at the two following localities: (1) within the Shuiwan locality, 30 paleomagnetic sites (a paleomagnetic site is eight independently oriented samples at various stratigraphic levels within a 1–2-m-thick stratigraphic interval) were sampled throughout the Mahalagou Formation (Table 2, LJ1–LJ8 and SW1–SW22), and (2) within the Xiejia locality, 41 paleomagnetic sites were sampled throughout the Chetougou and Xianshuihe Formations (Table 2, ROT1–ROT41). These sites supplement results from magnetostratigraphic sampling of the Xiejia section throughout Xining Basin Cenozoic stratigraphy, which we also compiled here for rotational analysis (Table 3; Dai et al., 2006). Bedding attitudes were consistently measured at each site and throughout the sections, providing accurate correction of the simple homoclinal structures affecting the stratigraphy (Fig. 2).

Analysis of Paleomagnetic Sites

For samples from the Shuiwan locality, stepwise (up to 18 steps) thermal demagnetization up to 670 °C was carried out in a magnetically shielded oven, and remanence measurements were performed on a 2G 755 DC SQUID cryogenic magnetometer at the Utrecht University Paleomagnetic Laboratory. Most samples from the Shuiwan locality displayed similar demagnetization behavior; after removal of a low-temperature component below 250–350 °C, characteristic remanent magnetizations (ChRM) decay linearly, reaching the origin on vector end-point diagrams between 600 °C and 620 °C (Fig. 4A). Further linear decay until 650–670 °C displayed by a few specimens suggests the combined presence of magnetite and hematite. In some samples, overlapping of components is indicated by nonunivectorial decay displayed by a great circle path on stereographic projection. This observation suggests the presence of a partial normal overprint in some samples from the Shuiwan locality (Fig. 4A). For the samples from paleomagnetic sites at the Xiejia locality, the same treatment applied on two pilot samples at each site yielded the same general demagnetization behavior as the Shuiwan samples (Fig. 4B). The rest of the Xiejia locality samples were thus demagnetized by a combination of thermal treatment up to 350 °C to remove the contribution of possible high-coercivity phases such as goethite, followed by alternating field (AF) demagnetization (8–11 steps from 5 to 120 mT)

TABLE 1. PALEOMAGNETIC RESULTS FROM THE NORTHEASTERN TIBETAN PLATEAU

			Site location		Observed direction					Reference pole				Rotation		
	Age		Lat.	Long.	I_s	D_s	α_{95}	*n*	Rev.	Age	Lat.	Long.	α_{95}	R	±	ΔR
Locality/Formation	(Ma)	Ref.	(°N)	(°E)	(°)	(°)	(°)		(±)	(Ma)	(°N)	(°E)	(°)	(°)		(°)
					Miocene-Quaternary Basin System											
Xunhua-Guide-Gonghe basin																
Amgigang	2.6–1.8	1	36.20	101.40	40.4	357.1	6.1	67m	+	10–0	86.3	172.0	2.6	–1.5	±	6.9
Gangjia	3.6–2.6	1	36.20	101.40	42.2	358.7	3.9	166m	+	10–0	86.3	172.0	2.6	–3.1	±	5.0
Herjia	7.8–3.6	1	36.10	101.40	42.7	3.3	1.5	635m	+	10–0	86.3	172.0	2.6	–1.1	±	3.1
Ashigong	11.5–7.8	1	36.20	101.60	44.6	4.4	3.0	157m	+	10–0	86.3	172.0	2.6	0.0	±	4.3
Low Garang	19–17.3	1	36.20	101.60	35.2	31.1	9.8	31m	+	25–15	81.4	149.7	4.5	22.5	±	10.8
Guidemen	20.8–19	1	36.20	101.60	22.5	43.9	10.4	36m	+	25–15	81.4	149.7	4.5	35.3	±	10.2
Xiongchen	N1	2	36.25	101.77	39.2	10.3	8.4	7	n/a	30–10	84.0	154.8	2.7	4.1	±	9.1
Xining	pre-N1	1	36.10	101.40	34.0	35.7	6.8	71m	+	40–20	82.8	158.1	3.8	27.9	±	7.7
Ryuieshan	K1	3	36.50	101.15	37.8	35.8	6.3	1	n/a	130–110	78.2	189.4	2.4	21.2	±	6.8
					Paleocene-Miocene Basin System											
Linxia Basin																
Linxia/HWJ	6–4.3	4	35.70	103.10	40.0	3.4	5.7	67m	+	10–0	86.3	172.0	2.6	–0.9	±	6.5
Linxia/LS	7.8–6	4	35.70	103.10	34.5	9.9	6.1	51m	+	15–5	85.0	155.7	3.1	4.8	±	6.7
Linxia/DX	13.8–13.1	4	35.70	103.10	43.0	13.7	4.6	84m	+	20–10	84.2	154.9	3.2	7.8	±	6.0
Linxia/SZ	14.7–13.1	4	35.70	103.10	33.4	13.7	10.1	41m	+	20–10	84.2	154.9	3.2	7.8	±	10.2
Linxia/ZZ	21.4–14.7	4	35.70	103.10	39.5	14.6	5.0	84m	+	25–15	81.4	149.7	4.5	6.3	±	7.1
Linxia/TL	29–21.4	4	35.70	103.10	37.2	14.3	4.7	91m	+	30–20	83.8	153.2	5.3	8.1	±	7.2
Lanzhou Basin																
DuitingouU	20–15	5	36.23	103.23	38.2	8.0	4.7	100m	+	25–15	81.4	149.7	4.5	–0.4	±	6.8
DuitingouL	25–20	5	36.23	103.23	37.8	10.0	4.7	55m	+	25–15	81.4	149.7	4.5	1.6	±	6.8
Minhe	K1	3	36.30	102.90	40.4	50.5	9.8	3	n/a	130–110	78.2	189.4	2.4	35.9	±	10.6
Lanzhou	K1	3	36.20	103.40	46.9	43.2	6.8	6	n/a	130–110	78.2	189.4	2.4	28.6	±	8.3
Longzhong-Jingning Basin																
LongxiB	N2	2	34.97	104.43	56.7	4.8	5.5	5	+	10–0	86.3	172.0	2.6	0.6	±	8.4
Wenfeng	N1	2	34.94	104.80	43.7	8.8	8.5	7	n/a	30–10	84.0	154.8	2.7	2.9	±	9.8
Heichuan	E2	2	35.16	103.97	51.5	32.8	5.4	12	n/a	50–30	81.3	162.4	3.3	23.2	±	7.7
LongxiA	K1	2	34.93	104.39	40.1	47.7	6.2	6	n/a	130–110	78.2	189.4	2.4	33.3	±	6.9
Dingxi	K1	6	35.57	104.31	51.4	29.2	8.3	4	n/a	130–110	78.2	189.4	2.4	14.7	±	11.0
LintaoA	K1	6	35.85	103.78	50.3	32.3	4.2	11	+	130–110	78.2	189.4	2.4	17.7	±	5.8
LintaoB	K1	6	35.83	103.77	48.9	39.3	7.5	4	n/a	130–110	78.2	189.4	2.4	24.7	±	9.5
West Qinling basins																
NiudingShan	E1–2	2	34.18	104.81	55.3	28.5	3.1	7	n/a	60–40	80.9	164.4	3.4	18.5	±	5.6
Nanyang	E1–2	2	33.97	104.67	57.6	38.1	3.0	4	n/a	60–40	80.9	164.4	3.4	28.1	±	5.7
Dangchang	K1	2	34.08	104.47	50.9	35.8	2.9	18	n/a	130–110	78.2	189.4	2.4	21.5	±	4.4
Xining Basin																
Xiejia/Xianshuihe	18–17	*	36.52	101.87	43.3	11.6	8.6	18m	+	25–15	81.4	149.7	4.5	3.0	±	10.7
Xiejia/Xianshuihe sites	18–17	*	36.52	101.87	45.0	5.1	12.9	8	+	25–15	81.4	149.7	4.5	–3.5	±	15.5
Xiejia/Chetougou	18–23	*	36.52	101.87	41.9	8.8	6.3	55m	+	25–15	81.4	149.7	4.5	0.2	±	8.3
Xiejia/Chetougou sites	18–23	*	36.52	101.87	39.7	3.4	9.2	14	+	25–15	81.4	149.7	4.5	–5.1	±	10.8
Xiejia/Xiejia	30–23	*	36.52	101.87	51.3	3.7	3.3	162m	+	30–20	83.8	153.2	5.3	–2.6	±	7.0
Xiejia/Mahalagou	41.5–30	*	36.52	101.87	42.6	2.8	3.3	179m	+	40–30	81.2	173.4	4.6	–7.9	±	5.9
North Xining/Mahalagou	41.5–30	2	36.65	101.78	40.2	12.8	5.3	17	+	40–30	81.2	173.4	4.6	2.0	±	7.3
Xiejia/Honggou	50–41.5	*	36.52	101.87	40.1	26.2	5.6	33m	+	50–40	81.1	150.4	5.3	17.2	±	8.2
East Xining/Honggou	50–41.5	2	36.58	101.89	33.7	26.8	11.9	9	+	50–40	81.1	150.4	5.3	17.8	±	12.8
Xining/Honggou	50–41.5	7	36.5	102.0	40.8	29.3	13.2	5	i	50–40	81.1	150.4	5.3	20.3	±	15.2
Shancheng/Minhe	K2	2	36.38	102.05	33.9	30.1	9.6	15	n/a	90–70	81.4	206.1	5.9	20.0	±	10.9
East Xining/Hekou	K1	2	36.58	101.89	33.6	37.7	4.5	20	+	130–110	78.2	189.4	2.4	23.0	±	4.9

Note: Locality/Formation—paleomagnetic locality and sampled formation; Age—age of sampled formation given by numbers when magnetostratigraphic control on age span is available or given by epoch notation otherwise (K1—Early Cretaceous; K2—Late Cretaceous; E1—Paleocene; E2—Eocene; E3—Oligocene; N1—Miocene; N2—Pliocene). Lat. and Long.—latitude and longitude; Is and Ds—inclination and declination in stratigraphic coordinates; α_{95}—angular radius of 95% confidence on mean direction; *n*—number of sites used to calculate mean direction (followed by "m" if number of characteristic remanent magnetization (ChRM) directions from magnetostratigraphic levels); Rev.—outcome of reversals test ("+" if positive, "i" if indeterminate, and "n/a" if not applicable); Reference pole—Eurasian paleomagnetic pole from Besse and Courtillot (2002); Age—pole age limits of average sliding window; α_{95}—pole 95% confidence limit; R—vertical-axis rotation (clockwise is positive); ΔR—95% confidence limit on R; Rotation and flattening are derived from observed direction minus expected direction at locality calculated from reference pole.

References (Ref.): *—this study; 1—Yan et al. (2006); 2—Dupont-Nivet et al. (2004); 3—Halim et al. (1998); 4—Fang et al. (2003); 5—Flynn et al. (1999); 6—Yang et al. (2002); 7—Cogné et al. (1999).

TABLE 2. PALEOMAGNETIC SITE-MEAN DIRECTIONS

Site ID#	Level (m)	*n*	Dg (°)	Ig (°)	Ds (°)	Is (°)	*k*	α_{95} (°)
Shuiwan locality								
Mahalagou Formation								
LJ001*	<–10	7	25.9	45.0	23.5	43.3	17.3	14.9
LJ002	<–10	11	12.2	43.6	10.3	41.3	168.6	3.5
LJ003	<–10	8	27.7	29.3	26.4	27.7	29.5	10.4
LJ005*	<–10	4	211.3	–38.6	210.1	–38.1	36.4	15.4
LJ006*	<–10	7	227.8	–28.9	226.2	–28.2	12.6	17.7
LJ008*	<–10	8	224.4	–34.8	222.9	–34.1	19.2	13.0
SW001*	–10	8	135.7	–45.6	139.9	–16.3	27.0	10.9
SW002	–11	5	225.0	–25.1	214.8	–14.0	48.5	11.1
SW003*	–12	4	217.4	–44.5	198.6	–27.7	15.0	24.6
SW004*	–15	8	208.6	–51.9	188.5	–31.1	16.3	14.2
SW005	–16	8	233.9	–37.9	214.4	–28.9	90.6	5.9
SW006	–17	5	272.5	–56.9	220.8	–60.2	44.9	11.5
SW007*	–20	7	241.5	–34.7	222.3	–29.8	18.5	14.4
SW008*	–21	8	229.1	–44.8	206.4	–32.5	9.8	18.6
SW009*	–23	4	263.6	–60.5	210.2	–58.0	20.5	20.8
SW010*	–25	3	321.0	–60.4	232.1	–85.0	1.2	99.9
SW011	11	6	36.6	43.4	18.6	26.4	179.3	5.0
SW012	15	7	20.0	53.5	2.2	30.1	43.8	9.2
SW013	18	8	34.7	43.6	17.2	25.9	54.5	7.6
SW014*	55	8	40.8	47.8	18.8	31.7	12.3	16.5
SW015	74	8	53.1	47.7	27.0	36.5	34.0	9.6
SW016	78	6	235.9	–39.5	214.8	–31.1	25.9	13.4
SW017*	82	4	16.3	57.3	358.1	32.6	20.8	20.6
SW018	84	5	235.2	–67.6	188.4	–51.4	29.6	14.3
SW019*	88	7	274.2	–57.4	220.8	–61.3	5.8	27.4
SW020*	101	8	249.1	–52.5	212.9	–46.9	9.6	18.8
SW021*	105	6	260.2	–68.7	194.4	–60.1	15.2	17.7
SW022*	114	4	273.2	–55.9	222.9	–60.1	22.5	19.8
Shuiwan geographic		11	42.6	46.0			18.9	10.8
Shuiwan stratigraphic		11			23.1	34.4	24.8	9.4
Shuiwan normal		6			17.0	31.6	63.0	8.5
Shuiwan reverse		5			211.3	–37.5	15.7	19.9
Indeterminate reversals test; critical angle = 21.2°; normal/reverse angle = 13.2°.								
Xiejia locality								
Xianshuihe Formation								
ROT27	782	8	17.5	53.6	9.8	50.1	160.9	4.4
ROT28	784	8	189.5	–53.3	182.6	–49.0	153.1	4.5
ROT30	789	5	20.4	24.3	17.7	21.3	40.5	12.2
ROT31	793	5	44.6	32.7	40.1	32.5	31.8	13.8
ROT32	798	8	183.3	–63.5	174.4	–58.5	166.5	4.3
ROT33	800	8	186.6	–54.1	179.8	–49.5	33.2	9.8
ROT34–37*	810	7	178.6	–49.0	173.7	–43.8	17.2	15.0
ROT39	818	7	343.3	47.0	340.2	40.8	143.0	5.1
ROT41	825	6	169.8	–50.9	165.5	–45.1	57.4	8.9
Xianshuihe geographic		8	11.2	49.1			19.4	12.9
Xianshuihe stratigraphic		8			5.1	45.0	19.4	12.9
Chetougou Formation								
ROT1	780	8	38.8	32.9	34.4	32.0	83.5	6.1
ROT10*	748	8	352.6	39.0	349.5	33.4	19.5	12.9
ROT11	747	7	355.1	27.1	353.1	21.7	41.1	9.5
ROT13*	705	5	193.3	–48.9	187.1	–45.0	15.3	20.2
ROT14	704	5	201.5	–48.7	194.7	–45.6	56.9	10.2
ROT16	700	7	210.1	–43.9	203.9	–41.8	27.8	11.6
ROT17	697	6	206.5	–42.0	200.8	–39.6	30.1	12.4
ROT19	683	7	184.6	–39.3	180.6	–34.7	75.3	7.0
ROT2	779	6	31.0	64.8	17.7	62.4	352.2	3.6
ROT20	680	8	166.5	–50.6	162.6	–44.6	78.6	6.3
ROT22	677	6	170.5	–25.0	168.8	–19.3	28.6	12.7
ROT24	671	8	204.7	–50.8	197.2	–48.0	47.7	8.1
ROT25	669	8	353.8	46.6	349.7	41.1	30.3	10.2
ROT4	776	3	20.0	50.1	13.0	46.9	135.4	10.6
ROT5	774	6	356.4	27.8	354.2	22.5	25.9	13.4
ROT6	766	7	333.2	43.4	331.4	36.8	47.9	8.8

(*continued*)

TABLE 2. PALEOMAGNETIC SITE-MEAN DIRECTIONS (*continued*)

Site ID#	Level (m)	*n*	Dg (°)	Ig (°)	Ds (°)	Is (°)	*k*	α_{95} (°)
Chetougou geographic		14	8.4	44.0			19.5	9.2
Chetougou stratigraphic		14			3.4	39.7	19.5	9.2
Xiejia sites normal		11			5.2	39.0	14.2	12.6
Xiejia sites reverse		11			182.7	–44.2	31.2	8.3
Positive reversals test class C; critical angle = 15.1°; normal/reverse angle = 5.5°.								

Note: Site ID#—paleomagnetic site identification; Ig and Dg—inclination and declination of site-mean direction in geographic coordinates (with no structural correction); Is and Ds—inclination and declination of site-mean direction in stratigraphic coordinates (after restoration of local bedding to horizontal); *k*—concentration parameter; α_{95}—angular radius of 95% confidence on mean direction; *n*—number of sample characteristic remanent magnetization (ChRM) directions averaged to calculate site-mean direction. Averages of site-mean directions are given by formation and by polarity (normal and reverse); reversals test—outcome of the reversals test of McFadden and McElhinny (1990).

*Discarded site-mean direction.

TABLE 3. FORMATION-MEAN DIRECTIONS FROM XIEJIA MAGNETOSTRATIGRAPHIC SECTION (DAI ET AL., 2006)

Formation	*n*	Dg (°)	Ig (°)	Ds (°)	Is (°)	*k*	α_{95} (°)
Xianshuihe geographic	18	19.2	48.4			15.1	9.2
Xianshuihe stratigraphic	18			11.6	43.3	17.1	8.6
Xianshuihe normal	5			24.5	42.9	38.8	12.4
Xianshuihe reverse	13			186.5	–43.1	15.0	11.1
Positive reversals test class C; critical angle = 15.9°; normal/reverse angle = 13.2°.							
Cwhetougou geographic	55	16.7	49.8			14.0	5.3
Chetougou stratigraphic	55			8.8	41.9	17.3	6.3
Chetougou normal	32			7.2	41.7	15.9	6.5
Chetougou reverse	23			191.9	–52.0	11.4	9.4
Positive reversals test class C; critical angle = 11.2°; normal/reverse angle = 10.8°.							
Xiejia geographic	162	22.9	54.6			12.1	3.3
Xiejia stratigraphic	162			3.7	51.3	12.1	3.3
Xiejia normal	75			4.1	50.2	12.6	4.8
Xiejia reverse	87			183.4	–52.2	11.6	4.7
Positive reversals test class B; critical angle = 6.6°; normal/reverse angle = 2.1°.							
Mahalagou geographic	179	64.2	44.1			10.1	3.5
Mahalagou stratigraphic	179			2.8	42.6	11.1	3.3
Mahalagou normal	55			359.4	41.5	12.5	5.7
Mahalagou reverse	124			184.4	–43.0	10.1	4.6
Positive reversals test class B; critical angle = 9.1°; normal/reverse angle = 4.0°.							
Honggou geographic	33	58.7	33.1			16.3	6.4
Honggou stratigraphic	33			26.2	40.1	21.2	5.6
Honggou normal	19			20.5	41.5	25.2	6.8
Honggou reverse	14			213.8	–37.8	19.8	9.2
Positive reversals test class C; critical angle = 11.2°; normal/reverse angle = 10.9°.							
Qijiachuan geographic	17	53.3	45.9			8.2	13.2
Qijiachuan stratigraphic	17			5.7	47.3	13.3	10.2
Qijiachuan normal	13			356.1	48.7	15.6	10.9
Qijiachuan reverse	4			211.9	–38.5	30.9	16.8
Negative reversals test; critical angle = 20.8°; normal/reverse angle = 27.5°.							

Note: Formation—name of formation over which characteristic remanent magnetization (ChRM) directions have been averaged. Ig and Dg—inclination and declination of formation-mean direction in geographic coordinates (with no structural correction); Is and Ds—inclination and declination of formation-mean direction in stratigraphic coordinates (after restoration of local bedding to horizontal); *k*—concentration parameter; α_{95}—radius of cone of 95% confidence about formation-mean direction; *n*—number of sample ChRM directions averaged to calculate formation-mean paleomagnetic direction. Average directions are also given for normal and reverse polarity within each formation. Reversals test—outcome of the application of the reversals test of McFadden and McElhinny (1990).

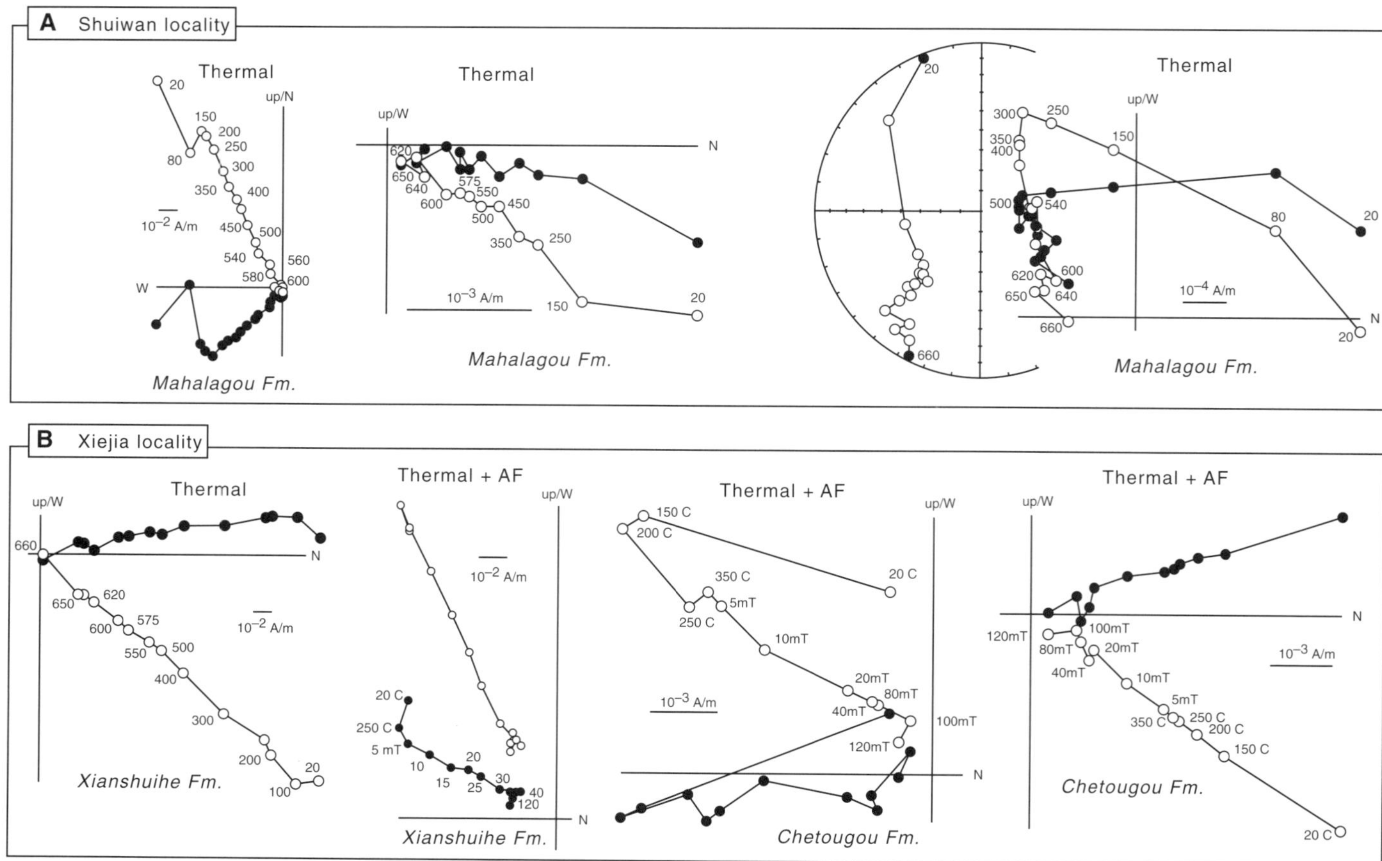

Figure 4. Characteristic vector end-point diagrams of demagnetizations. (A) Thermal demagnetization of the Shuiwan locality in the Mahalagou Formation. Right-side diagram shows overlapping components interpreted to result from partial normal overprint. (B) Thermal and alternating field (AF) demagnetizations of the Xiejia locality in the Xianshuihe and Chetougou Formations. Projection in the horizontal (vertical) plane is illustrated by black (open) dots, and adjacent numbers correspond to temperature step in °C or peak AF in mT.

using an inline 2G degausser and RF-SQUID magnetometer with sample autofeeder in a shielded room. For most samples, this AF treatment successfully demagnetized 80%–100% of the ChRM, further suggesting magnetite as the dominant carrier. All ChRM directions were assessed on vector end-point diagrams and calculated using least squares analysis (Kirschvink, 1980) to estimate the maximum angular deviation (MAD) on a minimum of four measurements. These principal component analyses were performed without forcing line fits through the origin. A few samples showing erratic demagnetization behavior resulting in ChRM directions with a MAD above 15° were systematically rejected. The remaining ChRM directions used in this analysis have straightforward demagnetization behavior and low MAD values, typically 10° to 5° (Fig. 4). On the remaining ChRM directions, Fisher (1953) statistics were used to calculate site-mean directions and their α_{95} (95% confidence angle) from the paleomagnetic sites at both localities (Table 2). To avoid any possible bias from isolated outlying directions, site-mean directions with α_{95} above 15° and/or dispersion parameter (k) below 20 were systematically rejected for further analysis, as well as one widely aberrant direction (SW001) from the Shuiwan section data set. After structural corrections, the remaining site-mean directions were compared on stereographic projections showing separate sets of normal- and reversed-polarity directions (Fig. 5).

Analysis of Magnetostratigraphic ChRM Directions

For rock magnetic analysis of magnetostratigraphic levels from the Xiejia section, see Dai et al. (2006). Of the ChRM directions reported in Dai et al. (2006), all directions with MAD above 15° were systematically rejected for the present analysis. In addition, to remove possible transitional directions likely found in these sediments, where low sedimentation rates record a relatively high of number of paleomagnetic reversals, we applied the recursive cut-off method developed by Vandamme (1994) on separate sets of reverse- and normal-polarity populations from each formation. On the remaining ChRM directions, Fisher (1953) statistics were used to calculate formation-mean directions for the formations sampled for magnetostratigraphy at the Xiejia section (Table 3). After structural corrections, the remaining ChRM directions were compared on stereographic projections showing separate sets of normal- and reversed-polarity directions (Fig. 6).

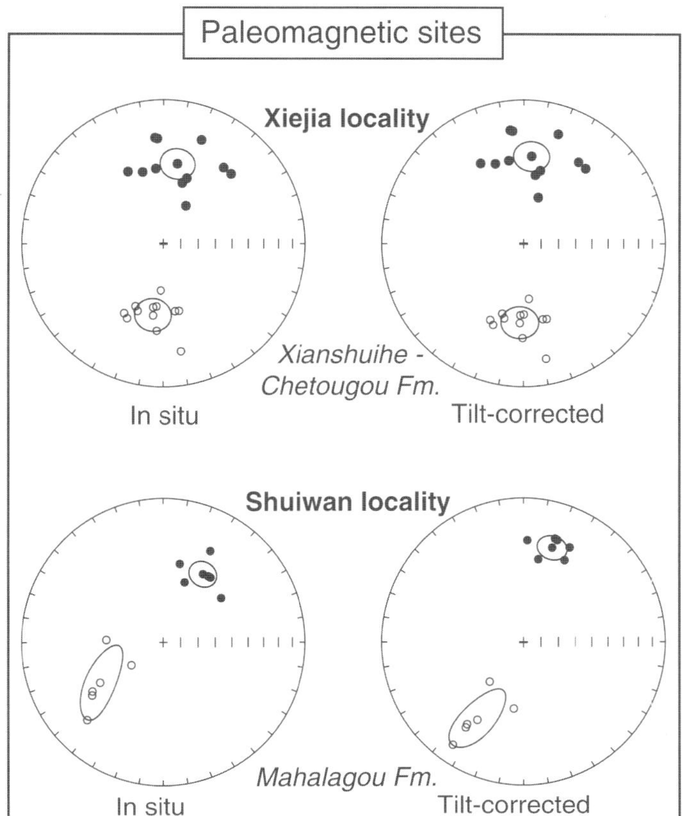

Figure 5. Stereographic projections in the lower (black dots) and upper (open dots) hemisphere of site-mean directions from the Xiejia and Shuiwan localities given before (in situ) and after (tilt corrected) tilt corrections. Averages of normal and reverse site-mean directions are given with 95% confidence ellipse (Tauxe, 1998).

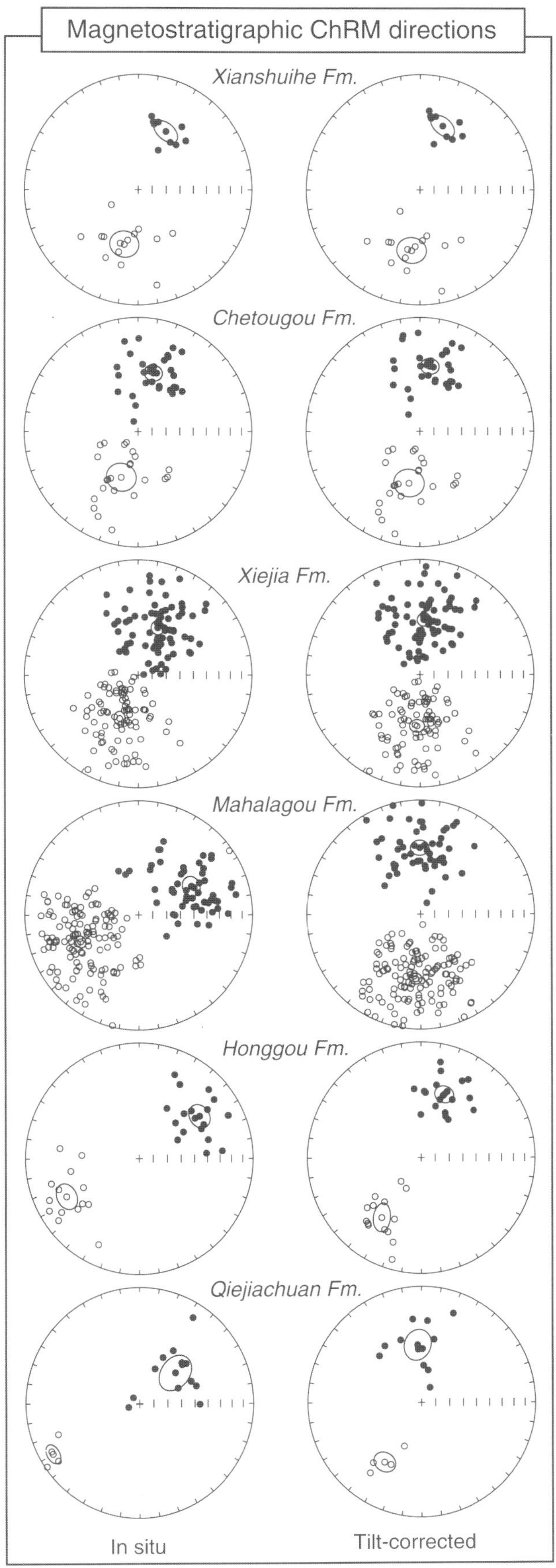

PALEOMAGNETIC RESULTS

Reliability Tests

Although prefolding magnetization is indicated by a positive regional fold test (McFadden, 1990) applied to the combined site-mean directions and formation-mean directions, the reversals test of McFadden and McElhinny (1990) yielded mixed results and shed light on the primary origin and the reliability of the observed directions critical for rotational analysis. The reversals test applied to site-mean directions was positive for the Xiejia locality sites but indeterminate for the Shuiwan locality

Figure 6. Stereographic projections in the lower (black dots) and upper (open dots) hemisphere of characteristic remanent magnetization (ChRM) directions from the magnetostratigraphy of different formations at the Xiejia section given before (in situ) and after (tilt corrected) tilt corrections. Averages of normal and reverse site-mean directions are given with 95% confidence ellipse (Tauxe, 1998).

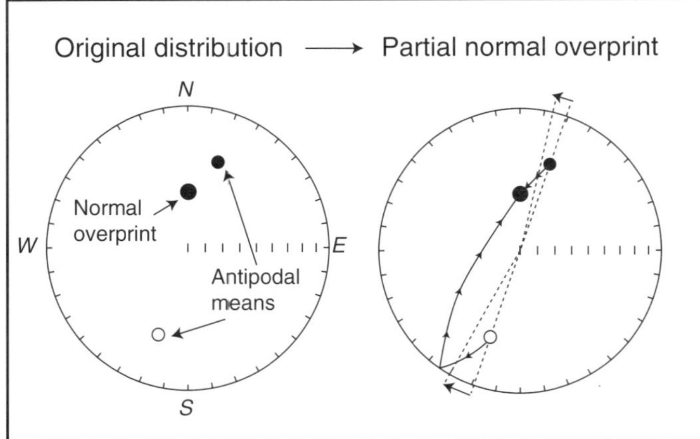

Figure 7. Illustration of the potential effect of an unresolved partial normal overprint on an originally antipodal distribution (left plot). The mean normal direction is biased toward a steeper and more northerly direction, while the mean reversed direction is biased toward a shallower and more westerly direction (right plot).

sites (Table 2). For magnetostratigraphic results from the Xiejia sections, the reversals test was applied to ChRM directions for each formation. All passed the reversals test, except for ChRM of the Qijiachuan Formation (Table 3). For both data sets that failed the reversals test, a clear bias of normal directions toward more northerly direction and reversed direction toward more westerly directions was observed. Based on this characteristic distribution of site-mean directions, we interpret the failure of the reversals test for these data sets to result from the partial contribution of an unresolved normal-polarity overprint as illustrated on Figure 7. Failure of the reversals test for these two data sets reflects a significant bias in directions that preclude using them for analysis of tectonic rotation. As a result, we rejected these two data sets from further analysis. For the remaining data sets from the Honggou to Xianshuihe Formations, positive reversals tests with nearly antipodal normal and reverse directions suggest that these directions have not been significantly biased, and thus they provide a set of reliable results for rotational analysis.

TABLE 4. INCREMENTAL MOVING AVERAGE OVER MAGNETOSTRATIGRAPHIC ChRM DIRECTIONS

Level (m)	*n*	Dm (°)	Im (°)	*k* (°)	α_{95} (°)	Age (Ma)	Dx (°)	±	δD (°)	R (°)	±	ΔR (°)
367.5	15	0.5	48.4	12.9	11.1	34.13	10.0	±	4.3	−9.4	±	13.9
364.6	15	2.5	49.0	11.9	11.6	34.20	10.0	±	4.3	−7.5	±	14.6
361.6	15	4.4	46.6	13.3	10.9	34.26	10.0	±	4.3	−5.6	±	13.2
358.5	15	5.3	47.3	14.5	10.4	34.33	10.0	±	4.3	−4.7	±	12.8
355.7	15	5.1	46.6	13.8	10.7	34.39	10.0	±	4.3	−4.8	±	13.0
353.0	15	5.6	43.3	13.0	11.0	34.45	10.0	±	4.3	−4.3	±	12.7
350.1	15	3.3	43.2	14.5	10.4	34.51	10.0	±	4.3	−6.6	±	12.0
347.8	15	2.0	42.2	13.5	10.8	34.56	10.0	±	4.3	−7.9	±	12.2
345.6	15	2.8	41.4	13.9	10.6	34.61	10.0	±	4.3	−7.2	±	11.9
343.9	15	1.3	40.0	12.6	11.2	34.64	10.0	±	4.3	−8.7	±	12.3
342.1	15	1.2	36.9	12.8	11.1	34.75	10.0	±	4.3	−8.7	±	11.7
340.0	15	2.8	37.2	12.5	11.2	34.90	10.0	±	4.3	−7.2	±	11.9
338.0	15	5.5	38.5	12.8	11.1	35.06	10.0	±	4.3	−4.5	±	11.9
336.0	15	5.6	39.0	12.7	11.2	35.21	10.0	±	4.3	−4.4	±	12.1
333.8	15	3.2	35.6	10.9	12.1	35.37	10.0	±	4.3	−6.7	±	12.5
331.8	15	2.6	32.7	11.2	12.0	35.54	10.0	±	4.3	−7.3	±	11.9
329.5	15	0.7	32.0	12.7	11.2	35.72	10.0	±	4.3	−9.2	±	11.1
327.2	15	−1.6	33.4	11.0	12.1	35.91	10.0	±	4.3	−11.5	±	12.1
324.6	15	0.6	34.0	10.7	12.3	36.13	10.0	±	4.3	−9.4	±	12.4
322.0	15	−3.3	33.0	9.5	13.1	36.34	10.0	±	4.3	−13.2	±	13.0
319.6	15	−2.0	34.0	8.9	13.6	36.57	10.0	±	4.3	−11.9	±	13.6
317.2	15	−4.3	36.0	8.5	13.9	36.74	10.0	±	4.3	−14.3	±	14.3
314.6	15	−0.5	37.9	8.1	14.3	36.90	10.0	±	4.3	−10.4	±	15.0
312.0	15	−0.7	38.4	8.0	14.4	37.07	10.0	±	4.3	−10.7	±	15.2
309.1	15	2.4	39.5	8.6	13.8	37.26	10.0	±	4.3	−7.5	±	14.9
306.1	15	−0.6	41.7	9.3	13.2	37.45	10.0	±	4.3	−10.5	±	14.7
303.2	15	−4.8	42.3	9.6	13.0	37.63	10.0	±	4.3	−14.8	±	14.6
300.3	15	−8.2	40.4	9.4	13.2	37.82	10.0	±	4.3	−18.2	±	14.4
297.4	15	−6.4	38.2	9.2	13.3	38.00	10.0	±	4.3	−16.3	±	14.1
294.6	15	−5.1	39.4	10.2	12.6	38.16	10.0	±	4.3	−15.0	±	13.5
291.7	15	−5.1	41.9	10.7	12.3	38.26	10.0	±	4.3	−15.1	±	13.7
288.5	15	−4.6	42.3	10.5	12.4	38.38	10.0	±	4.3	−14.6	±	13.9
285.2	15	−1.8	43.3	10.1	12.7	38.52	10.0	±	4.3	−11.7	±	14.5
282.3	15	−1.3	42.7	9.9	12.8	38.66	10.0	±	4.3	−11.2	±	14.4
279.3	15	2.2	47.5	10.9	12.2	38.81	10.0	±	4.3	−7.8	±	14.9
276.3	15	1.7	49.1	11.5	11.8	38.95	10.0	±	4.3	−8.3	±	14.9
272.5	15	6.7	49.0	11.1	12.0	39.13	10.0	±	4.3	−3.3	±	15.2
268.8	15	4.8	50.5	10.3	12.5	39.31	10.0	±	4.3	−5.2	±	16.3
264.9	15	5.5	50.1	10.1	12.6	39.49	10.0	±	4.3	−4.4	±	16.3

(*continued*)

Rotational Analysis

Based on the data sets accepted by the reliability tests, we examined the likelihood of vertical-axis rotation during deposition of sediments. When comparing results by formations, a marked decrease in declination is immediately apparent between the Honggou Formation, which has N-NE declination, and the Mahalagou and subsequent formations, which show northerly declinations (Fig. 6). To estimate this decrease in rotation, the net rotation values are formally derived following the methods of Beck (1980) and Demarest (1983) by comparing formation means to the expected direction calculated using the reference paleomagnetic pole for Eurasia (Besse and Courtillot, 2002) of the appropriate age (Table 1). This analysis suggests that a net clockwise rotation of ~20°, significant at the 95% confidence level, occurred sometime between deposition of the Honggou and Mahalagou Formations, as previously observed in published results from the Xining Basin (Dupont-Nivet et al., 2004). We note that inclination values are systematically lower than the ~60° expected from the apparent polar wander path of Eurasia of this age. This can be interpreted as resulting from flattening of inclination during depositional processes and subsequent compaction as observed in numerous red beds in Asia (Dupont-Nivet et al., 2002b; Gilder et al., 2001).

To better estimate the timing of the rotation, we focus on the declination values between the Honggou and Mahalagou Formations from the available magnetostratigraphic ChRM direction data set of the Xiejia section. The age of magnetostratigraphic levels is estimated by linear interpolation of assumed constant accumulation rate to the nearest correlated chron boundary (Dai et al., 2006). Throughout these formations from 34 to 48 Ma, we applied a sliding window, averaging Fisher (1953) statistics for 15 successive ChRM directions incrementally at each ChRM direction (Table 4). These moving averages of 15 ChRM directions included long time spans from the magnetostratigraphic data set (longer than 2 m.y.), likely averaging paleosecular

TABLE 4. INCREMENTAL MOVING AVERAGE OVER MAGNETOSTRATIGRAPHIC ChRM DIRECTIONS (*continued*)

Level (m)	*n*	Dm (°)	Im (°)	*k* (°)	α_{95} (°)	Age (Ma)	Dx (°)	±	δD (°)	R (°)	±	ΔR (°)
259.7	15	5.2	49.9	10.1	12.7	39.74	10.0	±	4.3	–4.8	±	16.3
254.4	15	8.4	50.6	10.1	12.7	40.00	10.0	±	4.3	–1.6	±	16.5
249.6	15	12.4	51.8	11.8	11.6	40.26	10.0	±	4.3	2.5	±	15.6
244.9	15	19.0	51.5	13.8	10.7	40.58	10.0	±	4.3	9.0	±	14.3
240.2	15	21.9	51.8	15.0	10.2	40.88	10.0	±	4.3	11.9	±	13.8
235.5	15	25.2	54.0	19.0	9.0	41.20	10.0	±	4.3	15.2	±	12.8
230.6	15	28.2	54.8	19.9	8.8	41.52	10.0	±	4.3	18.3	±	12.8
226.2	15	30.9	52.7	17.8	9.3	41.81	10.0	±	4.3	20.9	±	12.9
221.5	15	30.4	50.6	18.8	9.1	42.12	10.0	±	4.3	20.5	±	12.0
216.7	15	30.4	50.1	18.0	9.3	42.44	10.0	±	4.3	20.5	±	12.1
211.9	15	31.1	48.0	22.0	8.3	42.84	10.0	±	4.3	21.2	±	10.6
206.9	15	29.2	47.8	22.3	8.3	43.28	10.0	±	4.3	19.2	±	10.5
202.1	15	27.1	46.1	20.2	8.7	43.71	10.0	±	4.3	17.2	±	10.7
196.8	15	29.1	43.2	27.6	7.4	44.00	10.0	±	4.3	19.2	±	8.9
191.2	15	29.9	41.7	25.5	7.7	44.27	10.0	±	4.3	19.9	±	9.0
187.4	15	32.9	41.8	27.0	7.5	44.45	10.6	±	4.5	22.3	±	8.8
183.6	15	32.7	42.4	24.3	7.9	44.63	10.6	±	4.5	22.1	±	9.3
179.6	15	33.4	42.5	24.9	7.8	44.82	10.6	±	4.5	22.8	±	9.2
175.4	15	30.3	43.4	21.1	8.5	45.02	10.6	±	4.5	19.7	±	10.1
170.8	15	29.7	44.5	22.5	8.2	45.24	10.6	±	4.5	19.1	±	9.9
166.3	15	27.7	41.9	19.3	8.9	45.46	10.6	±	4.5	17.1	±	10.3
161.9	15	26.1	39.0	21.1	8.5	45.67	10.6	±	4.5	15.5	±	9.5
157.3	15	23.7	41.2	19.8	8.8	45.89	10.6	±	4.5	13.1	±	10.1
152.6	15	23.0	39.1	16.4	9.7	46.11	10.6	±	4.5	12.4	±	10.7
146.9	15	24.4	40.0	15.9	9.9	46.45	10.6	±	4.5	13.8	±	11.0
141.1	15	24.7	38.5	15.7	10.0	46.85	10.6	±	4.5	14.1	±	10.8
135.0	15	25.4	38.2	15.8	9.9	47.28	10.6	±	4.5	14.8	±	10.7
129.7	15	27.5	39.3	15.7	10.0	47.65	10.6	±	4.5	16.9	±	10.9
124.8	15	26.4	38.2	15.7	10.0	47.95	10.6	±	4.5	15.8	±	10.8

Note: Level—average stratigraphic level in meters according to levels given in Dai et al. (2006); *n*—number of characteristic remanent magnetization (ChRM) directions averaged (fixed to 15); Im and Dm—Fisher (1953) statistic mean inclination and declination of 15 ChRM directions within sliding window (in stratigraphic coordinates); *k*—concentration parameter; α_{95}— radius of cone of 95% confidence about mean direction; Age—age of the average stratigraphic level estimated using linear interpolation to the nearest correlated chron (Dai et al., 2006); D_x ± δD—expected declination and associated 95% confidence calculated from the apparent polar wander path of Besse and Courtillot (2002) of Eurasia at 40 and 50 Ma; R ± ΔR—rotation and associated 95% confidence derived following the methods of Beck (1980) and Demarest (1983).

variation. This procedure yielded window-mean directions with 95% confidence angles ranging from 7.4° to 14.4° (average = 10.9°). To estimate the timing of the rotation, the obtained window-mean declinations were plotted with respect to their magnetostratigraphically determined age averaged over the sliding window and compared to the expected declination calculated from the apparent polar wander path (Fig. 8). The record of clockwise rotation is clearly indicated by a progressive decrease in declinations occurring at ca. 41 Ma, in the lower part of the Mahalagou Formation. On either side of the transition, the nearest statistically different window-mean directions are found at 41.8 Ma (declination of 30.9° ± 9.3°) and 39.7 Ma (declination of 5.2° ± 12.9°), respectively, and most of the declination decrease is recorded within this time interval. In addition, the rotation (defined as the difference between the observed and expected declination; see Table 4) is statistically significant at 41.8 Ma (rotation of 20.9° ± 12.9°) and decreases systematically until 39.7 Ma, when it becomes insignificant (rotation of −4.8° ± 16.3°). As a result, we interpret the rotation to have occurred ca. 41 Ma, mostly between 41.8 Ma and 39.7 Ma.

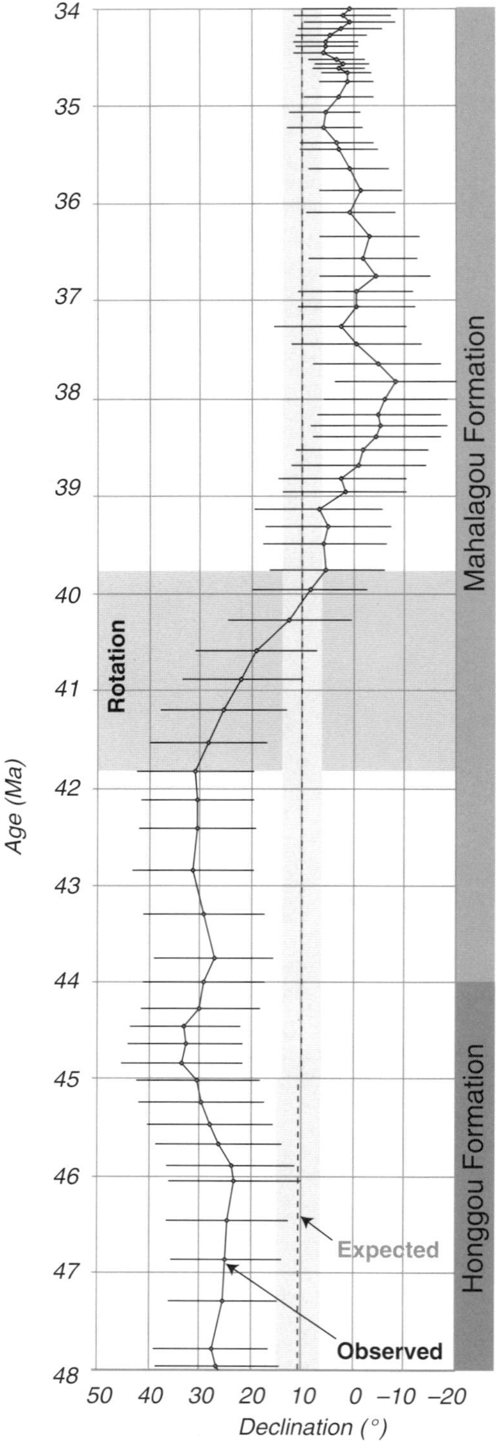

Figure 8. Moving average over the magnetostratigraphic characteristic remanent magnetization (ChRM) directions of the Xiejia section from 34 to 48 Ma (Dai et al., 2006). A sliding window was applied, averaging Fisher (1953) statistics for 15 successive ChRM directions incrementally at each ChRM direction (see Table 4). Mean declinations with 95% confidence are plotted against magnetostratigraphically determined age at the Xiejia section and compared to the expected declination (dotted line with 95% confidence in gray shaded area) from the poles of Eurasia (Besse and Courtillot, 2002). Most of the clockwise rotation occurred within the indicated horizontal shaded time interval.

DISCUSSION

Diachronous Rotations in Two Basin Systems

To understand the regional significance of our results, we examined existing paleomagnetic data over the northeastern Tibetan Plateau region (Table 1). Consistent with our results, previous studies from the Xining Basin show clockwise rotations of ~20°–30° in sediments from the Honggou Formation and older rocks, and no statistically significant rotations in younger deposits (Dupont-Nivet et al., 2004, Cogné et al., 1999). More regionally, data from the rest of the Paleocene-Miocene basin system are consistent in magnitude and timing with Xining Basin rotations (Fig. 9A). Clockwise rotations of ~15°–35° are systematically recorded in Paleogene and older rocks, and no significant rotation since at least 29 Ma is recorded by magnetostratigraphically dated sediments from the Linxia Basin (Fang et al., 2003). The sum of these results indicates that a regional tectonic mechanism generating widespread clockwise rotations was active during the early history of the Paleocene-Miocene basin system. Remarkably, the existing paleomagnetic data from the Gonghe-Guide-Xunhua basins (Fang et al., 2005; Yan et al., 2006) show a mid-Miocene (17–11 Ma) ~25° clockwise rotation that is restricted to the Miocene-Quaternary basin system located to the west of the Laji Shan range (except for the poorly dated Xiongchen locality; see Fig. 9B). The fact that this second rotation phase does not appear to have affected the Paleogene-Miocene basin system suggests a mid-Miocene regional westward shift of deformation and tectonic activity on the structural boundary between the two basin systems (i.e., the Laji Shan thrust belt; Fig. 1) during the mid-Miocene between ca. 17 and 11 Ma.

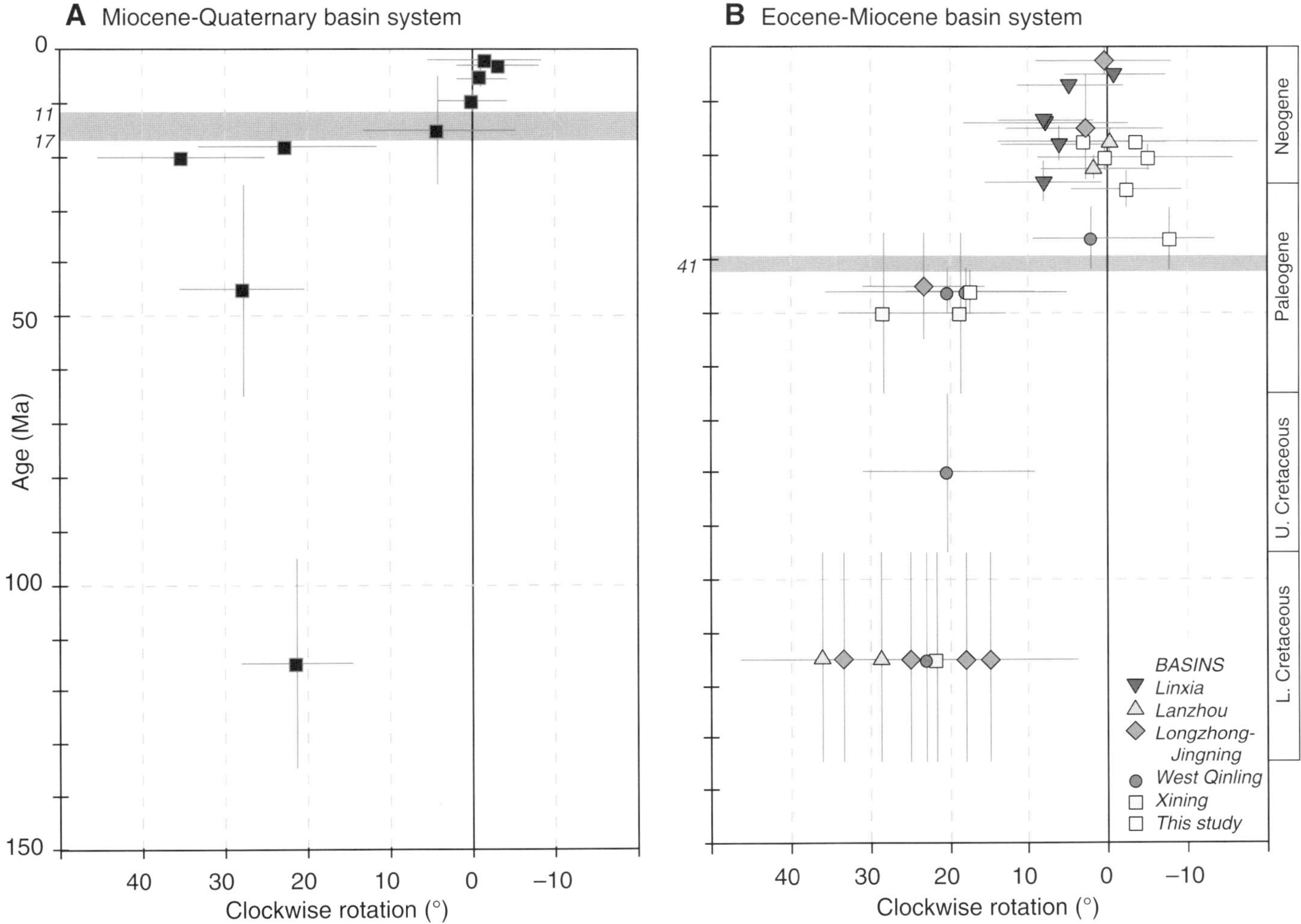

Figure 9. Clockwise rotation relative to the apparent polar wander path of Eurasia as a function of age of the sampled rocks within the northeastern Tibetan Plateau separated in (A) the Paleocene-Miocene basin system and (B) the Miocene-Quaternary basin system (data from Table 1). Thin black lines indicate 95% error bars in clockwise rotation and uncertainty in age (or age span for magnetostratigraphic results; see Table 1). Shaded horizontal areas indicate age range of tectonic rotation.

Regional Tectonic Context of the Early Eocene and Mid-Miocene Rotation Events

The timing brackets on the deformation events outlined by the two distinct rotation phases (at ca. 41 Ma and 17–11 Ma) support previous tectonic interpretations proposed for this region. The younger event that affected the Miocene-Quaternary basin system west of the Laji Shan range can be associated with documented onset of sediment accumulation in the Miocene to Quaternary compressional to transpressional setting (Fang et al., 2005; Yan et al., 2006). This deformation was coeval with widespread middle to late Miocene transpressional activity on various faults and associated range uplift in the northeastern Tibetan Plateau (e.g., George et al., 2001; Kirby et al., 2002; Pares et al., 2003). This activity is well documented in the Kunlun fault system, the Altyn Tagh fault system, and the Qilian Shan belt, possibly resulting from northeastward propagation of deformation of the Indian-Asia collision and Tibetan Plateau uplift (Fig. 1; Meyer et al., 1998; Tapponnier et al., 2001). The tectonic mechanism responsible for clockwise rotation of the Miocene-Quaternary basins must have been associated with crustal deformation on the bounding Laji Shan thrusts. We speculate, based on the absence of Miocene or younger rotation both in the Qaidam Basin to the west (Dupont-Nivet et al., 2002a, 2003) and in the Longzhong Basin to the east, that the Gonghe-Guide-Xunhua Basins may have rotated with crustal slivers in a regional N-S–trending right-lateral shear system driven by differential northward convergence at the eastern margin of the Qaidam Basin (Dupont-Nivet et al., 2004; England and Molnar, 1990).

Early Tertiary tectonic events related to the Paleocene-Miocene basin systems are still poorly understood and documented. Eocene deformation and subsidence in the Xining Basin and the central and northern Tibetan Plateau have been attributed to early propagation of deformation from the

India-Asia collision (Horton et al., 2003; Spurlin et al., 2005; Liu et al., 2003; Dai et al., 2006). Alternatively, the Paleogene may have been a major period of tectonic activity in the Qinling Shan, with widespread Eocene sedimentation in half-graben basins such as the Weihe graben in the Ordos block directly adjacent to the east of the Longzhong Basin (Fig. 1; Bellier et al., 1991; Ratschbacher et al., 2003; Zhang et al., 1998). This deformation has been attributed to far-field effects of back-arc rifting associated with Pacific-Eurasia convergence, which reached a minimum between 53.0 and 39.5 Ma (Allen et al., 1997; Northrup et al., 1995). Although both the Indian-Asia collision and the Pacific margin were located over 2000 km from the Qinling Shan region, it has been suggested that the Paleogene stress field in the Qinling Shan region resulted from the combined effects of Pacific subduction and the India-Asia collision (Ratschbacher et al., 2003). This geodynamic setting may have prompted opening of basins and Eocene clockwise rotations through large-scale N-S–trending dextral shear associated with the accommodation of the penetration of India into Asia (Fournier et al., 2004; Schellart and Lister, 2005). Further assessment of structural features associated with the Paleocene-Miocene basin system is necessary to establish the tectonic mechanism driving clockwise rotations.

ACKNOWLEDGMENTS

We thank R. Molina Garza and J. Geissman for detailed and thoughtful reviews. This work was co-supported by a European Union Marie Curie Fellowship and Netherlands Organization for Scientific Research (NWO) grants to G. Dupont-Nivet and S. Dai, and the Knowledge Innovation Program of the Chinese Academy of Sciences (Grant No. kzcx2-yw-104), the funds of the National Science Foundation of China (40571017, 40721061, 40334038), National Key Project for Basic Research (2005CB422000), and Key Project of Science and Technology of the Ministry of Education of China (grant no. 306016). We thank Meng QingQuan, Cheng Yu, and Tang Yuhu for field assistance.

REFERENCES CITED

Allen, M.B., Macdonald, D.I.M., Xun, Z., Vincent, S.J., and Brouet-Menzies, C., 1997, Early Cenozoic two-phase extension and late Cenozoic thermal subsidence and inversion of the Bohai Basin, northern China: Marine and Petroleum Geology, v. 14, p. 951–972, doi: 10.1016/S0264-8172(97)00027-5.

Argand, E., 1924, La tectonique de l'Asie: Compte-rendu du 13ème Congrès Géologique International, Brussel, v. 1, p. 171–372.

Avouac, J.P., and Tapponnier, P., 1993, Kinematic model of active deformation in central Asia: Geophysical Research Letters, v. 20, p. 895–898, doi: 10.1029/93GL00128.

Beck, M.E., 1980, Paleomagnetic record of plate-margin tectonic processes along the western edge of North America: Journal of Geophysical Research, v. 85, p. 7115–7131, doi: 10.1029/JB085iB12p07115.

Bellier, O., Vergely, P., Mercier, J.L., Ning, C.C., Deng, N.G., Yi, M.C., and Long, C.X., 1991, Analyse tectonique et sedimentaire dans les monts Li Shan (province du Shaanxi-Chine du Nord): Datation des regimes tectoniques extensifs dans le graben de la Weihe: Bulletin de la Société Géologique de France, v. 162, p. 101–112.

Besse, J., and Courtillot, V., 2002, Apparent and true polar wander and the geometry of the geomagnetic field in the last 200 million years: Journal of Geophysical Research, v. 107(B11), no. 2300, doi: 10.1029/2000JB000050.

Cogné, J.P., Halim, N., Chen, Y., and Courtillot, V., 1999, Resolving the problem of shallow magnetizations of Tertiary age in Asia: Insights from paleomagnetic data from the Qiangtang, Kunlun, and Qaidam blocks (Tibet, China), and a new hypothesis: Journal of Geophysical Research, v. 104, p. 17,715–17,734, doi: 10.1029/1999JB900153.

Dai, S., Fang, X., Dupont-Nivet, G., Song, C., Gao, J., Krijgsman, W., Langereis, C., and Zhang, W., 2006, Magnetostratigraphy of Cenozoic sediments from the Xining Basin: Tectonic implications for the northeastern Tibetan Plateau: Journal of Geophysical Research, v. 111, doi: 10.1029/2005JB004187.

Demarest, H.H., 1983, Error analysis of the determination of tectonic rotations from paleomagnetic data: Journal of Geophysical Research, v. 88, p. 4321–4328, doi: 10.1029/JB088iB05p04321.

Dupont-Nivet, G., Guo, Z., Butler, R.F., and Jia, C., 2002a, Discordant paleomagnetic direction in Miocene rocks from the central Tarim Basin: Evidence for local deformation and inclination shallowing: Earth and Planetary Science Letters, v. 199, p. 473–482, doi: 10.1016/S0012-821X(02)00566-6.

Dupont-Nivet, G., Butler, R.F., Yin, A., and Chen, X., 2002b, Paleomagnetism indicates no Neogene rotation of the Qaidam Basin in North Tibet during Indo-Asian Collision: Geology, v. 30, p. 263–266, doi: 10.1130/0091-7613(2002)030<0263:PINNRO>2.0.CO;2.

Dupont-Nivet, G., Butler, R.F., Yin, A., and Chen, X., 2003, Paleomagnetism indicates no Neogene rotation of the northeastern Tibetan Plateau: Journal of Geophysical Research, v. 108, doi: 10.1029/2003JB002399.

Dupont-Nivet, G., Horton, B.K., Zhou, J., Waanders, G.L., Butler, R.F., and Wang, J., 2004, Paleogene clockwise tectonic rotation of the Xining-Lanzhou region, northeastern Tibetan Plateau: Journal of Geophysical Research, v. 109, no. B04401, doi: 10.1029/2003JB002620.

Dupont-Nivet, G., Krijgsman, W., Langereis, C.G., Abels, H.A., Dai, S., and Fang, X., 2007, Tibetan plateau aridification linked to global cooling at the Eocene-Oligocene transition: Nature, v. 445, p. 635–638.

England, P., and Molnar, P., 1990, Right-lateral shear and rotation as the explanation for strike-slip faulting in eastern Tibet: Nature, v. 344, p. 140–142, doi: 10.1038/344140a0.

Fang, X., Garzione, C., Van der Voo, R., Li, J., and Fan, M., 2003, Flexural subsidence by 29 Ma on the NE edge of Tibet from the magnetostratigraphy of Linxia Basin, China: Earth and Planetary Science Letters, v. 210, p. 545–560.

Fang, X., Yan, M., Van der Voo, R., Rea, D.K., Song, C., Pares, J.M., Gao, J., Nie, J., and Dai, S., 2005, Late Cenozoic deformation and uplift of the NE Tibetan Plateau: Evidence from high-resolution magnetostratigraphy of the Guide Basin, Qinghai Province, China: Geological Society of America Bulletin, v. 117, p. 1208–1225, doi: 101130/B25727.1.

Fisher, R.A., 1953, Dispersion on a sphere: Proceedings of the Royal Society of London, Series A, Mathematical and Physical Sciences, v. 217, p. 295–305, doi: 10.1098/rspa.1953.0064.

Flynn, L.J., Downs, W., Opdyke, N.D., Huang, K., Lindsay, E., Ye, J., Xie, G., and Wang, X., 1999, Recent advances in the small mammal biostratigraphy and magnetostratigraphy of Lanzhou Basin: Chinese Science Bulletin, v. 44, supplement, p. 109–117.

Fournier, M., Jolivet, L., Davy, P., and Thomas, J.-C., 2004, Backarc extension and collision: An experimental approach to the tectonics of Asia: Geophysical Journal International, v. 157, p. 871–889.

George, A.D., Marshallsea, S.J., Wyrwoll, K.-H., Chen, J., and Lu, Y., 2001, Miocene cooling in the northern Qilian Shan, northeastern margin of the Tibetan Plateau, revealed by apatite fission-track and vitrinite-reflectance analysis: Geology, v. 29, p. 939–942, doi: 10.1130/0091-7613(2001)029<0939:MCITNQ>2.0.CO;2.

Gilder, S., Chen, Y., and Sen, S., 2001, Oligo-Miocene magnetostratigraphy and rock magnetism of the Xishuigou section, Subei (Gansu Province, western China) and implications for shallow inclinations in central Asia: Journal of Geophysical Research, v. 106, p. 30,505–30,522, doi: 10.1029/2001JB000325.

Halim, N., Cogné, J.P., Chen, Y., Atasiesi, R., Besse, J., Courtillot, V., Gilder, S., Marcoux, J., and Zhao, R.L., 1998, New Cretaceous and early Tertiary paleomagnetic results from Xining-Lanzhou Basin, Kunlun and Qiangtang blocks, China: Implications on the geodynamic evolution of Asia: Journal of Geophysical Research, v. 103, p. 21,025–21,045, doi: 10.1029/98JB01118.

Holt, W.E., Chamot-Rooke, N., Le Pichon, X., Haines, A.J., Shen-Tu, B., and Ren, J., 2000, Velocity field in Asia inferred from Quaternary fault slip rates and global positioning system observations: Journal of Geophysical Research, v. 105, p. 19,185–19,209, doi: 10.1029/2000JB900045.

Horton, B.K., Yin, A., Spurlin, M.S., Zhou, J., and Wang, J., 2003, Paleocene-Eocene syncontractional sedimentation in narrow, lacustrine-dominated basins of east-central Tibet: Geological Society of America Bulletin, v. 114, p. 771–786, doi: 10.1130/0016-7606(2002)114<0771:PESSIN>2.0.CO;2.

Horton, B.K., Dupont-Nivet, G., Zhou, J., Waanders, G.L., Butler, R.F., and Wang, J., 2004, Mesozoic-Cenozoic evolution of the Xining-Minhe and Dangchang Basins, northeastern Tibetan plateau: Magnetostratigraphic and biostratigraphic results: Journal of Geophysical Research, v. 109, no., B04402, doi: 10.1029/2003JB002913.

Kirby, E., Reiners, P.W., Krol, M.A., Whipple, K.X., Hodges, K.V., Farley, K.A., Tang, W., and Chen, Z., 2002, Late Cenozoic evolution of the eastern margin of the Tibetan Plateau: Inferences from Ar/Ar and (U-Th)/He thermochronology: Tectonics, v. 21, doi: 10.1029/2000TC001246.

Kirschvink, J.L., 1980, The least-square line and plane and the analysis of paleomagnetic data: Geophysical Journal of the Royal Astronomical Society, v. 62, p. 699–718.

Liu, Z., Zhao, X., Wang, C., Liu, S., and Yi, H., 2003, Magnetostratigraphy of Tertiary sediments from the Hoh Xil Basin: Implications for the Cenozoic history of the Tibetan Plateau: Geophysical Journal International, v. 154, p. 233–252, doi: 10.1046/j.1365-246X.2003.01986.x.

McFadden, P.L., 1990, A new fold test for palaeomagnetic studies: Geophysical Journal International, v. 103, p. 163–169, doi: 10.1111/j.1365-246X.1990.tb01761.x.

McFadden, P.L., and McElhinny, M.W., 1990, Classification of the reversal test in palaeomagnetism: Geophysical Journal International, v. 103, p. 725–729, doi: 10.1111/j.1365-246X.1990.tb05683.x.

Meyer, B., Tapponnier, P., Bourjot, L., Metivier, F., Gaudemer, Y., Peltzer, G., Shummin, G., and Zhitai, C., 1998, Crustal thickening in the Gansu-Qinghai, lithospheric mantle, oblique and strike-slip controlled growth of the Tibetan Plateau: Geophysical Journal International, v. 135, p. 1–47, doi: 10.1046/j.1365-246X.1998.00567.x.

Molnar, P., England, P., and Martinod, J., 1993, Mantle dynamics, uplift of the Tibetan Plateau, and the Indian monsoon: Reviews of Geophysics, v. 31, p. 357–396, doi: 10.1029/93RG02030.

Northrup, C.J., Royden, L.H., and Burchfiel, B.C., 1995, Motion of the Pacific plate relative to Eurasia and its potential relation to Cenozoic extension along the eastern margin of Eurasia: Geology, v. 23, p. 719–722, doi: 10.1130/0091-7613(1995)023<0719:MOTPPR>2.3.CO;2.

Pares, J.M., Van der Voo, R., Downs, W.R., Yan, M., and Fang, X., 2003, Northeastward growth and uplift of the Tibetan Plateau: Magnetostratigraphic insights from the Guide Basin: Journal of Geophysical Research, v. 108, no. 2017, doi: 10.1029/2001JB001349.

Qinghai Bureau of Geology and Mineral Resources, 1991, Regional geology of the Qinghai Province: Beijing, Geological Publishing House, 662 p.

Qiu, Z., Wang, B., Qiu, Z., Heller, F., Yue, L., Xie, G., and Wang, X., 2001, Land-mammal geochronology and magnetostratigraphy of mid-Tertiary deposits in the Lanzhou Basin, Gansu Province, China: Eclogae Geologicae Helvetiae, v. 94, p. 373–385.

Ratschbacher, L., Hacker, B.R., Calvert, A., Webb, L.E., Grimmera, J.C., McWilliams, M.O., Ireland, T., Dongg, S., and Hug, J., 2003, Tectonics of the Qinling (Central China): Tectonostratigraphy, geochronology, and deformation history: Tectonophysics, v. 366, p. 1–53, doi: 10.1016/S0040-1951(03)00053-2.

Ruddiman, W.F., Kutzbach, J.E., and Prentice, I.C., 1997, Testing climatic effects of orography and CO2 with general circulation and biome models, *in* Ruddiman, W.F., ed., Tectonic uplift and climate change: New York, Plenum Press, p. 203–235.

Schellart, W.P., and Lister, G.S., 2005, The role of the East Asian active margin in widespread extensional and strike-slip deformation in East Asia: Journal of the Geological Society, v. 162, p. 959–972, doi: 10.1144/0016-764904-112.

Spurlin, M.S., Yin, A., Horton, B.K., Zhou, J., and Wang, J., 2005, Structural evolution of the Yushu-Nangqian region and its relationship to syncollisional igneous activity, east-central Tibet: Geological Society of America Bulletin, v. 117, p. 1293–1317, doi: 10.1130/B25572.1.

Tapponnier, P., 2001, Oblique stepwise rise and growth of the Tibetan Plateau: Science, v. 394, p. 1671–1677, doi: 10.1126/science.105978.

Tapponnier, P., Xu, Z., Roger, F., Meyer, B., Arnaud, N., Wittlinger, G., and Yang, J., 2001, Oblique stepwise rise and growth of the Tibetan Plateau: Science, v. 294, p. 1671–1677.

Tauxe, L., 1998, Paleomagnetic Principles and Practice: Dordrecht, Kluwer Academic Publisher, 299 p.

Vandamme, D., 1994, A new method to determine paleosecular variation: Physics of the Earth and Planetary Interiors, v. 85, p. 131–142.

Wang, C., Zhao, X., Liu, Z., Lippert, P.C., Graham, S.A., Coe, R.S., Yi, H., Zhu, L., Liu, S., and Li, Y., 2008, Constraints on the early uplift history of the Tibetan Plateau: Proceedings of the National Academy of Sciences, v. 105, p. 4987–4992.

Yan, M., Van der Voo, R., Fang, X.-M., Pares, J.M., and Rea, D.K., 2006, Paleomagnetic evidence for a mid-Miocene clockwise rotation of about 25° of the Guide Basin area in NE Tibet: Earth and Planetary Science Letters, v. 241, p. 234–247, doi: 10.1016/j.epsl.2005.10.013.

Yang, T., Yang, Z., Sun, Z., and Lin, A., 2002, New Early Cretaceous paleomagnetic results from Qilian orogenic belt and its tectonic implications: Science in China, ser. D, v. 45, p. 565–576.

Yin, A., and Harrison, M.T., 2000, Geologic evolution of the Himalayan-Tibetan orogen: Annual Review of Earth and Planetary Sciences, v. 28, p. 211–280, doi: 10.1146/annurev.earth.28.1.211.

Zhai, Y., and Cai, T., 1984, The Tertiary System of Gansu Province, *in* Wang, Y.M., ed., Gansu Geology: People's Press of Gansu, p. 1–40, http://www.nau.edu/~qsp/will_downs/86.pdf.

Zhang, Y.Q., Mercier, J.L., and Vergely, P., 1998, Extension in the graben systems around the Ordos (China), and its contribution to the extrusion tectonics of South China with respect to Gobi-Mongolia: Tectonophysics, v. 285, p. 41–75, doi: 10.1016/S0040-1951(97)00170-4.

Manuscript Accepted by the Society 04 April 2008

Printed in the USA

The Geological Society of America
Special Paper 444
2008

The dynamic support and decoupling process of the Tibetan lithosphere based on the integration of flexural modeling with other geological and geophysical studies

Yu Jin
Chevron International Exploration & Production, Beijing 100004, China

Erchie Wang
Institute of Geology and Geophysics, China Academy of Sciences, Beijing 100029, China

Xiaodian Jiang
Department of Marine Geology, China University of Oceanography, Qingdao 266003, China

ABSTRACT

The mechanism for uplift of the Tibetan Plateau and the mechanism for maintaining the high gravitational potential of the plateau have been a focus of discussion for decades in the earth science community around the world. One theory based on gravity data, topographic data, and flexural modeling proposes that the Tibetan Plateau is partially supported by the lithosphere of the surrounding lowlands, including the Indian plate, the Tarim Basin, and the Sichuan Basin, in the form of regional compensation. We present two-dimensional (2-D) transects of the lowlands to the highlands around the plateau to review the lateral compensation and weakening of the Indian plate when it is subducting beneath the Tibetan Plateau in the south, the possible support of the Tarim lithosphere while it is underplating the northern edge of the Tibetan Plateau, and the possible lateral compensation style beneath the eastern plateau associated with the western foredeep of the Sichuan Basin.

Lateral variation of lithosphere strength is traditionally calculated based on the coherence theory in the spectral domain. This theory requires the study areas to be rectangular and the elastic thickness within each rectangular box to be constant. However, a rectangular geometry is hardly consistent with any natural shape of a tectonic unit. In this paper, we calculate lateral variation in lithosphere strength consistent with the natural shapes of tectonic units, such as the Himalayas, Western Kun Lun Shan, Central Tibetan Plateau, Qilian Shan, Tarim Basin, Tian Shan, and so on, based on three-dimensional (3-D) flexural modeling in the spatial domain.

Flexural modeling from gravity data around the Tibetan Plateau is mainly based on the way in which the Moho deflects from the deformed and weakened lithosphere in convergent plate boundaries. However, flexural modeling hardly differentiates among the variations in architecture of deformation in the upper crust, the lower

Jin, Y., Wang, E., and Jiang, X., 2008, The dynamic support and decoupling process of the Tibetan lithosphere based on the integration of flexural modeling with other geological and geophysical studies, *in* Burchfiel, B.C., and Wang, E., eds., Investigations into the Tectonics of the Tibetan Plateau: Geological Society of America Special Paper 444, p. 89–104, doi: 10.1130/2008.2444(06). For permission to copy, contact editing@geosociety.org.

crust, and the upper mantle lithosphere. In order to further understand the present architecture and the evolution of Tibetan lithosphere, an integrated study using gravity data, surface faulting, structural modeling, seismic imaging, deep seismic sounding, and earthquake events is also discussed.

Rheological modeling using a simple geothermal structure at various locations over the Tibetan Plateau reveals that the base of the strong upper crust of Tibet is at 30 to 35 km depth. The tendency for the strong upper crust to flow on the weak, ductile lower crust (or middle and lower crusts) depends on the development of the thickness of the weak lower crust. In other words, the weakness depends on the thickness difference between the base of the upper crust and the Moho depth. The larger the difference is, the easier it is for the upper crust to flow relative to the strong upper mantle lithosphere. Therefore, improvements in Moho depth calculations based on 3-D spatial flexural modeling will help to clarify the mobility of the lower-crustal flow of the Tibetan plateau.

Keywords: plateau, lithosphere, tectonics, gravity, seismic.

INTRODUCTION

The mechanical strength of Earth's lithosphere has been studied since 1970 due to the nature of the moving plates that govern the global tectonics of Earth, which are not rigid but are subject to deformation (Dorman and Lewis, 1970; Walcott, 1970; Watts and Cochran, 1974; McNutt and Parker, 1978; Watts et al., 1980, 1982; McNutt and Menard, 1982; Turcotte and Schubert, 1982; McNutt, 1983, 1984; Lyon-Caen and Molnar, 1983, 1984; Forsyth, 1985; Sheffels and McNutt, 1986; Bechtel et al., 1987, 1990; McNutt et. al., 1988; Zuber et al., 1989; Wolfe and McNutt, 1991; Gail and McNutt, 1992; Filmer et al., 1993; Jin et al., 1994, 1996; Lowry and Smith, 1994; Macario et al., 1995; Stewart and Watts, 1997; Burov and Molnar, 1998; Lorenzo et al., 1998; Rehbinder and Yakubenko, 1999; Simons et al., 2000; Buck, 2001; Simons and van der Hilst, 2002; Jin and Jiang, 2002; Braitenberg et al., 2003; Jiang et al., 2004; Jiang and Jin, 2005; Swain and Kirby, 2006).

Due to the lithology difference between continental crust and oceanic crust, the light, weak, granite-diorite–rich, buoyant upper and middle crust of the continental lithosphere tends to be detached or decoupled from the strong, heavy, gabbro-peridotite–rich lower crust or upper mantle lithosphere under tectonic deformation (Wernicke, 1981; Zuber et al., 1986; Buck, 1988; Kruse et al., 1991; Jin et al., 1994; Jin and Jiang, 2002, their color plate 2). The mechanical strength of a decoupled lithosphere is significantly less than the un-decoupled ones, such as the Indian Shield, the Canadian Shield, and the Australian continent (McNutt et al., 1988). Decoupled lithosphere is characterized by an elastically strong but brittle upper crust, which can have pervasive earthquakes due to faulting (Langin et al., 2003); it can also be characterized by a high resistivity zone (Wei et al., 2001), a ductile fluid-like middle crust, or lower crust that can be represented as a high conductive zone, and finally an elastically strong upper mantle lithosphere. The Tibetan lithosphere falls into the category of decoupled lithosphere.

The decoupling of the continental lithosphere is difficult to image due to the current limitation of imaging technologies, the insufficiency of imaging data (Alsdorf et al., 1998; Kumar et al., 2006), and the significant cost of seismic data acquisition. The continental shelf of West Africa is a good analog for the Tibetan Plateau in the sense of decoupling. The cross section outlining the crustal architecture of the continental shelf of West Africa from seismic imaging shows a good, visible decoupled portion of the lithosphere with brittle failure in the upper part, ductile flow in the middle, and a strong, competent crust at the base (Fig. 1).

In this paper, we will present a comprehensive review of flexural modeling over the Tibetan Plateau and its vicinity and integrate a wealth of information from other studies, such as tectonics, seismic imaging, earthquake seismology, and magnetotelluric data, in order to further understand the rheological decoupling process of the Tibetan lithosphere.

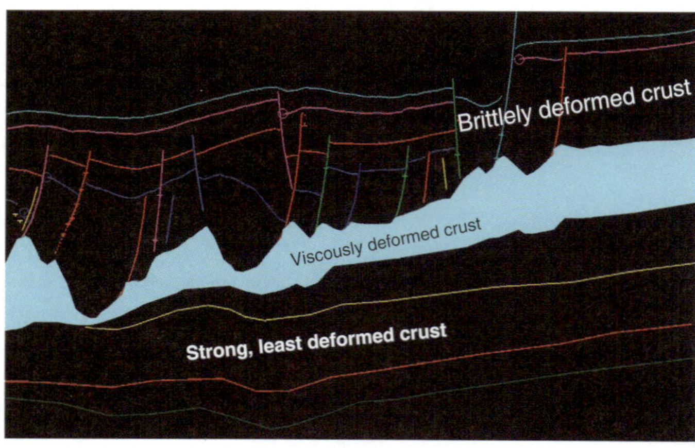

Figure 1. The cross section of a crustal deformation model showing rheological variation in depth. The model is based on a seismic imaging profile from the continental shelf of West Africa. The upper part of the section shows brittle deformation, the middle part shows viscous deformation, and the lower part shows the least deformation (or nonbrittle deformation). The colors of the horizons show the different time surfaces. The color of the faults show different fault surfaces and becomes significant for building up fault framework in three dimensions.

LATERAL SUPPORT AND LITHOSPHERE DECOUPLING DUE TO THE CONVERGENCE BETWEEN INDIA AND EURASIA

Even though the model from the West Africa continental shelf (Fig. 1) presents a good analog of the decoupling process for the Tibetan lithosphere, it is still difficult to visualize the architecture of the decoupled Tibetan lithosphere due the paucity of imaging technology widely applied to the upper crust mainly for hydrocarbon exploration and development. However, the thick crust of Tibet indicates that the lithosphere is decoupled, even using a rheological model bearing the simplest geothermal variation in depth, and a more realistic geothermal model should correlate well with the electrical conductivity model shown in Figure 3 of Wei et al. (2001). Figure 2 shows a rheological model of the yield strength envelope (Goetze and Evans, 1979; Kirby, 1980) for the Tibetan Plateau assuming that the crust of the central Tibetan Plateau is 60 km thick in average and has no stretching ($\beta = \delta = 1$) (Royden and Keen, 1980). With such a large thickness of the Tibetan crust, a cooling process for the lithosphere after a thermal impulse (McKenzie, 1978; Royden and Keen, 1980) does not need to exist for the Tibetan lithosphere to be decoupled. The lithosphere will stay decoupled by a weak zone from 30 to 60 km deep, or even shallower, based on conductance observations (Wei et al., 2001) or the frequency of seismicity (Langin et al., 2003) through geologic time (Fig. 2), unless the thick crust can be thinned to a normal crust thickness with an average of ~33 km.

The doubling of the Tibetan crust is still a puzzle, and there is a variety of thickening and crustal flow models (Dewey and Burke, 1973; England and McKenzie, 1982; Zhao and Morgan, 1987; England and Houseman, 1986; Molnar et al., 1993; Jin et al., 1994; Willett and Beaumont, 1994). However, there is one thing in common among all these models: the crustal doubling is derived from the convergence between India and Eurasia. The difference depends on whether the doubling is caused by pure shear deformation (Dewey and Burke, 1973; England and McKenzie, 1982; England and Houseman, 1986) or from a lower crust injection (Zhao and Morgan, 1987).

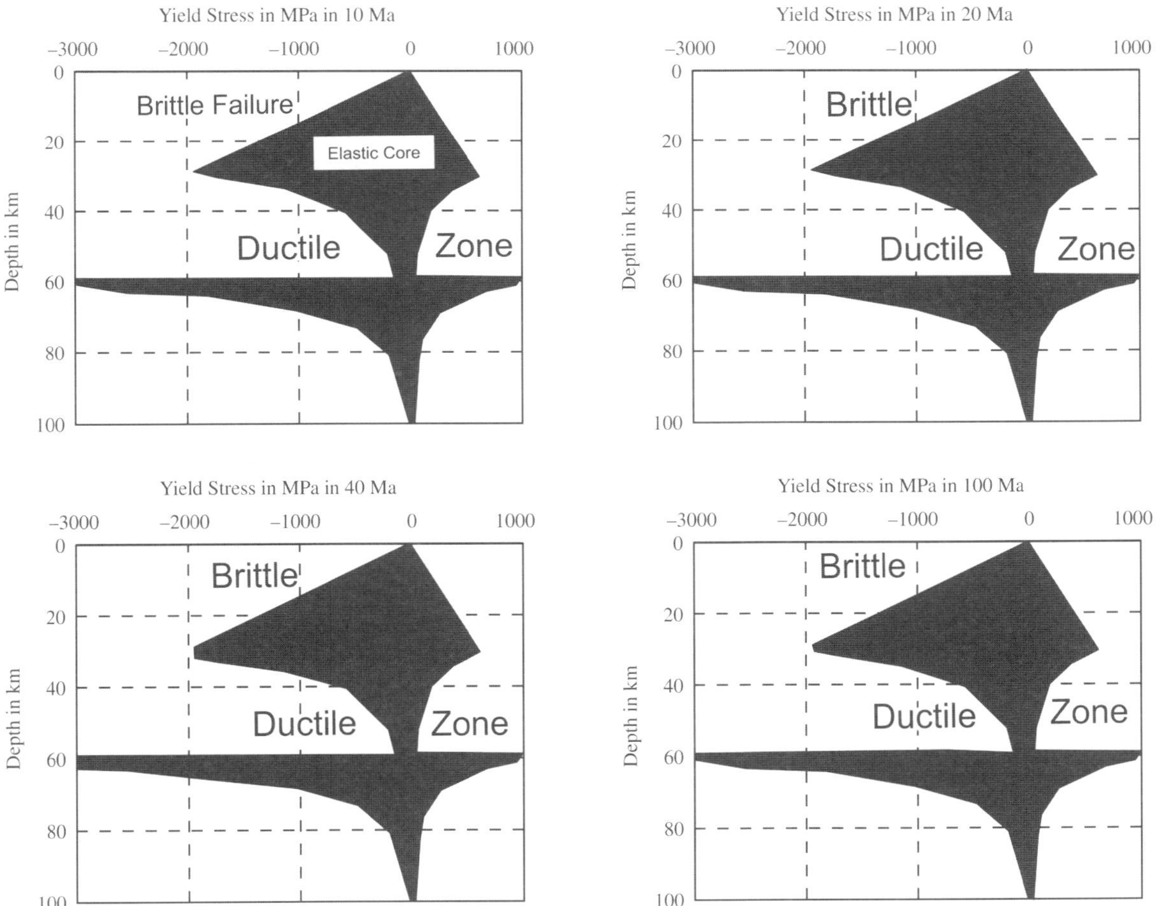

Figure 2. Yield strength envelope of Tibetan Plateau at 10 Ma, 20 Ma, 40 Ma, and 100 Ma, assuming that the crust of Tibet is 60 km and there is no stretching ($\beta = \delta = 1$). The negative values are the yield stresses under compression, and the positive values are the extensional stresses. The black area is the nonyielding, elastic core of the central Tibetan lithosphere. The linear, brittle failure boundary is based on Byerlee's law (1977) and the Mohr circle. The nonlinear viscous yielding boundary is based on the nonlinear viscous constitutive law (Turcotte and Schubert, 1982). The comprehensive formulation of this yield strength envelope can be found in Section 5 of Chapter 2 in Jin and Jiang (2002).

Based on rheology, the upper crust of Tibet and its upper mantle lithosphere should not be deformed at the same rate and in the same style due to its strength variation with depth (Fig. 2) or due to the overthickening of its crust (Jin et al., 1994). Figure 2 shows a possible rheological pattern of the Tibetan lithosphere that assumes the Tibetan crust is 60 km thick. With such a thick crust, the middle and lower crust of Tibet will yield, and the yielding will decouple the deformation style between the upper crust and the upper mantle lithosphere. Also, decoupling is time-invariant as long as the crust is doubled.

As we mentioned already, the exact process of crustal doubling in Tibet is still unknown. However, one consensus about the doubling is that it originated from the convergence between the Indian and Eurasian plates. While the Indian plate subducts beneath the Eurasian plate, the light, buoyant, quartz-rich Indian crust cannot dive into the mantle with the heavy, olivine-rich Indian mantle lithosphere, and so it becomes detached from the Indian lithosphere. Part of the detached India crust contributes to the formation of the Himalayas (Alsdorf et al., 1998) and to the weakening of the Indian Lithosphere (Lyon-Caen and Molnar, 1983; Jin et al., 1996).

The other part of the detached, light Indian crust may be injected to the middle and lower crusts of Tibet to decouple the Tibetan lithosphere (Zhao and Morgan, 1987). This weakening and injection process into the lithosphere caused by the detachment of the Indian crust in the form of imbricates can also be documented by flexural modeling (Jin et al., 1996). Figure 3 shows the location of the major flexural transect of Jin et al. (1996) as well as the locations of the International Deep Profiling of Tibet and the Himalayas (INDEPTH) deep sound-

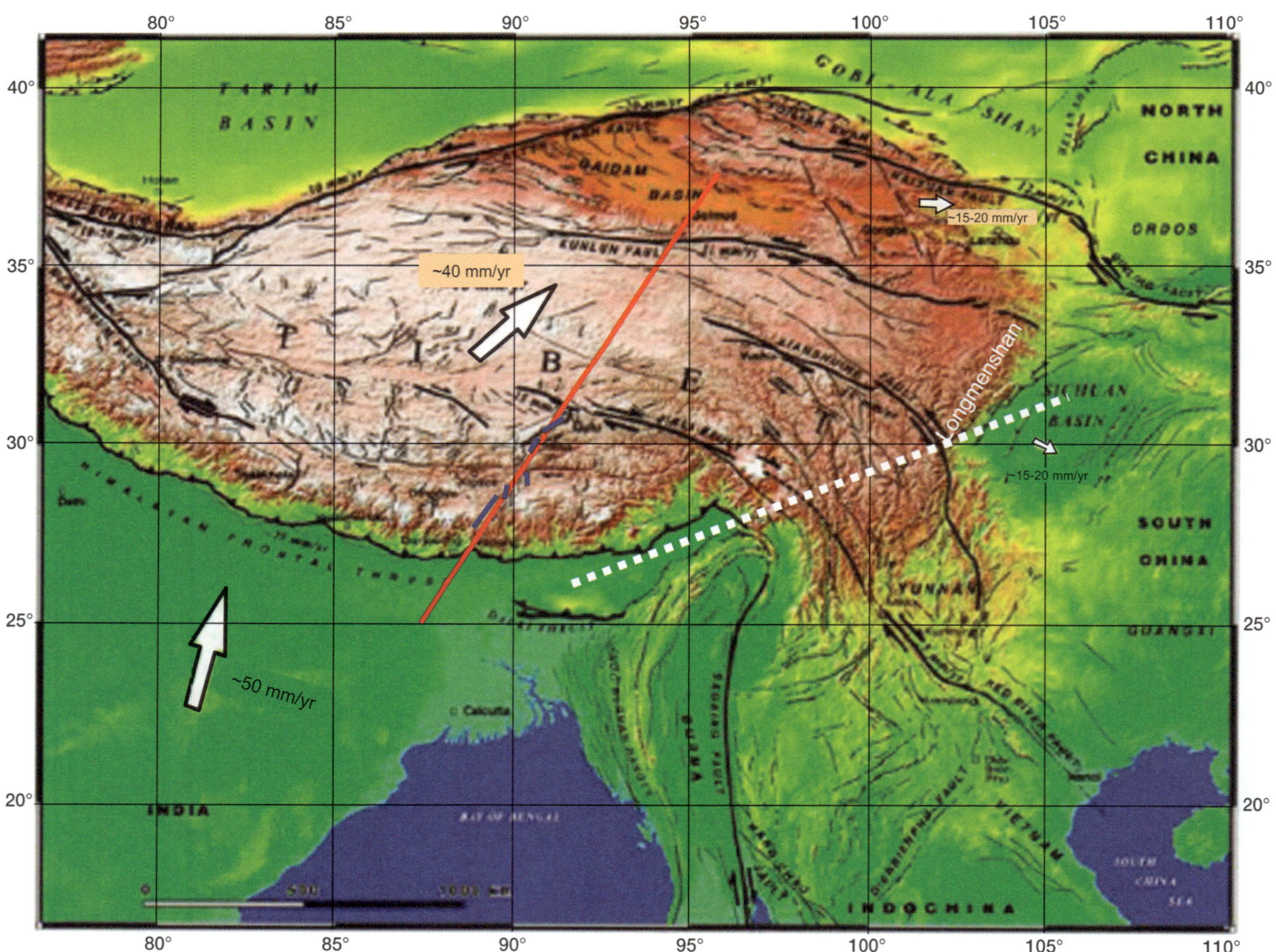

Figure 3. Gravity transect (red line) across the southern part of the Tibetan Plateau (Jin et al., 1996). The blue short segments along the southern part of the gravity transect are the International Deep Profiling of Tibet and the Himalayas (INDEPTH) deep sounding reflection seismic survey lines from Tib-1 to Tib-11 (Alsdorf et al., 1998). The white dashed line is a proposed seismic transect from India to the Sichuan Basin. The background is the Central Asia topography. The white arrows with black outlines show the crustal direction of movement, and sizes of the arrows are plotted in the relative scale of the plate moving velocities (Tapponnier et al., 2001).

ing reflection profiles along the transect. The cross section of the geotransect in Figure 3 and the major reflectors from the INDEPTH seismic imaging (Alsdorf et al., 1998) are shown in Figure 4 (thick black solid segments).

Two distinct features can be seen in the cross section. First, the effective elastic strength of the plate is weakened from 90 km in the lowland to 30 km at the mantle suture between India and Eurasia. Second, the major INDEPTH reflectors in black (Alsdorf et al., 1998) have very different structural attitudes on each side of the mantle suture. To the south of the mantle suture, the reflectors GDR and MHT significantly dip to the north; however, to the north of the mantle suture, the reflector YDR is almost horizontal. YDR is at a depth of ~15 km (plate 2 of Alsdorf et al., 1998) and needs to be further investigated in order to understand the origin of the reflectors. Nevertheless, the steep dips of GDR and MHT should only reflect the intraboundaries of various imbricates (Fig. 5) detached from the Indian lithosphere (Nelson et al., 1996; Makovsky et al., 1999).

One possible interpretation of the distinct difference between the north and south reflectors relative to the mantle suture is that the detached crustal imbricates from the Indian plate meet with the week, viscous middle and lower crusts of Tibet at the mantle suture between India and Eurasia (Fig. 5), the imbricates are melted and injected into the Tibetan middle and lower crusts, and the melts force the upper crust of Tibet upward hydraulically (Zhao and Morgan, 1987), so that the reflectors within the upper crust or at the boundary between the middle and lower crusts of Tibet are kept semihorizontal to the north of the mantle suture (Figs. 4 and 5).

The behavior of the weak middle and lower crusts of Tibet (Figs. 2 and 5) allows the upper crust of Tibet to flow independently from the upper mantle lithosphere of Tibet (Shen et al.,

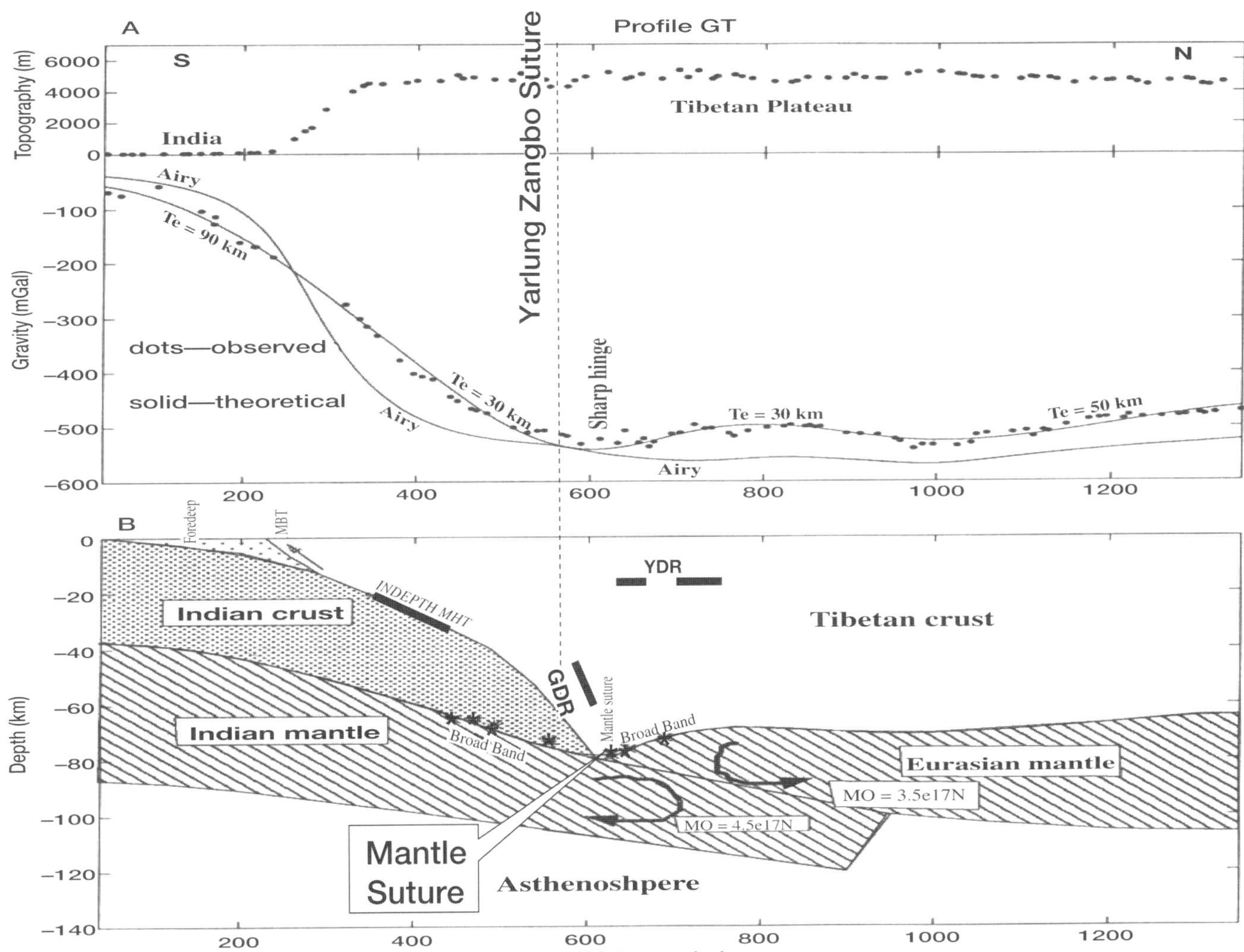

Figure 4. Lithosphere structure of Tibet based on gravity data and flexural modeling (Jin et al., 1996). The thick black solid segments of the lithosphere model in B are the major seismic reflectors imaged by INDEPTH Project (Makovsky et al., 1996; Alsdorf et al., 1998). MHT—Main Himalayan thrust; GDR—Gangdese deep reflection; YDR—Yamdrok-Damxung Reflector; Te—effective elastic thickness; M0—bending moment. The dots in the lower part of A are the observed Bouguer gravities.

2001; Chen et al., 2000). The analogue from West Africa in Figure 1 demonstrates this type of deformation, where the shallower crust is rafted on a weak, viscous zone, and the crust below the viscous zone is almost undeformed. One thing is certain: Tibet is in the compressional domain and West Africa is in the extensional domain. The authors only intend to use the decoupling case of West Africa as an analogue.

LATERAL SUPPORT OF THE TIBETAN PLATEAU FROM THE TARIM PLATE AND LITHOSPHERE COUPLING OR WELDING AT THE NORTHERN EDGE OF THE PLATEAU

The northern Tibetan Plateau is bounded by the Altyn Tagh and West Kunlun faults. A question that remains is whether the sinistral motion of the Altyn Tagh fault (Peltzer et al., 1989; Bendick et al., 2000; Jiang et al., 2004) penetrates the entire thickness of the lithosphere or whether the crustal motion is decoupled from the upper mantle lithosphere of Tibet.

A seismic tomographic study suggests that the Altyn Tagh fault is a localized lithospheric shear zone (Wittlinger et al., 1998). Flexural modeling along the new 2-D gravity surveys (Jiang et al., 2004) across the northern boundary of the Tibetan Plateau also shows that a local compensation model has to be invoked in order to obtain the best fit for the observed gravities along the Altyn Tagh transect. The consistency of these two independent studies is puzzling because the majority of the Tibetan upper crust is flowing on its weak lower crust (Chen et al., 2000; Shen et al., 2001), like the rheological analog in West Africa (Fig. 1), but the Altyn Tagh fault seems to cut through the whole lithosphere.

One possible answer is explained with rheological modeling. The gravity transect across the Altyn Tagh fault suggests that

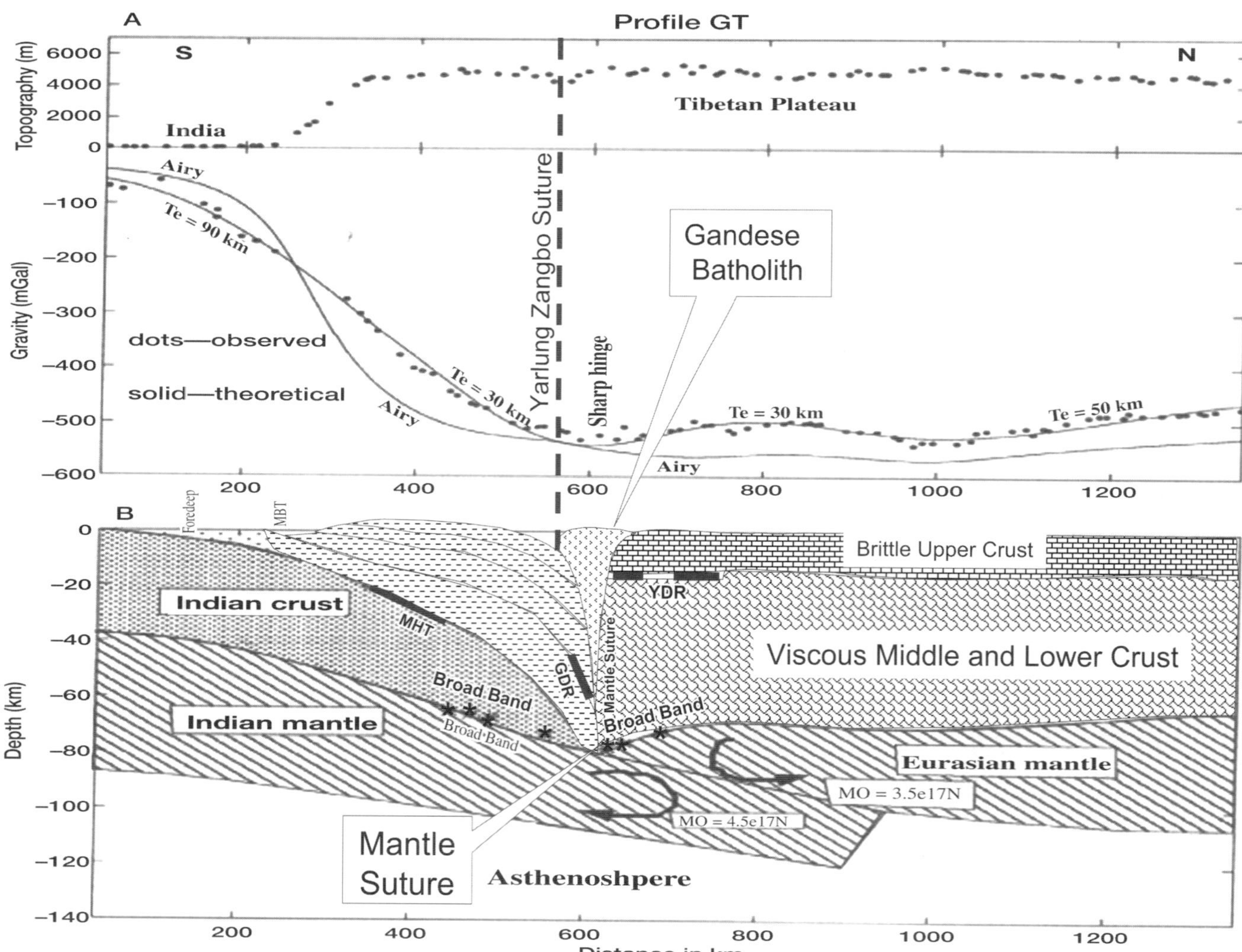

Figure 5. Lithospheric structure of Tibet and the subducting Indian plate based on an integrated study of gravity data, flexural modeling (Jin et al., 1996), deep sounding seismic reflection (thick black solid segments) (Zhao and Nelson, 1993), broadband seismology (black stars) (Yuan et al., 1997), and the geological mapping of Tibet (Pan et al., 2004). MHT—Main Himalayan Thrust; GDR—Gangdese Deep Reflection; YDR—Yamdrok-Damxung Reflector; Te—effective elastic thickness; M0—bending moment.

the crustal thickness underneath the fault (100 km location on Figure 10 of Jiang et al., 2004) is ~40 km. The rheological profile of a 40-km-thick crust with a normal geothermal variation at depth is significantly different from the rheological profile of a 60-km-thick crust. The latter tends to have a well-developed mushy, weak lower crust to allow the upper crust to cruise on it, even with a normal geotherm (Figs. 1 and 2). However, the former has a rather resistant lower crust, which tends to help the upper crust and the upper mantle lithosphere to weld together and deform at the same pace (Fig. 6).

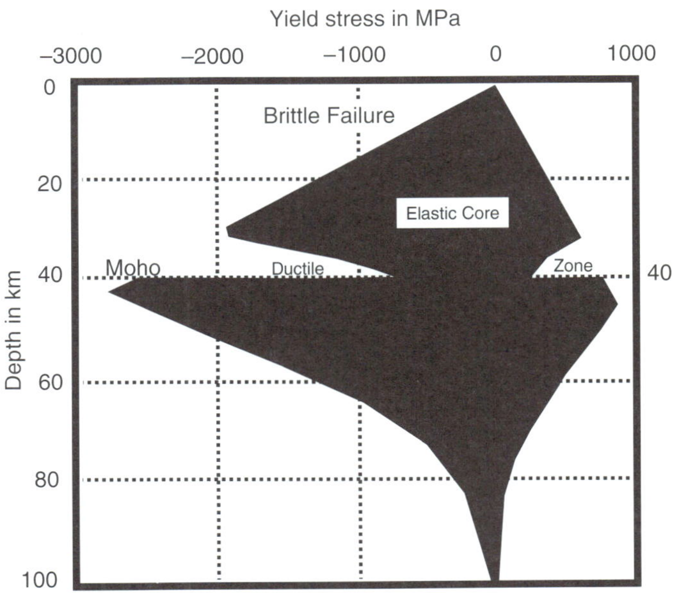

Figure 6. Yield strength envelope of the lithosphere underneath the Altyn Tagh fault. The black area is the nonyielding, elastic core of the lithosphere. The linear, brittle failure boundary is based on Byerlee's law (1977) and the Mohr circle. The nonlinear viscous yielding boundary is based on the nonlinear viscous constitutive law (Turcotte and Schubert, 1982). The comprehensive formulation of this yield strength envelope can be found in Section 5 of Chapter 2 in Jin and Jiang (2002).

LATERAL SUPPORT AND ELASTIC REBOUND OF THE LITHOSPHERE AT THE EASTERN EDGE OF THE TIBETAN PLATEAU

Flexural modeling across the eastern edge of the Tibetan Plateau from the Sichuan Basin (Fig. 7) also suggests that the eastern boundary of the plateau is still in regional compensation, with a continuous plate of about Te = 47 km (Te—effective elastic thickness) in the lowland (Sichuan) and Te = 36 km underneath the highland (Jiang and Jin, 2005). The question that remains for the flexural compensation here is that the foreland basin of Sichuan is not Cenozoic in age but Mesozoic (Burchfiel et al., 1995). The author recently went to Ya An, the foredeep area of Longmen Shan (Jiang and Jin, 2005) and consulted with

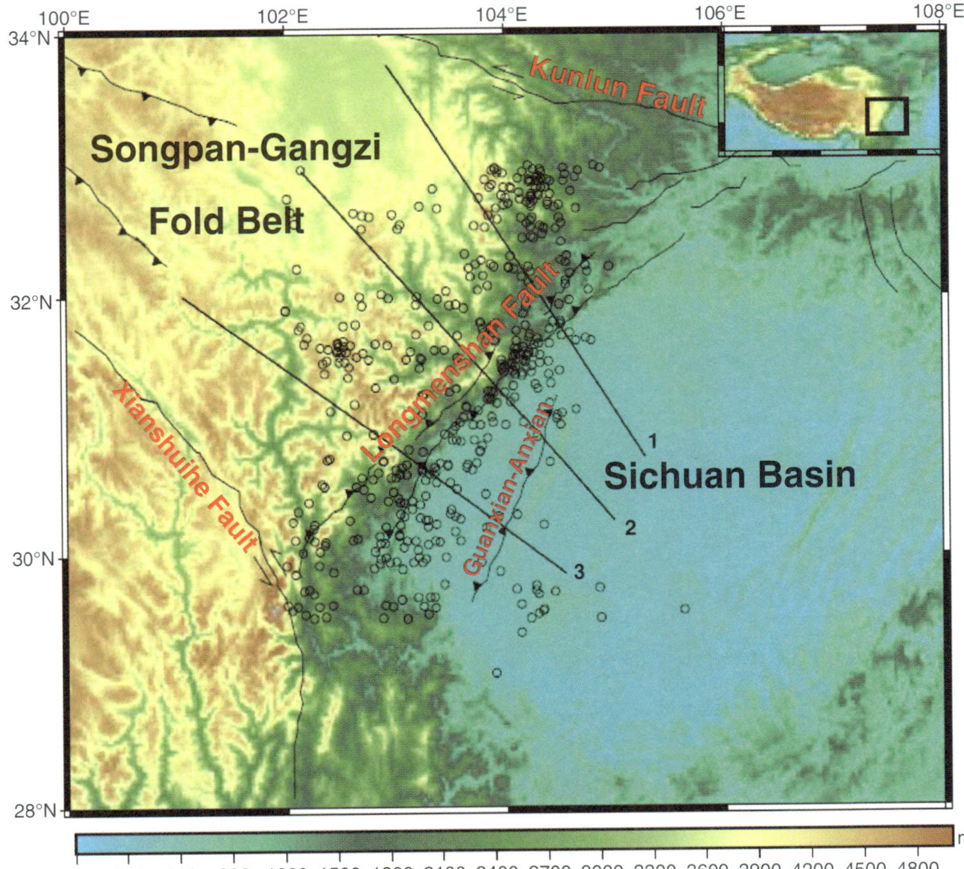

Figure 7. Current earthquake events (black circles) in the Longmenshan area, including the eastern Tibetan Plateau and the western Sichuan Basin. The seismic events are cataloged in Table 1 and are collected through the data center of Science Academy of China. Profiles 1, 2, and 3 are the seismic transects for flexural modeling (Jiang and Jin, 2005). The flexural modeling with some earthquake events at depth is shown in Figure 8.

the local geologists and geophysicists of Sichuan Petroleum Administration about Cenozoic deposition at the foredeep. The answer is positive. However, the author needs seismic imaging data to quantify the thickness and lithology of the Cenozoic sediments in the foredeep. The Sichuan Basin contains many kilometers of Jurassic-Cretaceous terrestrial sediments with fluvial mudstones, sandstones, and conglomerates and many kilometers of middle Lower Triassic-Permian shallow-marine carbonates (Ma et al., 2007). There was a maximum of 1500 m of Cenozoic sediments deposited during the Himalayan orogeny (Ma et al., 2007). At present, the maximum thickness of the foredeep sediments (Fig. 2 of Jiang and Jin, 2005) still exceeds 10 km; however, the hydrocarbon maturity threshold in the western foreland basins of Sichuan used to be 4 km deep, and now it is almost exposed at the surface (David Rowley, 2006, personal commun.). In the other words, the foredeep basin in front of Longmenshan (Fig. 7) has been experiencing a substantial elastic rebound, and at the same time, the frontal mass load (Longmenshan) of the western foredeep of the Sichuan Basin has been undergoing denudation. As we can see from an animation of global reconstruction (Scotese, 2006), the mass around the Longmenshan area has been moving toward the northeast since the collision between India and Eurasia. A foredeep basin in front of Longmenshan can only be well developed if the mass of Longmenshan moves toward the southeast.

Even though there is no clear evidence that the Longmenshan thrust (Figs. 7 and 8) is currently obducting onto the western part of the Sichuan Basin to form a Cenozoic foredeep, the fault should still be active based on current earthquake observations. Figure 7 shows the current earthquake events in the Longmenshan area and the western Sichuan Basin. If a 10-km-wide swath of earthquake focal depths is projected onto profile 2 (Fig. 7), we can see that some of the events are possibly derived from slip or rupture on the Longmenshan thrust and the frontal thrust, the Guanxian fault (Fig. 8). The focal solutions of these projected seismic events should be further investigated

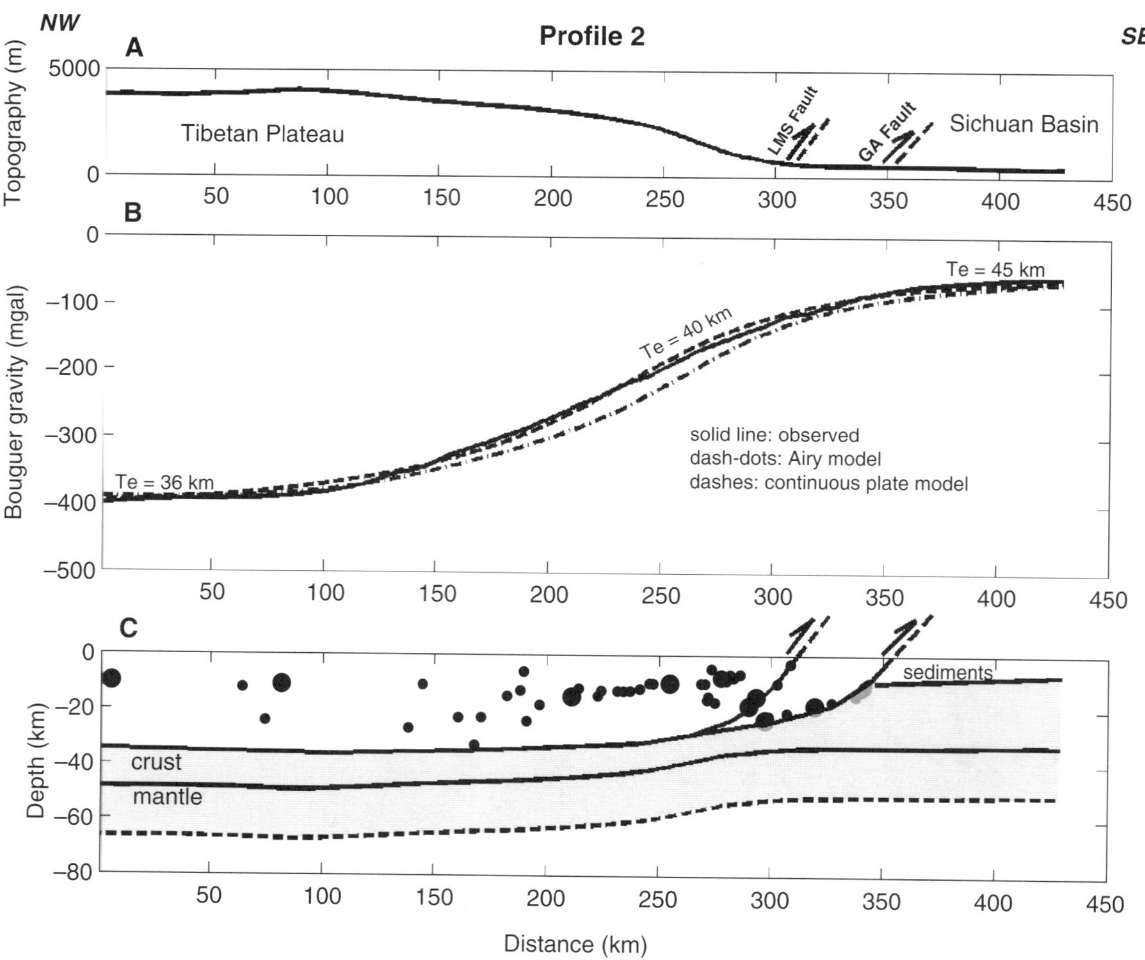

Figure 8. Cross section of flexural modeling across Longmenshan. The location of the profile is shown in Figure 7. (A) Topography: thrust fault symbols show the surface location of the Longmenshan thrust and the Guanxian thrust. (B) Observed and theoretical Bouguer gravity with various effective elastic thicknesses. (C) Possible Earth model derived from flexural modeling and earthquake focal depths. The swath to project the focal depth along the cross section is 10 km wide.

to see if the earthquakes are purely compressional or oblique. In Project INDEPTH III, 57 broadband and short-period seismic stations were deployed in central Tibet from August 1998 through May 1999 (Langin et al., 2003). These stations detected 267 local earthquakes with magnitudes from 1.7 to 5.8 during the observation period. The focal depths of most of the earthquakes were less than 20 km, with significant uncertainty bars (Fig. 9 of Langin et al., 2003). The determination of focal depths of small earthquakes is still challenging at the moment. On the other hand, weakening of the effective elastic strength of the lithosphere from the lowland to the highland should also be correlated to the pervasive faulting of the upper crust (Fig. 6 of Langin et al., 2003). The brittle failure of the upper crust is demonstrated well by the earthquake events (Fig. 8).

Flexural modeling in Figure 8 also shows that the brittle failure domain of the eastern Tibetan crust characterized by the earthquake events is located from a depth close to the surface all the way to ~35 km deep. However, the flexural modeling also shows that the Moho is at ~50 km depth in the eastern Tibetan Plateau. The 15 km of lower crust from 35 km depth to the Moho forms the mushy ductile zone, which permits the brittle upper crust to move relative to the strong upper mantle lithosphere. A rheological cross section in the eastern plateau with a crustal thickness of 50 km is shown in Figure 9. This scenario requires a considerably thick ductile zone like in the central Tibetan Plateau (Fig. 2) to decouple the lithosphere.

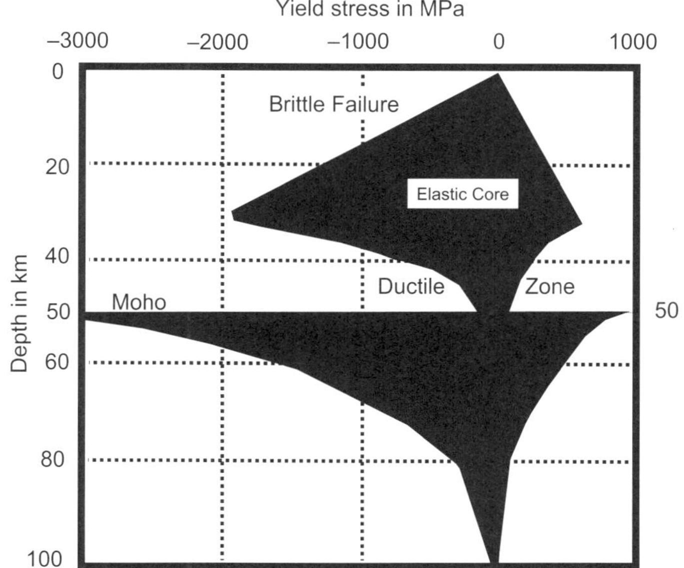

Figure 9. Rheological cross section of the lithosphere in the eastern Tibetan Plateau. The black area is the nonyielding, elastic core of the lithosphere. The linear, brittle failure boundary is based on Byerlee's law (1977) and the Mohr circle. The nonlinear viscous yielding boundary is based on the nonlinear viscous constitutive law (Turcotte and Schubert, 1982). The comprehensive formulation of this yield strength envelope may be found in Section 5 of Chapter 2 in Jin and Jiang (2002).

TWO MODES OF DEFORMATION FOR THE TIBETAN LITHOSPHERE

The mode of deformation of the Tibetan lithosphere has been proposed to be partitioned into two parts: deformation of the strong, brittle upper crust and deformation of the strong, elastic upper mantle lithosphere (Burchfiel and Royden, 1991; Jin et al., 1994; Royden, 1996; Shen et al., 2001). In other words, crustal deformation is decoupled from deformation of the upper mantle lithosphere by the soft, ductile lower crust of Tibet. Figure 1 shows a good analog for this type of decoupling with two modes of deformation. Rheologically speaking, there exist two separate strong cores on the yield strength envelopes (Figs. 2, 6, and 9). If deformation of the weak, viscous zone in the middle part of the crust (Fig. 1) or the middle part of the lithosphere of Tibet (Figs. 2, 6, and 9) is considered, then three modes of deformation should be defined instead of two.

The base of the strong but brittle upper crust (or the upper and middle crusts) of Tibet is ~30–35 km deep based on earthquake events (Fig. 8; Table 1) and rheological modeling with a simple thermal model (Figs. 2, 6, and 9). With an average elevation of 4–5 km over the Tibetan Plateau, it is straightforward to imagine a giant mass block of >2,000,000 km^2 with a semi-flat top and a semi-flat bottom moving in response to the convergence between India and Eurasia. The ease with which the giant mass block moves around on the viscous, weak lower crust depends on the extent that the viscous or ductile zone is developed. If the base of the strong upper crust of Tibet is approximately horizontal and ~35 km deep (Figs. 2, 6, 8, and 9), the isopach of the weak zone is defined by the Moho depth minus 35 km. The thicker the isopach is, the easier it is for the giant mass block to move. Also, the giant mass block is not rigid. It can adjust its geometry through internal brittle deformation (Fig. 8 or Fig. 1 as an analog).

The thickness of the viscous isopach is critical to the movement of the giant mass above, and the depth of the Moho is a critical measurement of the thickness of the isopach. In the center of Tibet, the Moho is between 65 and 75 km deep relative to sea level (Fig. 4). A ~30-km-thick, well-developed weak zone exists to allow the central piece of the giant mass to move with ease. However, the rapid transition of the Moho depth from 60 km to ~35 km at the northern edge of the Tibetan Plateau (Jiang et al., 2004) and from 50 to 35 km at the eastern edge of the Tibetan Plateau (Jiang and Jin, 2005) largely reduces the isopach thickness of the weak zone and hence hinders the freedom of movement of the giant mass of the Tibetan upper crust.

In contrast, the crustal mass near the eastern Himalayan syntaxis of the Tibetan Plateau (Fig. 3) is kept at a relatively high elevation at present. If the area is in local isostasy, the Moho is certainly deep enough to form a thick, viscous zone in the lower crust to carry the giant mass of the central Tibetan Plateau with high gravity potential (Molnar, 1988) to flow toward the south and southeast (Chen et al., 2000; Burchfiel, 2006; Cook, 2006). A seismic transect (white dash line in Fig. 3) from India to the Sichuan Basin via Bangladesh and Burma, integrating a survey

TABLE 1. SMALL TO LARGER EARTHQUAKE EVENTS THAT OCCURRED OVER THE EASTERN TIBETAN PLATEAU AND THE WESTERN SICHUAN BASIN SINCE EARLY 1980

Date	Time	Latitude (°N)	Longitude (°E)	Depth (km)	Magnitude
04/07/81	17-03-17.0	30.8	103.25	10	3.0ML
05/06/81	10-12-15.2	31.17	102.63	15	3.8Ms
04/05/82	05-52-04.4	31.3	102.82	15	4.0Ms
10/30/82	23-01-27.7	31.12	102.41	15	3.1Ms
12/26/82	15-27-39.2	30.28	103.59	10	3.3Ms
03/19/83	06-06-36.2	30.69	103.48	18	4.7ML
05/01/84	07-48-27.2	32.35	101.85	18	3.1ML
06/23/85	21-11-31.4	31.24	103.24	10	3.2ML
09/02/88	08-12-17.3	31.69	102.2	20	3.3ML
02/17/89	13-23-07.5	32.27	101.73	17	3.4ML
03/01/89	13-00-33.0	31.5	102.49	13	5.5Mb
09/22/89	02-25-50.4	31.58	102.51	12	6.5Ms
12/08/89	04-10-44.1	31.54	102.65	20	3.1ML
04/07/90	21-27-33.5	31.46	102.58	13	3.3ML
04/22/90	20-02-11.0	30.63	103.35	10	3.4ML
05/03/90	22-40-13.9	31.61	102.32	21	4.0Ms
06/25/90	18-00-27.1	31.54	102.49	11	3.4ML
07/12/90	05-50-08.6	31.59	102.48	6	3.4ML
10/01/90	18-02-24.2	30.29	103.94	6	4.6Mb
10/06/90	23-47-55.4	30.68	103.36	8	3.4ML
11/06/90	08-36-46.3	34	104.49	11	3.3ML
02/18/91	09-06-21.5	31.66	102.53	20	5.4Mb
11/06/91	09-57-10.4	31.6	102.39	13	3.7ML
02/20/92	04-42-29.1	32.31	101.67	10	3.4ML
05/30/92	02-26-15.6	31.37	102.99	16	4.5Mb
03/02/93	13-36-41.3	30.51	103.56	18	4.0Ms
12/30/93	04-34-18.7	30.73	102.89	16	3.2ML
05/16/95	20-18-33.0	31.75	102.12	18	3.7ML
08/01/98	20-08-46.9	30.7	103.48	26	3.8ML
11/01/98	18-26-26.8	30.82	103.12	11	3.0ML
06/13/99	03-34-42.3	31.23	103.26	15	3.1ML
06/16/99	04-10-35.2	30.94	102.8	19	2.8ML
09/08/99	01-06-40.3	30.47	103.82	4	3.9ML
11/12/99	05-34-11.8	30.66	103.7	16	3.6ML
11/16/99	12-27-10.3	30.21	103.83	10	3.6ML
10/11/99	20-07-59.5	30.81	103.77	21	2.9ML
01/28/01	20-14-38.6	31.09	103.24	7	3.2ML
02/07/01	16-20-35.4	30.35	103.51	16	2.9ML
09/22/01	22-34-51.4	32.01	101.97	33	4.3Mb
09/24/01	00-18-03.7	30.68	103.65	25	3.0ML
01/26/02	01-14-13.8	31.45	103.22	16	4.0ML
05/10/03	15-33-50.3	31.65	102.48	14	3.1ML
07/12/78	21-49-57.4	32	103.2	23	5.4Ms
10/19/81	11-38-09.7	32.97	102.15	10	3.5Ms
11/07/81	10-54-36.5	31.26	104	15	4.7Ms
02/19/82	09-51-18.9	31.08	104.48	15	3.7Ms
09/21/83	10-58-52.1	31.01	104.01	19	4.0ML
08/01/84	09-03-02.8	31.94	103.19	33	3.6ML
11/30/84	06-19-54.7	31.42	103.99	33	3.2ML
01/14/85	22-52-24.8	31.99	102.94	11	3.4ML
02/01/85	09-31-43.3	31.98	103.39	14	3.7ML
04/26/95	20-11-20.4	32.62	102.71	19	4.0Ms
08/07/85	21-34-04.2	31.66	103.92	7	3.2ML
01/30/87	12-16-39.1	31.93	103.5	14	2.9ML
05/15/87	03-41-27.8	31.7	103.57	16	4.1ML
10/10/87	12-26-59.8	32.64	102.63	12	3.3ML
10/16/87	15-22-03.7	31.72	103.89	4	3.6ML
12/27/87	19-18-10.9	31.8	103.78	11	4.0ML
12/28/87	21-00-01.5	31.76	103.8	14	3.8Ms
07/25/88	13-11-32.1	31.26	103.9	12	3.2ML
10/05/88	02-30-53.7	31.1	104.52	10	3.6ML
01/24/89	18-07-25.0	31.22	103.86	11	3.9ML
03/27/89	17-20-55.7	31.48	103.97	11	3.5ML
04/10/89	15-21-13.6	31.53	103.57	14	3.2ML
06/20/89	19-53-50.9	31.32	104.28	3	3.1ML
07/11/89	11-32-34.4	31.81	104.44	19	4.1Ms

(*continued*)

TABLE 1. SMALL TO LARGER EARTHQUAKE EVENTS THAT OCCURRED OVER THE EASTERN TIBETAN PLATEAU AND THE WESTERN SICHUAN BASIN SINCE EARLY 1980 (*continued*)

Date	Time	Latitude (°N)	Longitude (°E)	Depth (km)	Magnitude
09/08/91	21-23-29.7	31.01	103.71	12	3.2ML
01/09/92	00-08-05.0	31.78	103.31	5	3.3ML
03/30/93	09-18-10.9	31.18	103.93	20	3.3ML
04/22/93	07-32-56.1	31.59	103.36	24	3.1ML
07/11/93	15-15-23.3	31.45	103.98	10	4.5Ms
09/30/93	08-04-53.0	31.29	104.3	9	4.6Mb
06/23/95	11-51-29.8	32.02	102.88	27	3.4ML
09/09/95	01-19-24.6	31.31	104.08	10	4.2ML
10/31/95	05-07-55.2	31.94	103.56	6	3.0ML
01/25/96	02-27-35.5	31.53	103.23	25	3.2ML
05/09/96	15-03-40.1	31.12	103.76	17	3.5ML
10/18/96	12-28-46.2	31.71	103.19	24	3.4ML
02/17/97	19-03-07.7	30.92	104.29	14	3.0ML
06/13/97	13-38-53.6	31.03	104.17	17	2.8ML
10/12/97	16-59-36.9	31.26	103.5	14	2.9ML
04/17/98	12-58-02.9	31.78	103.91	17	3.0ML
04/17/98	13-24-02.0	31.77	103.88	23	2.9ML
02/01/99	02-19-23.5	31.29	103.55	15	3.3ML
03/23/99	02-47-58.4	31.18	104.41	16	3.4ML
09/23/99	07-57-08.9	31.21	103.88	21	3.6ML
02/04/00	16-29-47.5	31.27	103.88	15	2.8ML
02/15/00	10-17-02.5	31.27	103.85	15	2.9ML
04/21/00	17-36-27.2	31.06	104.47	20	2.9ML
11/04/00	21-41-06.8	31.52	103.24	8	3.5Ms
01/21/01	00-16-35.3	31.21	103.84	9	2.9ML
01/21/01	05-08-47.9	31.07	103.88	8	3.4ML
07/18/01	12-11-13.7	31.22	104.09	29	3.2ML
07/29/01	01-22-20.5	31.58	103.14	8	3.6ML
08/09/01	08-16-47.1	31.51	103.97	9	2.6ML
08/04/02	12-29-08.8	31.59	103.45	12	3.5ML
09/16/02	20-47-46.0	32.52	102.7	24	3.9Ms
11/16/02	16-18-03.3	31.01	103.78	23	3.3ML
05/05/03	19-00-02.3	31.19	104.05	13	3.1ML
05/11/79	17-30-54.0	32.09	104.56	0	4.0Ms
05/17/79	19-17-38.9	32.85	103.89	0	3.2Ms
05/21/79	21-32-02.3	32.56	104.21	0	4.3Ms
11/11/79	18-57-47.8	32.79	104.14	0	3.8Ms
04/06/80	23-53-59.5	32.46	104.25	10	3.3Ms
06/12/80	23-52-12.9	32.94	103.94	0	4.0Ms
09/01/81	01-22-54.0	32.7	104	15	3.7Ms
09/18/81	13-14-23.2	32.83	104	10	3.6ML
02/14/82	13-29-56.5	32.94	103.99	0	2.6Ms
03/21/82	11-37-00.0	32.54	104.33	0	3.2Ms
02/21/84	02-14-58.2	32.72	104.21	0	3.1ML
03/16/85	14-59-28.5	32.77	103.98	6	3.6ML
06/09/85	22-12-22.2	33.03	103.98	6	3.1ML
01/04/86	10-52-52.0	32.78	104.31	10	3.8ML
04/22/86	00-27-25.7	32.71	104.24	11	3.5ML
09/14/86	02-30-28.8	32.89	103.83	0	3.1ML
11/01/86	20-34-02.3	33.43	103.41	31	4.3ML
12/03/86	23-18-21.8	32.42	104.52	13	4.2ML
04/27/87	13-56-22.9	33.25	103.63	14	2.7ML
06/05/87	20-11-04.7	32.6	104.16	5	2.8ML
04/13/87	15-05-14.5	32.75	104.02	15	3.3ML
03/23/88	09-56-21.3	32.74	104.15	17	3.2ML
06/10/88	12-15-38.6	32.48	104.35	9	4.0ML
11/19/88	14-47-33.6	33.04	103.95	8	3.8ML
12/04/91	03-10-27.6	33.18	103.75	16	3.5ML
04/02/92	09-21-08.7	32.96	103.95	9	4.2ML
01/05/93	16-38-44.6	32.05	104.77	14	3.4ML
10/04/93	07-49-50.7	32.45	104.27	7	3.9ML
03/06/94	21-42-15.4	32.53	104.36	12	3.7ML
09/04/94	20-13-37.9	32.31	104.57	26	4.5Mb
09/05/94	00-17-02.9	32.18	104.6	25	3.7ML
09/16/94	04-38-53.7	32.19	104.76	23	4.2Ms
05/08/95	18-37-32.3	33.25	103.5	28	5.0Ms

(*continued*)

TABLE 1. SMALL TO LARGER EARTHQUAKE EVENTS THAT OCCURRED OVER THE EASTERN TIBETAN PLATEAU AND THE WESTERN SICHUAN BASIN SINCE EARLY 1980 (*continued*)

Date	Time	Latitude (°N)	Longitude (°E)	Depth (km)	Magnitude
02/19/96	08-50-01.8	32.6	104.26	9	3.0ML
03/27/96	19-37-27.7	32.71	104.04	5	3.3ML
04/23/96	13-35-28.0	32.52	104.47	5	4.0Ms
03/04/98	00-15-26.3	32.83	104.06	13	2.9ML
05/12/98	09-37-19.2	33.29	103.71	30	3.4ML
07/14/98	14-44-32.6	32.12	104.56	25	3.6ML
09/14/99	19-03-46.4	32.82	104.07	14	3.1ML
10/26/99	09-06-47.3	32.28	104.52	9	3.1ML
11/16/00	03-08-25.4	33.13	103.87	23	3.2ML
06/16/01	10-16-05.4	33.09	103.85	20	3.0ML
10/26/01	01-08-38.6	33.04	103.8	15	3.2ML
05/29/03	11-24-51.3	32.69	104.25	15	3.4ML

Note: The seismic data were collected by China University of Oceanography through Science Academy of China. ML—local magnitude; Ms—surface wave magnitude; Mb—body wave magnitude.

of gravity, structural, and sedimentary geologies as well as a passive seismic survey, such as the one Langin et al. (2003) published, should be conducted in order to understand the crustal deformation, the possible lower crustal flow, the Moho compensation style, and the strength of the lithosphere to the east of the eastern Himalayan syntaxis. The seismic imaging technology from acquisition to processing in the mountainous area has been improved significantly in recent years, and the author is currently designing a seismic survey in a mountainous area with a relief from 200 m to 1400 m. If financial support permits, the authors could conduct an active seismic survey through the transect from India to the Sichuan Basin.

CURRENT STATUS OF MECHANICAL STRENGTH MODELING OF THE LITHOSPHERE OVER THE TIBETAN PLATEAU AND ITS VICINITY

Mechanical modeling of the elastic strength of the Tibetan lithosphere and its vicinity can be dated back to the early work by Lyon-Caen and Molnar in 1983 and 1984, and also Royden's work in 1993. Jin et al. (1994) proposed a two-mode deformation model for the Tibetan lithosphere based on flexural modeling using spatially distributed gravity data modeled in the spectral domain (the rectangular boxes in Fig. 10). Jin et al. (1996) proposed the location of the mantle suture (Figs. 4 and 5) and the mantle front of the subducting Indian lithosphere based on 2-D flexural modeling using the integrated gravity data sets from South Asia and China (six red lines in Fig. 10).

In 1997 and 1998, Massachusetts Institute of Technology (MIT), China University of Oceanography, and the Bureau of Geophysical Prospecting of China National Petroleum Corporation started a campaign of two gravity surveys across the northern boundary of the Tibetan Plateau (Jiang et al., 2004). The first survey across the Altyn Tagh was conducted in 1997, and the second one across the West Kunlun was conducted in 1998. The flexural modeling based on gravity data from the first survey further suggested that the Altyn Tagh fault is a lithospheric-scale fault (Wittlinger et al., 1998; Jiang et al., 2004). The fault is not simply constrained within the crust because the weak, ductile zone is poorly developed (Fig. 6).

Most of the mechanical modeling of the lithosphere over the Tibetan Plateau and its vicinity is presented on 2-D cross sections (Lyon-Caen and Molnar, 1983; Lyon-Caen and Molnar, 1984; Royden, 1993; Jin et al., 1996; Jiang et al., 2004; Jiang and Jin, 2005). The disadvantage of the 2-D flexural model on a cross section is that the model ignores the fact that the lithosphere is a three-dimensional deflected surface under deformation. A good analog for the deformed surface of the lithosphere under tectonic stressing in geological history due the convergence between India and Eurasia is the Central Asia theoretical Moho depth based on 3-D flexural modeling in the spatial domain (Fig. 11) (Jin and Jiang, 2002). This figure shows that there is a large limitation in 2-D flexural modeling on an undulating 3-D surface.

The coherence technique (Forsyth, 1985) can measure the spatial variation of the mechanical strength of the lithosphere, and it is a good supplement to flexural modeling on a 2-D cross section. The technique also takes subsurface loading (McNutt, 1983) into consideration. However, the coherence technique is conducted in the spectral domain, where a rectangular box of gravity and topographic data is required for Fourier transforms (white boxes in Fig. 10). In each box, the flexural rigidity or the elastic thickness (Te) of the modeled lithosphere needs to be a constant. The mechanical strength of a tectonic unit in the lithosphere is dependent on its tectonic history, and no tectonic unit tends to be rectangular.

In order to address the limitation of 2-D flexural modeling on a cross section and the limitation of the coherence technique, Jin (1997) and Jin and Jiang (2002) calculated the spatial variation of the mechanical strength of the lithosphere over the Tibetan Plateau and its vicinity in the spatial domain by solving the governing equation of plate flexure (Equation 3.15 of Jin, 1997; Equation 4.45 of Jin and Jiang, 2002) using finite difference. Figure 12 shows the spatial strength (Te) variation of the lithosphere over the Tibetan Plateau and its vicinity. To some extent, the strength variation is consistent with the tectonic units in the area. The Indian plate is strong, with a Te value of ~80 km. The central Tibetan Plateau and overall Central Asia have an intermediate plate thickness of ~40 km. In the active mountain-building areas of Tian Shan, Qilian Shan, the northern and southern boundaries of the Tibetan Plateau, and the eastern Himalayan syntaxis, the lithosphere is weak, with a Te value around 20 km. Nevertheless, the variation in Te values needs to be further constrained by the results from the 2-D flexural modeling on the cross sections showed in Figure 10.

DISCUSSION AND CONCLUSION

The Tibetan Plateau has a unique average elevation over 4 km and a crustal thickness almost double the normal crustal thickness. The gravity potential energy of this elevated giant mass tends to drive the plateau to flow from the highland toward

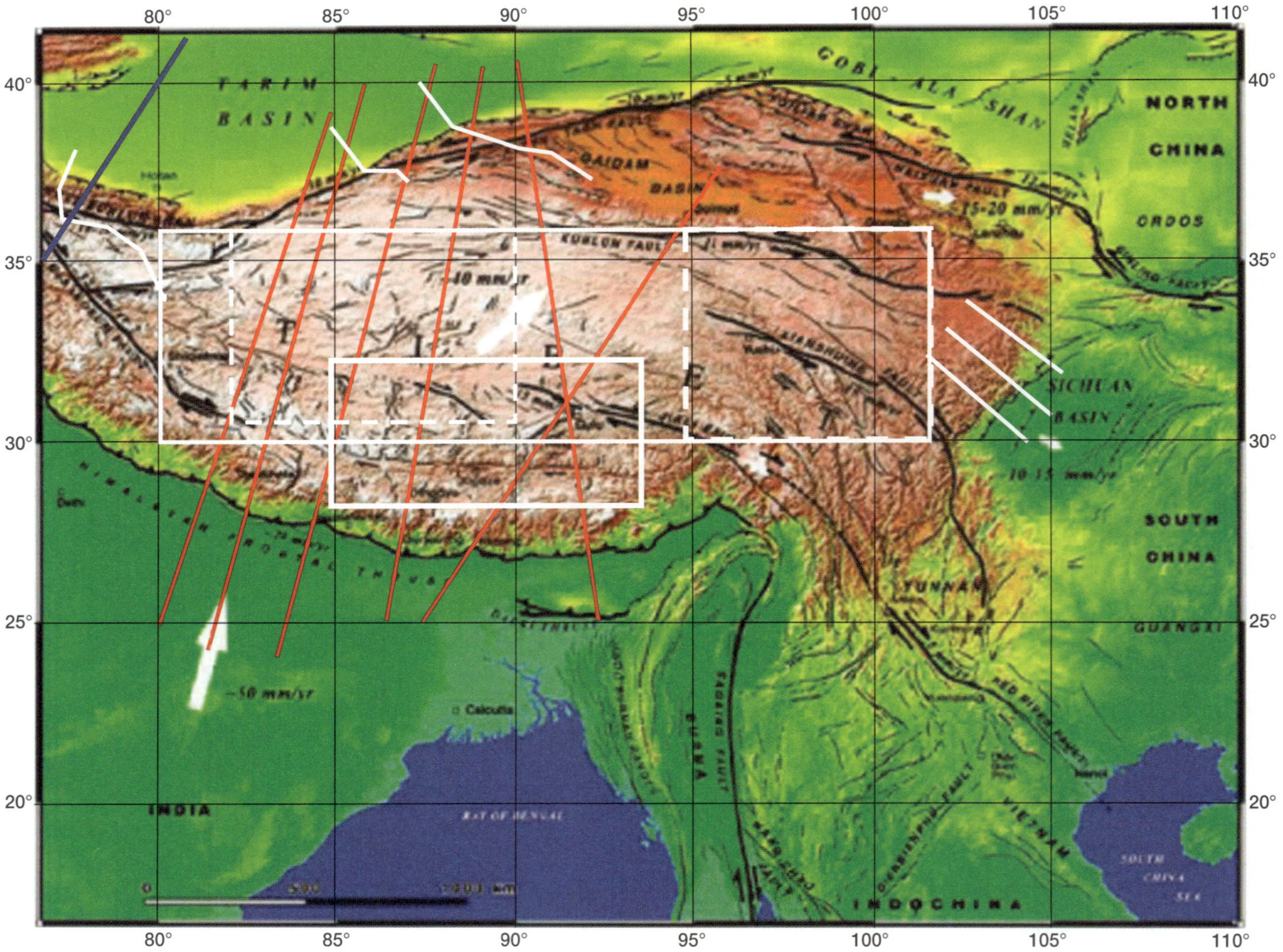

Figure 10. Summary map of flexural modeling for the lithosphere of the Tibetan Plateau and vicinity. The white rectangular boxes are the areas for the coherence modeling based on 5′ × 5′ Bouguer gravity data from China (Jin et al., 1994). The red transects are the two-dimensional (2-D) flexural modeling based on gravity data from South Asia, including Pakistan, India, Bangladesh, and Indochina from Lamont-Doherty Earth Observatory (LDEO) of Columbia University and the 5′ × 5′ gravity from China (Jin et al., 1996). The three white survey paths from the Tarim Basin to the northern Tibetan Plateau show the field gravity survey work conducted by Massachusetts Institute of Technology (MIT), Ocean University of China (OUC), and Bureau of Geophysical Prospecting of China National Petroleum Corporation (Jiang et al., 2004). The three white transects across the eastern edge of the Tibetan Plateau show the flexural modeling work based on the 5′ × 5′ gravity grid from China (Jiang and Jin, 2005). The blue transect across West Kunlun is from Lyon-Caen and Molnar (1984).

its surrounding lowlands. However, whether the giant mass flows or not depends on the buildup of the weak lower crust of Tibet. In other words, a thicker and weaker lower crust will make the stronger upper crust flow easier and vice versa.

Rheological modeling using a simple thermal structure for the Tibetan lithosphere (Figs. 2, 6, and 9) shows that the Tibetan Plateau should have a strong upper crust (or upper and middle crusts) of 30–35 km. In reality, the thermal structure of the Tibetan lithosphere is more complicated than the simple one-dimensional steady-state thermal model as shown in Figure 2 of Wei et al. (2001). The simple thermal model used here is just the lower threshold of the thermal structure for the Tibetan lithosphere. The more thermally active the plateau is, the more possible it is for lithospheric decoupling to occur. The thickness of the weak lower crust is the difference between the base of the upper crust and the Moho. A deeper Moho results in a weaker lower crust, which makes it easier for the upper crust to flow relative to the strong upper mantle lithosphere. The Moho depth variation can be obtained through 2-D (cross section), 3-D (spatial surface) flexural modeling, and deep seismic sounding (Li et al., 2006). Therefore, results from the flexural modeling indicate the way in which the upper crust flows in the Tibetan Plateau.

Flexural modeling plays a key role in defining the crustal flow of the Tibetan Plateau because the Moho subsurface

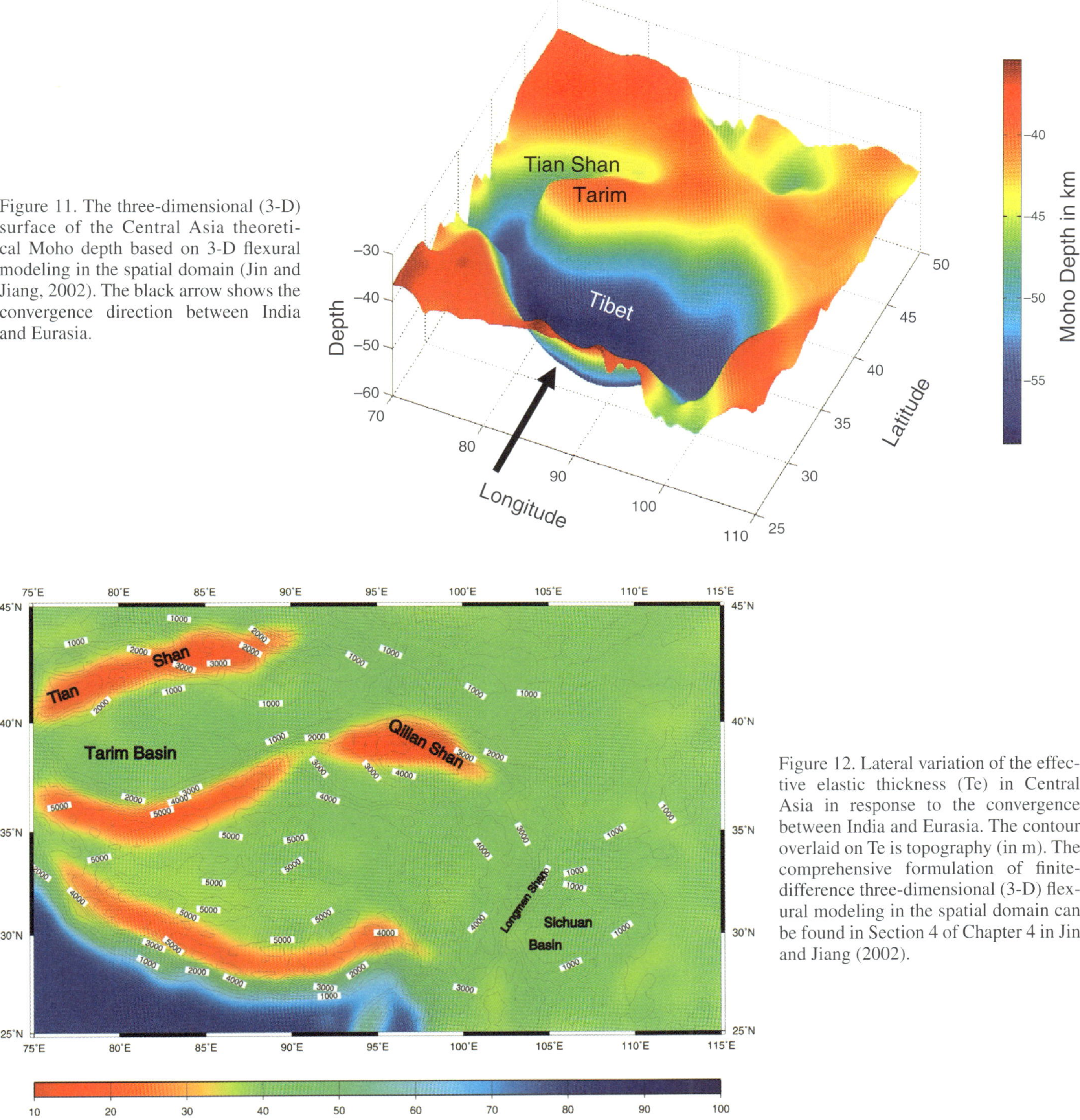

Figure 11. The three-dimensional (3-D) surface of the Central Asia theoretical Moho depth based on 3-D flexural modeling in the spatial domain (Jin and Jiang, 2002). The black arrow shows the convergence direction between India and Eurasia.

Figure 12. Lateral variation of the effective elastic thickness (Te) in Central Asia in response to the convergence between India and Eurasia. The contour overlaid on Te is topography (in m). The comprehensive formulation of finite-difference three-dimensional (3-D) flexural modeling in the spatial domain can be found in Section 4 of Chapter 4 in Jin and Jiang (2002).

denotes the lower boundary of the weak, ductile lower crust on which the upper crust flows. Figure 13 shows the Moho depth variation calculated from finite-difference 3-D flexural modeling in the spatial domain based on observed gravity and topographic data (Jin and Jiang, 2002). However, Moho geometry derived from 3-D modeling needs to be constrained by the results from the 2-D flexural modeling on the transects shown in Figure 10, and it also needs to be constrained in some key locations, such as the eastern and western Himalayan syntaxes and Qilian Shan, where no 2-D flexural modeling work has ever been conducted. Other independent work to constrain the Moho depth may include deep seismic sounding (DSS). Profiling from DSS can image the Moho refraction and wide-angle reflection with considerable accuracy (Li et al., 2006).

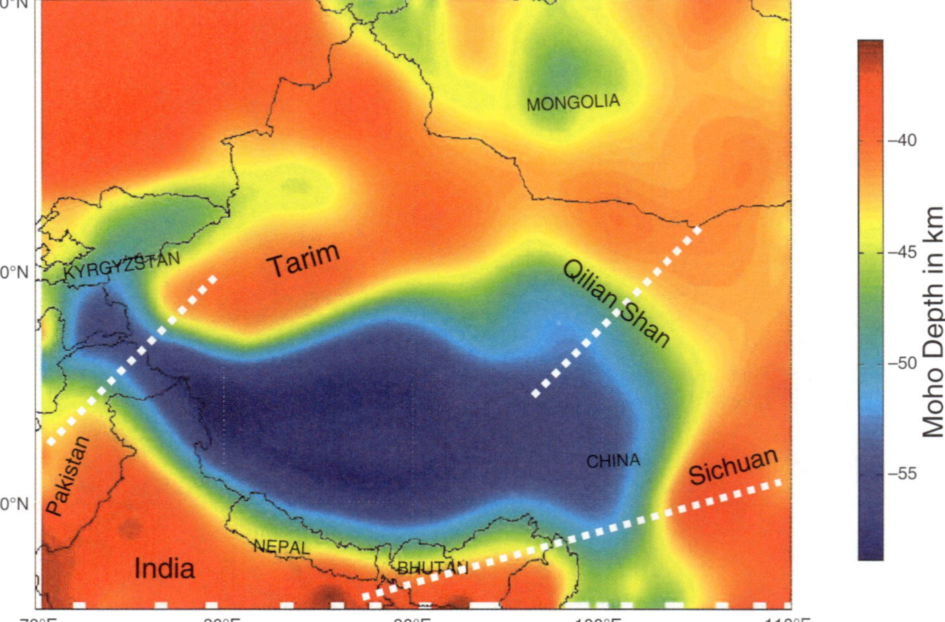

Figure 13. The Central Asia Moho depth map calculated from finite-difference three-dimensional (3-D) flexural modeling in the spatial domain (Jin and Jiang, 2002). The three white dashed lines are proposed transects of future 2-D flexural modeling to constrain the 3-D result.

One lesson from the gravity surveys of MIT and CUO in 1997 and 1998 (Jiang et al., 2004) was the lack of structural geologists or sedimentologists among the survey team, which resulted in no multidiscipline check between crustal geology and geophysical flexural modeling on the lithosphere along the survey lines. An integrated group composed of a structural geologist, a sedimentologist, a gravity survey team, a DSS team, and a passive seismic survey team should be formed if the proposed survey transects (Fig. 13) from India to Sichuan, from Pakistan to Tarim, and across Qilian Shan are to be conducted.

ACKNOWLEDGMENTS

We thank David Rowley for the significant discussion on the rebound of the oil maturity threshold of the Sichuan Basin during Cenozoic. We thank Walter Mooney and Dogan Seber for their comprehensive review of the paper. We also thank Xiaodian Jiang's team at China University of Oceanography for preparing the local seismic information in the eastern Tibetan Plateau. This research was supported by Institute of Geology and Geophysics, China Academy of Sciences, China, and China University of Oceanography.

REFERENCES CITED

Alsdorf, D., Makovsky, Y., Zhao, W., Rown, L.D., Nelson, K.D., Klemperer, S., Hauck, M., Ross, A., Cogan, M., Clark, M., Che, J., and Kuo, J., 1998, INDEPTH (International Deep Profiling of Tibet and the Himalayas) multi-channel seismic reflection data: Description and availability: Journal of Geophysical Research, v. 103, p. 26,993–26,999, doi: 10.1029/98JB01078.

Bechtel, T.D., Forsyth, D.W., and Swain, C.J., 1987, Mechanism of isostatic compensation in the vicinity of the East African Rift, Kenya: Geophysical Journal of the Royal Astronomical Society, v. 90, p. 445–465.

Bechtel, T.D., Forsyth, D.W., Sharpton, V.L., and Grieve, R.A., 1990, Variations in effective elastic thickness of the North American lithosphere: Nature, v. 343, p. 636–638, doi: 10.1038/343636a0.

Bendick, R., Biham, R., Freymueller, J., and Larson, K., 2000, Geodetic evidence for a slow slip rate in the Altyn Tagh system: Nature, v. 404, p. 69–72, doi: 10.1038/35003555.

Braitenberg, C., Wang, Y., Fang, J., and Hsu, H.T., 2003, Spatial variations of flexure parameters over the Tibet-Qinghai plateau: Earth and Planetary Science Letters, v. 205, p. 211–224, doi: 10.1016/S0012-821X(02)01042-7.

Buck, W.R., 1988, Flexural rotation of normal faults: Tectonics, v. 7, p. 7959–7973.

Buck, W.R., 2001, Accretional curvature of lithosphere at magmatic spreading centers and flexural support of axial highs: Journal of Geophysical Research, v. 106, no. B3, p. 3953, doi: 10.1029/2000JB900360.

Burchfiel, C., 2006, Early Cenozoic tectonics of the eastern Himalayan syntaxis, *in* Abstracts of the International Conference on Continental Dynamics and Environmental Change of the Tibetan Plateau: Xining, Qinghai, China, China Academy of Sciences and Geological Society of America.

Burchfiel, B.C., and Royden, L.H., 1991, Tectonics of Asia 50 years after the death of Emile Argand: Eclogae Geologicae Helvetiae, v. 84, no. 3, p. 599–629.

Burchfiel, C., Chen, Z., Liu, Y., and Royden, L.H., 1995, Tectonics of the Longmen Shan and adjacent regions: International Geology Review, v. 37, p. 661–735.

Burov, E.B., and Molnar, P., 1998, Gravity anomalies over the Ferghana valley (Central Asia) and intracontinental deformation: Journal of Geophysical Research, v. 103, p. 18,137–18,152, doi: 10.1029/98JB01079.

Byerlee, J.D., 1977, Friction of rocks, *in* Everden, J.F., ed., Experimental Studies of Rock Friction with Application to Earthquake Prediction: Menlo Park, California, U.S. Geological Survey, p. 55–77.

Chen, Z., Burchfiel, B.C., Liu, Y., King, R.W., Royden, L.H., Tang, W., Wang, E., Zhao, J., and Zhang, X., 2000, Global positioning system measurements from eastern Tibet and their implications for India/Eurasia intercontinental deformation: Journal of Geophysical Research, v. 105, no. B7, p. 16,215–16,228, doi: 10.1029/2000JB900092.

Cook, K., 2006, The role of lower crustal flow and lateral variations of crustal strength in shaping the eastern margin of Tibet, *in* Abstracts of the International Conference on Continental Dynamics and Environmental Change of the Tibetan Plateau: Xining, Qinghai, China, Chinese Academy of Sciences and Geological Society of America.

Dewey, J.F., and Burke, K., 1973, Tibetan, Variscan and Precambrian basement reactivation: Products of continental collision: The Journal of Geology, v. 81, p. 683–692.

Dorman, L.M., and Lewis, B.T.R., 1970, Experimental isostasy: Theory of the determination of the Earth's isostatic response to a concentrated load: Journal of Geophysical Research, v. 75, p. 3357–3365, doi: 10.1029/JB075i017p03357.

England, P., and Houseman, G., 1986, Finite strain calculations of continental deformation: 2. Comparison with the India-Asia collision: Journal of Geophysical Research, v. 91, p. 3664–3676, doi: 10.1029/JB091iB03p03664.

England, P., and McKenzie, D.P., 1982, A thin viscous sheet model for continental deformation: Geophysical Journal of the Royal Astronomical Society, v. 70, p. 295–321.

Filmer, P.E., McNutt, M.K., and Wolfe, C.J., 1993, Elastic thickness of the lithosphere in the Marquesas and Society Islands: Journal of Geophysical Research, v. 98, p. 19,565–19,577, doi: 10.1029/93JB01720.

Forsyth, D.W., 1985, Subsurface loading and estimates of the flexural rigidity of continental lithosphere: Journal of Geophysical Research, v. 90, p. 12,623–12,632, doi: 10.1029/JB090iB14p12623.

Gail, C., and McNutt, M.K., 1992, Geophysical constrains on the state of stress along the Marquesas Fracture Zone: Journal of Geophysical Research, v. 97, p. 4425–4438.

Goetze, C., and Evans, B., 1979, Stress and temperature in the bending lithosphere as constrained by experimental rock mechanics: Geophysical Journal of the Royal Astronomical Society, v. 58, p. 463–478.

Jiang, X.D., and Jin, Y., 2005, Mapping the deep lithospheric structure beneath the eastern margin of the Tibetan Plateau from gravity anomalies: Journal of Geophysical Research, v. 110, p. B07407, doi: 10.1029/2004JB003394.

Jiang, X.D., Jin, Y., and McNutt, M.K., 2004, Lithospheric deformation beneath the Altyn Tagh and West Kunlun faults from recent gravity survey: Journal of Geophysical Research, v. 109, p. B05406, doi: 10.1029/2003JB002444.

Jin, Y., 1997, State-of-stress and rheology of Tibet and its vicinity from gravity anomalies and numerical models [PhD Thesis]: Cambridge, Massachusetts Institute of Technology.

Jin, Y., and Jiang, X.D., 2002, Lithosphere Dynamics: Beijing, China, Publishing House of Sciences, 231 p.

Jin, Y., McNutt, M.K., and Zhou, Y.S., 1994, Evidence from gravity and topography data for folding of Tibet: Nature, v. 371, p. 669–674, doi: 10.1038/371669a0.

Jin, Y., McNutt, M.K., and Zhou, Y.S., 1996, Mapping the descent of Indian and Eurasian plates beneath the Tibetan Plateau from gravity anomalies: Journal of Geophysical Research, v. 101, p. 11,275–11,290, doi: 10.1029/96JB00531.

Kirby, S.H., 1980, Tectonic stresses in the lithosphere: Constraints provided by the experimental deformation of the rocks: Journal of Geophysical Research, v. 85, p. 6353–6363, doi: 10.1029/JB085iB11p06353.

Kruse, S., McNutt, M.K., Phipps-Morgan, J., Royden, L., and Wernicke, B., 1991, Lithospheric extension near Lake Mead, Nevada: A model for ductile flow in the lower crust: Journal of Geophysical Research, v. 96, p. 4435–4456, doi: 10.1029/90JB02621.

Kumar, P., Yuan, X., Kind, R., and Ni, J., 2006, Imaging the colliding Indian and Asian lithospheric plates beneath Tibet: Journal of Geophysical Research, v. 111, p. B06308, doi: 10.1029/2005JB003930.

Langin, W., Brown, L.D., and Sandvol, E.A., 2003, Seismicity of Central Tibet from Project INDEPTH III seismic recordings: Bulletin of the Seismological Society of America, v. 93, p. 2146–2159, doi: 10.1785/0120030004.

Li, S., Mooney, W.D., and Fan, J., 2006, Crustal structure of mainland China from deep seismic sounding data: Tectonophysics, v. 420, p. 239–252, doi: 10.1016/j.tecto.2006.01.026.

Lorenzo, J.M., O'Brien, G.W., Stewart, J., and Jandon, K., 1998, Inelastic yielding and forebulge shape across a modern foreland basin: Northwest shelf of Australia, Timor Sea: Geophysical Research Letters, v. 25, p. 1455–1458.

Lowry, A.R., and Smith, R.B., 1994, Flexural rigidity of the Basin and Range–Colorado Plateau–Rocky Mountain transition from coherence analysis of gravity and topography: Journal of Geophysical Research, v. 99, p. 20,123–20,140, doi: 10.1029/94JB00960.

Lyon-Caen, H., and Molnar, P., 1983, Constraints on the structure of the Himalayas from an analysis of gravity anomalies and a flexural model of the lithosphere: Journal of Geophysical Research, v. 88, p. 8171–8192, doi: 10.1029/JB088iB10p08171.

Lyon-Caen, H., and Molnar, P., 1984, Gravity anomalies and the structure of western Tibet and the southern Tarim Basin: Geophysical Research Letters, v. 11, p. 1251–1254, doi: 10.1029/GL011i012p01251.

Ma, Y.S., Guo, X., Guo, T., Huang, R., Cai, X., and Li, G., 2007, The Puguang gas field: New giant gas discovery in the mature Sichuan Basin, southwest China: American Association of Petroleum Geologists (AAPG) Bulletin, v. 91, p. 627–643, doi: 10.1306/11030606062.

Macario, A., Malinverno, A., and Haxby, W.F., 1995, On the robustness of elastic thickness estimates obtained using the coherence methods: Journal of Geophysical Research, v. 100, p. 15,163–15,172, doi: 10.1029/95JB00980.

Makovsky, Y., Klemperer, S.L., Liyan, H., and Deyuan, L., 1996, Structural elements of the southern Tethyan Himalaya crust from wide-angle seismic data: Tectonics, v. 15, no. 5, p. 997–1005, doi: 10.1029/96TC00310.

Makovsky, Y., Klemperer, S.L., Ralschbacher, L., and Alsdorf, D., 1999, Mid-crustal reflector on INDEPTH wide angle profiles: An ophiolitic slab beneath the India-Asia suture in southern Tibet: Tectonics, v. 18, p. 793–808, doi: 10.1029/1999TC900022.

McKenzie, D.P., 1978, Some remarks on the development of sedimentary basins: Earth and Planetary Science Letters, v. 40, p. 25–32, doi: 10.1016/0012-821X(78)90071-7.

McNutt, M.K., 1983, Influence of plate subduction on isostatic compensation in Northern California: Tectonics, v. 2, p. 399–415, doi: 10.1029/TC002i004p00399.

McNutt, M.K., 1984, Lithosphere flexure and thermal anomalies: Journal of Geophysical Research, v. 89, p. 11,180–11,194, doi: 10.1029/JB089iB13p11180.

McNutt, M.K., and Menard, H.W., 1982, Constraints on yield strength in the oceanic lithosphere derived from observations of flexure: Geophysical Journal of the Royal Astronomical Society, v. 71, p. 363–394.

McNutt, M.K., and Parker, R.L., 1978, Isostasy in Australia and the evolution of the compensation mechanism: Science, v. 199, p. 773–775, doi: 10.1126/science.199.4330.773.

McNutt, M.K., Diament, M., and Kogan, M.G., 1988, Variation of elastic plate thickness at continental thrust belts: Journal of Geophysical Research, v. 93, p. 8825–8838, doi: 10.1029/JB093iB08p08825.

Molnar, P., 1988, A review of geophysical constraints on the deep structure of the Tibetan Plateau, the Himalayas and the Karakorum and their tectonics: Royal Society of London Philosophical Transactions, ser. A., v. 326, p. 33–88.

Molnar, P., England, P., and Martinod, J., 1993, Mantle dynamics, uplift of Tibetan Plateau, and the Indian Monsoon: Reviews of Geophysics, v. 31, p. 357–396, doi: 10.1029/93RG02030.

Nelson, K.D., Zhao, W., Brown, L.D., Kuo, J., Che, J., Liu, X., Klemperer, S.L., Makovsky, Y., Meissner, R., Mechie, J., Kind, R., Wenzel, F., Ni, J., Nabelek, J., Chen, L., Handong, T., Wenbo Wei, Jones, A.G., Booker, J., Unsworth, M., Kidd, W.S.F., Hauck, M., Alsdorf, D., Ross, A., Cogan, M., Wu, C., Sandvol, E., and Edwards, M., 1996, Partially molten middle crust beneath southern Tibet: Synthesis of project INDEPTH results: Science, v. 274, p. 1684–1688, doi: 10.1126/science.274.5293.1684.

Pan, G.T., Ding, J., Yao, D.S., and Wang, L.Q., 2004, Geological Map of Qinghai-Xizang Plateau and Adjacent Areas: Chengdu, China, Chengdu Cartographic Publishing House.

Peltzer, G., Tapponnier, P., and Armijo, R., 1989, Magnitude of late Quaternary left-lateral displacements along the northern edge of Tibet: Science, v. 246, p. 1285–1289, doi: 10.1126/science.246.4935.1285.

Rehbinder, G., and Yakubenko, P.A., 1999, Displacement and flexural stresses of a loaded elastic plate on a viscous liquid: Journal of Geophysical Research, v. 104, p. 10,827, doi: 10.1029/1998JB900065.

Royden, L.H., 1993, The tectonic expression of slab pull at continental convergent boundaries: Tectonics, v. 12, p. 303–325, doi: 10.1029/92TC02248.

Royden, L., 1996, Coupling and decoupling of crust and mantle in convergent orogens: Implications of strain partitioning in the crust: Journal of Geophysical Research, v. 101, p. 17,679–17,705, doi: 10.1029/96JB00951.

Royden, L., and Keen, C.E., 1980, Rifting processes and thermal evolution of the continental margin of eastern Canada determined from subsidence curves: Earth and Planetary Science Letters, v. 51, p. 343–361, doi: 10.1016/0012-821X(80)90216-2.

Scotese, C. R., 2006, Maps and Drift Animation: A PALEOMAP Production (computer animation).

Sheffels, B., and McNutt, M.K., 1986, Role of subsurface loads and regional compensation in the isostatic balance of the Transverse Range, California:

Evidence for intracontinental subduction: Journal of Geophysical Research, v. 91, p. 6419–6431, doi: 10.1029/JB091iB06p06419.

Shen, F., Royden, L.H., and Burchfiel, B.C., 2001, Large-scale crustal deformation of the Tibetan Plateau: Journal of Geophysical Research, v. 106, no. B4, p. 6793–6816, doi: 10.1029/2000JB900389.

Simons, F.J., and van der Hilst, R.D., 2002, Age-dependent seismic thickness and mechanical strength of the Australian lithosphere: Geophysical Research Letters, v. 29, p. 1529, doi: 10.1029/2002GL014962.

Simons, F.J., Zuber, M.T., and Korenaga, J., 2000, Isostatic response of the Australian lithosphere: Estimation of effective elastic thickness and anisotropy using multi-taper spectral analysis: Journal of Geophysical Research, v. 105, p. 19,163–19,184, doi: 10.1029/2000JB900157.

Stewart, J., and Watts, A.B., 1997, Gravity anomalies and spatial variation of flexural rigidity at mountain ranges: Journal of Geophysical Research, v. 102, no. B3, p. 5327–5352, doi: 10.1029/96JB03664.

Swain, C.J., and Kirby, J.F., 2006, An effective elastic thickness map of Australia from wavelet transforms of gravity and topography using Forsyth's method: Geophysical Research Letters, v. 33, no. 2, p. L02314, doi: 10.1029/2005GL025090.

Tapponier, P., Xu, Z., Roger, F., Meyer, B., Arnaud, N., Wittlinger, G., and Yang, J., 2001, Oblique stepwise rise and growth of the Tibet Plateau: Science, v. 294, 1671–1677, doi: 10.1126/science.105978.

Turcotte, D.L., and Schubert, G., 1982, Geodynamics: Application of Continuum Physics to Geological Problems: New York, John Wiley and Sons, p. 104–131.

Walcott, R.I., 1970, Flexural rigidity, thickness and viscosity of the lithosphere: Journal of Geophysical Research, v. 75, p. 3941–3954, doi: 10.1029/JB075i020p03941.

Watts, A.B., and Cochran, J.R., 1974, Gravity anomalies and flexure of the lithosphere along the Hawaiian-Emperor seamount chain: Geophysical Journal of the Royal Astronomical Society, v. 38, p. 119–141.

Watts, A.B., Bodine, J.H., and Steckler, M.S., 1980, Observations of flexure and the state of stress in the oceanic lithosphere: Journal of Geophysical Research, v. 85, p. 6369–6376, doi: 10.1029/JB085iB11p06369.

Watts, A.B., Karner, G.D., and Steckler, M.S., 1982, Lithosphere flexure and the evolution of sedimentary basins: Royal Society of London Philosophical Transactions, ser. A., v. 305, p. 249–281, doi: 10.1098/rsta.1982.0036.

Wei, W., Unsworth, M., Jones, A., Booker, J., Tan, H., Nelson, D., Chen, L., Li, S., Solon, K., Bedrosian, P., Jin, S., Deng, M., Ledo, J., Kay, D., and Boberts, B., 2001, Detection of widespread fluids in the Tibetan crust by magnetotelluric studies: Science, v. 292, p. 716–718, doi: 10.1126/science.1010580.

Wernicke, B., 1981, Low-angle normal faults in the Basin and Range Province—Nappe tectonics in an extending orogen: Nature, v. 291, p. 645–648, doi: 10.1038/291645a0.

Willett, S.D., and Beaumont, C., 1994, Subduction of Asian lithospheric mantle beneath Tibet inferred from models of continental collision: Nature, v. 369, p. 642–645, doi: 10.1038/369642a0.

Wittlinger, G., Tapponnier, P., Poupinet, G., Jiang Mei, Shi Danian, G. Herquuel, and Masson, F., 1998, Tomographic evidence for localized lithospheric shear along the Altyn Tagh fault: Science, v. 282, p. 74–76.

Wolfe, C.J., and McNutt, M.K., 1991, Compensation of Cretaceous seamounts of the Darwin Rise, northwest Pacific Ocean: Journal of Geophysical Research, v. 96, p. 2363–2374, doi: 10.1029/90JB01893.

Yuan, X., Ni, J., Kind, R., Mechie, J., and Sandvol, E., 1997, Lithospheric and upper mantle structure of southern Tibet from a seismological passive source experiment: Journal of Geophysical Research, v. 102, no. B12, p. 27,491–27,500, doi: 10.1029/97JB02379.

Zhao, W.J., Nelson, K.D., and Project INDEPTH Team, 1993, Deep seismic reflection evidence for continental underthrusting beneath southern Tibet: Nature, v. 366, p. 557–559, doi: 10.1038/366557a0.

Zhao, W.L., and Morgan, J., 1987, Injection of Indian crust into Tibetan lower crust: A two-dimensional finite element model study: Tectonics, v. 6, p. 489–504, doi: 10.1029/TC006i004p00489.

Zuber, M.T., Pamentier, E.M., and Fletcher, R.C., 1986, Extension of continental lithosphere: A model for two scales of Basin and Range deformation: Journal of Geophysical Research, v. 91, p. 4826–4838, doi: 10.1029/JB091iB05p04826.

Zuber, M.T., Bechtel, T.D., and Forsyth, D.W., 1989, Effective elastic thickness of the lithosphere and mechanisms of isostatic compensation in Australia: Journal of Geophysical Research, v. 94, p. 9353–9367, doi: 10.1029/JB094iB07p09353.

Manuscript Accepted by the Society 04 April 2008

Printed in the USA

The Geological Society of America
Special Paper 444
2008

Exhumation of old rocks during the Zagros collision in the northwestern part of the Zagros Mountains, Iran

Alireza Nadimi*
Geology Department, Payame Noor University of Isfahan, Kohandej, Isfahan, POB 81456-617, Iran

Hassan Nadimi
Physics Department, Payame Noor University of Isfahan, Kohandej, Isfahan, POB 81456-617, Iran

ABSTRACT

Arabia-Eurasia convergence is accommodated in the Zagros Mountains of southwestern Iran and in the seismic belts of the central Caspian, Alborz, and Kopeh Dagh of northern Iran. The Zagros is a NW-trending fold-and-thrust belt made up of a 6–15-km-thick sedimentary pile, which overlies the Precambrian metamorphic basement. During the Zagros orogeny, some of the Precambrian basement and Lower Paleozoic strata were exhumed from depth and are now exposed in the Golpayegan region in the northwestern part of the Zagros Mountains. The tectonic evolution of the Golpayegan region and the exhumation of the old rocks are interpreted as the product of three major sequential geotectonic events. (1) Major thrusts formed during shortening and exposure of the basement rocks in the Aligudarz block. The rock units are strongly imbricated and sheared, which suggest a high amount of cumulative shortening in the northern Zagros. (2) NE-SW–trending extensional faults (e.g., Eastern Mute, Western Mute, and Mahallat faults) formed during lateral extensional movement after middle Miocene time. In this event, a set of NE-SW–trending horsts and grabens was formed. In the horsts (e.g., Hassan-Robat, Mute, and Mahallat horsts), the Precambrian basement rocks and Lower Paleozoic strata are exposed. (3) Strike-slip movements began that remain active today. Strike-slip motions are well documented for the late Pliocene–Quaternary period. In the Golpayegan region, the Shazand and Dehagh faults cut through the NE-trending normal faults and through Quaternary deposits. Drainages are displaced by ~4 km of dextral movement along the Shazand fault and near the city of Golpayegan.

Keywords: collision, exhumation, strike-slip, Zagros, Sanandaj-Sirjan zone, Golpayegan.

*geotecton@yahoo.com

Nadimi, A., and Nadimi, H., 2008, Exhumation of old rocks during the Zagros collision in the northwestern part of the Zagros Mountains, Iran, *in* Burchfiel, B.C., and Wang, E., eds., Investigations into the Tectonics of the Tibetan Plateau: Geological Society of America Special Paper 444, p. 105–122, doi: 10.1130/2008.2444(07).

INTRODUCTION

Intense crustal deformation can be described by the dynamics of orogenic wedge development in many orogens. The internal parts of ancient orogens, in particular, contain key information concerning the initiation, evolution, and destruction of such orogenic wedges (Dahlen, 1990; Boyer, 1995; Mitra and Sussman, 1997). During the orogeny and superposed extensional tectonics, high-pressure metamorphic rocks and associated older rocks and basement complexes of the deep levels of the lithosphere are exhumed and become exposed at Earth's surface. High-pressure metamorphic rocks form an important component in geophysical and geochemical models of the lithosphere and have a crucial role in the construction of geodynamic models concerned with convergent plate margins, especially foreland fold-and-thrust belts (Karabinos, 1988; Kurz and Froitzheim, 2002). Special mechanisms are necessary to bring these rocks to shallow crustal levels (exhumation: England and Molnar, 1990; Ring et al., 1999) and allow access for structural and petrological investigation.

The appearance of high-pressure metamorphic rocks and basement complexes at the surface requires mechanisms that reverse the kinematic path traced by these units, i.e., from subduction to exhumation (Kurz and Froitzheim, 2002). Several mechanisms may contribute to removal of material above the rocks within a subduction zone, and to emplacement of these rocks at shallower lithospheric levels. Qualitative kinematic models that have been developed from field observations basically describe three mechanisms that contribute to the exhumation of deep crustal rocks: (1) erosion (e.g., Hacker et al., 1995; Jamieson et al., 2002); (2) extension (e.g., Ring et al., 2001); and (3) extrusion (e.g., Michard et al., 1993; Chemenda et al., 1995). Erosion alone is clearly not sufficient to exhume rocks buried to mantle depth. Thus, additional tectonic processes are required (Kurz and Froitzheim, 2002).

During the Zagros orogeny, some of the older rocks and Precambrian basement were exhumed from depth and exposed in the northwestern part of the Zagros Mountains (Nadimi and Nadimi, 2006a). Here, Precambrian metamorphic basement and Infracambrian–Lower Cambrian sedimentary formations crop out at the surface. These rocks were emplaced at shallow depths by faulting during forearc extension. In this paper, we attempt to demonstrate, based on available geologic data, the tectonic evolution of the region and exhumation of these old rocks during the Zagros orogeny.

ZAGROS FOLD-AND-THRUST BELT

The Zagros fold-and-thrust belt, which extends NW-SE for ~2000 km, lies on the northeastern margin of the Arabian plate (Fig. 1). Like other fold-and-thrust belts, it shows intense shortening close to the suture, which becomes less intense toward the foreland. The Zagros orogen resulted from the convergence between the Iran block and the Arabian plate as the Neotethyan realm (rifting of which started in the Late Permian–Early Triassic, when the Central Iranian block separated from Gondwana and Arabia) disappeared through a succession of subduction-obduction-collision stages (Agard et al., 2005).

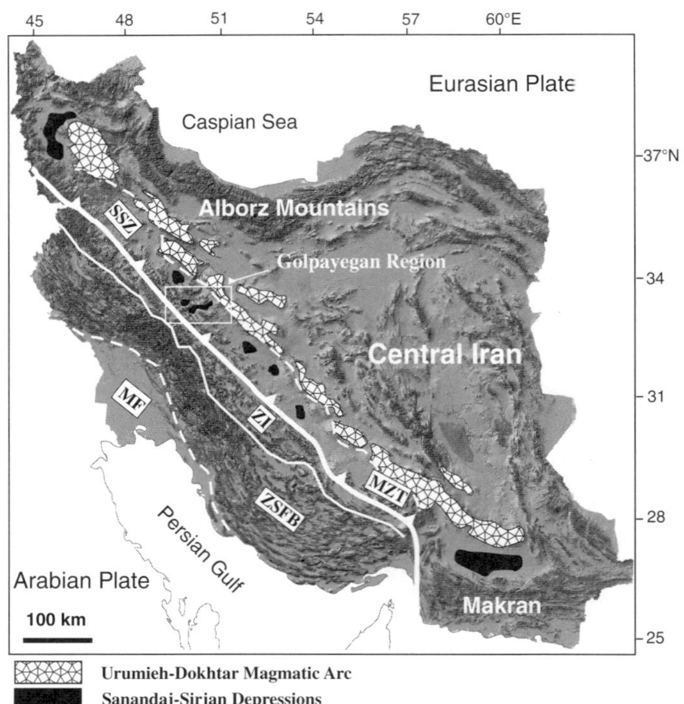

Figure 1. Topography, and principal active fold-and-thrust belt of the Arabia-Eurasia collision in Iran. The structural zones of the Zagros orogen are drawn by a thick line. Abbreviations: SSZ—Sanandaj-Sirjan zone, ZI—Zagros imbricate belt, ZSFB—Zagros simply folded belt, MF—Mesopotamian foreland basin, and MZT—Main Zagros thrust. Rectangle indicates the Golpayegan region.

The Zagros is an example of a young and active orogenic belt where geomorphology and seismicity provide invaluable constraints on the style of deformation in ways not possible with ancient, inactive orogens (Jackson, 1980; Sattarzadeh et al., 2000). The Arabia-Eurasia collision was preceded by Campanian-Maastrichtian ophiolite obduction onto the Arabian passive continental margin (Stoneley, 1990; Yilmaz, 1993). Continental collision probably began in the Oligocene at the northern promontory of the Arabian plate (Yilmaz, 1993) and in the Miocene to the SE (Stoneley, 1981), forming the Bitlis-Zagros suture (Blanc et al., 2003). Based on thermochronologic data and plate reconstructions, the Zagros collision began in middle to late Miocene time (Axen et al., 2001; McQuarrie et al., 2003; Guest et al., 2006). The collisional process is still active, with an N-S–directed convergence rate of ~20 ± 2 mm/yr, according to recent global positioning system (GPS) data (Vernant et al., 2003). The original Zagros suture is known as the Main Zagros thrust (Blanc et al., 2003; Agard et al., 2005). Part of this fault zone is active in a right-lateral sense, where it is known as the Main Recent fault (Blanc et al., 2003).

The Zagros fold-and-thrust belt is located between the Tertiary Urumieh-Dokhtar magmatic arc to the northeast and the Arabian plate margin to the southwest. The belt is subdivided into four parallel NW-SE–trending structural zones that include, from northeast to southwest, the Sanandaj-Sirjan metamorphic zone, the imbricated belt, the folded belt, and the Mesopotamian foreland basin with buried folds, which extends to the SE into the Persian Gulf (Fig. 1) (Falcon, 1974; Colman-Sadd, 1978).

The Sanandaj-Sirjan Zone

The Sanandaj-Sirjan zone is a narrow belt that lies to the southwest of the Urumieh-Dokhtar magmatic arc. The Sanandaj-Sirjan zone is separated from the NW-trending high mountains of the imbricated belt on the southwest by the Main Zagros thrust. The development of the 1500-km-long, 150- to 200-km-wide Sanandaj-Sirjan zone was related to the destruction of the Tethys Ocean during Cretaceous and Tertiary time as a result of convergence and continental collision between the African-Arabian and the Eurasian plates (Agard et al., 2005; Ghasemi and Talbot, 2006). The Sanandaj-Sirjan zone consists of mainly Jurassic, interbedded phyllites and meta-volcanic rocks that are characterized by a moderate metamorphic overprint except close to large-scale Mesozoic calc-alkaline plutons (Agard et al., 2005). These metamorphic rocks are unconformably overlain by Barremian–Aptian Orbitolina limestones typical of strata that cover central Iran (Stöcklin, 1968). During most of the late half of the Mesozoic, the Sanandaj-Sirjan zone was an active Andean-like margin characterized by calc-alkaline magmatic activity that progressively migrated northward (Berberian and King, 1981; Sengor, 1990). It behaves today as a fairly rigid block, moving northward at 14 mm/yr with respect to stable Eurasia (Vernant et al., 2004). Older magmatic activity in the Sanandaj-Sirjan zone includes Late Triassic and Early Jurassic tholeiitic mafic volcanic rocks (Alavi and Mahdavi, 1994), which are interpreted as remnants of Tethyan oceanic crust, and Late Proterozoic to early Paleozoic mafic rocks, which formed during earlier extensional events (Berberian and King, 1981; Rachidnejad-Omran et al., 2002).

A NW-trending series of depressions, which includes the Lake Urumieh, Tuzlu Gol, Khomein, Golpayegan, Gawkhuni, and Jaz Murian depressions, separates the Sanandaj-Sirjan zone from the Urumieh-Dokhtar magmatic arc and the rest of the Central Iranian Plateau to the northeast (Fig. 1) (Tillman et al., 1981; this study). Most of the depressions are surrounded by faults.

The Imbricated Belt

The Zagros imbricate belt (or High Zagros), to the southwest of the Main Zagros thrust, is a narrow NW-SE–trending thrust belt up to 80 km wide. It is bounded to the SW by the High Zagros fault, which is currently seismically active along a few segments (Berberian, 1995). This zone forms the topographically highest part of the Zagros, with summits over 4000 m above sea level (Blanc et al., 2003). The belt consists of a zone of thrust faults that transported numerous slices of metamorphosed and nonmetamorphosed Phanerozoic stratigraphic units of the African-Arabian passive continental margin, as well as its "obducted" ophiolites, from the collisional suture zone on the northeast toward interior parts of the Arabian craton to the southwest. The tectonic slices consist of Mesozoic limestones, radiolarites, obducted ophiolite remnants, and Eocene volcanics and flyschs (Agard et al., 2005).

The total thickness of the sedimentary rocks within the imbricate belt increases from the Main Zagros thrust to the High Zagros fault, as determined from gravimetric and magnetic data (Bosold et al., 2005). Such thickening is typical for the inner, more complex forelands of orogenies. It is usually a result of sedimentary thickening of the synorogenetic sequence in the downwarped synclines that develop in front of approaching thrust belts, and of section duplication caused by thrust imbrications, which increase in intensity toward the High Zagros fault (Bosold et al., 2005).

The Simply Folded Belt

The Zagros simply folded belt consists of a thick sedimentary sequence that overlies the highly metamorphosed rocks of the Precambrian basement. The belt represents the passive margin of the Arabian plate folded during Cenozoic time. Anticlines of the belt form the principal hydrocarbon traps of the Zagros. The main reservoir is the Asmari Limestone, and the Gachsaran Formation forms the regional seal (Beydoun et al., 1992).

The Zagros simply folded belt is often cited as one of the regions in the world showing the best examples of large-scale detachment folds (Colman-Sadd, 1978) and their regular and well-rounded geometry. This is due to a very efficient décollement level in the Hormoz Salt, located at the base of the thick (up to 10 km) sedimentary sequence (Sherkati et al., 2005).

Structurally, the Zagros simply folded belt is characterized by well-known NW-SE–trending folds. The high aspect ratio of the folds suggest that they are related to faulting (forced folds) and are not buckle folds (Cosgrove and Ameen, 2000). The fold geometries suggest that several décollement zones are present in the sedimentary cover (Blanc et al., 2003).

Evolution of the Zagros Basin

The Zagros Basin is defined by a 6–15-km-thick succession of cover sedimentary rocks deposited over an extraordinarily wide and long region along the N-NE margin of the Arabian plate. Since the end of Precambrian time, this region has evolved within a number of different tectonic settings. The Zagros Basin was part of the stable supercontinent of Gondwana in Paleozoic times, it was a passive margin in Mesozoic times, and it became a convergent orogen in Cenozoic times (Bahroudi and Koyi, 2004).

The basement at depth under the Zagros Basin is traditionally considered to be a continuation of the Precambrian shield exposed in Arabia (Falcon, 1969; Berberian and King, 1981;

Al Laboun, 1986; Alsharhan and Nairn, 1997). Most of the Phanerozoic cover sediments surrounding the Arabian Shield display two dominant structural elements, with a variety of trends. These are inferred to be caused by deep basement faults and can be divided into transverse and longitudinal sets relative to the NW-SE trend of the Zagros fold-and-thrust belt (Berberian, 1995). The longitudinal NW-SE faults are almost perpendicular to the NE-trending shortening direction and presumably are NW-trending high-angle reverse faults with dip-slip displacement coeval with formation of the Zagros fold-and-thrust belt (Berberian, 1995). Reliable fault-plane solutions and focal depths of the earthquakes in the Zagros using long-period World Wide Standardized Seismograph Network (WWSSN) P-waveforms show that most earthquakes occur on high-angle thrust faults (40°–50°) at centroid depths of 8–12 km in the uppermost part of the metamorphic basement beneath the Lower Cambrian Hormoz Salt and near the top of the sedimentary cover on a large number of faults distributed across the belt that are concealed by the folded shallow sedimentary cover (Jackson, 1980; Jackson and McKenzie, 1984; Ni and Barazangi, 1986).

THE GOLPAYEGAN REGION

The Golpayegan region in the Sanandaj-Sirjan zone is surrounded by the Zagros fold-and-thrust belt within the Alpine-Himalayan orogenic system of western Asia. It is located between the Urumieh-Dokhtar magmatic arc to the northeast and the Zagros imbricate belt to the southwest (Figs. 1 and 2). Being situated to the northeast of the Zagros Neotethyan suture and its parallel Cenozoic magmatic arc (Urumieh-Dokhtar magmatic arc), the Golpayegan region is an area of continuous intracontinental deformation formed in response to the ongoing convergence between the Arabian plate (Gondwana) and the Central Iranian terrane.

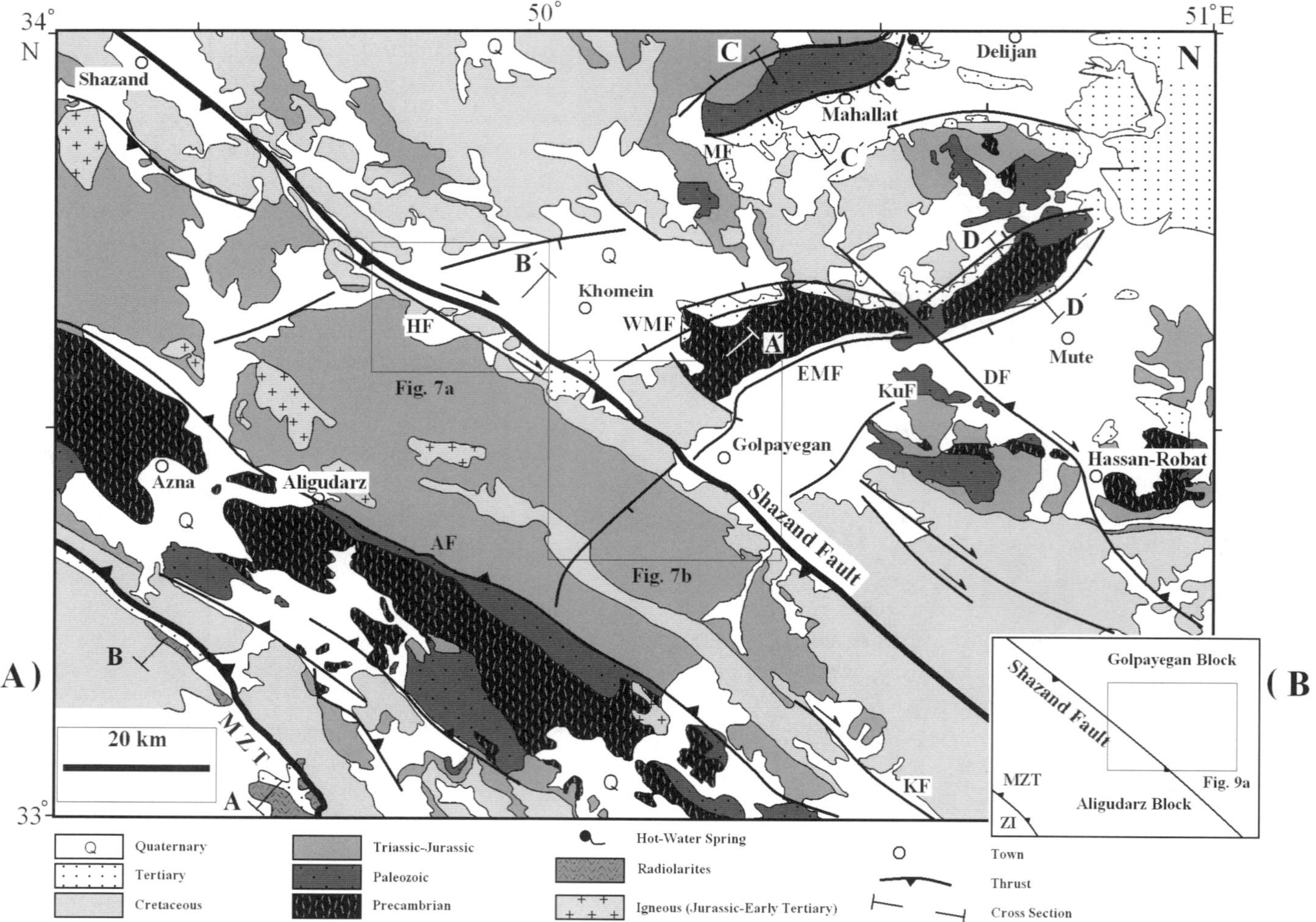

Figure 2. (A) Geological map of the Golpayegan region. The major faults of the Golpayegan region are shown by thick lines. Abbreviations: MZT—Main Zagros thrust, DF—Dehagh fault, KF—Khansar fault, AF—Aligudarz fault, MF—Mahallat fault, WMF—Western Mute fault, EMF—Eastern Mute fault, HF—Hosseinabad fault, and KuF—Kunjijan fault. (B) Location of the structural blocks of the Golpayegan region, including the Golpayegan and Aligudarz blocks and the Zagros imbricate zone (ZI).

Mesozoic and Tertiary sedimentary rocks overlying Paleozoic and older sedimentary rocks predominate in the study area, and they are largely covered by Quaternary rocks. Old metamorphic rocks of the Sanandaj-Sirjan zone are exposed in the Mute and Aligudarz areas (Fig. 2). The rocks in the Mute area consist of gneiss, marble, amphibolite, mica schist, phyllite, quartzite, decimeter-thick to several-meters-long quartz veins, and rare meter-thick magnetite-rich horizons. Thiele (1966) and Thiele et al. (1968) assigned a Precambrian age to the metamorphic rocks and to the Mute gold ore formation based on the assumption that overlying unmetamorphosed formations were Late Precambrian, Infracambrian, and Cambrian in age. Previous $^{40}Ar/^{39}Ar$ dating has indicated that gold ore formation at Mute was early to middle Eocene in age (Moritz et al., 2006). In the Hassan-Robat area, the crystalline and intensely deformed metamorphic basement of Precambrian age makes up the main body of the mountains and is in fault contact with younger rock units (Fig. 3). Field evidence shows that the basement in the area is pre-Triassic in age. Subhorizontal mylonitic foliation in the Precambrian metamorphic rocks is present in the Mute high mountains. The Precambrian metamorphic rocks of Aligudarz consist of marble, amphibole, meta-dolomite, quartzite, mica schist, and meta-rhyolite.

Sedimentary structures and the spatial distribution of the Mesozoic and Cenozoic rocks indicate that differential vertical movements across major faults controlled the depositional patterns of most, if not all, of the mapped stratigraphic units (Tillman et al., 1981). Lower Cretaceous sedimentary rocks unconformably overlie the older Mesozoic units and form the mountains ranges (Fig. 3F). The Lower Cretaceous gray limestones are intensity foliated and folded and show a weak recrystallization. Tertiary sedimentary deposits locally overlie the Lower Cretaceous units (Tillman et al., 1981). Eocene volcanic rocks of the Urumieh-Dokhtar magmatic arc occur in the northeastern part of the region (Moritz et al., 2006). Middle Eocene conglomerates and sandy conglomerates are present on the northeast side of the Golpayegan region and unconformably overlie all the older units (Fig. 3G). In the south, near Khomein, the sediments overlie folded Lower Cretaceous limestone with angular unconformity. These rocks unconformably overlie the Precambrian metamorphic rocks and Precambrian–Lower Paleozoic sedimentary rocks on the northwestern and northeastern side of Mute high mountains, and they are in fault contact with Upper Precambrian–Lower Paleozoic sedimentary rocks around Mahallat.

Miocene sedimentary rocks (a part of the Upper Oligocene–Lower Miocene Qom Formation) commonly have a 2-m-thick basal conglomerate member overlain by a 50- to 100-m-thick white to pink, thick-bedded, very fossiliferous clastic limestone (Fig. 3H; Tillman et al., 1981). These rocks overlie the Precambrian metamorphic rocks on the northwestern side of the Mute high mountains and in the Hassan-Robat area and overlie Upper Paleozoic and Mesozoic rocks with angular unconformity on the southeastern side of the Mute high mountains (Fig. 4).

During Pliocene time, conglomerates were deposited within local closed basins. The Pliocene conglomerates are overlain by fine silts and clay interbedded with and capped by alluvial fans and gravels (locally cemented), which represent a continuous unit at about the same elevation throughout much of the region (Tillman et al., 1981).

Structural Blocks of the Golpayegan Region

The Golpayegan region consists, from northeast to southwest, of two major structural domains: the Golpayegan block and Aligudarz block, which are separated by the Shazand fault (Fig. 2). The Golpayegan block is located between the Urumieh-Dokhtar magmatic arc to the northeast and the Shazand fault to the southwest. The Golpayegan block consists of, from east to west, two NE-trending depressions, the Golpayegan and Khomein depressions, separated by the Hassan-Robat, Mute, and Mahallat high mountains.

The Aligudarz block is surrounded by the Shazand fault to the northeast and the Main Zagros thrust to the southwest. Rock units in the block are strongly folded, imbricated, sheared, and thrusted, and the structures trend NW-SE. Most folds in the Aligudarz block are asymmetric with inclined axial planes and have axes trending NW-SE parallel with the Zagros folds. In southwestern part of the region, tectonic mélanges composed of fragments of the normal stratigraphic units and/or dismembered ophiolite complexes are well developed.

Two first-order active fault systems, a NW-SE–trending (e.g., Shazand fault) and a NE-SW–trending (e.g., Eastern Mute fault) system, are identified within the region (Fig. 2A). The NW-SE–trending faults are the main structures in the northeastern Zagros Mountains, especially in the Golpayegan region. These faults are oblique-slip faults and have both reverse (thrust) and right-lateral strike-slip motions. In the Aligudarz block, a combination of strike-slip (right-lateral) and reverse (thrust) movements associated with these faults has generated a complex pattern of regional deformation involving crustal shortening, horizontal blocks rotation, and localized uplift.

The NE-SW–trending faults are second-order structures in the northeastern Zagros Mountains. These faults are distributed in different sizes in the Golpayegan block. The faults have normal dip-slip motion, and, in some places, a few have a left-lateral strike-slip component. In the Golpayegan block, normal movement on the faults has generated a NE-SW–trending horst-and-graben system (Fig. 5). In the horsts, the Precambrian basement and Lower Paleozoic units are separated by high-angle normal faults from younger formations. The horsts correlate with the Hassan-Robat, Mute, and Mahallat high mountains that separate the Golpayegan and Khomein grabens.

The NE-SW–trending faults are major extensional fractures that formed parallel to the shortening direction of the Zagros after middle Miocene time. Tillman et al. (1981) mapped numerous small NE-SW–trending normal faults (<10 km length) between the Golpayegan and Gawkhuni depressions. Most of the NE-trending normal faults in the area dip northwest and have down-to-the-west displacements and clearly crosscut and offset Miocene and younger exposures (Tillman et al., 1981).

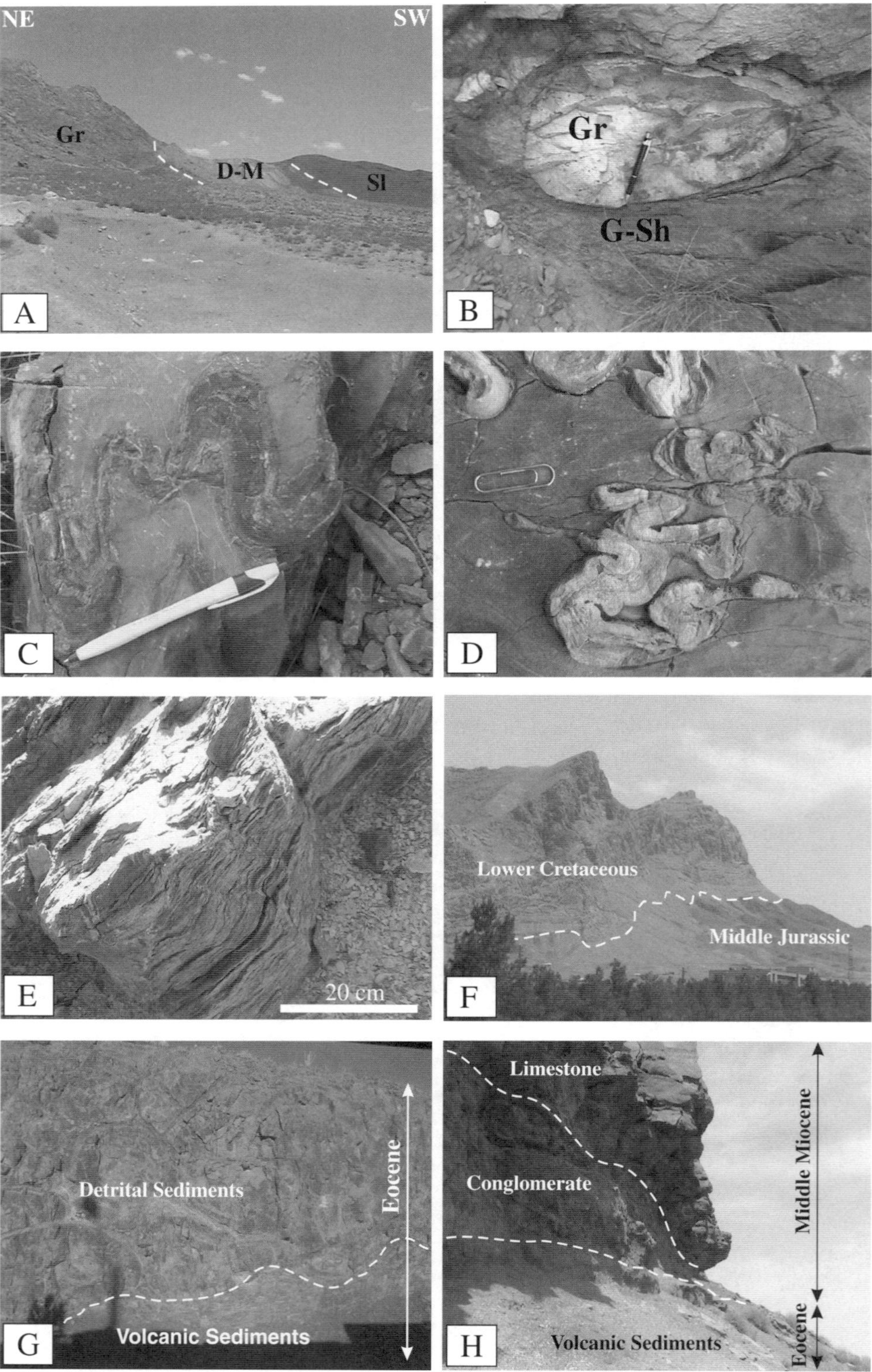

Figure 3. (A) Outcrop views of the Precambrian basement in the Hassan-Robat area. The major rock units of the area are drawn by thick lines. Abbreviations: Gr—granite (Precambrian), D-M—marble and dolomite-marble (Precambrian), and Sl—slate (middle Jurassic). (B) A granitic fragment (Gr) in the folded and sheared graphitic shales (G-Sh) with Triassic age showing that the granite is older than Triassic. (C–D) Ductile deformation in the Precambrian marbles in the Hassan-Robat area. (E) Subhorizontal mylonitic foliation in the Precambrian metamorphic rocks. (F) Unconformity between the Lower Cretaceous and Middle Jurassic units. (G) Volcanic and detrital sediments of Eocene age in the Mahallat area. (H) Unconformity between middle Miocene and Eocene units.

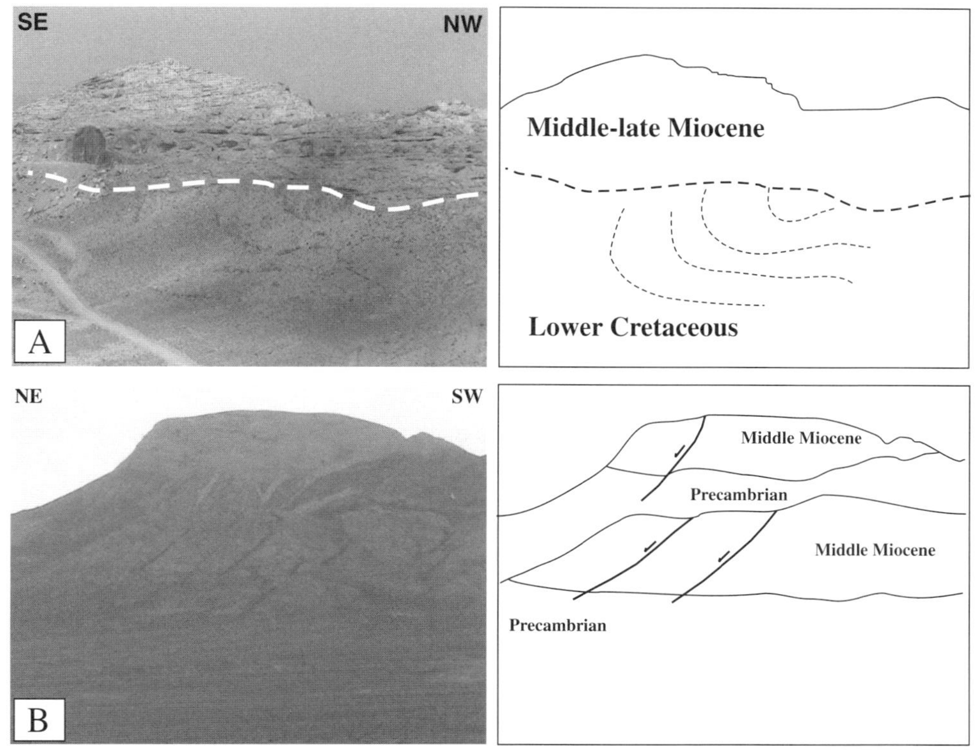

Figure 4. (A) Outcrop views of the angular unconformity between the folded Lower Cretaceous limestones and shales and Miocene limestones. (B) Location of the middle Miocene limestone on the Precambrian basements in Hassan-Robat area.

Major Faults of the Golpayegan Region

The observations used in this paper to infer the presence of active faults were taken from the results of numerous studies, both within Iran and elsewhere, which have explored the link between slip on active faults and the development of landscapes.

Shazand Fault

The Shazand fault is one of the major NW-SE–trending faults of the northeastern side of Zagros Mountains and Main Zagros thrust (Fig. 6). The fault marks the boundary between the mountains of the Aligudarz block to the southwest and the Golpayegan and Khomein depressions of the Golpayegan block to the northeast (Fig. 2A). The Shazand fault is 186 km long and trends NW-SE from northwest of Shazand in the northwestern Golpayegan block toward southwestern Isfahan. Earthquakes indicate the presence, depth, and geometry of faults in the Golpayegan region and are useful in explaining the geometries of shallower structures (Fig. 6B). Field observations and earthquake data show that the Shazand fault is a major fault (not inferred) and that is dips southwest, in contrast to the description given by Thiele et al. (1968) and Tillman et al. (1981). The fault strikes N50°W and it dips 40–50°SW. The fault has right-lateral strike-slip and reverse movements. Structural and morphotectonic evidence shows that the Shazand fault and its parallel branches, the Hosseinabad and Dera faults, play an important role in active tectonics of the block (Fig. 7).

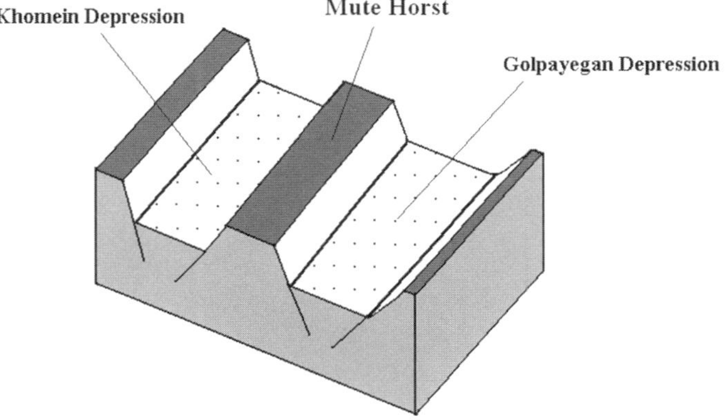

Figure 5. Schematic model showing the horst-and-graben structures in the Golpayegan block.

The Hosseinabad fault is 37 km long and lies south of the Shazand fault. Its strikes N55°W and dips 80°W. The fault displaces the Khomein River dextrally by ~4.2 km and cuts the Quaternary sediments and alluvial fans. The fault has reverse and right-lateral motions. The parallel Dera fault is smaller than the Hosseinabad fault.

Along the Shazand fault, Quaternary sediments and alluvial fans and rivers are cut, and the Golpayegan depression is also uplifted (Fig. 7). The Khomein and Golpayegan Rivers are displaced dextrally about ~4 km along the Shazand fault near Golpayegan and south of Khomein.

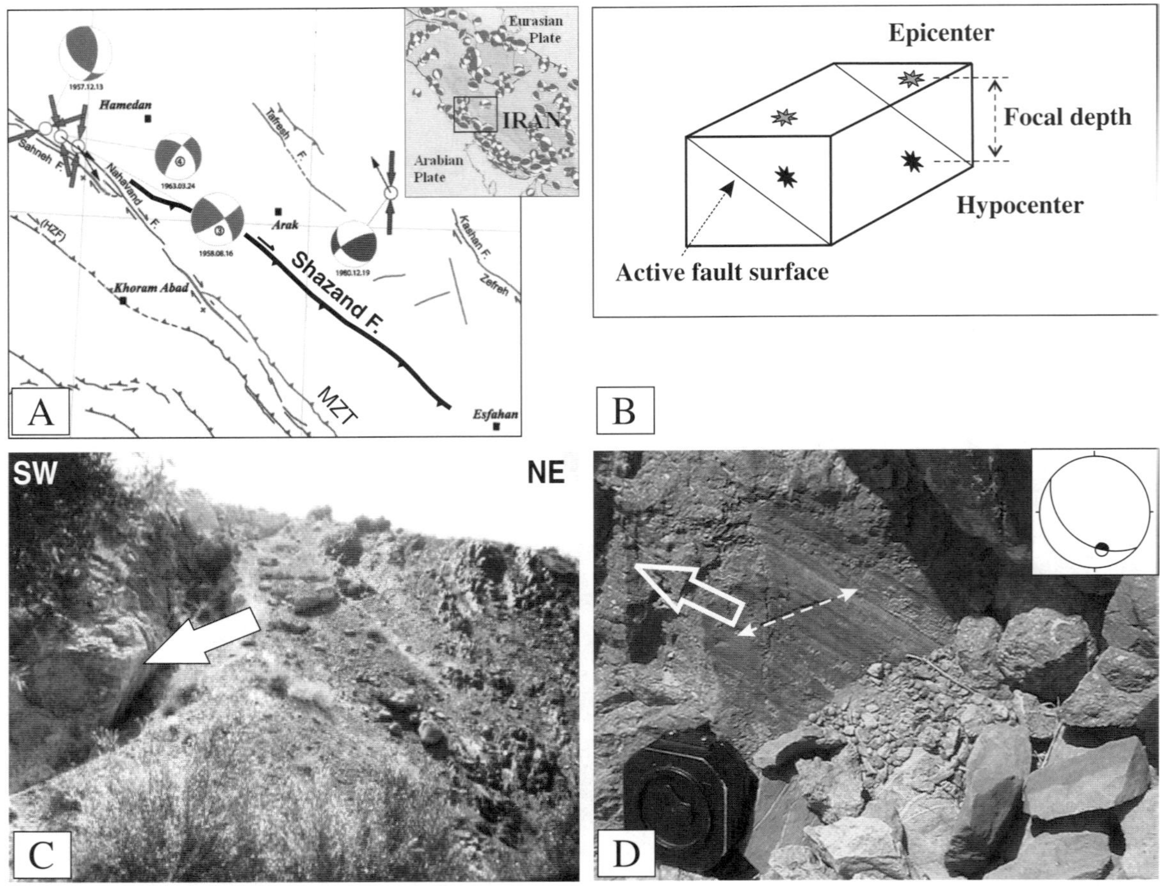

Figure 6. (A) Location of the Shazand fault in the northwestern Zagros. MZT and HZF are the Main Zagros thrust and High Zagros fault, respectively (modified from Hessami et al., 2003). (B) Schematic model of an active fault and its earthquakes. The earthquake hypocenters are located on the active fault plane at different depths. In the oblique fault, the earthquakes are located on a band. The depths can be used for measurement of the active fault-plane dip. (C) The Shazand fault plane in the southwest Khomein. (D) Close-up of the Shazand fault plane in the south Golpayegan region. Strong stretching lineations in the Cretaceous rocks show reverse and right-lateral strike-slip motions.

Dehagh Fault

The Dehagh fault trends NW-SE and can be traced from the Mute horst in the Golpayegan block toward the village of Hassan-Robat and the town of Dehagh and then toward Isfahan city (Fig. 2A). The fault is 200 km long, strikes N45–50°W, and dips 71°NE. It has dextral strike-slip movement. In the Golpayegan block, the fault cuts the Precambrian rocks of the Mute horst and has 2.2 km of offset (Fig. 8). Structural analyses of the areas around the village of Hassan-Robat show that the maximum principal stress direction is N10°E.

Eastern Mute Fault

The Eastern Mute fault trends NE-SW and can be followed from Dotu Mountain southwest of Golpayegan in the Aligudarz block to northeast of the village of Mute, to the northeast (Figs. 2A and 7B). In the Aligudarz block, the Golpayegan River lies in the valley along the trace of the Eastern Mute fault. The fault strike changes from N50°E to N70°E, and its dip is 50–60°SE. The Eastern Mute fault marks the boundary between the Mute horst to the northwest and the Golpayegan depression to the southeast. The fault can be traced for ~90 km and cuts middle Miocene strata, indicating that the fault was initiated after middle Miocene time. The Eastern Mute and Western Mute faults are extensional faults that formed by longitudinal extensional stress in the Zagros collision zone after the middle Miocene. The Eastern Mute fault is active and has cut young sediments and alluvial fans. Along the fault and along the Golpayegan River, there are fault scarps. The Eastern Mute fault near Golpayegan forms a 16-m-high scarp on the surface of the first (late Pleistocene–Holocene) river terrace (Figs. 8D and 8E). The fault scarps all face southeast, and the northwest side of the scarp is topographically higher, showing an active component of normal dip-slip motion along the fault. The Eastern Mute fault is displaced near Hekal Mountain southwest of Golpayegan ~500 m left-laterally, showing that the Eastern Mute fault has a small component of left-lateral strike-slip motion (Figs. 8C and 8F). The Eastern Mute fault is displaced right-laterally by the Shazand and Dehagh faults (Fig. 9A).

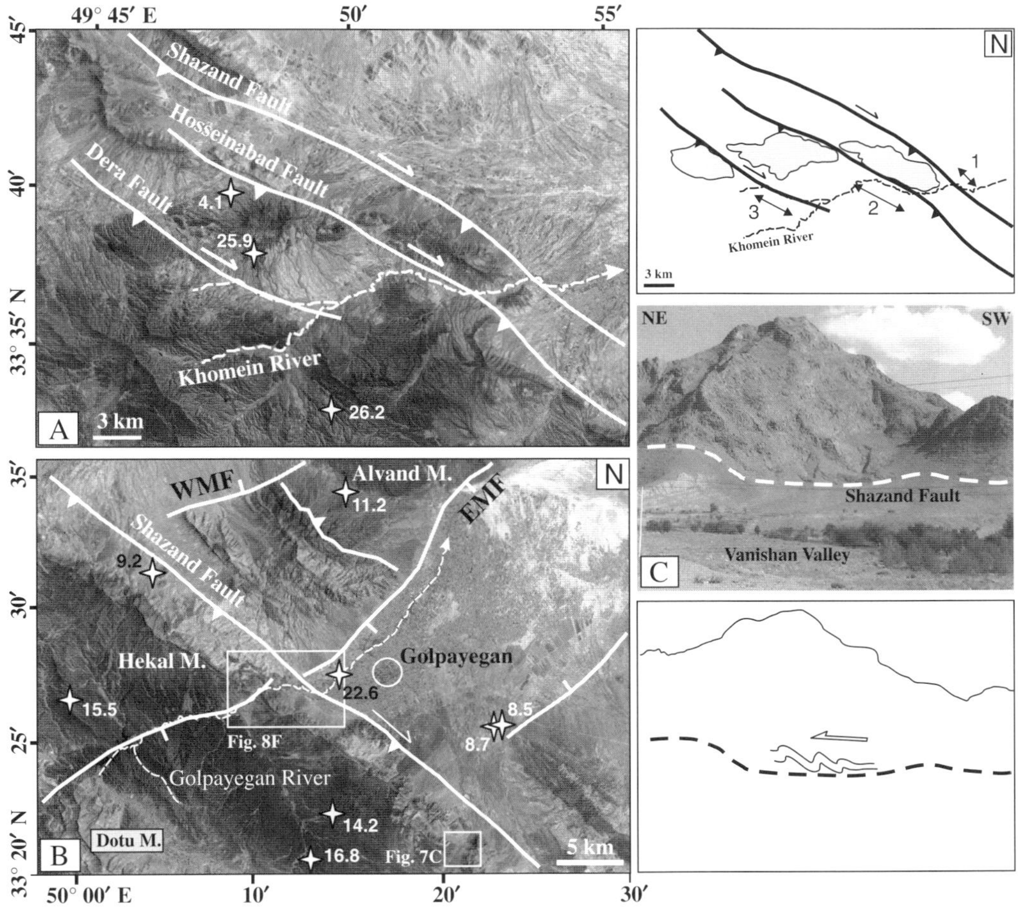

Figure 7. (A) Landsat Thematic Mapper (TM) image of south Khomein. The Khomein River is directed E-NE from the north Aligudarz Mountains. Lines show the Shazand, Hosseinabad, and Dera faults. These faults cut the river and alluvial deposits. 1, 2, and 3 are the displacements of the river by the faults. The Shazand and Hosseinabad faults cut the young sediments in the right part of the figure. Stars and numbers are epicenters and focal depths of earthquakes. The epicenters with different depths are located on active faults of the area. (B) Landsat TM image of the Golpayegan area. The Golpayegan River is directed NE from the southern mountains. The Shazand fault has displaced the Eastern Mute fault (EMF) right-laterally. WMF—Western Mute fault. Earthquake epicenters are located on the active faults. Locations of the figures are shown in Figure 2A (earthquake data are from www.iiees.ac.ir). (C) Outcrop views of the Shazand fault in Vanishan Valley in southeast Golpayegan. Location is shown in B.

Western Mute Fault

The Western Mute fault trends NE-SW (Figs. 2A and 9A). It can be followed from east of Khomein in the Golpayegan block to northeast of the village of Mute. The Western Mute fault forms the western boundary of the Mute horst and is located between the Mute horst and Khomein depression. The Western Mute fault is ~65 km long and has normal displacement. Its strikes N70°E and dips 70°NW. The fault cuts middle Miocene rock strata. The scarps along the fault face northwest, and the southeast side is topographically higher, indicating normal displacement.

Mahallat Faults

Mahallat horst is bounded by two NE-SW–trending faults (Fig. 2A). The Eastern Mahallat fault can be traced for ~40 km. This fault separates Eocene conglomerates from the Upper Precambrian–Lower Paleozoic strata. Field observations show that the fault has normal dip-slip motion, and, at the surface, it dips southeast to east. The fault scarps also show that the fault has normal displacement (Fig. 9C). There are some hot-water springs along the fault, especially northeast of the town of Mahallat.

The Western Mahallat fault can be traced for ~25 km. The fault is displaced by the Eastern Mahallat fault on the northeast side of the Mahallat horst. The scarps along the fault indicate normal dip-slip motion.

Kunjijan Fault

The Kunjijan fault forms the eastern border of the Golpayegan depression (Fig. 9A). It can be traced for 26 km and has normal displacement. The fault strikes N40°E and dips 34°NW.

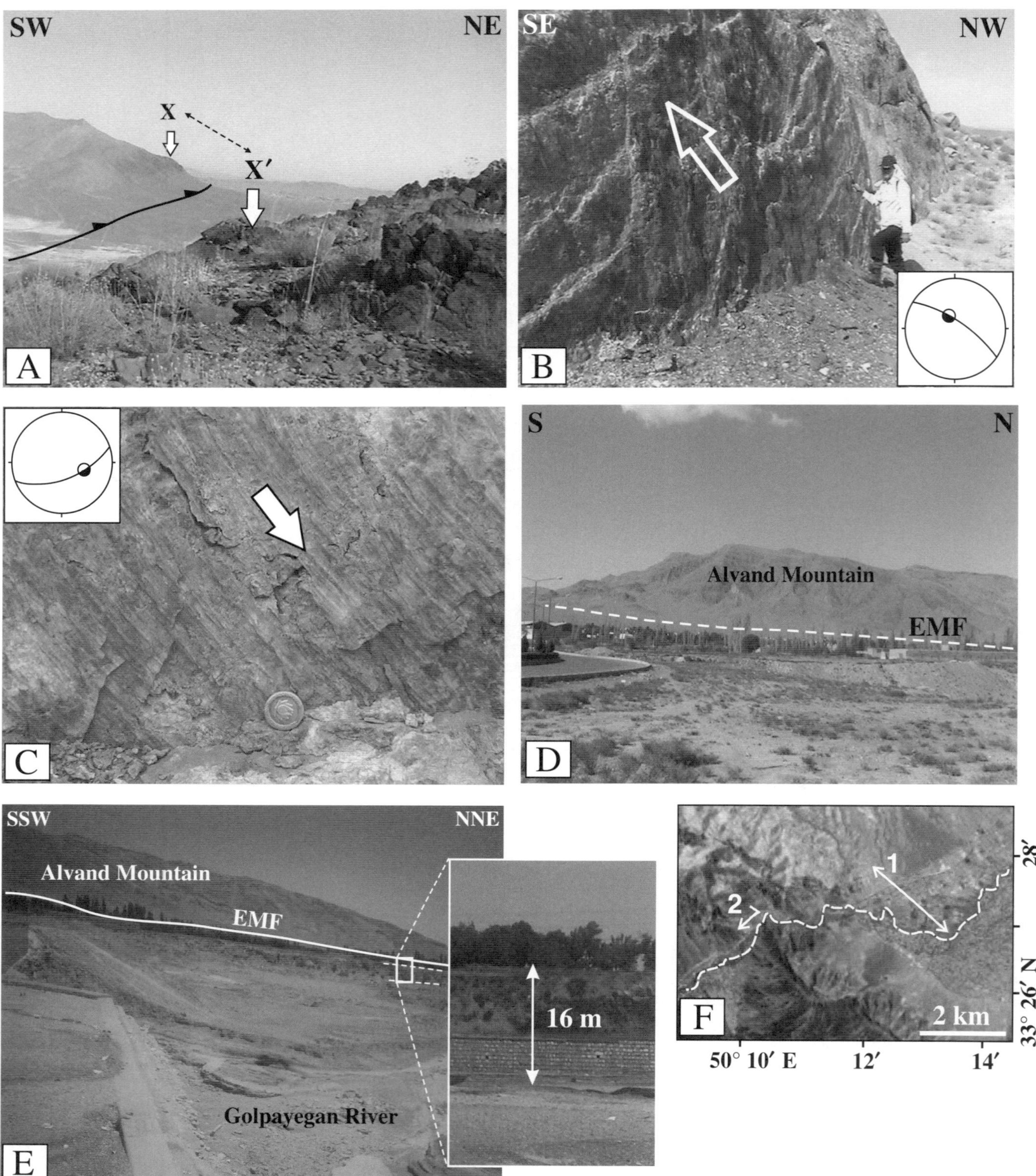

Figure 8. (A) Outcrop views of the Dehagh fault in the Hassan-Robat area. The fault displaces the Cretaceous rocks right-laterally. X–X′ is 2.2 km in length. (B) The Dehagh fault plane in east Mute. The lineations show reverse and right-lateral strike-slip motions. (C) Close-up of the Eastern Mute fault plane in west Golpayegan. Lineations show normal and left-lateral strike-slip motions. (D) Outcrop views of the Eastern Mute fault (EMF) close to Golpayegan. (E) Outcrop views of the Eastern Mute fault and its scarps. (F) Landsat TM image of the Golpayegan River offset. 1 is right-lateral displacement of the river by the Shazand fault; 2 is left-lateral offset of the Hekal Mountain by the Eastern Mute fault. Location is shown in Figure 7B.

Figure 9. (A) Landsat TM image of the Golpayegan block and the active faults. Thick lines show major faults of the Golpayegan block. Abbreviations: SF—Shazand fault, DF—Dehagh fault, WMF—Western Mute fault, EMF—Eastern Mute fault, and KuF—Kunjijan fault. Stars and numbers are epicenters and focal depths of earthquakes. The epicenters with different depths are located on active faults of the area. (B–C) The Mahallat fault scarps in the young sediments.

SECTIONS ACROSS THE GOLPAYEGAN REGION

Several sections described here show large differential vertical movements along the faults. Figure 10 shows four cross sections through the Golpayegan region based on a map that combines a 1:250,000 geological map (Thiele et al., 1968), satellite image interpretation, and regional field work within and around the Golpayegan and Aligudarz blocks. The Sanandaj-Sirjan and Zagros seismicity record is an important constraint on the structure, and it was used in construction of the cross sections.

1. The section line 10A summarizes the structure across both the Golpayegan block and Aligudarz block, including the Main Zagros thrust zone. Important structures recognized at the shallow level include tectonic slices along the contacts, drag folds, and boudinage. There are two major outcrops of Precambrian metamorphic basement, one in the SW part of the Aligudarz block and the other in the NE part of the Golpayegan block. In the SW, the Precambrian basement rocks were exposed during thrusting along imbricate structures. However, in the NE, the basement rocks were exposed during exhumation of deep crustal levels.

2. The section line 10B characterizes the structure across the Aligudarz block and the Main Zagros thrust. In the SW, the Main Zagros thrust zone includes imbricates of Paleozoic-Mesozoic strata and an exposure of Precambrian metamorphic basement. The exposures of crystalline basement suggest that there may be a major detachment in the zone at the level of the Precambrian basement. Some SW-dipping faults in this region give the zone the overall appearance of a flower structure. The intensity of deformation increases to the SW and toward the Main Zagros thrust zone. Within the Aligudarz block, major thrusts verge to the SW, and in the Golpayegan block, they verge to the NE. The Cretaceous and older units are strongly folded, forming overturned folds in the Golpayegan block and Aligudarz block. Fold wavelengths decrease toward the NE.

3. Figures 10C and 10D are cross sections through the Mahallat and Mute horsts. The sections are across the normal-faulted exhumed units in the Golpayegan block, and they show

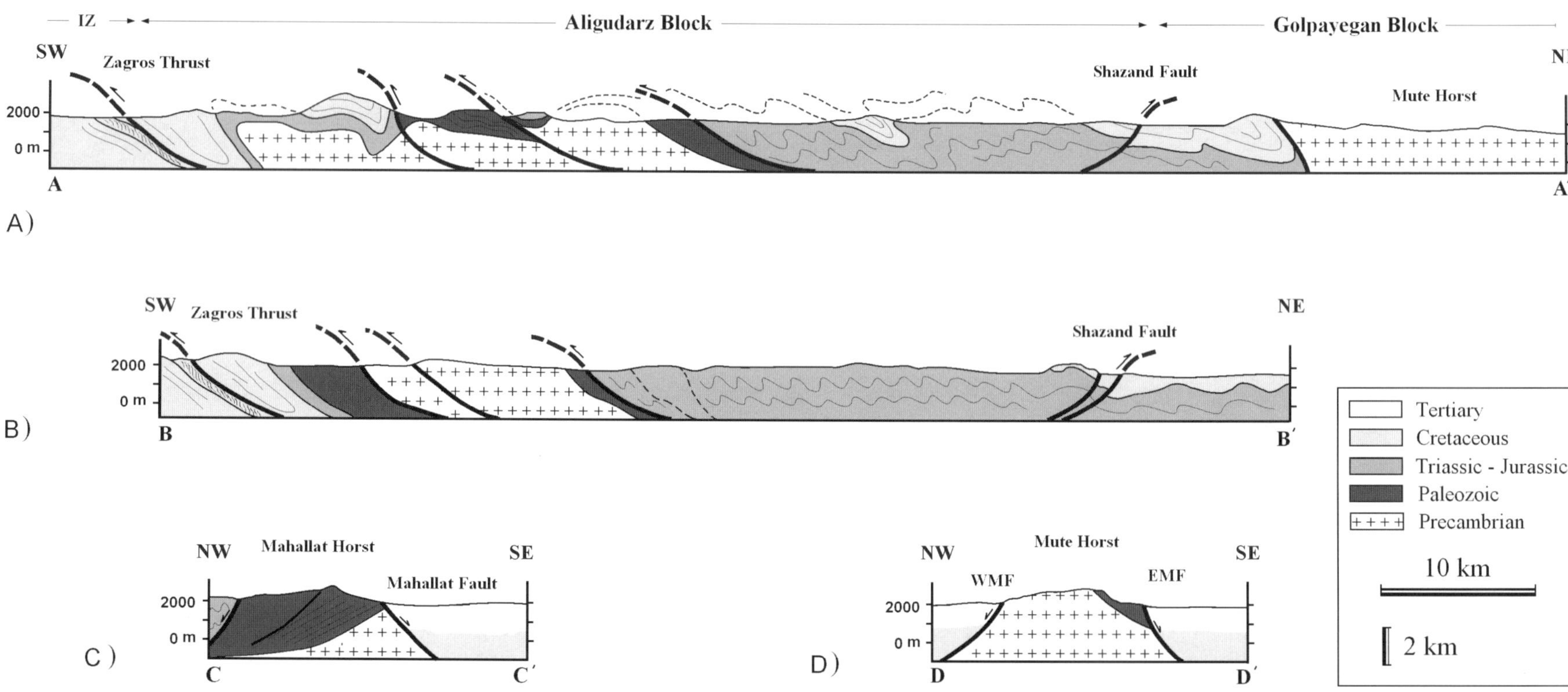

Figure 10. Simplified section across the Golpayegan region. (A) Cross section through the Golpayegan region and the Zagros thrust. This section shows the Precambrian basement and Lower Paleozoic outcrops at the Golpayegan block and Aligudarz block. (B) Cross section through the Aligudarz block and the Zagros thrust. This section show that the old rocks have outcropped only in the SW part of the section. (C) Cross section through the Mahallat horst. (D) Cross section through the Mute horst. Both sections show the fault contact of the horst and adjacent areas. Locations of the sections are shown in Figure 2A.

the NW-SE extension of the area. These faults cut Oligocene–middle Miocene strata and show that the faults were formed after the middle Miocene.

DISCUSSION

Here, we attempted to demonstrate, based on available geologic data, the tectonic evolution of the Golpayegan region of the Sanandaj-Sirjan zone and northwestern part of the Zagros orogen during convergence of the Arabian-Eurasian plates and exhumation of the old rocks during the deformation (Fig. 11). We focused our study on large-scale ductile and brittle structures formed after obduction and late Pliocene–Quaternary strike-slip movements. Both ductile and brittle structures can be interpreted within a single, continuous extensional and exhumation event, which started with ductile deformation and gradually changed into low-grade, brittle deformation.

Tectonic evolution of the Golpayegan region and exhumation of the old rocks are interpreted as the product of three major sequential geotectonic events: (1) formation of major thrusts during shortening and exposure of the basement rocks in the Aligudarz block, (2) formation of the extensional fractures during lateral extensional deformation and exhumation of the basement and old rocks in the Golpayegan block, and (3) the late Pliocene–Quaternary strike-slip movements.

Thrusting in the Aligudarz Block

Like other fold-and-thrust belts, the Aligudarz block shows intense shortening close to the suture. The tectonics of the Aligudarz block, southwestern part of Sanandaj-Sirjan zone, are characterized by numerous thrusts transporting rock units from northeast to southwest (Fig. 11). The Pliocene-Quaternary–age faults are seismic and active. The thrusts appear to form small- and large-scale duplexes and imbricate systems of several generations.

The Aligudarz block shows a maximum intensity of deformation. Rocks are strongly imbricated and sheared. In most areas, tectonic mélanges composed of fragments of the normal stratigraphic units and/or dismembered ophiolite complexes are well developed. The deformed structures in the Aligudarz block result from the convergence between the Eurasian and Arabian plates. This convergence started with subduction and later obduction of the Neotethyan upper mantle and oceanic crust. The thrusts and imbricated structures formed at diverse scales in the Aligudarz block and the Zagros orogen. Paul et al. (2006), based on gravity and seismological data, proposed a scenario where the Zagros wedge underthrusts the Sanandaj-Sirjan zone along a crustal-scale fault linked at the surface to the Main Zagros thrust. During this event, the Precambrian metamorphic rocks and old units were thrusted to shallower crustal levels and exposed in the Aligudarz

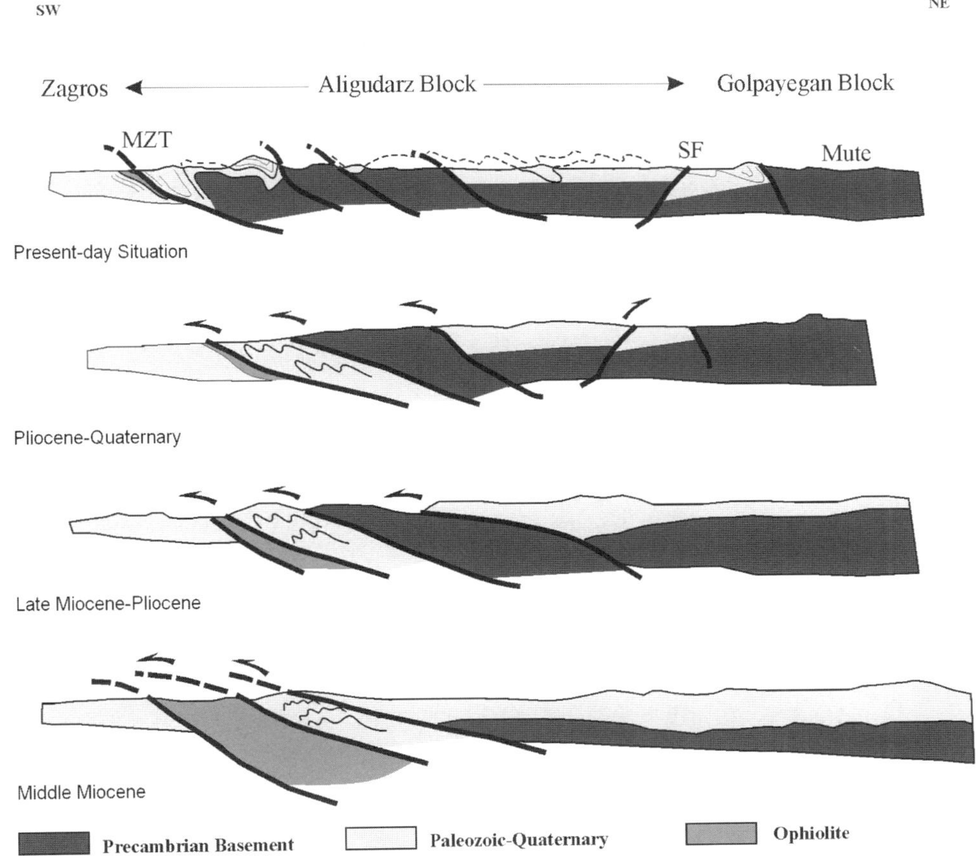

Figure 11. Progression of convergence and exhumation of the old rocks in the Golpayegan block and thrusting of the rocks in the Aligudarz block in the Golpayegan region and the Zagros thrust zone. In the NE part, from down to up, the old rock was exhumed during lateral extension movements. In the SW part, from down to up, the old rocks and other rocks unit were thrusted southwestward. MZT and SF are the Main Zagros thrust and Shazand fault, respectively.

block. Thrusting and compressional deformation continues today. Present-day convergence in the Zagros Mountains has been explained by the effects of ridge push and mantle-plume activity in the Red Sea–Afar region (Zeyen et al., 1997) and by slab pull toward the north beneath Turkey (Molinaro et al., 2005).

Development of the Extensional Structures

Geological evidence from this study and previous studies (Tillman et al., 1981; Moritz et al., 2006) indicates that extensional tectonism occurred in the Sanandaj-Sirjan zone during Tertiary time. There is some sedimentary and structural evidence for uplift the region. Molinaro et al. (2005) suggested that the lithospheric mantle was thinned below the crustal root and proposed that the thinning was related to recent slab break-off. Thermal uplift and surface erosion generated by this event probably modified the orogenic wedge taper in the mountain range and triggered the involvement of deeper levels of the crust in the deformation. The sudden and widespread occurrence of Pliocene-Pleistocene Bakhtiari conglomerates would be the sedimentary marker of this major change of tectonic regime (Molinaro et al., 2005). Structural studies show that there are two senses of extension during Tertiary time. Tillman et al. (1981) and Moritz et al. (2006) inferred that the NW-SE–trending normal faults and the NE-SW–trending extension was active during pre-Miocene time, and, in this study, we discuss NW-SE–trending extension that developed after middle Miocene time.

Evidence for Exhumation

During the extensional tectonism, Precambrian metamorphic rocks were exhumed and emplaced at a shallower lithospheric level. As summarized by Ring et al. (1999), exhumation of metamorphic rocks occurs as a result of ductile thinning, erosion, and normal faulting. The subhorizontal foliation in the metamorphic rocks at Mute (Moritz et al., 2006), south and west Golpayegan, and east Hassan-Robat (this study) confirms that ductile thinning was involved in the exhumation of the metamorphic rocks (Ring et al., 1999). The foliated metamorphic rocks were cut by later extensional structures. In the north, at Mute, the Eocene gold-bearing ore bodies formed along the extensional structures (Moritz et al., 2006).

Exhumation of metamorphic domes in several orogens is accompanied by the formation of sedimentary basins (e.g., Echtler and Malavieille, 1990; Cassard et al., 1993; Neubauer et al., 1995; Bonev et al., 2006). In the Golpayegan block, coarse clastic Eocene sedimentary rocks, including conglomerate alternating with sandstone, sandy shale, and volcanic rocks, overlie older formations with a distinct unconformity (Thiele et al., 1968) and were defined as syntectonic conglomerates by Alavi (1994). These sedimentary rocks are distributed in the northeastern parts of the study area and near the Urumieh-Dokhtar magmatic arc. The rocks indicate that extensional tectonism and exhumation in the Urumieh-Dokhtar magmatic arc during the Eocene was associated with formation of sedimentary basins around the uplifted areas.

Based on geological observations, two sets of normal faults and two extensional episodes are present. The first episode occurred before Miocene time and formed the NW-SE–trending normal faults during brittle deformation of metamorphic basement rocks. These faults may have formed at the surface of the uplifted region, and they have low distribution and frequency. The second episode occurred after middle Miocene time and formed the NE-SW–trending normal faults during lateral extensional movements (Fig. 12). These faults are the main set of normal faults, and they have high distribution and frequency.

A schematic structural model for the sequential formation of the faults in the Golpayegan block is shown in Figure 13. In the Golpayegan block, NE-SW shortening generated lateral extension stresses in a collision forearc. During the lateral extension of the region, some NE-SW–trending sigmoidal fractures formed. The borders of the fractures are correlated with normal faults. Formation of the faults and the lateral extension created a set of horsts and grabens parallel to the shortening direction. In the horsts (e.g., Hassan-Robat, Mute, and Mahallat), Precambrian basement and Lower Paleozoic strata were exhumed during the extension. The bounding faults of the exhumed rocks cut the Eocene strata in the Mahallat and Hassan-Robat horsts, on the northwestern side of Mute horst, and the Oligocene–middle Miocene Qom Formation and Quaternary sediments on the southeastern side of Mute horst. This shows that exhumation of the basement rocks in the Golpayegan block started after middle Miocene time and is continuing today.

Recent folding and uplifting of the belt are evident in the deeply incised Golpayegan and Khomein River valleys cutting the folds of the Aligudarz block mountains, in raised beaches, by the height of Quaternary alluvial terraces of the Golpayegan and Khomein depressions, and by the uplift of historical canals. During uplifting of the area, a 10-m-high scarp on the surface of the first river terrace was formed. Syntectonic conglomerates of Pliocene and Pliocene-Quaternary age in the Sanandaj-Sirjan zone correlate well with the tectonic events that affected the Sanandaj-Sirjan zone.

Strike-Slip Movements

Prior to the middle Miocene, the Arabian plate interacted with the Iranian block along the Main Zagros thrust. Since the change in the direction of relative plate motion in the late Pliocene, the Arabian plate has been moving northward, oblique to the Zagros chain (Bachmanov et al., 2004). During the convergence and crustal shortening, strike-slip movements started. In the region, active tectonic movements include right-lateral slip at the rate of ~2 mm/yr (Nadimi and Nadimi, 2006b) along the NW-SE–trending faults, and active thrusting accompanied by folding. The right-lateral strike-slip fault activity during Pliocene-Quaternary time is observable in the rock units and sediments.

Figure 13 shows the younger deformations in the Golpayegan region. The strike-slip faults cut the older faults. Conjugate NW- and NE-trending faults formed active blocks that are bounded by seismic faults. During movement and rotation of the blocks, the faults cut and rotated the Quaternary alluvial fans and rivers and uplifted the region. Drainages are dextrally displaced by ~4 km along the Shazand fault and near the town of Golpayegan.

During the strike-slip movements, some older faults (e.g., Shazand fault) that had reverse or thrust motions during the first geotectonic event changed their motions to oblique-slip or strike-slip motions. The Shazand fault cut the Eastern Mute normal fault and seems to be younger than Eastern Mute fault. However, the Shazand fault, after starting strike-slip motion, displaced the Eastern Mute fault right-laterally, and only its strike-slip motion is younger than its reverse motion.

CONCLUSION

Continental collision between the Iranian and Arabian plates resulted in the formation of the Zagros fold-and-thrust belt and its associated foreland basin. The convergence is accommodated in the NW Zagros by a combination of shortening on NW-SE–trending folds and thrusts, and right-lateral strike-slip faults. The existence of a component of strike-slip movement early in the convergence history points to a highly oblique convergence setting. The tectonic evolution of the Golpayegan region and exhumation of the old rocks are interpreted as the product of three major sequential geotectonic events: (1) formation of major thrusts during shortening and the exposure of the basement in the Aligudarz block, (2) formation of extensional fractures during lateral extensional movement and exhumation of the basement rocks in the Golpayegan block, and (3) strike-slip movements that started in late Pliocene–Quaternary time.

1. Thrusting and overthrusting of Sanandaj-Sirjan zone units in our study area suggest a large magnitude of cumulative shortening in northern Zagros. The Aligudarz block shows a maximum intensity of deformation, as the rocks are strongly folded, imbricated, and sheared. In the block and during thrusting, the Precambrian metamorphic rocks were exposed in NW-trending belts parallel to Zagros.

2. Ductile and brittle structures and sedimentary formations can be interpreted to have formed within a single, continuous extensional and exhumation event. Based on geological evidence,

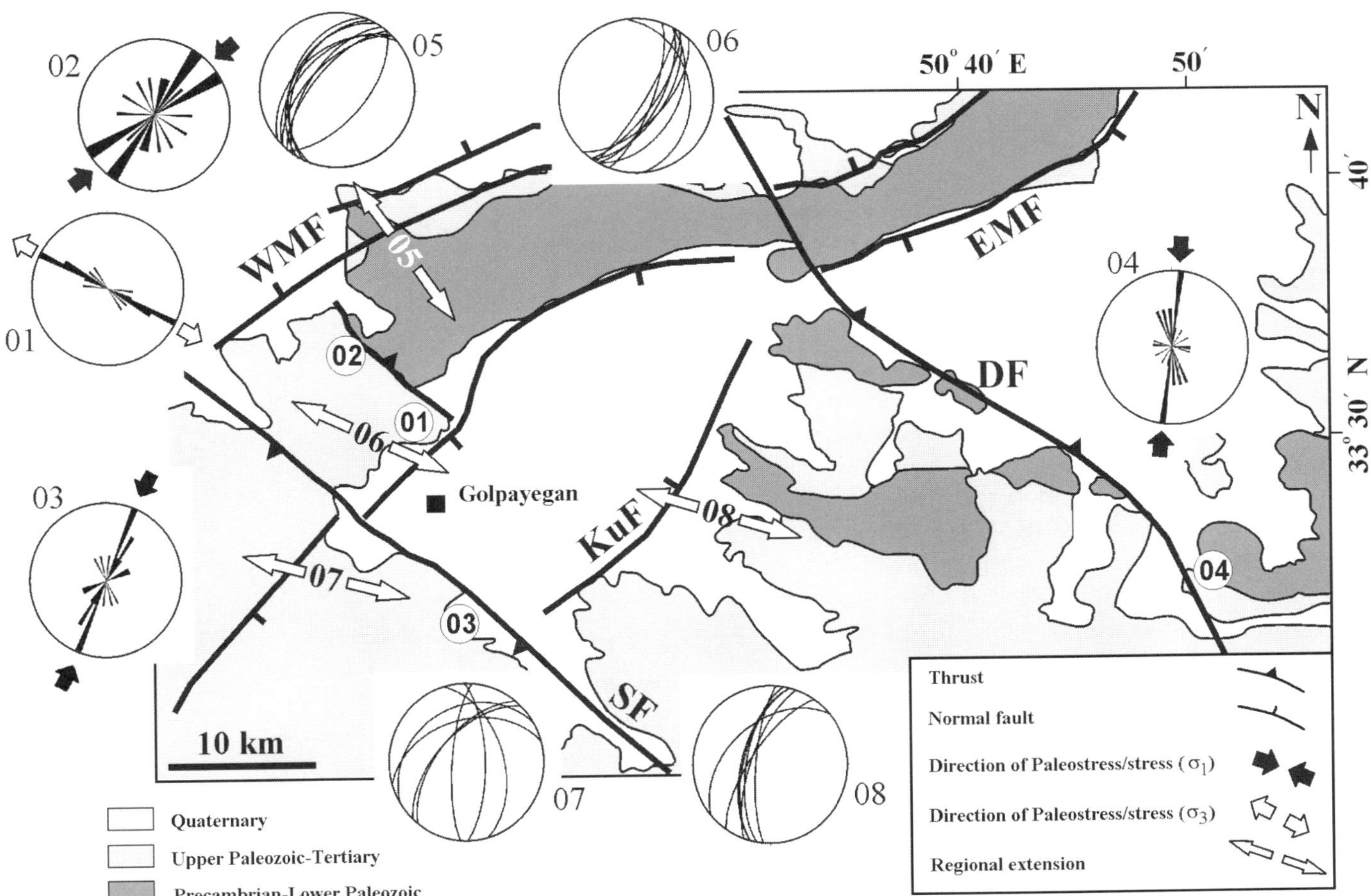

Figure 12. Geological map of the Golpayegan region and results of stress inversion. Rose diagrams show orientations of the local σ_1 and σ_3 axes deduced from microtectonic measurements. Stereoplots show measured fault planes. Arrows are trends of the inferred horizontal σ_3 axes (σ_1 axes are vertical or near-vertical). Stereoplots and rose diagrams show the lateral extensional direction in the region. The major faults are shown by thick lines, and names are abbreviated: SF—Shazand; DF—Dehagh; WMF—Western Mute; EMF—Eastern Mute; KuF—Kunjijan faults.

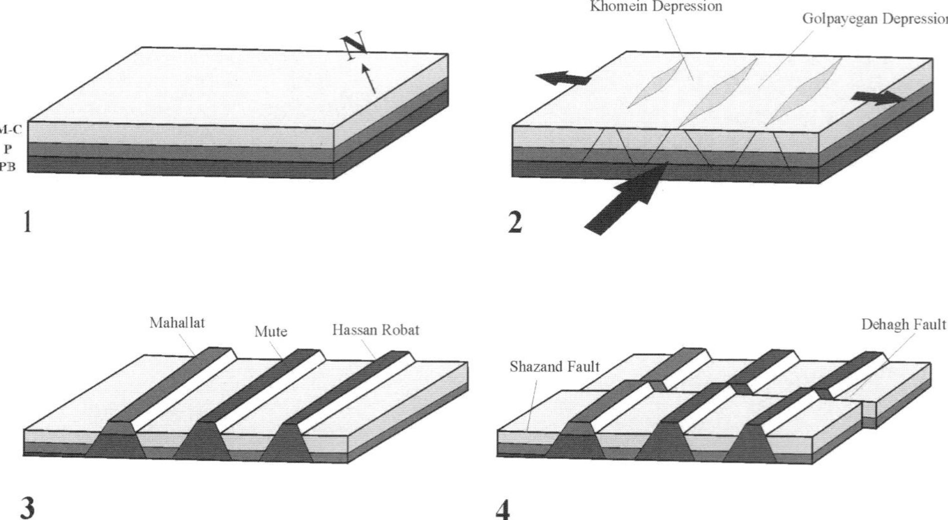

Figure 13. Schematic sequential model showing evolution of the extensional structures in the Golpayegan block. Abbreviations: M-C—Mesozoic-Cenozoic strata, P—Paleozoic strata, and PB—Precambrian basement. 1—The region before generation of lateral extension fractures. 2—NE-SW shortening generated lateral extension stresses, and during the lateral extension, some NE-SW trending sigmoidal fractures and faults formed. 3—Formation of the faults and the lateral extension created a set of horsts and grabens parallel to the shortening direction. 4—During the convergence and crustal shortening, strike-slip movements started and the strike-slip faults cut the older faults and horst and graben structures.

the event can be subdivided into two extensional episodes in the Golpayegan region. The first episode occurred before Miocene time and formed the NW-SE–trending normal faults in the metamorphic basement rocks. The second episode occurred during NE-SW–trending shortening in the collisional forearc area after middle Miocene time and formed the NE-SW–trending normal faults (e.g., Eastern Mute fault, Western Mute fault, and Mahallat faults). Exhumation of the Precambrian basement and accompanying old rocks within the Golpayegan block was related to the development of major extensional structures that generated a system of NE-SW–trending horsts and grabens parallel to the shortening direction. In the horsts (e.g., Hassan-Robat, Mute, and Mahallat), the exhumed basement rocks are bordered by NE-SW–trending normal faults.

3. Strike-slip motions are well documented for Pliocene–Quaternary time. In the region, the Shazand and Dehagh faults cut through the Eastern Mute fault and Western Mute fault and through Quaternary deposits. Drainages are dextrally displaced by ~4 km along the Shazand fault and near the town of Golpayegan.

ACKNOWLEDGMENTS

We thank many of our friends and colleagues of the Zamin Forozan Sofeh Co., Isfahan, Iran, and Erchie Wang of the Institute of Geology and Geophysics, Beijing, China, who helped us in various ways. B. Clark Burchfiel, Gary Axen, Charles M. Verdel, and an anonymous reviewer provided helpful reviews, and their assistance is gratefully acknowledged.

REFERENCES CITED

Agard, P., Omrani, J., Jolivet, L., and Mouthereau, F., 2005, Convergence history across Zagros (Iran): Constraints from collisional and earlier deformation: International Journal of Earth Sciences, v. 94, p. 401–419, doi: 10.1007/s00531-005-0481-4.

Alavi, M., 1994, Tectonics of the Zagros orogenic belt of Iran: New data and interpretations: Tectonophysics, v. 229, p. 211–238, doi: 10.1016/0040-1951(94)90030-2.

Alavi, M., and Mahdavi, M.A., 1994, Stratigraphy and structures of the Nahavand region in western Iran and their implications for the Zagros tectonics: Geological Magazine, v. 131, p. 43–47.

Al Laboun, A.A., 1986, Stratigraphy and hydrocarbon potential of the Paleozoic succession in both the Tabuk and Widyan Basins, Arabia, *in* Halbouty, M.T., ed., Future Petroleum Provinces of the World: American Association of Petroleum Geologists Memoir 40, p. 399–425.

Alsharhan, A.S., and Nairn, A.E.M., 1997, Sedimentary Basins and Petroleum Geology of the Middle East: Amsterdam, Elsevier Science, 843 p.

Axen, G.J., Lam, P.S., Grove, M., Stockli, D.F., and Hassanzadeh, J., 2001, Exhumation of the west-central Alborz Mountains, Iran, Caspian subsidence, and collision-related tectonics: Geology, v. 29, p. 559–562, doi: 10.1130/0091-7613(2001)029<0559:EOTWCA>2.0.CO;2.

Bachmanov, D.M., Trifonov, V.G., Hessami, K.T., Kozhurin, A.I., Ivanova, T.P., Rogozhin, E.A., Hademi, M.C., and Jamali, F.H., 2004, Active faults in the Zagros and central Iran: Tectonophysics, v. 380, p. 221–241, doi: 10.1016/j.tecto.2003.09.021.

Bahroudi, A., and Koyi, H.A., 2004, Tectono-sedimentary framework of the Gachsaran Formation in the Zagros foreland basin: Marine and Petroleum Geology, v. 21, p. 1295–1310, doi: 10.1016/j.marpetgeo.2004.09.001.

Berberian, M., 1995, Master "blind" thrust faults hidden under the Zagros folds: Active tectonics and surface morphotectonics: Tectonophysics, v. 241, p. 193–224, doi: 10.1016/0040-1951(94)00185-C.

Berberian, M., and King, G.C.P., 1981, Towards a paleogeography and tectonic evolution of Iran: Canadian Journal of Earth Sciences, v. 18, p. 210–265.

Beydoun, Z.R., Hughes Clarke, M.W., and Stoneley, R., 1992, Petroleum in the Zagros Basin: A Late Tertiary foreland basin overprinted onto the outer edge of a vast hydrocarbon-rich Palaeozoic–Mesozoic passive margin shelf, *in*

MacQueen, R.W., and Leckie, D.A., eds., Foreland Basins and Fold Belts: American Association of Petroleum Geologists Memoir 55, p. 309–339.

Blanc, E.J.P., Allen, M.B., Inger, S., and Hassani, H., 2003, Structural styles in the Zagros simple folded zone, Iran: Journal of the Geological Society of London, v. 160, p. 401–412.

Bonev, N., Burg, J.-P., and Ivanov, Z., 2006, Mesozoic-Tertiary structural evolution of an extensional gneiss dome—The Kesebir-Kardamos dome, Eastern Rhodope (Bulgaria-Greece): International Journal of Earth Sciences, v. 95, p. 318–340, doi: 10.1007/s00531-005-0025-y.

Bosold, A., Schwarzhans, W., Julapour, A., Ashrafzadeh, A.R., and Ehsani, S.M., 2005, The structural geology of the High Central Zagros revisited (Iran): Petroleum Geoscience, v. 11, p. 225–238.

Boyer, S.E., 1995, Sedimentary basin taper as a factor controlling the geometry and advance of thrust belts: American Journal of Science, v. 295, p. 1220–1254.

Cassard, D., Feybesse, J.L., and Lescuyer, J.L., 1993, Variscan crustal thickening and late overstacking during the Namurian-Westphalian in the western Montagne Noire (France): Tectonophysics, v. 222, p. 33–53, doi: 10.1016/0040-1951(93)90188-P.

Chemenda, A.I., Mattauer, M., Malavielle, J., and Bokun, A., 1995, A mechanism for syncollisional rock exhumation and associated normal faulting: Result from physical modeling: Earth and Planetary Science Letters, v. 132, p. 225–232, doi: 10.1016/0012-821X(95)00042-B.

Colman-Sadd, S.P., 1978, Fold development in Zagros simply folded belt, southwest Iran: American Association of Petroleum Geologists Bulletin, v. 62, no. 6, p. 984–1003.

Cosgrove, J.W., and Ameen, M.S., 2000, A comparison of the geometry, spatial organization and fracture patterns associated with forced folds and buckle folds, *in* Cosgrove, J.W., and Ameen, M.S., eds., Forced Folds and Fractures: Geological Society of London Special Publication 169, p. 7–21.

Dahlen, F.A., 1990, Critical taper model of fold-and-thrust belts and accretionary wedges: Annual Review of Earth and Planetary Sciences, v. 18, p. 55–99, doi: 10.1146/annurev.ea.18.050190.000415.

Echtler, H., and Malavieille, J., 1990, Extensional tectonics, basement uplift and Stephano-Permian collapse basin in a late Variscan metamorphic core complex (Montagne Noire, Southern Massif Central): Tectonophysics, v. 177, p. 125–138, doi: 10.1016/0040-1951(90)90277-F.

England, P., and Molnar, P., 1990, Surface uplift, uplift of rocks and exhumation of rocks: Geology, v. 18, p. 1173–1177, doi: 10.1130/0091-7613(1990)018<1173:SUUORA>2.3.CO;2.

Falcon, N.L., 1969, Problems of the relationship between surface structure and deep displacements illustrated by the Zagros Range, *in* Kent, P.E., Satterthwaite, G.E., and Spencer, A.M., eds., Time and Place in Orogeny: Geological Society of London Special Publication 3, p. 9–22.

Falcon, N.L., 1974, Southern Iran: Zagros Mountains, *in* Spencer, A.M., ed., Mesozoic–Cenozoic Orogenic Belts: Data for Orogenic Studies: Geological Society of London Special Publication 4, p. 9–22.

Ghasemi, A., and Talbot, C.J., 2006, A new tectonic scenario for the Sanandaj-Sirjan zone (Iran): Journal of Asian Earth Sciences, v. 26, p. 683–693, doi: 10.1016/j.jseaes.2005.01.003.

Guest, B., Axen, G.J., Lam, P.S., and Hassanzadeh, J., 2006, Late Cenozoic shortening in the west-central Alborz Mountains, northern Iran, by combined conjugate strike-slip and thin-skinned deformation: Geosphere, v. 2, no. 1, p. 35–52, doi: 10.1130/GES00019.1.

Hacker, B.R., Ratschbacher, L., Webb, L., and Shuwen, D., 1995, What brought them up? Exhumation of Dabie Shan ultrahigh-pressure rocks: Geology, v. 23, p. 743–746, doi: 10.1130/0091-7613(1995)023<0743:WBTUEO>2.3.CO;2.

Hessami, K., Jamali, F., and Tabassi, H., 2003, Major Active Faults of Iran: Seismology Research Center, Tehran, International Institute of Earthquake Engineering and Seismology (IISSE), scale 1:2,500,000.

Jackson, J.A., 1980, Reactivation of basement faults and crustal shortening in orogenic belts: Nature, v. 283, p. 343–346, doi: 10.1038/283343a0.

Jackson, J.A., and McKenzie, D.P., 1984, Active tectonics of the Alpine–Himalayan belt between Western Turkey and Pakistan: Geophysical Journal of the Royal Astronomical Society, v. 77, p. 185–284.

Jamieson, R.A., Beaumont, C., Nguyen, M.H., and Lee, B., 2002, Interaction of metamorphism, deformation and exhumation in large convergent orogens: Journal of Metamorphic Geology, v. 20, p. 9–24, doi: 10.1046/j.0263-4929.2001.00357.x.

Karabinos, P., 1988, Heat transfer and fault geometry in the Taconian thrust belt, western New England, *in* Mitra, G., and Wojtal, S.F., eds., Geometries and Mechanism of Thrusting, with Special Reference to the Appalachians: Geological Society of America Special Paper 222, p. 35–45.

Kurz, W., and Froitzheim, N., 2002, The exhumation of eclogite-facies metamorphic rocks—A review of models confronted with examples from the Alps: International Geology Review, v. 44, p. 702–743, doi: 10.2747/0020-6814.44.8.702.

McQuarrie, N., Stock, J.M., Verdel, C., and Wernicke, B.P., 2003, Cenozoic evolution of Neotethys and implications for the causes of plate motions: Geophysical Research Letters, v. 30, p. 2036, doi: 10.1029/2003GL017992.

Michard, A., Chopin, C., and Henry, C., 1993, Compression versus extension in the exhumation of the Dora-Maria coesite-bearing unit, Western Alps, Italy: Tectonophysics, v. 221, p. 173–193, doi: 10.1016/0040-1951(93)90331-D.

Mitra, G., and Sussman, A.J., 1997, Structural evolution of connecting splay duplexes and their implications for critical taper; an example based on geometry and kinematics of the Canyon Range culmination, Sevier belt, central Utah: Journal of Structural Geology, v. 19, p. 503–521, doi: 10.1016/S0191-8141(96)00108-3.

Molinaro, M., Zeyen, H., and Laurencin, X., 2005, Lithospheric structure beneath the south-eastern Zagros Mountains, Iran: Recent slab break-off?: Terra Nova, v. 17, p. 1–6, doi: 10.1111/j.1365-3121.2004.00575.x.

Moritz, R., Ghazban, F., and Singer, B.S., 2006, Eocene gold ore formation at Muteh, Sanandaj-Sirjan tectonic zone, western Iran: A result of late-stage extension and exhumation of metamorphic basement rocks within the Zagros orogen: Economic Geology and the Bulletin of the Society of Economic Geologists, v. 101, p. 1497–1524.

Nadimi, A., and Nadimi, H., 2006a, The Zagros collision: Exhumation of old rocks in the northwestern part of Zagros Mountain, Iran [abs.], *in* Burchfiel, B.C., ed., International Conference on Continental Dynamics and Environmental Change of the Tibetan Plateau: PR China, Institute of Geology and Geophysics, Chinese Academy of Science Publication, Abstracts, p. 3.

Nadimi, A., and Nadimi, H., 2006b, Active tectonics of the South Shahreza, N. Zagros Mountains, Iran, *in* 6th International Conference on the Geology of the Middle East: UAE, United Arab Emirates University Publication, Abstracts, p. 272.

Neubauer, F., Dallmeyer, R.D., Dunkl, I., and Schirnik, D., 1995, Late Cretaceous exhumation of the metamorphic Gleinalm dome, eastern Alps: Kinematics, cooling history and sedimentary response in a sinistral wrench corridor: Tectonophysics, v. 242, p. 79–98, doi: 10.1016/0040-1951(94)00154-2.

Ni, J., and Barazangi, M., 1986, Seismotectonics of the Zagros continental collision zone and a comparison with the Himalayas: Journal of Geophysical Research, v. 91, p. 8205–8218, doi: 10.1029/JB091iB08p08205.

Paul, A., Kaviani, A., Hatzfeld, D., Vergne, J., and Mokhtari, M., 2006, Seismological evidence for crustal-scale thrusting in the Zagros mountain belt (Iran): Geophysical Journal International, v. 166, p. 227–237, doi: 10.1111/j.1365-246X.2006.02920.x.

Rachidnejad-Omran, N., Emami, M.H., Sabzehei, M., Rastad, E., Bellon, H., and Piqué, A., 2002, Lithostratigraphie et histoire Paléozoïque à Paléocène des complexes métamorphiques de la région de Muteh, zone de Sanandaj-Sirjan (Iran méridional): Comptes Rendus Geoscience, v. 334, p. 1185–1191, doi: 10.1016/S1631-0713(02)01861-8.

Ring, U., Brandon, M.T., Willett, S.D., and Lister, G.S., 1999, Exhumation processes, *in* Ring, U., Brandon, M.T., Lister, G.S., and Willett, S.D., eds. Exhumation Processes: Normal Faulting, Ductile Flow, and Erosion: Geological Society of London Special Publication 154, p. 1–27.

Ring, U., Layer, P.W., and Reischmann, T., 2001, Miocene high-pressure metamorphism in the Cyclades and Crete, Aegean Sea, Greece: Evidence for large-magnitude displacement on the Cretan detachment: Geology, v. 29, p. 395–398, doi: 10.1130/0091-7613(2001)029<0395:MHPMIT>2.0.CO;2.

Sattarzadeh, Y., Cosgrove, J.W., and Vita-Finzi, C., 2000, The interplay of faulting and folding during the evolution of the Zagros deformation belt, *in* Cosgrove, J.W., and Ameen, M.S., eds., Forced Folds and Fractures: Geological Society of London Special Publication 169, p. 187–196.

Şengör, A.M.C., 1990, A new model for the late Palaeozoic–Mesozoic tectonic evolution of Iran and implications for Oman, *in* Robertson, A.H.F., Searle, M.P., and Ries, A.C., eds., The Geology and Tectonics of the Oman Region: Geological Society of London Special Publication 49, p. 797–831.

Sherkati, S., Molinaro, M., Frizon de Lamotte, D., and Letouzey, J., 2005, Detachment folding in the Central and Eastern Zagros fold-belt (Iran): Salt mobility, multiple detachments and late basement control: Journal of Structural Geology, v. 27, p. 1680–1696, doi: 10.1016/j.jsg.2005.05.010.

Stöcklin, J., 1968, Structural history and tectonics of Iran: A review: The American Association of Petroleum Geologists Bulletin, v. 52, p. 1229–1258.

Stoneley, R., 1981, The geology of the Kuh-e Dalneshin area of southern Iran, and its bearing on the evolution of southern Tethys: Journal of the Geological Society of London, v. 138, p. 509–526, doi: 10.1144/gsjgs.138.5.0509.

Stoneley, R., 1990, The Arabian continental margin in Iran during the Late Cretaceous, *in* Robertson, A., Searle, M., and Ries, A., eds., The Geology and Tectonics of the Oman Region: Geological Society of London Special Publication 49, p. 787–795.

Thiele, O., 1966, Zum Alter der Metamorphose in Zentraliran: Mitteilungen der Geologischen Gesellschaft [Wien], v. 58, p. 87–101.

Thiele, O., Alavi, M., Assefi, R., Hushmand-zadeh, A., Seyed-Emami, K., and Zahedi, M., 1968, Explanatory Text of the Golpayegan Quadrangle Map, 1:250,000: Geological Survey of Iran Geological Quadrangle E7, 24 p.

Tillman, J.E., Poosti, A., Rossello, S., and Eckert, A., 1981, Structural evolution of Sanandaj-Sirjan Ranges near Esfahan, Iran: American Association of Petroleum Geologists Bulletin, v. 65, p. 674–687.

Vernant, P., Nilforoushan, F., Hatzfeld, D., Abbassi, M.R., Vigny, C., Masson, F., Nankali, H., Martinod, J., Ashtiani, A., Bayer, R., Tavakoli, F., and Chéry, J., 2004, Present-day crustal deformation and plate kinematics in the Middle East constrained by GPS measurements in Iran and northern Oman: Geophysical Journal International, v. 157, p. 381–398, doi: 10.1111/j.1365-246X.2004.02222.x.

Yilmaz, Y., 1993, New evidence and model on the evolution of the southeast Anatolian orogen: Geological Society of America Bulletin, v. 105, p. 251–271, doi: 10.1130/0016-7606(1993)105<0251:NEAMOT>2.3.CO;2.

Zeyen, H., Volker, F., Wehrle, V., Fuchs, K., Sobolev, S.V., and Altherr, R., 1997, Styles of continental rifting: Crust-mantle detachment and mantle plumes: Tectonophysics, v. 278, p. 329–352, doi: 10.1016/S0040-1951(97)00111-X.

Manuscript Accepted by the Society 04 April 2008

Printed in the USA